MULTIPLE DRUG RESISTANCE IN CANCER

This volume is dedicated to my parents John and Katie,
to my wife Honor and to Aedin, Peter and Isolde.

Multiple Drug Resistance in Cancer

Cellular, Molecular and Clinical Approaches

Edited by

M. CLYNES

National Cell & Tissue Culture Center/Bioresearch Ireland,
Dublin City University, Glasnevin, Dublin, Ireland

Reprinted from Cytotechnology, volume 12, 1993.

SPRINGER SCIENCE+BUSINESS MEDIA, B.V.

A C.I.P. Catalogue record for this book is available from the Library of Congress.

ISBN 978-94-010-4355-7 ISBN 978-94-011-0826-3 (eBook)
DOI 10.1007/978-94-011-0826-3

Printed on acid-free paper

CONTENTS

Introduction vii

P-glycoprotein structure and evolutionary homologies 1
J.M. Croop

Molecular analysis of the multidrug transporter 33
U.A. Germann

Molecular cytogenetics of multiple drug resistance 63
P.V. Schoenlein

Antibodies in the study of multiple drug resistance 91
Y. Heike, T. Tsuruo

Non-P-glycoprotein multidrug resistance in cell lines which are defective in the cellular accumulation of drug 109
M.S. Center

Topoisomerase I in multiple drug resistance 127
A. Pessina

Topoisomerase II in multiple drug resistance 137
G.A. Hofmann, M.R. Mattern

Glutathione-related enzymes, glutathione and multidrug resistance 155
J.A. Moscow, K.H. Dixon

Pharmacologic circumvention of multidrug resistance 171
J.M. Ford, W.N. Hait

Multidrug resistance (MDR) genes in haematological malignancies 213
K. Nooter, P. Sonneveld

Human cell lines as models for multidrug resistance in solid tumours 231
M. Clynes, M. Heenan, K. Hall

Studies on low-level MDR cells 257
G. Belvedere, E. Dolfini

Differing patterns of cross-resistance resulting from exposures to specific antitumour drugs or to radiation *in vitro* 265
B.T. Hill

The use of reverse transcriptase-polymerase chain reaction (RT-PCR) to investigate specific gene expression in multidrug-resistant cells 289
L. O'Driscoll, C. Daly, M. Saleh, M. Clynes

Mathematical models for multidrug resistance and its reversal 315
S. Michelson

The biology of radioresistance: similarities, differences and interactions with drug resistance 325
S.N. Powell, E.H. Abraham

Kinetic resistance to anticancer agents 347
M.-T. Dimanche-Boitrel, C. Garrido, B. Chauffert

Cytochromes P450 and drug resistance 358
J. Doehmer, A.R. Goeptar, N.P.E. Vermeulen

Role of matrix metalloproteinases in invasion and metastasis: biology, diagnosis and inhibitors 367
S. McDonnell, B. Fingleton

Clinical significance of cellular resistance in tumours to cytotoxic chemotherapy and radiotherapy 385
M. Pomeroy, M. Moriarty

Key word index 393

Cytotechnology **12**: vii–viii, 1993.

Introduction

This volume brings together reviews by different experts on a wide range of research topics relevant to understanding how tumours withstand the toxic effects of chemotherapy. Some types of cancer can now be cured by chemotherapy, but for the majority of disseminated cancers where surgery alone is insufficient, neither chemotherapy nor any other available treatment can provide long-term cure for a large number of affected patients.

Since most types of cancer show some response to combination chemotherapy, and since a few forms of cancer can be completely cured by this approach, it is reasonable to hope that an understanding of why some cancers show resistance — intrinsic or acquired — to a range of drugs could lead to the design of better treatment and consequently to complete cures for more cancer patients.

Human tumour cells grown in culture are useful model systems for investigating drug resistance mechanisms and for testing ways of restoring sensitivity to multidrug-resistant cells. Much of this work has been done with established cell lines made resistant by exposure *in vitro* to progressively increasing concentrations of chemotherapeutic drugs such as adriamycin. These are excellent experimental systems, especially since the parental unselected cells are available as controls. Because, however, such cell lines and their environment are not exactly equivalent to tumours *in vivo*, in particular in relation to tissue organization and kinetics, *in vitro* studies need to be seen not as ends in themselves, but rather as the base necessary for design of confirmatory work on human cancer tissue. In general, however, studies with multidrug-resistant cell lines appear to yield results and concepts which are relevant to human cancers *in vivo*.

Much of the research on multiple drug resistance in cancer is based on the assumption that it is a genetic phenomenon resulting in biochemical-physiological changes at the cellular level. The bulk of experimental evidence confirms that this is a very important (and possibly but not certainly the most important) element in determining resistance. There is also some evidence that genetically-resistant cells may stably express mechanisms which are inducible, although perhaps to a lesser degree, in sensitive cancer cells as part of their biochemical defence mechanisms against xenobiotics. Most of the chapters in this volume focus on cellular mechanisms of multidrug resistance.

Overexpression of the membrane drug-transport protein, P-glycoprotein is a proven mechanism of resistance. **Ursula Germann** discusses current knowledge on structure-function relationships in this molecule, while **James Croop** describes the structure and evolutionary homologies of the genes for P-glycoprotein and related molecules, and **Patricia Schoenlein** outlines how the amplification of *mdr* genes in resistant cells has been studied. **Melvin Center** reviews recent developments indicating that other, as yet poorly-defined, transport molecules and mechanisms, including some of relevance to intracellular transport and distribution of drugs, may be important in resistance determination. The role of topoisomerases in toxicity of and resistance to some drugs is increasingly of interest, and this complex field is very clearly reviewed by **Augusto Pessina** (Topoisomerase I) and **Glenn Hoffman** and **Michael Mattern** (for Topoisomerase II). The question of whether glutathione-related enzymes play an important role either in multidrug-resistance or in alkylating agent-resistance has been controversial, and the evidence is critically presented here in the review by **Jeffrey Moscow and**

Katherine Dixon. The possible role of other enzymes of xenobiotic metabolism, such as cytochrome P450 enzymes, is discussed by **Johannes Doehmer, Arnold Goeptar and Nico Vermeulen**.

Quite a lot of information is now available on the occurrence of gene products related to the mechanisms discussed above. This information, for tissue-specific cell lines and from clinical studies is reviewed by **Kees Nooter and Pieter Sonneveld** (for hematological malignancies) and by **Martin Clynes, Mary Heenan** and **Keara Hall** (for solid tumours). Cell lines displaying a very low level of resistance, which may however be a good *in vitro* reproduction of lower-level clinical resistance, are described by **George Belevedere** and **Ersilia Dolfini**. **Bridget Hill** in her contribution has reviewed the different cross-resistance patterns seen in different multidrug-resistant lines, and in particular in lines made chemoresistant by exposure to radiation.

Identification in tissues or cells, at the highest possible level of specificity and sensitivity, of gene products related to resistance determination is of central importance to the study of multidrug resistance. **Yuji Heike** and **Takashi Tsuruo** describe some of the antibodies being used to study P-glycoprotein-mediated resistance, and **Lorraine O'Driscoll, Carmel Daly** and **Martin Clynes** review the more sensitive methods available for detecting and quantifying specific mRNAs.

The ultimate goal of MDR research is of course not merely to identify mechanisms but to build on this understanding to design new effective treatments. **James Ford** and **William Hait** describe the pharmacological approaches that are being used to reverse or circumvent multidrug resistance, in particular strategies to antagonise the action of P-glycoprotein. **Seth Michelson** in another chapter, presents mathematical models for P-glycoprotein-mediated resistance and its pharmacological reversal.

While most research on MDR has concentrated on intrinsic cellular properties, tumour cells *in vivo* are organized as tissues, and other factors including oxygen supply, penetration of drug and cell cycle/cell kinetic parameters may also play an important role in clinical resistance. Interesting cellular models for "kinetic" resistance to drugs are described here by **Marie-Thérèse Dimanche-Boitrel, Carmen Garrido and Bruno Chauffert**. Cells which have become resistant to particular chemotherapeutic drug combinations may be sensitive to other forms of therapy; the similarities and differences between mechanisms of chemo- and radioresistance are discussed by **Simon Powell** and **Edward Abraham**. Also, since cancer invasion, metastasis and induction of angiogenesis are all important steps in spread and pathology of tumours *in vivo*, therapy aimed at blocking these steps could be a useful addition to the options available to attack multidrug resistant tumours; **Susan McDonnell** and **Barbara Fingleton** review the role of matrix metalloproteases (one possible set of targets for pharmacological intervention) in cancer invasion and metastasis. Finally, **Maeve Pomeroy and Michael Moriarty** view the problem of multidrug resistance and current research in the area, from a clinician's viewpoint.

The authors exchanged abstracts prior to writing their chapters, to avoid significant overlap, but in general each author had the freedom to expand his or her topic as they felt to be most relevant. I feel that this has led to a more valuable expression of opinion and ideas which might not have been achieved if we had tried to completely eliminate overlap or to define the limits of each chapter too tightly. The chapters review the literature up to mid-1993.

I hope that this collection of up-to-date and critical reviews, and the ideas presented and questions posed, will be helpful to researchers working in clinical, molecular, pharmacological and cellular aspects of multiple drug resistance, and to others who want to find out what is currently happening in this field. Especially, I hope that it will contribute in some way to accelerating progress towards curing more cancer patients.

Martin Clynes

Cytotechnology **12**: 1–32, 1993.

P-glycoprotein structure and evolutionary homologies

James M. Croop
Division of Pediatric Oncology, The Dana-Farber Cancer Institute and The Children's Hospital, Harvard Medical School, Boston, MA, USA

Key words: drug resistance, MDR, multidrug resistance, P-glycoprotein

Abstract

Analysis of multidrug resistant cell lines has led to the identification of the P-glycoprotein multigene family. Two of the three classes of mammalian P-glycoproteins have the ability to confer cellular resistance to a broad range of structurally and functionally diverse cytotoxic agents. P-glycoproteins are integral membrane glycoproteins comprised of two similar halves, each consisting of six membrane spanning domains followed by a cytoplasmic domain which includes a nucleotide binding fold. The P-glycoprotein is a member of a large superfamily of transport proteins which utilize ATP to translocate a wide range of substrates across biological membranes. This superfamily includes transport complexes comprised of multicomponent systems, half P-glycoproteins and P-glycoprotein-like homologs which appear to require 12 α-helical transmembrane domains and two nucleotide binding folds for substrate transport. P-glycoprotein homologs have been isolated and characterized from a wide range of species. Amino acid sequences, the similarities between the halves and intron/exon boundaries have been compared to understand the evolutionary origins of the P-glycoprotein.

Introduction

Resistance to chemotherapeutic agents remains the major obstacle in the successful therapy of human cancer. Many tumors are intrinsically resistant to many of the most potent cytotoxic agents used in cancer therapy. Other tumors, initially sensitive, recur and are resistant not only to the initial therapeutic agents but also to other drugs which the tumor has not previously been exposed. This resistance of human tumors to multiple chemotherapeutic agents has been difficult to study in the clinical setting. However, significant progress has been made in characterizing the phenomenon of resistance to multiple cytotoxic agents *in vitro* (Croop *et al.*, 1988; Endicott and Ling, 1989; Gottesman and Pastan, 1989; Roninson, 1991). Cell lines have been selected for resistance to anticancer agents, many of which are antibiotics originating in microorganisms or alkaloids from plants, by slowly increasing the concentration of the cytotoxic agent in a step-wise fashion. Derivative lines, many thousand fold resistant to the selecting agent, have been generated which are cross resistant to a broad spectrum of structurally and functionally diverse cytotoxic compounds. These multidrug resistant cell lines have provided an experimental model for the analysis of cellular resistance to multiple cytotoxic agents. An understanding of the biological basis of multidrug resistance has emerged which points to a unifying mechanism responsible for cross resistance to specific cytotoxic agents. Molecular analysis of the multidrug resistant cell lines has identified a gene family whose members are capable of conferring resistance to multiple cytotoxic agents and are included in a large superfamily of transport proteins.

Multidrug resistant cell lines were initially generated using anthracylines, colchicine, vinca alkaloids, or actinomycin D (Nielsen and Skovsgaard, 1992). The cell lines usually displayed the highest level of resistance to the selecting agent with lesser degrees of resistance to the other cytotoxic agents in these categories (Table 1). Although the compounds appear quite diverse in their structure and mode of action, many are natural products and can usually be categorized as small lipophilic cations (Beck, 1990). In addition, a broad range of compounds, which often share structural characteristics with these substrates, have been shown to be capable of reversing multidrug resistance (Ford and Hait, 1990). Several initial observations provided clues for understanding the mechanism of multidrug resistance. First, multidrug resistant cell lines accumulated less of the cytotoxic agents than nonresistant lines. The decreased accumulation was associated with an energy dependent outward efflux of the compounds (Dano, 1973; Skovsgaard, 1978). Secondly, Juliano and Ling (1976) identified a 170 kD membrane glycoprotein, termed P-glycoprotein (Pgp), overexpressed in multidrug resistant cell lines. The amount of P-glycoprotein in the plasma membrane appeared to correlate with the level of drug resistance. Finally, karyotypic analysis of multidrug resistant cell lines indicated the presence of double minute chromosomes or homogeneously staining regions, abnormalities consistently associated with gene amplification (Biedler and Riehm, 1970). These observations, consistent with overexpression of a gene product responsible for decreased accumulation of cytotoxic agents, directed the means for elucidating the mechanism of multidrug resistance.

Table 1. Compounds included in the multidrug resistant phenotype or which the P-glycoprotein transports

Doxorubicin	Actinomycin D
Daunorubicin	Puromycin
Mitoxantrone	Emetine
Colchicine	Ethidium bromide
Vinblastine	Gramacidin D
Vincristine	Mithramycin
Taxol	Lipophilic cations
Etoposide	Lipophilic peptides
Teniposide	^{99m}Tc-SESTAMIBI
Rhodamine 123	Verapamil

Substrates included in the multidrug resistance phenotype or known to be transported by P-glycoprotein.

Isolation cDNAs encoding the mammalian P-glycoprotein

Using a variety of strategies, genes overexpressed in multidrug resistant cell lines have been isolated and characterized. A series of cDNAs displaying a high level of nucleic acid similarity, each encoding a polypeptide with features of the P-glycoprotein have been identified. These cDNAs have been classified based on sequence similarity and have been designated members of the *mdr* or *pgp* multigene family (Juranka *et al.*, 1989) (Table 2).

Specific DNA sequences amplified in multidrug resistant cell lines were first identified using the technique of DNA renaturation in agarose gels (Roninson *et al.*, 1984). DNA sequences amplified in two independently derived hamster multidrug resistant cell lines were identified with a total length of at least 150 kbp. A 1.1 kbp BamHI fragment amplified in both cell lines was excised

Table 2. Classification and nomenclature of the mammalian P-glycoproteins

Class	I	II	III
Human	*mdr1*		*mdr2/3*
Mouse	*mdr3* *mdr1a*	*mdr1* *mdr1b*	*mdr2* *mdr2*
Hamster	*pgp1*	*pgp2*	*pgp3*

Classes I and II convey multidrug resistance.

from the gel and cloned. This probe indicated that the 1.1 kbp BamHI fragment was amplified in both cell lines and that the level of drug resistance correlated with the level of amplification. The fragment did not, however, hybridize to a mRNA species. Using the 1.1 kbp BamHI fragment, a contiguous 140 kbp domain amplified in the two hamster multidrug resistant cell lines was isolated (Gros *et al.*, 1986c). Probes spanning the amplified domain indicated that a mRNA species of approximately 5 kbp was overexpressed in the multidrug resistant cell lines. The hamster probes were used to isolated two distinct cDNAs from a nondrug resistant, mouse pre-B cell library (Gros *et al.*, 1986a). One of the murine cDNAs, *mdr1*, was capable of conferring the complete multidrug resistant phenotype when transfected into drug sensitive cell lines (Gros *et al.*, 1986a; Croop *et al.*, 1987). The second cDNA, *mdr2*, although highly homologous to the first, did not convey drug resistance (Gros *et al.*, 1988). A third murine cDNA, *mdr3*, also capable of conveying multidrug resistance, was subsequently isolated from the original pre-B cell library using probes from the previously cloned cDNAs (Devault and Gros, 1990).

Two distinct human genomic clones were isolated by cross hybridization with probes from the amplified hamster domain (Roninson *et al.*, 1986). One of these clones, human MDR1, identified amplified DNA sequences and a 4.5 kbp mRNA species overexpressed in multidrug resistant cell lines. MDR1 was subsequently used to screen a cDNA library derived from a multidrug resistant cell line to isolate a series of overlapping human cDNA clones which encoded sequences capable of conveying multidrug resistance (Chen *et al.*, 1986; Ueda *et al.*, 1987). Similarly, using an independently derived hamster *mdr* probe (Van der Bliek *et al.*, 1986), a novel set of cDNAs with high homology to human *mdr1* were isolated from a human liver cDNA library and designated *mdr3* (Van der Bliek *et al.*, 1987, 1988b). As with mouse *mdr2*, transfection studies indicated that human *mdr3*, was not capable of conveying multidrug resistance.

Using the C219 monoclonal antibody directed against the P-glycoprotein (Kartner *et al.*, 1985), a 600 basepair clone, λCHP1, was isolated from an expression library generated from a multidrug resistant hamster cell line (Riordan *et al.*, 1985). Molecular analysis of multidrug resistant cell lines with λCHP1 indicated that the gene encoding the P-glycoprotein was amplified and overexpressed a mRNA of approximately 4.7 kbp. Subsequently, three members of the hamster *pgp* multigene family have been isolated using clones derived from λCHP1 (Endicott *et al.*, 1991).

Analysis of RNA transcripts overexpressed in multidrug resistant cell lines similarly identified cDNAs encoding P-glycoprotein. Differential screening of cDNA libraries from drug sensitive and drug resistant cell lines identified a series of 5 genes overexpressed in the multidrug resistant cell lines (Van der Bliek *et al.*, 1986). All five genes where present on a single amplicon. Only the cDNA which cross hybridized with λCHP1, however, was consistently overexpressed in a variety of multidrug resistant cell lines (De Bruijn *et al.*, 1986). Similarly, using C_0t fractionated genomic DNA to generate a probe from amplified sequences, cDNAs encoding the P-glycoprotein were identified in a library generated from a multidrug resistant cell line (Scotto *et al.*, 1986).

Cross hybridization (Gros *et al.*, 1986c; Ueda *et al.*, 1986) and sequence comparisons indicate that the cDNAs isolated from the human, murine and hamster libraries are members of a multigene family which encodes the P-glycoprotein. Comparisons of intron sequences (Ng *et al.*, 1989), untranslated sequences (Endicott *et al.*, 1987; Hsu *et al.*, 1989) and coding regions (Hsu *et al.*, 1989; Endicott *et al.*, 1991) have identified three classes of mammalian P-glycoproteins as indicated in Table 2. The mammalian P-glycoprotein multigene families are clustered in tandem on a single chromosome, chromosome 7 in humans (Chin *et al.*, 1989; Raymond *et al.*, 1990; Lincke *et al.*, 1991). Classes I and II are capable of conveying multidrug resistance, while class III cannot. Systematic searches have never identified a human class II, suggesting that classes I and II represent a gene duplication occurring after the separation of murine and human evolution (Chin *et al.*, 1989; Ng *et al.*, 1989).

The predicted structure of the P-glycoprotein

The isolation of cDNAs encoding the P-glycoprotein enabled analysis of the structure of the polypeptide. The nucleic acid sequences of the cDNAs which encode the mammalian P-glycoproteins predict an integral membrane glycoprotein with remarkably similar structures (Fig. 1). The length of the mammalian P-glycoproteins vary from 1276 to 1281 amino acids predicting a molecular weight of approximately 140 kD. This is consistent with purified or labelled P-glycoprotein which migrates at a molecular weight of 130—180 kD in polyacrylamide gels. Although P-glycoprotein mobility in polyacrylamide gels is sensitive to the conditions employed (Greenberger *et al.*, 1988b), approximately 10—15 kD of the observed molecular weight is accounted for by N-linked glycosylation (Greenberger *et al.*, 1988a). Indeed, the predicted amino acid structure of the P-glycoprotein indicates 7—10 potential N-linked glycosylation sites [N-X-S/T].

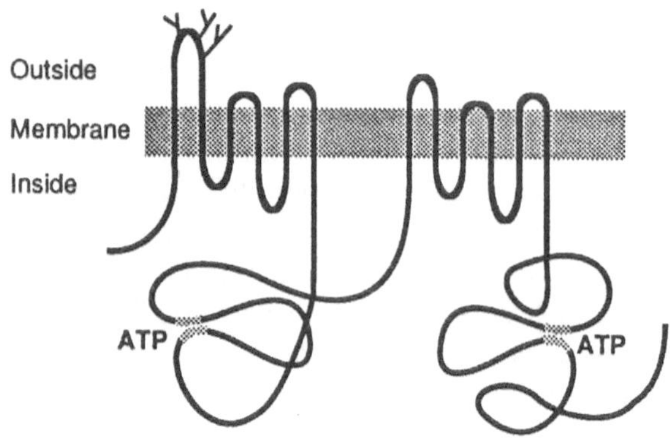

Fig. 1. Predicted structure and membrane orientation of the P-glycoprotein. Cluster of glycosylation sites in first external domain and nucleotide binding folds identified.

Examination of the amino acid sequence identifies a number of significant structural features (Figs. 1 and 2). Independent hydrophobicity analyses using the methods of Kyte-Doolitte (Gros *et al.*, 1986b) and Eisenberg (Chen *et al.*, 1986) predict a polypeptide which traverses the plasma membrane 12 times. A short, highly charged, cytoplasmic domain precedes three membrane loops which are followed by a large cytoplasmic domain. A cluster of the potential N-linked glycosylation sites are consistently located in the first external loop in virtually all P-glycoproteins. The large cytoplasmic domain includes the paired consensus sequences G-x(4)-G-K-[ST] and [RK]-x(3)-G-x(3)-L-[Hydrophobic](3) separated by approximately

```
Mdr5' M..E.................F...........K.K..........FRY..W.D.L.M..GT..A..HG..LPL.M..FG.MTD.F...............   100
Mdr3' ........R..S..........................P..SF.....LN..EWPY..VG..CA..NG..QP......S.....F...............   764

      ..............LEE.M..YAYYY.G.G.GVL..AYIQVS.W.LAAGRQI.KIR..FFHAI..QE.GWFD.....E..LNTRLTDD.SKI..GIGDK.   200
      ...................N.FSL.FL..G...F.T.F.QGFTFGKAGEILT.R.R.M.F..MLRQD.SWFD...N.TG.L.TRLA.D...V.GA....L   864

      GMFFQ...TF...FI.G..F..GWKLTLVI.A.SP..GLS...WAK.L..F..KEL.AYAKAGAVAEE...AIRTVIAFGGQ.KELERY...LE.AK..G   300
      A...QN.ANLGTG.I.S....YGWQLTLLL....P.I...G..EMK.L.G.A..DKK..E..GKIATEAIEN.RT.VSLT.E.KFE.MY...L..PYRN.   964

      IKKA..A.IS.G.A.LL.YASYALAFWYG..LV.S.EY..G...TVFF..L.G.FS.G...P.I..FANARGAA...F.IIDN.P.IDS.S..G.KPD..   400
      ...AH..GITF...QA..YFSYA.CFRFG.YL.....M.F..V.LVFSA.V.GA.A.G..SSFAPDYAKA..SA.......E..P.IDSYS..GL.P...   1064

      .GNL.F....F.YFSR....ILKGLNLKV.SGQTVALVG.SGCGK.TT.QL.QRLYDP..G....DGQDIR..NV..LREIIGVVSQEPVLF.TTIAENI   500
      EG...F....FNYPTR...PVLQGL..EVKKGQTL.LVGSSGCGKSTVVQLLERFY.P.AG.V.LDG.E...LN.Q..RA.LGIVSQEP.LFDCSIAENI   1164

      .YG..R..VTM.EI.KAVKEANAY.FIMKLP..FDTLVG.RGAQLSGGQKQRIAIARALVRNPKILLLDEATSALDTESEA.VQ.ALDKAR.GRTTIVIA   600
      AYGDNSR.V...EI..AA..ANIH.FI..LP.KY.T.VGDKGTQLSGGQKQR.AI.RAL.RQP..LLLDEATSALDTESEKVVQEALDKAREGRTC.VIA   1264

      HRLST.RNAD.IAGF..GVIVE.G.H.ELM...G.Y..LV..QT.G.....     651
      HRLSTIQNADLIVV..NG.VKEHGTHQQLLAQKGIYFSM...QAG......     1315
```

Fig. 2. Consensus sequence for mammalian P-glycoproteins. Mammalian P-glycoproteins listed in Table 3 were aligned using the UWGCG Pileup program (gap weight of 3.0 and a gap length weight of 0.1) and the identical amino acids listed using the single letter code. The proximal and distal halves were aligned using the Gap program (gap weight 5 and gap length weight 0.3). Dots represent nonidentical residues or gap. Transmembrane domains identified under the single lines and the nucleotide binding domains under the double lines (Endicott *et al.*, 1991). Shaded amino acids represent identity consensus of all P-glycoproteins described in Table III except *pfmdr1*, *STE6*, *mdl*, *ltpgpA*, *mrp* and *CFTR*. Numbering includes gaps.

120 amino acids which function as a nucleotide binding fold in a wide range of ATPases (Walker *et al.*, 1982). The cytoplasmic domain is followed by three additional membrane loops and another large cytoplasmic domain with the consensus sequence for a second nucleotide binding fold. Experimental evidence indicates that the P-glycoprotein binds and hydrolyzes ATP (Ambudkar *et al.*, 1992; Sarkadi *et al.*, 1992) which is required for drug transport (Horio *et al.*, 1988). In addition, epitope mapping of the monoclonal antibody C219 confirms the cytoplasmic localization of the hydrophilic domain (Georges *et al.*, 1990).

This model of the P-glycoprotein suggests that it is comprised of two halves, each with a similar structure. A comparison of the amino acid sequence of the carboxy and amino halves indicates approximately 40% identical and an additional 25% conserved amino acid homology which has suggested duplication of an ancestral gene (Fig. 3). The highest similarity between the moieties surrounds the nucleotide binding folds in the large cytoplasmic domains with almost 60% amino acid identity with an additional 20% conserved substitutions. When the two halves of the P-glycoprotein are aligned, a short segment termed the "linker" region (Van der Bliek *et al.*, 1987) which does not readily align with the corresponding region in the proximal half of the polypeptide appears to bridge the amino and carboxy halves. The linker regions of the class I and class II P-glycoproteins, which are capable of conveying drug resistance, each contain consensus sequences for cAMP- and cGMP-dependent protein kinase phosphorylation sites, while the class III members do not (Hsu *et al.*, 1989). Analysis of P-glycoprotein in vesicles from multidrug resistant cell lines indicates that phosphorylation occurs exclusively on serine residues including at least three in the "linker" region (Chambers *et al.*, 1993). Similar sites are found in the first cytoplasmic domain of the two human P-glycoproteins but not in the murine or hamster homologs. In addition, a large number of potential casein kinase II, tyrosine kinase and

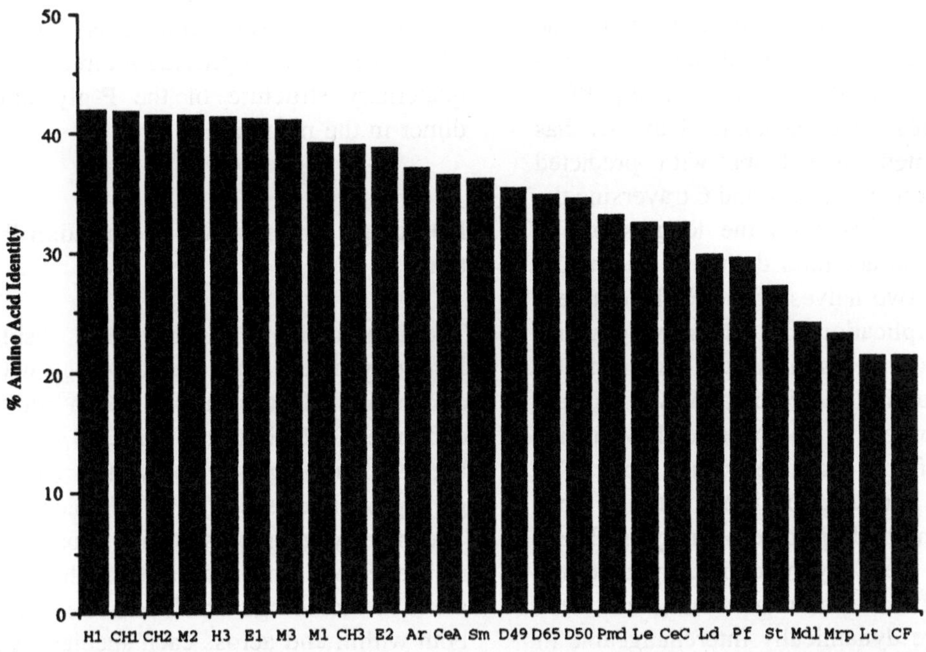

Fig. 3. Amino acid identity between the amino and carboxy half of the P-glycoprotein homologs. The two halves of the P-glycoproteins aligned using the UWGCG Gap program. Abbreviations for described homologs: H, human; CH, Chinese Hamster; M, mouse; E, *E. histolytica*; Ar, *atpgp1*; Ce, *C. elegans*; Sm, *S. mansoni*; D, *Drosophila*; Pmd, *pmd$_1$*; Le, *lemdr1*; Ld, *ldmdr1*; Pf, *pfmdr1*; St, *STE6*; Lt, *ltpgpA*; CF, *CFTR*.

protein kinase C phosphorylation sites span the length of all the members of the P-glycoprotein family.

Recently, the topography of the P-glycoprotein predicted by the hydrophobicity analysis has been questioned. An extracellular glycosylation site has been identified between predicted transmembrane domains 8 and 9 utilizing an *in vitro* translation assay (Zhang and Ling, 1991). Similarly, this region would be placed on the extracellular side of the plasma membrane based upon its localization within the lumen of the endoplasmic reticulum (Skatch *et al.*, 1993). These observations suggest that this domain is extracellular, not on the cytoplasmic side as indicated by the hydropathy analysis. Models placing predicted transmembrane domain 8 on the outside of the cell with predicted transmembrane 9 or 10 traversing the membrane back to the cytoplasmic side have been proposed to account for these observations. Only four loops would traverse the membrane in the carboxy half of the P-glycoprotein, i.e., predicted domains 7, 9 or 10, 11 and 12. Likewise, analysis of the amino half of the P-glycoprotein with the *in vitro* translation assay indicates that predicted transmembrane domain 3 is intracellular and predicted transmembrane domain 5 extracellular (Zhang *et al.*, 1993). This would suggest that the amino half also has only four transmembrane loops with predicted transmembrane domains 1, 2, 4 and 6 traversing the plasma membrane. Based on the lengths of the intracellular and extracellular domains, this model indicates that the two halves of the P-glycoprotein are not tandem duplications but rather topographical mirror images. Interestingly, the experimental data for both the amino and carboxy halves indicate that only a portion of the polypeptides have moieties comprised of four transmembrane loops. The remainder have a structure consistent with the model predicted by the hydrophobicity analysis. These data suggest that the P-glycoprotein may have two or more conformational structures. Whether these configurations are dynamically interchangeable in the plasma membrane, required for drug transport, or are associated with discrete functions remains to be clarified.

A growing body of evidence suggests that the P-glycoprotein may function as a dimer. Freeze fracture analysis of multidrug resistant cell lines demonstrates an increase in the density of intramembrane particles with a diameter predicting a molecular mass of 340 kD (Sehested *et al.*, 1989). The particles cluster in the presence of verapamil, although not doxorubicin, suggesting they may represent localization of the P-glycoprotein. These results are complemented by radiation inactivation analysis of membrane preparations from multidrug resistance cell lines indicating that the P-glycoprotein has a molecular mass of 250 kD, about twice the size of the unglycosylated polypeptide (Boscoboinik *et al.*, 1990). Finally, immunoprecipitation of P-glycoprotein from multidrug resistant cells incubated with a cross-linking agent identifies a minor band at 340 kD in addition to the major 180 kD species (Naito and Tsuruo, 1992). The 340 kD band is photolabelled with ^{3}H-Azidopine, but only the 180 kD species is detected when the cells are treated with a reversible cross-linking reagent. These observations provide indirect evidence for a quaternary structure of the P-glycoprotein as a dimer in the membrane.

Homology amongst the mammalian P-glycoproteins

The mammalian P-glycoproteins display a high level of primary structure homology with 69–94% amino acid identity and up to an additional 10% conserved substitutions (Figs. 4A and 5). The highest homologies exist within each of the P-glycoprotein classes (87–94% identical, Fig. 5). The family members which convey multidrug resistance are more similar to each other than they are to the homologs which do not convey resistance both within and across each species. A lower than

Fig. 4. Amino acid identity between each of the P-glycoprotein homologs and (A) human *Mdr1* or (B) *Drosophila Mdr49*. P-glycoprotein homologs in Table 3 were aligned using the UWGCG Pileup program and identical residues determined using the Distances program. Abbreviations as in Fig. 3.

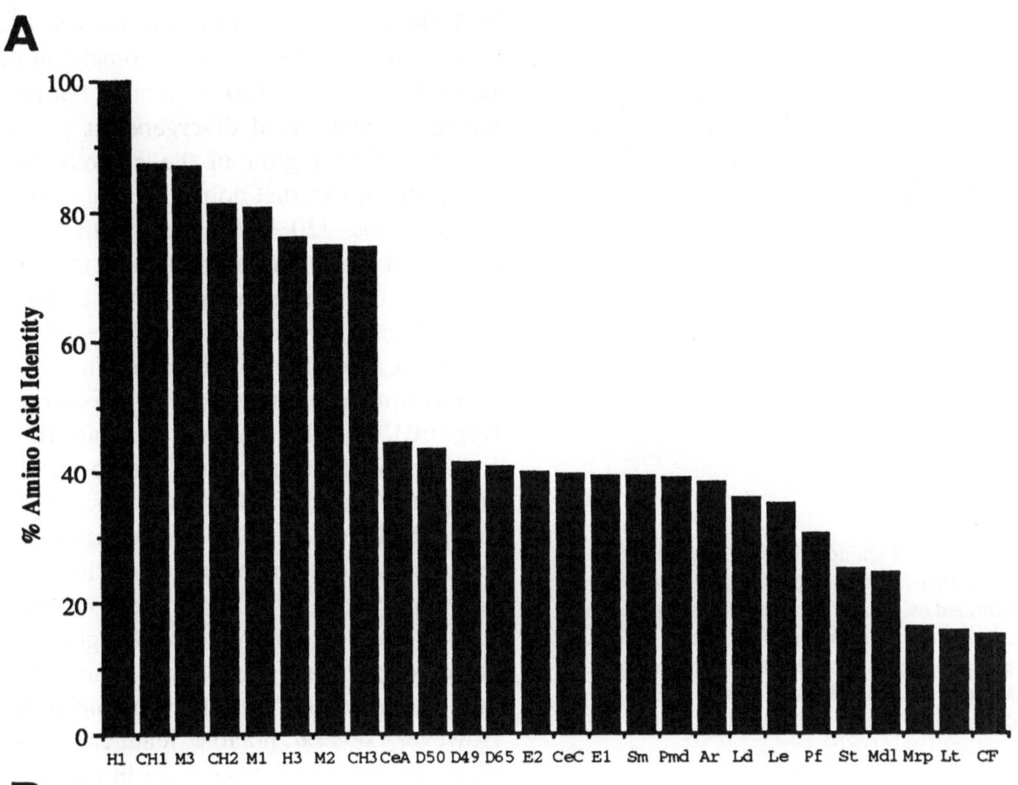
A
% Amino Acid Identity
100
80
60
40
20
0
H1 CH1 M3 CH2 M1 H3 M2 CH3 CeA D50 D49 D65 E2 CeC E1 Sm Pmd Ar Ld Le Pf St Mdl Mrp Lt CF

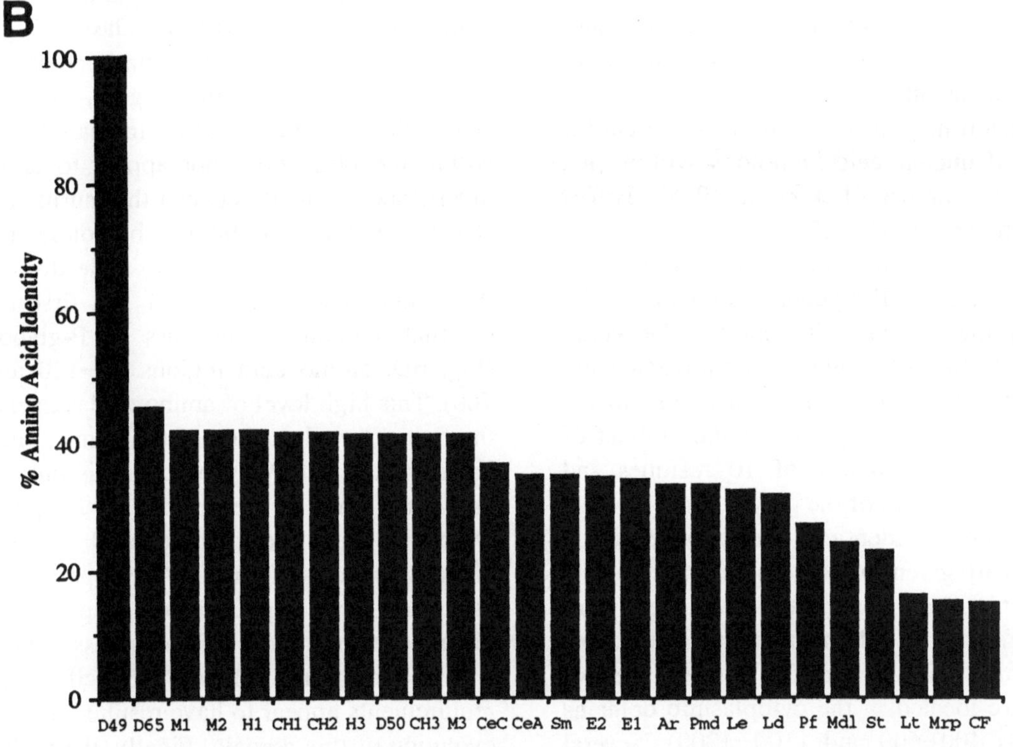
B
% Amino Acid Identity
100
80
60
40
20
0
D49 D65 M1 M2 H1 CH1 CH2 H3 D50 CH3 M3 CeC CeA Sm E2 E1 Ar Pmd Le Ld Pf Mdl St Lt Mrp CF

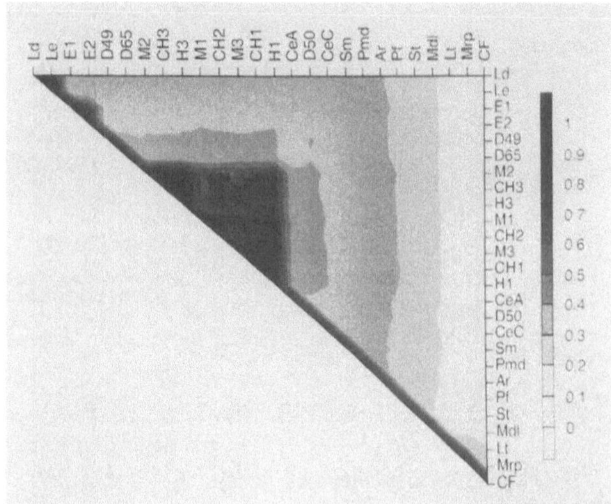

Fig. 5. Graphical analysis of amino acid identities among all P-glycoproteins. Percent identical residues of any two P-glycoproteins can be determined by comparing level of gray at the x and y axis intercept with the scale at the right. Scale indicates percent amino acid identity as a shade of gray from 0 to 1.0. Percent identical residues between any two P-glycoproteins determined as described in Fig. 4. Abbreviations as in Fig. 3.

expected silent nucleotide replacement has suggested that hamster class I and II may not have evolved independently from each other and that gene conversion may have occurred to account for high levels of nucleic acid homology within each of the species (Van der Bliek *et al.*, 1988b; Hsu *et al.*, 1989; Endicott *et al.*, 1991).

A graphical representation of the amino acid similarity among the P-glycoproteins can be displayed using the Plotsimilarity analysis (Devereux *et al.*, 1984). In this analysis, amino acids are compared at each position in a series of aligned polypeptides. A score is given for the amount of similarity within a window of 10 residues and plotted along the length of the polypeptide. A value of 1.5 is given for identical amino acids, while lesser values are given for homologies based on the Dayhoff amino acid substitution table. Plotsimilarity analysis of all mammalian P-glycoproteins (Fig. 6A) indicates that the highest levels of homology are located in the cytoplasmic domains (amino acids 400–600 and 1100–1300). Several notable regions account for the majority of the difference between the family members. The region with the largest amount of amino acid divergence is located in the first external domain, in the region of amino acids 80–120. A similar, but not quite as extensive, amount of divergence is present in the corresponding region in the carboxy half of the polypeptide, external domain four encompassed by amino acids 720–760. Finally, the cytoplasmic domain at the amino terminus and the corresponding region at the beginning of the carboxy half of the polypeptide (amino acids 640–700) are quite divergent among the P-glycoproteins.

Plotsimilarity analyses of P-glycoprotein class I (Fig. 6B), class II (Fig. 6C) and class III (Fig. 6D) indicates a similar pattern of amino acid homology along the polypeptide. As expected, based on higher homologies within each class of P-glycoprotein, amino acid homology along the entire length of the polypeptide is higher with values closer to 1.5. In addition, the regions of highest homology and divergence are located in similar regions found in the comparison of all the mammalian homologs. However, several notable features are observed. First, the degree of divergence in the first external domain present in class I and class II P-glycoproteins is almost as high as found across all of the classes (Fig. 6A). In this region, around amino acids 80–120, the general increased homology within the class does not appear to apply. This divergence is found between the murine and hamster as well as the human homologs (data not shown). Second, there is very little divergence in the amino acid sequence of the first or fourth external domain in the class III P-glycoproteins (Fig. 6D, amino acid regions 80–120 and 720–760). This high level of amino acid conservation in the first external domain of the human, mouse and hamster class III P-glycoproteins is quite remarkable since this is the region of the polypeptide which displays the highest level of divergence amongst the class I and II P-glycoproteins. This is also the same region which displays the most divergence among the different classes within each species (data not shown). Specific functional requirements appear to have resulted in differential evolution of this domain. Finally, the high level of homology along the entire length of the P-glyco-

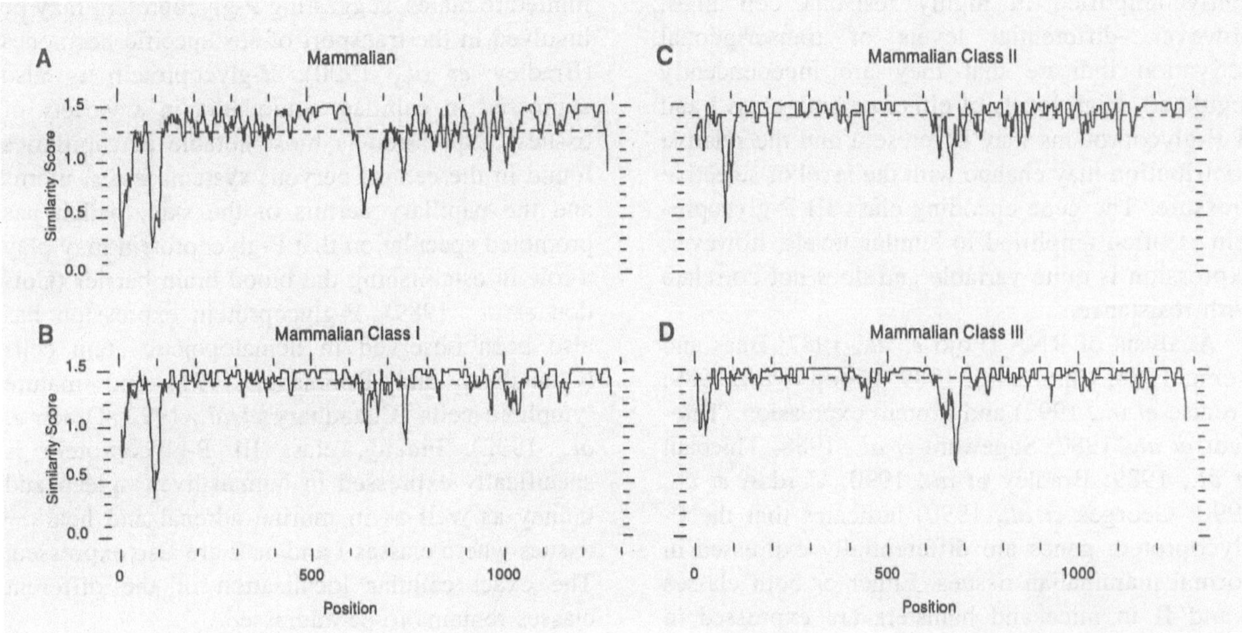

Fig. 6. Plotsimilarity analysis of mammalian (A), class I (B), II (C), and III (D) P-glycoproteins. Mammalian P-glycoproteins listed in Table 3 aligned with UWGCG Pileup program and the running average of the amino acid similarities plotted along the length of the polypeptides as described in text. Dashed line represents average similarity. Position in amino acids.

proteins, exemplified by scores approaching 1.5, is indicative that the analogous amino and carboxy halves of different P-glycoproteins display considerably higher similarity than the two halves of any individual P-glycoprotein, both within and across the different classes (compare Figs. 4A and 5 with Fig. 3).

P-glycoprotein expression

The identification of molecular and immunological probes for the different classes of P-glycoprotein provided a means to precisely examine their expression in multidrug resistant cell lines and normal tissues. Most multidrug resistant cell lines generated by stepwise selection overexpress P-glycoprotein at levels proportional to the amount of drug resistance. Characterization of human multidrug resistant cell lines indicates that at relatively low levels of resistance, *mdr1* transcripts are overexpressed prior to DNA amplification (Shen *et al.*, 1986; Lemontt *et al.*, 1988). As resistance increases with selection to higher drug concentrations, an increased number of *mdr1* gene copies are present. RNA overexpression, however, remains increased out of proportion to the amount of gene amplification, often exceeding a 10-fold excess in transcripts (Shen *et al.*, 1986; Van der Bliek *et al.*, 1988a). Although the *mdr3* gene is often amplified at levels similar to *mdr1*, RNA expression is usually not detected.

Expression of both class I and II P-glycoproteins correlates with multidrug resistance in murine cell lines (Hsu *et al.*, 1989; Raymond *et al.*, 1990). At relatively low levels of resistance, murine lines which overexpress class I P-glycoprotein have increased *mdr* transcripts prior to gene amplification. Highly resistant lines express more class I RNA than would be expected from the amount of DNA amplification, similar to the human cell lines. In multidrug resistant cell lines which overexpress class II P-glycoprotein, however, both RNA expression and gene amplification are proportional and roughly parallel the level of resistance. The genes for both class I and II P-glycoproteins are consistently amplified in highly resistant cell lines, however, differential levels of transcriptional activation indicate that they are independently regulated. High levels of either or both class I and II P-glycoproteins may be present and the relative contribution may change with the level of selective pressure. The gene encoding class III P-glycoprotein is often amplified to similar levels, however, expression is quite variable and does not correlate with resistance.

Analysis of RNA (Fojo *et al.*, 1987; Baas and Borst, 1988; Chin *et al.*, 1989; Croop *et al.*, 1989; Trezise *et al.*, 1992) and protein expression (Thiebaut *et al.*, 1987; Sugawara *et al.*, 1988; Thiebaut *et al.*, 1989; Bradley *et al.*, 1990; Cordon *et al.*, 1990; Georges *et al.*, 1990) indicates that the P-glycoprotein genes are differentially expressed in normal mammalian tissues. Either or both classes I and II in mice and hamsters are expressed in tissues where human *mdr1* is expressed. P-glycoprotein is consistently expressed at high levels at a variety of secretory surfaces including the bile canaliculi (Buschman *et al.*, 1992), the proximal tubules of the kidney, and intestinal and colonic epithelium. P-glycoprotein is also detected at the epithelial surfaces of the pancreas, bronchial mucosa, ovarian follicles and prostate. A dramatic increase in P-glycoprotein expression occurs in the murine uterus during gestation (Arceci *et al.*, 1988; Arceci *et al.*, 1990). The low levels normally present in the epithelium lining the uterine cavity and secretory glands gradually increase under the combined control of estrogen and progesterone to the highest levels found in mouse tissues. Expression of the cystic fibrosis gene (see below) appears to complement that of *mdr* in a variety of secretory epithelial tissues, including the uterus. The high levels of expression of the cystic fibrosis gene found in the nonpregnant uterine epithelium decrease to very low levels during pregnancy (Trezise *et al.*, 1992).

P-glycoprotein expression is high in other specialized tissues in addition to those with epithelial secretory surfaces. Very high levels of expression are found in the adrenal gland, primarily in the cortical regions. Observations in hamsters indicate that the high level of adrenal expression is limited to males, suggesting P-glycoprotein may be involved in the transport of sex specific hormones (Bradley *et al.*, 1990). P-glycoprotein is also expressed in capillary endothelia in a variety of tissues. Expression is most notable in capillaries found in the central nervous system, testes, uterus and the papillary dermis of the skin, which has prompted speculation that P-glycoprotein may play a role in establishing the blood brain barrier (Cordon *et al.*, 1989). P-glycoprotein expression has also been observed in hematopoietic stem cells (Chaudhary and Roninson, 1991) and mature lymphoid cells (Chaudhary *et al.*, 1992; Drach *et al.*, 1992). Finally, class III P-glycoprotein is specifically expressed in human liver, spleen and kidney as well as in murine adrenal and heart — tissues where classes I and/or II are also expressed. The exact cellular localization of the different classes remains to be addressed.

Unfortunately, analysis of P-glycoprotein expression in normal tissues has not provided significant insight on its normal physiological role. In addition, differences in molecular probes, immunological reagents and fixation protocols have produced a number of discrepancies between laboratories. Differences in expression between individual samples have also been noted. Finally, the sensitivity of the various techniques and reagents remain to be clarified and a general agreement on the best methodology for P-glycoprotein detection is lacking (Herzog *et al.*, 1992). Although quantitative analysis of RNA expression by the polymerase chain reaction appears to be the most sensitive method, the inclusion of connective tissue in the samples and potential lack of correlation between RNA and protein expression in cells not selected for drug resistance (Hill *et al.*, 1990; Wu *et al.*, 1992) may confound the analysis. Characterization of protein expression using immunological methods is reliable for detecting protein, but sensitivity at low levels of expression has not been clarified (Toth *et al.*, 1992). Differences in protein function due to differential glycosylation and phosphorylation are only now beginning to be addressed. It is clear that issues remain in characterizing P-glycoprotein expression and function in both normal and malignant tissues.

P-glycoprotein homologs

P-glycoprotein homologs have been isolated from a wide range of species. Analysis of cDNA and genomic clones indicate a high level of amino acid similarity and structural conservation, based on hydrophobicity analyses, present across large evolutionary distances. High levels of similarity even remain in the nucleic acid sequence in many of the distant homologs. Isolation of multiple P-glycoproteins and cross hybridization of genomic DNA to molecular probes indicate the presence of a multigene family in most organisms (Juranka *et al.*, 1989). Although specific functions have not yet been identified for many of the P-glycoprotein homologs, the variability in the number of family members and the divergence in the primary sequence suggest that this transport protein has readily adapted to a range of functional capacities. Recent observations on the ability of the human P-glycoprotein to function as an ATP (Abraham *et al.*, 1993) and a volume regulated chloride channel (Gill *et al.*, 1992) implies multiple mechanistic properties associated with the polypeptide and have challenged the traditional distinctions between transporters and channels (Jan and Jan, 1992). Figures 3, 5 and 7 show the homologies between the amino acid sequences of the P-glycoprotein homologs and should be referred to for the comparisons described in the text.

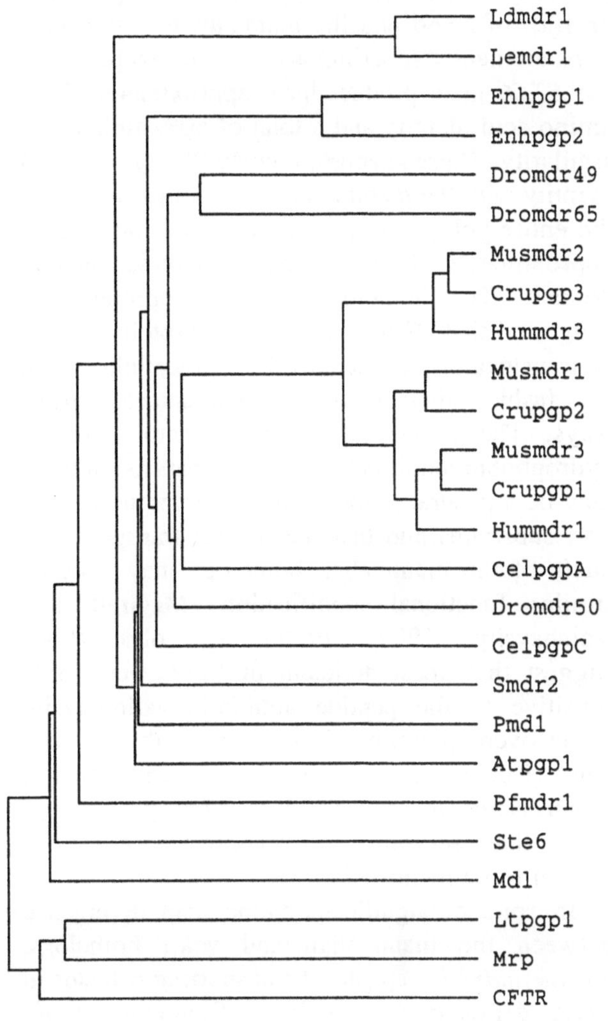

Fig. 7. Dendrogram of P-glycoproteins. Pair wise clustering of P-glycoproteins listed in Table 3 aligned with UWGCG Pileup program. Clusters described in the text: 1) Mammalian, 2) *Ltpgp1*, *Mrp*, *CFTR*, 3) *Pfmdr1*, *Ste6*, *Mdl*, and 4) remainder of homologs. See Fig. 5 for amino acid identity between homologs.

Saccharomyces cerevisiae

The yeast *Saccharomyces cerevisiae STE6* gene encodes a P-glycoprotein homolog (Kuchler *et al.*, 1989; McGrath and Varshavsky, 1989). *STE6* is one of several nonlethal "sterile" mutations in *MAT***a** cells which inhibit haploid *S. cerevisiae* mating and the formation of diploid cells. *MAT***a** cells secrete **a** factor, a 12 amino acid, peptide pheromone which induces conjugation. The **a** factor lacks a hydrophobic N-terminal signal sequence and sites for N-linked glycosylation suggesting that it is not exported using traditional secretory pathways. However, **a** factor does undergo post-translational farnesylation and methylation which render it lipophilic. Yeast carrying the *STE6* mutation produce **a** factor but are unable to secrete the peptide to the extracellular space while overexpression of *STE6* results in a striking increase in the amount of **a** factor released from the cells (Kuchler *et al.*, 1989).

Analysis of the predicted structure of the *STE6* protein product reveals a polypeptide with 1290 amino acids comprised of 12 membrane spanning domains and two cytoplasmic domains containing nucleotide binding folds. The hydrophobicity profiles of *STE6* and the mammalian P-glycoproteins are nearly superimposable. The two halves of the *STE6* gene product have approximately 27% amino acid identity and a total of 50% amino acid similarity. There is approximately 25% amino acid identity with the mammalian P-glycoproteins along the entire polypeptide, with the highest homology, approximately 40%, in the nucleotide binding domain. The homology to P-glycoprotein has suggested that *STE6* is a transport protein capable of translocating **a** factor to the extracellular space in a fashion distinct from classical secretory pathways. The requirement that **a** factor undergo hydrophobic modification implies that lipophilicity may be a general property of P-glycoprotein transport substrates and that the natural transport products of the mammalian homologs may undergo similar functional modification (McGrath and Varshavsky, 1989). Preliminary observations suggest that yeast deficient in *STE6* were more sensitive to the peptide antibiotic valinomycin, while overexpression of the gene results in resistance (Kuchler *et al.*, 1989). Cross hybridization of *S. cerevisiae* genomic DNA by *STE6* probes suggests that additional P-glycoprotein homologs remain to be identified.

In spite of significant amino acid divergence between the mammalian and yeast homologs, murine *mdr3* is capable of transporting **a** factor in *STE6* deficient *S. cerevisiae* (Raymond *et al.*, 1992). Although export is less efficient than *STE6*, functional homology across a large evolutionary distance appears to remain. Similar experiments with human *mdr1* indicate that the human homolog is not capable of **a** factor transport even though homology between the mammalian P-glycoproteins is much higher to each other than to *STE6* (Kuchler and Thorner, 1992). Expression of human *mdr1* in yeast did, however, convey resistance to valinomycin. These functional assays for the murine and human homologs expressed in *S. cerevisiae* should provide a useful system for rapid mutational analysis of the mammalian homologs. Such an analysis has provided new insights on the structural and functional requirements of *STE6* transport of **a** factor. Mutations generated in either the proximal or distal nucleotide binding domains of *STE6* inhibit export of **a** factor suggesting that each half of the polypeptide is required for specific functional activities (Berkower and Michaelis, 1991). Similarly, expression of either half of *STE6* is not sufficient for export of **a** factor. Co-expression of the two halves, however, does result in export of **a** factor. This observation suggests that the two halves can self assemble in the plasma membrane to form a functional *STE6* transporter. This probably reflects the underlying structural organization and functional interactions required of the hydrophobic and nucleotide binding domains required for P-glycoprotein and related membrane transport proteins (Fig. 8 and see below).

Schizosaccharomyces pombe

A P-glycoprotein homolog has been isolated from a leptomycin B resistant strain of *Schizosaccharomyces pombe* (Nishi *et al.*, 1992). Leptomycin B is an antifungal antibiotic isolated from a strain of *Streptomyces* which induces morphological abnormalities in a variety of fungi including the fission yeast *S. pombe*. A leptomycin B resistant strain of *S. pombe* identified after chemical mutagenesis has been utilized to characterize the genes responsible for resistance. A genomic library from the resistant line was cloned into a multicopy expression vector and introduced into a susceptible stain. Three distinct clones were identified which were capable of conveying leptomycin B resistance. Genetic analysis of the clone which conveyed the highest level of resistance suggested that it was not a mutant gene, but rather a wild-type gene able to confer resistance when carried on a multicopy plasmid. Overexpression of this plasmid resulted in resistance to leptomycin B, cycloheximide, valinomycin and staurosporine B but not actinomycin D. Disruption of the wild-type gene indicated that it was nonessential, however, the null mutant was hypersensitive to the agents to which it conveyed resistance as well as actinomycin D. Analysis of the clone indicated that an open reading frame consisting of 1362 amino acids with similarity to the predicted amino acid sequence of the P-glycoprotein was responsible for resistance. The gene

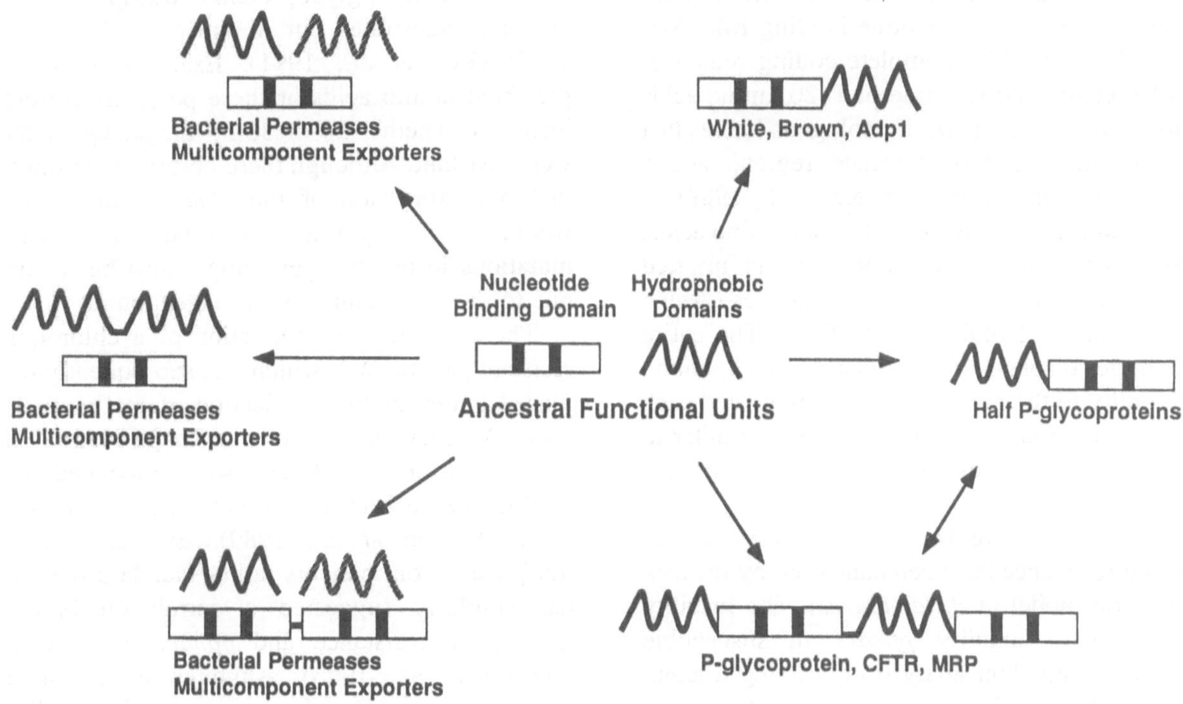

Fig. 8. Schematic representation of the evolutionary origins and organization of the functional units utilized in P-glycoprotein transport.

was designated *pmd1*⁺ for *S. pombe mdr1*-like gene. The primary structure predicted a polypeptide with two similar halves, each containing six transmembrane domains and a nucleotide binding domain. *Pmd1*⁺ has higher amino acid identity to the mammalian P-glycoproteins (37—40%) than to the yeast homolog *STE6* (24%).

Plasmodium falciparum

The rapid emergence of *Plasmodium falciparum* resistant to antifolate antimalarials has led to a dependence upon quinoline-containing compounds for the treatment and prophylaxis of malaria. Parasites resistant to these newer agents, however, are increasingly prevalent in most endemic areas of the world, posing a serious threat to control of the disease. Analysis of resistant *P. falciparum* initially suggested that the organisms displayed a phenotype similar to mammalian multidrug resistant cell lines. Isolates found in the wild were shown to be resistant to multiple antimalarials such as chloroquine, mefloquine and quinine (Martin *et al.*, 1987). In addition, *P. falciparum* isolates resistant to chloroquine were shown to efflux radiolabelled chloroquine 40—50 times more rapidly than susceptible strains, even though the initial uptake rates were similar (Krogstad *et al.*, 1987). Finally, verapamil and diltiazem, reversal agents of mammalian multidrug resistance, as well as vinblastine and doxorubicin all were shown to increase the accumulation of chloroquine in the resistant but not the susceptible isolates (Krogstad *et al.*, 1987; Martin *et al.*, 1987). These observations suggested that a P-glycoprotein homolog may have been involved in the resistance phenotype of *P. falciparum*.

Evidence for P-glycoprotein homologs in *P. falciparum* was obtained by screening genomic libraries using degenerate oligonucleotide probes to sequences in the highly conserved nucleotide binding fold (Foote *et al.*, 1989) and analysis of polymerase chain reaction products which were designed to span the nucleotide binding fold (Wilson *et al.*, 1989). The complete coding region of *pfmdr1* reveals a polypeptide of 1420 amino acids in length consisting of two homologous halves that each contain six transmembrane regions and a nucleotide binding domain (Foote *et al.*, 1989). A stretch of approximately 50 additional amino acids, not present in other P-glycoproteins, is inserted between the two consensus sequences which comprise the nucleotide binding folds. The entire polypeptide displays approximately 30% amino acid identity to the mammalian homologs and each of the two homologous halves are more similar to the mammalian homologs than they are to each other.

Characterization of the role of *pfmdr1* in antimalarial resistance has been hampered by the lack of a genetic model in *P. falciparum.* The inability to transfect or amplify *pfmdr1* in susceptible isolates has limited the analysis of defining whether the gene plays a direct role in drug resistance. Conflicting evidence has further confounded its role in drug resistant *P. falciparum.* Over half of the analyzed chloroquine resistant isolates have amplified *pfmdr1* (Triglia *et al.*, 1991; Wilson *et al.*, 1993). Pulsed field electrophoresis indicate that *pfmdr1* is amplified as a head to tail, tandem array of repeats on chromosome five where the gene is normally located. Amplification of *pfmdr1* in resistant isolates correlates with overexpression of two stage specific transcripts (Volkman *et al.*, 1993). A 8.5-kbp transcript is expressed in both ring and trophozoite stages, while a 7.5-kbp transcript is only expressed in trophozoites. Similarly, quantitative immunoblotting indicates that the protein is not overexpressed in resistant isolates unless the gene has been amplified (Cowman *et al.*, 1991). However, the gene product does localize to the digestive vacuole, where chloroquine normally accumulates, suggesting that it may modulate intracellular chloroquine concentrations.

Analysis of the *pfmdr1* gene in chloroquine resistant isolates indicated the presence of two alleles correlating with resistance (Foote *et al.*, 1990). Amino acid substitutions occur at locations where independent mutations affect the ability of the mammalian P-glycoprotein to transport specific cytotoxic substrates out of the cell (Choi *et al.*, 1988; Gros *et al.*, 1991). Examination of the predicted amino acids at these positions correctly identified whether 34 of 36 *P. falciparum* isolates were resistant. Although these observations suggest that overexpression of the *pfmdr1* gene product would not be required for chloroquine resistance, mutations in another gene might also be required due to the low frequency of alleles found.

The original characterization of a chloroquine resistant isolate, W2, which had subsequently been placed under stepwise selection of mefloquine *in vitro*, W2mef, indicated that amplification and overexpression of *pfmdr1* was associated with mefloquine resistance but not chloroquine resistance (Wilson *et al.*, 1989). Evaluation of *P. falciparum* from patients failing standard therapies has similarly suggested a correlation between mefloquine resistance and *pfmdr1* amplification (Wilson *et al.*, 1993). Although there was no correlation between the presence of the alleles described above and chloroquine resistance, amplification of the *pfmdr1* gene was found in 10 of the 11 isolates. All 10 of the isolates which had amplified *pfmdr1* were resistant to all compounds tested. The only isolate which did not display amplification of *pfmdr1*, however, was sensitive to mefloquine and halofantrine.

A role for *pfmdr1* in conveying resistance to antimalarials in *P. falciparum* could not be found utilizing a genetic cross between chloroquine resistant and sensitive parasites (Wellems *et al.*, 1990). The cross was made between a chloroquine sensitive line, HB3, containing a single copy of *pfmdr1*, and a resistant line, Dd2, a derivative of W2mef, which contains four copies. Gametocytes from the two lines were produced *in vitro*, fed to mosquitoes which subsequently fed on a chimpanzee. Individual progeny of the cross were isolated from the blood by limiting dilution. Eight chloroquine resistant and eight sensitive isolates were identified and characterized *in vitro*. Each of the progeny clones displayed a sensitive or resistant phenotype identical to one of the parents. There were no phenotypes of intermediate resistance. Neither the number of copies of Dd2 *pfmdr1*, or even the presence of Dd2 *pfmdr1* correlated with chloroquine resistance. Resistant and sensitive isolates where just as likely to receive *pfmdr1* from either parent. Similarly, there were no significant differences among the parent or progeny clones in the level of mefloquine resistance. Analysis of the sensitive HB3 parent, however, has indicated that it has one of the alleles associated with chloroquine resistance and in addition, may be somewhat resistant to mefloquine (Foote *et al.*, 1990). The analysis of this cross could be complicated if the HB3 *pfmdr1* gene has the ability to convey resistance. Although this cross does not appear to implicate *pfmdr1*, it does suggest, however, that a single genetic locus is responsible for chloroquine resistance.

Leishmania

Leishmania is an intracellular parasite responsible for a range of clinical syndromes including cutaneous, mucosal and visceral infiltration. The treatment of choice remains pentavalent antimonials. One mechanism of resistance in *Leishmania* selected with heavy metals, methotrexate, primaquine, or terbinafine correlates with amplification of H circles (Ellenberger and Beverley, 1989; Ouellette *et al.*, 1990). H circles are extrachromosomal genetic elements comprised of a 30 kbp inverted repeat separated by unique DNA segments which encode at least 20 polyadenylated RNAs. Amplification occurs from pre-existing circles or the *de novo* generation of H circles from a chromosomal copy. The chromosomal copy contains only a single copy of the duplicated region indicating that the circles cannot be formed by a simple excision (Ouellette *et al.*, 1991). A P-glycoprotein homolog, *ltpgpA*, was first identified in amplified H-circles from *L. tarentolae* using cross hybridization to a mammalian *mdr1* probe (Ouellette *et al.*, 1990). Cross hybridization of genomic DNA with molecular probes suggests that there are at least five P-glycoprotein homologs. Sequence analysis of *ltpgpA* indicates a continuous open reading frame of 1548 amino acids encoding a polypeptide with two similar halves, each containing six transmembrane domains and a nucleotide binding fold. Homology to mammalian P-glycoproteins extends only to the 200 amino acids surrounding the nucleotide binding domain as indicated by only 15% amino acid identity with the entire polypeptide. There does not appear to be significant homology between the two halves of *ltpgpA* outside of the nucleotide binding domain with only 21% amino acid identity between the halves. In addition, the amino terminus is much longer than the corresponding region in the mammalian homologs. These differences, as well as closer homology to other P-glycoprotein-like transport proteins (Figs. 5 and 7, see below) have suggested that *ltpgpA* may be distinct from the other P-glycoproteins (Cole *et al.*, 1992).

Isolation and expression of four DNA fragments spanning the H region of *L. major* have implicated *lmpgpA*, the homolog of *ltpgpA*, as conveying resistance to the heavy metals arsenite, antimony and lead (Callahan and Beverley, 1991). Introduction of the H region fragment which includes *lmpgpA* and further amplification by co-selection with the neomycin resistance gene confers heavy metal resistance to the parasite. There was no increase in resistance to cytotoxic agents included in the mammalian multidrug resistant phenotype such as vinblastine and puromycin. Resistance was associated with a two fold decrease in arsentite accumulation. Similarly, when clones which included *ltpgpA* were introduced into the parasites, a similar pattern of heavy metal resistance was obtained although the level was approximately 2—4-fold less than with the *lmpgpA* constructs. Similarly, a P-glycoprotein-like transport protein in yeast, HMT1, has recently been shown to convey resistance to heavy metals (Ortiz *et al.*, 1992). Analysis of an H circle in *L. tarentolae* amplified during selection with arsenite revealed a novel fusion of *ltpgpA* and a second P-glycoprotein homolog designated *ltpgpB* which resides just outside the chromosomal location of the H region (Ouellette *et al.*, 1991). The exact alignment of the sequences of the P-glycoprotein homologs suggests that the H circle was formed by intrachromosomal homologous recombination between *ltpgpA* and *ltpgpB*.

Selection of *L. donovani* with increasing concentrations of vinblastine have successfully generated parasite lines cross resistant to vinblastine, daunorubicin and puromycin (Henderson *et al.*, 1992). The resistant parasites display markedly decreased uptake of radiolabelled puromycin. Drug resistance, however, was not reversed with either verapamil or quinidine even though the parasites displayed a multidrug resistant phenotype similar to mammalian multidrug resistant cells. Probes were generated to determine whether amplification of a leishmanial P-glycoprotein homolog correlated with drug resistance. Two polymerase chain reaction products designed to span a nucleotide binding fold were isolated and characterized. The product with approximately 50—55% amino acid identity to the murine nucleotide binding domain indicated that homologous genomic sequences were amplified 50-fold in the multidrug resistant parasites. Hybridization with the product which displayed 35—40% amino acid identity and a probe from *ltpgpA* revealed no evidence of amplification. Northern blot analysis indicated that a 12.5 kbp mRNA was overexpressed in the resistant parasites. A 5.7 kbp genomic clone from *L. donovani* with the complete coding sequence of *ldmdr1* was isolated using the polymerase chain reaction product, transfected into *Leshmania* and amplified by co-selection with the neomycin resistance gene. The transfected parasites displayed the same multidrug resistant phenotype as the original resistant organisms. Similar results have been obtained by selecting *L. enriettii* with vinblastine (D. Wirth, personal communication). An analogous gene, *lemdr1* (Genbank accession L080-91), has been isolated from *L. enriettii* which is amplified on a 35—40 kbp extrachromosomal circle in the resistant organisms. Analysis of the predicted amino acid sequences reveal polypeptides with a hydrophobicity profile similar to the mammalian P-glycoproteins, 83% amino acid identity with each other but only 24% identity to *ltpgpA*. There is approximately 37% amino acid identity to the mammalian homologs indicating that *ldmdr1* and *lemdr1*, which convey multidrug resistance to hydrophobic substrates, are much closer evolutionarily to the mammalian P-glycoproteins than to *ltpgpA* (Figs. 4, 5 and 7).

Drosophila

Three members of the *Drosophila mdr* multigene family have been isolated and characterized. The P-glycoprotein homologs have been designated relative to their chromosomal location, *Mdr49* on chromosome two, *Mdr65* on chromosome three (Wu *et al.*, 1991) and *Mdr50* also on chromosome two, but far from *Mdr49* (M. Dean, personal communication). Each of the *Drosophila* genes are localized at different chromosomal locations, in contrast to the mammalian multigene family which is clustered on a single chromosome. *Mdr49* and *Mdr65* were isolated from a *Drosophila* head cDNA library using a murine *mdr1* cDNA probe which included the sequences for the nucleotide binding fold. *Mdr50* was isolated using a probe generated from a polymerase chain product encompassing an nucleotide binding domain. The predicted amino acid sequence for all three P-glycoprotein homologs indicate that the polypeptides have two homologous halves, each with six transmembrane domains and a nucleotide binding fold. The *Drosophila* homologs display approximately 40—45% amino acid identity to the mammalian homologs. However, the nucleic acid and amino acid sequences of the *Drosophila* P-glycoproteins are as divergent from each other as they are from the mammalian homologs, in contrast to the mammalian *mdr* families which display a high level of intraspecies similarity (compare Figs. 4A and B). Preliminary observations indicate that *Mdr49* and *Mdr65* are differentially expressed at the secretory surfaces of tissues analogous to mammalian structures where P-glycoprotein is expressed (R. Arceci and J. Croop, unpublished data).

Mdr49 is located in a chromosomal region which has been the subject of previous genetic analysis (Wu *et al.*, 1991). The precise genetic location of *Mdr49* relative to the deficiencies with known chromosomal breakpoints in the region was determined using probes spanning the cDNA. Three deficiencies were identified which had chromosomal breakpoints within the *Mdr49* gene. vg^{135} had a breakpoint in the proximal coding region, vg^{C} near the center of the gene and vg^{138} probably just distal to the coding region. In addition, the *b8* line deleted the entire gene. Each of the deficiencies is a large chromosomal deletion which is homozygous lethal. However, the identification of deleted and disrupted *Mdr49* genes in the previously generated deficiencies allowed genetic crosses resulting in flies lacking a wild-type *Mdr49* gene. Utilizing the balancer chromosome, *CyO*, which gives flies curly wings, vg^{135}/*CyO* and vg^{C}/*CyO* flies were mated with *b8/CyO* flies. The flies with straight wings would be the vg^{135}/*b8* and vg^{C}/*b8* progeny with *Mdr49* deleted on the *b8* chromosome and disrupted on the other. Molecular analysis indicated loss of chromosomal material and truncated mRNA species consistent with the predicted genetic localization.

Functional analysis of the mutant *Mdr49* fly lines indicated that they were fertile and viable under laboratory conditions. The lack of a lethal phenotype was consistent with previous saturation mutagenesis in the region which had not identified any vital genes in the location of *Mdr49*. The effect of losing the *Mdr49* gene in flies stressed with a cytotoxic agent was ascertained by determining the survival of flies developing on colchicine. Fertilized eggs were placed on media containing increasing concentrations of colchicine and the number of adult flies emerging determined for 19 randomly chosen control lines and the flies bearing a disrupted *Mdr49*. The vg^{C}/*b8* flies were approximately two fold more sensitive to colchicine than the controls and could easily be distinguished from survival of any of the other lines at 10 μM colchicine. Neither the vg^{135}/*b8* or the heterozygous vg^{135}, and vg^{C} and *b8* lines showed an increased sensitivity to colchicine. Thus the loss of the wild-type *Mdr49* in and of itself did not cause the heightened colchicine sensitivity during development. It is possible, however, that the remaining portion of the disrupted *Mdr49* gene gives rise to a truncated or fusion protein or RNA species which is deleterious to development on colchicine. The vg^{C} RNA transcript from the flies with increased sensitivity includes almost half of the *Mdr49* coding sequence while the vg^{135} transcript is very short and unlikely to provide deleterious information. Isolation and expression of the *Mdr49* gene from the vg^{C} chromosome should help elucidate whether these possibilities are responsible for the increased sensitivity to colchicine in the flies with a disrupted *Mdr49*.

Entameoba histolytica

Entameoba histolytica is an enteric protozoan responsible for amoebic dysentery in man. Although drug resistance has not been a major therapeutic problem, case reports of failed drug treatment have been described. *E. histolytica* mutated with ethylmethane-sulphonate and selected in emetine have resulted in a resistant clone with a stable phenotype. Probes generated from polymerase chain reaction products predicted to span the nucleotide binding fold identified four distinct DNA fragments containing P-glycoprotein-like sequences (Samuelson *et al.*, 1990). The probes identified 4.5–5 kbp mRNAs overexpressed in the emetine resistant line. When the resistant line was selected with increasing concentrations of emetine, higher levels of mRNA expression were observed. The sequences for two P-glycoprotein homologs, *Enhpgp1* and *Enhpgp2*, have been characterized in detail (Descoteaux *et al.*, 1992). The two homologs are 1302 and 1310 amino acids in length displaying a higher level of amino acid identity to each other, 67%, than to other P-glycoprotein homologs. This is in contrast to *Drosophila*, *Leishmania* and *C. elegans* where similarities between P-glycoproteins within the same organism are no greater than the similarities to homologs in other species. Each of the *E. histolytica* homologs display approximately 38–40% amino acid identity with the mammalian P-glycoproteins. The predicted structure is very similar to the mammalian P-glycoprotein except that the amino terminus is not as hydrophilic. Two pseudogenes of the P-glycoprotein were also identified.

Arabidopsis thaliana

A P-glycoprotein homolog has been identified in the plant *A. thaliana* (Dudler and Hertig, 1992). Degenerate oligonucleotides corresponding to highly conserved amino acid sequences in the nucleotide binding fold were used to screen a genomic library. Two distinct classes of genomic sequences were obtained, and one *atpgp1*, further characterized. The predicted polypeptide was 1286 amino acids in length consisting of two similar halves, each comprised six transmembrane domains and a cytoplasmic domain with the consensus sequences for a nucleotide binding domain. The two halves displayed 37% amino acid identity with each other and 36—39% amino acid identity with the mammalian P-glycoproteins. Northern blot analysis indicated that *atpgp1* was expressed in all parts of the plant examined, including the leaves, roots, flower buds and open flowers. Highest levels of expression of the 4.5 kbp transcript were found in the peduncles.

Caenorhabditis elegans

Four homologs have been identified in the *C. elegans pgp* multigene family (Lincke *et al.*, 1992). Three sets of genomic clones, which each mapped to a different chromosome, were isolated using a human *mdr1* probe. Further analysis of the clones indicated that four P-glycoprotein gene homologs were present in the three genetic loci. cDNAs for two of the homologs, *Celpgpa* and *Celpgpc*, have been characterized in detail. They encode polypeptides 1321 and 1254 amino acids in length. Each is comprised of two homologous halves containing six transmembrane domains and a cytoplasmic region with a nucleotide binding fold. The hydrophobicity plots of both homologs are virtually superimposable on that of human *mdr1*. The two homologs, however, have only 40% amino acid identity with each other and 38—40% amino acid identity with the mammalian P-glycoproteins. The two genes have 14 and 13 exons, respectively. Although the *C. elegans* homologs share only one intron/exon junction, there are four and five intron positions, respectively, conserved with the mammalian *mdr* genes. Partially processed RNA transcripts for one of the P-glycoprotein homologs increased with growth of the nematodes at heat shock temperatures over an 8-h period. There were no increases with arsenite or actinomycin D, however, prolonged exposure to emetine appears to have resulted in increased expression. Transformation of nematodes with expression constructs which fuse the promotor regions of the two *gpg* genes with *lacZ* coding region indicate that expression of both genes is restricted to the intestinal cells (Lincke *et al.*, 1993).

Other P-glycoprotein homologs

A *P-glycoprotein* homolog has been isolated from *Escherichia coli* and designated *Mdl* (Genbank accession L08627). It has a similar structure to the P-glycoproteins except that the amino end appears truncated with only five transmembrane domains. There is approximately 20—25% amino acid identity with the other P-glycoprotein homologs. Partial sequences for two P-glycoproteins have been isolated from winter flounder using hamster *pgp* probes and have highlighted the difficulty in relating specific homologs to the mammalian classes (Chan *et al.*, 1992). Finally, a P-glycoprotein homolog, *Smdr2*, has been isolated from the parasite *Schistosoma mansoni* using a polymerase chain reaction product generated to span a nucleotide binding domain (I. Bosch and C. Schoemaker, personal communication).

Evolutionary homologies

A comparison of the amino acid sequence of the P-glycoprotein to the protein sequences in the genetic databases indicates homology to a rapidly expanding group of polypeptides in both prokaryotes and eukaryotes. The region of homology is highest in the approximately 200 amino acids which surround the nucleotide binding domains. The vast majority of these polypeptides transport molecules across biological membranes and have been designated the ATP binding cassette (ABC) transporters (Hyde *et al.*, 1990) or traffic ATPases (Ames *et al.*, 1992). A signature pattern for this superfamily of proteins, distinguishing them from other ATPases, consists of the amino acids [LIVMFY]-S-[SG]-G-x(3)-[RKA]-[LIVMYA]-x-LIVMF]-[AG] situated between the paired consensus sequences of the nucleotide binding domain (Bairoch, 1992). This consensus is manifested as the sequence LSGGQKQ-

RIAIA in virtually all of the P-glycoproteins (Fig. 2). In most members of this large group of polypeptides, however, the similarity is more extensive and in many, extends to predicted transmembrane domains. These transport proteins can be categorized into four major subgroups: multicomponent transport systems which include a small hydrophilic polypeptide with a nucleotide binding domain which either 1) shuttle substrates into bacteria, termed the periplasmic permease systems or 2) export substrates and toxins out of microorganisms, 3) polypeptides approximating one half of a P-glycoprotein with a hydrophobic and hydrophilic domain and 4) transport proteins with a structure similar to the P-glycoprotein comprised of two similar halves each containing transmembrane domains and a nucleotide binding domain (Fig. 8). A variety of configurations are capable of transporting similar substrates in different organisms.

Bacterial permeases

The bacterial periplasmic permeases provide an energy dependent transport system for the uptake of nutrients in Gram negative bacteria (Ames *et al.*, 1990; Higgins, *et al.*, 1990). These multicomponent systems provide a mechanism for specific sugars, metals, amino acids, peptides, or cofactors which have traversed the outer membrane into the periplasmic space to cross the inner cell membrane (Table 3 and Fig. 8). The permeases are characteristically comprised of four individual polypeptides encoded in one or two operons. Each individual system includes a periplasmic substrate-binding protein which is specific for the substrate. Although there is little primary amino acid homology amongst the substrate-binding proteins, a striking similarity in the tertiary structure has been deduced by x-ray crystallography (Adams and Oxender, 1989). When bound to the substrate, the binding proteins undergo a conformational change which allows interaction with the membrane bound components of the system. The membrane bound components of the transport system are comprised of two hydrophobic polypeptides and a third member with a nucleotide binding domain. This latter polypeptide displays a high level of similarity to the analogous member of each permease as well as the entire cytoplasmic domain of the P-glycoprotein. The nucleic acid sequence encoding the region around the nucleotide binding domain even displays nucleic acid homology to the mammalian P-glycoproteins in some of the permeases such as *potA* of the spermidine and putrescine transport system, and *braG* of the branched chain amino acid transport system (data not shown). The two hydrophobic polypeptides are each comprised of five or six putative transmembrane domains with approximately 25—30% identical and an additional 25% conserved amino acid homology with each other. In a high affinity *mycoplasma* transport system, the hydrophobic proteins appear to be fused together as a single polypeptide which is consistent with the proposed function of the hydrophobic domains as pseudodimers (Higgins *et al.*, 1990). Some of the hydrophobic domains in different permeases display homology to each other and in some cases (*upgE*, *araJ*, *oppB*) even very faint homology to the P-glycoprotein (data not shown). Although the domain with homology to the cytoplasmic region of P-glycoprotein does not include a large number of hydrophobic residues, it appears to be membrane bound on the cytoplasmic side of the inner membrane (Ames, 1986). In several of the permease systems this domain with the nucleotide binding fold is duplicated (*mglA*, *rbsA*, *araG*) and in the *opp* system there are two similar polypeptides with nucleotide binding domains, both of which are required for transport. ATP hydrolysis is tightly coupled to the transport of the substrates and presumably results in conformational changes which allow translocation of the substrate across the membrane. The entire permease complex, comprised of polypeptides containing twelve membrane spanning regions associated with two nonhydrophobic polypeptides which include nucleotide binding domains, suggests a functional unit quite similar to the predicted structure of the P-glycoprotein.

Multicomponent export systems

Hydrophilic polypeptides with homology to the cytoplasmic region of the P-glycoprotein act as

Table 3. Polypeptides with amino acid similarity to the P-glycoprotein. Annotated BLAST search of the combined protein databases using human *Mdr1* as the query sequence. Results separated into categories described in the text. Original BLAST retrieved 225 entries. Duplicates, similar operons in different species and strains, partial sequences and certain open reading frames (ORF) with unknown functions removed. *Lemdr1* was not in protein database but included in all other analyses. References are available from database designations

P-glycoproteins		Smallest Poisson Probability P(N)
SP:MDR1_HUMAN	Human Mdr1	0.0
SP:MDR1_CRIGR	Hamster Pgp1	0.0
SP:MDR3_MOUSE	Mouse Mdr3	0.0
SP:MDR1_MOUSE	Mouse Mdr1	0.0
SP:MDR2_CRIGR	Hamster Mdr2	0.0
PIR:JH0502	Rat Mdr1b	0.0
SP:MDR3_CRIGR	Hamster Mdr3	0.0
SP:MDR2_MOUSE	Mouse Mdr2	0.0
SP:MDR3_HUMAN	Human Mdr3	0.0
PIR:S19979	C. elegans CelpgpA	7.4e–217
GP:ENHPGP2A_1	E. histolytica Enhpgp2	4.1e–208
PIR:A41249	Drosophila Mdr49	8.0e–197
PIR:S19980	C. elegans CelpgpC	4.1e–184
GPU:DROMDR50A_1	Drosophila Mdr50	6.8e–181
PIR:S21957	A. thaliana Atpgp1	4.6e–180
GP:ENHPGP1A_1	E. histolytica Enhpgp1	2.0e–179
PIR:B41249	Drosophila Mdr65	9.7e–165
PIR:S20548	S. pombe PMD1+	7.9e–154
GP:LEILDMDR1A_1	Leishmania Lemdr1	1.8e–121
GPU:ECOMDL_1	E. coli Mdl	2.8e–79
SP:STE6_YEAST	S. cerevisiae STE6	6.5e–67
SP:MDR_PLAFF	P. falciparum Pfmdr1	2.6e–60
GPU:HUMMRPX_1	Human MRP	2.6e–23
SP:MDR_LEITA	Leishmania LtpgpA	3.6e–16
SP:CFTR_HUMAN	Human CFTR	1.7e–15

(continued)

members of multicomponent transport systems involved in the export of a variety of substrates from microorganisms (Table 3 and Fig. 8). *Streptomyces* which produce doxorubicin and daunorubicin express two genes located in tandem which convey resistance to these antibiotics in heterologous hosts (Guilfoile and Hutchinson, 1991). One of the polypeptides is hydrophilic and contains a nucleotide binding domain, *drrA*, and the other hydrophobic with six potential transmembrane domains, *drrB*. Another resistance operon in *Streptomyces* encodes *oleC*, a polypeptide comprised of a nucleotide binding domain, and a second hydrophobic polypeptide with six putative transmembrane domains. Together these polypeptides encode resistance to oleandomycin. Similarly, polypeptides containing two nucleotide binding domains linked in tandem have been identified in *Streptomyces* which produce macrolide antibiotics (Schoner *et al.*, 1992). These polypeptides act as specific resistance determinants in *Streptomyces* producing spiramycin, *srmB*, carbomycin, *carA*, and tylosin, *tlrC*, and appear to transport the antibiotics out of the organisms. Other potential members of the transport systems have not yet been identified.

Bacterial operons responsible for the export of antibiotics have also been identified. Host resistance to the peptide, antibiotic microcin B17 is encoded in the *mbcF* gene in *Enterobacteriaceae*. The MbcF polypeptide includes a single nucleotide

Table 3. (continued)

Half P-glycoproteins		Smallest Poisson Probability P(N)
SP:MSBA_ECOLI	MsbA bacterial growth >32°C	5.2e–85
SP:NDVA_RHIME	NdvA beta-(1→2)Glucan export	3.6e–68
SP:HETA_ANASP	HetA heterocyst differentiation	1.2e–65
GP:YSPHMT1M_1	HMT1 heavy metal resistance	2.4e–65
PIR:S18855	ClyI-B toxin export	2.7e–62
SP:HLYB_ECOLI	HlyB toxin export	3.7e–62
SP:LKTB_PASHA	LtkB toxin export	1.0e–61
SP:CHVA_AGRTU	ChvA Beta-(1→2)Glucan export	5.4e–61
SP:CYAB_BORPE	CyaB toxin export	2.8e–60
PIR:B41538	PSF-2 MHC peptide transporter	1.5e–50
PIR:A39203	ComA bacterial competence	2.2e–44
PIR:A41538	PSF-1 MHC peptide transporter	1.4e–42
GP:LACLCN_2	LcnC bacteriocin export	1.0e–38
SP:CVAB_ECOLI	CvaB colicin export	2.3e–36
GPU:ECOAAT_1	MdrA tRNA transferase *E. coli*	2.8e–28
GP:BACSPA_2	SpaB bacteriocin export	1.6e–21
SP:PRTD_ERWCH	PrtD protease export	8.5e–20
GP:LLNISI_4	NisT lantibiotic nisin export	8.5e–18
GP:PDCBCTCN_3	PA-1 bacteriocin export	8.4e–15
SP:ADP1_YEAST	Adp1 ORF with EGF motif	3.4e–06
SP:PMP7_RAT	Pmp70 peroxisome protein	8.0e–06
SP:WHIT_DROME	White Drosophila	8.4e–06
GP:HUMPMP_1	Pmp70 peroxisome protein	0.00086
GP:PSESYRD_1	SyrD syringomycin resistance	0.0051
SP:BROW_DROME	Brown Drosophila	0.037

(continued)

binding domain and interacts with a hydrophobic polypeptide to export microcin B17 from the cytoplasmic to the periplasmic space (Garrido *et al.*, 1988). The biosynthesis operon of the peptide, lantibiotic epidermin from *Staphylococci* similarly encodes a polypeptide with a nucleotide binding domain presumably responsible for export (Schnell *et al.*, 1992). Polypeptides consisting of duplicated nucleotide binding domains linked together encode an inducible mechanism of erythromycin resistance, *msrA* and virginiamycin resistance, *vgA*, in *Staphylococci*. MsrA has been shown to function by decreasing accumulation of the antibiotic (Ross *et al.*, 1990). The hydrophilic and hydrophobic domains of these multicomponent units presumably interact to transport the toxic antibiotics out of the organisms. Although, the structural organization of the efflux complexes have not yet been clarified, the similarity of the components to those found in bacterial permeases suggest that the two may be similar. Understanding the mechanism of substrate influx versus toxin efflux will certainly provide insight into the transport mechanisms of the P-glycoprotein.

Multicomponent transport systems responsible for the transport of substrates which form the bacterial capsule and are responsible for microbial virulence have similarly been identified. The *bexA* gene encodes a hydrophilic polypeptide containing a nucleotide binding domain and *bexB* a polypeptide with six potential transmembrane domains in *H. influenzae* type b (Kroll *et al.*, 1988). Both are essential for polysaccharide export to the capsule and disruption of this complex results in loss of capsule formation. The polysaccharide capsule of *N. meningitides* requires the function of a similar operon, *ctr*. Similarly, *kpsT* and *kpsM* encode analogous proteins which are required for polysialic

Table 3. (continued)

Bacterial Permeases		Smallest Poisson Probability P(N)
SP:PROV_ECOLI	ProV glycine beteine uptake	4.7e–17
SP:HISP_SALTY	HisP histidine uptake	8.5e–16
SP:OPPF_BACSU	OppF oligopeptide uptake	3.9e–15
SP:POTA_ECOLI	PotA spermidine uptake	4.8e–15
SP:P29_MYCHR	P29 hypothetical mycoplasma	7.5e–15
GP:M225334S2_1	LacK lactose uptake	2.5e–13
SP:CYSA_ECOLI	CysA sulfate uptake	9.6e–13
GP:S77552_3	NocP octopine uptake	2.6e–12
SP:SFUC_SERMA	SfuC iron uptake	1.3e–11
SP:UGPC_ECOLI	UgpC glycerol-3-phosphate uptake	2.6e–11
PIR:A39741	CycV cytochrome c biogenesis	2.9e–11
SP:BRAF_PSEAE	BraF branched-chain AA uptake	7.2e–11
SP:NOSF_PSEST	NosF copper uptake	1.7e–10
SP:GLNO_BACST	GlnQ glutamine uptake	5.0e–10
SP:ARAG_ECOLI	AraG arabinose uptake	5.4e–10
SP:LIVG_ECOLI	LivG leucine uptake	1.0e–09
SP:PSTB_ECOLI	PstB phosphate uptake	2.0e–09
SP:PHNC_ECOLI	PhnC alkylphosphonate uptake	2.1e–09
SP:FECE_ECOLI	FecE iron (III) uptake	5.8e–09
SP:DCID_BACSU	DciD dipeptide uptake	1.0e–08
SP:MALK_ECOLI	MalK maltose uptake	1.1e–08
GP:S83898_1	MsmK multiple sugar uptake	1.1e–08
SP:MGLA_ECOLI	MglA galactose uptake	1.5e–08
SP:FEPC_ECOLI	FepC ferric enterobactin uptake	8.6e–07
SP:MODC_ECOLI	Modc molybdenum uptake	6.4e–06
SP:OPPD_SALTY	OppD oligopeptide uptake	8.0e–06
SP:RBSA_ECOLI	RbsA ribose uptake	1.2e–05
GP:LACLEUILV_5	Ilv5 branched chain AA uptake	4.6e–05
SP:BTUD_ECOLI	BtuD vitamin B12 uptake	0.44

(continued)

acid incorporation in the capsule of *E. coli* K1, the causative agent of neonatal septicemia and meningitis. These transport systems appear represent a generalized mechanism to translocate substrates across the cytoplasmic membrane for insertion in the capsule. Knowledge of these systems in pathogenic bacteria provide a new target for antibiotics by inhibition of substrate transport, analogous to the action of reversal agents in multidrug resistant cells.

Several systems which may not transport substrates also include hydrophilic domains with homology to P-glycoprotein. The nodulation protein NodI from *Rhizobium*, which is required for the bacteria to form root nodules on leguminous host plants, includes a similar nucleotide binding domain (Rossen *et al.*, 1984). The *ftsE* gene is required by *E. coli* for cellular division. Within this operon is an open reading frame for a hydrophilic polypeptide which contains the consensus sequence for a nucleotide binding domain (Gill *et al.*, 1986). The ABC excinuclease of *E. coli*, an ATP dependent DNA repair enzyme, is made up of three subunits, including UvrA with a structure comprised of a duplication with nucleotide binding domains in each half (Husain *et al.*, 1986). Similarly, elongation factor-3 of *S. cerevisiae*, which appears to transduce energy for stimulation of aminoacyl tRNA binding to the ribosome, has a duplicated structure with a nucleotide binding domain in each

Table 3. (continued)

Multicomponent exporters		Smallest Poisson Probability P(N)
GP:STMCARA_1	CarA carbomycin resistance	1.1e–10
PIR:S21597	SrmB spiromycin resistance	5.9e–09
GP:STMDAUN_2	DrrA daunorubicin resistance	1.9e–08
GP:S110314_1	VgA viriniamycin resistance	7.9e–07
GP:STMOLEC_4	OleC oleandomycin resistance	2.0e–06
SP:MSRA_STAEP	MsrA erthromycin resistance	4.8e–05
SP:TLRC_STRFR	TlrC tylosin resistance	0.00068
GP:STAEPIDNA_1	EpiY epidermin resistance	0.037
SP:BEXA_HAEIN	BexA polysaccharide to capsule	0.051
SP:YCP5_SYNY3	Cpn60 chaperone ORF	0.071
GP:STAMSRA_2	MsrA erythromycin resistance	0.12
SP:XPSU_ECOLI	KpsU polysialic acid to capsule	0.53
SP:MCBF_ECOLI	MbcF microcin B17 resistance	0.85
PIR:S15223	CtrD polysaccharide to capsule	0.87
Others		
GP:BACSPOOK_7	SpoOK sporulation & competence	3.8e–15
GPU:ODNCPOSATO_1	Chloroplast ATPase ORF	1.2e–09
SP:YNTR_RHIME	NtrA/RpoN sigma factor	1.6e–08
SP:MBPX_MARPO	MbpX chlorplast ORF	2.2e–08
GPU:STRSTRA_1	FimA fimbral protein	6.0e–08
SP:NODI_BRAJA	NodI nodulation protein	1.7e–06
SP:FTSE_ECOLI	FtsE cell division protein	3.3e–06
SP:EF3_YEAST	EF-3 elongation factor 3	6.5e–06
PIR:PS0228	Infl initiation factor 1	3.4e–05
SP:CHVD_AGRTU	ChvD nodulation protein	4.7e–05
SP:UVRA_ECOLI	UvrA excinuclease subunit A	0.12

half (Qin *et al.*, 1990). These diverse examples suggest that the functional unit used to bind and hydrolyze ATP in the transport systems can be utilized for ATP hydrolysis in a variety of cellular functions. This observation has further suggested that the hydrophobic domains of the transport complexes are responsible for substrate specificity and translocation.

Half P-glycoproteins

Polypeptides approximating one half of a P-glycoprotein, consisting of a series of transmembrane loops followed by a hydrophilic region containing a potential nucleotide binding domain, have been identified in both prokaryotes and eukaryotes (Table 3 and Fig. 8). A series of prokaryotic transport proteins which export virulence toxins (HlyB, LtkB, CyaB, ClyI-B, CyaB), polysaccharides (NdvA, ChvA), antibiotics (CvaB, SpaB, LcnC, SyrD, NisT), protease activities (PrtD), heavy metals (Hmt), and the toxic compound aminotriazole (Atr1) have been identified. In addition, prokaryotic polypeptides which are involved in the formation of a polysaccharide layer during heterocyst differentiation (HetA), growth in temperatures greater than 32°C (MsbA), competence for genetic transformation (ComA) and association with a tRNA protein transferase (MdrA) have been shown to have a similar structure. The hydrophobicity plots of each of these polypeptides are very similar to either half of the P-glycoprotein. A comparison of the primary structure indicates an identical amino acid similarity to the P-glycoprotein ranging from 25–35% with an additional 20%

conserved residues spanning the entire polypeptide. Similarity is highest in the hydrophilic domain with 40–45% identical amino acids. Significant nucleic acid similarity is even present between the mammalian P-glycoproteins and several of the prokaryotic sequences including *hlyB*, *cyaB*, *msbA*, *lcnC*, *chvA* in the range of 45% surrounding the nucleotide binding domain.

Recently, two mammalian polypeptides have been implicated in the transport of short peptides from the cytosol into the endoplasmic reticulum where they are bound by the major histocompatibility class I molecules for antigen presentation (Bahram *et al.*, 1991). Each of these polypeptides appears to represent half of a P-glycoprotein with eight putative transmembrane domains followed by a cytoplasmic domain containing the consensus sequence for a nucleotide binding fold. Comparison of the amino acid sequences of the two peptide transporters indicates approximately 40% amino acid identity with an additional 20% conserved residues. Similarity is highest surrounding the nucleotide binding domain and decreases in the proximal regions. Comparison of the amino acid sequence of the human peptide transporters to human *mdr1* indicates approximately 30% identical and an additional 25% conserved amino acid residues. Both of the peptide transporters are slightly more similar to the carboxy half of the human P-glycoprotein. The genes encoding these peptide transporters are located within 10 kbp of each other in the major histocompatibility class II region. The two polypeptides appear to assemble and function as a heterodimer with both transporters required for the formation of stable class I molecules and antigen presentation (Kelly *et al.*, 1992). Again, the predicted functional structure appears similar to the proposed structure of the P-glycoprotein.

A second group of mammalian half P-glycoproteins are associated with peroxisomes. A partially deleted half P-glycoprotein gene has been identified in some patients with adrenoleukodystrophy (ALD) (Mosser *et al.*, 1993). The principal abnormality in ALD is the accumulation of very-long-chain fatty acids due to impaired oxidation in peroxisomes. The candidate gene is suspected of translocating a component of the oxidative metabolic pathway into the organelle. This gene has highest similarity, 39% amino acid identity, with another 70 kD peroxisomal membrane protein, *Humpmp*, of unknown function (Kamijo *et al.*, 1992). It, too, however, is presumed to transport substrates into the peroxisome since the nucleotide binding domain is localized to the cytoplasmic side of the organelle. Both peroxisome transporters have approximately 20–25% amino acid identity with the peptide transporters as well as human *mdr1*, most of which surrounds the nucleotide binding domain.

The *Drosophila* white and brown gene products are structurally related to the P-glycoprotein. Each consists of a nucleotide binding fold followed by a series of transmembrane domains (O'Hare *et al.*, 1984; Dreesen *et al.*, 1988). The products of these genes have been proposed to transport pigment precursors in the eye, possibly functioning as heterodimers (Dreesen *et al.*, 1988). Similarly, an open reading frame identified during the sequencing of the yeast genome, Adp1, predicts a polypeptide with a similar structure which includes an epidermal growth factor motif. The transposition of the transmembrane and nucleotide binding domains relative to the organization found in the P-glycoprotein and half P-glycoprotein-like polypeptides would suggest that the functional units found in the bacterial permeases originally had significant latitude as building blocks in forming transport proteins (Fig. 8).

P-glycoprotein-like transporters

Analysis of the gene responsible for cystic fibrosis indicates that it has amino acid homology and a similar structure to the P-glycoprotein (Riordan *et al.*, 1989). The amino acid sequence of the cystic fibrosis transmembrane conductance regulator (CFTR) predicts a polypeptide 1480 amino acids long with two repeat motifs, each which consists of six membrane spanning domains and a hydrophilic region containing the consensus sequences for a nucleotide binding fold. The two halves of the polypeptide display only 21% amino acid identity with each other and 15% amino acid identity with the mammalian P-glycoproteins. The only homology surrounds the nucleotide binding domain. A large, highly charged cytoplasmic domain of 241 amino acids, termed the R group, contains a cluster of consensus sequences for protein kinase A and protein kinase C phosphorylation sites and links the two halves of the molecule at the analogous position of the "linker" region in the P-glycoprotein.

A number of cell lines displaying a cross resistant phenotype similar to multidrug resistant cell lines but which do not overexpress the P-glycoprotein have been isolated by stepwise selection in doxorubicin (Mirski *et al.*, 1987; McGrath and Center, 1988; Baas *et al.*, 1990; Slapak *et al.*, 1990). Recently, a multidrug resistance-associated protein, MRP, has been identified in one such cell line by differential hybridization (Cole *et al.*, 1992). Molecular analyses indicate that the gene encoding MRP is amplified and overexpressed in the resistant cell line. Although the line displays resistance to multiple agents, cyclosporin A does not reverse the resistance and major differences in net drug accumulation do not appear to be part of the resistant phenotype. Analysis of the predicted amino acid sequence of MRP indicates homology to the P-glycoprotein with two similar halves each containing a nucleotide binding domain. The hydrophobicity analysis, however, was used to assign eight transmembrane domains in the proximal half of the polypeptide and four in the distal half. Other predictions are possible including up to 12 transmembrane domains in the proximal and six in the distal halves. This cluster of potential transmembrane domains at the amino terminus is distinct from all of the other P-glycoprotein homologs. Although the proximal and distal halves are similar, the homology is primarily located around the nucleotide binding domain and the two halves are not as similar to each other as most P-glycoprotein homologs (Fig. 3). MRP displays highest similarity to LtpgpA and CFTR with approximately 30 and 24% amino acid identity, respectively, with only about 15% amino acid identity to any of the P-glycoproteins (Fig. 5). These observations have suggested that MRP, LtpgpA and CFTR are more closely related to each other than the other P-glycoprotein homologs and represent a separate grouping which originated from different precursors or diverged before the separation of plants and animals (Cole *et al.*, 1992). Of note, however, the nucleic acid sequence surrounding the nucleotide binding domain is more similar to the P-glycoproteins than LtpgpA or CFTR (data not shown).

The predicted structure of the P-glycoprotein is similar to a variety of channels and transport proteins but lacks substantial similarity to their amino acid sequences. A consensus structure consisting of 12 α-helical, transmembrane spanning domains has been proposed for a wide range of transport systems which require of energy coupling (Malone, 1990). Primary sequence comparisons have identified a putative "major facilitator superfamily" comprised of drug resistance proteins, sugar facilitators, Krebs cycle intermediates which symport H^+, phosphate ester-phosphate antiporters and oligosaccharide-H^+ symporters with this structure (Marger and Saier, 1993). In addition, this generalized structure is found in adenylyl cyclase which is comprised of two similar halves each containing six transmembrane domains and a large hydrophilic region with a cyclic nucleotide binding domain (Kupinski *et al.*, 1989). Voltage-gated sodium (Noda *et al.*, 1986) and calcium channels (Tanabe *et al.*, 1987) are comprised of four similar groups each including six transmembrane spanning domains. These hydrophobic regions are interspersed with cytoplasmic domains predicting a structure not dissimilar to a pair of linked P-glycoproteins. The P-glycoprotein, however, does not display significant homology at the amino acid level to these transport proteins and channels. Whether the lack of amino acid homology with structural homology represents divergence from the ancestral origins of the transport proteins or convergence to motifs which perform similar functional activities is unknown.

Evolutionary origins of the P-glycoprotein

The P-glycoprotein is the member of a large superfamily of transport proteins which move a wide range of substrates across plasma membranes. The general structure of 12 membrane spanning domains and two hydrophilic moieties which include nucleotide binding domains appears similar across large evolutionary distances. However, the composition of the functional units comprising the structure varies and a range of combinations can direct transport in and out of the cell as well as across different organelles. Cytotoxic compounds are exported not only by P-glycoproteins, but also by multicomponent transport systems and half P-glycoproteins in a variety organisms.

Although the functional domains of the P-glycoprotein are observed in both prokaryotes and eukaryotes, the precise evolutionary origins have been a matter of speculation. The observation that the two halves of the P-glycoprotein displayed a high level of amino acid homology originally suggested that the polypeptide represented gene duplication of a primordial gene. Identification of the intron/exon boundaries of the human *mdr1* (Chen *et al.*, 1990), *mdr3* (Lincke *et al.*, 1991) and mouse *mdr1* (Raymond and Gros, 1989) P-glycoproteins have provided more information on the possible evolutionary mechanisms involved. Each of the mammalian homologs contains 28 exons with introns placed at homologous sites along the gene. The most notable feature is that the majority of the boundaries in the proximal and distal halves of the gene do not occur at the corresponding locations in the two halves. Only two intron boundaries are located at identical nucleotides in the two halves and occur within the nucleotide binding domain. Another intron boundary in a hydrophobic region occurs at the same amino acid. The lack of similarity between the halves suggested that the P-glycoprotein may have originated from two genes with common ancestors evolving separately prior to fusion (Chen *et al.*, 1990). However, since both moieties have a similar number of exons, over 50% nucleic acid identity in the coding regions and the exons appear to encode functional domains, the theory for gene duplication has similarly been advanced (Raymond and Gros, 1989). Further insight has been provided with determination of the intron/exon boundaries of the *A. thaliana* homolog (Dudler and Hertig, 1992). Nine introns have been identified, eight in the proximal half and one in the distal half. None of the introns in the proximal half match the position of the intron in the distal half. Four of the introns correspond to similar positions in the mammalian gene including two in the proximal half aligning with the introns which are identical in the nucleotide binding domain. In addition, however, three introns in one half of the *A. thaliana* homolog correspond to introns in the opposite half of the mammalian gene. Preliminary observations in the *C. elegans* (Lincke *et al.*, 1992) and *Drosophila* homologs (J. Croop, unpublished data) suggest a similar conservation of introns in one half of the homolog with introns in the contralateral half of the mammalian gene. These latter observations suggest that a duplicated primordial gene could have differentially lost introns to account for the variability between the proximal and distal halves of the mammalian P-glycoprotein. This would also explain the similarity found in intron location across different halves in the evolutionarily divergent homologs (Dudler and Hertig, 1992). Such a scenario, however, might be expected to lead to exons with larger than average size in the mammalian gene, which is not the case (Chen *et al.*, 1990).

Examination of the similarity among the currently described P-glycoproteins reveals several distinct groupings which presumably have evolutionary significance. The mammalian P-glycoproteins can be identified as a single group with over 65% identity amongst all of the homologs as indicated in Figs. 5 and 7. Within each species, the mammalian P-glycoproteins display very high levels of similarity in the range of 72–85%. A similar level of homology is present between the strain specific *lemdr1* and *ldmdr1*, but is not found among the homologs in other organisms. The two *E. histolytica* approach this level of homology with 68% identity, but the homologs in *Drosophila* and *C. elegans* are as divergent from each other as they are from the other P-glycoproteins (Fig. 5). Interestingly, most of the homologs, except as noted above, display a slightly higher homology to a mammalian P-glycoprotein than any other homolog (data not shown). The recently described clustering of homologies among *ltpgp1*, *Mrp* and *CFTR* can similarly be seen in Figs. 5 and 7 (Cole *et al.*, 1992). However, note that the amino acid identities among these three homologs are only 20–30% compared to approximately 15% identity with each of the other P-glycoprotein homologs. The majority of the remaining homologs display over 30% amino acid identity to one another. The exceptions are *pfmdr1*, *Ste6* and *Mdl* which appear to represent a separate grouping with only 20–30% amino acid identity with the other homologs. These four groups are similarly distinguished in the pairings found in Fig. 7.

Direct comparisons of the P-glycoprotein homologs to human *mdr1* support these groupings. Figure 4A clearly indicates the high level of homology to the mammalian P-glycoproteins, a lesser but similar level of homology to each of the remaining homologs with the exception of *Ste6* and *Mdl* and *Mrp*, *ltpgp1* and *CFTR* which segregate into two discrete groups. The *pfmdr1* homolog is intermediate in homology to *Ste6* and *Mdl* and the majority of homologs. An similar result is obtained comparing *Drosophila Mdr49* to the other homologs (Fig. 4B). In addition, a BLAST search of the genetic databases (Altshul *et al.*, 1990) to identify proteins with homology to human *mdr1* indicates a similar clustering of the homologs. Table 3 indicates that the probability approaches zero for a chance occurrence to have identified human *mdr1* and the other mammalian homologs as similar. The other homologs are clustered as described above with *pfmdr1*, *Ste6*, and *Mdl* and *ltpgp1*, *Mrp*, and *CFTR*, and the remainder of the homologs defining three distinct levels in the probability of being similar to human *mdr1*. These distinctions are highlighted by the fact that many of the half P-glycoproteins are more likely than *pfmdr1*, *Ste6*, *Mdl*, *ltpgp1*, *Mrp* or *CFTR* to have matched human *mdr1*. Furthermore, BLAST searches of the genetic databases with nucleic acid sequences from the nucleotide binding domains of the mammalian P-glycoproteins identify all of the homologs except *Ste6*, *Mdl*, *ltpgp1* and *CFTR* (data not shown). However, a number of half P-glycoproteins and multicomponent transport genes are identified. Finally, BLAST searches with the nucleic acid sequences from the carboxy half of the mammalian P-glycoproteins consistently match many more of the homologs as well as other polypeptides with evolutionary homology than searches with the amino end.

Characterization of the differences in homology between the proximal and distal halves of the P-glycoprotein homologs is also consistent with the assignment of these groupings. The mammalian P-glycoproteins display the highest levels of homology between the two halves with approximately 40% identical amino acid residues. The majority of the remaining homologs are 30–40% identical, while *ltpgp1*, *Mrp* and *CFTR* are 21–23% identical. The two halves of *pfmdr1*, *Ste6* and *Mdl* range from 24 to 29% identity. Differential levels of similarity between the two halves suggest that the structural or functional requirements have evolved over time. Unfortunately, the normal physiological roles for most P-glycoproteins are unknown so it is difficult determine the significance of these differences. Since, in general, the carboxy and amino halves are more similar to the respective halves among P-glycoprotein homologs (data not shown), either similarity remains to the original ancestral moieties which fused to form the P-glycoprotein or divergence between the two halves of a duplicated gene occurred prior to further radiation. Whether or not these and similar comparisons will further delineate distinct groupings or a graded continuum of differences between the homologs awaits the identification of additional P-glycoproteins.

Acknowledgements

I would like to thank Dr. Robert Arceci for critically reviewing the figures, and along with Drs. Chris Slapak and Ting Wu for reviewing the manuscript. This work was supported by NIH Grant CA48162 and American Cancer Society Grant JFRA-420.

References

Abraham EH, Prat AG, Gerweck L, Seneveratne T, Arceci R, Kramer R, Guidotti G and Cantiello HF (1993) The multidrug resistance (*mdr1*) gene product functions as an ATP channel. Proc Natl Acad Sci USA 90: 312–316.

Adams MD and Oxender DL (1989) Bacterial perplamic binding protein tertiary structures. J Biol Chem 264: 15739–15742.

Altshul SF, Gish W, Miller W, Meyers E and Lipman DJ (1990) Basic local alignment search tool. J Mol Biol 215: 403–410.

Ambudkar SV, Lelong IH, Zhang J, Cardarelli CO, Gottesman MM and Pastan I (1992) Partial purification and reconstitution of the human multidrug-resistance pump: characterization of the drug-stimulatable ATP hydrolysis. Proc Natl Acad Sci USA 89: 8472–8476.

Ames G (1986) Bacterial periplasmic transport systems: structure, mechanism and evolution. Annu Rev Biochem 55: 397–425.

Ames GF, Mimura CS, Holbrook SR and Shyamala V (1992) Traffic ATPases: a superfamily of transport proteins operating from *Escherichia coli* to humans. Adv Enzymol Related Areas Mol Biol 65: 1–47.

Ames GF, Mimura CS and Shyamala V (1990) Bacterial periplasmic permeases belong to a family of transport proteins operating from *Escherichia coli* to human: Traffic ATPases. Fems Microbiol Rev 6: 429–446.

Arceci RJ, Baas F, Raponi R, Horwitz SB, Housman D and Croop J (1990) Multidrug resistance gene expression is controlled by steroid hormones in the secretory epithelium of the uterus. Mol Reprod Dev 25: 101–109.

Arceci RJ, Croop JM, Horwitz SB and Housman D (1988) The gene encoding multidrug resistance is induced and expressed at high levels during pregnancy in the secretory epithelium of the uterus. Proc Natl Acad Sci USA 85: 4350–4354.

Baas F and Borst P (1988) The tissue dependent expression of hamster P-glycoprotein genes. FEBS Lett 229: 329–332.

Baas F, Jongsma AP, Broxterman HJ, Arceci RJ, Housman D, Scheffer GL, Riethorst A, Van GM, Nieuwint AW and Joenje H (1990) Non-P-glycoprotein mediated mechanism for multidrug resistance precedes P-glycoprotein expression during in vitro selection for doxorubicin resistance in a human lung cancer cell line. Cancer Res 50: 5392–5398.

Bahram S, Arnold D, Bresnahan M, Strominger JL and Spies T (1991) Two putative subunits of a peptide pump encoded in the human major histocompatibility complex class II region. Proc Natl Acad Sci USA 88: 10094–10098.

Bairoch A (1992) PROSITE: a dictionary of sites and patterns in proteins. Nucleic Acids Res 20, Suppl: 2013–2018.

Beck WT (1990) Multidrug resistance and its circumvention. Eur J Cancer 26: 513–515.

Berkower C and Michaelis S (1991) Mutational analysis of the yeast a-factor transporter STE6, a member of the ATP binding cassette (ABC) protein superfamily. EMBO J 10: 3777–3785.

Biedler JL and Riehm H (1970) Cellular resistance to actinomycin D in Chinese hamster cells in vitro: cross-resistance, radioautographic and cytogenetic studies. Cancer Res 30: 1174–1184.

Boscoboinik D, Debanne MT, Stafford AR, Jung CY, Gupta R and Epland RM (1990) Dimerization of the P-glycoprotein in membranes. Biochim Biophys Acta 1027: 225–258.

Bradley G, Georges E and Ling V (1990) Sex-dependent and independent expression of the P-glycoprotein isoforms in chinese hamster. J Cell Physiol 145: 398–408.

Buschman E, Arceci R, Croop J, Mingxin C, Arias I, Housman D and Gros P (1992) Isoform specific antibodies identify the bilie canalicular membrane as the primary site of expression of the P-glycoprotein encoded by mouse *mdr2*. J Biol Chem 267: 18093–18099.

Callahan HL and Beverley S (1991) Hewavy metal resistance: a new role for P-glycoproteins in *Leishmania*. J Biol Chem 266: 18427–18430.

Chambers TC, Poh J, Raynor RL and Kuo JF (1993) Identification of specific sites in human P-glycoprotein phosphorylated by protein kinase C. J Biol Chem 268: 4592–4595.

Chan KM, Davies PL, Childs S, Veinot L and Ling V (1992) P-glycoprotein genes in the winter flounder, *Pleuronectes americanus*: isolation of two type of genomic clones carrying 3' terminal exons. Biochim Biophys Acta 1171: 65–72.

Chaudhary PM, Mechetner EB and Roninson IB (1992) Expression and activity of the multidrug resistance P-glycoprotein in human peripheral blood lymphocytes. Blood 80: 2735–2739.

Chaudhary PM and Roninson IB (1991) Expression and activity of P-glycoprotein, a multidrug efflux pump, in human hematopoietic stem cells. Cell 66: 85–94.

Chen CJ, Chin JE, Ueda K, Clark DP, Pastan I, Gottesman MM and Roninson IB (1986) Internal duplication and homology with bacterial transport proteins in the mdr1 (P-glycoprotein) gene from multidrug-resistant human cells. Cell 47: 381–389.

Chen CJ, Clark D, Ueda K, Pastan I, Gottesman MM and Roninson IB (1990) Genomic organization of the human multidrug resistance (MDR1) gene and origin of P-glycoproteins. J Biol Chem 265: 506–514.

Chin JE, Soffir R, Noonan KE, Choi K and Roninson IB (1989) Structure and expression of the human MDR (P-glycoprotein) gene family. Mol Cell Biol 9: 3808–3820.

Choi K, Chen CJ, Kriegler M and Roninson IB (1988) An altered pattern of cross-resistance in multidrug-resistant human cells results from spontaneous mutations in the mdr1 (P-glycoprotein) gene. Cell 53: 519–529.

Cole SPC, Bhardwaj G, Gerlach JH, Mackie JE, Grant CE, Almquist KC, Stewart AJ, Kurz EU, Duncan AMV and Deeley RG (1992) Overexpression of a transporter gene in a multidrug-resistant human lung cancer cell line. Science 258: 1650–1654.

Cordon CC, O'Brien JP, Boccia J, Casals D, Bertino JR and Melamed MR (1990) Expression of the multidrug resistance gene product (P-glycoprotein) in human normal and tumor tissues. J Histochem Cytochem 38: 1277–1287.

Cordon CC, O'Brien JP, Casals D, Rittman GL, Biedler JL, Melamed MR and Bertino JR (1989) Multidrug-resistance gene (P-glycoprotein) is expressed by endothelial cells at blood-brain barrier sites. Proc Natl Acad Sci USA 86: 695–698.

Cowman AF, Karcz S, Galatis D and Culvenor JG (1991) A P-glycoprotein homologue of Plasmodium falciparum is localized to the digestive vacuole. J Cell Biol 113: 1033–1042.

Croop J, Guild B, Gros P and Housman D (1987) Genetics of multidrug resistance: relationship of a cloned gene to the complete multidrug resistance phenotype. Cancer Res 47: 5982—5988.

Croop JM, Gros P and Housman DE (1988) Genetics of multidrug resistance. J Clin Invest 81: 1303—1309.

Croop JM, Raymond M, Haber D, Devault A, Arceci RJ, Gros P and Housman DE (1989) The three mouse multidrug resistance (*mdr*) genes are expressed in a tissue-specific manner in normal mouse tissues. Mol Cell Biol 9: 1346—1350.

Dano K (1973) Active outward transport of daunomycin in resistant Ehrlich ascites tumor cells. Biochim Biophys Acta 323: 446—483.

De Bruijn MH, Van der Bliek AM, Biedler JL and Borst P (1986) Differential amplification and disproportionate expression of five genes in three multidrug-resistant Chinese hamster lung cell lines. Mol Cell Biol 6: 4717—4722.

Descoteaux S, Ayala P, Orozco E and Samuelson J (1992) Primary sequence of two P-glycoprotein genes of *Entamoeba histolytica*. Mol Biochem Parasit 54: 201—211.

Devault A and Gros P (1990) Two members of the mouse *mdr* gene family confer multidrug resistance with overlapping but distinct drug specificities. Mol Cell Biol 10: 1652—1663.

Devereux J, Haeberli P and Smithies O (1984) A comprehensive set of sequence analysis programs for the VAX. Nuc Acids Res 12: 387—395.

Drach D, Zhao S, Drach J, Mahadevia R, Gattringer C, Huber H and Andreeff M (1992) Subpopulations of normal peripheral blood and bone marrow cells express a functional multidrug resistant phenotype. Blood 80: 2729—2734.

Dreesen TD, Johnson DH and Henikoff S (1988) The brown protein of Drosophila melanogaster is similar to the white protein and to components of active transport complexes. Mol Cell Biol 8: 5206—5215.

Dudler R and Hertig C (1992) Structure of an *mdr*-like gene from *Arabidopsis thaliana*. J Biol Chem 267: 5882—5888.

Ellenberger TE and Beverley SM (1989) Multiple drug resistance and conservative amplification of the H region in Leishmania major. J Biol Chem 264: 15094—15103.

Endicott J, Juranka P, Farida S, Gerlach J, Deuchars K and Ling V (1987) Simultaneous expression of two P-glycoprotein genes in drug-sensitive Chinese hamster ovary cells. Mol Cell Biol 7: 4075—4081.

Endicott JA and Ling V (1989) The biochemistry of P-glycoprotein-mediated multidrug resistance. Annu Rev Biochem 58: 137—171.

Endicott JA, Sarangi F and Ling V (1991) Complete cDNA sequences encoding the Chinese hamster P-glycoprotein gene family. Dna Seq 2: 89—101.

Fojo AT, Ueda K, Slamon DJ, Poplack DG, Gottesman MM and Pastan I (1987) Expression of a multidrug-resistance gene in human tumors and tissues. Proc Natl Acad Sci USA 84: 265—269.

Foote SJ, Kyle DE, Martin RK, Oduola AM, Forsyth K, Kemp DJ and Cowman AF (1990) Several alleles of the multidrug-resistance gene are closely linked to chloroquine resistance in Plasmodium falciparum. Nature 345: 255—258.

Foote SJ, Thompson JK, Cowman AF and Kemp DJ (1989) Amplification of the multidrug resistance gene in some chloroquine-resistant isolates of P. falciparum. Cell 57: 921—930.

Ford JM and Hait WN (1990) Pharmacology of drugs that alter multidrug resistance in cancer. Pharmacol Rev 42: 155—199.

Garrido M, Herrero M, Kolter R and Moreno F (1988) The export of the DNA replication inhibitor Microcin B17 provides immunity for the host cell. EMBO J 7: 1853—1862.

Georges E, Bradley G, Gariepy J and Ling V (1990) Detection of P-glycoprotein isoforms by gene-specific monoclonal antibodies. Proc Natl Acad Sci USA 87: 152—156.

Gill DR, Hatful GF and Salmond GP (1986) A new cell division operon in *Escherichia coli*. Mol Gen Genet 86: 134—145.

Gill DR, Hyde SC, Higgins CF, Valverde MA, Mintenig GM and Sepulveda FV (1992) Separation of drug transport and chloride channel functions of the human multidrug resistance P-glycoprotein. Cell 71: 23—32.

Gottesman M and Pastan I (1989) The multidrug resistance transporter, a double-edged sword. J Biol Chem 263: 12163—12166.

Greenberger LM, Lothstein L, Williams SS and Horwitz SB (1988a) Distinct P-glycoprotein precursors are overproduced in independently isolated drug-resistant cell lines. Proc Natl Acad Sci USA 85: 3762—3766.

Greenberger LM, Williams SS, Georges E, Ling V and Horwitz SB (1988b) Electrophoretic analysis of P-glycoproteins produced by mouse J774.2 and Chinese hamster ovary multidrug-resistant cells. J Natl Cancer Inst 80: 506—510.

Gros P, Ben Neriah Y, Croop JM and Housman DE (1986a) Isolation and expression of a complementary DNA that confers multidrug resistance. Nature 323: 728—731.

Gros P, Croop J and Housman D (1986b) Mammalian multidrug resistance gene: complete cDNA sequence indicates strong homology to bacterial transport proteins. Cell 47: 371—380.

Gros P, Croop J, Roninson I, Varshavsky A and Housman DE (1986c) Isolation and characterization of DNA sequences amplified in multidrug-resistant hamster cells. Proc Natl Acad Sci USA 83: 337—341.

Gros P, Dhir R, Croop J and Talbot F (1991) A single amino acid substitution strongly modulates the activity and substrate specificity of the mouse *mdr1* and *mdr3* drug efflux pumps. Proc Natl Acad Sci 88: 7289—7293.

Gros P, Raymond M, Bell J and Housman D (1988) Cloning and characterization of a second member of the mouse mdr gene family. Mol Cell Biol 8: 2770—2778.

Guilfoile PG and Hutchinson CR (1991) A bacterial analog of the *mdr* gene of mammalian tumor cells is present in *Streptomyces peucetius*, the producer of daunorubicin and doxorubicin. Proc Natl Acad Sci USA 88: 8553—8557.

Henderson DM, Sifri CD, Rodgers M, Wirth DF, Hendrickson N and Ullman B (1992) Multidrug resistance in Leishmania donovani is conferred by amplification of a gene homologous to the mammalian mdr1 gene. Mol Cell Biol 12: 2855—2865.

Herzog CE, Trepel JB, Mickley LA, Bates SE and Fojo AT (1992) Various methods of analysis of mdr-1/P-glycoprotein in human colon cancer cell lines. J Natl Cancer Inst 84: 711–716.

Higgins CF, Hyde SC, Mimmack MM, Gileadi U, Gill DR and Gallagher MP (1990) Binding protein-dependent transport systems. J Bioenerg Biomembr 22: 571–592.

Hill BT, Deuchars K, Hosking LK, Ling V and Whelan RD (1990) Overexpression of P-glycoprotein in mammalian tumor cell lines fractionated X irradiation in vitro. J Natl Cancer Inst 82: 607–612.

Horio M, Gottesman MM and Pastan I (1988) ATP-dependent transport of vinblastine in vesicles from human multidrug-resistant cells. Proc Natl Acad Sci USA 85: 3580–3584.

Hsu S, Lothstein L and Horwitz S (1989) Differential over-expression of the three mdr gene family members in multidrug resistant J774.2 mouse cells. J Biol Chem 264: 12053–12062.

Husain I, Van Houten B, Thomas D and Sancar A (1986) Sequences of *Escherichia coli uvrA* gene and protein reveal two potential ATP binding sites. J Biol Chem 261: 4895–4901.

Hyde S, Emsley P, Hartshorn M, Mimmack M, Gileadi U, Pearce S, Gallagher M, Gill D, Hubbard R and Higgins C (1990) Structural model of ATP-binding proteins associated with cystic proteins associated with cystic fibrosis, multidrug resistance and bacterial transport. Science 346: 362–365.

Jan LY and Jan YN (1992) Tracing the roots of ion channels. Cell 69: 715–718.

Juliano RL and Ling V (1976) A surface glycoprotein modulating drug permeability in Chinese hamster ovary cell mutants. Biochim Biophys Acta 455: 152–162.

Juranka PF, Zastawny RL and Ling V (1989) P-glycoprotein: multidrug-resistance and a superfamily of membrane-associated transport proteins. FASEB J 3: 2583–2592.

Kamijo K, Kamijo T, Ueno I, Osumi T, Osumi T and Hashimoto T (1992) Nucleotide sequence of the human 70 kDa peroxisomal membrane protein: a member of ATP-binding cassette transporters. Biochim Biophys Acta 1129: 323–327.

Kartner N, Evernden-Porrelle D, Bradley G and Ling V (1985) Detection of P-glycoprotein in multidrug-resistant cell lines by monoclonal antibodies. Nature 316: 820–823.

Kelly A, Powis S, Kerr L, Mockridge I, Elliott T, Bastin J, Uchanska-Ziegler B, Ziegler A, Trowsdale J and Townsend A (1992) Assembly and function of the two ABC transporter proteins encoded in the human major histocompatibility complex. Nature 355: 641–644.

Krogstad DJ, Gluzman IY, Kyle DE, Oduola AM, Martin SK, Milhous WK and Schlesinger PH (1987) Efflux of chloroquine from Plasmodium falciparum: mechanism of chloroquine resistance. Science 238: 1283–1285.

Kroll JS, Hopkins I and Moxon ER (1988) Capsule loss in H. influenzae type b occurs by recombination-mediated disruption of a gene essential for polysaccharide export. Cell 53: 347–358.

Kuchler K, Sterne R and Thorner J (1989) *Saccharomyces cerevisiae* STE6 gene product: a novel pathway for protein export in eukaryotic cells. EMBO J 13: 3973–3984.

Kuchler K and Thorner J (1992) Functional expression of human *mdr1* in the yeast *Saccharomyces cerevisiae*. Proc Natl Acad Sci USA 89: 2302–2306.

Kupinski J, Coussen F, Bakalyar HA, Tang W, Feinstein PG, Orth K, Slaughter C, Reed R and Gilman A (1989) Adenylyl cyclase amino acid sequence: possible channel- or transporter-like structure. Science 244: 1558–1564.

Lemontt JF, Azzaria M and Gros P (1988) Increased mdr gene expression and decreased drug accumulation in multidrug-resistant human melanoma cells. Cancer Res 48: 6348–6353.

Lincke CR, Smit JJ, Van der Bliek AM, Van der Velde-Koerts T and Borst P (1991) Structure of the human *MDR3* gene and physical mapping of the human *MDR* locus. J Biol Chem 266: 5303–5310.

Lincke CR, Broeks A, The I, Plasterk RHA and Borst P (1993) The expression of two P-glycoprotein (*pgp*) genes in transgenic *Caenorhabditis elegans* is confined to intestinal cells. EMBO J 12: 1615-1620.

Lincke CR, The I, Van GM and Borst P (1992) The P-glycoprotein gene family of *Caenorhabditis elegans*. Cloning and characterization of genomic and complementary DNA sequences. J Mol Biol 228: 701–711.

Malone PC (1990) A consensus structure for membrane transport. Res Micro 141: 374–383.

Marger MD and Saier MH (1993) A major superfamily of transmembrane facilitators that catalyse uniport, symport and antiport. Trend Bio Sci 18: 13–20.

Martin SK, Oduola AM and Milhous WK (1987) Reversal of chloroquine resistance in Plasmodium falciparum by verapamil. Science 235: 899–901.

McGrath JP and Varshavsky A (1989) The yeast STE6 gene encodes a homologue of the mammalian multidrug resistance P-glycoprotein. Nature 340: 400–404.

McGrath T and Center MS (1988) Mechanisms of multidrug resistance in HL60 cells: evidence that a surface membrane protein distinct from P-glycoprotein contributes to reduced cellular accumulation of drug. Cancer Res 48: 3959–3963.

Mirski SE, Gerlach JH and Cole SP (1987) Multidrug resistance in a human small cell lung cancer cell line selected in adriamycin. Cancer Res 47: 2594–2598.

Mosser J, Douar A, Sarde C, Kioschis P, Feil R, Moser H, Poustka A, Mandel J and Aubourg P (1993) Putative X-linked adrenoleukodystrophy gene shares unexpected homology with ABC transporters. Nature 361: 726–730.

Naito M and Tsuruo T (1992) Functionally active homodimer of P-glycoprotein in multidrug resistant tumor cells. Biochem Biophys Res Comm 185: 284–290.

Ng WF, Sarangi F, Zastawny RL, Veinot DL and Ling V (1989) Identification of members of the P-glycoprotein multigene family. Mol Cell Biol 9: 1224–1232.

Nielsen D and Skovsgaard (1992) P-glycoprotein as multidrug transporter: a critical review of current multidrug resistant cell lines. Biochim Biophys Acta 1139: 169–183.

Nishi K, Yoshida M, Nishimura M, Nishikawa M, Nishiyama M, Horinouchi S and Beppu T (1992) A leptomycin B resistance gene of Schizosaccharomyces pombe encodes a protein similar to the mammalian P-glycoproteins. Mol Microbiol 6: 761–769.

Noda M, Ikeda T, Kayano T, Suzuki H, Takeshima H, Kurasaki M, Takahashi H and Numa S (1986) Existence of distinct sodium channel messenger RNAs in rat brain. Nature 320: 188–192.

O'Hare K, Murphy C, Levis R and Rubin G (1984) DNA sequence of the *white* locus of *Drosophila melanogaster*. J Mol Biol 180: 437–455.

Ortiz DF, Keppel L, Speiser DM, Scheel G, McDonald G and Ow DW (1992) Heavy metal tolerance in the fission yeast requires an ATP-binding-cassette-type vacuolar membrane transporter. EMBO J 11: 3491–3499.

Ouellette M, Fase-Fowler F and Borst P (1990) The amplified H circle of methotrexate-resistant *Leishmania tarentolae* contains a novel P-glycoprotein gene. EMBO J 9: 1027–1033.

Ouellette M, Hettema E, Wust D, Fase FF and Borst P (1991) Direct and inverted DNA repeats associated with P-glycoprotein gene amplification in drug resistant Leishmania. EMBO J 10: 1009–1016.

Qin S, Xie A, Bonato MC and McLaughlin CS (1990) Sequence analysis of the translational elongation factor 3 from *Saccharomyces cerevisiae*. J Biol Chem 265: 1903–1912.

Raymond M and Gros P (1989) Mammalian multidrug-resistance gene: correlation of exon organization with structural domains and duplication of an ancestral gene. Proc Natl Acad Sci USA 86: 6488–6492.

Raymond M, Gros P, Whiteway M and Thomas DY (1992) Functional complementation of yeast *ste6* by a mammalian multidrug resistance mdr gene. Science 256: 232–234.

Raymond M, Rose E, Housman D and Gros P (1990) Physical mapping, amplification and overexpression of the mouse *mdr* gene family in multidrug resistant cells. Mol Cell Biol 10: 1642–1651.

Riordan J, Rommens J, Kerem B, Alon N, Rozmahel R, Grzelczak Z, Zielenski J, Lok S, Plavsic M, Chou J, Drumm M, Iannuzzi M, Collins F and Tsui L (1989) Identification of the cystic fibrosis gene: cloning and characterization of complementary DNA. Science 245: 1066–1073.

Riordan JR, Deuchars K, Kartner N, Alon N, Trent J and Ling V (1985) Amplification of P-glycoprotein genes in multidrug-resistant mammalian cell lines. Nature 316: 817–819.

Roninson IB (ed) (1991) Molecular and Cellular Biology of Multidrug Resistance in Tumor Cells. Plenum Press, New York.

Roninson IB, Abelson H, Housman DE, Howell N and Varshavsky A (1984) Amplification of specific DNA sequences correlates with multidrug resistance in Chinese hamster cells. Nature 309: 626–628.

Roninson IB, Chin JE, Choi KG, Gros P, Housman DE, Fojo A, Shen DW, Gottesman MM and Pastan I (1986) Isolation of human mdr DNA sequences amplified in multidrug-resistant KB carcinoma cells. Proc Natl Acad Sci USA 83: 4538–4542.

Ross JI, Eady EA, Cove JH, Cunliffe WJ, Baumberg S and Wootton JC (1990) Inducible erythromycin resistance in staphylococci is encoded by member of the ATP-binding transport super-gene family. Mol Microbiol 4: 1207–1214.

Rossen L, Johnston AW and Downie JA (1984) DNA sequence of the *Rhizobium* leguminosarum nodulation genes nodAB and C required for root hair curling. Nucl Acids Res 12: 9497–9508.

Samuelson J, Ayala P, Orozco E and Wirth D (1990) Emetine-resistant mutants of *Entamoeba histolytica* overexpress mRNAs for multidrug resistance. Mol Biochem Para 38: 281–290.

Sarkadi B, Price EM, Boucher RC, Germann UA and Scarborough GA (1992) Expression of the human multidrug resistance cDNA in insect cells generates a high activity drug-stimulated membrane ATPase. J Biol Chem 267: 4854–4858.

Schnell N, Engelke G, Augustin J, Rowenstein R, Ungermann V, Goetz F and Entian KD (1992) Analysis of the genes involved in the biosynthesis of lantibiotic epidermin. Eur J Biochem 204: 57–68.

Schoner B, Geistlich M, Rosteck P, Rao RN, Seno E, Reynolds P, Cox K, Burgett S and Hershberger C (1992) Sequence similarity between macrolide resistance determinants and ATP-binding transport proteins. Gene 115: 93–96.

Scotto KW, Biedler JL and Melera PW (1986) Amplification and expression of genes associated with multidrug resistance in mammalian cells. Science 232: 751–755.

Sehested M, Simpson D, Skovsgaard T and Buhl-Jenson P (1989) Freeze-fracture study of plasma membranes in wild type and daunorubicin-resistant Ehrlich ascites tumor and P388 leukemia cells. Virchows Arch [b] 56: 327–335.

Shen DW, Fojo A, Chin JE, Roninson IB, Richert N, Pastan I and Gottesman MM (1986) Human multidrug-resistant cell lines: increased mdr1 expression can precede gene amplification. Science 232: 643–645.

Skach WR, Calayag MC and Lingappa VR (1993) Evidence of an alternative model of human P glycoprotein structure and biogenesis. Biol Chem 268: 6903-6908.

Skovsgaard T (1978) Mechanism of cross-resistance between vincristine and daunorubicin in Ehrlich ascites tumor cells. Cancer Res 38: 4722–4727.

Slapak CA, Daniel JC and Levy SB (1990) Sequential emergence of distinct resistance phenotypes in murine erythroleukemia cells under adriamycin selection:decreased anthracycline uptake precedes increased P-glycoprotein expression. Cancer Res 50: 7895–7901.

Sugawara I, Kataoka I, Morishita Y, Hamada H, Tsuruo T, Itoyama S and Mori S (1988) Tissue distribution of P-glycoprotein encoded by a multidrug-resistant gene as revealed by a monoclonal antibody, MRK 16. Cancer Res 48: 1926–1929.

Tanabe T, Takeshima H, Mikami A, Flockerzi V, Takahashi R, Kangawa K, Kojima M, Matsuo H, Hirose T and Numua S (1987) Primary structure of the receptor for calcium channel blockers from skeletal muscle. Nature 328: 313–318.

Thiebaut F, Tsuruo T, Hamada H, Gottesman MM, Pastan I and Willingham MC (1987) Cellular localization of the multidrug-resistance gene product P-glycoprotein in normal human tissues. Proc Natl Acad Sci USA 84: 7735–7738.

Thiebaut F, Tsuruo T, Hamada H, Gottesman MM, Pastan I and

Willingham MC (1989) Immunohistochemical localization in normal tissues of different epitopes in the multidrug transport protein P170: evidence for localization in brain capillaries and cross-reactivity of one antibody with a muscle protein. J Histochem Cytochem 37: 159–164.

Toth K, Vaughan MM, Slocum HK, Fredericks WJ, Chen YF, Arredondo MA, Harstrick A, Karakousis C, Baker RM and Rustum YM (1992) Comparison of an immunoperoxidase "sandwich" staining method and western blot detection of P-glycoprotein in human cell lines and sarcomas. Am J Pathol 140: 1009–1016.

Trezise AEO, Romano PR, Gill DR, Hyde SC, Sepulveda FV, Buchwald M and Higgins CF (1992) The multidrug resistance and cystic fibrosis genes have complementary epithelial expression. EMBO J 11: 4291–4303.

Triglia T, Foote SJ, Kemp DJ and Cowman AF (1991) Amplification of the multidrug resistance gene pfmdr1 in Plasmodium falciparum has arisen as multiple independent events. Mol Cell Biol 11: 5244–5250.

Ueda K, Cardarelli C, Gottesman MM and Pastan I (1987) Expression of a full-length cDNA for the human "MDR1" gene confers resistance to colchicine, doxorubicin and vinblastine. Proc Natl Acad Sci USA 84: 3004–3008.

Ueda K, Cornwell MM, Gottesman MM, Pastan I, Roninson IB, Ling V and Riordan JR (1986) The mdr1 gene, responsible for multidrug-resistance, codes for P-glycoprotein. Biochem Biophys Res Commun 141: 956–962.

Van der Bliek A, Van der Velde-Koerts T, Ling V and Borst P (1986) Overexpression and amplification of five genes in a multidrug resistant chinese hamster ovary cell line. Mol Cell Biol 6: 1671–1678.

Van der Bliek AM, Baas F, Ten Houte de Lange T, Kooiman PM, Van der Velde-Koerts T and Borst P (1987) The human mdr3 gene encodes a novel P-glycoprotein homologue and gives rise to alternatively spliced mRNAs in liver. EMBO J 6: 3325–3331.

Van der Bliek AM, Baas F, Van der Velde-Koerts T, Biedler JL, Meyers MB, Ozols RF, Hamilton TC, Joenje H and Borst P (1988a) Genes amplified and overexpressed in human multidrug-resistant cell lines. Cancer Res 48: 5927–5932.

Van der Bliek AM, Kooiman P, Schneider C and Borst P (1988b) Sequence of *mdr3* cDNA encoding a human P-glycoprotein. Gene 71: 401–411.

Volkman SK, Wilson C and Wirth D (1993) Stage-specific transcripts of the *Plasmodium falciparum pfmdr1* gene. Mol Biochem Para 57: 203–212.

Walker JE, Saraste M, Runswick MJ and Gay NJ (1982) Distantly related sequences in the a and b subunits of ATP synthetase, myosin, kinases and other ATP-requiring enzymes and a common nucleotide binding fold. EMBO J 1: 945–951.

Wellems TE, Panton LJ, Gluzman IY, Do RVE, Gwadz RW, Walker JA and Krogstad DJ (1990) Chloroquine resistance not linked to mdr-like genes in a Plasmodium falciparum cross. Nature 345: 253–255.

Wilson C, Seranno A, Wasley B, Bogenshutz M, Shankar A and Wirth D (1989) Amplification of a gene related to mammalian *mdr* genes in drug resistant *Plasmodium falciparum*. Science 244: 1184–1186.

Wilson C, Volkman SK, Thaithong S, Martin RK, Kyle DE, Milhous WK and Wirth D (1993) Amplification of *pfmdr1* associated with mefloquine and halofantrine resistance in *Plamsodium falciparum* from Thailand. Mol Biochem Para 57: 151–160.

Wu C-t, Budding M, Griffin M and Croop J (1991) Isolation and characterization of *Drosophila* multidrug resistance gene homologues. Mol Cell Biol 11: 3940–3948.

Wu L, Smythe AM, Stinson SF, Mullendore LA, Monks A, Scudiero DA, Paull KD, Koutsoukos AD, Rubinstein LV, Boyd MR and Shoemaker RH (1992) Multidrug-resistant phenotype of disease oriented panels of human tumor cell lines used for anticancer drug screening. Cancer Res 52: 3029–3034.

Zhang JT, Duthie M and Ling V (1993) Membrane topology of the N-terminal half of the hamster P-glycoprotein molecule. J Biol Chem 268: 15101–15110.

Zhang JT and Ling V (1991) Study of membrane orientation and glycosylated extracellular loops of mouse P-glycoprotein by in vitro translation. J Biol Chem 266: 18224–18232.

Address for offprints: James M. Croop, Division of Pediatric Oncology, The Dana-Farber Cancer Institute and The Children's Hospital, Harvard Medical School, Boston, MA, USA.

Cytotechnology **12**: 33–62, 1993.

Molecular analysis of the multidrug transporter

Ursula A. Germann
Laboratory of Cell Biology, National Cancer Institute, National Institutes of Health, Building 37, Room 1B22, Bethesda, MD 20892, USA

Key words: ATPase, ATP-binding cassette superfamily, drug transport, multidrug resistance, P-glycoprotein

Abstract

The multidrug resistance gene product, P-glycoprotein or the multidrug transporter, confers multidrug resistance to cancer cells by maintaining intracellular levels of cytotoxic agents below a killing threshold. P-glycoprotein is located within the plasma membrane and is thought to act as an energy-dependent drug efflux pump. The multidrug transporter represents a member of the ATP-binding cassette superfamily of transporters (or traffic ATPases) and is composed of two highly homologous halves, each of which harbors a hydrophobic transmembrane domain and a hydrophilic ATP-binding fold. This review focuses on various biochemical and molecular genetic approaches used to analyze the structure, function, and mechanism of action of the multidrug transporter, whose most intriguing feature is its ability to interact with a large number of structurally and functionally different amphiphilic compounds. These studies have underscored the complexity of this membrane protein which has recently been suggested to assume alternative topological and quaternary structures, and to serve multiple functions both as a transporter and as a channel. With respect to the multidrug transporter activity of P-glycoprotein, progress has been made towards the elucidation of essential amino acid residues and/or polypeptide regions. Furthermore, the drug-stimulatable ATPase activity of P-glycoprotein has been established. The mechanism of drug transport by P-glycoprotein, however, is still unknown and its physiological role remains a matter of speculation.

Introduction

This review focuses on a molecular genetic and biochemical analysis of the multidrug resistance (*mdr*) gene product P-glycoprotein ("P" for permeability), that confers multidrug resistance to mammalian cells by acting as a multidrug transporter (Juliano and Ling, 1976; Roninson, 1991; Gottesman and Pastan, 1993). P-glycoproteins of this class include the human *MDR*1 (Chen *et al.*, 1986), the mouse *mdr*3 (or *mdr*1a) (Hsu *et al.*, 1989; Devault and Gros, 1990) and *mdr*1 (or *mdr*1b) (Gros *et al.*, 1986b), the hamster *pgp*1 and *pgp*2 (Gros *et al.*, 1986c; Endicott *et al.*, 1987), and the rat *pgp*1 and *pgp*2 (or *mdr*1b) gene products (Silverman *et al.*, 1991; Deuchars *et al.*, 1992).

The multidrug transporter is one of the most intriguing members of the superfamily of ATP-binding cassette (ABC) transporters (Hyde *et al.*, 1990) or traffic ATPases (Mimura *et al.*, 1991). This superfamily consists of membrane-associated proteins of different origins including bacteria, archebacteria, yeast, plants, insects, animals, and man (for reviews see Ames and Lecar, 1992; Ames *et al.*, 1992; Higgins, 1992, 1993; and J.M. Croop, this issue). ABC transporters play an important role in various biological processes such as the uptake of nutrients, the extrusion of noxious compounds, the secretion of toxins, the transport of ions and peptides, or cell signaling. While many ABC transporters are integral membrane proteins localized at the cell surface, some of them are present in intracellular membranes. Characteristic of this transporter superfamily is that all members share the same overall architecture: they are built from one, two, or multiple sets of a hydrophobic transmembrane domain (which most often contains six predicted membrane-spanning regions) and a hydrophilic nucleotide binding fold. In the majority of prokaryotic traffic ATPases the transmembrane domains and the nucleotide binding folds are encoded by separate open reading frames within an operon, and noncovalently linked complexes (e.g., dimers, trimers, or tetramers) form functional transporters (Ames and Lecar, 1992; Ames *et al.*, 1992; Higgins 1992, 1993). Eukaryotic ABC transporters are usually encoded by a single gene (Ames and Lecar, 1992; Ames *et al.*, 1992; Higgins, 1992, 1993) and consist of either one or two basic units, each composed of a hydrophobic membrane-integral part and a nucleotide binding fold. Of the eukaryotic members identified so far, the mammalian ABC transporters localized in the plasma membrane are predicted to contain a total of 12 membrane spanning regions and two nucleotide binding folds, whereas those localized in intracellular membranes are predicted to contain only six transmembrane regions and one nucleotide binding fold, and probably function as dimers (Higgins, 1993).

The putative ATP binding folds harbor several amino acid sequence motifs that are highly conserved among all ABC transporters (Ames *et al.*, 1992; Higgins, 1992). Two core consensus motifs, a G-$(X)_4$-G-K-(T)-$(X)_6$-I/V sequence ("Walker motif A") and a hydrophobic pocket sequence R/K-$(X)_3$-G-$(X)_3$-L-$(\text{hydrophobic})_4$-D ("Walker motif B") are generally found in any ABC transporter and many other nucleotide binding proteins (Walker *et al.*, 1982). They are directly involved in the binding and hydrolysis of nucleotides. Two further short stretches, the so-called "center region" located approximately in the middle of the intracytoplasmic loop between the two Walker motifs, and the "linker peptide" L-S-G-G-$(X)_3$-R-hydrophobic-X-hydrophobic-A immediately preceding the "Walker motif B", are highly conserved among traffic ATPases and are not present in other nucleotide binding proteins (Shyamala *et al.*, 1991). The functions of these latter motifs is still unknown. Models based on calculations of secondary structures as well as on the resolved tertiary structures of the three nucleotide binding proteins adenylate kinase, EF-Tu, and p21-ras predict that these two motifs are part of a loop that may undergo conformational changes upon hydrolysis of bound nucleotides, and thus promote contact between the transmembrane domain and the nucleotide binding fold (Hyde *et al.*, 1990; Mimura *et al.*, 1991).

The amino acid sequence identity within the membrane associated domains of different ABC transporters is generally low. Highest sequence homology can be observed in loop regions that connect putative membrane spanning regions (Manavalan *et al.*, 1993). These intra- or extracytoplasmic loop regions are also predicted to be of similar size (Manavalan *et al.*, 1993). Despite their overall hydrophobic character, however, very few sequences are conserved within the putative membrane-integral segments.

ABC transporters share an overall sequence identity of approximately 30% in their nucleotide binding folds and it is believed that most of their transport activities require ATP hydrolysis as a source of energy (Ames *et al.*, 1992; Higgins, 1992). The discovery of the ABC superfamily of transporters has important implications for the *mdr* subfamily both from an evolutionary, and biochemical point of view. While the analysis of the overall structural organization of ABC transporter genes has already helped to determine major functional domains, a direct DNA and protein sequence comparison may allow the identification of subgroups within this superfamily that have similar biological activities. Furthermore, such a comparative analysis may also implicate highly homologous regions within the polypeptide chains of functionally similar proteins that may be important for their biochemical properties. As will be discussed in a later section in more detail, some similarities in function have been associated with P-glycoprotein and the yeast STE6 protein (Raymond *et al.*, 1992), and with P-glycoprotein and the cystic fibrosis transmembrane conductance regulator (CFTR) gene product (Valverde *et al.*, 1992).

Mammalian multidrug transporters are single-chain proteins with two homologous halves each of which contains six predicted membrane-spanning regions followed by a nucleotide binding fold (for reviews see e.g., Endicott and Ling, 1989; Roninson, 1991; Gottesman and Pastan, 1993). A generally accepted working model for the topology of the plasma membrane-associated multidrug transporter based entirely on hydropathy calculations is presented in Fig. 1A (Chen *et al.*, 1986; Gros *et al.*, 1986b). The polypeptide chain consists of approximately 1280 amino acid residues and the orientation of the molecule is such that both the amino- and carboxy-terminus are located on the cytoplasmic side of the plasma membranes. Hence, in this model the two putative nucleotide binding regions are located intracellularly which is in agreement with their postulated role to fulfill the energy requirements of the drug transport process. These ATP-binding/utilization sites represent the most conserved regions between the various multidrug transporter isoforms, underscoring their pivotal role for function. Mammalian P-glycoproteins are glycosylated, usually within the first extracellular loop that harbors several consensus N-linked glycosylation sites which are, however, not conserved between different P-glycoprotein isoforms (Devault and Gros, 1990). An amino acid sequence comparison between various multidrug transporters reveals that besides the nucleotide binding sites, regions of high sequence identity include the first and second intracytoplasmic loop both in the amino- and in the carboxy-terminal half (Gottesman and Pastan, 1988). The amino- and carboxy-termini, the first extracytoplasmic loop, and the connector region which links the two halves of the protein represent the most divergent regions (Gottesman and Pastan, 1988).

Gene transfer experiments involving *mdr* cDNAs under the control of several eukaryotic promoters have clearly shown that the expression of the multidrug transporter is sufficient to confer the multidrug resistance phenotype to cultured cells (Gros *et al.*, 1986a; Ueda *et al.*, 1987; Devault and Gros, 1990). Although the two forms of multidrug transporter in rodents exhibit a high degree of amino acid sequence identity (e.g., 71% overall sequence identity for the mouse *mdr*1 and *mdr*3 gene products), they appear to confer a somewhat different pattern of drug resistance to cultured cells upon transfection (Devault and Gros, 1990). As will be described later, the differences in the drug substrate specificity of mouse multidrug transporter isoforms can, however, not be easily ascribed to a few nonconserved amino acid residues only.

The overexpression of P-glycoprotein causes cells to become resistant to a great variety of structurally and functionally unrelated amphiphilic cytotoxic drugs. These include anticancer drugs such as anthracyclines, epipodophyllotoxins, and *Vinca* alkaloids, as well as many other cytotoxic agents some of which are listed in Table 1. According to one current working hypothesis, the multidrug transporter mediates multidrug resistance by acting as an energy-dependent pump that concomitantly decreases drug uptake and increases drug efflux (reviewed in Gottesman and Pastan, 1993). In this model, the active removal of the drugs from the cell membrane prevents their intracellular accumulation and cytotoxic effects. This review will focus on recent studies on the biosynthesis, structure, function, and mechanism of action of the multidrug transporter. Furthermore, the chloride and ATP channel activities that have recently been associated with this intriguing transport system will be discussed, and efforts undertaken by different laboratories to elucidate the physiological role of P-glycoprotein will be described.

In vitro model systems to study the multidrug transporter

During the past two decades, many useful model systems have been developed to study the P-glycoprotein-mediated mechanism of multidrug resis-

Table 1. Compounds which interact with the multidrug transporter

Anticancer drugs	Other cytotoxic agents	Reversing agents	Cyclic and linear peptides
vinblastine	colchicine	verapamil	gramicidin D
vincristine	podophyllotoxin	nifedipine	valinomycin
daunomycin	puromycin	azidopine	*N*-Acetyl-leucyl-leucyl-norleucine
doxorubicin	emetine	quinidine	yeast **a**-factor pheromone
mitoxantrone	ethidium bromide	amiodorone	
etoposide		reserpine	
teniposide		cyclosporine A	
actinomycin D		rapamycin	
mitomycin C		FK506	
mithramycin		progesterone	
taxol		tamoxifen	
topotecan		forskolin	

tance. More than two decades ago Kessel *et al.* (1968) and Biedler and Riehm (1970) noticed that cultured cells selected for resistance to single cytotoxic agent such as actinomycin D, or a *Vinca* alkaloid, respectively, display cross-resistance to a wide range of apparently dissimilar drugs. Based on these initial observations many different laboratories have established a large number of mostly human and rodent multidrug resistant cell lines, usually by exposing the drug-sensitive parental cells to stepwise increasing, but sublethal concentrations of drug (reviewed in Beck and Danks, 1991; Sugimoto and Tsuruo, 1991). Occasionally, a pretreatment with a mutagen or tumor promoter was applied to facilitate the initial step(s) of selection (Akiyama *et al.*, 1985; Shen *et al.*, 1986; Gupta *et al.*, 1988). Despite slightly varying cross resistance patterns, most of these cultured multidrug resistant cell lines produce elevated levels of P-glycoprotein as first described by Juliano and Ling (1976). Thus, these multidrug resistant cell lines represent very useful model systems to study various aspects of the P-glycoprotein-mediated mechanism of multidrug resistance. They were initially used to characterize the multidrug resistance phenotype in great detail, to define drug resistance patterns, and to identify putative reversing agents (Ford and Hail, this volume; Belvedere and Dolfini, this volume; Clynes et al., this volume; Nooter and Sonneveld, this volume). Various drug-sensitive parental cells and their multidrug resistant derivatives have also been used to determine the kinetics of uptake, efflux, and accumulation of drugs, and to analyze the energy requirements of the drug transport process (Michelson, this volume). Extensive studies on the intracellular localization, biosynthesis, maturation (involving posttranslational modifications such as glycosylation and phosphorylation), stability, degradation, structure, and topology of P-glycoprotein, and initial characterizations of its drug and nucleotide interaction sites have been performed in multidrug resistant cell lines (this chapter). Furthermore, highly multidrug resistant cells have been used successfully to raise antibodies against various extracellular and intracellular epitopes of P-glycoprotein (Heike and Tsuruo, this volume). Mammalian multidrug resistant cell lines also serve as a valuable source for the preparation of plasma membrane vesicles to analyze the P-glycoprotein-mediated drug transport process (this chapter). Moreover, membranes enriched in P-glycoprotein have allowed its partial purification and reconstitution into proteoliposomes for further biochemical analysis (this chapter).

Whereas low-level drug-resistant cultured cells are characterized by the expression of elevated levels of *mdr* mRNA and P-glycoprotein, a molecular genetic analysis of increasingly multidrug resistant cell lines has revealed that *mdr* genes may be amplified during the drug selection process (Schoenlein, this volume). Thus, in highly multidrug resistant cells large copy numbers of *mdr* genes accumulate either in homogeneously staining regions or on extrachromosomal elements (episome, double minute chromosome) (reviewed in Schoenlein, 1993). The amplification of *mdr* genes has greatly facilitated their molecular cloning. Once the cDNAs encoding various P-glycoproteins became available, other *in vitro* model systems were established to perform a molecular analysis of the multidrug transporter. Via DNA- and recombinant retrovirus-mediated transfections it was confirmed that expression of the multidrug transporter was sufficient to confer multidrug resistance to drug-sensitive cells (Gros *et al.*, 1986a; Ueda *et al.*, 1987; Pastan *et al.*, 1988; Devault and Gros, 1990). Stable transfections have facilitated analysis of the functional activity of many different P-glycoprotein mutants. Direct selection of transfected cells by growing them in the presence of a drug substrate for P-glycoprotein (e.g., colchicine, vinblastine, vincristine, see Table 1) has been used to identify fully functional multidrug transporter mutants (e.g., Currier *et al.*, 1989). Selection systems involving a co-transfected dominant selectable marker (e.g., a neomycin resistance gene or an adenosine deaminase gene) present on different plasmids, or present within the same expression vector as the *mdr* gene, but controlled by a different promoter, have proven to be helpful in the enrichment of transfected cell populations in cells expressing P-glycoprotein mutants that are somewhat hampered in their functional activity (Azzaria *et al.*, 1989; Currier *et al.*, 1989; Kane *et al.*, 1989; Currier *et al.*, 1992).

Several heterologous expression systems have also been developed to study the recombinant multidrug transporter. That which most closely approximates the highly multidrug resistant mammalian systems is a baculovirus-mediated expression system in which recombinant human P-glycoprotein is produced in large amounts in *Spodoptera frugiperda* Sf9 insect cells after infection with an *Autographa californica* nuclear polyhedrosis virus carrying *MDR*1 cDNA under the control of the polyhedrin promoter (Germann *et al.*, 1990). Insect cells are able to perform many of the higher eukaryotic posttranslational modifications (reviewed in O'Reilly *et al.*, 1992) and synthesize high levels of an underglycosylated form of P-glycoprotein that is inserted into both intra- and extracellular membranes and assumes a similar structure and function as its native mammalian counterpart (Germann *et al.*, 1990; Sarkadi *et al.*, 1992). Heterologous expression of recombinant human and mouse P-glycoprotein has also been established in *Saccharomyces cerevisiae* (Saeki *et al.*, 1991; Kuchler and Thorner, 1992; Raymond *et al.*, 1992). Like the insect cells, yeast also produces an underglycosylated, but functionally active form of multidrug transporter and thus represents a very useful organism for the further examination of the biological role and mechanism of action of P-glycoprotein. Expression of full-length recombinant P-glycoprotein in *Escherichia coli* has proven to be difficult, which is unfortunate, since this system might offer clear advantages over other systems with respect to the ease and speed of production and analysis of multidrug transporter mutants. Even relatively low amounts of recombinant multidrug transporter seem to interfere with bacterial cell growth and appear to be toxic (U.A. Germann, G.L. Evans, I. Pastan, M.M. Gottesman, unpubl.). Despite numerous attempts in many different laboratories, the successful heterologous expression of a biologically active, full-length mammalian multidrug transporter in bacteria has thus far not been published, although the presence of structural and/or functional homologues in prokaryotes, e.g., *drr*AB (Guilfoile and Hutchinson, 1991) and *bmr* (Neyfakh *et al.*, 1991) implicates that this approach may be feasible. Preliminary experiments, however, suggest that expression of the murine *mdr*1 gene product under control of an inducible T7 promoter in an *OmpT* minus *E. coli* host strain may be possible and there is some evidence for transport of lipophilic cations in this system (E. Bibi, F. Talbot, P. Gros and R. Kaback, pers. comm.). Mouse P-glycoprotein has also been expressed after injection of *mdr*1b mRNA in *Xenopus* oocytes (Castillo *et al.*, 1990). Again, this recombinant P-glycoprotein exhibits structural and functional features similar to the native counterpart produced in murine cells. *Xenopus* oocytes may provide a model to analyze the recently suggested P-glycoprotein-associated channel functions (see below) in more detail. Unfortunately, this expression system appears to be erratic and sometimes fails to provide consistent data for so far unknown reasons (S.B. Horwitz, pers. comm). As will be described in the following paragraph, the *Xenopus* expression system has also been particularly useful in studying the biogenesis of the multidrug transporter, providing data complementary to those obtained from a cell free *in vitro* transcription-translation-translocation system.

Biosynthesis of P-glycoprotein

Immunocytochemical localization studies have revealed that in highly multidrug resistant cells the majority of P-glycoprotein is localized in the plasma membrane at the cell surface which is consistent with its proposed function as a multidrug transporter (Willingham *et al.*, 1987). Some P-glycoprotein has also been detected in association with intracellular organelles that are involved in the processing of glycosylated, multispanning integral membrane proteins, e.g., the Golgi apparatus. Both in a baculovirus-mediated and in a yeast heterologous expression system large amounts of recombinant P-glycoprotein are not only present at the cell surface, but also in intracellular membranes such as Golgi-like vesicles or the nuclear envelope (Germann *et al.*, 1990; Kuchler and Thorner, 1992). It is currently not known whether these internal locations of underglycosylated forms of P-glycoprotein are of any biological significance, i.e., whether P-glycoprotein may assume some so far unknown intracellular transport function within insect or yeast cells. At the present time, it seems likely that these alternative locations reflect an inefficiency of the cellular translocation machinery in processing the overexpressed recombinant multidrug transporter. It is interesting to note, however, that a difference in intracellular drug distribution has often been observed in multidrug resistant cells, which could be accounted for by P-glycoprotein in endocytic vesicles, even though immunocytochemical evidence for such localization is lacking thus far.

Studies on the biogenesis and maturation of polytopic membrane proteins have shown that internal hydrophobic signal sequences usually act in a polar fashion from the amino-terminus to direct the insertion of the polypeptide chain into the endoplasmic reticulum (Friedlander and Blobel, 1985; Wessels and Spiess, 1988). Information on the biosynthesis of P-glycoprotein is so far very limited, but as it becomes more and more obvious that some of the so-called "nonfunctional" mutants may not be processed correctly, this field of research is expected to gain more attention in the near future. Recently, a cell free *in vitro* transcription-linked translation system including dog pancreas microsomal membranes has allowed definition of at least two signal recognition particle (SRP) and SRP-docking protein-dependent signal sequences, one in each half of P-glycoprotein (Zhang and Ling, 1991). In general, stringent control mechanisms may be necessary to ensure a proper membrane insertion of the nascent polypeptide chain which most likely is crucial for the functional activity of P-glycoprotein.

Recently, such coupled transcription-translation-translocation analyses and supplementary studies in *Xenopus* oocytes have revealed alternative topological forms of P-glycoprotein (Zhang and Ling, 1991; Skach *et al.*, 1993; Zhang *et al.*, 1993). Skach *et al.* (1993) used a series of full-length, truncated, and chimeric *MDR*1 cDNA expression vectors to demonstrate biogenesis of a human P-glycoprotein variant with only 10 membrane-spanning regions (Fig. 1B). For putative transmembrane domains 1 through 4, the topological structure of this P-glycoprotein variant is consistent with the originally predicted and so far generally accepted 12-transmembrane domain model (Fig. 1A). The putative cytosolic loop between the transmembrane domains 8 and 9, however, was found to be protected from proteinase K digestion presumably because it resides in the endoplasmic reticulum lumen. Thus, an alternative topology of the carboxy-terminal half of P-glycoprotein was proposed to position this region to the noncytosolic side of the membrane (Fig. 1B). Similarly, Zhang and Ling (1991) postulated an alternative topological variant of P-glycoprotein that contains only four membrane-spanning regions in the carboxy-terminal half (Fig. 1C). Their model is based on the finding of a novel *in vitro* site of glycosylation present between the putative transmembrane domains 8 and 9, which in the 12-transmembrane domain model would be

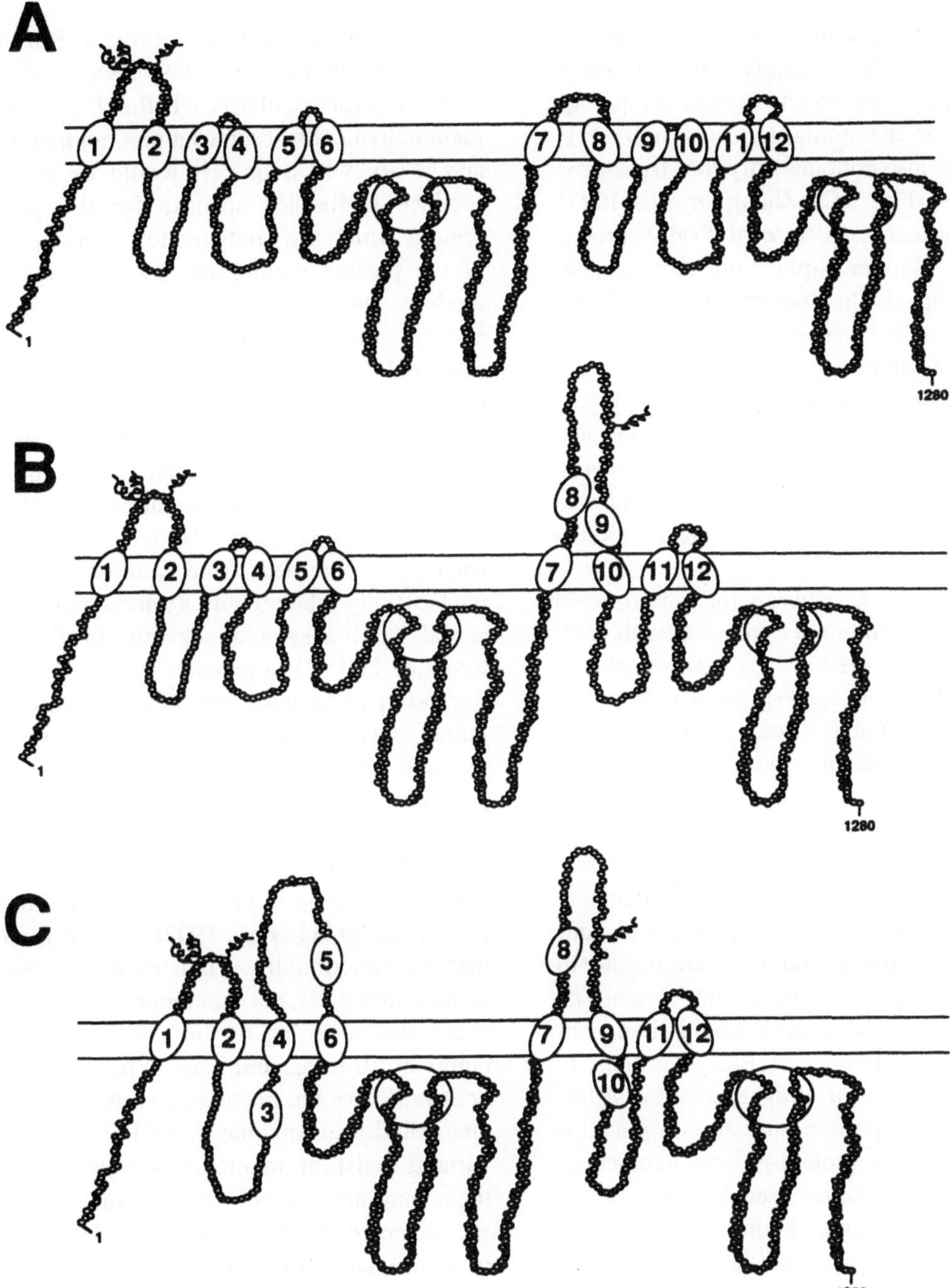

Fig. 1. Alternative topological models of the human multidrug transporter. (A) A 12-transmembrane domain model is predicted by a computer-assisted hydropathy profile and amino acid sequence comparison of P-glycoprotein with bacterial transport proteins (Chen *et al.*, 1986; Gros *et al.*, 1986). Putative transmembrane regions are shown as numbered ellipses. Two ATP-binding sites, based on homology with other ABC transporters, are circled, putative N-linked carbohydrates are represented as wiggly lines. (B) Alternative 10-transmembrane domain model proposed by Skach *et al.* (1993) based on studies on the biogenesis of human P-glycoprotein performed *in vitro* and *in vivo* in *Xenopus* oocytes. (C) Eight-transmembrane domain model proposed by Zhang and co-workers (Zhang and Ling, 1991; Zhang *et al.*, 1993) based on the finding of an additional glycosylation site *in vitro* between the putative transmembrane domains 8 and 9 in model and on studies on the biogenesis of the amino-terminal half of hamster P-glycoprotein. Adapted from reference (Germann *et al.*, 1993).

located on the cytoplasmic site of the plasma membrane (Fig. 1A). Interestingly, a more recent *in vitro* analysis has suggested an alternative topological orientation of the amino-terminal half of P-glycoprotein that also contains only four transmembrane segments (Fig. 1C; Zhang *et al.*, 1993). Although the primary structure of P-glycoprotein appears to be a tandem duplication, these studies taken together suggest that the amino-and carboxy-terminal halves of P-glycoprotein may not necessarily fold in the same way during biogenesis and that the two halves may represent structurally (and perhaps functionally) distinct membrane-associated moieties. It is interesting to note, however, that in both of these studies the alternate topological forms represent less than 50% of the total P-glycoprotein analyzed.

The main difference between these two novel 10- and eight-transmembrane domain models (Fig. 1B,C) and the initially predicted 12-transmembrane domain model (Fig. 1A) for P-glycoprotein is that two, or four hydrophobic stretches, respectively, protrude from the plasma membrane. If these hydrophobic regions were really proven to be exposed from the cell surface in a fraction of multidrug transporter molecules present in mammalian cells, it would be tempting to speculate that they might be involved in interactions with amphiphilic drug substrates and thus contribute to drug substrate recognition, binding, or release. The binding and hydrolysis of ATP could then considerably alter the topology of P-glycoprotein as part of its mechanism of drug extrusion. Alternatively, these hydrophobic stretches might also offer putative sites for protein-protein interaction, such as multimer formation (see below). At the present time it is unknown whether these alternate topological forms of P-glycoprotein might reflect functional intermediates or different functional phenotypes (such as transporter or channel, see below).

In principle, all three hypothetical eight-, ten-, and 12-transmembrane domain models are in agreement with existing cellular epitope localization data obtained with various antibodies specifically recognizing the amino- and carboxy-terminal ends, the two ATP-binding sites, and the first and fourth extracellular loop of P-glycoprotein (Kartner *et al.*, 1985; Yoshimura *et al.*, 1989; Georges *et al.*, 1990, 1993). In order to clarify whether P-glycoprotein in mammalian cells assumes a predominant topology, and if so, which one, it would be necessary to produce antibodies specific for the polypeptide regions which are predicted to be on opposite sides of the plasma membranes by the three different models. Depending on the outcome of such an analysis, the assumed topology of this and maybe other ABC transporters might eventually have to be revised.

Apart from these studies performed in cell free systems, no other experiments have thus far indicated that the carboxy-terminal half of P-glycoprotein is glycosylated. The analysis of proteolytic fragments obtained from multidrug transporter photoaffinity labeled with azidopine *in vivo* did not reveal a glycosylated segment in the carboxy-terminal half (Bruggemann *et al.*, 1989, 1992). Similarly, when wild-type human multidrug transporter or various P-glycoprotein mutants that lack putative glycosylation sites were treated with *N*-glycanase (a mixture of peptide-*N*-glycosidase F and endoglycosidase F), no carbohydrates were detected other than the ones present in the first extracellular loop in the amino-terminal half of the protein (Schinkel *et al.*, 1993). Thus it is suggested that the functionally active multidrug transporter located in the plasma membrane either contains 12 membrane-spanning regions, or that it assumes an eight- or 10-transmembrane domain structure, but does not carry the postulated carbohydrate moiety. One could also imagine that all of these topologic variants exist in mammalian cells, but some of them may never reach the cell surface since they are subjected to intracellular degradation. Obviously additional studies are needed to determine the predominant and functionally active topological form of the multidrug transporter. In the absence of any other clear evidence for the co-existence of the eight- and 10-transmembrane domain variants in multidrug resistant mammalian cells, the 12-transmembrane domain model continues to be the preferred topological model for P-glycoprotein (Fig. 1A). It is this model that will be referred to throughout this review.

The quaternary structure of P-glycoprotein represents another unsolved puzzle. Since 12 transmembrane regions and two nucleotide binding folds constitute a minimal functional unit for many mammalian ABC transporters, it has been assumed that the multidrug transporter predominantly acts in a monomeric state. It is possible, however, that P-glycoprotein fulfills its function as a homodimer or even as a homotetramer. Indeed, according to several radiation inactivation and freeze fracture studies, the estimated size of the multidrug transporter in mammalian plasma membranes is larger than the one predicted for a monomer suggesting an association of multiple subunits to form a functional protein complex (Wright *et al.*, 1985; Arsenault *et al.*, 1988; Boscoboinik *et al.*, 1990; Weinstein *et al.*, 1990). Chemical cross-linking studies also support the idea of functionally active homodimers (Naito and Tsuruo, 1992). A more recent thorough analyses of radiolabeled P-glycoprotein expressed in multidrug resistant hamster cells involving sucrose gradient velocity sedimentation of postnuclear lysates has indicated that a large fraction of P-glycoprotein assembles to oligomers in an early biosynthetic compartment (M.S. Poruchynsky and V. Ling, pers. comm.). Oligomers were also obtained after treatment of cells *in situ* with a crosslinker prior to any detergent solubilization. Interestingly, P-glycoprotein monomers, dimers, and oligomers are all labeled by photoactive drug- or ATP-analogs, and are post-translationally modified by phosphorylation, although to different extents (M.S. Poruchynsky and V. Ling, pers. comm.). These studies may imply that the formation and dissociation of P-glycoprotein oligomers may contribute to its function as a mediator of multidrug resistance, but additional experiments are required to support this hypothesis. Detergent-solubilized P-glycoprotein purified from plasma membranes of multidrug resistant KB-V1 cells seems to be predominantly monomeric (S. Ambudkar, I. Pastan, M.M. Gottesman, pers. comm.).

Pulse-chase labeling experiments have indicated that P-glycoprotein is synthesized as a nonglycosylated precursor (apparent molecular weight of approximately 130,000–140,000), which is slowly ($t_{1/2}$ = 1–2 h in human cells, $t_{1/2}$ = 20–30 min in mouse cells) processed to its mature form (apparent molecular weight ranging from 145,000 to 180,000) (Greenberger *et al.*, 1988; Richert *et al.*, 1988; Ichikawa *et al.*, 1991). P-glycoprotein appears to be a very stable protein *in vitro* with a half-life of 48–72 h in cultured human multidrug resistant cells and approximately 18 h in mouse cells, where the rate of degradation is similar among multidrug transporter isoforms (Richert *et al.*, 1988; Cohen *et al.*, 1990). Interestingly, distinct isoforms of P-glycoprotein differ in the degree and sites of glycosylation. To date, the composition of any P-glycoprotein carbohydrate moiety has not been elucidated. Several experiments suggest that glycosylation is not required to provide a functionally active multidrug transporter. Treatment with tunicamycin does not decrease the levels of drug resistance in multidrug resistant cells, and multidrug resistant sublines can be selected from glycosylation-defective parental cells (Ling *et al.*, 1983; Ichikawa *et al.*, 1991). As mentioned above, insect cells and yeast cells were found to produce underglycosylated forms of recombinant multidrug transporter. Despite the partial lack of carbohydrates, however, the majority of these recombinant P-glycoproteins are translocated to the cell surface of the appropriate host cell and appear to be functionally similar to their native counterparts expressed in mammalian cells (Germann *et al.*, 1990; Saeki *et al.*, 1991; Kuchler and Thorner, 1992; Raymond *et al.*, 1992; Sarkadi *et al.*, 1992). A more conclusive recent study involved a series of P-glycoprotein mutants that are partially or fully carbohydrate-deficient (Schinkel *et al.*, 1993). This study has demonstrated that the lack of *N*-glycosylation in the putative first extracellular loop of the human *MDR*1 gene product affects neither the level of drug resistance, nor the pattern of cross-resistance. Since the absence of these three putative *N*-glycosylation sites drastically decreased the number of drug-resistant clones upon transfection, however, it has been suggested that *N*-glycosylation may contribute to the correct folding and/or proper routing and/or stabilization of P-glycoprotein *en route* to or within the plasma membrane.

P-glycoproteins overexpressed in multidrug

resistant cell lines are posttranslationally modified by phosphorylation (e.g., Carlsen *et al.*, 1977; Garman *et al.*, 1983; Roy and Horwitz, 1985; Hamada *et al.*, 1987; Richert *et al.*, 1988; Myers *et al.*, 1989; Schurr *et al.*, 1989; Chambers *et al.*, 1990; Ma *et al.*, 1991; Yu *et al.*, 1991; Bates *et al.*, 1992). This covalent modification of P-glycoprotein is also observed in cells stably transfected with *mdr*1 cDNA (Schurr *et al.*, 1989) or in the heterologous baculovirus-mediated expression system (Germann *et al.*, 1990). Treatment of multidrug resistant cells with protein kinase inhibitors (e.g., H-87, H9, K525a, or staurosporine) may circumvent drug resistance (Miyamoto *et al.*, 1990; O'Brian *et al.*, 1991; Miyamoto *et al.*, 1993; Sampson *et al.*, 1993). Reversal of the multidrug resistance phenotype may be due to changing the phosphorylation status of P-glycoprotein, although in some cases (e.g., staurosporine) competitive binding to the multidrug transporter may cause increased intracellular drug accumulation (Sato *et al.*, 1990) (I.H. Lelong, I. Pastan, M.M. Gottesman, unpubl.) and some kinase inhibitors (e.g., staurosporine) also influence *MDR*1 mRNA levels (Chaudhary and Roninson, 1992; Sampson *et al.*, 1993). Levels of cAMP dependent protein kinase have also been shown to modulate *mdr* mRNA levels in Chinese hamster and mouse cells (Abraham *et al.*, 1987, 1990; Chin *et al.*, 1992).

The multidrug transporter has been suggested to undergo rapid cycles of phosphorylation and dephosphorylation (Ma *et al.*, 1991) and protein phosphorylation differences may play a role in the altered multidrug resistance phenotype. Preliminary studies suggest an involvement of the membrane-associated protein phosphatases 1 and 2A (Chambers *et al.*, 1992). The protein kinase(s) responsible for P-glycoprotein phosphorylation has(ve) also not been clearly identified yet. Possible candidates include protein kinase C (PKC) (Chambers *et al.*, 1990; Chambers *et al.*, 1992), cAMP-dependent protein kinase (PKA) (Mellado and Horwitz, 1987), a novel kinase that is calcium-independent and phospholipid-dependent, but so far has not been characterized in great detail (Staats *et al.*, 1990), or other novel, membrane-bound kinases (S. Ambudkar, I.H. Lelong, I. Pastan, and M.M. Gottesman, pers. comm.). It is possible that depending on the species, cell type, and/or state of differentiation several kinases contribute to P-glycoprotein phosphorylation. Alternatively, a cascade involving different kinases might be regulating P-glycoprotein phosphorylation.

Many multidrug resistant cell lines express increased levels of protein kinases, most prominently PKC (Fine *et al.*, 1988; Posada *et al.*, 1989; O'Brian *et al.*, 1991). Activation of PKC *in vivo* with 12-*O*-tetradecanoylphorbol-13-acetate (TPA) stimulates P-glycoprotein phosphorylation in several multidrug resistant cell lines and decreases drug accumulation (Hamada *et al.*, 1987; Chambers *et al.*, 1990), while inhibition of PKC can partially reverse the multidrug resistance phenotype (Posada *et al.*, 1989; Ma *et al.*, 1991; Chambers *et al.*, 1992). Furthermore, transfection of the P-glycoprotein overexpressing human breast cancer cell line MCF-7 with PKCalpha enhances P-glycoprotein phosphorylation as well as levels of drug resistance (Yu *et al.*, 1991). Hence, many studies indicate a role for PKC, in particular PKCalpha in regulating multidrug resistance, but not necessarily other PKC isoenzymes (Ahmad *et al.*, 1992; Blobe *et al.*, 1993). Interestingly, in a recent report, PKC activation by TPA and diacylglycerol (DAG) in normal human lymphocytes has been associated not only with increased multidrug transporter activity, but also increased levels of *MDR*1 mRNA and P-glycoprotein expression, suggesting a PKC-mediated signal transduction pathway for the upregulation of *MDR*1 gene expression. In other studies, SW620 human colon carcinoma cells were treated with sodium butyrate resulting in a decreased phosphorylation of P-glycoprotein that correlated with lowered transport activity for some drugs (Bates *et al.*, 1992). Although this lowered functional activity could be partially restored with TPA, a direct involvement of PKC was not demonstrated.

Taken together, many of these observations have to be interpreted with caution and the exact role of P-glycoprotein phosphorylation must await *in vitro* evidence for a functional effect on purified, reconstituted P-glycoprotein, and direct *in vivo* evidence based on mutagenesis. The identification of actual phosphorylation sites within the multidrug transporter may help in this respect. Early studies have implicated multiple phosphorylation sites within P-glycoprotein, mainly serine residues (Hamada *et al.*, 1987, Ma *et al.*, 1991). Recently, a partial amino acid sequence analysis of human P-glycoprotein revealed that three Ser residues clustered within the linker region are specifically phosphorylated by PKC *in vitro*, and most likely also *in vivo* (Chambers *et al.*, 1993). The human multidrug transporter may, therefore, contain a putative small regulatory domain between its two homologous halves, at a location analogous to the phosphorylatable R-domain that regulates the chloride channel function of the CFTR gene product (Riordan *et al.*, 1989). One approach to assess the role of P-glycoprotein phosphorylation in intact cells could involve site-directed mutagenesis analysis of these identified sites of phosphorylation, e.g., the substitution by a nonphosphorylatable amino acid such as Ala, or replacement by Asp, that resembles a permanently phosphorylated Ser-like residue.

P-glycoprotein: a multidrug transporter

It is generally believed that P-glycoprotein serves as an energy-dependent drug efflux pump. The definite proof of this hypothesis, however, still awaits its purification to homogeneity and its subsequent reconstitution into artificial plasma membranes, in which state its biological activity could be examined. Such studies may also help to completely elucidate the mechanism of action of P-glycoprotein. Currently, the multidrug transport activity may be viewed in an abstract sense as a series of conformational changes of the P-glycoprotein polypeptide chain that is triggered through drug interactions (drug binding, drug transport through the membrane lipid bilayer, and drug release), and nucleotide interactions (ATP binding, ATP hydrolysis, and release of ADP). Major open questions include the number, location, and nature of the drug interaction sites, the relative importance of the two putative ATP binding sites, and the mechanism and stoichiometry of the energy transduction and the drug transport process. In the following section, the most promising approaches to study the multidrug transporter structure, function, and mechanism of action will be reviewed.

Identification of functional domains by a mutational analysis

Both naturally occurring and genetically engineered P-glycoprotein mutants have helped to identify amino acid residues and protein domains which directly contribute to its function as a multidrug transporter. As described in the introduction, P-glycoprotein represents a tandemly duplicated molecule and consists of two highly homologous halves. Transfection experiments with truncated *MDR*1 cDNAs encoding either half of the multidrug transporter have demonstrated that both intact halves are required to confer multidrug resistance to drug-sensitive cells, and neither half can function independently as a transporter (Currier *et al.*, 1989) and (L.E. Airan, U.A. Germann, S.J. Currier, I. Pastan, M.M. Gottesman, unpubl.). Preliminary experiments suggest that the two halves of P-glycoprotein do not reconstitute to a functional multidrug transporter when expressed concomitantly in the same cell (L.E. Airan, U.A. Germann, S.J. Currier, I. Pastan, M.M. Gottesman, unpubl.). It would be interesting, however, to determine the multidrug transporter activity of P-glycoprotein mutants that consist of a duplicated amino- or carboxy-terminal half, respectively.

An analysis of various genetically engineered insertion and deletion mutants of the multidrug transporter has revealed that some small changes are tolerated, but most of the polypeptide chain is indispensable (Currier *et al.*, 1989). Minor carboxy-terminal deletions of three, seven and 23 amino acids residues preserve a biologically active multidrug transporter suggesting that the very carboxy-terminal end of P-glycoprotein is not essential for drug binding and transport activity. This hypothesis is supported by the fact that small epitope additions (Loo and Clarke, 1993), or carboxy-terminal translational fusions with another protein, e.g., human adenosine deaminase (Germann *et al.*, 1989), or bacterial β-galactosidase (Shimabuku *et al.*, 1992) do not interfere with the ability of P-glycoprotein to confer multidrug resistance. A larger deletion of 53 amino acids from the carboxy terminus, however, eliminates the multidrug transport function (Currier *et al.*, 1989). To date, no deletions or other mutations of the amino-terminus have been described in great detail. A few naturally occurring alternatively spliced *pgp*1 gene transcripts have been detected in a highly actinomycin D resistant Chinese hamster cell line, but they have so far not been proven to produce a functional multidrug transporter (Devine *et al.*, 1991). The amino-terminal end represents one of the least conserved regions between different multidrug transporters and, therefore, may not be very critical for its drug binding and/or transport ability. This region may nevertheless be essential e.g., for its biosynthesis, maturation, and/or translocation to the plasma membrane as may be suggested by the finding that a chimeric P-glycoprotein with an amino-terminal protein extension was not able to confer multidrug resistance to drug-sensitive cells upon transfection (U.A. Germann, I. Pastan, and M.M. Gottesman, unpubl.).

As pointed out in the introduction, both the amino- and carboxy-terminal half of all P-glycoproteins harbor a highly homologous ATP binding/utilization domain that is characterized by several conserved elements such as the "Walker motifs A and B" (Walker *et al.*, 1982) , the "center region", and the "linker peptide" (Shyamala *et al.*, 1991). Different laboratories have shown by site-directed mutagenesis that both of these two predicted nucleotide binding regions are essential for the multidrug transport function. Discrete substitutions of the conserved Gly with Ala, or the conserved Lys with Arg within the "Walker motif A" in either the amino- or carboxy-terminal half of the mouse *mdr*1b gene product were found to abrogate its ability to confer the multidrug resistance phenotype to drug sensitive cells (Azzaria *et al.*, 1989). Interestingly, these mutations did not interfere with photoaffinity labeling by 8-azido-ATP, even when present in both halves of P-glycoprotein, suggesting that ATP binding was not impaired, but that a subsequent step in the drug transport process was defective. More radical mutations in the nucleotide binding region(s) in either half of the human *MDR*1 gene product in which the conserved Lys residues were replaced by Met residues reduced ATP binding (B. Schott, B.M. Morse, I.B. Roninson, pers. comm.). Surprisingly, transfectants selected for expression of a neomycin resistance gene within the same plasmid as the mutated *MDR*1 cDNA, but controlled by a different promoter, showed low-level resistance to vinblastine. Subsequent selection of the corresponding transfectants for increased resistance to vinblastine led to increased expression of these mutant forms of P-glycoprotein (B. Schott, B.M. Morse, I.B. Roninson, pers. comm.). Taken together, these results suggest that both homologous ATP binding domains are crucial for a fully functional P-glycoprotein and the two halves do not act independently, but cooperate. This latter conclusion is supported by a genetically engineered human P-glycoprotein mutant that carries a carboxy-terminal type ATP-binding fold (including both Walker motifs and the loop in between) in both halves of the molecule (P. Wu, S.J. Currier, I. Aksentijevich, I. Pastan, M.M. Gottesman, U.A. Germann, unpubl.). In contrast to wild-type *MDR*1 cDNA, the transfection of such a modified *MDR*1 cDNA does not allow direct selection of drug-resistant clones. Unless this P-glycoprotein mutant is affected in its stability or translocation to the plasma membrane, this result may point to a functional polarity within the multidrug transporter and indicate that both ATP-binding folds are required to provide a fully active multidrug transporter. It is tempting to speculate that the amino- and carboxy-terminal ATP-binding folds of human P-glycoprotein may either serve different functions, or that they need specific interactions with the appropriate transmembrane domain to fulfill an equivalent function. In this respect it is interesting to note that the amino-terminal half of P-glycoprotein, but not the carboxy-terminal half has the capacity to independently catalyze ATP hydrolysis (Shimabuku *et al.*, 1991, 1992).

Interestingly, an analysis involving chimeric mouse *mdr*1/*mdr*2 gene products has revealed that the appropriate *mdr*2 ATP-binding domains in either half of P-glycoprotein complement the drug transport activity of the *mdr*1b gene product (Buschman and Gros, 1991). A series of stable transfections with similar human *MDR*1/*MDR*2 chimeras confirmed this finding (P. Wu, S.J. Currier, I. Aksentijevich, I. Pastan, M.M. Gottesman, U.A. Germann, unpubl.) The human *MDR*2/3 (Van der Bliek *et al.*, 1987, 1988) and the mouse *mdr*2 (Gros *et al.*, 1988; Buschman *et al.*, 1992) gene products have so far not been directly associated with the drug resistance phenomenon (Schinkel *et al.*, 1991). Since their highly homologous (approximately 88% identical) ATP-binding folds are functionally interchangeable with the appropriate human *MDR*1 or mouse *mdr*1 ATP-binding folds, however, it is quite likely that their ATP-binding domains support some so-far unknown energy-dependent transport activity.

In contrast, chimeras in which either the amino-terminal or the carboxy-terminal ATP-binding fold was replaced by the appropriate ATP-binding fold of the ABC transporter CFTR were found to be unable to confer drug resistance to drug-sensitive cells upon transfection (P. Wu, S.J. Currier, I. Aksentijevich, I. Pastan, M.M. Gottesman, U.A. Germann, unpubl.). Provided that additional experiments show these chimeras to represent stable proteins expressed at the cell surface, these data suggest that either the approximately 30% identical *MDR*1 and CFTR ATP-binding folds are functionally unrelated, or, more likely, that they are protein-specific and that some highly divergent regions between the two Walker motifs may harbor essential amino acid residues that direct this specificity.

Although the two halves of P-glycoprotein seem to interact with each other for functional activity, the short stretch (approximately 25 amino acids) which connects them and carries potential *in vivo* phosphorylation sites (Chambers *et al.*, 1993) exhibits some small degree of flexibility. Apart from very few identical amino acids this connecting region is not conserved between different forms of P-glycoprotein. Transfection studies involving chimeras have demonstrated that the highly divergent *mdr*2 connecting region can substitute for the *mdr*1 connecting region without perturbing the multidrug transporter function of the murine P-glycoprotein (Buschman and Gros, 1991). Furthermore, the connecting region of the human *MDR*1 gene product tolerates small insertions of 17 to 18 amino acids as long as they assume a predicted random coil secondary structure (L.E. Airan, U.A. Germann, S.J. Currier, I. Pastan, M.M. Gottesman, unpubl.). In contrast, a polypeptide insertion of the same size with a predicted alpha-helical secondary structure interferes with the ability to confer multidrug resistance to drug sensitive cells (L.E. Airan, U.A. Germann, S.J. Currier, I. Pastan, M.M. Gottesman, unpubl.). Although suggested to be somewhat critical for a functional multidrug transporter, the precise role of this connecting region remains to be elucidated.

A functional analysis of chimeric genes obtained by exchanging homologous domains of the mouse *mdr*1 and *mdr*2 genes has implied that potential drug substrate interaction sites may reside within the transmembrane domains of P-glycoprotein (Buschman and Gros, 1991). The approach to use chimeras between the mouse *mdr*1 and *mdr*3 genes to explore the basis of the different cross resistance patterns conferred by the encoded P-glycoproteins suggested that drug substrate specificity is achieved by complex interactions of structural elements present in both homologous halves (Dhir and Gros, 1992). Indeed, several point mutations scattered throughout the primary structure, but within predicted transmembrane domains or in close proximity thereof have been shown to influence the cross resistance pattern conferred by P-glycoprotein, presumably because they are involved in recognition, binding, or release of drug substrates (Fig. 2). A Gly185 to Val185 substitution occurred spontaneously near the putative transmembrane 3 in P-glycoprotein when multidrug resistant human KB carcinoma cells were selected in stepwise increasing amounts of colchicine in the growth medium after an initial treatment with ethyl methanesulfonate (Choi *et al.*, 1989; Kioka *et al.*, 1989). Subsequent gene transfer studies showed that the Val185 mutant P-glycoprotein confers preferential resistance to colchicine and etoposide, but decreased resistance to *Vinca* alkaloids and actinomycin D in comparison with the Gly185 wild-type P-glycoprotein (Safa *et al.*, 1990). In one study the Val185 mutant P-glycoprotein was demonstrated to bind photoactive analogs of colchicine less well and photoactive analogs of vinblastine better than the Gly185 wild-type P-glycoprotein, suggesting that this single amino acid substitution near the putative transmembrane domain 3 may interfere with drug substrate release, rather than initial drug substrate binding (Safa *et al.*, 1990). In another study vinblastine was found to be a more effective inhibitor of photoaffinity labeling of the Val185 mutant P-glycoprotein than of the Gly185 wild-type P-glycoprotein and it was concluded that the Val185 mutation may affect the initial interactions of the drug with the multidrug transporter (Bruggemann *et al.*, 1992). Recent measurements of the kinetics of drug uptake by transfected NIH3T3 cells expressing approximately equal amounts of either the Val185 mutant or the Gly185 wild-type P-glycoprotein have, however, indicated that a possible explanation may not be that simple, since complex effects were observed due to this single amino acid substitution (W. Stein, C.O. Cardarelli, M.M. Gottesman and I. Pastan, pers. comm.).

Another spontaneously selected P-glycoprotein mutant with two adjacent amino acid substitutions in the putative transmembrane domain 6, namely Gly338 to Ala338, and Ala339 to Pro339, has been found in multidrug resistant hamster cells selected in the presence of high concentrations of actinomycin D (Devine *et al.*, 1992). This double mutant P-glycoprotein confers a preferential resistance to actinomycin D, mainly due to the Pro substitution of Ala339 (P.W. Melera, pers. comm.).

Several genetically engineered discrete amino acid changes also alter the substrate specificity of P-glycoprotein drastically. A second substitution of Asn183 by Ser183 next to the Val185 substitution within the first intracytoplasmic loop close to the transmembrane domain 3 of the human multidrug transporter (see above) causes a relative increase in resistance to actinomycin D and vinblastine, without decreasing the relative colchicine resistance (Currier *et al.*, 1992). Hence, nonidentical, but overlapping sites within P-glycoprotein may be involved in determining its drug transport specificity.

A single substitution within the putative transmembrane domain 11, by which a Ser at position 941 in the mouse *mdr*1, or at position 939 in the mouse *mdr*3 gene product is replaced by a Phe, reduces the capacity to confer colchicine and adriamycin resistance drastically, while the capacity to confer vinblastine resistance is only slightly affected (Gros *et al.*, 1991). Furthermore, this Ser to Phe mutation also decreases the capacity of verapamil and progesterone to modulate the multidrug transporter activity (Kajiji *et al.*, 1993). Since the mutant P-glycoproteins are much less efficiently labeled by photoaffinity labeling reagents azidopine and iodoarylazidoprazosin, it is suggested that the Ser to Phe substitutions within the putative transmembrane domain 11 interfere with the initial binding of drugs to P-glycoprotein (Kajiji *et al.*, 1993).

A site-directed mutagenesis of 13 Pro residues into Ala within or close to the predicted transmembrane domains of the human *MDR*1 gene product has revealed another two amino acid residues that are important for its multidrug transporter function. Mutation of either Pro223 in the amino-terminal half within the putative transmembrane domain 4, or of Pro866 at an identical position in the carboxy-terminal half within the putative transmembrane domain 10 reduces the ability of the multidrug transporter to confer resistance to colchicine, adriamycin, or actinomycin D, while the ability to confer resistance to vinblastine is maintained (Loo and Clarke, 1993).

Taken together, these studies confirm that genetic variations may play a role in determining the cross-resistance pattern that accompanies the multidrug resistance phenotype. This is in agreement with the distinct cross-resistance patterns conferred by different multidrug transporter isoforms and a series of genetically engineered mouse *mdr1*/*mdr*3 chimera (Devault and Gros, 1990; Dhir and Gros, 1992). The single amino acids which have been identified so far to influence the substrate specificity of P-glycoproteins are widely separated from each other in the primary structure of P-glycoprotein, but they are all in close proximity or within putative transmembrane regions. It is tempting to speculate that in a three-dimensional model the majority of the transmembrane segments participate in the transport process and are part of a pore- or channel-like structure through which the drugs are pumped out of the cell. Since various drug substrates are affected differentially by the described mutations, however, their initial P-glycoprotein binding sites may not necessarily be identical.

Interestingly, deletion mutations that may constrain the P-glycoprotein tertiary structure also appear to alter the relative drug resistance pattern. A variant of an expressed human *MDR*1 mRNA lacking exon 20 was detected in an Adriamycin-resistant subline of human myelogenous leukemia cells K562/ADM (Y. Sugimoto, T. Tsuruo, pers. comm.). Stable transfections were performed to assess the functional activity of this P-glycoprotein mutant in which approximately half (amino acids 800—827) of the putative 4th intracytoplasmic loop is missing. These experiments revealed that the exon 20-deletion mutant of the *MDR*1 cDNA is able to confer multidrug resistance to drug-sensitive cells, but the cross resistance pattern is slightly changed when compared to wild-type (Y. Sugimoto, T. Tsuruo, pers. comm.). In other studies, a genetically engineered deletion of 20 amino acids comprising two of the three putative glycosylation sites in the first extracytoplasmic loop of the human multidrug transporter was shown to not only cause an overall decrease of the drug transport activity, but to also modulate the substrate specificity, most strikingly for rhodamine 123 (Schinkel *et al.*, 1993).

Several genetically engineered P-glycoprotein mutations within or close to membrane-spanning regions resulted in a protein unable to confer multidrug resistance. These include a series of mouse *mdr1*/*mdr2* chimera in which transmembrane segments were exchanged (Dhir and Gros, 1992). Furthermore, substitution of Pro709 by Ala located on the cytoplasmic side of the putative transmembrane segment 7 was found to induce a structural perturbation that was inconsistent with multidrug transporter activity (Loo and Clarke, 1993). Within the first intracytoplasmic loop that connects the putative membrane-spanning regions 2 and 3 a small segment of four *MDR*1 amino acids (residues 165, 166, 168, and 169) was found to be crucial for multidrug transport activity (Currier *et al.*, 1992). Substitution of these four amino acids by the appropriate *MDR*2 amino acids resulted in a P-glycoprotein mutant that was incapable of conferring drug resistance to drug-sensitive cells. Furthermore, a chimeric P-glycoprotein was analyzed in which the full-length putative first intracytoplasmic *MDR*1 loop between amino acids 140 and 229 was substituted by the appropriate *MDR*2 region. This chimera, containing a total of 17 amino acid differences, was found to be located at the cell surface and to be labeled by azidopine, but to be unable to confer the multidrug resistance phenotype to drug-sensitive cells. Surprisingly, a change of only the four *MDR*2 amino acid residues 165, 166, 168, and 169 in this chimera back to *MDR*1 residues fully restored multidrug transporter activity, underscoring their importance for this function. At the present time it remains to be elucidated whether all four of these *MDR*1 amino acid residues or only one, two, or three of them are essential.

Biochemical determination of drug- and ATP-binding sites

Initially, biochemical evidence for the drug binding capacity by P-glycoprotein was gained indirectly by demonstrating that radioactively labeled cytotoxic drugs (e.g., vinblastine) or agents that reverse multidrug resistance (e.g., verapamil) bind to plasma membranes isolated from multidrug resistant cells, but not to plasma membranes from drug-sensitive cells (Cornwell *et al.*, 1986a). Later on, the use of radioactive photoaffinity labeling reagents and P-glycoprotein-specific antibodies has unequivocally allowed the identification of the multidrug transporter as a drug- and ATP-binding protein (Cornwell *et al.*, 1991).

P-glycoprotein is labeled by a great variety of photoactive analogs of drugs and reversing agents. These include e.g., vinblastine analogs (Cornwell *et al.*, 1986b), colchicine analogs (Safa *et al.*, 1989, 1990), verapamil analogs (Safa, 1988), iodomycin, a daunomycin derivative (Busche *et al.*, 1989), azidopine (Safa *et al.*, 1987; Akiyama *et al.*, 1988; Bruggemann *et al.*, 1989; Yoshimura *et al.*, 1989; Greenberger *et al.*, 1990), forskolin derivatives (Morris *et al.*, 1991), arylazidoprazosin (Greenberger *et al.*, 1991), cyclosporin A (Foxwell *et al.*, 1989) and others (reviewed in Beck and Qian, 1992; Safa, 1993). These photoaffinity labeling reactions are specific as demonstrated by competition experiments involving an excess of nonradioactive drugs and/or reversing agents that are associated with the multidrug resistance phenotype. While unlabeled drug and/or reversing agents inhibit these photoaffinity labeling reactions in a dose-dependent manner, the relative efficiency of blocking is quite variable. For instance, photolabeling of P-glycoprotein by many different compounds is much more effectively inhibited by *Vinca* alkaloids than by colchicine (Safa, 1993). It is, therefore, suggested that either these drugs bind to separate, but overlapping or coupled sites, or that P-glycoprotein harbors a common drug-acceptor site that displays variable affinity for different drugs.

Many efforts have been undertaken to identify the location of these photoaffinity labeling sites within the primary structure of P-glycoprotein in order to determine putative drug interaction sites. The most common approach involves 1) labeling of multidrug resistant cells with a radioactive photoaffinity labeling reagent, 2) extracting labeled P-glycoprotein and purifying it by immunoprecipitation or gel electrophoresis under denaturing conditions, 3) subjecting the labeled P-glycoprotein to either proteolytic digestion or chemical cleavage, and 4) subsequently identifying the radioactively labeled P-glycoprotein fragments with a series of monoclonal or polyclonal antibodies raised against different linear epitopes of P-glycoprotein (e.g., Bruggemann *et al.*, 1989; Greenberger, 1993). Initial studies revealed an azidopine labeling site in the carboxy-terminal half of mouse P-glycoprotein (Greenberger *et al.*, 1990), and at least one azidopine labeling site in either half of the human multidrug transporter (Bruggemann *et al.*, 1989; Yoshimura *et al.*, 1989; Bruggemann *et al.*, 1992). The exact azidopine labeling sites have not been identified yet, but for the human multidrug transporter one site comprises the putative transmembrane domains 3–6 and the other one is between the putative transmembrane domains 7–12 (see Fig. 2; Bruggemann *et al.*, 1989). Both of these sites are equally labeled by azidopine (Bruggemann *et al.*, 1992). Mouse P-glycoprotein was later demonstrated to contain at least two labeling sites for arylazidoprazosin (Greenberger *et al.*, 1991). As for the azidopine labeling sites of the human P-glycoprotein, there is one arylazidoprazosin labeling site present in each half of the mouse P-glycoprotein, namely one within the putative transmembrane domains 6 or within the immediately following cytoplasmic region in the amino-terminal half, the other one at an analogous site in the carboxy-terminal half within or near the transmembrane domain 12 (see Fig. 2; Greenberger *et al.*, 1991; Greenberger, 1993). Interestingly, overlapping, if not the same sites within the human multidrug transporter are labeled by a forskolin derivative (see Fig. 2; D.I. Morris, L.M. Greenberger, E.P. Bruggemann, C.O. Cardarelli, M.M. Gottesman, I. Pastan, K.B. Seamon, pers. comm.). No exact labeling sites have yet been determined for any of these photoaffinity labeling reagents, thus it is not known whether a few P-glycoprotein amino acids are specifically involved in these reactions. At the present time it is also unclear whether these major sites are the only labeling sites, or whether there exist other minor labeling sites. Nevertheless it has been well established that both homologous halves of P-glycoprotein interact with various photoactive drug analogs. This finding suggests two possible models for the mechanism of action of P-glycoprotein. Either the two halves of P-glycoprotein act independently and represent two different drug binding and transport sites, or the two halves interact with each other and form a single pore- or channel-like structure through which the drugs are exported. Recent studies have indicated that the amino- and carboxy-terminal site are equally labeled by azidopine and also equally inhibited by vinblastine in a dose-dependent manner (Bruggemann *et al.*, 1992). Since this was observed for both the wild-type *MDR*1 gene product and the previously described P-glycoprotein mutant, that carries a Gly185 to Val185 substitution resulting in a decreased overall vinblastine transport, these data support a single transport channel model in which the two halves of P-glycoprotein interact and cooperate for function. Provided that the photoaffinity labels mark the site(s) of transport, this hypothesis still allows various drugs to enter and initially bind to P-glycoprotein at nonidentical sites.

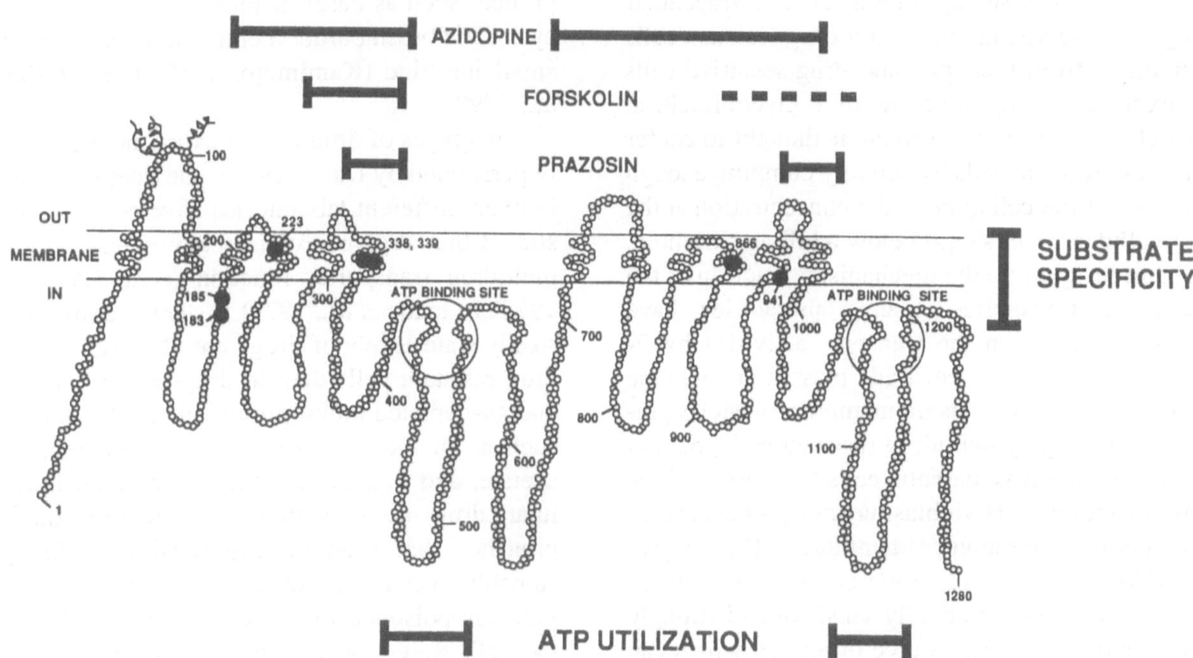

Fig. 2. Two-dimensional model of the multidrug transporter inserted into the plasma membrane (see Fig. 1A) showing functional domains. Heavy bars indicate the photoaffinity labeling sites for azidopine (Bruggemann *et al.*, 1989, 1992), a forskolin derivative (D.I. Morris, L.M. Greenberger, E.P. Bruggemann, C. Cardarelli, M.M. Gottesman, I. Pastan, K.B. Seamon, pers. commun.), and iodoarylazidoprazosin (Greenberger, 1993), the drug substrate specificity sites at positions 183 (Currier *et al.*, 1992), 185 (Choi *et al.*, 1989; Kioka *et al.*, 1989), 223 (Loo *et al.*, 1993), 338, 339 (Devine *et al.*, 1992), 866 (Loo and Clarke, 1993) and 941 (Gros *et al.*, 1991), and the nucleotide binding domains (Walker *et al.*, 1982). Adapted from reference (Chin *et al.*, 1993).

As discussed earlier the P-glycoprotein-mediated mechanism of multidrug resistance is highly dependent on intracellular ATP levels and the multidrug transporter bears two homologous consensus nucleotide binding sites that are essential for its functional activity. The multidrug transporter has been demonstrated to be affinity labeled by 8-azido-ATP (Cornwell *et al.*, 1987, 1991; Sarkadi *et al.*, 1992). Both ATP and GTP, as well as the nonhydrolyzable ATP-analogs AMP-PNP compete for this reaction. Consistent with this finding is that the multidrug transporter is able to use both ATP and GTP as a source of energy (Lelong *et al.*, 1992). The 8-azido-ATP labeling is unaffected, however, by ADP, ribose-5-phosphate, or drug substrates (Cornwell *et al.*, 1987). 2-Azido-ATP as well as 7-chloro-4-nitrobenzo-2-oxa-1,3-diazole (NBD-Cl) were also shown to label P-glycoprotein and ATP acts as competitive inhibitor for these reactions (Al-Shawi and Senior, 1993). General topological models of ABC transporter ATP-binding domains obtained by computer-assisted modeling (Hyde *et al.*, 1990; Mimura *et al.*, 1991) and the structure-function analysis of the nucleotide binding domains described in the previous section may provide an idea about the location of the ATP-binding sites within P-glycoprotein, but the precise amino acids that are involved in ATP-binding have so far not been identified experimentally. Moreover, it is unknown whether both the amino- and the carboxy-terminal ATP-binding sites are functionally equivalent, or whether ATP binds to one of the two sites preferentially.

Drug transport and ATPase activity

Generally, lower steady state levels of intracellular drug are observed in intact multidrug resistant cells that differ from their parental drug-sensitive cells by expressing large amounts of P-glycoprotein at the cell surface. P-glycoprotein is thought to confer drug resistance to cells by actively pumping enough drug out of the cell to keep the concentration at the intracellular target site(s) below a killing threshold. To gain insight into the mechanism of action of the multidrug transporter, many laboratories have initiated studies on the transport activities by P-glycoprotein using purified plasma membrane vesicles. Inside-out plasma membrane vesicles prepared from highly multidrug resistant cells, but not from drug-sensitive parental cells have allowed the demonstration of ^{3}H-vinblastine transport against an apparent concentration gradient due to P-glycoprotein (Horio *et al.*, 1988, 1991; Lelong *et al.*, 1992). This transport is osmotically sensitive and strongly depends on a constant source of energy which can be provided by either ATP or GTP, but not by their nonhydrolyzable analogs, or ADP, AMP, CTP, UTP, ITP, NAD^+, NADH, $NADP^+$, and NADPH (Horio *et al.*, 1988; Lelong *et al.*, 1992). The transport is also blocked by the ATPase inhibitor vanadate in a noncompetitive manner (Horio *et al.*, 1988). P-glycoprotein is clearly the mediator of this ^{3}H-vinblastine transport, since it is inhibited by unlabeled vinblastine and other drugs to which multidrug resistant cells are cross-resistant, including vincristine, daunorubicin, actinomycin D, and colchicine (Horio *et al.*, 1988). Furthermore, agents that reverse P-glycoprotein-mediated multidrug resistance (e.g., verapamil) block ^{3}H-vinblastine transport (Horio *et al.*, 1988). Unfortunately, due to technical difficulties caused by the general hydrophobic character of P-glycoprotein substrates, it has so far been impossible to clearly establish the mechanism by which drug substrates and reversing agents inhibit ^{3}H-vinblastine transport. The published data suggest, however, that many of these compounds act in a competitive or mixed manner, and thus compete with ^{3}H-vinblastine for binding to P-glycoprotein by interacting with the same or overlapping site(s) (Horio *et al.*, 1991). Similar results were obtained by studying inside-out vesicles from normal tissues that are rich in P-glycoprotein, such as basal membranes from rat hepatocytes, and brush border membrane vesicles from rat small intestine (Kamimoto *et al.*, 1989; Hsing *et al.*, 1992).

Two types of drug uptake experiments, originally performed by Danø (1973), and later on repeated in many different laboratories have clearly demonstrated that P-glycoprotein requires energy for its multidrug transporter function (e.g., Skovsgaard, 1978a,b; Inaba *et al.*, 1979). As noted above, lower steady-state levels of drugs are observed in multidrug resistant cells than in drug-sensitive cells. In energy-depleted multidrug resistant cells, however, such as in the presence of azide, cyanide, iodoacetate, and 2-deoxy-D-glucose, the total intracellular drug accumulation is increased and approaches that seen in drug-sensitive cells, presumably because the energy-requiring drug efflux pump is poisoned by lowering the ATP levels in the cell. Re-energizing the energy-depleted drug-containing cells (i.e., by adding glucose) reactivates the pump and causes the intracellular drug concentration to decrease again toward values characteristic for multidrug resistant cells. Hence, it can be concluded that cellular ATP levels influence the efficacy of the multidrug transporter. This is in agreement with the previously described ATP-dependence that was noticed in drug transport experiments performed *in vitro* using isolated plasma membrane vesicles (Horio *et al.*, 1988).

Surprisingly, only recent studies have conclusively demonstrated that P-glycoprotein exhibits an ATPase activity that is stimulated by adding drug substrates. Initial attempts to purify P-glycoprotein by affinity chromatography resulted in protein preparations with a low capacity ATPase activity (1–3 nmol ATP/min/mg), which was not drug-dependent and could also not account for the high levels of ATP consumption due to drug transport in multidrug resistant cell membranes (Hamada and Tsuruo, 1988a,b). In contrast, an approximately 1000-fold higher specific ATPase activity (3–5 μmol ATP/min/mg) was observed for recombinant P-glycoprotein overproduced in insect cells after infection with a lytic *MDR*1-baculovirus (Sarkadi *et al.*, 1992). Interestingly this P-glycoprotein-specific ATPase activity is enhanced approximately 3- to 5-fold by the addition of small amounts of drug substrates or multidrug resistance reversing agents. Different drugs show different concentrations for half-maximal activation, and drugs which do not represent substrates for the multidrug transporter, such as 5-fluorouracil, do not stimulate the ATPase activity. This drug-induced ATPase activity of recombinant P-glycoprotein is dependent on magnesium ions and is inhibited by vanadate. While GTP is also a substrate, ADP or AMP are not hydrolyzed. Similarly, Chinese hamster ovary cell plasma membranes highly enriched in P-glycoprotein were also demonstrated to exhibit a high drug-stimulated ATPase activity with a specific activity of 9 μmol ATP/min/mg measured at 37°C and at the optimal pH 7.4 in the presence of 10 μM verapamil (Al-Shawi and Senior, 1993). Besides MgATP, 2'-dATP was found to be a good substrate, in contrast to CaATP, GTP and ITP that are hydrolyzed weakly. As found for the recombinant human P-glycoprotein ADP and AMP are not utilized as substrates. ADP and 5'-adenylyl imidodiphosphate were identified as competitive, and vanadate, and fluoroaluminate as noncompetitive inhibitors. Both 2-azido-ATP and NBD-Cl completely inhibit the hamster P-glycoprotein ATPase activity through covalent binding of approximately 1 mol reagent/mol protein. A single Lys residue at the active site was suggested to be involved in NBD-Cl-binding. *N*-ethylmaleimide was also found to act as a strong inhibitor, and since ATP protects markedly against such inhibition, the reactive Cys residue(s) were suggested to be located within the nucleotide binding site(s).

To date, the precise catalytic mechanism of ATP hydrolysis is still unclear. As discussed earlier, it is also unknown whether both of the putative ATP-binding sites actually represent catalytic sites, and if so, whether they interact with each other. In this respect an analysis of immunoprecipitated P-glycoprotein-β-galactosidase fusion protein and deletions thereof has provided some interesting data (Shimabuku *et al.*, 1991, 1992). This analysis revealed that the amino-terminal half of the human multidrug transporter differs from the carboxy-terminal half, since it has the capacity to catalyze ATP hydrolysis independently. Much lower rates of ATP hydrolysis, however, were observed for the amino-terminal half P-glycoprotein-β-galactosidase than for the full-length P-glycoprotein-β-galactosidase fusion protein.

A more complete understanding of the mechanism of action by the multidrug transporter still awaits its purification to homogeneity and subsequent functional reconstitution into proteoliposomes. Considerable progress has been made in this area of research since the first attempts to isolate P-glycoprotein from multidrug resistant mammalian cells by lectin-agarose affinity or immunoaffinity chromatography (Riordan and Ling, 1979; Hamada and Tsuruo, 1988a,b). Various laboratories have used plasma membrane vesicles prepared from multidrug resistant cells as their starting material and succeeded in the partial purification of P-glycoprotein that retains higher levels of ATPase activity (Ambudkar *et al.*, 1992; Doige *et al.*, 1992). Moreover, reconstitution of a drug-stimulatable ATPase activity into proteoliposomes has been achieved using partially purified P-glycoprotein from highly multidrug resistant human KB carcinoma cells (Ambudkar *et al.*, 1992). Partially pure P-glycoprotein preparations exhibit a high specific ATPase activity (5—12 μmol ATP/min/mg), which approaches the ATPase activity of well-characterized ion-transporting pumps. Interestingly, drug substrate stimulation of this ATPase activity, however, has only been observed after reconstitution of the multidrug transporter into phospholipid vesicles. This underscores the need for proper lipid-P-glycoprotein interactions in a functional transporter which has been implied by many other studies as well (e.g., Callaghan *et al.*, 1992; Saeki *et al.*, 1992; Sinicrope *et al.*, 1992; Doige *et al.*, 1993).

Models for the mechanism of drug transport

The precise mechanism by which P-glycoprotein in multidrug resistant cells maintains intracellular drug levels below a killing threshold has not yet been completely elucidated and is quite controversial. Early drug uptake experiments and kinetic analyses have suggested that decreased drug influx, or increased drug efflux represent potential mechanisms underlying P-glycoprotein-mediated multidrug resistance (Danø, 1973; Ling and Thompson, 1974). Initially, the ability of P-glycoprotein to act as an energy-dependent drug efflux pump has been emphasized (e.g., Gottesman and Pastan, 1988). Recently, the idea has evolved that P-glycoprotein itself may be responsible for both a decreased influx and increased efflux of drugs (Gottesman and Pastan, 1993). Thus, in an attempt to account for both of these activities, models for the mechanism of action of P-glycoprotein have been proposed in which the multidrug transporter removes drugs directly from the plasma membrane before they enter the cytosol (Gros *et al.*, 1986b; Raviv *et al.*, 1990; Gottesman and Pastan, 1993). The model for the mechanism of action of P-glycoprotein that is illustrated in Fig. 3 incorporates this feature.

The compounds which interact with P-glycoprotein are likely to enter the cell by diffusion (W. Stein, I. Pastan, M.M. Gottesman, pers. comm.). Although multidrug transporter substrates differ in structure and function, may either be protonated or uncharged at neutral pH, and range in molecular weight from approximately 100 to 2000 (mostly >400), they are generally amphiphilic and have high affinity coefficients for lipids (e.g., Friche *et al.*, 1990; Beck and Qian, 1992). Thus, they may partition into and out of the membrane in a rapid equilibrium, but move slowly, in a rate-limiting step, through the bilayer of nonpolar fatty acid side chains. This increased solubility in the lipid over the aqueous phase may render drugs a favorable target for an energized multidrug transporter which according to the model presented in Fig. 3 may recognize its substrates while they are still present within the lipid bilayer and, thus, remove them directly from the plasma membrane. It is important

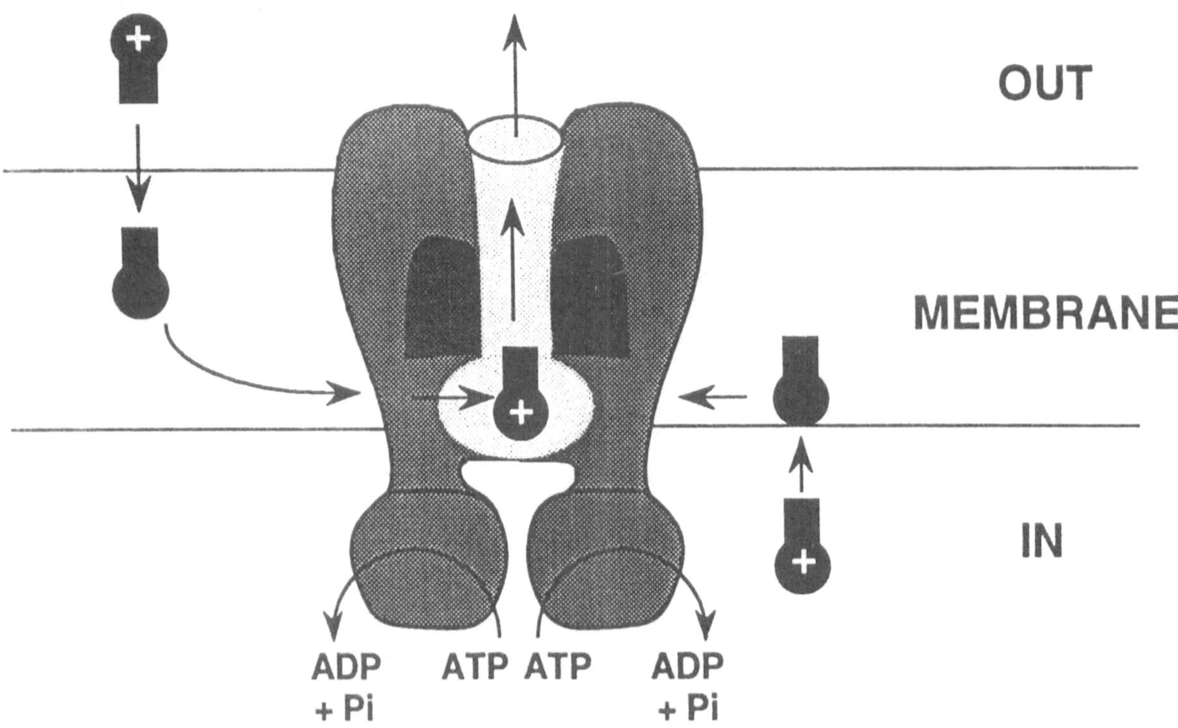

Fig. 3. Hypothetical model for the mechanism of drug transport by P-glycoprotein. Amphiphilic drug substrates enter a cell by diffusion through the plasma membrane and encounter the energized multidrug transporter while they are present within the lipid bilayer. P-glycoprotein directly removes the drugs from the membrane using energy provided by two essential ATP binding/utilization sites. The two homologous halves of the multidrug transporter interact in this drug transport process and form a single channel-like structure through which the drugs are exported. Drug substrates may also be extracted from the cytoplasm in a similar way. Adapted from reference (Gottesman *et al.*, 1991a).

to note, however, that some very hydrophobic agents, (e.g., camptothecin) which are basically water-insoluble, do not interact with P-glycoprotein. This indicates that a low degree of water solubility is required for a drug to be a substrate for the multidrug transporter.

Besides kinetic data (Skovsgaard, 1978a,b; Inaba *et al.*, 1979; W. Stein, I. Pastan, M.M. Gottesman, pers. comm.), phenomenological and experimental evidence supports the hypothesis that drug substrates are directly removed from the plasma membrane. Hydrophobic peptides to which multidrug resistant cells are cross-resistant (e.g., gramicidin D, valinomycin, or N-acetyl-leucyl-leucyl-norleucin) are thought to mainly accumulate in the plasma membrane of cells which do not express multidrug transporter (Sharma *et al.*, 1992; M. Horio, M.M. Gottesman, I. Pastan, pers. comm.). This hypothesis was corroborated using either fluorescence excitation spectroscopy or confocal fluorescence microscopy to analyze the accumulation of rhodamine 123 in drug-sensitive and multidrug resistant cells (Kessel, 1989; Weaver *et al.*, 1991). The mitochondrial dye and multidrug transporter substrate rhodamine 123 (Neyfakh *et al.*, 1989) was found to be localized in a more hydrophobic environment in drug-sensitive cells when compared to multidrug resistant cells, presumably in plasma membranes and eventually intracellular membranous structures. These data suggest that P-glycoprotein displaces rhodamine 123 from the plasma membranes of multidrug resistant cells.

Similar conclusions were also drawn from a study in which the intracellular localization of doxorubicin was investigated by measuring the transfer of the energy of doxorubicin to iodinated naphthalene azide (Raviv *et al.*, 1990). Whereas in drug-sensitive cells large amounts of this hydrophobic label were found to be associated with many different membrane proteins, generally lower amounts of iodinated naphthalene azide were found in multidrug resistant cells, and only in association with the multidrug transporter. Another study involved various fluorescent calcium and pH indicators either in the form of free acids or in the form of more hydrophobic acetoxymethylester derivatives (Homoloya *et al.*, 1993). It was found that only the uncharged hydrophobic derivatives are substrates of the multidrug transporter and that their extrusion from multidrug resistant cells is blocked by cytotoxic drugs and agents that reverse multidrug resistance. Since the acetoxy methylester derivatives are readily cleaved by highly active cytoplasmic esterases and converted into free acids upon entry into the cytosol, the low accumulation of these compounds in multidrug resistant cells suggests that they are extruded without ever entering the cytosol. Hence, acetoxy methylester derivatives of fluorescent calcium and pH indicators appear to be removed by P-glycoprotein directly from the plasma membrane.

According to the model presented in Fig. 3, the overall increase in drug efflux by the multidrug transporter results from a similar extraction of drugs that partition into the membrane from the cytoplasm. It is possible, however, that direct transport out of the cytoplasm may also contribute to drug efflux. A second characteristic feature of the model shown in Fig. 3 is that the two homologous halves of P-glycoprotein cooperate and build a single channel-like structure through which drug substrates are transported and released upon ATP hydrolysis. This hypothesis is consistent with the previously described data from different mutational analyses of P-glycoprotein and from photoaffinity labeling studies. With respect to initial drug binding, however, the model described above still allows multiple interaction sites that may differ for various substrates.

The broad substrate specificity of P-glycoprotein remains an enigma and any detailed model of mechanism will need to account for it. Other unsolved problems include the mechanism of energy transduction and its stoichiometry. It has been speculated that a flow of protons and/or other ions may represent a driving force (W. Stein, I. Pastan, M.M. Gottesman, pers. comm.), but definite answers require further investigation of purified functional protein. As mentioned at the beginning of this section, the mechanism of action of P-glycoprotein is still a matter of debate. The model discussed here represents only one out of many suggested working hypotheses and the interested reader is referred to other reviews for the description of other models (e.g., Endicott and Ling, 1989; Roninson, 1991; Schinkel and Borst, 1991; Higgins and Gottesman, 1992; Gottesman and Pastan, 1993).

P-glycoprotein: a multifunctional transporter

As discussed so far, a primary function of P-glycoprotein is to pump a great variety of amphiphilic drug substrates out of cells, some of which are listed in Table 1. There is increasing evidence that P-glycoprotein represents an unusual membrane transport protein and may serve multiple functions. The following section gives a brief overview about recently discovered novel aspects of this intriguing molecule.

Interactions with nondrug substrates

In Table 1, only few of the many compounds are summarized that have been demonstrated to interact with P-glycoprotein. These do not only include anticancer drugs, but other cytotoxic agents as well, including cytotoxic small peptides, e.g., valinomycin, gramicidin D, or the calpain and cathepsin inhibitory hydrophobic tripeptide *N*-acetyl-leucyl-leucyl-norleucine (Sharma *et al.*, 1992). Thus, P-glycoprotein has been suggested to play a role in the cellular export of small peptides and/or proteins that lack cleavable hydrophobic signal sequences. This hypothesis is supported by the finding that wild-type, but not mutant mammalian *mdr*1 products may complement the function of the *STE6* gene product in a deficient yeast strain and allow secretion of the farnesylated yeast **a**-factor mating peptide, although with low efficiency (Raymond *et al.*, 1992; B. Ni, I. Pastan, M.M. Gottesman, pers. comm.). In addition to structural similarities, P-glycoprotein may, therefore, also share analogous functional properties with homologous members of the ABC transporter superfamily such as the Ste6 protein of yeast (Michaelis, 1993), the hemolysin transport protein HlyB of *Escherichia coli* (Koronakis and Hughes, 1993), and the putative transport proteins Tap-1 and Tap-2 associated with the endoplasmic reticulum in mammalian cells and involved in antigen presentation (Townsend and Trowsdale, 1993).

Channel activity of P-glycoprotein

Studies by Valverde *et al.* (1992) have recently suggested that P-glycoprotein may be associated with a volume-regulated, ATP-dependent chloride channel activity in addition to its ATP-dependent multidrug transporter activity. Hence, P-glycoprotein may be not only be structurally, but also functionally related to yet another ABC transporter, CFTR, which is thought to play a critical role in fluid secretion across the apical membrane of certain epithelia by functioning as a chloride channel (Anderson *et al.*, 1991b). The chloride channels associated with these two proteins differ, however, in their electrophysiological properties. CFTR is a low conductance, time- and voltage-independent chloride channel that requires ATP hydrolysis for opening, and is activated by protein kinase A-dependent phosphorylation in response to changing cAMP levels (Anderson *et al.*, 1991a). The chloride channel associated with P-glycoprotein is outwardly-rectifying, voltage-dependent, and requires ATP-binding, or binding of a non-hydrolyzable analog thereof (Gill *et al.*, 1992). Hence, besides changes in the cell volume, ATP might be involved in the regulation of this P-glycoprotein-associated chloride channel activity, which appears to be distinct and separable from its multidrug transporter activity (Gill *et al.*, 1992). To date, it has, however, not clearly been established whether P-glycoprotein itself acts as chloride channel, or whether P-glycoprotein represents a modulator of this function. Measurements of single chloride channel currents together with functional reconstitution experiments are expected to resolve this matter of debate in the near future.

Interestingly, using *in situ* hybridization both *mdr* and *CFTR* genes have been demonstrated to be expressed in rat epithelial cells, but they do not appear to be coexpressed in the same cell (Trezise *et al.*, 1992). It may be speculated that their gene products serve analogous functions in different epithelial cells, or at different stages of development in the same epithelial cell. Furthermore, coordinate control of expression of these genes with respect to cell type- and cell stage-specificity may be suggested by these data. The idea that P-glycoprotein may be bifunctional and act both as transporter and as channel is very intriguing. Conversely, however, an energy-dependent transport activity has so far not been identified for the CFTR gene product.

Another channel function, namely that of an ATP conducting channel has recently been suggested to be associated with P-glycoprotein expressed in highly multidrug resistant hamster Chinese hamster ovary and human lung tumor cell lines (Abraham *et al.*, 1993). Release of ATP was found to be proportional to the P-glycoprotein content in these cells both by measuring whole-cell and single-channel currents. Direct experimental evidence to demonstrate that indeed P-glycoprotein itself fulfills the unusual role to allow steady release of ATP from cells is lacking thus far, but the suggested hypothesis provides a challenge for further studies.

Studies and speculations on the normal function of P-glycoprotein

To date, the physiological role of P-glycoprotein remains an enigma. Generally, P-glycoproteins appear to be expressed in a temporally regulated cell- and tissue-specific manner (Chaudhary and Roninson, 1991; Cordon-Cardo, 1991; Gottesman *et al.*, 1991b). In human adults, high levels of *MDR*1 mRNA are expressed in the adrenal glands, kidney, liver, small intestine, and colon, whereas low levels are detected in most other tissues (Fojo *et al.*, 1987). Consistent with these results highest levels of the *MDR*1 gene product have been detected immunohistochemically on the surface of adrenal cortical cells, on the luminal surface of the brush border of the proximal renal tubule, on the luminal surface of biliary hepatocytes, and on the mucosal surface of the columnar epithelial cells of the small and large intestine (Thiebaut *et al.*, 1987). Lower amounts of human P-glycoprotein have been found on the luminal surface of some small pancreatic ductules, and in certain specialized capillary endothelial cells of the brain and testis (Thiebaut *et al.*, 1989). P-glycoprotein has also been detected in CD34 positive bone marrow cells (Chaudhary and Roninson, 1991), in circulating lymphocytes (Chaudhary *et al.*, 1992), and in the trophoblast of the placenta (Sugawara *et al.*, 1988). In addition, P-glycoprotein expression is increased in the secretory and gestational endometrium (Axiotis *et al.*, 1991). Many of the corresponding tissues in mouse and hamster also express either one or both of the multidrug resistance-associated P-glycoproteins (Arceci *et al.*, 1988; Croop *et al.*, 1989).

While the tissue distribution of P-glycoprotein may help to identify its physiological role as a transporter, the specific, but complex pattern of P-glycoprotein expression is not necessarily consistent with a single class of physiological substrates. Hence, it has been hypothesized that P-glycoprotein may serve diverse transport functions. P-glycoprotein in the epithelial cells of excretory organs and in the endothelial cells of capillary blood vessels at blood-tissue barrier sites has been proposed to assume a general protective function and to serve in the transepithelial and transendothelial transport of xenobiotics ingested with food, and of endogenous metabolites (Kamimoto *et al.*, 1989; Tatsuta *et al.*, 1992; reviewed Cordon-Cardo, 1991; Gottesman *et al.*, 1991b). The presence of P-glycoprotein in the adrenal gland, in the endometrium of the gravid uterus, and the placenta is consistent with a potential involvement in steroid transport (secretion or protection of membranes from toxic effects of dissolved steroids). Interestingly, stably *mdr*-transfected tissue culture cells which express P-glycoprotein in a polar fashion on the apical surface show a several-fold increased net basal to apical transport of various drugs and other putative multidrug transporter substrates due to a concomitant increase of the basal to apical and decrease of the apical to basal transport (Horio *et al.*, 1989; Ueda *et al.*, 1992). Thus, this model system has proven particularly useful in the identification of steroids that are transported by human P-glycoprotein (Ueda *et al.*, 1992). Cortisol, a glucocorticoid and the main steroid produced in the human adrenal, as well as aldosterone, the main mineral-corticoid produced in the human adrenal, are both substrates of P-glycoprotein (Ueda *et al.*, 1992). Also estriol, the main estrogen produced during pregnancy in the fetoplacental unit has been found to be a multidrug transporter substrate (K. Ueda, Y. Tanigawara, pers. comm.). P-glycoprotein furthermore directly transports dexamethasone, but not corticosterone, deoxycorticosterone, estradiol, progesterone, or testosterone (K. Ueda, Y. Tanigawara, pers. comm.), even though some of these hydrophobic steroids (e.g., corticosterone and progesterone) have been demonstrated to interact with mouse P-glycoprotein (Wolf and Horwitz, 1992). The role of P-glycoprotein in steroid secretion is furthermore supported by the finding that inhibitors of the multidrug transporter block steroid secretion in mouse adrenal Y1 cells (Chin *et al.*, 1992). Moreover, the inactivation of one copy of the *mdr*1b gene interferes with the regulation of steroid secretion in mouse adrenal Y-1 cells (S. Altuvia, I. Pastan, M.M. Gottesman, pers. comm.). P-glycoprotein in the hematopoietic stem cells has been postulated to be involved in the export of a so far unidentified regulatory molecule and, thus, to modulate their differentiation and proliferation (Chaudhary and Roninson, 1991). Many of these proposed transport activities have yet to be proven. The possibility of P-glycoprotein-associated channel functions raises the intriguing hypothesis that certain tissues may require P-glycoprotein for either transport, or channel activity, or even both of these functions.

Acknowledgements

I thank many scientific colleagues and collaborators for stimulating discussions and for sharing unpublished results. I am especially grateful for their permission to cite unpublished data. I also thank Michael Gottesman, Ira Pastan, Yoshikazu Sugimoto and Greg Evans for critical reading of the manuscript.

References

Abraham I, Hunter RJ, Sampson KE, Smith S, Gottesman MM and Mayo JK (1987) Cyclic AMP-dependent protein kinase regulates sensitivity of cells to multiple drugs. Mol Cell Biol 7: 3098–3106.

Abraham I, Chin K-V, Gottesman MM, Mayo JK and Sampson KE (1990) Transfection of a mutant regulatory subunit gene of cAMP-dependent protein kinase causes increased drug sensitivity and decreased expression of P-glycoprotein. Exp Cell Res 189: 133–141.

Abraham EH, Prat AG, Gerweck L, Seneveratne T, Arceci RJ, Kramer R, Guidotti G and Cantiello HF (1993) The multidrug resistance (*mdr*1) gene product functions as an ATP channel. Proc Natl Acad Sci USA 90: 312–316.

Ahmad S, Trepel JB, Ohno S, Suzuki K, Tsuruo T and Glazer RI (1992) Role of protein kinase-C in the modulation of multidrug resistance – expression of the atypical gamma-isoform of protein kinase-C does not confer increased resistance to doxorubicin. Mol Pharmacol 42: 1004–1009.

Akiyama S-i, Fojo A, Hanover JA, Pastan I and Gottesman MM (1985) Isolation and genetic characterization of human KB cell lines resistant to multiple drugs. Somatic Cell Mol Genet 11: 117–126.

Akiyama S-i, Cornwell MM, Kuwano M, Pastan I and Gottesman MM (1988) Most drugs that reverse multidrug resistance also inhibit photoaffinity labeling of P-glycoprotein by a vinblastine analog. Mol Pharmacol 33: 144–147.

Al-Shawi MK and Senior AE (1993) Characterization of the adenosine triphosphatase activity of Chinese hamster P-glycoprotein. J Biol Chem 268: 4197–4206.

Ambudkar SV, Lelong IH, Zhang J, Cardarelli CO, Gottesman MM and Pastan I (1992) Partial purification and reconstitution of the human multidrug resistance pump: characterization of the drug-stimulatable ATP hydrolysis. Proc Natl Acad Sci USA 89: 8472–8476.

Ames GF-L and Lecar H (1992) ATP-dependent bacterial transporters and cystic fibrosis: analogy between channels and transporters. FASEB J 6: 2660–2666.

Ames GF-L, Mimura CS, Holbrook SR and Shyamala V (1992) Traffic ATPases: a superfamily of transport proteins operating from *Escherichia coli* to humans. Adv Enzymol 65: 1–47.

Anderson MP, Berger HA, Rich DP, Gregory RJ, Smith AE and Welsh MJ (1991a) Nucleotide triphosphates are required to open the CFTR channel. Cell 67: 775–784.

Anderson MP, Gregory RJ, Thompson S, Souza DW, Paul S, Mulligan RC, Smith AE and Welsh MJ (1991b) Demonstration that CFTR is a chloride channel by alteration of its anion selectivity. Science 253: 202–205.

Arceci RJ, Croop JM, Horwitz SB and Housman DE (1988) The gene encoding multidrug resistance is induced and expressed at high levels during pregnancy in the secretory epithelium of the uterus. Proc Natl Acad Sci USA 85: 4350–4354.

Arsenault AL, Ling V and Kartner N (1988) Altered plasma membrane ultrastructure in multidrug-resistant cells. Biochim Biophys Acta 938: 315–321.

Axiotis CA, Guarch R, Merino MJ, Laporte N and Neumann RD (1991) P-glycoprotein expression is increased in human secretory and gestational endometrium Lab Invest 65 577–581

Azzaria M, Schurr E and Gros P (1989) Discrete mutations introduced in the predicted nucleotide-binding sites of the *mdr1* gene abolish its ability to confer multidrug resistance Mol Cell Biol 9 5289–5297

Bates SE, Currier SJ, Alvarez M and Fojo AT (1992) Modulation of P-glycoprotein phosphorylation and drug transport by sodium butyrate Biochemistry 31 6366–6372

Beck WT and Danks MK (1991) Characteristics of multidrug resistance in human tumor cells In IB Roninson (ed) Molecular and Cellular Biology of Multidrug Resistance in Tumor Cells (pp 3–46) New York Plenum Publishing Corporation

Beck WT and Qian X-D (1992) Photoaffinity substrates for P-glycoprotein Biochem Pharmacol 43 89–93

Biedler JL and Riehm H (1970) Cellular resistance to actinomycin D in Chinese hamster cells *in vitro* cross-resistance, radioautographic, and cytogenetic studies Cancer Res 30 1174–1184

Blobe GC, Sachs CW, Khan WA, Fabbro D, Stabel S, Wetsel WC, Obeid LM, Fine RL and Hannun YA (1993) Selective regulation of expression of protein kinase-C (PKC) isoenzymes in multidrug resistant MCF-7 cells — functional significance of enhanced expression of PKC-alpha J Biol Chem 268 658–664

Boscoboinik D, Debanne MT, Stafford AR, Jung CY, Gupta RS and Epand RM (1990) Dimerization of the P glycoprotein in membranes Biochim Biophys Acta 1027 225–228

Bruggemann EP, Germann UA, Gottesman MM and Pastan I (1989) Two different regions of P-glycoprotein are photoaffinity labeled by azidopine J Biol Chem 264 15483–15488

Bruggemann EP, Currier SJ, Gottesman MM and Pastan I (1992) Characterization of the azidopine and vinblastine binding site of P glycoprotein J Biol Chem 267 21020–21026

Busche R, Tummler B, Riordan JR and Cano Gauci DF (1989) Preparation and utility of a radioiodinated analogue of daunomycin in the study of multidrug resistance Mol Pharmacol 35 414–421

Buschman E and Gros P (1991) Functional analysis of chimeric genes obtained by exchanging homologous domains of the mouse *mdr1* and *mdr2* genes Mol Cell Biol 11 595–603

Buschman E, Arceci RJ, Croop JM, Che MX, Arias IM, Housman DE and Gros P (1992) *mdr2* encodes P-glycoprotein expressed in the bile canalicular membrane as determined by isoform-specific antibodies J Biol Chem 267 18093–18099

Callaghan R, Vangorkom LCM and Epand RM (1992) A comparison of membrane properties and composition between cell lines selected and transfected for multi drug resistance Br J Cancer 66 781–786

Carlsen SA, Till JE and Ling V (1977) Modulation of drug permeability in Chinese hamster ovary cells possible role for phosphorylation of surface glycoproteins Biochim Biophys Acta 467 238–250

Castillo G, Vera J, Huang Yang C-P and Horwitz SB (1990) Functional expression of murine multidrug resistance in *Xenopus* oocytes Proc Natl Acad Sci USA 87 4737–4741

Chambers TC, McAvoy EM, Jacobs JW and Eilon G (1990) Protein kinase C phosphorylates P-glycoprotein in multidrug resistant human KB carcinoma cells J Biol Chem 265 7679–7686

Chambers TC, Zheng B and Kuo JF (1992) Regulation by phorbol ester and protein kinase-C inhibitors, and by a protein phosphatase inhibitor (okadaic acid), of P-glycoprotein phosphorylation and relationship to drug accumulation in multidrug-resistant human-KB cells Mol Pharmacol 41 1008–1015

Chambers TC, Pohl J, Raynor RL and Kuo JF (1993) Identification of specific sites in human P-glycoprotein phosphorylated by protein kinase C J Biol Chem 268 4592–4595

Chaudhary PM and Roninson IB (1991) Expression and activity of P-glycoprotein, a multidrug efflux pump, in human hematopoietic stem cells Cell 66 85–94

Chaudhary PM, Mechetner EB and Roninson IB (1992) Expression and activity of the multidrug resistance P-glycoprotein in human peripheral blood lymphocytes Blood 80 2735–2739

Chaudhary PM and Roninson IB (1992) Activation of *MDR1* (P-glycoprotein) gene expression in human cells by protein kinase-C agonists Oncol Res 4 281–290

Chen C-J, Chin JE, Ueda K, Clark DP, Pastan I, Gottesman MM and Roninson IB (1986) Internal duplication and homology with bacterial transport proteins in the *mdr1* (P-glycoprotein) gene from multidrug-resistant human cells Cell 47 381–389

Chin K-V, Chauhan SS, Abraham I, Sampson KE, Krolczyk AJ, Wong M, Schimmer B, Pastan I and Gottesman MM (1992) Reduced mRNA levels for the multidrug-resistance genes in cAMP-dependent protein kinase mutant cell lines J Cell Physiol 152 87–94

Chin K-V, Pastan I and Gottesman MM (1993) Function and regulation of the human multidrug resistance gene Adv Canc Res 60 157–180

Choi K, Chen C-J, Kriegler M and Roninson IB (1989) An altered pattern of cross-resistance in multidrug-resistant human cells results from spontaneous mutations in the *mdr1* (P-glycoprotein) gene Cell 53 519–529

Cohen D, Yang C-PH and Horwitz SB (1990) The products of the *mdr1a* and *mdr1b* genes from multidrug resistant murine cells have similar degradation rates Life Sci 46 489–495

Cordon Cardo C (1991) Immunohistochemical analysis of P-glycoprotein expression in normal and tumor tissues in humans In IB Roninson (ed) Molecular and Cellular Biology of Multidrug Resistance in Tumor Cells (pp 303–318) New York Plenum Publishing Corporation

Cornwell MM, Gottesman MM and Pastan I (1986a) Increased vinblastine binding to membrane vesicles from multidrug resistant KB cells J Biol Chem 262 7921–7928

Cornwell MM, Safa AR, Felsted RL, Gottesman MM and Pastan I (1986b) Membrane vesicles from multidrug-resistant human cancer cells contain a specific 150–170kDa protein detected by photoaffinity labeling. Proc Natl Acad Sci USA 83: 3847–3850.

Cornwell MM, Tsuruo T, Gottesman MM and Pastan I (1987) ATP-binding properties of P-glycoprotein from multidrug resistant KB cells. FASEB J 1: 51–54.

Cornwell MM, Pastan I and Gottesman MM (1991) Binding of drugs and ATP by P-glycoprotein and transport of drugs by vesicles from human multidrug-resistant cells. In: IB Roninson (ed) Molecular and Cellular Biology of Multidrug Resistance in Tumor Cells (pp. 229–242). New York: Plenum Publishing Corporation.

Croop JM, Raymond M, Haber D, Devault A, Arceci RJ, Gros P and Housman DE (1989) The three mouse multidrug resistance (*mdr*) genes are expressed in a tissue-specific manner in normal mouse tissues. Mol Cell Biol 9: 1346–1350.

Currier SJ, Ueda K, Willingham MC, Pastan I and Gottesman MM (1989) Deletion and insertion mutants of the multidrug transporter. J Biol Chem 264: 14376–14381.

Currier SJ, Kane SE, Willingham MC, Cardarelli CO, Pastan I and Gottesman MM (1992) Identification of residues in the first cytoplasmic loop of P-glycoprotein involved in the function of chimeric human *MDR* 1-*MDR*2 transporters. J Biol Chem 267: 25153–25159.

Danø K (1973) Active outward transport of daunomycin in resistant Ehrlich ascites tumor cells. Biochem Biophys Acta 323: 466–483.

Deuchars KL, Duthie M and Ling V (1992) Identification of distinct P-glycoprotein gene sequences in rat. Biochim Biophys Acta 1130: 157–165.

Devault A and Gros P (1990) Two members of the mouse *mdr* gene family confer multidrug resistance with overlapping but distinct drug specificities. Mol Cell Biol 10: 1652–1663.

Devine SE, Hussain A, Davide JP and Melera PM (1991) Full length and alternatively spliced pgp1 transcripts in multidrug-resistant Chinese hamster lung cells. J Biol Chem 266: 4545–4555.

Devine SE, Ling V and Melera PW (1992) Amino acid substitutions in the 6th transmembrane domain of P-glycoprotein alter multidrug resistance. Proc Natl Acad Sci USA 89: 4564–4568.

Dhir R and Gros P (1992) Functional analysis of chimeric proteins constructed by exchanging homologous domains of 2 P-glycoproteins conferring distinct drug resistance profiles. Biochemistry 31: 6103–6110.

Doige CA, Yu XH and Sharom FJ (1992) ATPase activity of partially purified P-glycoprotein from multidrug-resistant Chinese hamster ovary cells. Biochim Biophys Acta 1109: 149–160.

Doige CA, Yu X and Sharom FJ (1993) The effects of lipids and detergents on ATPase-active P-glycoprotein. Biochim Biophys Acta 1146: 65–72.

Endicott JA, Juranka PF, Sarangi F, Gerlach JH, Deuchars KL and Ling V (1987) Simultaneous expression of two P-glycoprotein genes in drug-sensitive Chinese hamster ovary cells. Mol Cell Biol 7: 4075–4081.

Endicott JA and Ling V (1989) The biochemistry of P-glycoprotein-mediated multidrug resistance. Annu Rev Biochem 58: 137–171.

Fine RL, Patel J and Chabner BA (1988) Phorbol esters induce multidrug resistance in human breast cells. Proc Natl Acad Sci USA 85: 582–586.

Fojo AT, Ueda K, Slamon DJ, Poplack DG, Gottesman MM and Pastan I (1987) Expression of a multidrug-resistance gene in human tumors and tissues. Proc Natl Acad Sci USA 84: 265–269.

Foxwell BMJ, Mackie A, Ling V and Ryffel B (1989) Identification of the multidrug resistance-related P-glycoprotein as a cyclosporin binding protein. Mol Pharmacol 36: 543–546.

Friche E, Jensen PB, Roed H, Skovsgaard T and Nissen NI (1990) In vitro circumvention of anthracycline-resistance in Ehrlich ascites tumour. Biochem Pharm 39: 1721–1726.

Friedlander M and Blobel G (1985) Bovine opsin has more than one signal sequence. Nature 318: 338–343.

Garman D, Albers L and Center MS (1983) Identification and characterization of a plasma membrane phosphoprotein which is present in Chinese hamster lung cells resistant to Adriamycin. Biochem Pharmacol 32: 3633–3637.

Georges E, Bradley G, Gariepy J and Ling V (1990) Detection of P-glycoprotein isoforms by gene-specific monoclonal antibodies. Proc Natl Acad Sci USA 87: 152–156.

Georges E, Tsuruo T and Ling V (1993) Topology of P-glycoprotein as determined by epitope mapping of MRK-16 monoclonal antibody. J Biol Chem 268: 1792–1798.

Germann UA, Gottesman MM and Pastan I (1989) Expression of a multidrug resistance-adenosine deaminase fusion gene. J Biol Chem 264: 7418–7424.

Germann UA, Willingham MC, Pastan I and Gottesman MM (1990) Expression of the human multidrug transporter in insect cells by a recombinant baculovirus. Biochemistry 29: 2295–2303.

Germann UA, Pastan I and Gottesman MM (1993) P-glycoproteins: mediators of multidrug resistance. Semin Cell Biol 4: 63–76.

Gill DR, Hyde SC, Higgins CF, Valverde MA, Mintenig GM and Sepúlveda FV (1992) Separation of drug transport and chloride channel functions of the human multidrug resistance P-glycoprotein. Cell 71: 23–32.

Gottesman MM and Pastan I (1988) The multidrug-transporter: a double-edged sword. J Biol Chem 263: 12163–12166.

Gottesman MM, Schoenlein PV, Currier SJ, Bruggemann EP and Pastan I (1991a) Biochemical basis for multidrug resistance in cancer. In: G Pretlow and P Pretlow (eds) Biochemical and Molecular Aspects of Selected Cancers (pp. 339–371). San Diego, CA: Academic Press Inc.

Gottesman MM, Willingham MC, Thiebaut F and Pastan I (1991b) Expression of the *MDR*1 gene in normal human tissues. In: IB Roninson (ed) Molecular and Cellular Biology of Multidrug Resistance in Tumors (pp. 279–289). New York: Plenum Publishing Corporation.

Gottesman MM and Pastan I (1993) Biochemistry of multidrug resistance mediated by the multidrug transporter. Annu Rev

Biochem 62: 385–427.

Greenberger LM, Lothstein L, Williams SS and Horwitz SB (1988) Distinct P-glycoprotein precursors are overproduced in independently isolated drug-resistant cell lines. Proc Natl Acad Sci USA 85: 3762–3766.

Greenberger LM, Yang C-PH, Gindin E and Horwitz SB (1990) Photoaffinity probes for the α1-adrenergic receptor and the calcium channel bind to a common domain in P-glycoprotein. J Biol Chem 265: 4394–4401.

Greenberger LM, Lisanti CJ, Silva JT and Horwitz SB (1991) Domain mapping of the photoaffinity drug-binding sites in P-glycoprotein encoded mouse *mdr*1b. J Biol Chem 266: 20744–20751.

Greenberger L (1993) Major photoaffinity drug labeling sites for iodoaryl azidoprazosin in P-glycoprotein are within, or immediately C-terminal to, transmembrane domains 6 and 12. J Biol Chem 268: 11417–11425.

Gros P, Ben Neriah Y, Croop J and Housman DE (1986a) Isolation and characterization of a complementary DNA that confers multidrug resistance. Nature 323: 728–731.

Gros P, Croop J and Housman D (1986b) Mammalian multidrug resistance gene: complete cDNA sequence indicates strong homology to bacterial transport proteins. Cell 47: 371–380.

Gros P, Croop J, Roninson IB, Varshavsky A and Housman DE (1986c) Isolation and characterization of DNA sequences amplified in multidrug-resistant hamster cells. Proc Natl Acad Sci USA 83: 337–341.

Gros P, Raymond M, Bell J and Housman DE (1988) Cloning and characterization of a second member of the mouse *mdr* gene family. Mol Cell Biol 8: 2770–2778.

Gros P, Dhir R, Croop J and Talbot F (1991) A single amino acid substitution strongly modulates the activity and substrate specificity of the mouse *mdr*1 and *mdr*3 drug efflux pumps. Proc Natl Acad Sci USA 88: 7289–7293.

Guilfoile PG and Hutchinson CR (1991) A bacterial analog of the *mdr* gene of mammalian tumor cells present in *Streptomyces peucetius*, the producer of daunorubicin and doxorubicin. Proc Natl Acad Sci USA 88: 8553–8557.

Gupta RS, Murray W and Gupta R (1988) Cross resistance pattern towards anticancer drugs of a human multidrug-resistant cell line. Br J Cancer 58: 441–447.

Hamada H, Hagiwara K-I, Nakajima T and Tsuruo T (1987) Phosphorylation of the M_r 170,000 to 180,000 glycoprotein specific to multidrug-resistant tumor cells: effects of verapamil, trifluoperazine, and phorbol esters. Cancer Res 47: 2860–2865.

Hamada H and Tsuruo T (1988a) Characterization of the ATPase activity of the M_r 170,000 to 180,000 membrane glycoprotein (P-glycoprotein) associated with multidrug resistance in K562/ADM cells. Cancer Res 48: 4926–4932.

Hamada H and Tsuruo T (1988b) Purification of the 170- to 180-kilodalton membrane glycoprotein associated with multidrug resistance — 170- to 180-kilodalton membrane glycoprotein is an ATPase. J Biol Chem 263: 1454–1458.

Higgins CF (1992) ABC transporters: from microorganisms to man. Annu Rev Cell Biol 8: 67–113.

Higgins CF and Gottesman MM (1992) Is the multidrug transporter a flippase? Trends Pharmacol Sci 17: 18–21.

Higgins CF (1993) The ABC transporter channel superfamily-an overview. Semin Cell Biol 4: 1–5.

Homoloya L, Holló Z, Germann UA, Pastan I, Gottesman MM and Sarkadi B (1993) Fluorescent cellular indicators are extruded by the multidrug resistance protein. J Biol Chem 268: 21493–21496.

Horio M, Gottesman MM and Pastan I (1988) ATP-dependent transport of vinblastine in vesicles from human multidrug-resistant cells. Proc Natl Acad Sci USA 85: 3580–3584.

Horio M, Chin K-V, Currier SJ, Goldenberg S, Williams C, Pastan I, Gottesman MM and Handler J (1989) Transepithelial transport of drugs by the multidrug transporter in cultured Madin-Darby canine kidney cell epithelia. J Biol Chem 264: 14880–14884.

Horio M, Lovelace E, Pastan I and Gottesman MM (1991) Agents which reverse multidrug-resistance are inhibitors of ^{3}H-vinblastine transport by isolated vesicles. Biochim Biophys Acta 1061: 106–110.

Hsing S, Gatmaitan Z and Arias IM (1992) The function of Gp170, the multidrug resistance gene product, in the brush border of rat intestinal mucosa. Gastroenterology 102: 879–885.

Hsu SI, Lothstein L and Horwitz SB (1989) Differential overexpression of three *mdr* gene family members in multidrug-resistant J774.2 mouse cells. J Biol Chem 264: 12053–12062.

Hyde SC, Emsley P, Hartshorn MJ, Mimmack MM, Gileadi U, Pearce SR, Gullagher MP, Gill DR, Hubbard RE and Higgins CF (1990) Structural and functional relationships of ATP-binding proteins associated with cystic fibrosis, multidrug resistance, and bacterial transport. Nature 346: 362–365.

Ichikawa M, Yoshimura A, Furukawa T, Sumizawa T, Nakazima Y and Akiyama S-I (1991) Glycosylation of P-glycoprotein in a multidrug-resistant KB cell line, and in human tissues. Biochim Biophys Acta 1073: 309–315.

Inaba M, Kobayashi H, Sukurai Y and Johnson RK (1979) Active efflux of daunorubicin and adriamycin in sensitive and resistant sublines of P388 leukemia. Cancer Res 39: 2200–2203.

Juliano RL and Ling V (1976) A surface glycoprotein modulating drug permeability in Chinese hamster ovary cell mutants. Biochim Biophys Acta 455: 152–162.

Kajiji S, Talbot F, Grizzuti K, Van Dyke-Phillips V, Agresti M, Safa AR and Gros P (1993) Functional analysis of P-glycoprotein mutants identifies predicted transmembrane domain 11 as a putative drug binding site. Biochemistry 32: 4185–4194.

Kamimoto Y, Gatmaitan Z, Hsu J and Arias IM (1989) The function of Gp170, the multidrug resistance gene product, in rat liver canalicular membrane vesicles. J Biol Chem 264: 11693–11698.

Kane SE, Reinhard DH, Fordis CM, Pastan I and Gottesman MM (1989) A new vector using the human multidrug resistance gene as a selectable marker enables overexpression of foreign genes in eukaryotic cells. Gene 84: 439–446.

Kartner N, Evernden-Porelle D, Bradley G and Ling V (1985) Detection of P-glycoprotein in multidrug-resistant cell lines by monoclonal antibodies. Nature 316: 820–823.

Kessel D, Botterill V and Wodinsky I (1968) Uptake and retention of daunomycin by mouse leukemic cells as factors in drug response. Cancer Res 28: 938–941.

Kessel D (1989) Exploring multidrug resistance by using rhodamine 123. Cancer Commun 1: 145–149.

Kioka N, Tsubota J, Kakehi Y, Komano T, Gottesman MM, Pastan I and Ueda K (1989) P-glycoprotein gene (*MDR*1) cDNA from human adrenal: normal P-glycoprotein carries Gly^{185} with an altered pattern of multidrug resistance. Biochem Biophys Res Commun 162: 224–231.

Koronakis V and Hughes C (1993) Bacterial signal peptide-independent protein export: HlyB-directed secretion of hemolysin. Semin Cell Biol 4: 7–15.

Kuchler K and Thorner J (1992) Functional expression of human *mdr*1 in the yeast *Saccharomyces cerevisiae*. Proc Natl Acad Sci USA 89: 2302–2306.

Lelong IH, Padmanabhan R, Lovelace E, Pastan I and Gottesman MM (1992) ATP and GTP as alternative energy sources for vinblastine transport by P-170 in KB-V1 plasma membrane vesicles. FEBS Lett 304: 256–260.

Ling V and Thompson LH (1974) Reduced permeability in CHO cells as a mechanism of resistance to colchicine. J Cell Physiol 83: 103–116.

Ling V, Kartner N, Sudo T, Siminovitch Land Riordan JR (1983) The multidrug resistance phenotype in Chinese hamster ovary cells. Cancer Treat Rep 67: 869–874.

Loo TW and Clarke DM (1993) Functional consequences of proline mutations in the predicted transmembrane domain of P-glycoprotein. J Biol Chem 268: 3143–3149.

Ma L, Marquardt D, Takemoto L and Center MS (1991) Analysis of P-glycoprotein phosphorylation in HL60 cells isolated for resistance to vincristine. J Biol Chem 266: 5593–5599.

Manavalan P, Smith AE and McPherson JM (1993) Sequence and structural homology among membrane-associated domains of CFTR and certain transport proteins. J Protein Chem 12: 279–290.

Mellado W and Horwitz SB (1987) Phosphorylation of the multidrug resistance associated glycoprotein. Biochemistry 26: 6900–6904.

Michaelis S (1993) STE6, the yeast a-factor transporter. Semin Cell Biol 4: 17–27.

Mimura CS, Holbrook SR and Ames GF-L (1991) Structural model of the nucleotide-binding conserved component of periplasmic permeases. Proc Natl Acad Sci USA 88: 84–88.

Miyamoto K-I, Wakusawa S, Nakamura S, Koshiura R, Otsuka K, Naito K, Hagiwara M and Hidaka H (1990) Circumvention of multidrug resistance in P388 murine leukemia cells by a novel inhibitor of cyclic AMP-dependent protein kinase, H-87. Cancer Lett 51: 37–42.

Miyamoto K, Inoko K, Ikeda K, Wakusawa S, Kajita S, Hasegawa T, Takagi K and Koyama M (1993) Effect of staurosporine derivatives on protein kinase activity and vinblastine accumulation in mouse leukaemia P388/ADR cells. J Pharm Pharmacol 45: 43–47.

Morris DI, Speicher LA, Ruoho AE, Tew KD and Seamon KB (1991) Interaction of forskolin with the P-glycoprotein multidrug transporter. Biochemistry 30: 8371–8379.

Myers MB, Rittmann-Grauer L, O'Brian JP and Safa AR (1989) Characterization of monoclonal antibodies recognizing a Mr 180,000 P-glycoprotein: differential expression of the M_r 170,000 P-glycoprotein in multidrug-resistant human tumor cells. Cancer Res 49: 3209–3214.

Naito M and Tsuruo T (1992) Functionally active homodimer of P-glycoprotein in multidrug-resistant tumor cells. Biochem Biophys Res Commun 185: 284–290.

Neyfakh AA, Serpinska AS, Chervonsky AV, Apasov SG and Kazarov AR (1989) Multidrug-resistance phenotype of a subpopulation of T-lymphocytes without drug selection. Exp Cell Res 185: 496–505.

Neyfakh AA, Bidnenko VE and Chen LB (1991) Efflux-mediated multidrug resistance in *Bacillus subtilis*: similarities and dissimilarities with the mammalian system. Proc Natl Acad Sci USA 88: 4781–4785.

O'Brian CA, Ward NE, Liskamp RM, de Bont DB, Earnest LE, van Boom JH and Fan D (1991) A novel N-myristylated synthetic octapeptide inhibits protein kinase C activity and partially reverses murine fibrosarcoma cell resistance to Adriamycin. Invest New Drugs 9: 169–179.

O'Reilly DR, Miller LK and Luckow VL (1992) Baculovirus Expression Vectors: A Laboratory Manual. New York: WH Freeman and Company.

Pastan I, Gottesman MM, Ueda K, Lovelace E, Rutherford AV and Willingham MC (1988) A retrovirus carrying an *MDR*1 cDNA confers multidrug resistance and polarized expression of P-glycoprotein in MDCK cells. Proc Natl Acad Sci USA 85: 4486–4490.

Posada JA, McKeegan EM, Worthington KF, Morin MJ, Jaken MS and Tritton TR (1989) Human multidrug resistant KB cells overexpress protein kinase C: involvement in drug resistance. Cancer Commun 1: 285–292.

Raviv Y, Pollard HB, Bruggemann EP, Pastan I and Gottesman MM (1990) Photosensitized labeling of a functional multidrug transporter in living drug-resistant tumor cells. J Biol Chem 265: 3975–3980.

Raymond M, Gros P, Whiteway M and Thomas DY (1992) Functional complementation of yeast *ste*6 by a mammalian multidrug resistance *mdr* gene. Science 256: 232–234.

Richert ND, Aldwin L, Nitecki D, Gottesman MM and Pastan I (1988) Stability and covalent modification of P-glycoprotein in multidrug-resistant KB cells. Biochemistry 27: 7607–7613.

Riordan JR and Ling V (1979) Purification of P-glycoprotein from plasma membrane vesicles of Chinese hamster ovary cell mutants with reduced colchicine permeability. J Biol Chem 254: 12701–12705.

Riordan JR, Rommens JM, Kerem B-S, Alon N, Rozmahel R, Grzelczak Z, Zielenski J, Lok S, Plavsic N, Chou J-L, Drumm ML, Iannuzzi MC, Collins FS and Tsui L-C (1989) Identification of the cystic fibrosis gene: cloning and characterization of complimentary DNA. Science 245: 1066–1073.

Roninson IB (1991) Molecular and Cellular Biology of Multidrug Resistance in Tumor Cells. New York: Plenum Publishing Corporation.

Roy SN and Horwitz SB (1985) A phosphoglycoprotein associated with taxol resistance in J774.2 cells. Cancer Res 45: 3856–3863.

Saeki T, Shimabuku AM, Azuma Y, Shibano Y, Komano T and Ueda K (1991) Expression of human P-glycoprotein in yeast cells – effects of membrane component sterols on the activity of P-glycoprotein. Agric Biol Chem 55: 1859–1865.

Saeki T, Shimabuku AM, Ueda K and Komano T (1992) Specific drug binding by purified lipid-reconstituted P-glycoprotein – dependence on the lipid composition. Biochim Biophys Acta 1107: 105–110

Safa AR, Glover CJ, Sewell JL, Meyers MB, Biedler JL and Felsted RL (1987) Identification of the multidrug-resistance-related membrane glycoprotein as an acceptor for calcium channel blockers. J Biol Chem 262: 7884–7888.

Safa AR (1988) Photoaffinity labeling of the multidrug-resistance-related P-glycoprotein with photoactive analogs of verapamil Proc Natl Acad Sci USA 85 7187–7191

Safa AR, Mehta ND and Agresti M (1989) Photoaffinity labeling of P-glycoprotein in multidrug-resistant cells with photoactive analogs of colchicine. Biochem Biophys Res Commun 161. 1402–1408

Safa AR, Stern RK, Choi K, Agresti M, Tamai I, Mehta ND and Roninson IB (1990) Molecular basis of preferential resistance to colchicine in multidrug-resistant human cells conferred by Gly to Val-185 substitution in P-glycoprotein Proc Natl Acad Sci USA 87 7225–7229.

Safa AR (1993) Photoaffinity labeling of P-glycoprotein in multidrug-resistant cells Cancer Invest 11. 46–56

Sampson KE, Wolf CL and Abraham I (1993) Staurosporine reduces P-glycoprotein expression and modulates multidrug resistance. Cancer Lett 68 7–14

Sarkadi B, Price EM, Boucher RC, Germann UA and Scarborough GA (1992) Expression of the human multidrug resistance cDNA in insect cells generates a high activity drug-stimulated membrane ATPase J Biol Chem 267 4854–4858

Sato W, Yusa K, Naito M and Tsuruo T (1990) Staurosporine, a potent inhibitor of C-kinase, enhances drug accumulation in multidrug-resistant cells Biochem Biophys Res Commun 173: 1252–1257.

Schinkel AH and Borst P (1991) Multidrug resistance mediated by P-glycoproteins. Semin Cancer Biol 2. 213–226.

Schinkel AH, Roelofs MEM and Borst P (1991) Characterization of the human *MDR*3 P-glycoprotein and its recognition by P-glycoprotein-specific monoclonal antibodies Cancer Res 51 2628–2635

Schinkel A, Kemp S, Dolle M, Rudenko G and Wagenaar E (1993) *N*-Glycosylation and deletion mutants of the human *MDR*1 P-glycoprotein. J Biol Chem 268: 7474–7481

Schoenlein PV (1993) Role of amplification in drug resistance In: LJ Goldstein and R Ozols (eds) Drug Resistance III Norwell, MA: Kluwer Academic Publishers, (in press)

Schurr E, Raymond M, Bell JC and Gros P (1989) Characterization of the multidrug resistance protein expressed in cell clones stably transfected with the mouse *mdr*1 cDNA. Cancer Res 49: 2729–2734.

Sharma RC, Inoue S, Roitelman J, Schimke RT and Simoni RD (1992) Peptide transport by the multidrug resistance pump. J Biol Chem 267: 5731–5734.

Shen D-w, Cardarelli C, Hwang J, Cornwell M, Richert N, Ishii S, Pastan I and Gottesman MM (1986) Multiple drug resistant human KB carcinoma cells independently selected for high-level resistance to colchicine, Adriamycin or vinblastine show changes in expression of specific proteins. J Biol Chem 261: 7762–7770.

Shimabuku AM, Saeki T, Ueda K and Komano T (1991) Production of a site specifically cleavable P-glycoprotein-β-galactosidase fusion protein. Agric Biol Chem 55: 1075–1080.

Shimabuku AM, Nishimoto T, Ueda K and Komano T (1992) P-glycoprotein-ATP hydrolysis by the N-terminal nucleotide-binding domain. J Biol Chem 267. 4308–4311.

Shyamala V, Baichwald V, Beall E and Ames GF-L (1991) Structure-function analysis of the histidine permease and comparison with cystic fibrosis mutations. J Biol Chem 266: 18714–18719.

Silverman JA, Raunio H, Gant TW and Thorgeirsson SS (1991) Cloning and characterization of a member of the rat multidrug resistance *(mdr)* gene family. Gene 106: 229–236.

Sinicrope FA, Dudeja PK, Bissonnette BM, Safa AR and Brasitus TA (1992) Modulation of P-glycoprotein-mediated drug transport by alterations in lipid fluidity of rat liver canalicular membrane vesicles. J Biol Chem 267: 24995–25002.

Skach WR, Calayag MC and Lingappa VR (1993) Evidence for an alternate model of human P-glycoprotein structure and biogenesis J Biol Chem 268: 6903–6908.

Skovsgaard T (1978a) Mechanisms of resistance to daunorubicin in Ehrlich ascites tumor cells. Cancer Res 38: 1785–1791.

Skovsgaard T (1978b) Mechanism of cross-resistance between vincristine and daunorubicin in Ehrlich ascites tumor cells. Cancer Res 38: 4722–4727.

Staats J, Marquardt D and Center MS (1990) Characterization of a membrane-associated protein kinase of multidrug-resistant HL60 cells which phosphorylates P-glycoprotein. J Biol Chem 265: 4084–4090.

Sugawara I, Kataoka I, Morishita Y, Hamada H, Tsuruo T, Itoyama S and Mori S (1988) Tissue distribution of P-glycoprotein encoded by a multidrug resistance gene revealed by monoclonal antibody MRK16. Cancer Res 48: 1926–1929

Sugimoto Y and Tsuruo T (1991) Development of multidrug resistance in rodent cell lines. In: IB Roninson (ed) Molecular and Cellular Biology of Multidrug Resistance in Tumor Cells (pp 57–70). New York: Plenum Publishing Corporation.

Tatsuta T, Naito M, Oh-hara T, Sugawara I and Tsuruo T (1992) Functional involvement of P-glycoprotein in blood-brain-barrier J Biol Chem 267: 20383–20391.

Thiebaut F, Tsuruo T, Hamada H, Gottesman MM, Pastan I and Willingham MC (1987) Cellular localization of the multidrug resistance gene product P-glycoprotein in normal human tissues. Proc Natl Acad Sci USA 84: 7735–7738.

Thiebaut F, Tsuruo T, Hamada H, Gottesman MM, Pastan I and Willingham MC (1989) Immunohistochemical localization in normal tissues of different epitopes in the multidrug transport protein, P170: evidence for localization in brain capillaries and cross-reactivity of one antibody with a muscle protein. J Histochem Cytochem 37: 159–164.

Townsend A and Trowsdale J (1993) The transporters associated with antigen presentation. Semin Cell Biol 4: 53–61.

Trezise AEO, Romano PR, Gill DR, Hyde SC, Sepúlveda FV, Buchwald M and Higgins CF (1992) The multidrug resistance and cystic fibrosis genes have complementary patterns of epithelial expression. EMBO J 11: 4291–4303.

Ueda K, Cardarelli C, Gottesman MM and Pastan I (1987) Expression of a full-length cDNA for the human "*MDR*1" (P-glycoprotein) gene confers multidrug resistance to colchicine, doxorubicin, and vinblastine. Proc Natl Acad Sci USA 84: 3004–3008.

Ueda K, Okamura N, Hirai M, Tanigawara Y, Saeki T, Kioka N, Komano T and Hori R (1992) Human P-glycoprotein transports cortisol, aldosterone, and dexamethasone, but not progesterone. J Biol Chem 267: 24248–24252.

Valverde MA, Diáz M, Sepúlveda FV, Gill DR, Hyde SC and Higgins CF (1992) Volume-regulated chloride channels associated with the human multidrug resistance P-glycoprotein. Nature 355: 830–833.

Van der Bliek AM, Baas F, Ten Houte de Lange T, Kooiman PM, Van der Velde-Koerts T and Borst P (1987) The human *mdr*3 gene encodes a novel P-glycoprotein homologue and gives rise to alternatively spliced mRNAs in liver. EMBO J 6: 3325–3331.

Van der Bliek AM, Kooiman PM, Schneider C and Borst P (1988) Sequence of *mdr*3 cDNA, encoding a human P-glycoprotein. Gene 71: 401–411.

Walker JE, Saraste M, Runswick MJ and Gay NJ (1982) Distantly related sequences in the a- and b-subunits of ATP synthase, myosin, kinases and other ATP-requiring enzymes and a common nucleotide binding fold. EMBO J 1: 945–951.

Weaver JL, Ine PS, Aszalos A, Schoenlein PV, Currier SJ, Padmanabhan R and Gottesman MM (1991) Laser scanning and confocal microscopy of daunorubicin, doxorubicin and rhodamine 123 in multidrug-resistant cells. Exp Cell Res 196: 323–329.

Weinstein RS, Kuszak JR, Kluskens LF and Coon JS (1990) P-glycoproteins in pathology: the multidrug resistance gene family in humans. Human Path 21: 34–48.

Wessels HP and Spiess M (1988) Insertion of a multispanning membrane protein occurs sequentially and requires only one signal sequence. Cell 55: 61–70.

Willingham MC, Richert ND, Cornwell MM, Tsuruo T, Hamada H, Gottesman MM and Pastan I (1987) Immunocytochemical localization of P170 at the plasma membrane of multidrug-resistant human cells. J Histochem Cytochem 35: 1451–1456.

Wolf DC and Horwitz SB (1992) P-glycoprotein transports corticosterone and is photoaffinity-labeled by the steroid. Int J Cancer 52: 141–146.

Wright LC, Dyne M, Holmes KT and Mountford CE (1985) Phospholipid and ether linked phospholipid content alter with cellular resistance to vinblastine. Biochem Biophys Res Commun 133: 539–545.

Yoshimura A, Kuwazuru Y, Sumizawa T, Ichikawa M, Ikeda S, Ueda T and Akiyama S-I (1989) Cytoplasmic orientation and two-domain structure of the multidrug transporter, P-glycoprotein, demonstrated with sequence-specific antibodies. J Biol Chem 264: 16282–16291.

Yu G, Ahmad S, Aquino A, Fairchild CR, Trepel JB, Ohno S, Suzuki K, Tsuruo T, Cowan JH and Glazer RI (1991) Transfection with protein kinase Cα confers increased multidrug resistance to MCF-7 cells expressing P-glycoprotein. Cancer Commun 3: 181–188.

Zhang J-Tand Ling V (1991) Study of membrane orientation and glycosylated extracellular loops of mouse P-glycoprotein by *in vitro* translation. J Biol Chem 266: 18224–18232.

Zhang J-T, Durbie M and Ling V (1993) Membrane topology on the N-terminal half of the hamster P-glycoprotein molecule. J Biol Chem 268: 15101–15110.

Cytotechnology **12**: 63–89, 1993.

Molecular cytogenetics of multiple drug resistance

Patricia V. Schoenlein
Department of Cellular Biology and Anatomy, Medical College of Georgia, Augusta, Georgia 30912, USA

Key words: cytogenetics, gene amplification, multidrug resistance, P-glycoproteins, physical mapping

Abstract

The refractory nature of many human cancers to multi-agent chemotherapy is termed multidrug resistance (MDR). In the past several decades, a major focus of clinical and basic research has been to characterize the genetic and biochemical mechanisms mediating this phenomenon. To provide model systems in which to study mechanisms of multidrug resistance, *in vitro* studies have established MDR cultured cell lines expressing resistance to a broad spectrum of unrelated drugs. In many of these cell lines, the expression of high levels of multidrug resistance developed in parallel to the appearance of cytogenetically-detectable chromosomal anomalies resulting from gene amplification. This review describes cytogenetic and molecular-based studies that have characterized DNA amplification structures in MDR cell lines and describes the important role gene amplification played in the cloning and characterization of the mammalian multidrug resistance genes (*mdr*). In addition, this review discusses the genetic selection generally used to establish the MDR cell lines, and how drug selections performed in transformed cell lines generally favor the genetic process of gene amplification, which is still exploited to identify drug resistance genes that may play an important role in clinical MDR.

Introduction

When cultured cells are exposed to a cytotoxic agent, such as a chemotherapeutic drug, individual clones can be selected that express either resistance to only the selecting drug (single-agent drug resistance) or resistance to multiple drugs that may be structurally and functionally unrelated. Such concomitant cross-resistance occurs frequently in cultured cell lines and is termed the multidrug resistance (MDR) phenotype. The MDR phenotype is also encountered in the clinical setting where many human cancers are refractory to multi-agent chemotherapy. A variety of physiological factors such as the degree of vascularity and oxygenation of the tumor can effect a poor response to multi-agent chemotherapy. However, a variety of cell-specific mechanisms appear to mediate MDR in human cancer and a large number of studies have focused on elucidating the physiological, molecular and genetic details of clinical MDR with the goal of improving treatment protocols.

Based on *in vitro* studies of multidrug resistant cultured cell lines, several cell-mediated MDR mechanisms have been proposed including changes in detoxifying systems, intracellular pH, DNA repair processes, drug target proteins, general membrane permeability, drug transport systems and drug efflux systems. As reviewed extensively in other chapters of this volume, the expression of MDR *in vitro* has been correlated specifically to alterations in the glutathione redox cycle (Kramer *et al.*, 1988), the glutathione S-transferase levels (Baptist *et al.*, 1986; Morrow and Cowan, 1990; Black *et al.*, 1990), the expression and/or activity of topoisomerase I and topoisomerase II (Ferguson *et al.*, 1988; Beck *et al.*, 1987; Danks *et al.*, 1988, 1989), and in the expression of a small family of related genes referred to as the multidrug resistance (*mdr*) genes that encode membrane-localized drug transporter proteins (for reviews see Endicott and Ling, 1989; Gottesman and Pastan, 1993).

The mechanisms mediating multidrug resistance in human tumors are not well defined with the exception of the role of one of the membrane-localized transporter proteins referred to as P-glycoprotein or P170 (for reviews, see Chapters 4 & 14 of this volume). A strong correlation has been established between P-glycoprotein gene expression and the refractory nature to chemotherapy of many multidrug resistant human cancers (reviewed in Goldstein *et al.*, 1992; Gottesman *et al.*, 1991). In addition, a recent study has clearly correlated the expression of P170 in *de novo* acute lymphoblastic leukemia (ALL) to a higher relapse rate, a shorter remission duration and a shorter median survival, emphasizing the value of P-glycoprotein as a prognostic indicator (Goasguen *et al.*, 1993). These results support earlier clinical studies of childhood neuroblastoma and sarcoma that also correlated the expression of P-glycoprotein to a poor prognosis (Chan *et al.*, 1990; Chan *et al.*, 1991).

P-glycoprotein is encoded by the human *MDR*1 gene which belongs to a small mammalian family of homologous multidrug resistance genes: the *MDR*1 and *MDR*2 (also referred to as *mdr*3) genes in human; the *mdr*1a, *mdr*1b and *mdr*3 genes in mouse; and the *pgp*1, *pgp*2 and *pgp*3 genes in hamster. The *mdr* genes have been grouped into three classes based on sequence similarities in their 3' noncoding regions (Ng *et al.*, 1989; summarized in Table 1) and this classification has been supported by the specificity of monoclonal antibodies for gene-specific protein epitopes (Georges *et al.*, 1990). The role of the different *mdr* genes in the

Table 1. Classification of multidrug resistance (*MDR)* genes[a]

	mdr genes		
Species	Class I	Class II	Class III
Hamster	*pgp*1	*pgp*2	*pgp*3
Mouse	*mdr*3(*mdr*1a)	*mdr*1(*mdr*1b)	*mdr*2
Human	*MDR*1		*MDR*2(mdr3)

[a]Adapted from Veinot and Ling(1993);
[b]The mammalian multidrug resistance genes are also designated *PGY* by the Committee on Standardized Human Gene Nomenclature.

multidrug resistance phenotype was determined most directly through gene transfer experiments of constitutively expressed *mdr* cDNAs that were shown to encode plasma membrane-localized P-glycoproteins. The class I and II P-glycoprotein transporters confer an MDR phenotype (Gros *et al.*, 1986a; Ueda *et al.*, 1987a; Guild *et al.*, 1988; Pastan *et al.*, 1988; Devault and Gros, 1990); whereas, the class III *mdr* genes do not appear to mediate drug-resistance (Schinkel *et al.*, 1991). The *mdr*-encoded MDR phenotype typically includes resistance to many different chemotherapeutic agents: vinca alkaloids (vinblastine and vincristine), anthracyclines (doxorubicin), epipodophyllotoxins (etoposide), antibiotics (actinomycin D), mitomycin C, taxol, topotecan, mithramycin, as well as some cytotoxic natural products, e.g., colchicine, puromycin emetine and gramicidin D. The P-glycoproteins efflux drugs from cells in an ATP-dependent manner and the expression of these transporters may also result in decreased drug influx (for reviews see Stein *et al.*, 1993; Gottesman and Pastan, 1993).

As determined by DNA sequence analyses of the human and rodent *mdr* cDNA clones (Chen *et al.*, 1986; Gros *et al.*, 1986c; Gerlach *et al.*, 1986), the P-glycoprotein isoforms are comprised of two homologous halves, each half containing 6 putative transmembrane domains and one hydrophillic ATP binding site. The P-glycoproteins are further classified as members of a superfamily of transport proteins, referred to as the ATP binding cassette (ABC) family, that share a similar protein architecture and are characterized by extensive amino acid sequence similarity where the ATP binding site(s) is located. ABC transporters are ubiquitous in prokaryotic and eukaryotic organisms and mediate a variety of biological functions. Some of the ABC transport proteins identified, especially in bacteria consist of only " half molecules", i.e., six putative transmembrane regions and one ATP binding site. The structure and function of the ABC transporter proteins has been extensively studied and reviewed (Hyde *et al.*, 1990; Ames and Lecar, 1992, Ames *et al.*, 1992; Higgins, 1992 and 1993; J.M. Croop this volume).

The initial identification of the multidrug resistance (*mdr*) gene family and subsequent studies characterizing the encoded P-glycoprotein drug transporters in mammalian somatic cells resulted, in part, from observations of abnormal karyotypes of metaphase chromosomes in multidrug resistant cultured cell lines. These aberrant chromosomal structures were frequently the result of localized gene amplification events that occurred at the mammalian *mdr* (*PGY*) gene locus resulting in an increase in the copy number and expression of the respective gene(s). The aim of this chapter is threefold: 1) to discuss the molecular cytogenetic studies characterizing amplicons in MDR expressing cultured cells; 2) to describe how gene amplification facilitated the identification of the mammalian *mdr* gene family and continues to play a role in identifying new genes involved in the development of the MDR phenotype; and 3) to discuss the phenomenon of genomic instability, a characteristic of transformed and tumorigenic cells that results in chromosomal abnormalities such as gene amplification.

Isolation of mammalian multidrug resistant cell lines

To identify multidrug resistance mechanisms operative in cancer, the most common approach has been to select multidrug resistant mammalian cell lines *in vitro* that exhibit patterns of drug resistance similar to the drug resistance profiles of human cancers. Typically, in these types of studies, cultured cells are selected for resistance to a single anti-cancer drug with a "classic" step-wise selection protocol that favors the selection of cells harboring amplified drug resistance genes. The drug is increased in small, sublethal increments allowing the selection of a rare number of mutant cells that gradually adapt to the cytotoxic agent. Step-wise drug selections usually are carried out over a period of several months to several years to generate cells expressing a high-level of drug resistance. Cell populations are often subcloned at independent steps of the drug selection to generate sublines expressing independent genetic changes resulting in drug resistance.

A multidrug resistance phenotype frequently develops during step-wise drug selections even though a cell line has been exposed to only one cytotoxic agent. The emergence of the MDR phenotype is identified by clonogenic and/or growth assays which compare the relative resistance (RR) of the drug resistant sublines to the parental subline for a broad spectrum of cytotoxic agents. The RR is usually defined as the ratio of the LD_{10} or LD_{50} (lethal dose allowing only 10% or 50% cell survival, respectively, also reported as the ED_{50} — effective dose resulting in 50% cell kill) of the resistant relative to the parental cell line. Table 2 summarizes data from one of the earliest reports describing the MDR phenotype in cultured cells (Biedler and Riehm, 1979). In this study, Chinese hamster lung (CHL) cells selected for resistance to actinomycin D developed cross-resistance to a variety of structurally and functionally unrelated drugs. Similar drug selections by a number of different laboratories were performed to establish MDR cell lines that have served as *in vitro* model systems to study clinically-relevant multidrug resistance mechanisms, e.g., the selection of multidrug resistant rodent cells (Riehm and Biedler, 1971; Dano, 1972; Ling and Thompson, 1974; Bech-Hansen *et al.*, 1976; Ling and Baker, 1978; Meyers and Biedler, 1981; Biedler and Peterson, 1981; Kartner *et al.*, 1983; Howell *et al.*, 1984); human leukemia and sarcoma cells Beck *et al.*, 1979; Siegfried *et al.*, 1983; Bhalla *et al.*, 1985; Tsuruo *et al.*, 1986); and human KB epidermoid carcinoma cell lines (Akiyama *et al.*, 1985; Shen *et al.*, 1986b). The growth characteristics, morphology

Table 2. Multidrug resistance profile of actinomycin D-selected DC-3F/AD IV cells[a]

Drug	Relative resistance [b]
Actinomycin D	376
Mithramycin	670
Vinblastine	239
Vincristine	189
Puromycin	84
Daunorubicin	29
Colcemid	18
Mitomycin C	3.1

[a]Table modified from Melera and Biedler (1991) and Biedler and Riehm (1979);
[b]Relative resistance is expressed as the ratio of 50% inhibitory concentrations of drug for resistant to control DC-3F (Chinese hamster lung cell line), obtained in a 3-day growth assay.

and karyotypes of these cell lines and of a large number of other human and rodent MDR cell lines have been recently reviewed by Beck and Danks (1991) and Sugimoto and Tsuruo (1991).

A general phenotype of independently-derived MDR cell lines was the overproduction of a glycosylated integral plasma membrane protein species of approximately 170 kDa molecular weight (gp 170,000) (Juliano and Ling, 1976; Beck *et al.*, 1979; Kartner *et al.*, 1983). This P-glycoprotein species was referred to as P-(permeability)glycoprotein (Pgp) because in biochemical studies its increased expression was correlated to the emergence of high levels of multidrug resistance and a reduced permeability of MDR cells to several unrelated drugs (Juliano and Ling, 1976). P-glycoprotein was initially identified when membrane polypeptides were analyzed by Juliano and Ling (1976) in a series of MDR Chinese hamster ovary (CHO) sublines that were selected in increasing colchicine concentrations. In subsequent studies, P-glycoproteins of molecular weights 150 kDa (gp 150,000) and 180 kDa (gp 180,000) were detected in MDR Chinese hamster lung cells (Biedler and Peterson, 1981) and in human leukemic cells (Beck and Cirtain, 1982), respectively. Further studies using specific antibodies raised against Pgp in Western blot analyses of MDR rodent and mammalian cell lines demonstrated the frequent association between the expression of P-glycoprotein(s) and the development of the MDR phenotype (reviewed in Endicott and Ling, 1989). The identification of specific membrane-localized proteins associated with the MDR phenotype supported earlier studies of MDR cells that had characterized either a decreased intracellular accumulation of drugs (Kessel *et al.*, 1968), a decreased membrane permeability to drugs (Biedler and Riehm, 1970), or an increased energy-dependent drug efflux (Dano, 1973) in multidrug-resistant cells. The P-glycoprotein(s) identified in all of these cell lines were later shown to be encoded by members of the multidrug resistance gene (*mdr*) family that were amplified in many of the MDR cell lines (discussed below).

In addition to the the increased synthesis of P-glycoproteins, other biochemical changes also occurred in the MDR cell lines as a result of *in vitro* drug selections (reviewed in Meyers and Biedler, 1991). For example, alterations of plasma membrane glycopeptides and gangliosides were characterized in MDR Chinese hamster cells (Peterson *et al.*, 1983). In another study of human MDR cell lines, several protein changes were characterized using two dimensional gel analyses (Richert *et al.*, 1985; Shen *et al.*, 1986b). These human MDR cell lines were independently derived from a subclone (referred to as KB-3–1) of the adenocarcinoma cell line HeLa in increasing concentrations of colchicine, vinblastine and Adriamycin (Fig. 1). With the exception of overexpression of P-glycoprotein, the protein changes observed in these cell lines were not consistent, e.g., decreased amounts of proteins of approximately 72–75 kDa (referred to as gp 72 and gp 75) only in the colchicine and adriamycin-selected cell lines; increased levels of a 48 kDa protein only in the Adriamycin and vinblastine-selected KB cell lines, increased levels of a 21 kDa protein only in the colchicine-selected KB cell line; and increased levels of a 32 kDa protein only in the vinblastine and colchicine-selected cell lines. In general, very few consistent protein changes other than P-glycoprotein expression were observed between MDR cell lines, suggesting that additional alterations in protein profiles between MDR and parental cell lines reflect secondary effects of metabolic changes that occur during drug selection. Interestingly, in

Colchicine resistant

⇓ EMS
.005 μg/ml
KB-Ch^{R}-8
⇓ EMS
.01 μg/ml
KB-Ch^{R}-8-5
⇓ EMS
.03 μg/ml
.1 μg/ml
KB-Ch^{R}-8-5-11
0.3 μg/ml
⇓ 0.5
0.7
1.0
KB-Ch^{R}-8-5-11-24
⇓
KB-C1
⇓
KB-C1.5
⇓
KB-C2.5
⇓
KB-C4
⇓
KB-C6

Vinblastine resistant

⇓ EMS
.0015 μg/ml
KB-VBL^{R}-5
⇓ EMS
.003 μg/ml
KB-VBL^{R}-5-2
⇓
KB-V.01
2 steps, .01 μg/ml
⇓ increments
KB-V.03
⇓
KB-V.05
⇓ 5 steps, .05 μg/ml
increments
KB-V.3
⇓
KB-V.8
⇓
KB-V1

Adriamycin resistant

⇓ EMS
.01 μg/ml
KB-ADR^{R}-5A
⇓
KB-A.02
6 steps, .005 μg/ml
⇓ increments
KB-A.05
⇓
KB-A.1
4 steps, .1 μg/ml
⇓ increments
KB-A.5
⇓
KB-A.7
⇓
KB-A.9
⇓
KB-A1

Fig. 1. Derivation of the MDR KB cells. The human HeLa subline, KB-3–1 was mutagenized with ethyl methanesulfonate (EMS) and selected in increasing amounts of the appropriate drug. Independently isolated subclones designated Ch^{R} (colchicine-resistant), ADR^{R} (adriamycin-resistant) and VBL^{R} (vinblastine-resistant) or resistant population, designated C (colchicine-resistant); V (vinblastine-resistant); A (adriamycin resistant) were isolated at a number of drug concentrations (the number immediately following the letter refers to the selecting concentrations of drug in ug/ml). KB-Ch^{R}-8–5-11–24 represents a cell line that was stored at -80°C immediately after adaptation to 1.0 μg/ml colchicine. KB-Ch^{R}-8–5-11–24 cells that continued to be passaged in culture were designated KB-C1. (Adapted from Akiyama *et al.*, 1985; Shen *et al.*, 1986b).

some cases the increased levels of specific proteins resulted from the coamplification of genes linked to the *mdr* locus, e.g., the increased synthesis of a protein of approximately 19 kDa in several vincristine-selected MDR cell lines (Meyers *et al.*, 1985) and the increased synthesis of the 21 kDa protein, termed C21, in the colchicine-selected KB cell lines (Shen *et al.*, 1986b). Both of these proteins have been identified as calcium-binding proteins that are encoded by a sorcin gene present in rodents (Meyers *et al.*, 1987) and in humans (Van der Bliek *et al.*, 1986a, 1986b). In rodent and human cells, the sorcin gene is linked to the *mdr* locus gene as determined by physical mapping studies (see below).

Gene amplification in MDR cell lines

Phenotypic and cytogenetic characteristics of many of the independently-derived MDR cell lines suggested that they harbored amplified copies of either identical or similar multidrug resistance gene(s). First, many of the MDR cell lines that were independently established developed similar cross resistance profiles, even though they often showed the greatest level of resistance (termed primary resistance) to the cytotoxic agent used for the drug selection. Second, chromosomal abnormalities commonly associated with gene amplification such as homogeneously staining regions (HSRs), abnormal banding regions (ABRs) or double minute chromosomes (DMs) were commonly observed in many of the highly resistant MDR cells (Biedler and Riehm, 1970; Riordan *et al.*, 1985; Fojo *et al.*, 1985; for reviews see Bradley *et al.*, 1988; Melera and Biedler, 1991; Gudkov *et al.*, 1991).

The distinct cytogenetic features of HSRs and ABRs were first identified *in vitro* in antifolate-resistant Chinese hamster cells and in cultured human neuroblastoma cells by Biedler and Spengler (1976a) and Biedler *et al.* (1980), respectively. In Fig. 2, the characteristic features of the HSR can be seen in the karyotype of a MDR human neuroblastoma cell line that contains amplified copies of the human *MDR*1 gene and of the human oncogene *MYCN*. After trypsin-Giemsa banding of chromosomes in metaphase spreads, HSRs lack the cross-striational arrays of darkly to lightly stained regions that result in a discernible longitudinal differentiation in normal chromosomes and appear unstained, while ABRs stain with distinct, but abnormal banding patterns. DM structures appear as extrachromosomal circular, paired chromatin structures (Kaufman *et al.*, 1979). ABRs and DM structures can be seen in Fig. 3 (see below). Frequently, the degree of multidrug resistance correlated to the length of the HSR or the number of DMs observed within the MDR cells.

The third characteristic shared by many of the MDR cell lines was based on the selection of drug-sensitive revertants. In the absence of drugs, cell lines became concomitantly sensitive to the same spectrum of drugs to which they originally expressed resistance and frequently lacked the obvious chromosomal aberrations seen in the MDR cell lines from which they were derived (Dahllof *et al.*, 1984; Meyers *et al.*, 1985; Roy and Horwitz, 1985; Fojo *et al.*, 1985; Lothstein and Horwitz, 1986). In some cases, cytogenetic analysis was performed during the selection of the drug-sensitive revertants and demonstrated decreased numbers of DM structures per cell and the shortening of HSRs concomitantly with decreased expression of resistance. These three characteristics, coupled with the fact that a high level of drug resistance in MDR cell lines could only be achieved through step-wise drug selections, provided strong indirect evidence that an unstable gene dosage mechanism, such as gene amplification, encoded the MDR phenotype and encouraged the use of molecular biology methods to clone and characterize the amplified drug resistance gene(s).

Cloning and identification of amplified multidrug resistance (MDR) gene sequences

A variety of approaches have been used to clone amplified DNAs in MDR cell lines. The observation of amplified gene structures present in MDR cell lines allowed the application of a differential cloning technique devised by Roninson (1983 and 1987) to identify and selectively clone amplified DNA sequences that were present in two independently derived MDR hamster cell lines (Roninson *et al.*, 1984), the LZ cell line (Adriamycin-selected) (Howell *et al.*, 1984) and the CH^RC5 cell line (Ling and Thompson, 1974). These cloned genomic DNAs were then used to facilitate the following studies: 1) the identification and characterization of the multidrug resistance gene, designated *mdr*1, within an approximately 120 kb contiguous region of rodent genomic DNA that was selectively expressed and amplified in a number of independently-selected MDR rodent cell lines (Roninson *et al.*, 1984; Gros *et al.*, 1986b); 2) the cloning and analysis of *mdr* cDNAs corresponding to over-expressed *mdr* mRNAs in rodent MDR cell lines (Gros *et al.*, 1986a); and 3) the identification, isolation and characterization of homologous human *mdr* genomic DNA sequences (Roninson *et al.*, 1986) that were subsequently used to isolate human *mdr* cDNAs (Ueda *et al.*, 1987b). A similar approach, also based on the amplification of specific genomic DNA sequences, used differential hybridization of Cot-fractionated DNA to clone

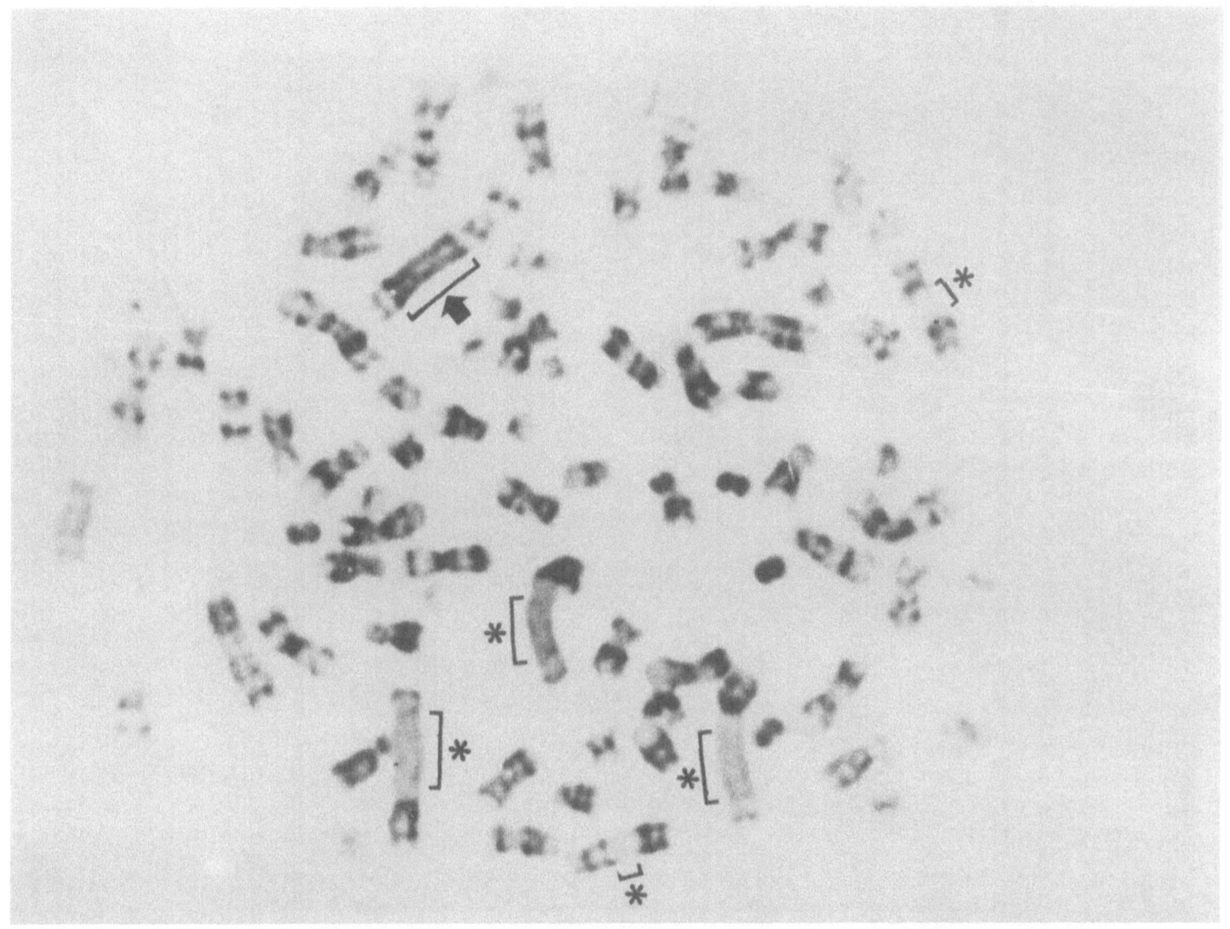

Fig. 2. Metaphase cell of the 124-fold actinomycin D-resistant human neuroblastoma cell line, BE (2)-C/ACT (0.2), stained by trypsin-Giemsa banding methods. The multidrug-resistant cells are characterized by a 50-fold increase in *MDR1* gene copy number manifested cytogenetically as a darkly staining HSR on chromosome 11p (arrow). The *MDR1* HSR stains more intensely than do 5 HSRs (asterisks), on chromosomes 4q and 6p, comprising the 150-fold amplified MYCN genes. HSRs are indicated by brackets. (Photograph kindly provided by JL Biedler and B Spengler).

mdr genes in the MDR Djungarian Hamster Cell lines (Gudkov *et al.*, 1987).

Cloning strategies have also targeted the increased expression of drug resistance genes. In these studies *mdr* cDNAs were isolated as a result of P-glycoprotein monoclonal antibody recognition of cDNA libraries (Kartner *et al.*, 1985; Riordan *et al.*, 1985; & Gerlach *et al.*, 1986) or as a result of cDNA library screening based on differential expression of mRNAs between parent (drug-sensitive) and MDR cell lines (Van der Bliek *et al.*, 1986a; Scotto *et al.*, 1986). Analysis of the cloned *mdr* cDNAs and/or genomic *mdr* DNAs have resulted in the identification and characterization of the multigene family of multidrug resistance genes in mammalian cells (Table 1). The genetic and biochemical methodologies used to clone and characterize the structure and function of the different *mdr* genes and full-length *mdr* cDNAs are reviewed in other chapters of this volume and in other published comprehensive reviews (Endicott and Ling, 1989; Roninson, *et al.*, 1991; Gottesman

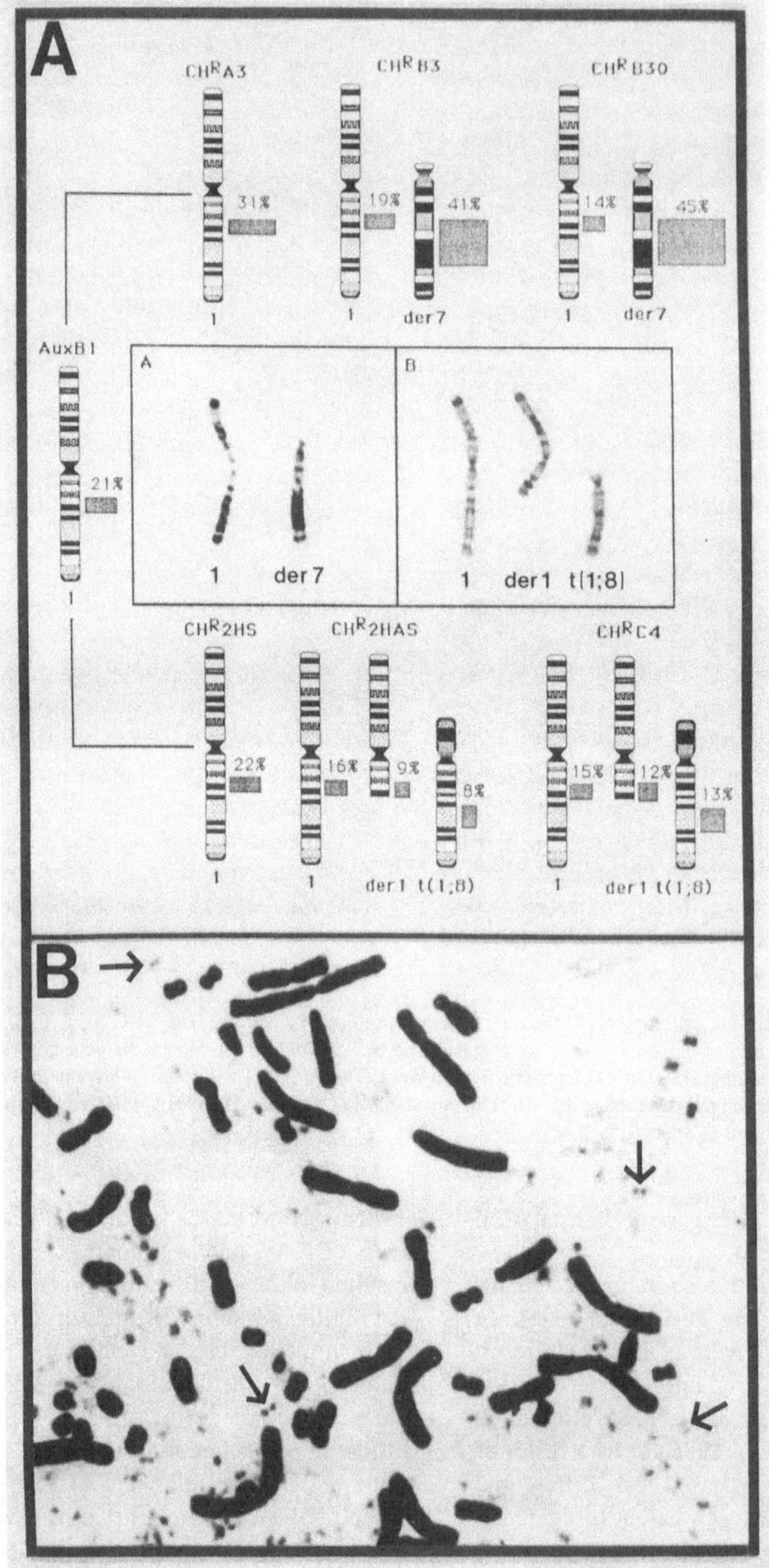
A
CH^R A3
CH^R B3
CH^R B30
31%
19%
41%
14%
45%
1
1
der7
1
der7
AuxB1
21%
1
A
1
der 7
B
1
der1 t(1;8)
CH^R 2HS
CH^R 2HAS
CH^R C4
22%
16%
9%
8%
15%
12%
13%
1
1
der1 t(1;8)
1
der1 t(1;8)
B

and Pastan, 1993).

With the availability of *mdr* clones, *in situ* hybridizations using *mdr* DNAs as probes confirmed that the gene amplification structures (HSRs, ABRs and DMs) in many of the MDR cell lines contained amplified *mdr* genes (Riordan *et al.*, 1985; Jongsma *et al.*, 1987; for review see Melera and Biedler, 1991; Gudkov *et al.*, 1991; Trent *et al.*, 1993). Figure 3 provides a representative example of *mdr* gene amplification structures as defined by cytogenetics and *in situ* hybridization methods. In this study, Trent and colleagues (1993) identified complex karyotypes with the formation of ABRs (Fig. 3A) and DM structures (Fig. 3B) during the progression of *mdr*1 gene amplification at the *mdr* (*PGY)* locus in MDR CHO sublines. The ideogram in Fig. 3 demonstrates results from *in situ* hybridization studies using the hamster *mdr1* cDNA probe pCHP1 (Riordan *et al.*, 1985) to localize chromosomes carrying single and multiple copies of the *mdr*1 gene in two series of MDR cell lines, designated the C4 and C5 panels. The two independent step-wise drug selections were originally performed by Ling and Thompson (1974) with clonal selections of the AuxB1 parent during growth in increasing concentrations of colchicine. Following their derivation, Ling and co-workers retained frozen samples of these cell lines which were used by Trent and coworkers to study the genesis of *mdr* gene amplification which is discussed in more detail in a later section of this review.

Characterization of MDR1-MDR2-containing circular DNA amplicons in human MDR adenocarcinoma cell lines

Once *mdr* cDNAs and genomic DNAs were cloned, Southern blot analyses of genomic DNA and Northern blot analyses of mRNA isolated from a variety of rodent and human MDR cell lines were performed. These studies confirmed the presence of *mdr* gene amplification in highly MDR cell lines and demonstrated the selective increased expression of an approximately 4.5–5.0 kb *mdr*-encoded mRNA species (Riordan *et al.*, 1985; Shen *et al.*, 1986a; Gros *et al.*, 1986b; Scotto *et al.*, 1986; Gros *et al.*, 1986b; De Bruijn *et al.*, 1986). The characterization of *mdr* gene amplification and expression in the human MDR KB cell lines, which have been previously described in Fig. 1, provide a representative example of such molecular biology studies. Table 3 summarizes published correlations between the MDR phenotype, the *MDR*1 gene copy number and/or the level of *MDR1* mRNA expression in some of these cell lines (Akiyama *et al.*, 1985; Fojo *et al.*, 1985; Shen *et al.*, 1986a and 1986b; Roninson *et al.*, 1986).

Studies of the MDR KB cells lines also demonstrate the phenomenon of differential amplification and expression of *mdr* genes which has been commonly observed in both rodent (Horwitz *et al.*, 1988; Hsu *et al.*, 1989; see below) and human MDR cell lines. Of the two human *mdr* genes, only

Fig. 3. (*facing page*) Direct evidence for the generation of double minutes (DMs) and abnormal banding regions (ABRs) in colchicine resistant, *mdr*-amplified CHO cell lines. The normal location of the *mdr*l gene in CHO is 1q26. These studies depict the alteration of this region associated with increasing drug resistance, and ultimately the generation of cytological evidence of gene amplification (DMs and ABRs). (Photograph kindly provided by J. Trent and adapted from Trent *et al.* 1993, with permission).

A. Ideograms documenting chromosome 1 abnormalities as well as ABR formation in two different panels of colchicine-selected resistant cells. The C5 series and its chromosome rearrangements are shown (top), and the C4 series is shown on the bottom. The AuxB1 parental drug sensitive cell line demonstrated a total of 21% autoradiographic grains to chromosome 1q localizing *PGY* sequences to 1q26. As illustrated in the C5 panel (top), increasing autoradiographic grain concentration was observed initially at the site of the single copy gene, and subsequently at the site of an ABR on der(7). Inset (A) illustrates the apparently normal G-banded chromosome (1) and the ABR-bearing der(7) chromosome observed in the $CH^RB3 \rightarrow CH^RB30$ members of the C5 panel which were known to harbor *mdrl* gene amplification. The C4 panel (bottom) showed no increase in autoradiographic grain distribution at the initial steps of selection (associated with no observed increase in gene copy number). However, by the second step of selection, a specific translocation involving the chromosome band encoding-*PGY* was observed with a net increase in the number of autoradiographic grains. Inset B illustrates the t(1;8) and der(1) chromosomes in addition to an apparently normal copy of chromosome 1 retained in the resistant sublines.

B. Evidence for DMs in the C5-panel shown in (A). G-banded metaphase from the C5-subline demonstrating numerous DMs (arrows). Note: This cell has been initially overexposed to allow better resolution of DMs.

Table 3. Amplification of MDR sequences in KB sublines and relative drug resistance[a]

Cell line	Relative resistance to			*MDR*1 copy number	*MDR*1 mRNA expression	Double-minute stucture
	Colchicine	Adriamycin	Vinblastine			
Colchicine selection						
KB-3-1	1	1	1	1	<.1[c]	–
KB-Ch^{R}-8	2.1[b]	1.1[b]	1.2[b]	1[c]	1[c]	–
KB-Ch^{R}-8-5	3.8[b]	3.2[b]	6.3[b]	1[c]	3[c]	–
KB-Ch^{R}8-5-11	40[b]	23[b]	51[b]	13[e]	80[c]	–
KB-Ch^{R}-8-5-11-24	128[b]	26[b]	20[b]	20[e]	158[e]	ND
KB-C1	263[d]	162[d]	96[d]	13[c]	120[e]	–
KB-C1.5	324[d]	ND	142[d]	14[c]	340[c]	+
KB-C2.5	487[c]	141[c]	206[c]	ND	ND	+
KB-C4	1750[c]	254[c]	159[c]	30[e]	378[e]	+
KB-C6	2100[d]	320[d]	370[d]	80[g]	820[g]	
KB-C1-R1	6[c]	4[c]	3[c]	1[c]	1[c]	–
Vinblastine selection						
KB-V1	171[d]	422[d]	213[d]	100[f]	320[c]	+
Adriamycin selection						
KB-A1	19[d]	97[d]	43[d]	70[f]	270[c]	+

[a]The *MDR1* gene copy numbers are given relative to parental KB-3-1 cells. In all colchicine-selected (Fig. 1) cell lines the *MDR2* copy number was determined to be approximately equivalent to that of *MDR1*. In KB-V1 and in KB-A1, the *MDR2* gene copy is similar to the unamplified level in KB-3-1. KB-V1 and KB-A1 were selected in multiple steps to resist vinblastine and Adriamycin at 1 μg/ml, respectively (refer to Fig. 1). Relative resistance is expressed as the LD_{10} of the resistant line divided by the LD_{10} of the parental KB-3-1 cells; [b]From Akiyama *et al.* (1985); [c]Shen *et al.* (1986a); [d]Shen *et al.* (1986b); [e]Choi *et al.* (1988); [f]Roninson *et al.* (1986); [g]Shen *et al.* (1988) N.D., Not determined. (Adapted from Roninson *et al.*, 1991)

the *MDR1* gene is consistently amplified and overexpressed in correlation to the increasing levels of drug resistance in human MDR cell lines. For example, in the vinblastine-selected cell line KB-V1 and the adriamycin-selected KB-A1 cell line, the *MDR*2 gene which lies immediately adjacent to the *MDR*1 gene on chromosome 7, band q21.1 (Fojo *et al.*, 1986; Callen *et al.*, 1987; reviewed in Trent and Callen, 1991) is not amplified (Chin *et al.*, 1989). In contrast, in all of the colchicine-selected sublines the *MDR*2 gene is coamplified with *MDR*1; however, a concomitant increase in *MDR*2 mRNA expression does not result (Chin *et al.*, 1989). These results provided indirect evidence that *MDR*1 gene expression confers the MDR phenotype in human cell lines.

Although DMs were observed in the highly MDR KB cell lines (KB-C1.5, KB-C2.5, KB-C4 and KB-C6), obvious amplification structures were not detected in KB-Ch^{R}-8-5-11 or KB-C1 with conventional cytogenetics (Fojo *et al.*, 1985; Schoenlein *et al.*, 1992) even though hybridization studies determined the presence of amplified *MDR*1 and *MDR*2 genes in both of these MDR cell lines (Table 3). In addition, the cell lines KB-V1 and KB-A1 that contain a very high copy of *MDR*1 (Table 3) showed very few DM structures. This discrepancy between the detection of amplified *MDR1* and/or *MDR*2 genes via hybridization studies and the lack of obvious DNA amplification structures was resolved when submicroscopic circular DNA structures (termed episomes) of <1,000 kilobase pairs (kbp) in size, were shown to harbor the amplified *mdr* genes in these cell lines (Ruiz *et al.*, 1989; Schoenlein *et al.*, 1992). Episomes were first identified with electron microscopy (Maurer *et al.*, 1987) and unidirectional and pulsed field gel electrophoresis (PFGE) techniques (Carroll *et al.*, 1987, Von Hoff *et al.*, 1988) in studies aimed at characterizing amplicon structures formed during the amplification of other mammalian drug resistance genes, i.e., the amplification of the dihydrofolate reductase gene *dhfr* and the *CAD* gene. Similar electrophoretic techniques (recently reviewed, Schoenlein *et al.*, 1993) were employed to identify *MDR*1-containing episomes present in the MDR KB cell lines. With unidirectional electrophoresis, episomes which are supercoiled circular DNAs can be detected and resolved in agarose gels in relation to the logarithm of their molecular weight. Episomes present in high copy can be visualized with ethidium bromide staining of the gel (e.g., MDR KB cell lines KB-A1 and KB-V1; Fig. 4).

PFGE techniques (Fig. 5) were used to approximate the size of the episomes present in KB-Ch^R-8–5-11 (890 kbp), KB-C1 (890 kbp), KB-V1 75C kbp) and KB-A1 (750 kbp; PV Schoenlein, unpublished). Similarly, the size of the DM structures in the highly MDR KB cell lines were determined to be approximately 1.8 and 3.6 megabase pairs (Mbp) (Fig. 5). For these studies, high molecular weight DNA isolated from the MDR cell lines was exposed to 3,000 cGy of radiation, which will linearize circular DNA molecules (amplicons) ranging in size from 200 kbp to 3.6 Mbp through the introduction of one double-strand break per circular DNA. Following linearization, the DNA amplicons were separated according to size using PFGE. The fractionated DNA was transferred to nitrocellulose and hybridized to an *MDR*1 cDNA probe.

Further PFGE studies of the colchicine-selected MDR KB cell lines that were established at different steps of the drug selection provided strong evidence that the *mdr*-containing DM structures were generated from the multimerization of the 890 kb *mdr* episome. This direct relationship between episome and DM structures was determined by restriction analysis of amplicon DNA using enzymes that digest mammalian DNA infrequently, i.e., *Not*I (Schoenlein *et al.*, 1992). Subsequent hybridization studies have determined that the 890 kb episome contains the *MDR*2 gene, the human sorcin gene and at least two other linked genes of unknown function (PV Schoenlein, Y Sugimoto, T Tsuruo, I Pastan and M Gottesman, unpublished).

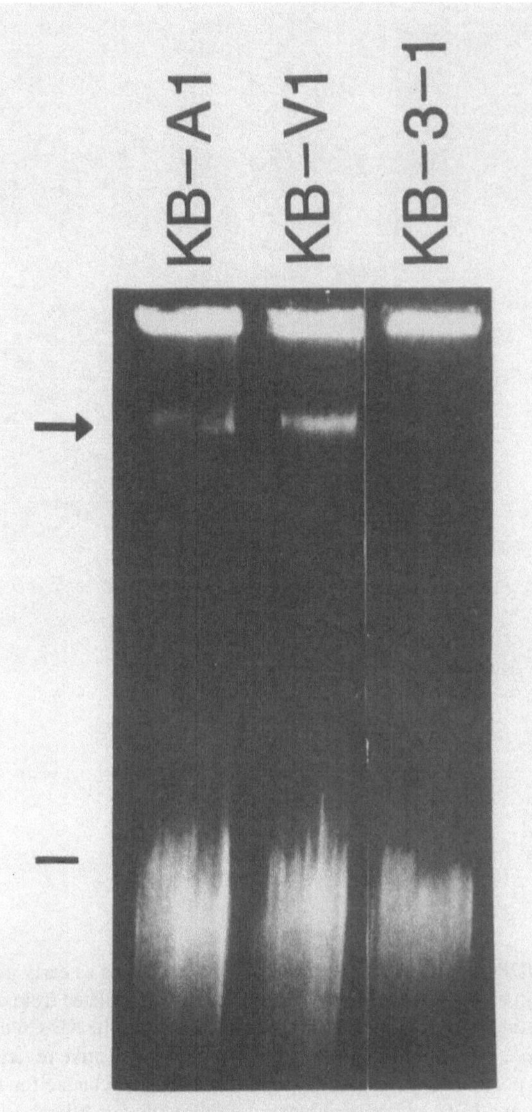

Fig. 4. *MDR*1-containing episomes present in multidrug resistant cell lines KB-V1 and KB-A1. An EtBr-stained agarose gel following unidirectional electrophoresis of agarose-embedded DNA. The supercoiled *MDR*1-containing circular DNAs (arrow) migrate below the origin (loading well) and above the randomly fragmented DNA (solid line). Taken from Schoenlein *et al.* (1993), with permission.

Other studies have documented the prevalence of *mdr*-containing extrachromosomal DNAs in rodent MDR cultured cell lines ranging in size from several hundred kbp to several Mbp (Meese *et al.*, 1992; Stahl *et al.*, 1992; see below).

As discussed earlier, many of the MDR cell lines revert to a normal (drug-sensitive) phenotype

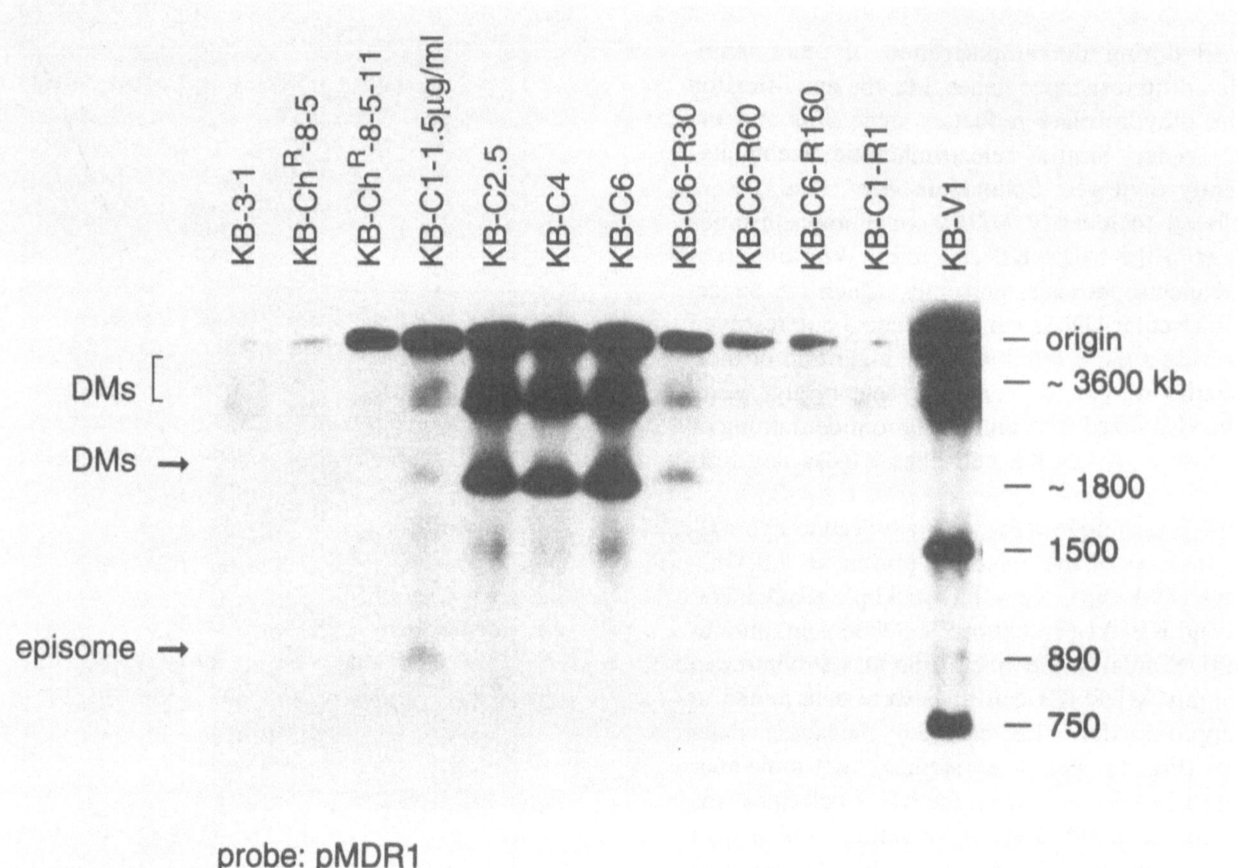

Fig. 5. MDR1-containing episomes and DM structures as early and late gene amplification events during drug selection (taken from Schoenlein *et al.* 1992, with permission). DNA was isolated from colchicine-selected cell lines growing in the presence of the selecting concentration of colchicine (refer to Fig. 1) and from the KB-C6 cell line after passage in medium without colchicine for 30–160 days (Shen *et al.*, 1988). KB-C1-R1 (lane 11) is a drug-sensitive revertant cell line (see Table 3 for drug-sensitive profile) subcloned from a KB-C1 population passaged in medium without colchicine for several months (Shen *et al.*, 1986a,b). Irradiated DNA was electrophoresed on a 1.2% agarose gel under the following conditions: 150 V, with a 120-s pulse for 22 h, followed by a 240-s pulse for 30 h, fractionated DNA was transferred to a nitrocellulose membrane, which was hybridized to a *MDR*1 cDNA probe, stripped and reprobed with a *MDR*2-specific cDNA probe. Both probes gave the same hybridization signals except in the lane containing DNA from KB-V1; with the *MDR*2 probe only a very light signal, equivalent to single copy, was detected at the origin of the gel. The DM structures of approximate size of 3.6 Mbp are located in the zone of compression under these gel conditions.

when they are passaged in the absence of drug. Such phenotypic reversion is a genotype that is characteristic of unstable forms of gene amplifications, such as DMs and HSRs. The lack of stability of circular *mdr* amplicons is clearly demonstrated in Fig. 5 where the loss of DM structures specifically carrying the *MDR*1 gene occurs with the emergence of a more drug-sensitive phenotype (lanes 8—11). In earlier studies, Gottesman and colleagues used conventional Southern analysis and *in situ* hybridization studies of whole cell populations to correlate the loss of *MDR*1 gene copies and gene expression (respectively) to the loss of the MDR phenotype in these drug-sensitive revertant cell lines. (Shen *et al.*, 1988).

The MDR phenotype mediated by amplification and expression of transfected P-glycoprotein genes

In independent studies, the amplification and expression of the human *MDR*1 gene was shown to confer the MDR phenotype to drug-sensitive mouse NIH 3T3 cells (Shen *et al.*, 1986c). In these studies NIH 3T3 cells were cotransfected with high-molecular-weight DNA isolated from the human MDR cell line KB-C1.5 (Fig. 1) and plasmid pSV2 neo DNA which provides a dominant selectable marker, resistance to the antibiotic G418. Initially, transfectants were selected in G418 to decrease selection of spontaneously occurring colchicine-resistant variants of the drug-sensitive NIH 3T3 cells. These primary transformants were then selected step-wise in increasing concentrations of colchicine to obtain a high level of MDR expression. Their MDR phenotype was confirmed with clonogenic assays that compared cross-resistance profiles to several unrelated drugs. From one of the highly MDR primary transformants NIH-T1-C1 (adapted to 1.0 μg/ml colchicine), high-molecular weight DNA was isolated and used to independently transfect drug-sensitive NIH 3T3 cells and secondary MDR transformants were selected in a similar fashion. From this second round of transfection, a highly MDR secondary transformant NIH-T2-C1 (adapted to growth in 1.0 μg/ml colchicine) was established. Subsequent to the selection of the MDR primary and secondary transformants, Southern and Northern blot hybridizations were used to determine that the human *MDR*1 gene was transferred from KB-C1.5 to NIH-T1-C1 and then to NIH-T2-C1. In each of the transformants described and in sublines established at earlier steps in the colchicine selection that were less multidrug resistant, *MDR*1 gene amplification and expression was demonstrated. These studies provided strong evidence that *MDR*1 gene expression, independent of *MDR*2 gene expression, conferred a MDR phenotype in human cells.

Similar transfection studies were performed in mouse LTA cells using genomic DNA from a colchicine-selected MDR hamster cell line CH^RC5 (Deuchars *et al.*, 1987). In independent MDR transfectants, the hamster *pgp*1 gene (class 1) was amplified and expressed as determined by Southern blot hybridizations and Western blot analyses. In concurrent studies, Gros and colleagues performed similar chromosome-mediated gene transfer of MDR. In these studies, murine LTA cells were used as the drug-sensitive recipients for transfection of high molecular weight genomic DNA that was isolated from an adriamycin-selected MDR hamster cell line LZ (Gros *et al.*, 1986d). Although these genetic studies provided strong evidence that *mdr* genes conferred the MDR phenotype, definitive proof came from transfection and/or infection studies of drug sensitive cell lines with expression vectors containing full-length *mdr* cDNAs (JM Croop, this volume).

Amplicons containing five linked genes in MDR cell lines

Non-*mdr* cDNAs were also cloned as a result of the cloning strategies based on differential gene expression in MDR and parent cell lines (Van der Bliek *et al.*, 1986; Scotto *et al.*, 1986) and these cDNAs facilitated studies aimed at characterizing the multigene amplification structures in some of the highly MDR cell lines. P Borst and colleagues analyzed the molecular cytogenetic characteristics of the colchicine-selected series of MDR CHO cell lines selected by Ling and Thompson (1974) and shown in the top panel of Fig. 3. A differential hybridization approach was used in which cDNA clone banks were constructed and screened to identify cDNAs corresponding to specific mRNA species that were expressed at higher levels in the colchicine-selected MDR CHO cell lines as compared to AuxB1 parental cells. The mRNA used to generate the clone banks of cDNA was isolated from the CH^RC5 cell line whose progenitor is the cell line CH^RB3 (Fig. 3A, top panel). During the initial step wise drug selection, CH^RC5 was selected in a higher concentration of colchicine subsequent to the derivation of CH^RB3.

Six gene classes (class 1–6) that were overexpressed in the CH^RC5 cells were identified and the cDNAs representing these gene classes were used to further characterize the genetics of MDR in the MDR CHO cell lines (reviewed in Borst and

Van der Bliek, 1991). Molecular studies with the cloned cDNAs demonstrated that only class 2, consisting of the multidrug resistance genes, *pgp*1 (class 2c), *pgp*2 (class 2b), and *pgp*3 (class 2a), was consistently overexpressed and amplified in independently-derived MDR cell lines (De Bruijn *et al.*, 1986; Jongsma *et al.*, 1987), providing strong evidence that the other genes are not required for the MDR phenotype. Differential amplification of the *mdr* genes within class 2 was often observed. For example, during the selection of the CH^RC5 MDR cell line from its progenitor CH^RB3, class 2c (*pgp*1) and classes 3–6 were further amplified; whereas, classes 2a (*pgp*3), 2b (*pgp*2) and class 1 were not coamplified (De Bruijn *et al.*, 1986; Jongsma *et al.*, 1987). Similar differential gene amplification patterns were observed when gene amplification profiles of the six gene classes were characterized in independently-derived MDR Chinese hamster lung cell lines.

The six gene classes were mapped to approximately 1,500 kbp of linked genomic DNA in hamster cells at the native *mdr* locus in the following relative linear order: class 1-class 2-class 3/class 6-class 5-class 4 (reviewed in Borst and Van der Bliek, 1991). For these mapping studies, PFGE was used to separate large fragments of genomic DNA generated from restriction enzyme digests, followed by Southern transfer of the fractionated DNA to membrane and hybridizations to the different cDNA probes. In addition, hybridization of the cDNA probes to metaphase chromosomes in situ of the Ch^RC5 cell line showed that DNAs of classes 2, 4 and 5 hybridize to HSRs located on chromosome 1 at approximately q26 (Jongsma *et al.*, 1987). These early data are consistent with the recent studies by Trent and colleagues (1993) (described above; Fig. 3). One of the coamplified genes (class 4) has been shown to encode a phosphorylated cytosolic calcium-binding protein referred to as sorcin/VP19 or CP22 (Van der Bliek *et al.*, 1986b; Meyers *et al.*, 1987); the identity and function of the other genes has not been reported.

The overall structure of the human and mouse *mdr* (*PGY*) locus appears to be similar to that in hamster. The murine *mdr* genes have been localized as a cluster on chromosome 5 (Martinsson and Levan, 1987) and the genes of class 3–5 are linked to the *mdr* gene cluster. In initial studies aimed at characterizing gene amplification patterns, Stahl *et al.* (1988) demonstrated coamplification of classes 3–5 with the *mdr* genes of class 2 in five independently-derived MDR mouse murine SEWA cell lines. These results provided strong evidence that gene classes 2–5 are syntenic to each other. In two other independently-derived MDR SEWA cell lines, differential amplification of gene classes 3–5 and of the *mdr* genes that comprise class 2 was observed. Similar differential amplification of *mdr* genes was observed in murine J774.2 MDR sublines (Hsu *et al.*, 1989): *mdr*1a was amplified in all cases, *mdr*1b was frequently coamplified, and *mdr*2 was infrequently amplified or amplified to a lesser degree. Class 1 was not coamplified or differentially amplified in any of the independently derived MDR SEWA cell lines that were examined. Recently, in MDR SEWA murine cells, a circular DNA amplicon has been characterized and shown to correspond, without any detectable changes in restriction enzyme map, to approximately 2,200 kbp of parental DNA. Within this amplicon and at the native murine *mdr* (*PGY*) locus, class 2 (comprised of the three mouse *mdr* genes *mdr*1a, *mdr*1b and *mdr*2), class 4 and class 5 were physically mapped with PFGE techniques, confirming the linkage of these five genes in murine cells.

In similar mapping studies in the human MDR ovarian carcinoma cell line 2780 AD, coamplification of the class 2 genes occurred with classes 4–6, strongly suggesting the synteny of these genes in human cells (Van der Bliek *et al.*, 1988). In addition, unpublished PFGE analysis from my laboratory has established a restriction map of the human *mdr* locus, spanning approximately 1,200 kbp of genomic DNA. This restriction map confirms linkage of the sorcin gene (class 4) to the *MDR*1 and *MDR*2 genes. Two unidentified genes map between the sorcin gene and the human *mdr* locus. As described in an earlier section, these five linked genes are all coamplified on the circular DNAs present in the colchicine-selected MDR cell lines (see above; Fig. 5).

Studies of the *mdr*-amplicons have clearly demonstrated that the amplified state of a gene does not always mediate a parallel increase in gene expression. In both rodent (Raymond *et al.*, 1990) and human MDR cell lines (Chin *et al.*, 1989; Van der Bliek *et al.*, 1988), coamplified *mdr* genes can be differentially expressed, i.e., the very low, barely detectable *MDR*2 expression levels in KB-C1 (described earlier). In addition, coamplification of the genes linked to the *mdr* locus does not necessarily mediate increased expression, i.e., the 30 fold amplification of the class 3 gene in the MDR rodent cell line DC-3F/DMXX was not accompanied by an increase in the mRNA level (De Bruijn *et al.*, 1986). Interestingly, when Tsuruo and co-workers selected for revertants of the MDR human myelogenous leukemia cell line K562/ADM, the level of *MDR*1 gene amplification was unaltered and only expression of the 4.5 kb *MDR*1 mRNA was decreased (Sugimoto *et al.*, 1987).

In recent studies Biedler *et al.* (1993) have suggested that amplified Pgp genes can be down-regulated at the level of transcription when cultured MDR cell lines are exposed to the chemosensitizer verapamil, and possibly other chemosensitizers. Verapamil is a calcium channel antagonist that appears to directly inhibit outward transport of drugs in MDR cells expressing P-glycoprotein by interacting with membrane-localized P-glycoprotein (reviewed in Chapter 16, this volume). Studies by Biedler and colleagues also suggest that verapamil activates mechanisms that lead to the loss of amplified MDR genes. The possibility that verapamil activates mechanisms that alter *mdr* gene expression or copy number in MDR cell lines is interesting in regards to the collateral sensitivity of MDR cell lines to the cytotoxic actions of verapamil and other modulators previously described (Warr *et al.*, 1986; Cano-Gauci and Riordan, 1987; Biedler *et al.*, 1988; Beck *et al.*, 1988b). The mechanism(s) underlying collateral sensitivity have not been identified and their elucidation may provide insight into the genetics and cell biology of MDR development, and possibly lead to improved circumvention of clinical MDR, a topic discussed in detail in Chapter 16 of this volume.

Cross-resistance profiles of MDR cell lines

In both rodent and human cells, the relative resistance of the MDR cell line is usually highest to the selecting agent. In rodent cells, the cross resistance profile depends to some degree on the P-glycoprotein isoform that is expressed because the *mdr*1a and *mdr*1b genes encode MDR phenotypes with distinct, but overlapping cross resistance profiles (Devault and Gros, 1990). However, the different cross-resistance profiles of human MDR lines should not result from isoform distinctions since *MDR*2 does not appear to encode drug resistance (Schinkel *et al.*, 1991).

In one case, during colchicine selection of the MDR cell line KB-Ch^{R}-8–5-11–24 (also referred to as KB-C1; Fig. 1), base pair mutations were found to occur in the human *MDR*1 gene encoding an amino acid change at position 185 of the *MDR*1 coding sequence (Glycine to Valine). These mutations result in preferential colchicine resistance with a diminished resistance to vinblastine (Choi *et al.*, 1988); e.g., the relative resistance to colchicine and vinblastine of cell line KB-C1 to its progenitor KB-Ch^{R}-8–5-11 (Table 3). These base pair mutations were not detected in any of the *MDR*1 copies present in KB-Ch^{R}-8–5-11; therefore, subsequent to the cloning of KB-Ch^{R}-8–5-11, these mutations probably arose spontaneously in an *MDR*1 gene harbored on one of the 890 kb circular DNAs within this cell line (see above, Fig. 5). The mutated *MDR*1-circular DNA appears to have been preferentially selected during subsequent steps in the colchicine selection because the same base pair mutations are present in the *MDR*1 gene copies (Choi *et al.*, 1988) harbored on the circular DNAs identified in cell lines KB-C1.5, KB-C2.5, KB-C4 and KB-C6 (Schoenlein *et al.*, 1992).

In another study, during actinomycin D-selection of a hamster cell line, a change in the cross resistance profile occurred as a result of a double mutation identified in the putative transmembrane domain six of the *MDR*1 protein at positions 338 and 339 (Gly to Ala and Ala to Pro, respectively) (Devine *et al.*, 1992). These mutations appeared to decrease the transport of drugs other than actinomycin D (P Melera, personal communication) giving the appearance of preferential actinomycin D resistance. Genetically engineered mutations in *mdr* cDNAs also have been shown to affect the pattern of multidrug resistance (reviewed in Chapter 2 of this volume). Although these types of genetic changes help define the structure-function interactions of P-glycoproteins, their frequency of occurrence and contribution to the different MDR phenotypes in human cancers is not known. It is quite possible that spontaneous mutations occur infrequently, if at all, *in vivo* and that the spontaneous mutations that arise during *in vitro* drug selections are a consequence of gene amplification events. For example, amplified genes, particularly genes harbored on extrachromosomal elements, may not be repaired as efficiently as native chromosomal DNA.

Mechanisms mediating early events in *MDR* gene amplification

Studies aimed at understanding the processes mediating gene amplification should provide valuable insight into tumor cell biology because gene amplification is an obvious manifestation of the increased chromosomal instability characteristic of cancer cells. Such studies have identified several different mechanism(s) that appear to mediate gene amplification such as direct excision (Wahl, 1989), chromosome breakage (Ma *et al.*, 1993; Wahl *et al.*, 1993), recombination (Schimke, 1988; Stark and Wahl, 1984; Hamlin *et al.*, 1984; Smith *et al.*, 1990; Trask and Hamlin, 1989; Stark *et al.*, 1989; Ma *et al.*, 1993) and premature condensation of chromosomes found within micronuclei (Sen *et al.*, 1989). However, the molecular details of these mechanisms and how they contribute to the size, composition and structural variability of DNA amplification structures awaits further studies.

The mechanism(s) of *mdr* gene amplification and maintenance of the amplified *mdr* copies appears to depend to some degree on the cell line being analyzed, suggesting that the inherent "plasticity" of the cell line may play an important role in the evolution of *mdr* gene amplification. In a recent study, Horwitz and colleagues provide evidence that a LINE-1 DNA element, which belongs to a family of short interspersed repeat DNAs in mammalian cells, may mediate unequal sister chromatid exchange during amplification of the *mdr* locus in the murine macrophage-like cell line J774.2 (Cohen *et al.*, 1992). Previous studies have demonstrated that the mammalian HSAG middle repetitive DNA elements stimulate *in vitro* gene amplification in both rodent and human cell lines (reviewed in McArthur *et al.*, 1993). Together, these studies raise the interesting possibility that repetitive DNA sequences in mammalian DNA may play an active role in the genomic instability of tumor cells.

The amplification of the *mdr* (*PGY)* locus in Chinese hamster ovary (CHO) cells, appears to proceed through a recombination-based process generating an interstitial chromosomal duplication, followed by a series of unequal homologous sister chromatid exchanges. This mechanism was proposed by Trent and colleagues (1993) who studied *mdr*1a gene amplification in MDR CHO sublines established at early steps of step-wise colchicine selection. The series of MDR sublines and the cytologic techniques used to characterize them, i.e., *in situ* hybridization are shown in Fig. 3 (C4 and C5 panel). Gene amplification via this type of mechanism was initially described for the early steps of *CAD* and *dhfr* gene amplification (Smith *et al.*, 1990; Trask and Hamlin,1989). A characteristic early step in this amplification mechanism is the appearance of large distinctive chromosomal structures containing an array of amplified genes attached to one end of the chromosome arm that harbors the single copy gene. Trent and colleagues (1993) observed such distinctive chromosomal structures in the MDR CHO cell lines that were established at early steps of the colchicine selection. These structures were localized at one end of the long arm of chromosome 1 where single copy *mdr* genes have been previously localized in unselected cells (Jongsma *et al.*, 1987; Melera and Biedler, 1991).

Figure 3A (top panel) schematically shows the initial increase in *mdr* gene copy that was detected during colchicine selection of the C5 series of MDR sublines. As depicted in the figure, increasing autoradiographic grains were detected at the site of the single copy gene on the long arm of chromosome 1 in CH^{R}A3, prior to the formation of the ABR on der(7). In the C4 series (Fig. 3A, bottom panel) an initial translocation led to a variety of derivative chromosome 1's all of which had alterations near the *PGY* locus. Karyotypes of all the der(1) chromosomes and the overall complexity of karyotypic changes that occurred at early steps of the colchicine selection for both the C4 and C5 series of MDR cell lines, including the formation of DMs at later steps in the selection (Fig. 3 B), have been extensively characterized (Trent *et al.*, 1993). In addition to these studies, Trent and colleagues have recently identified linear extrachromosomal DNA elements harboring amplified *mdr* genes in an MDR mouse cell line, raising the possibility that linear DNA structures are unstable intermediates in the dynamic process of gene amplification (Meese *et al.*, 1992). These studies, which are a small representative of a large number of gene amplification studies over the past several decades (reviewed in a multiauthor volume, Kellems 1993), emphasize the active role that gene amplification structures play in the already increased genomic fluidity of transformed (neoplastic) cells (see Conclusions and Perspectives).

Kopnin and colleagues have also addressed *mdr* gene amplification mechanisms at very early steps of *in vitro* drug selections in Djungarian hamster DM-15 (Gudkov *et al.*, 1991) and murine P388 cell lines (Kopnin *et al.*, 1992). In both cell lines, at early steps of drug selection the formation of extrachromosomal structures, also referred to as small chromatin bodies (SCBs) or mini-chromosomes appeared to mediate *mdr* gene amplification. In the DM-15 cell lines, prior to SCB formation, an additional chromosome 4 or a segment of chromosome 4 is consistently detected with karyotype analysis. Shortly after the appearance of the additional chromosome 4 structure, an HSR is localized to one end of any one of the three chromosome 4's present in the cells. Sokova (1986) has previously localized the Djungarian hamster *mdr* (*PGY*) locus to this chromosome. Therefore, this chronology of amplification events is similar in sequence to *mdr* gene amplification in CHO cells as described by Trent and co-workers (see above) where an amplified array of genes is attached to one end of the chromosome carrying the single copy gene through mechanisms involving recombination and unequal sister chromatid exchange. In the MDR P388 cells, no other gene amplification structures were observed prior to the formation of *mdr*-containing SCBs. Therefore, formation of *mdr*-extrachromosomal DNAs may be the initial step in *mdr* gene amplification in the MDR P388 cell lines. Such extrachromosomal DNAs could be generated as a result of a recombination-based mechanism or through deletion of small genomic regions via a direct excision type mechanism. It cannot be ruled out that, in the MDR P388 cell lines, linear arrays of amplified *mdr* genes may have formed at very early stage(s) but escaped detection. However, it should be noted that early stages of *dhfr* and *CAD* gene amplification were analyzed in concurrent studies and did not show the formation of extrachromosomal SCBs as observed in the MDR P388 cell lines (Gudkov *et al.*, 1991; Kopnin *et al.*, 1992).

Kopnin and colleagues have also identified the formation of SCBs as an early step in an *in vivo* system where MDR expressing tumors derived from weekly intraperitoneal transplantations in mice were analyzed for *mdr* gene amplification following chemotherapy (Demidova *et al.*, 1991). In this system, amplified copies of the *mdr* locus (30–50 fold) were present on SCBs, prior to the formation of large newly formed "chromosome-like" structures. In four of five sublines derived *in vivo*,the copy number of the *mdr*1a, *mdr*1b and *mdr*2 gene was similar. Whether the appearance of extrachromosomal SCBs or chromosome-like structures represent a specific mechanism of gene amplification that more accurately depicts gene amplification processes in human cancers must be addressed in future studies. Interestingly, in over 200 cases of a variety of human cancers where biopsies have been analyzed for gene amplification, double minute chromosomes frequently appear to harbor the amplified gene(s) (reviewed by Benner *et al.*, 1991). These observations suggest that *in vivo* amplification mechanisms may favor either the formation of extrachromosomal amplification structures or the maintenance of cells harboring amplified genes on circular DNAs.

*MDR*1 gene expression and amplification in human cancer

In a large-scale study, Goldstein and colleagues addressed the possible role of P-glycoprotein in MDR in human cancer. To correlate the expression of P-glycoprotein to the MDR phenotype(s) operative *in vivo*, the *MDR*1 RNA expression levels was analyzed in >400 human tumor specimens representing a variety of human cancers (Goldstein *et al.*, 1989). Other studies, often aimed at specific types of cancers, have also correlated the levels of either *MDR*1 gene expression or P-glycoprotein to the MDR nature of different human malignancies. These studies suggest that the multidrug resistant phenotype of some cancers is an inherent property of the neoplasm because the tumor develops from cells that normally express the *MDR*1 gene, e.g., the adrenal cortical cells, renal proximal tubule epithelium, biliary hepatocytes, mucosal cells of the small and large intestine, capillary endothelial cells of testis and brain (Thiebaut *et al.*, 1987; Thiebaut *et al.*, 1989; reviewed in Cordon-Cardo *et al.*, 1991). However, *MDR*1 gene expression was detected in multidrug resistant tumors derived from tissues that do not normally express the *MDR*1 gene. In some of these cancers, *MDR*1 gene expression may occur unfortuitously as a consequence of the malignant transformation of these cells. In addition, *MDR*1 gene expression appears to be either turned on or greatly increased following chemotherapy. In these cases of "acquired" MDR, *MDR*1 gene expression is thought to result from the selection of tumor cells that have undergone a spontaneous mutation resulting in increased expression of the *MDR*1 gene. Alternatively, the acquired expression of the *MDR*1 gene may be an adaptation of the tumor cells to the presence of the chemotherapeutic agents, i.e., *in vitro* studies by Kohno and colleagues (1989) which found that the *MDR*1 promoter could be activated by the addition of anticancer agents including vincristine, daunomycin and adriamycin. In general, the transcriptional "activation" of the *MDR*1 gene in human cancers is not well-defined. Even in cultured cell lines, factors regulating *MDR*1 transcription remain poorly understood (reviewed in Chapter 3, this volume).

Although increased transcription of the human *MDR*1 gene in cultured cell lines often results from gene amplification, the *in vivo* amplification of the *MDR*1 gene in cancer is rarely reported, possibly because it occurs infrequently or because it is difficult to detect. Due to the heterogeneity of tumor cell populations, if only a small percentage of cancer cells harbor amplified *MDR*1 genes, particularly on extrachromosomal circular DNAs, it would be very difficult to detect these structures with current technologies. Two recent studies report the amplification of the multidrug resistance gene *MDR1* in clinical samples of mammary carcinoma from unrelated patients (Lonn *et al.*, 1992; Lonn *et al.*, 1993). In these studies, the cells were obtained by fine-needle biopsies and the number of amplified *MDR1* genes was determined by PCR. Neither cytogenetic nor molecular approaches were used in these studies to confirm the presence of specific amplification structures harboring the extra *MDR1* gene copies, most probably because of the limiting amount of tumor material. However, the evidence in these studies does suggest a localized gene amplification event as opposed to the occurrence of aneuploidy. Further, the *MDR1* gene amplification in these cancers appeared to be the result of tumorigenic progression as opposed to a response to cytotoxic agents.

Atypical multidrug resistance and gene amplification

There are a variety of MDR cell lines whose drug resistance mechanism(s) have not been defined and appear to be distinct from the P-glycoprotein mediated MDR phenotype. The characteristics of these cell lines has been published in a number of studies, particularly with respect to their drug resistance profile or the identification of a specific size of protein species that appears concomitantly with the development of the MDR phenotype

(Sugawara *et al.*, 1988; Norris *et al.*, 1989; Yang and Trujillo, 1990; Zwelling *et al.*, 1990; Chen *et al.*, 1990). In some cases it has been determined that the drug resistance mechanism(s) operative in these MDR cell lines affect the intracellular concentration of drugs (McGrath and Center, 1987; Slovak *et al.*, 1988; Slapak *et al.*, 1990; Eijdems *et al.*, 1992; Marquardt and Center, 1992). However, there is very little information describing the molecular nature of these non-P-glycoprotein mediated MDR mechanisms, also referred to as atypical multidrug resistance (at-MDR), and their role, if any, in the expression of chemotherapeutic resistance in human cancers. Many of these MDR cell lines were selected by the "classical" step-wise drug selection (described above) that favors gene amplification events. Therefore, amplified gene sequences may be present in these MDR cell lines that could potentially be used as genetic and molecular tools to identify specific drug resistance genes, as exemplified by the *in vitro* studies that identified and characterized the *mdr*-encoded mechanism of multidrug resistance.

For example, in one study Beck *et al.* (1987) described an abnormally banded region (ABR) on chromosome 13q in the at-MDR cell line CEM/-VM1. This cell line was derived from a human leukemic cell line in increasing concentrations of the epipodophyllotoxin teniposide and showed cross-resistance to a wide variety of natural product antitumor drugs (Danks *et al.*, 1987). However, CEM/VM1 is not resistant to the *Vinca* alkaloids nor is this cell line impaired in its ability to accumulate radiolabeled epipodophyllotoxin. These results suggest that a non-P-glycoprotein mechanism appears to mediate the MDR which may be encoded by an amplified gene other than *mdr* located on the ABR on chromosome 13q. In another study of an at-MDR cell line (derived from a human colon carcinoma cell), cytogenetic analysis identified a homogeneously staining region on the short arm of chromosome 7 that was associated with the development of drug resistance (Dalton *et al.*, 1988).

The applicability of using gene amplification as a tool to identify novel multidrug resistance genes is well represented in a recent study by Cole and colleagues (1992). The MDR cell line H69AR whose drug resistance mechanism was characterized in this study was derived from a step-wise doxorubicin selection of the small cell lung carcinoma cell line NCI-H69 and showed obvious gene amplification structures. A putative multidrug resistance gene sequence was isolated from a cDNA library that was constructed from H69AR mRNA and screened by differential hybridization with total cDNA prepared from H69 and H69AR mRNA. Southern (DNA) blot and Northern (RNA) blot analyses of H69 and H69AR DNA confirmed that this cDNA sequence represented an amplified gene in H69AR that was overexpressed. This gene, designated the multidrug resistance protein (MRP) gene, is overexpressed in inherently resistant NSCLC cell lines, suggesting that MRP may play a role in multidrug resistance in human cancer. Interestingly, DNA sequence analysis of the *MRP* gene demonstrates that MRP is a member of the adenosine triphosphate (ATP)-binding cassette (ABC) superfamily of transport proteins that includes the multidrug transporter P-glycoprotein and the cystic fibrosis transmembrane conductance regulator (CFTR).

Conclusion and perspectives

The genomic instability of transformed and tumorigenic cells is manifested as chromosomal aberrations such as gene amplification, chromosomal translocations, interstitial deletions and alterations in chromosomal ploidy (Yunis, 1983; Bishop, 1987). These aberrations can result in a variety of abnormal phenotypes, e.g., increased proliferation and tumorigenesis, as well as the development of multidrug resistance. In human tumors, discernible gene amplification structures harboring oncogenes have been frequently documented. Their functional significance in some cases of cancer is indisputable because a clear correlation has been established between the amplification of specific oncogenes and the increased proliferative ability of tumor cells, as well as a poor prognosis for the patient; e.g., *MYCN* gene amplification in neuroblastomas (Brodeur *et al.*, 1984; Seeger *et al.*, 1985) and *neu* (*c-erbB*-2) (Slamon *et al.*, 1987), *MYC* and *Int*-2 gene amplification in primary human breast cancers (Donovan-Peluso *et al.*, 1991; Tsuda *et al.*, 1989).

The biological significance of cytologically detectable chromosomal abnormalities in mammalian cells was first indicated *in vitro* in antifolate resistant Chinese hamster lung cells when Biedler and Spengler (1976b) demonstrated a correlation between the presence of HSRs and the increased synthesis of dihydrofolate reductase (DHFR), the enzyme which encodes resistance to the folate antagonist methotrexate. Soon after these observations, Schimke and colleagues using molecular biological approaches, including *in situ* hybridization methods, provided firm evidence that amplified copies of the DHFR gene were actually present on HSRs (Alt *et al.*, 1978; Nunberg *et al.*, 1978; Dolnick *et al.*, 1979) and/or on DM structures (Kaufman *et al.*1979). Taken together, these pioneering studies established an important association between cytogenetic (chromosomal) abnormalities and the increased copy number and expression of drug resistance genes in cultured cell lines.

To understand the processes mediating genomic instability in neoplastic cells, mammalian drug resistant cell lines containing amplified drug resistance genes have provided model systems (comprehensively reviewed in a multiauthor volume, Kellems, 1993). By comparing the relative occurrence of amplification of drug resistance genes, *in vitro* studies have demonstrated an increased gene amplification frequency in transformed cells as compared to normal diploid fibroblasts (Sager, 1982; Wright *et al.*, 1990; Tlsty *et al.*, 1989 and 1990). Interestingly, once amplification structures have been formed they contribute to genomic instability. For example, several studies have provided evidence that episomes are precursors of DMs during the amplification process (Carroll *et al.*, 1988; 1987; Von Hoff *et al.*, 1988; Ruiz and Wahl, 1990; VanDevanter *et al.*, 1990; Schoenlein *et al.*, 1992) and that DMs integrate into chromosomes to form HSRs and ABRs (Biedler, 1982; Carroll *et al.*, 1988; Ruiz and Wahl, 1990; Von Hoff *et al.*, 1990).

The exact mechanisms initiating gene amplification have not been elucidated. However, the maintenance of genomic integrity in normal mammalian cells appears to be similar to the process in yeast (reviewed in Hartwell and Weinert, 1989) which is a highly regulated process involving active "check-points" of the cell-cycle. These check-points mediate cell-cycle arrest when suboptimal growth conditions occur that could lead to DNA damage. In mammalian cells, one determinant in the proposed "check-point" pathway is the tumor suppressor gene product p53. When cells lacking functional P53 are exposed to drugs (Livingston *et al.*, 1992; Yin *et al.*, 1992) or gamma irradiation (Kuerbitz *et al.*, 1992), they show an increased gene amplification frequency and an inability to growth-arrest. Interestingly, *p53* gene mutations and/or loss of *p53* heterozygosity occur in a variety of human cancers (for reviews and references see Lane and Benchimol, 1990; Levine *et al.*, 1991).

In vitro studies have also utilized gene amplification as a genetic tool to identify, clone and characterize over twenty single-agent drug resistance genes (reviewed in Kellems, 1993), the recently identified multidrug resistance protein (MRP) gene (Cole *et al.*, 1992) and a small family of multidrug resistance (*mdr*) genes (for reviews see, Endicott and Ling, 1989, Gros *et al.*, 1991; Roninson *et al.*, 1991; Borst and Van der Bliek, 1991; Gottesman and Pastan, 1993). The stepwise drug selections used to establish the drug resistant cell lines from which these genes were identified and cloned usually resulted in gene amplification in the absence of directed mutagenesis. However, in the case of the human adenocarcinoma MDR KB cell lines from which the human *MDR*1 gene was cloned (Fig. 1, Table 3), treatment with the mutagen ethyl methanesulfonate (EMS) was required to increase the frequency of obtaining drug resistant clones in the initial steps of the drug selections (Akiyama *et al.*, 1985; Shen *et al.*, 1986b). In contrast, EMS mutagenesis was not required for later steps in the selection of the MDR KB cell lines. Interestingly, in the colchicine-selected sublines once an MDR subline was established that contained an *mdr* amplicon, i.e., KB-Ch^{R}-8–5–11, EMS treatment was not necessary in later steps in the drug selection. This observation has also been made in the vinblastine-selected sublines (PV

Schoenlein, unpublished) and leads one to speculate that EMS mutagenesis led to a more "plastic" genome in the KB cell lines with either specific mutations in *p53* or other genes whose products are involved in maintaining chromosomal integrity within cells.

The actual role of gene amplification in conferring drug resistance *in vivo*, particularly clinical MDR, awaits further investigation. Although *MDR*1-gene amplification has not been reported as a major mechanism to achieve increased expression of P170 in human cancer, the possible regulation of MDR expression by amplified oncogenes in human cancers must be considered. As further novel, clinically relevant MDR mechanisms are identified and characterized, it is probable that the process of gene amplification will contribute to the development of some clinical MDRs. If gene amplification is found to play an appreciable role in the MDR of human cancer, either determinants mediating genomic instability in cancer cells or the amplification structures themselves should be considered as future targets of clinical therapy.

Acknowledgements

I would like to thank Michael M Gottesman, John T Barrett and Robert Wrenn for critical reading of this manuscript. My appreciation is also extended to June Biedler and Michael Gottesman for helpful discussions, and to June Biedler, Barbara Spengler and Jeffrey Trent for providing relevant photographs used in this text. In addition, I would like to thank Martin Clynes for his continued patience.

References

Akiyama S-i, Fojo A, Hanover JA, Pastan I, and Gottesman MM (1985) Isolation and genetic characterization of human KB cell lines resistant to multiple drugs. Somat Cell Mol Genet 11: 117–126.

Alt FW, Kellems RE, Bertino JR, and Schimke RT (1978) Selective multiplication of dihydrofolate reductase genes in methotrexate-resistant variants of cultured murine cells. J Biol Chem 253: 1357–1370.

Ames GF-L and Lecar H (1992) ATP-dependent bacterial transporters and cystic fibrosis: analogy between channels and transporters. FASEB J 6: 2660–2666.

Ames GF-L, Mimura CS, Holbrook SR and Shyamala V (1992) Traffic ATPases: a superfamily of transport proteins operating from *Escherichia coli* to humans. Adv Enzymol 65: 1–47.

Baptist G, Tulpule A, Sinha BK, Katki AG, Myers CE and Cowan KH (1986) Overexpression of a novel anionic glutathione S-transferase in multidrug resistant human breast cancer cells. J Biol Chem 261: 15544–15549.

Beck WT, Mueller TJ and Tanzer LR (1979) Altered surface membrane glycoproteins in Vinca alkaloid-resistant human leukemic lymphoblasts. Cancer Res 39: 2070–2076.

Beck WT and Cirtain MC (1982) Continued expression of vinca alkaloid resistance by CCRF-CEM cells after treatment with tunicamycin or pronase. Cancer Res 42: 184–189.

Beck WT and Danks MK (1991) Characteristics of multidrug resistance in human tumor cells. In: IB Roninson (ed.) Molecular and Cellular Biology of Multidrug Resistance in Tumor Cells (pp. 3–46) Plenum Publishing Corporation, New York.

Beck WT, Cirtain MC, Danks MK, Felsted RL, Safa AR, Wolverton JS, Suttle DP, and Trent JM (1987) Pharmacologic, molecular and cytogenetic analysis of "atypical" multidrug resistant human leukemic cells. Cancer Res 47: 5455–5460.

Beck WT, Cirtain MC, Wolverton JS, Safa AR and Felsted RL (1988) Collateral sensitivity of human multidrug resistant (MDR) leukemic cell lines to verapamil. Proc Am Assoc Cancer Res 29: 307.

Bech-Hansen NT, Till JE and Ling V (1976) Pleiotropic phenotype of colchicine resistant CHO cells. Cross resistance and collateral sensitivity. J Cell Physiol 88: 23–32.

Benner SE, Wahl GM and Von Hoff DD. (1991) Double minute chromosomes and homogeneously staining regions in tumors taken directly from patients versus in human tumor cell lines. Anti-Cancer Drugs 2: 11–25.

Bhalla K, Hindenburg A, Taub RN, and Grant S (1985) Isolation and characterization of an anthracycline-resistant human leukemic cell line. Cancer Res 45: 3657–3662.

Biedler JL and Riehm H (1970) Cellular resistance to actinomycin D in Chinese hamster cells in vitro: Cross-resistance, radioautographic and cytogenetic studies. Cancer Res 30: 1174–1184.

Biedler JL and Riehm H (1979) Cellular resistance to Actinomycin D in Chinese hamster cells *in vitro*. Cancer Res 30: 1174–1184.

Biedler JL and Spengler BA (1976a) Metaphase chromosome anomaly: Association with drug resistance and cell-specific products. Science 191: 185–187.

Biedler JL and Spengler BA (1976b) A novel chromosome abnormality in human neuroblastoma and antifolate-resistant Chinese hamster cell lines in culture. J Natl Cancer Inst 57: 683–695.

Biedler JL, Melera PW and Spengler BA. (1980) Specifically altered metaphase chromosomes in antifolate-resistant Chinese hamster cells that overproduce dihydrofolate reductase. Cancer Genet Cytogenet 2: 47–60.

Biedler JL and Peterson RHF (1981) Altered plasma membrane glycoconjugates of Chinese hamster cells with acquired resistance to actinomycin D, daunorubicin and vincristine In AC Sartorelli, JS Lazo, and JR Bertino (eds.) Molecular Actions and Targets for Cancer Chemotherapeutic Agents (pp. 453–482) Academic Press, New York.

Biedler JL (1982) Evidence for transient existence of amplified DNA sequences in antifolate-resistant, vincristine resistant, and human neuroblastoma cells In: RT Schimke (ed) Gene Amplification (pp. 39–45) Cold Spring Harbor Laboratory, Cold Spring Harbor, N.Y.

Biedler JL, Meyers MB and Spengler BA (1988) Collateral sensitivity of multidrug-resistant Chinese hamster cells to calcium channel blockers: Correlation with P-glycoprotein content. Proc Am Assoc Cancer Res 29: 295.

Biedler JL, Chang T-d, Druskin H, Meyers MB and Spengler BA (1993) Modulation of multidrug resistance gene expression by chemosensitizing agents. Proc Am Assoc Cancer Res 34: 578.

Bishop MJ (1987) Molecular genetics of cancer. Science 235: 305–311.

Black SM, Beggs JD, Hayes JD, Bartoszek A, Muramatsu M, Sakai M and Wolf CR (1990) Expression of human glutathione S-transferases in *Saccharomyces cerevisiae* confers resistance to the anticancer drugs adriamycin and chlorambucil. Biochem J 268: 309–315.

Borst P and Van der Bliek AM (1991) Amplification of several different genes in multidrug-resistant chinese hamster cell lines In: IB Roninson (ed.) Molecular and Cellular Biology of Multidrug Resistance in Tumor Cells (pp 107–115) Plenum Publishing Corporation, New York.

Bradley G, Juranda PF and Ling V (1988) Mechanism of multidrug resistance. Biochim Biophy Acta 948: 87–128.

Brodeur GM, Seeger RC, Schwab M, Varmus HE and Bishop JM (1984) Amplification of N-myc in untreated human neuroblastomas correlates with advanced disease stage. Science 224: 1121–1124.

Callen DF, Baker E, Simmers RN, Seshadri R and Roninson IB (1987) Localization of the human multiple drug resistance gene, *MDR*1 to 7q21.1. Hum Genet 77: 142–144.

Cano-Gauci DF and Riordan JR (1987) Action of calcium antagonists on multidrug resistant cells Biochem Pharmacol 36: 2115–2123.

Carroll SM, Gaudray P, DeRose ML, Emery JF, Meinkoth JL, Nakkim E, Subler M, Von Hoff DD, and Wahl GM (1987) Characterization of an episome produced in hamster cells that amplify a transfected CAD gene at high frequency: functional evidence for a mammalian replication origin Mol Cell Biol 7: 1740–1750

Carroll SM, DeRose ML, Gaudray P, Moore CM, Needham-VanDevanter DR, Von Hoff DD and Wahl GM (1988) Double minute chromosomes can be produced from precursors derived from a chromosomal deletion Mol Cell Biol 8 1525–1533

Chan HSL, Thorner PS, Haddad G and Ling V (1990) Immunohistochemical detection of P-glycoprotein· prognostic correlation in soft tissue sarcoma of childhood. J Clin Oncol 8 689–704.

Chan HSL, Haddad G, Thorner PS, DeBoer G, Lin YP, Ondrusek N, Yeger H and Ling V (1991) P-glycoprotein expression as a predictor of the outcome of therapy for neuroblastoma N Engl J Med 325: 1608–1614.

Chen Y-N, Mickley LA, Schwartz AM, Acton EM, Hwang J and Fojo AT (1990) Characterization of adriamycin-resistant human breast cancer cells which display overexpression of a novel resistance-related membrane protein. J Biol Chem 265 10073–10080.

Chen C-j, Chin JE, Ueda K, Clark DP, Pastan I, Gottesman MM and Roninson IB (1986) Internal duplication and homology with bacterial transport proteins in the *mdr*1 (P-glycoprotein) gene from multidrug-resistant human cells. Cell 47. 381–389.

Chin JE, Soffir R, Noonan KE, Choi K and Roninson IB (1989) Structure and expression of the human MDR (P-glycoprotein gene family.) Mol Cell Biol 9· 3808–3820.

Choi K, Chen C-j, Kriegler M and Roninson IB (1988) An altered pattern of cross-resistance in multidrug-resistant human cells results from spontaneous mutations in the *mdr*1 (P-glycoprotein) gene. Cell 53: 519–523.

Cohen D, Higman SM, Hsu, S I-H and Horwitz SB (1992) The involvement of a LINE-1 element in a DNA rearrangement upstream of the *mdr*1a gene in a taxol multidrug-resistant murine cell line J Biol Chem 267: 20248–20254

Cole SPC, Bhardwaj G, Gerlack JH, Mackie JE, Grant CE, Almquist KC, Stewart AJ, Kurz EU, Duncan AMV, and Deeley RG (1992) Overexpression of a transporter gene in a multidrug-resistant human lung cancer cell line. Science 258: 1650–1654.

Cordon-Cardo C (1991) Immunohistochemical analysis of P-glycoprotein expression in normal and tumor tissues in humans In: IB Roninson (ed.) Molecular and cellular biology of multidrug resistance in tumor cells (pp. 303–318) Plenum Publishing Corporation, New York.

Dahllof B, Martinsson T and Levan G (1984) Resistance to actinomycin D and to vincristine induced in a SEWA mouse tumor cell line with concomitant appearance of double minutes and a low-molecular-weight protein Exp Cell Res 152 415–426.

Dalton WS, Cress AE, Alberts DS and Trent JM (1988) Cytogenetic and phenotypic analysis of a human colon carcinoma cell line resistant to mitoxantrone. Cancer Res 48 1882–1888.

Danks MK, Yalowich JC and Beck WT (1987) Atypical multiple drug resistance in a human leukemic cell line selected for resistance to teniposide (VM-26) Cancer Res 47 1297–1301

Danks MK, Schmidt CA, Cirtain MC, Suttle DP and Beck WT (1988) Altered catalytic activity of and DNA cleavage by DNA topoisomerase II from human leukemic cells selected for resistance to VM-26 Biochemistry 27: 8861–8869

Danks MK, Schmidt CA, Deneka DA and Beck WT (1989) Increased ATP requirement for activity of and complex formation by DNA topoisomerase II from human leukemic CCRF-CEM cells selected for resistance to VM-26. Cancer

Commun 1: 101–109.

Dano K (1972) Cross resistance between vinca alkaloids and anthracyclines in Ehrlich ascites tumor *in vivo*. Cancer Chemotherapy Reports 56: 701–708.

Dano K (1973) Active outward transport of daunomycin in resistant Ehrlich ascites tumor cells. Biochim Biophys Acta 323: 466–483.

De Bruijn MHL, Van der Bliek AM, Biedler JL and Borst P (1986) Differential amplification and disproportionate expression of five genes in three multidrug-resistant Chinese hamster lung cell lines. Mol Cell Biol 6: 4717–4722.

Demidova NS, Chernova OB, Siyanova EY, Goncharova AS and Kopnin BP (1991) Newly formed chromosome-like structures in independent mouse P388 sublines with developed *in vivo mdr*1 gene amplification. Somat Cell Mol Genet 17: 581–590.

Deuchars KL, Du R-P, Naik M, Evernden-Porelle D, Kartner N, Van der Bliek AM and Ling V (1987) Expression of hamster P-glycoprotein and multidrug resistance in DNA-mediated transformants of mouse LTA cells. Mol Cell Biol 7: 718–724.

Devault A and Gros P (1990) Two members of the mouse *mdr* gene family confer multidrug resistance with overlapping but distinct drug specificities. Mol Cell Biol 10: 1652–1663.

Devine SE, Ling V, and Melera PW (1992) Amino acid substitutions in the 6th transmembrane domain of P-glycoprotein alter multidrug resistance. Proc Natl Acad Sci USA 89: 4564–4568.

Dolnick BJ, Berenson RJ, Bertino JR, Kaufman RJ, Nunberg JH and Schimke RT (1979) Correlation of dihydrofolate reductase elevation with gene amplification in a homogeneously staining chromosomal region in L5178Y cells. J Cell Biol 83: 394–402.

Donovan-Peluso M, Contento AM, Tobon H, Ripepi B and Locker J (1991) Oncogene amplification in breast cancer. Am J Pathol 138: 835–845.

Eijdems EWHM, Borst P, Jongsma APM, DeJong SD, DeVries EGE, VanGroenigen M, Versantvoort CHM, Nieuwint AWM and Baas F (1992) Genetic transfer of ono-P-glycoprotein-mediated multidrug resistance (MDR) in somatic cell fusion: Dissection of a compound MDR phenotype. Proc Natl Acad Sci USA 89: 3498–3502.

Endicott JA and Ling V (1989) The biochemistry of P-glycoprotein-mediated multidrug resistance. Annu Rev Biochem 58: 137–171.

Ferguson PJ, Fisher MH, Stephenson J, Li D-, Zhou B-s and Cheng Y-c (1988) Combined modalities of resistance in etoposide-resistant human KB cell lines. Cancer Res 48: 5946–5964.

Fojo AT, Whang-Peng J, Gottesman MM and Pastan I (1985) Amplification of DNA sequences in human multidrug-resistant KB carcinoma cells. Proc Natl Acad Sci USA 82: 7661–7665.

Fojo AT, Lebo R, Simizu N, Chin JE, Roninson IB, Merlino GT, Gottesman MM and Pastan I (1986) Localization of multidrug resistance-associated DNA sequences to human chromosome 7. Somat Cell Mol Genet 12: 415–420.

Georges E, Bradley G, Gariepy J and Ling V (1990) Detection of P-glycoprotein isoforms by gene-specific monoclonal antibodies. Proc Natl Acad Sci USA 87: 152–146.

Gerlach JH, Endicott JA, Juranka PF, Henderson G, Sarangi F, Deuchars KL and Ling V (1986) Homology between P-glycoprotein and a bacterial haemolysin transport protein suggests a model for multidrug resistance. Nature 324: 485–489.

Goasguen JE, Dossot J-M, Fardel O, Le Mee F, Le Gall E, Leblay R, LePrise PY, Chaperon J and Fauchet R (1993) Expression of the multidrug resistance-associated P-glycoprotein (P-170) in 59 cases of *de novo* acute lymphoblastic leukemia: Prognostic Implications. Blood 81: 2394–2398.

Goldstein LJ, Galski H, Fojo AT, Willingham MC, Lai S-L, Gazdar A, Pirker R, Green A, Crist JW, Brodeur GM, Lieber M, Cossman J, Gottesman MM and Pastan I (1989) Expression of a multidrug resistance gene in human cancers. J Natl Canc Inst 81: 116–124.

Goldstein LJ, Pastan I, and Gottesman MM (1992). Multidrug resistance in human cancer. In Critical Reviews in Oncology/Hematology 12: 243–253.

Gottesman MM, Goldstein LJ, Fojo AT, Galski H and Pastan I (1991) Expression of the multidrug resistance gene in human cancer. In: IB Roninson (ed.) Molecular and Cellular Biology of Multidrug Resistance in Tumor Cells (pp. 291–301) Plenum, New York.

Gottesman MM and Pastan I (1993) Biochemistry of multidrug resistance mediated by the multidrug transporter. In: Annual Review of Biochemistry. Vol. 62. Richardson, Abelson, Meister and Walsh (eds.) Annual Reviews Inc. Palo Alto.

Gros P, Ben Neriah Y, Croop J and Housman DE (1986a) Isolation and characterization of a complementary DNA that confers multidrug resistance. Nature 323: 728–731.

Gros P, Croop J, Roninson IB, Varshavsky A and Housman DE (1986b) Isolation and characterization of DNA sequences amplified in multidrug-resistant hamster cells. Proc Natl Acad Sci USA 83: 337–341.

Gros P, Ben Neriah Y, Croop J and Housman DE (1986d) Chromosome-mediated gene transfer of multidrug resistance. Mol Cell Biol 6: 3785–3790.

Gros P, Croop J and Housman D (1986c) Mammalian multidrug resistance gene: complete cDNA sequence indicates strong homology to bacterial transport proteins. Cell 47: 371–380.

Gros P, Raymond M and Housman, D (1991) Cloning and characterization of mouse *mdr* genes. In: I. Roninson (ed.) Molecular and Cellular Biology of Multidrug Resistance in Tumor Cells (pp. 73–88) Plenum Publishing Corporation, New York.

Gudkov AV, Chernova OV, Kazarov AR and Kopnin BP (1987) Cloning and characterization of DNA sequences amplified in multidrug resistant Djungarian hamster and mouse cells. Somat Cell Mol Genet 13: 609–619.

Gudkov AV, Chernova OB and Kopnin BP (1991) Karyotype and amplicon evolution during stepwise development of multidrug resistance in Djungarian hamster cell lines. In: IB Roninson (ed.) Molecular and Cellular Biology of Multidrug

Resistance in Tumor Cells (pp. 147–168) Plenum Press, New York.
Guild BC, Mulligan RC, Gros P and Housman DE (1988) Retroviral transfer of a murine cDNA for multidrug resistance confers pleiotropic drug resistance to cells without prior drug selection. Proc Natl Acad Sci USA 85: 1595–1599.
Hamlin JL, Milbrandt JD, Heintz NH and Azizkhan JC (1984) DNA sequence amplification in mammalian cells. Int Rev Cytol 90: 31–82.
Hartwell LH and Weinert TA (1989) Checkpoints: controls that ensure the order of cell cycle events. Science 246: 629–634.
Higgins CF (1992) ABC transporters: from microorganisms to man. Annu Rev Cell Biol 8: 67–113.
Higgins CF (1993) The ABC transporter channel superfamily — an overview. Semin Cell Biol 4: 1–5.
Horwitz SB, Goei S, Greenberger L, Lothstein L, Mellado W, Samar SN, Yang C-PH and Zeheb R (1988) Multidrug resistance in the mouse macrophage-like cell line J774.2 In: PV Wooley and KD Tew (eds.) Mechanisms of Drug Resistance in Neoplastic Cells (pp. 223–242) Academic Press, New York.
Hsu SI, Lothstein L and Horwitz SB (1989) Differential overexpression of three *mdr* gene family members in multidrug-resistant J774.2 mouse cells. J Biol Chem 264: 12053–12062.
Howell N, Belli TA, Zaczkiewics LT, and JA Belli (1984) High level, unstable Adriamycin resistance in a Chinese hamster mutant cell line with double minute chromosomes. Cancer Res 44: 4023–4030.
Hyde SC, Emsley P, Hartshorn MJ, Mimmack MM, Gileadi U, Pearce SR, Gullagher MP, Gill DR, Hubbard RE and Higgins CF (1990) Structural and functional relationships of ATP-binding proteins associated with cystic fibrosis, multidrug resistance, and bacterial transport. Nature 346: 362–365.
Jongsma APM, Spengler BA, Van der Blick AM, Borst P and Biedler JL (1987) Chromosomal localization of three genes coamplified in the multidrug-resistant CH^RC5 Chinese hamster ovary cell line. Cancer Res 47: 2875–2878.
Juliano RL and Ling V (1976) A surface glycoprotein modulating drug permeability in Chinese hamster ovary cell mutants. Biochim Biophys Acta 455: 152–162.
Kartner N, Riordan JR, and Ling V (1983) Cell surface P-glycoprotein associated with multidrug resistance in mammalian cell lines. Science 221: 1285–1288.
Kartner N, Evernden-Porelle D, Bradley G and Ling V (1985) Detection of P-glycoprotein in multidrug-resistant cell lines by monoclonal antibodies. Nature 316: 820–823.
Kaufman RJ, Brown PC and Schimke RT (1979) Amplified dihydrofolate reductase genes in unstable methotrexate-resistant cells are associated with double minute chromosomes. Proc Natl Acad Sci USA 76: 5669–5673.
Kellems RE (1993) Gene Amplification in Mammalian Cells. 543 pp. Marcel Dekker, Inc., New York, N.Y.
Kessel D, Botterill V and Wodinsky I (1968) Uptake and retention of daunomycin by mouse leukemic cells as factors in drug response. Cancer Res 28: 938–941.
Kohno K, Sato S-i, Takano H, Matsuo K-i and Kuwano M (1989) The direct activation of human multidrug resistance gene (MDR1) by anticancer agents. Biochem Biophys Res Commun 165: 1415–1421.
Kopnin BP, Sokova OI and Demidova NS (1992) Regularities of karyotypic evolution during stepwise amplification of genes determining drug resistance. Mutation Res 276: 163–177.
Kramer RA, Zakher J and Kim G (1988) Role of the glutathione redox cycle in acquired and *de novo* multidrug resistance. Science 241: 694–697.
Kuerbitz SJ, Plunkett BS, Walsh WV and Kastan MB (1992) Wild-type p53 is a cell cycle checkpoint determinant following irradiation. Proc Natl Acad Sci USA, in press.
Lane DP and Benchimol S (1990) p53: oncogene or anti-oncogene? Genes Dev 4: 1–8.
Levine AJ, Momand J and Finlay CA (1991) The p53 tumour suppressor gene. Nature 351: 453–456.
Ling V and Thompson LH (1974) Reduced permeability in CHO cells as a mechanism of resistance to colchicine. J Cell Physiol 83: 103–116.
Ling V and Baker RM (1978) Dominance of colchicine resistance in hybrid CHO cells. Somat Cell Genet 4: 193–200.
Lonn U, Lonn S, Nylen U and Stenkvist B (1992) Appearance and detection of multiple copies of the *mdr*-1 gene in clinical samples of mammary carcinoma. Int J Cancer 51: 682–686.
Lonn U, Lonn S and Stenkvist B (1993) Reduced occurrence of mdr-1 amplification in stage-IV breast-cancer patients treated with tamoxifen compared with other endocrine treatments. Int J Cancer 53: 574–578.
Lothstein L and Horwitz SB (1986) Expression of phenotypic traits following modulation of colchicine resistance in J774.2 cells. J Cell Physiol 127: 253–260.
Livingstone LR, White A, Sprouse J, Livanos E, Jacks T and Tlsty TD (1992) Altered cell cycle arrest and gene amplification potential accompany loss of wild-type p53. Cell 70: 923–935.
Ma C, Martin S, Trask and Hamlin JL (1993) Sister chromatid fusion initiates amplification of the dihydrofolate reductase gene in Chinese hamster cells. Genes and Develop 7: 605–620.
Martinsson T and Levan G (1987) Localization of the multidrug resistance-associated 170 kDa P-glycoprotein gene to mouse chromosome 5 and to homogeneously staining regions in multidrug-resistant mouse cells by in situ hybridization. Cytogenet Cell Genet 45: 99–101.
Marquardt D and Center MS (1992) Drug transport mechanisms in HL60 cells isolated for resistance to adriamycin: evidence for nuclear drug accumulation and redistribution in resistant cells. Cancer Res 52: 3157–3163.
Maurer BJ, Lai E, Hamkalo BA, Hood L and Attardi G (1987) Novel submicroscopic extrachromosomal elements containing amplified genes in human cells. Nature (London) 327: 434–437.
McArthur JG, Beitel LK and Stanners CP (1993) The stimulation of gene amplification in mammalian cells by HSAG middle repetitive elements. In: RE Kellems (ed.) Gene

Amplification in Mammalian Cells (pp. 485—499) Marcel Dekker, Inc. New York.

McGrath T and Center M (1987) Adriamycin resistance in HL60 cells in the absence of detectable P-glycoprotein. Biochem Biophys Res Commun 145: 1171—1176.

Meese EU, Horwitz SB and Trent JM (1992) Evidence for linear extrachromosomal elements mediating gene amplification in the multidrug-resistant J774.2 murine cell line. Cancer Genet Cytogenet 59: 20—25.

Melera PW and Biedler JL (1991) Molecular and cytogenetic analysis of multidrug resistance-associated gene amplification in Chinese hamster, mouse sarcoma, and human neuroblastoma cells. In: IB Roninson (ed.) Molecular and Cellular Biology of Multidrug Resistance in Tumor Cells (pp. 243—257) Plenum Press, New York.

Meyers MB and Biedler JL (1981) Increased synthesis of a low molecular weight protein in vincristine-resistant cells. Biochem Biophys Res Commun 99: 228—235.

Meyers MB, Spengler BA, Chang TD, Melera PW and Biedler JL (1985) Gene amplification-associated cytogenetic aberrations and protein changes in vincristine-resistant Chinese hamster, mouse, and human cells. J Cell Biol 100: 588—597.

Meyers MB, Schneider KA, Spengler BA, Chang T-D and Biedler JL (1987) Sorcin (V19), a soluble acidic calcium-binding protein overproduced in multidrug-resistant cells. Biochem Pharmacol 36: 2373—2380.

Meyers MB and Biedler JL (1991) Protein changes in multidrug resistant cells. In: IB Roninson (ed.) Molecular and Cellular Biology of Multidrug Resistance in Tumor Cells (pp. 243—261) Plenum Press, New York.

Meyers MB, Schneider KA, Spengler BA, Chang T-D and Biedler JL (1987) Sorcin (V19), a soluble acidic calcium-binding protein overproduced in multidrug-resistant cells. Identification of the protein by anti-sorcin antibody. Biochem Pharmacol 36: 2373—2380.

Morrow CS and Cowan KH (1990) Glutathione S-transferases and drug resistance. Cancer Cells 2: 15—22.

Ng WF, Sarangi F, Zastawny RL, Veinot-Drebot L (1989) Identification of members of the P-glycoprotein multigene family. Mol Cell Biol 9: 1224—1232.

Norris MD, Haber M, King M and Davey R. A (1989) A typical multidrug resistance in CCRF-CEM cells selected for high level methotrexate resistance: reactivity to monoclonal antibody C219 in the absence of P-glycoprotein expression. Biochem Biophys Res Commun 165: 1435—1441.

Nunberg JH, Kaufman J, Schimke RT, Urlaub G and Chasin LA (1978) Amplified dihydrofolate reductase genes are localized to a homogeneously staining region of a single chromosome in a methotrexate-resistant Chinese hamster ovary cell line. Proc Natl Acad Sci USA 75: 5553—5556.

Pastan I, Gottesman MM, Ueda K, Lovelace E, Rutherford AV and Willingham MC (1988) A retrovirus carrying an MDR1 cDNA confers multidrug resistance and polarized expression of P-glycoprotein in MDCK cells. Proc Natl Acad Sci USA 85: 4486—4490.

Peterson RHF, Meyers MB, Spengler BA and Biedler JL (1983) Alteration of plasma membrane glycopeptides and gangliosides of Chinese hamster cells accompanying development of resistance to daunorubicin and vincristine. Cancer Res 43: 222—228.

Raymond M, Rose E, Housman DE and Gros P (1990) Physical mapping, amplification, and overexpression of the mouse *mdr* gene family in multidrug-resistant cells. Mol Cell Biol 10: 1642—1651.

Richert N, Akiyama S-I, Shen S-W, Gottesman MM and Pastan I (1985) Multiply drug resistant human KB carcinoma cells have decreased amounts of a 75 and 72 kDa glycoprotein. Proc Natl Acad Sci USA 82: 2330—2333.

Riordan JR, Deuchars K, Kartner N, Alon N, Trent J and Ling V (1985) Amplification of P-glycoprotein genes in multidrug-resistant mammalian cell lines. Nature 316: 817—819.

Roy SM and Horwitz SB (1985) A phosphoglycoprotein associated with taxol resistance J774.2 cells. Cancer Res 45: 3856—3863.

Riehm H and Biedler JL (1971) Cellular resistance to daunomycin in Chinese hamster cells *in vitro*. Cancer Res 31: 409—412.

Riordan JR, Duechars K, Kartner N, Alon N, Trent, J and Ling V (1985) amplification of P-glycoprotein genes in multidrug-resistant mammalian cell lines. Nature 316: 817—819.

Roninson IB (1983) Detection and mapping of homologous, repeated and amplified DNA sequences by DNA renaturation in agarose gels. Nucleic Acids Res 11: 5413—5431.

Roninson IB (1987) Use of in-gel DNA renaturation for detection and cloning of amplified genes. In: MM Gottesman (ed.) Methods in Enzymology (pp. 332—371) academic Press, San Diego, California.

Roninson IB, Abelson H, Housman DE, Howell N, and Varshavsky A (1984) Amplification of specific DNA sequences correlates with multidrug resistance in Chinese hamster cells. Nature (London) 309: 626—628.

Roninson IB, Chin JE, Choi K, Gros P, Housman DE, Fojo A, Shen D-w, Gottesman MM and Pastan I (1986) Isolation of the human *mdr* DNA sequences amplified in multidrug-resistant KB carcinoma cells. Proc Natl Acad Sci USA 83: 4538—4542.

Roninson IB,Pastan I and Gottesman MM (1991) Isolation and characterization of the human *MDR* (P-Glycoprotein) Genes. In: IB Roninson (ed.) Molecular and Cellular Biology of Multidrug Resistance in Tumor Cells (pp. 91—104) Plenum Press, New York.

Ruiz JC, Choi K, Von Hoff DD, Roninson IB and Wahl GM (1989) Autonomously replicating episomes contain mdr1 genes in a multidrug-resistant human cell line. Mol Cell Biol 9: 109—115.

Ruiz JC and Wahl GM (1990) Chromosomal destabilization during gene amplification. Mol Cell Biol 10: 3056—3066.

Sager R (1982) The role of genomic rearrangements in tumor cell heterogeneity. In: AH Owens, Jr., DS Coffey and SB Baylin (eds.) Tumor Cell Heterogeneity: Origins and Implications (pp. 411—423) Academic Press, New York.

Schimke RT (1988) Gene amplification in cultured cells. J Biol Chem 263: 5989—5992.

Schoenlein PV, Shen D-w, Barrett JT, Pastan I and Gottesman

MM (1992) Double minute chromosomes carrying the human MDR1 and MDR2 genes are generated from the dimerization of submicroscopic circular DNAs in colchicine-selected KB carcinoma cells. Mol Biol Cell 3: 507–520.

Schoenlein PV, VanDevanter DR and Gottesman MM (1993) Extrachromosomal elements in mammalian cells In: KW Adolph (ed.) Methods in Molecular Genetics 2 (in press).

Scotto KW, Biedler JL and Melera PW (1986) Amplification and expression of genes associated with multidrug resistance in mammalian cells. Science 232: 751–755.

Seeger RC, Brodeur GM, Sather H, Dalton A, Siegel SE, Wong KY and Hammond D (1985) Association of multiple copies of the N-*myc* oncogene with rapid progression of neuroblastomas. N Engl J Med 313: 1111–1116.

Sen S, Hittelman WN, Teeter LD and Tien Kuo M (1989) Model for the formation of double minutes from prematurely condensed chromosomes of replicating micronuclei in drug-treated Chinese hamster ovary cells undergoing DNA amplification. Cancer Res 49: 6731–6737.

Shen D-w, Fojo A, Chin JE, Roninson IB, Richert N, Pastan I and Gottesman MM (1986a) Human multidrug-resistant cell lines: Increased *mdr*1 expression can proceed gene amplification. Science 232: 643–645.

Shen D-w, Cardarelli C, Hwang J, Cornwell M, Richert N, Ishii S, Pastan I and Gottesman MM (1986b) Multiple drug-resistant human KB carcinoma cells independently selected for high-level resistance to colchicine, adriamycin, or vinblastine show changes in expression of specific proteins J Biol Chem 261: 7762–7770.

Shen D-w, Fojo A, Roninson IB, Chin JE, Soffir R, Pastan I and Gottesman MM (1986c) Multidrug resistance in DNA-mediated transformants is linked to transfer of the human *mdr*1 gene. Mol Cell Biol 6: 4039–4045.

Shen D-w, Pastan I and Gottesman MM (1988) In situ hybridization analysis of acquisition and loss of the human multidrug resistance gene. Cancer Res 48: 4334–4339.

Schinkel AH, Roelofs MEM and Borst P (1991) Characterization of the human *MDR*3 P-glycoprotein and its recognition by P-glycoprotein-specific monoclonal antibodies. Cancer Res 51: 2628–2635.

Siegried, JA, Tritton, TR, and Sartorelli, AC (1983). Comparison of anthracycline concentrations in S180 cell lines of varying sensitivity. Eur J Cancer Clin Oncol 19: 1133–1141.

Slamon DJ, Clark GM, Wong SG, Levin WJ, Ullrich A and McGuire WL (1987) Human breast cancer: correlation of relapse and survival with amplification of the HER-2/*neu* oncogene. Science 235: 177–182.

Slapak CA, Daniel JC and Levy SB (1990) Sequential emergence of distinct resistance phenotypes in murine erythroleukemia cells under adriamycin selection: decreased anthracycline uptake precedes increased P-glycoprotein expression. Cancer Res 50: 7895–7901.

Slovak ML, Hoeltge GA, Dalton WS and Trent JM (1988) Pharmacological and biological evidence for differing mechanisms of doxorubicin resistance in two human tumor cell lines. Cancer Res 48: 2793–2797.

Smith KA, Gorman PA, Stark MB, Groves RP and Stark GR (1990) Distinctive chromosomal structures are formed very early in the amplification of CAD genes in Syrain hamster cells. Cell 63: 1219–1227.

Sokova OI (1986) Nomenclature of metaphase and prometaphase Djungarian hamster chromosomes. Cytology 28: 211–214.

Stahl F, Marinsson T, Dahllof B and Levan G (1988) Amplification and overexpression of the P-glycoprotein genes and differential amplification of three other genes in SEWA murine multidrug-resistant cells. Hereditas 198: 251–258.

Stahl F, Wettergren Y and Levan G (1992) Amplicon structure in multidrug resistant murine cells: a nonrearranged region of genomic DNA corresponding to large circular DNA. Mol Cell Biol 12: 1179–1187.

Stark GR and Wahl GM (1984) Gene amplification. Ann Rev Biochem 53: 447–491.

Stark GR, Debatisse M, Giulotto E, and Wahl GM (1989) Recent progress in understanding mechanisms of mammalian DNA amplification. Cell 57: 901–908.

Stein W, Pastan I and Gottesman MM (1992) Trends in Biochemical Sci (submitted).

Sugimoto Y, Roninson IB and Tsuruo T (1987) Decreased expression of the amplified *mdr*1 gene in revertants of multi-drug-resistant human myelogenous leukemia K562 occurs without loss of amplified DNA. Mol Cell Biol 7: 4549–4552.

Sugimoto Y and Tsuruo T (1991) Development of multidrug resistance in rodent cell lines. In: IB Roninson (ed.) Molecular and Cellular Biology of Multidrug Resistance in Tumor Cells (pp. 57–65) Plenum Press, New York.

Sugawara I, Ohkochi E, Hamada H, Tsuruo T and Mori, S (1988) Cellular and tissue distribution of MRK20 murine monoclonal antibody-defined 85-kDa protein in adriamycin-resistant cancer cell lines. Jpn J Cancer Res 79: 1101–1110.

Thiebaut F, Tsuruo T, Hamada H, Gottesman M M, Pastan I and Willingham M C (1987). Cellular localization of the multidrug resistance gene product P-glycoprotein in normal human tissues. Proc Natl Acad Sci USA 84: 7735–7738.

Thiebaut F, Tsuruo T, Hamada H, Gottesman MM, Pastan I and Willingham MC (1989) Immunohistochemical localization in normal tissues of different epitopes in the multidrug transport protein, P170: evidence for localization in brain capillaries and cross-reactivity of one antibody with a muscle protein. J. Histochem. Cytochem 37: 159–164.

Tlsty TD, Margolin BH and Lum K (1989) Differences in the rates of gene amplification in nontumorigenic and tumorigenic cell lines as measured by Luria-Delbruck fluctuation analysis. Proc Natl Acad Sci USA 86: 9441–9445.

Tlsty TD (1990) Normal diploid human and rodent cells lack a detectable frequency of gene amplification. Proc Natl Acad Sci USA 87: 3132–3136.

Trask BJ and Hamlin JL (1989) Early dihydrofolate reductase gene amplification events on CHO cells usually occur on the same chromosome arm as the original locus. Genes Dev 3: 1913–1925.

Trent JM and Callen DF (1991) Chromosome localization of P-glycoprotein genes in drug-sensitive and – resistant human cells. In: IB Roninson (ed.) Molecular and Cellular Biology of Multidrug Resistance in Tumor Cells (pp. 169–188) Plenum Press, New York.

Trent JM, Meese EU, Meltzer PS, Slovak ML, Thompson F and Ling V (1993) Molecular cytogenetic characterization of the sequential development of multidrug resistance in two panels of colchicine-resistant cell lines. In: RE Kellems (ed.) Gene Amplification in Mammalian Cells (pp. 71–85) Marcel Dekker Inc., New York.

Tsuda H, Hirohashi S, Shimosato Y, Hirota T, Tsugane S, Yamamoto H, Miyajima N, Toyoshima K, Yamamoto T, and Yokota J (1989) Correlation between long-term survival in breast cancer patients and amplification of two putative oncogene-coamplification units: hst-1/int-2 and c-erbB-2/ear-1. Cancer Res 11: 3104–3108.

Tsuruo T, Iida-Saito H, Kawabata H, Ohara T, Hamada H and Utakoji T, (1986) Characteristics of resistance to adriamycin in human myelogenous leukemia K562 resistant to adriamycin and in isolated clones. Jpn J Cancer Res 77: 682–689.

Ueda K, Cardarelli C, Gottesman MM and Pastan I (1987a) Expression of a full-length cDNA for the human "*MDR*1" (P-glycoprotein) gene confers multidrug resistance to colchicine, doxorubicin, and vinblastine. Proc Natl Acad Sci USA 84: 3004–3008.

Ueda K, Clark DP, Chen C-J, Roninson IB, Gottesman MM and Pastan I (1987b) The human multidrug resistance (*MDR*1) gene: cDNA cloning and transcription initiation. J Biol Chem 262: 505–508.

Van der Bliek AM, Van der Velde-Koerts T, Ling V and Borst P (1986a) The overexpression and amplification of five genes in a multidrug-resistant Chinese hamster ovary cell line. Mol Cell Biol 6: 1671–1678.

Van der Bliek AM, Meyers MB, Biedler JL, Hes E and Borst P (1986b) A 22 kDa protein (Sorcin/V19), encoded by an amplified gene in multidrug resistant cells, is homologous to the calcium-binding light chain of calpain. EMBO J 5: 3201–3208.

Van der Bliek AM, Baas F, Van der Velde-Koerts T, Biedler JL, Meyers MB, Oxols RF, Hamilton TC, Joenje H and Borst P (1988) Genes amplified and overexpressed in human multidrug-resistant cell lines. Cancer Res 48: 5927–5932.

VanDevanter DR, Piaskowski BD, Casper JT, Douglass EE and Von Hoff, DD (1990) Ability of circular extrachromosomal DNA molecules to carry amplified *MYCN* proto-oncogenes in human neuroblastomas *in vivo*. J Natl Cancer Inst 82: 1815–1821.

Veinot L and Ling V (1993) Amplification of the P-glycoprotein gene family and multidrug resistance. In: RE Kellems (ed.) Gene Amplification in Mammalian Cells (pp. 287–299) Marcel Dekker Inc., New York.

Von Hoff DD, Needham-VanDevanter DR, Yucel J, Windle BE and Wahl GM (1988) Amplified human *MYC* oncogenes localized to replicating submicroscopic circular DNA molecules. Proc Natl Acad Sci USA 85: 4804–4808.

VonHoff DD, Forseth B, Clare CN, Hansen KL and VanDeVanter D (1990) Double minutes arise from circular extrachromosomal sites in human HL-60 leukemia cells. J Clin Invest 85: 1887–1895.

Wahl GM (1989) The importance of circular DNA in mammalian gene amplification. Cancer Res 49: 1333–1340.

Wahl GM, Carroll SM and Windle BE (1993) Cytogenetic and molecular dynamics of mammalian gene amplification: evidence supporting chromosome breakage as an initiating event. In: RE Kellems (ed.) Gene Amplification in Mammalian Cells (pp. 513–533) Marcel Dekker, Inc. New York.

Warr JR, Brewer F, Anderson M and Fergusson J (1986) Verapamil hypersensitivity of vincristine resistant Chinese hamster ovary cell lines. Cell Biol Int Rep 10: 389–399.

Wright JA, Smith HS, Watt FM, Hancock MC, Hudson DL and Stark GR (1990) DNA amplification is rare in normal human cells. Proc Natl Acad Sci USA 87: 1791–1795.

Yang L-Y and Trujillo JM (1990) Biological characterization of multidrug-resistant human colon carcinoma sublines induced/selected by two methods. Cancer Res 50: 3218–3225.

Yin Y, Tainsky MA, Bischoff FZ, Strong LC and Wahl GM (1992) Wild-type p53 restores cell cycle control and inhibits gene amplification in cells with mutant p53 alleles. Cell 70: 937–948.

Yunis JJ (1983) The chromosomal basis of human neoplasia. Science 221: 227–235.

Zwelling L, Slovak ML, Doroshow JH, Hinds M, Chan D, Parker E, Mayes J, Sie KL, Meltzer PS and Trent JM (1990) HT1080/DR4: a P-glycoprotein-negative human fibrosarcoma cell line exhibiting resistance to topoisomerase II-reactive drugs despite the presence of a drug-sensitive topoisomerase II. J Natl Cancer Inst 82: 1553–1561.

Cytotechnology **12**: 91–107, 1993.

Antibodies in the study of multiple drug resistance

Yuji Heike[1] and Takashi Tsuruo[2]
[1]*Pharmacology Division, National Cancer Center Research Institute, Tsukiji 5-1-1 Chuo-ku, Tokyo 104, Japan;* [2]*Institute of Molecular and Cellular Biosciences, University of Tokyo, Yayoi, Bunkyo-ku, Tokyo 113, Japan*

Key words: C219, diagnosis of multidrug resistance, monoclonal antibody, multidrug resistance, MRK16, P-glycoprotein, therapy of multidrug resistance

Abstract

Multidrug resistance (MDR) is a major problem in cancer chemotherapy. As P-glycoprotein is the key molecule in MDR, many investigators have constructed anti-P-glycoprotein monoclonal antibodies (MAbs). Those antibodies, including MRK16 and C219, were used for elucidation of the mechanism of MDR and for overcoming of MDR. This article describes the characterization of the antibodies against the P-glycoprotein and other proteins of multidrug-resistant tumor cells, and discusses the therapeutic implication of the antibodies.

Abbreviation: ADCC — antibody-dependent cell-mediated cytotoxicity

Introduction

Resistance of tumors against multiple drugs is a major problem in cancer chemotherapy. P-glycoprotein, which transports various cytotoxic drugs outside the cells, is one of the key molecules in multidrug resistance [1—3]. P-glycoprotein has been proven to be a protein which binds anticancer drugs, to be an ATPase, and to be localized at the plasma membrane of resistant cells [4,5]. Transfection of cloned *MDR*-1 sequence, the genes encoding human P-glycoprotein, conferred multidrug resistance (MDR) on sensitive cells [6—8]. The amount of P-glycoprotein expression in tumor samples was found to be elevated in intrinsically drug-resistant cancers of the colon, kidney and adrenal as well as in some tumors that had acquired drug resistance after chemotherapy [9—13]. Because P-glycoprotein appears to be involved in both acquired and intrinsic MDR in human cancers, the investigation of its mechanism and the means of selective killing of tumor cells expressing P-glycoprotein is very important for successful cancer therapy.

To achieve this, many investigators, including us, have constructed anti-P-glycoprotein MAbs (Table 1) [14—21]. Using these MAbs, characterization and analysis of P-glycoprotein have been carried out. These studies have revealed the importance in P-glycoprotein on the mechanism of drug resistance of several cancers. Recently, several experiments were carried out for overcoming MDR in test animals using these MAbs. In this chapter, we summarize the characteristics of anti-P-glycoprotein MAbs and their application especially for diagnostic and therapeutic purposes.

Table 1. Anti-P-glycoprotein MAbs reported previously

Antibody	Investigator
MRK16	Hamada, H. *et al.* 1986 [14]
MRK17	
C219	Norbert, K. *et al.* 1985 [16]
JSB-1	Scheper, R.J. *et al.* 1988 [17]
HYB-241	Meyers, M.B. *et al.* 1989 [18]
HYB-612	
HYB-195	
265/F4	Lathan, B. *et al.* 1985 [19]
4E3	Robert, J.A. *et al.* 1993 [20]
UIC2	Mechetner, E.B. *et al.* 1992 [21]
MAb57	Cenciarelli, C. *et al.* 1991 [43]
G8-2	Yanovich, S. *et al.* 1989 [44]
32G7	Danks, M.K. *et al.* 1985 [45]

Anti-P-glycoprotein MAb

In this section, we describe anti-P-glycoprotein MAbs reported previously, mainly two major MAbs, MRK16 and C219 (Table 1).

MRK16

This Mab was established by the immunization of BALB/c mice with intact multidrug resistant human myelogenous leukemia cell line, K562/ADM [14]. This MAb has IgG2a subclass and recognizes an extracellular epitope of the molecule of human *MDR*-1 product. Its epitope includes at least two (the first and fourth) of the six predicted extracellular domains, which make up the extracellular peptide loops, and the antibody recognizes the epitope situated in close proximity in the native protein [22]. MRK16 increases intracellular accumulation of vincristine and actinomycin-D, and enhanced vincristine cytotoxicity in K562/ADM cells [14] (Fig. 1A), and also increases intracellular concentration of vincristine in 2780AD cells [14, 15] (Fig. 1B). It induces ADCC reaction of human and mouse effector cells against P-glycoprotein positive MDR cells [23,24]. As MRK16 was the first antibody reported which recognized the extracellular portion of human *MDR*-1 product, many investigators have used this MAb for diagnostic [25—28] and therapeutic purposes [23,24,29].

Recently, we succeeded in constructing the recombinant chimeric antibody MH-162, in which the antigen-recognizing variable regions of MRK16 are joined to the constant regions of human antibodies. The apparent avidity of the chimeric antibody to the K562/ADM cell antigen is similar to that of the all-mouse antibody MRK16 as determined by enzyme-linked immunosorbent assay and also by competition inhibition cell binding radioimmunoassay. MH-162 augments the killing activity of human effector cells against drug-resistant tumor cells more effectively than the all-mouse monoclonal MRK16 [30].

C219

Ling and his coworkers established the MAb C219 (IgG2a) directed against SDS-solubilized plasma membranes of multidrug-resistant Chinese hamster ovary and human leukemic MDR cells [16]. C219 detects a cytoplasmic sequence six residues away from the consensus sequence which has been proposed to be the ATP-binding domain of the polypeptide [31]. As this domain is highly conserved in all P-glycoprotein isoforms, human and mouse, C219 not surprisingly recognizes all types of the P-glycoproteins thus far characterized, and the antibody stains the P-glycoprotein of both human and mouse tumor cell lines [31,32].

Other MAbs

Besides MRK16 and C219, other MAbs against P-glycoprotein have also been reported (Table 1).

MRK17 (an IgG1 isotype antibody) was established simultaneously with MRK16 by us [14]. Like MRK16, it recognized the extracellular domain of P-glycoprotein. The antibody, however, had no effect on drug accumulation and no potentiation of drug cytotoxicity, but has been found to inhibit the growth of MDR cells. The chimeric antibody MH-171, which was constructed from MRK17 by the same procedure as MH-162, also augments the killing activity of human effector cells [33].

JSB-1 was established by Scheper *et al.* in 1988 [17]. This MAb was obtained after immunization of mice with colchicine resistant Chinese hamster

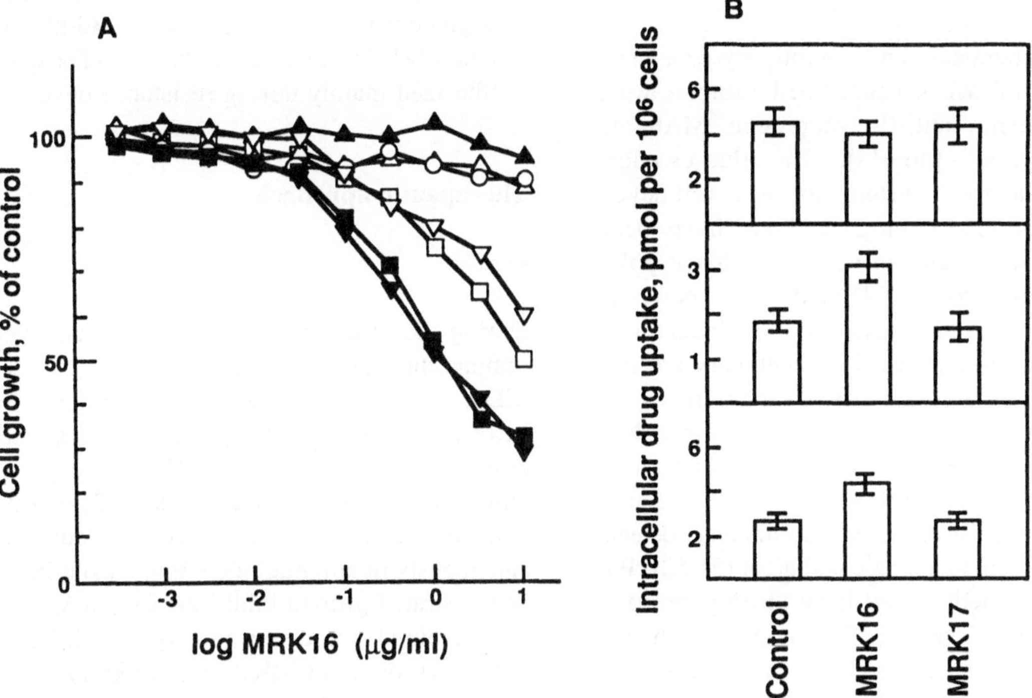

Fig. 1. (A) Modulation of drug cytotoxicity by MRK16.

The effect of MRK16 on the growth of K562/ADM cells was examined without drugs (○); with adriamycin at the concentration of IC_{20} (150 ng/ml) (Δ) or IC_{50} (500 ng/ml) (▲); with vincristine at the concentration of IC_{20} (250 ng/ml) (∇) or IC_{50} (500 ng/ml) (▼); with actinomycin D at the concentration of IC_{20} (30 ng/ml) (□) or IC_{50} (100 ng/ml) (■). (B) The effect of MAbs on the cellular uptake of the drugs; adriamycin (top), vincristine (middle), actinomycin D (bottom).

ovary cell line, CHrC5. The isotype is IgG1, and it recognizes the cytoplasmic epitope closely linked to one recognized by C219 [34–38].

265/F4 was obtained by the immunization of mice with resistant Chinese hamster ovary CHOC5 cells [19]. It detected a 170 kDa protein in CHO resistant cell membranes [32,39].

HYB-612, HYB-241 and HYB-195 were established by the immunization of mice with vincristine-resistant human neuroblastoma cells (SH-SY5Y/VCR) [18,40–42]. MAb 4E3 was established by immunization with human MDR uterine adenocarcinoma cell line (ME180/Dox500) [20]. This MAb recognized the extracellular epitope of human P-glycoprotein. Other MAbs, MAb57 [43], G8-2 [44] and 32G7 [45] have also been reported to react with P-glycoprotein.

Recently, Mechetner and Roninson developed the MAb UIC2 that recognizes an epitope of the extracellular domain of P-glycoprotein [21]. This MAb has IgG2a subclass. Differing from MRK16, it recognizes another epitope of extracellular domain; notably the antibody shows a different reaction against a mutant form of P-glycoprotein having a deletion the first extracellular loop. UIC2 inhibits the efflux of drugs from MDR cells and increase the accumulation of the drugs as well as enhancing the cytotoxicity of anticancer drugs against MDR cancer cells.

Diagnostic approach

In this section, we summarized the application of anti-P-glycoprotein MAbs for diagnostic purpose.

Immunohistochemistry and immunocytochemistry

Immunohistochemical and immunocytochemical examinations of MDR cancer cell samples were carried out using anti-P-glycoprotein MAbs in many laboratories [46–48]. The fluorescence-activated cell analysis system has been well developed, and immunostaining methods have been improved for clear diagnosis. Although the polymerase chain reaction (PCR) method is a powerful tool for detecting small amounts of P-glycoprotein [49,50], immuno- and cyto-histochemistry are still useful methods for diagnosis as they involve a simpler technique and are more economical. Numerous reports have been published and are mentioned in Chapters 4 and 9 of this volume in detail.

The fluorescence-activated cell analysis detects the P-glycoprotein at a single cell level [51,52]. We have used this method and have already reported the biphasic expression of P-glycoprotein in primary cultured human renal cell carcinoma cells [53].

Unfortunately, in clinical samples, it is usually difficult to detect P-glycoprotein positive cells by the simple flow cytometry analysis, because of the low proportion of P-glycoprotein positive cells. Some means of augmenting the sensitivity of the analysis such as panning methods or use of immunobeads are under consideration [54].

P-glycoprotein family

Recently, it was found that mammalian P-glycoproteins are encoded by small gene families. In humans, two genes *MDR*-1 and *MDR*-2 (also called *MDR*-3) are present [55–59]. Although human *MDR*-1 confers multidrug resistance *in vitro* and *in vivo*, the contribution of *MDR*-3 gene products on multidrug resistance has not yet been clarified [2,60–62]. Distinction between the subfamilies of P-glycoprotein is important for estimating drug sensitivities of the cells. MRK16, MRK17, JSB-1, HYB-241, UIC-2 and 4E3 antibodies react to only *MDR*-1 product [14,17,19–21]. The diagnosis using these MAbs is useful for evaluating sensitivity against anti-cancer drugs.

Moreover, existence of subfamilies in *MDR*-1 gene products has been reported, the one is a 170-kDa glycoprotein, and the other a 180-kDa glycoprotein [18]. In most cases, the 170-kDa species is synthesized mainly during resistance development.

Therapeutic approach

In vivo

During the past 5 years, our group has developed immunotherapeutic approaches for overcoming MDR cancers using anti-P-glycoprotein MAbs, MRK16 and MRK17 [14]. We have evaluated the effect of MRK16 and MRK17 on the growth of human MDR ovarian cancer cells 2780AD [23]. 2780AD cells (10^7/mouse) were inoculated subcutaneously in athymic mice. In the experiment, we demonstrated growth inhibition by the MAbs (Fig. 2A). Animals received intravenous injection of various amounts of MRK16 or MRK17 on days 2 and 7 after tumor inoculation. Subcutaneous tumors never developed in the mice treated with 10 µg or more per animal of MRK16. When the mice were treated with 1 µg per animal of MRK16 or 1 or 10 µg per animal of MRK17, the development of tumors was substantially retarded as compared to control mice where palpable tumors developed by day 20.

In the next experiment, the therapy with antibodies was started after the tumors became palpable (150–400 mm^3 in volume) (Fig. 2B). One group of animals received 6 mg/kg of adriamycin intravenously three times as control therapy. Adriamycin did not induce significant inhibition of 2780AD tumor growth, but it did significantly inhibit growth of the parent tumor A2780 *in vivo* [76% inhibition was observed on day 45 (data not shown)]. Animals received various amounts of MRK16 or MRK17 intravenously on days 20, 23 and 26, and the average tumor volume was measured. One or two days after the first injection of MRK16, the size of the tumor masses remarkably decreased in all five mice. After the third MRK 16 injection, the average tumor mass was substantially reduced, and in four out of five mice, the tumor mass had completely disappeared. Among these

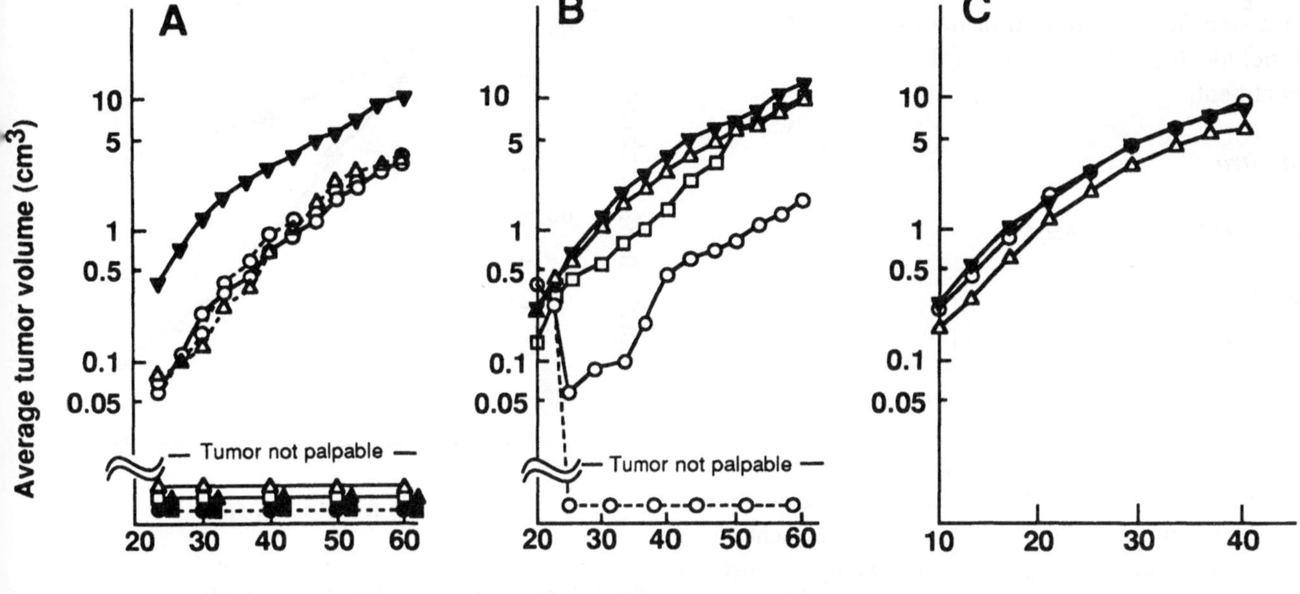

Fig. 2. In vivo effects of MRK16, MRK17 and adriamycin on the growth of drug-resistant tumor xenografts.

All animals were given injections of 10^7 2780AD cells s.c. on day 0. (A) On days 2 and 7 after tumor cell inoculation, groups of animals were treated i.v. with 0.001 (○), 0.01 (Δ), 0.03 (□), 0.1 (●), 0.3 (▲) or 1 mg (■) of MRK16 (———) or MRK17 (-------) antibody. Untreated animals (▼) served as controls. (B) When the tumors became palpable on day 20 (150–400 mm^3 in volume; as determined from the formula: $V=1/2\ ab^2$, where a is the longest diameter and b is the shortest diameter of the tumor), mice were randomised into 4 groups. One group was untreated (▼), a second group was treated with 1 mg of MRK16 on days 20, 23, and 26 (○), a third group was treated with 1 mg of MRK17 on days 20, 23 and 26 (Δ), and the fourth group was treated with 6 mg/kg of adriamycin on days 20, 24 and 28, an optimal dose and schedule for adriamycin treatment (□). (C) Drug-sensitive parent A2780 cells were inoculated s.c. and MRK16 (○) and MRK17 (Δ) were given at 1 mg per animal on days 2 and 7 after tumor inoculation. As in A and B, (▼) represents the growth of untreated control tumors.

four mice, a palpable tumor recurred in one mouse on day 42; the other 3 mice remained tumor-free up to day 60. No significant inhibition of tumor growth was observed in animals which received 1 mg of MRK17 intravenously on days 20, 23 and 26. These MAbs had no effect on the *in vivo* growth of drug-sensitive parent tumor A2780 cells, suggesting that the effect of these antibodies was specific for the drug-resistant tumor cells over-expressing the P-glycoprotein.

Several mechanisms, such as direct cytotoxicities of the MAbs or immune response induced by MAbs, may be involved in the growth inhibition of the xenografts. Complement-dependent cytotoxicity (CDC) for 2780AD cells was clearly demonstrated *in vitro* using MRK16 and rabbit serum as complement. Antibody-dependent cell-mediated cytolysis (ADCC) by MRK16 was more clearly demonstrated than by MRK17. In the growth inhibition experiments in several other tumor xenograft models, IgG2a subclass of the mouse-derived anti-cancer MAb induced more efficient cell-mediated cytotoxicity as compared to other mouse IgG subclasses [63]. The finding is compatible with our observation in which MRK16 (IgG2a) was more effective in treatment of established tumors than MRK17 (IgG1). The benefit of MRK16 use *in vivo* in MDR cancer treatment has also been verified in the model of xenograft of MDR human colon cancer (HT29mdr1) [64].

Rittmann *et al.* reported that pretreatment of human MDR tumor BRO/pFRmdr1.6 with HYB-

241 prior to administration of vinca alkaloid results in a significant inhibition of the tumor. [65]. They conclude the effectiveness of HYB-241 in MDR treatment.

In vitro

1. Direct effects of MAbs

MRK16

We reported that MRK16 inhibited P-glycoprotein associated drug efflux from K562/ADM thus increasing the intracellular accumulation of vincristine and actinomycin D [14] (Fig. 1B). The antibody enhances vincristine and actinomycin D cytotoxicities in K562/ADM cells in a dose-dependent manner (Fig. 1A). Similar enhancement of vincristine and actinomycin D cytotoxicity has also been observed in MDR 2780AD cells. However, MRK16 had no effect on parent K562 or A2780 cells. Enhanced cytotoxicities of vincristine and actinomycin D in K562/ADM cells by MRK16 seem to be mediated by the enhanced intracellular accumulation of these antitumor agents. However, the effects observed with MRK16 were partial, and the antibody did not restore the drug levels to those observed in sensitive cells. Differing from MRK16, MRK17 had no effect on intracellular accumulation of adriamycin, vincristine or actinomycin D. The antibody had slight growth inhibitory effects against P-glycoprotein positive cells (Fig. 3).

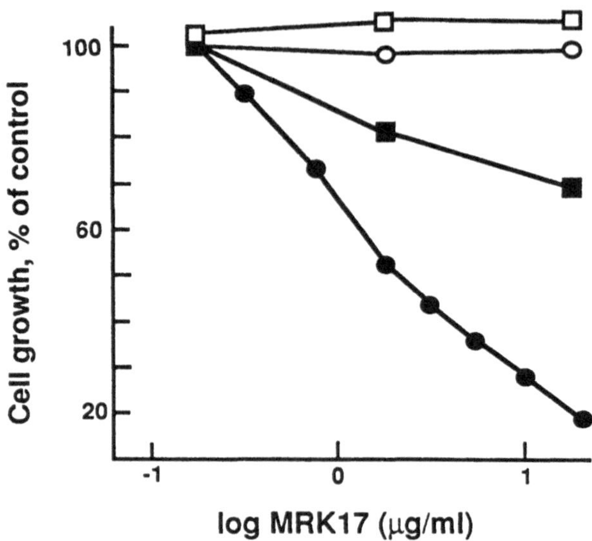

Fig. 3. Specific growth inhibitory effect of MRK17 on multidrug-resistant human cell lines.

K562 cells (○), K562/ADM cells (●), A2780 (□) or 2780AD cells (■). Cell growth is expressed as a percentage of the growth without the MAb.

Cyclosporin augmented the antitumor effects of vincristine and adriamycin synergistically with MRK16 by increasing the intracellular concentrations of anti-cancer drugs [66]. The combined effects of MRK16 and cyclosporin on the reversal of vincristine resistance in K562/ADM cells are shown in Table 2. MRK16 remarkably increased the cellular accumulation of cyclosporine in K562/ADM cells. The increased cyclosporine in the cells enhanced the cellular accumulation of vincristine and adriamycin (Fig. 4) and enhanced the cytotoxicities of both drugs in the cells (Table 2). These are the mechanisms explaining the synergistic modulation of MDR by MRK16 and cyclosporine in K562/ADM cells. The synergistic effect in K562/ADM cells was most prominent at 3 μM cyclosporine, followed by 1 and 10 μM cyclosporine (Table 2). Combining MRK16 with cyclosporine reduced the dose of cyclosporine needed for overcoming MDR. The synergistic modulation of MDR was observed in the combinations uses of cyclosporine and verapamil and of verapamil and quinine. All of these drugs competitively inhibit the transport of anti-tumor against through P-glycoprotein.

UIC2

UIC2 inhibited the function of P-glycoprotein and increased the accumulation of anticancer drugs such as vinblastine, vincristine, colchicine, taxol and VP-16 [21] (Fig. 5). UIC2 had maximal inhibitory effect at 10 μg/ml against P-glycoprotein positive cells. The inhibition by UIC2 has been reported to be qualitatively different from that of the other anti P-glycoprotein MAbs. Further examinations focusing on clinical application are warranted.

Table 2. Synergistic reversal of vincristine and adriamycin resistance in K562/ADM cells by combined treatment with MRK16 and cyclosporine

Cell line	Modifier	Concentration of modifier (mM)	IC_{50} of vincristine, nM		IC_{50} of adriamycin, nM	
			without MRK16	with 10 μM/ml MRK16	without MRK16	with 10 μM/ml MRK16
K562	Cyclosporine	0	3.3(1)	2.6(1.3)	23(1)	24(1)
		1	1.8(1.8)	1.7(1.9)	20(1.2)	20(1.2)
		3	1.6(2.1)	1.6(2.1)	17(1.4)	15(1.5)
		10	1.3(2.5)	1.2(2.8)	15(1.5)	15(1.5)
K562/ADM	Cyclosporine	0	639(1)	440(1.5)	2970(1)	3270(0.9)
		1	460(1.4)	99(6.5)	3210(0.9)	700(4.2)
		3	155(4.1)	7.7(83)	560(5.3)	59(50)
		10	5.2(123)	1.9(336)	34(87)	20(149)
K562	Verapamil	0	4.3(1)	3.9(1.1)	14(1)	11(1.3)
		1	2.4(1.8)	3.9(1.1)	7.2(1.9)	8.7(1.6)
		3	2.0(2.2)	2.0(2.2)	6.6(2.1)	6.9(2)
		10	2.2(2)	2.1(2)	6.1(2.3)	7.2(1.9)
K562/ADM	Verapamil	0	689(1)	484(1.4)	248(1)	2950(0.8)
		1	505(1.1)	530(1.3)	1020(2.4)	990(2.5)
		3	102(6.8)	60(11)	268(9.3)	230(11)
		10	21(33)	12(57)	130(19)	95(26)
K562	FK-506	0	3.2(1)	3.2(1)	15(1)	20(0.8)
		1	2.9(1.1)	2.6(1.2)	17(0.9)	16(0.9)
		3	2.4(1.3)	2.1(1.5)	11(1.4)	8.1(1.9)
		10	1.8(1.8)	1.7(1.9)	9(1.7)	7(2.1)
K562/ADM	FK-506	0	712(1)	531(11.3)	2490(1)	2550 (1)
		1	703(1)	531(11.3)	1590(1.6)	1910(1.3)
		3	537(1.3)	613(1.2)	455(5.5)	551(4.5)
		10	20(36)	17(42)	44(57)	42(59)

Numbers in parentheses = fold decreases in IC_{50} values as compared to those obtained in the absence of a modifier.

Other MAbs

HYB-241, HYB-612 and HYB-195 have been reported to increase the accumulation of vincristine and actinomycin D in MDR cells. No evidence of cytotoxicity against MDR cancers, however, has been reported [67,68].

MAb 4E3 produced a mild potentiation of vinblastine and actinomycin D cytotoxicities in MDR cells at high concentrations (50 μg/ml or 100 μg/ml) [20]. 265/F4 was reported to have selective growth-inhibitory effects against MDR CHO-cells [69].

2) Immunotoxin

Pastan *et al.* constructed an immunotoxin composed of MRK16 and *Pseudomonas* exotoxin [29]. The effects of the immunotoxin (MRK16-PE) on the viability of the cells having different levels of resistance were examined [29] (Fig. 6). The MDR cell lines with high levels of P-glycoprotein were very sensitive to the MRK16-PE conjugate. The competition test by 100- to 1000-fold excess of MRK16 revealed substantial reversal of cell killing.

In the report, MRK16 and *Pseudomonas* exotoxin were coupled by chemical procedure. Recently the recombinant immunotoxin, in which the cell recognition domain of the toxin (*Pseudomonas* toxin or diphtheria toxin) had been removed, was constructed by recombinant technique [70–74]. Immunotoxins prepared by this method have low toxicities. The application of this technique to produce anti-P-glycoprotein immunotoxins is warranted.

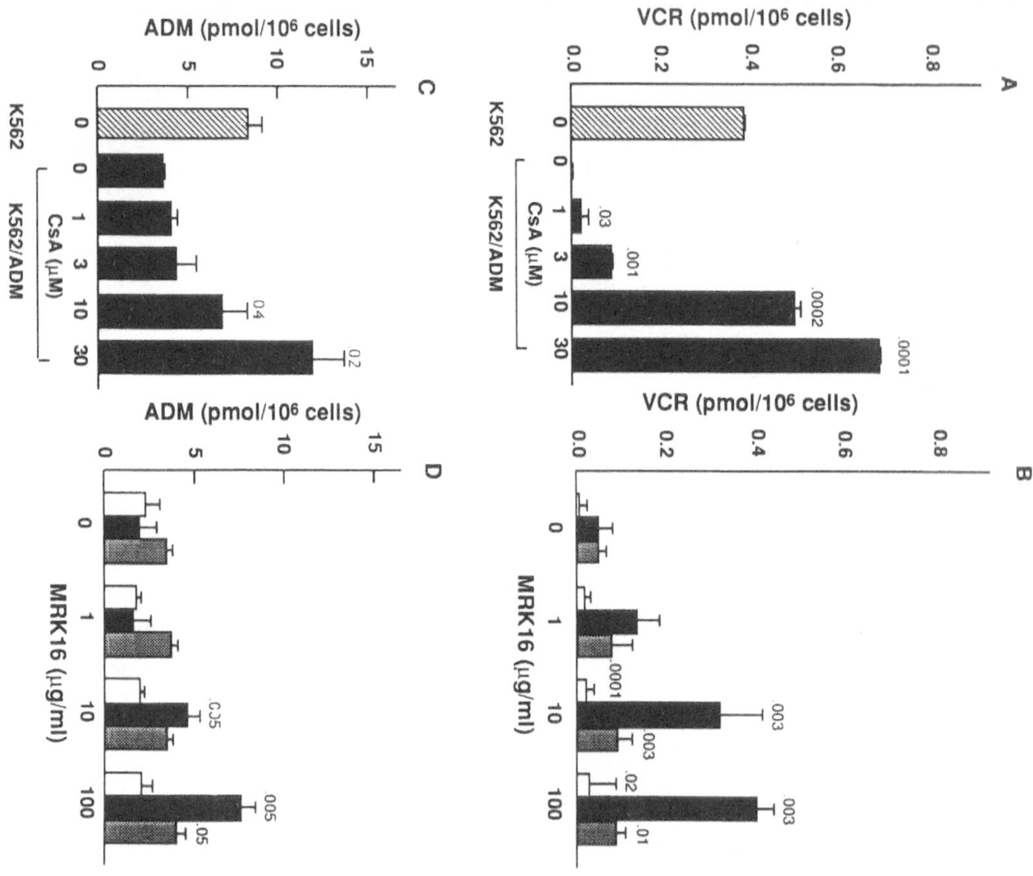

Fig. 4. Increment of intracellular accumulation of vincristine (VCR) and adriamycin (ADM) in K562 and K562/ADM cells by cyclosporine A.

Cellular accumulation of [^{3}H]VCR (A) and [^{14}C]ADM (C) in K562 and K562/ADM cells was examined in the absence or presence of indicated concentrations of cyclosporine (CsA). Cellular accumulation of [^{3}H]VCR (B) and [^{14}C]ADM (D) in K562/ADM cells was examined in the absence or presence of indicated concentrations of MRK16 alone (□), MRK16 combined with 3 mM CsA (■), and MRK16 combined with 3 mM verapamil (▨).

3) ADCC activity

We have already reported that MRK16 effectively prevented tumor development in athymic mice inoculated subcutaneously with the drug-resistant human ovarian cancer cells 2780AD [23] (Fig. 2). The effects of MRK16 *in vivo* were due to ADCC reactions. Both MRK16 and MRK17 augments the cytotoxicities of human effector cells against P-glycoprotein positive cancer cells (24). Interestingly, MRK16 (IgG2a subclass) augmented both lymphocyte- and monocyte-mediated cytolysis, but MRK17 (IgG1 subclass) augmented only monocyte-mediated cytolysis (Fig. 7 and Table 3).

Mouse-human chimeric antibodies, MH-162 or MH-171, are composed of the antigen-recognizing variable regions of MRK16 or MRK17, respectively, and constant regions of human IgG1 anti-

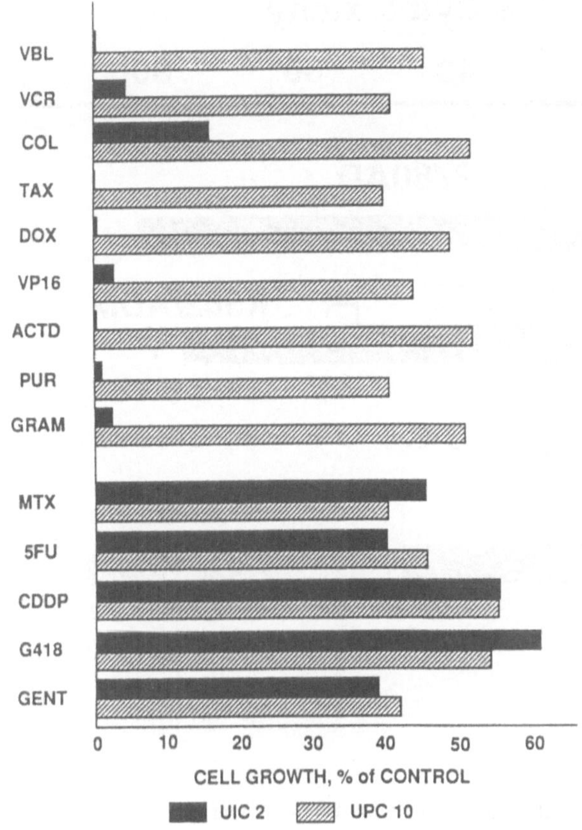

Fig. 5. Potentiation of cytotoxic effects of various drugs by MAb UIC2.

Bars represent the viability of BALB/c 3T3-1000 cells treated with indicated drugs (as determined by the MTT assay) in the presence of MAb UIC2 at 20 μg/ml (■) or control at 20 μg/ml (▨). Cell viability is expressed relative to that of control cells grown with control antibody in the absence of drugs. (From Mechetner *et al.* [21])

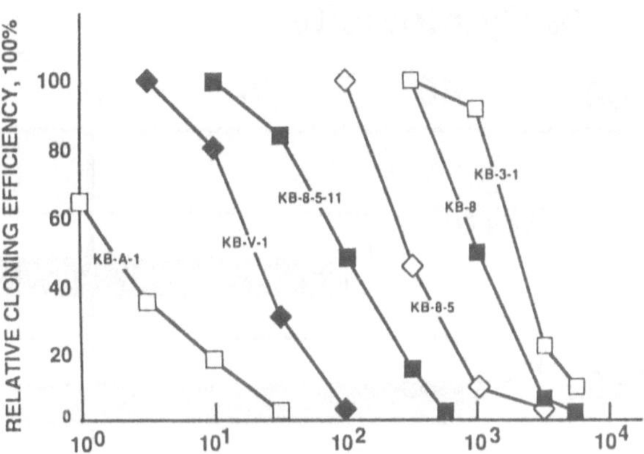

Fig. 6. Sensitivity of cells expressing various levels of drug resistance to MRK16-PE.

Cells expressing various levels of P-glycoprotein were incubated for 10 days with MRK16-PE at the indicated concentrations. After 10 days, the viable cells were stained and counted. (From FitzGerald *et al.* [29])

body. They were more effective than MRK16 and MRK17 in mediating ADCC of human effector cells [30,33] (Fig. 8). The effector cell analysis revealed that the chimeric antibodies augmented the lymphocyte-mediated cytolysis more effectively than all-murine MAbs [75]. There are no differences between MH-162 and MH-171 in ADCC activity of human effector cells. A major limitation in the clinical use of murine-derived MAbs is the immune response elicited against foreign protein, which may render the antibody ineffective and may also harm the patients. The chimeric immunoglobulins circumvent the problems of antigenicity of murine-derived MAbs as most immunoglobulin antigenicity resides in the constant domain.

MAb57 (IgG2a subclass) was proven to be useful in selectively destroying MDR target cells in an ADCC assay system [43].

4) *Bispecific antibodies*

Van Dijk *et al.* reported that the bispecific antibodies composed of MRK16 and anti-CD3 MAb (OKT3) lysed MDR ovarian cancer cells by CD3 positive human effector cells [76]. We developed the bispecific $F(ab')_2$ composed of the Fab fragments from MRK16 and OKT3 [53]. Some advantages were obtained by the deletion of the Fc portion from bispecific antibodies as described below. The bispecific $F(ab')_2$ enhanced the binding and cytotoxicity of human effector cells against drug resistant renal cancer cells (Fig. 9). The bispecific $F(ab')_2$ increased the direct interaction between CD3 positive effector cells and MDR cancer cells to render the latter more sensitive to killing by the effector cells. The detailed mechanisms of cell-killing by bispecific $F(ab')_2$ are not clear. The induction of several adhesion molecules and cytokines after stimulation of CD3 molecules

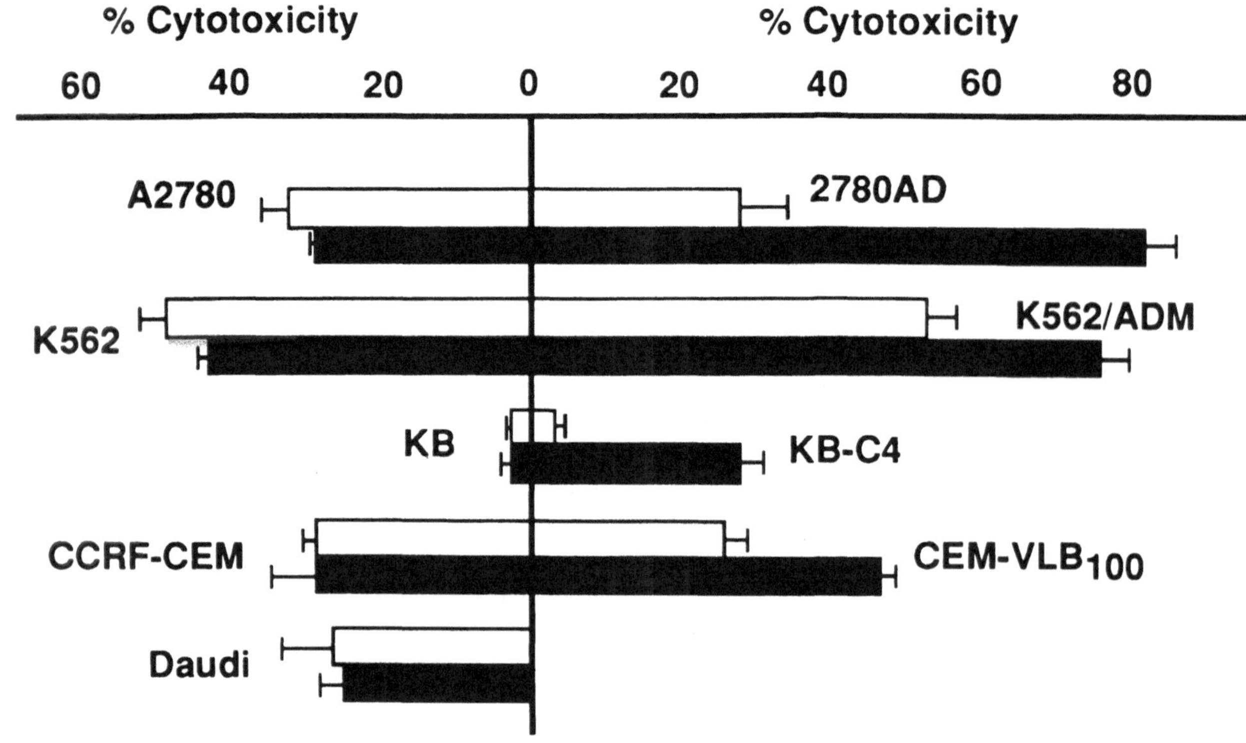

Fig. 7. Spectrum of target cells in ADCC reaction.

Four pairs of sensitive cells and their resistant cells were used as target cells. Target cells with (■) or without (□) 10^{-2} μg/ml of MRK16 were added. Effector cells were lymphocytes, and the effector-target ratio was 20. MRK16 augmented the killing activity of lymphocyte against MDR cancer cells.

are postulated. Advantages of the $F(ab')_2$ lacking Fc portion over the whole MAbs are: i) as $F(ab')_2$ does not bind to FcR+ cells in the reticuloendothelial system, its rapid clearance can be avoided and it remains in circulation for a long time. ii) undesirable immunoreaction caused by the murine Fc portion which is strongly antigenic in humans can be avoided. Nitta *et al.* has reported that the combination therapy with a bispecific $F(ab')_2$ composed of anti-glioma X anti-CD3 antibodies and lymphokine-activated killer cells is promising [77]. The augmentation of the cytotoxicity of effector cells by the bispecific $F(ab')_2$ could be beneficial for overcoming MDR cancers because of its unique mechanism and low toxicity.

Table 3. ADCC induced by MRK16 and MRK17 by human effector cells

Target cells	Effector cells	% Cytolysis		
		MAb(–)	MRK16	MRK17
A2780	Lymphocyte	0.0 ± 0.8	1.6 ± 1.4	0.8 ± 1.6
	Monocyte	2.4 ± 1.7	3.0 ± 2.2	3.3 ± 2.3
2780AD	Lymphocyte	9.4 ± 4.0	32.2 ± 5.3	10.1 ± 5.3
	Monocyte	9.3 ± 1.6	40.1 ± 3.7	32.2 ± 3.5

Effector target ratio: 40. The concentration of MAb was 10^{-2} μg/ml.

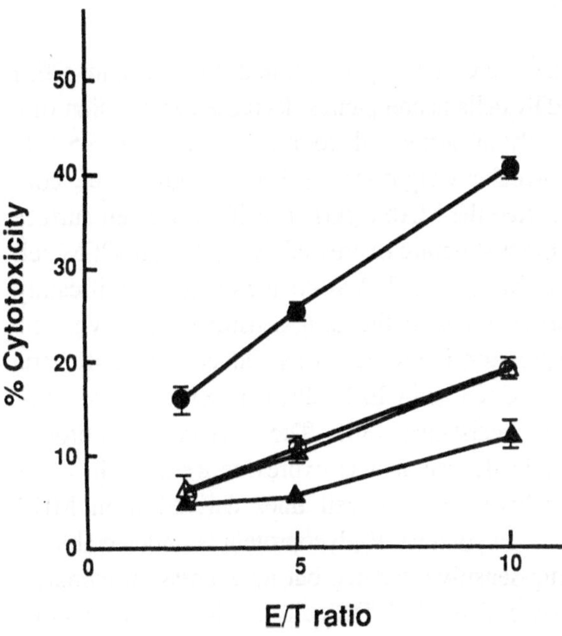

Fig. 8. ADCC with MH-162 or MRK16 for 2780AD cells.

ADCC was carried out using 1 μg/ml of MH162 (●), 1 μg/ml of MRK16 (○), 1 μg/ml of antiphosphorylchorine mouse-human chimeric antibody (Δ), or without monoclonal antibody (▲). Effector cells were peripheral mononuclear cells.

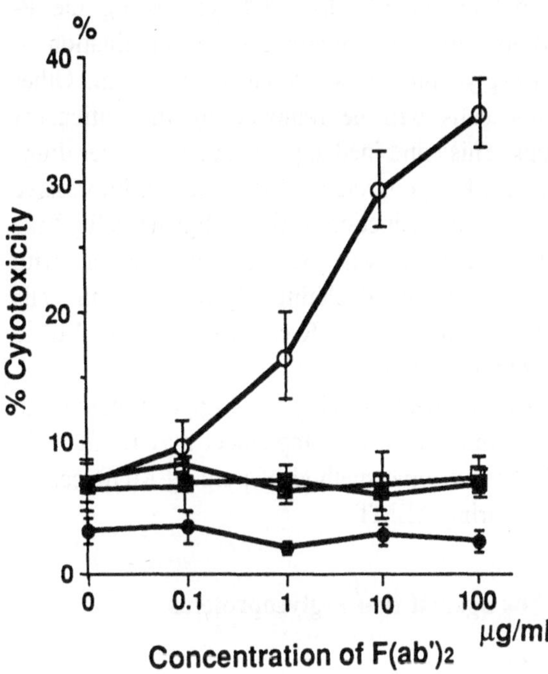

Fig. 9. Effects of the bispecific $F(ab')_2$ on the cytolytic activity of PBMCs.

Effector cells were fresh PBMCs. The effector-target ratio was 10. Target cells were P-glycoprotein positive human renal cancer cells with bispecific $F(ab')_2$ (○), P-glycoprotein positive renal cancer cells with $F(ab')_2$ of MRK16 (□), P-glycoprotein positive renal cancer cells with $F(ab')_2$ of OKT3 (■), P-glycoprotein negative renal cancer cells with bispecific $F(ab')_2$ (●).

5) *Purging of MDR cancer cells by MAbs*

Recently, high dose chemotherapy has been carried out against leukemia, lymphoma and some solid tumors. The development of supportive therapies, especially therapies with hematopoietic growth factor (ex. G-CSF), has made high dose chemotherapy easier and safer. Autologous bone marrow transplantation (ABMT) is a promising supportive therapy during a course of high dose chemotherapy. ABMT after high-dose chemotherapies, however still has a critical problem. As repeated chemotherapy leads to development of resistance to a number of drugs [78–80], the possibility of contamination by MDR cancer cells at the time of bone marrow harvest can not be neglected. Finding methods to remove P-glycoprotein positive MDR tumor cells is very important. Angelo *et al.* reported selective elimination of MDR cells by using MRK16 and a sheep antimouse immunoglobulin antibody conjugated to the ribosome-inactivating protein saporin 6 [81]. In cell suspensions composed of 90% normal bone marrow cells and 10% MDR-positive cells, treatment with MRK16 followed by the antimouse immunotoxin causes the elimination of 99% MDR positive cell. They report that a 2-log elimination of P-glycoprotein positive cells was obtained by treatment with MRK16 and antimouse immunotoxin. They also report that human normal hematopoietic precursors (granulocytic colony-forming units, erythroid, and megakaryocytic-forming units) are not affected by the MRK16 plus immunotoxin treatment.

Aihara *et al.* also demonstrated the possibility of selective killing of MDR cells from bone marrow using anti-P-glycoprotein MAb 17F9 and rabbit complement [82]. Selective removal of MDR cells, 3-log reduction of malignant leukemic cells from mixtures of MDR cancer cells and normal bone marrow cells has been achieved without affecting normal precursor cells. Cells expressing the P-glycoprotein were removed by a combination of anti-P-glycoprotein MAb and complement. Other tumor cells will be removed by the anticancer drugs. This combined approach of anticancer drugs and anti-P-glycoprotein MAb yields an impressive nearly 5-log reduction in the malignant cells. Like 17F9, MRK16 was used in combination with complement for depleting MDR myeloma cells from a mixture of MDR cancers and normal bone marrow cells [83].

The approach utilizing anti-P-glycoprotein MAb and immunotoxin or complement may result in an effective ex vivo method of purging MDR cancer cells during ABMT.

MAbs against non-P-glycoproteins

HOT12

Besides P-glycoprotein, a low molecular weight cytoplasmic protein (Mr 19,000–22,000) is over-expressed in some MDR cells [84]. The protein, revealed to be a calcium-binding protein has been designated CP22 or sorcin (soluble resistance-related calcium binding protein). We raised monoclonal antibody HOT12 against the Mr 22,000 protein purified from human leukemia cell line K562/ADM. HOT12 is also reactive with the Mr 22,000 protein from Chinese hamster ovary cells (CHRC5), suggesting that the protein is conserved among species. Although the Mr 22,000 protein was found to be highly expressed in drug-resistant cell lines, it is not necessarily related directly to drug resistance. For example, the weakly resistant K562/VCR expressed high amounts of the protein compared to the highly resistant cell lines such as 2780AD and KB-C4. The protein remained highly expressed in revertant cell lines, indicating that overexpression of the protein was not sufficient to maintain the MDR phenotype. From the data now available, the role of the Mr 22,000 protein in multidrug resistance is not clear. The amplification of the gene for the Mr 22,000 protein and overproduction of the protein could be merely consequences of the amplification of P-glycoprotein genes as suggested by Van dB *et al.* [85].

MAb56

The increased P-glycoprotein in KB and CEM MDR cells accompanies decreased expression of a family of surface glycoproteins, of 70- to 95-kDa molecular weight [86–88]. Cianfriglia *et al.* constructed the MAb (IgG1) specific for a cell surface protein structure expressed by the human CEM cell line, MAb56 [89]. A 90-kDa protein is significantly associated with the drug-sensitive phenotype. Its expression is progressively reduced in MDR variants of KB and CEM cells, parallel to the extent of drug resistance [86]. The surface glycoprotein MC56 determinant is expressed *de novo* in drug-sensitive revertant cell lines derived from MDR cells. In contrast, P-glycoprotein is undetectable on drug-sensitive parents, but its expression increases progressively in MDR cells parallel to the acquired level of drug resistance. In MDR cell lines showing high levels of MDR and the strong expression of P-glycoprotein, the amount of MC56 cellular determinant is practically undetectable. In revertant cell lines, both MC56 and P-glycoprotein are expressed but at an intermediate level according to the relative degree of drug resistance. The expression of the MC56 molecule on a variety of human cells and tissues might become a candidate as a marker for diagnosis for drug sensitivity. The correlation between MC56 and MDR in clinical samples is unclear at present.

Conclusion

P-glycoprotein is now known to be a key molecule of MDR. As P-glycoprotein is known to be expressed by both acquired and intrinsic MDR cells in human cancers clinically, the development of diagnostic and therapeutic approaches targeting P-glycoprotein is important. MAbs are the strong arms for studies towards these purposes. During the last decade, the mechanisms of MDR have been well clarified by these MAbs, and many attempts at diagnosis and therapy of MDR cancers have been made. The results revealed the clear contribution of P-glycoprotein in MDR. Various chemical compounds (including many dihydropyridine-type calcium channel blockers such as verapamil, R-verapamil, diltiazem, quinidine, trifluoperamine and cyclosporine) have been found to reverse MDR *in vitro* [90–94], and in a few instances *in vivo* [90,91,93–96]. Unfortunately, most of these agents have undesirable side-effects *in vivo*, such as to cause cardiovascular disorders at the concentrations found necessary to reverse multidrug resistance *in vitro* [97,98].

As P-glycoprotein is located on the membranes of MDR cancer cells, it is considered as a suitable antigen for immunotherapy. Many studies, utilizing immunotoxins, induction of host defence by MAbs, chimeric Abs and bispecific Abs have been carried out. *In vitro* and in some cases *in vivo*, the results have revealed some promising results. However, problems may arise because P-glycoprotein is constitutively expressed in some normal human organs and tissues including adrenal, kidney, liver, colon [2,99,100] and capillary endothelial cells in the brain [101–103]. In these tissues P-glycoprotein might act to transport normal metabolites or cytotoxic compounds into the blood, urine, bile, or lumen of the colon [100–103]. In therapies targeting the P-glycoprotein, the expression of the protein in normal tissues seems to be the most serious obstacle.

There may be some problems involved in the clinical application of anti-P-glycoprotein MAbs. However, it is clear that the MAbs are of considerable interest as candidates for therapeutic agents against MDR human cancers. At present, clinical phase I trials of MRK16 in the treatment of resistant tumors are under consideration.

References

1. Bell DR, Gerlach JH, Kartner N, Buich RN and Ling V (1985) Detection of P-glycoprotein in ovarian cancer: a molecular marker associated with multidrug resistance. J Clin Oncol 3: 311–315.
2. Fojo AT, Ueda K, Slamon DJ, Poplack DG, Gottesman MM and Pastan I (1987) Expression of a multidrug-resistance gene in human tumors and tissues. Proc Natl Acad Sci USA 84: 265–269.
3. Tsuruo T, Sugimoto Y, Hamada H, Roninson I, Okumura M, Adachi K, Morishima Y and Ohno R (1987) Detection of multidrug resistance markers, P-glycoprotein and *mdr*-1 mRNA, in human leukemia cells. Jpn J Cancer Res 78: 1415–1419.
4. Hamada H and Tsuruo T (1988) Purification of the 170- to 180-kilodalton membrane glycoprotein associated with multidrug resistance: the 170- to 180-kilodalton membrane glycoprotein is an ATPase. J Biol Chem 263: 1454–1458.
5. Hamada H and Tsuruo T (1988) Characterization of the ATPase activity of the 170- to 180-kilodalton membrane glycoprotein (P-glycoprotein) associated with multidrug resistance. Cancer Res 48: 4926–4932.
6. Shen D-W, Fojo A, Roninson IB, Chin JE, Soffir R, Pastan I and Gottesman MM (1986) Multidrug resistance in DNA mediated transformants is linked to transfer of the human *mdr*-1 gene. Mol Cell Biol 6: 4039–4045.
7. Gros P, Neriah UB, Croop JM and Housman DE (1986) Isolation and expression of a complementary DNA (mdr) that confers multidrug resistance. Nature 323: 728–731.
8. Ueda K, Cardarelli C, Gottesman MM and Pastan I (1987) Expression of a full-length cDNA for the human "MDR1" gene confers resistance to chochicine, doxorubicin and vinblastine. Proc Natl Acad Sci USA 84: 3004–3008.
9. Tsuruo T, Iida-Saito H, Kawabata H, Oh-hara T, Hamada H and Utakoji T (1986) Characteristics of resistance to adriamycin in human myelogenous leukemia K562 resistant to adriamycin and in isolated clones. Jpn J Cancer Res 77: 682–692.
10. Akiyama S, Fojo A, Hanover J and Pastan I (1985) Isolation and genetic characterization of human KB cell lines resistant to multiple drugs. Somatic Cell Mol Genet 11: 117–126.
11. Beck WT, Mueller TJ and Tanzer LR (1979) Altered surface membrane glycoproteins in vinca alkaloid-resistant human leukemic lymphoblasts. Cancer Res 39: 2070–2076.
12. Weinstein RS, Jakate SM, Dominguez JM, Lebovitz MD, Koukoulis GK, Kuszak JR, Klusens LF, Grogan TM, Saclarides TJ and Roninson IB (1991) Relationship of the expression of the multidrug resistance gene product (P-glycoprotein) in human colon carcinoma to local tumor aggressiveness and lymph node metastasis. Cancer Res 51: 2720–2726.
13. Rochlitz CF, Lobeck H, Peter S, Reuter J, Mohr B, de KE, Huhn D and Herrmann R (1992) Multiple drug resistance gene expression in human renal cell cancer is associated with the histologic subtype. Cancer 69: 2993–2998.
14. Hamada H and Tsuruo T (1986) Functional role for the 170- to 180-kDa glycoprotein specific to drug-resistant tumor cells as revealed by monoclonal antibodies. Proc Natl Acad Sci USA 83: 7785–7789.
15. Broxterman HJ, Kuiper CM, Schuurhuis GJ, Tsuruo T, Pinedo HM and Lankelma J (1988) Increase of daunorubicin and vincristine accumulation in multidrug resistant human ovarian carcinoma cells by a monoclonal antibody reacting with P-glycoprotein. Biochem Pharmacol 37: 2389–2393.
16. Norbert K, Deanna E-P, Grace B and Victor L (1985) Detection of P-glycoprotein in multi-drug resistant cell line by monoclonal antibodies. Nature 316: 820–823.

17. Scheper RJ, Bulte JW, Brakkee JG, Quak JJ, Van der Schoot E, Balm AJ, Meijer CJ, Broxterman HJ, Kuiper CM, Lankelma J and Pinedo HM (1988) Monoclonal antibody JSB-1 detects a highly conserved epitope on the P-glycoprotein associated with multidrug resistance. Int J Cancer 42: 389–394.
18. Meyers MB, Rittmann GL, O'Brien JP and Safa AR (1989) Characterization of monoclonal antibodies recognizing a Mr 180,000 P-glycoprotein: differential expression of the Mr 180,000 and Mr 170,000 P-glycoprotein in multidrug-resistant human tumor cells. Cancer Res 49: 3209–3214.
19. Lathan B, Edwards DP, Dressler LG, Von Hoff DD and McGuire WL (1985) Immunological detection of Chinese hamster ovary cells expressing a multidrug resistance phenotype. Cancer Res 45: 5064–5069.
20. Robert JA, Kimberly S, Johannes B, Alfred S, Frank B and James C (1993) Monoclonal antibody to an external epitope of the human *mdr*-1 P-glycorprotein. Cancer Res 53: 310–317.
21. Mechetner EB and Roninson IB (1992) Efficient inhibition of P-glycoprotein-mediated multidrug resistance with a monoclonal antibody. Proc Natl Acad Sci USA 89: 5824–5828.
22. Georges E, Tsuruo T and Ling V (1993) Topology of P-glycoprotein as determined by epitope mapping of MRK16 monoclonal antibody. J Biol Chem 268: 1792–1798.
23. Tsuruo T, Hamada H, Sato S and Heike Y (1989) Inhibition of multidrug-resistant human tumor growth in athymic mice by anti P-glycoprotein monoclonal antibodies. Jpn J Cancer Res 80: 627–631.
24. Heike Y, Hamada H, Inamura N, Sone S, Ogura T and Tsuruo T (1990) Monoclonal anti-P-glycoprotein antibody-dependent killing of multidrug-resistant tumor cells by human mononuclear cells. Jpn J Cancer Res 81: 1155–1161.
25. Becker I, Becker KF, Meyermann R and Hollt V (1991) The multidrug-resistance gene *MDR*-1 is expressed in human glial tumors. Acta Neuropathol Berl 82: 516–519.
26. Carulli G, Petrini M, Marini A, Ambrogi F, Ucci G, Riccardi A, Luoni R and Grassi B (1990) P-glycoprotein expression in multiple myeloma. Haematologica 75: 288–290.
27. Kim R, Hirabayashi N, Nishiyama M, Aogi K and Toge T (1991) Clinical significance of P-glycoprotein expression analyzed by inununohistochemical staining in cancer tissues. Jpn J Surg 21: 590–593.
28. Nabors MW, Griffin CA, Zehnbauer BA, Hruban RH, Phillips PC, Grossman SA, Brem H and Colvin OM (1991) Multidrug resistance gene (*MDR*-1) expression in human brain tumors. J Neurosurg 75: 941–946.
29. FitzGerald DJ, Willingham MC, Cardarelli CO, Hamada H, Tsuruo T, Gottesman MM and Pastan I (1987) A monoclonal antibody Pseudomonas toxin conjugate that specifically kills multidrug resistant cells. Proc Natl Acad Sci USA 84: 4288–4292.
30. Hamada H, Miura K, Ariyoshi K, Heike Y, Sato S, Kameyama K, Kurosawa Y and Tsuruo T (1990) Mouse-human chimeric antibody against the multidrug transporter P-glycoprotein. Cancer Res 50: 3167–3171.
31. Georges E, Bradley G, Gariepy J and Ling V (1990) Detection of P-glycoprotein isoforms by gene-specific monoclonal antibodies. Proc Natl Acad Sci USA 87: 152–156.
32. Barrand MA and Twentyman PR (1992) Differential recognition of *mdr*-1b gene products in multidrug resistant mouse tumour cell lines by different monoclonal antibodies. Br J Cancer 65: 239–245.
33. Ariyoshi K, Hamada H, Naito M, Heike Y Seimiya H Maezawa K and Tsuruo T (1992) Mouse-human chimeric antibody MH-171 against the multidrug transporter P-glycoprotein. Jpn J Cancer Res 83: 515–521.
34. Finstad CL, Yin BW, Gordon CM, Federici MG, Welt S, Lloyd KO (1991) Some monoclonal antibody reagents (C219 and JSB-1) to P-glycoprotein contain antibodies to blood group A carbohydrate determinants: a problem of quality control for immunohistochemical analysis. J Histochem Cytochem 39: 1603–1610.
35. Pavelic ZP, Sever Z, Fontaine RN, Baker VV, Reising J, Denton DM, Pavelic L and Khalily M (1991) Detection of P-glycoprotein with JSB-1 monoclonal antibody in B-5 fixed and paraffin-embedded cell lines and tissues. Sel Cancer Ther 7: 49–58.
36. deVries E, Meijer C, Timmer BH, Berendsen HH, deLeij L, Scheper RJ and Mulder NH (1989) Resistance mechanisms in three human small cell lung cancer cell lines established from one patient during clinical follow-up. Cancer Res 49: 4175–4158.
37. Ramael M, van den Bossched J, Buysse C, van Meerbeeck J, Segers K, Vermeire P and van Marck E (1992) Immunoreactivity for P-170 glycoprotein in malignant mesothelioma and in non-neoplastic mesothelium of the pleura using the murine monoclonal antibody JSB-1. J Pathol 167: 5–8.
38. Chitnis M, Hegde U, Chavan S, Juvekar A and Advani S (1991) Expression of the multidrug transporter P-glycoprotein and *in vitro* chemosensitivity: correlation with *in vivo* response to chemotherapy in acute myeloid leukemia. Sel Cancer Ther 7: 165–173.
39. Volm M, Bak MJ, Efferth T, Lathan B and Mattern J (1988) Immunocytochemical detection of a resistance-associated glycoprotein in tissue culture cells, ascites tumors and human tumor xenografts by Mab 265/F4 Anticancer Res 8: 531–535.
40. Nabors MW, Griffin CA, Zehnbauer BA, Hruban RH Phillips PC, Grossman SA, Brem H and Colvin OM (1991) Multidrug resistance gene (*MDR*-1) expression in human brain tumors. J Neurosurg 75: 941–946.
41. Radosevich JA, Robinson PG, Rittmann GL, Wilson B, Leung JP, Maminta ML, Warren W, Rosen ST and Gould VE (1989) Immunohistochemical analysis of pulmonary and pleural tumors with the monoclonal antibody HYB-612 directed against the multidrug resistance (*MDR*-1) gene product, P-glycoprotein. Tumour Biol 10: 252–257.

42. Ito Y, Tanimoto M, Kumazawa T, Okumura M, Morishima Y, Ohno R and Saito H (1989) Increased P-glycoprotein expression and multidrug-resistant gene (*mdr*-1) amplification are infrequently found in fresh acute leukemia cells sequential analysis of 15 cases at initial presentation and relapsed stage. Cancer 63: 1534–1538.
43. Cenciarelli C, Currier SJ, Willingham MC, Thiebaut F, Germann UA, Rutherford AV, Gottesman MM, Barca S, Tombesi M, Morrone S, Santoni A, Mariani M, Ramoni C, Dupuis ML and Cianfriglia M (1991) Characterization by somatic cell genetics of a monoclonal antibody to the *MDR*-1 gene product (P-glycoprotein): determination of P-glycoprotein expression in multi-drug-resistant KB and CEM cell variants. Int J Cancer 47: 533–543.
44. Yanovich S, Hall RE and Gewirtz DA (1989) Characterization of a K562 multidrug-resistant cell line. Cancer Res 49: 4499–4503.
45. Danks MK, Metzger DW, Ashmun RA and Beck WT (1985) Monoclonal antibodies to glycoproteins of vinca alkaloid-resistant human leukemic cells. Cancer Res 45: 3220–3224.
46. Campos L, Guyotat D, Archimbaud E, Calmard OP, Tsuruo T, Troncy J, Treille D and Fiere D (1992) Clinical significance of multidrug resistance P-glycoprotein expression on acute nonlymphoblastic leukemia cells at diagnosis. Blood 79: 473–476.
47. Robey CS, Bruner JM and Cafferty LL (1991) P-glycoprotein expression in gastroesophageal adenocarcinomas, their metastases and surrounding mucosa: a mapping study. Mod Pathol 4: 694–697.
48. Schneider J, Bak M, Efferth T, Kaufmann M, Mattern J and Volm M (1989) P-glycoprotein expression in treated and untreated human breast cancer. Br J Cancer 60: 815–818.
49. Murphy LD, Herzog CE, Rudick JB, Fojo AT and Bates SE (1990) Use of the polymerase chain reaction in the quantitation of *mdr*-1 gene expression. Biochemistry 29: 10351–10356.
50. Rochlitz CF, de Kant E, Neubauer A, Heide I, Bohmer R, Oertel J and Huhn D (1992) PCR-determined expression of the *MDR*-1 gene in chronic lymphocytic leukemia. Ann Hematol 65: 241–246.
51. Cumber PM, Jacobs A, Hoy T, Fisher J, Whittaker JA, Tsuruo T and Padua RA (1990) Expression of the multiple drug resistance gene (*mdr*-1) and epitope masking in chronic lymphatic leukemia. Br J Haematol 76: 226–230.
52. Benson MC, Giella J, Whang IS, Buttyan R, Hensle TW, Karp F and Olsson C (1991) Flow cytometric determination of the multidrug resistant phenotype in transitional cell cancer of the bladder: implications and applications. J Urol 146: 982–986.
53. Heike Y, Okumura K and Tsuruo T (1992) Augmentation by bispecific $F(ab')_2$ reactive with P-glycorpotein and CD3 of cytotoxicity of human effector cells on P-glycoprotein positive human renal cancer cells. Jpn J Cancer Res 83: 366–372.
54. Padmanabhan R, Tsuruo T, Kane SE, Willingham MC, Howard BH, Gottesman MM and Pastan I (1991) Magnetic-affinity cell sorting of human multidrug-resistant cells. J Natl Cancer Inst 83: 565–569.
55. Chen C, Chin JE, Ueda K, Clark DP, Pastan I, Gottesman MM and Roninson IB (1986) Internal duplication and homology with bacterial transport proteins in the *mdr*-1 (P-glycoprotein) gene from multidrug-resistant human cells. Cell 47: 381–389.
56. Van der Bliek AM, Kooiman PM, Schneider C and Borst P (1988) Sequence of *mdr*3 cDNA encoding a human P-glycoprotein. Gene 71: 401–411.
57. Schinkel AH, Roelofs MEM and Borst P (1991) Characterization of the human *MDR*3 P-glycoprotein and its recognition by P-glycoprotein-specific monoclonal antibodies. Cancer Res 51: 2628–2635.
58. Roninson IB, Chin JE, Choi K, Gros P, Housman DE, Fojo A, Shen E-W, Gottesman MM and Pastan I (1986) Isolation of human mdr DNA sequences amplified in multidrug-resistant KB carcinoma cell. Proc Natl Acad Sci USA 83: 4538–4542.
59. Schinkel AH, Roelofs EM and Borst P (1991) Characterization of the human MDR3 P-glycoprotein and its recognition by P-glycoprotein-specific monoclonal antibodies. Cancer Res 51: 2628–2635.
60. Chin JE, Soffir R, Noonan KE, Choi K and Roninson IB (1988) Structure and expression of the human *MDR* (P-glycoprotein) gene family. Mol Cell Biol 9: 3808–3820.
61. Theiebaut F, Tsuruo T, Hamada H, Gottesman MM, Pastan I and Willingham MC (1987) Cellular localization of the multidrug resistance gene product in normal human tissues. Proc Natl Acad Sci USA 84: 7735–7738.
62. Croop JM, Raymond M, Hber D, Devault A, Arceci RJ, Gros P and Housman DE (1989) The three mouse multidrug resistance (mdr) genes are expressed in a tissue specific manner in normal mouse tissues. Mol Cell Biol 9: 1346–1350.
63. Adams DO, Hall T, Steplewski Z and Koprowski H (1984) Tumors undergoing rejection induced by monoclonal antibodies of the IgG2a isotype contain increased numbers of macrophages activated for a distinctive form of antibody-dependent cytolysis. Proc Natl Acad Sci USA 81: 3506–3510.
64. Pearson JW, Fogler WE, Volker K, Usui N, Goldenberg SK, Gruys E, Riggs CW, Komschlies K, Wiltrout RH, Tsuruo T, Pastan I, Gottesman MM and Longo DL (1991) Reversal of drug resistance in a human colon cancer xenograft expressing *MDR*-1 complementary DNA by *in vivo* administration of MRK16 monoclonal antibody. J Natl Cancer Inst 83: 1386–1391.
65. Rittmann GL, Yong MA, Sanders V and Mackensen DG (1992) Reversal of vinca alkaloid resistance by anti-P-glycoprotein monoclonal antibody HYB-241 in a human tumor xenograft. Cancer Res 52: 1810–1816.
66. Naito M, Tsuge H, Kuroko C, Koyama T, Tornida A, Tatsuta T, Heike Y and Tsuruo T (1993) Enhancement of cellular accumulation of cyclosporine by anti-P-glycoprotein monoclonal antibody MRK16 and synergistic modula-

tion of multidrug resistance. J Natl Cancer Inst 85: 311–317.

67. Meyers MB, Rittmann GL, O'Brien JP and Safa AR (1989) Characterization of monoclonal antibodies recognizing a Mr 180,000 P-glycoprotein: differential expression of the Mr 180,000 and Mr 170,000 P-glycoproteins in multidrug-resistant human tumor cells. Cancer Res 49: 3209–3214.
68. Rittmann GL, Yong MA, Sanders V and Mackensen DG (1992) Reversal of vinca alkaloid resistance by anti-P-glycoprotein monoclonal antibody HYB-241 in a human tumor xenograft. Cancer Res 52: 1810–1816.
69. Efferth T, Lathan B and Volm M (1991) Selective growth-inhibition of multidrug-resistant CHO-cells by the monoclonal antibody 265/F4. Br J Cancer 64: 87–89.
70. Ashorn P, Moss B, Weinstein JN, Chaudhary VK, FitzGerald DJ, Pastan I and Berger EA (1990) Elimination of infectious human immunodeficiency virus from human T-cell cultures by synergistic action of CD4-Pseudomonas exotoxin and reverse transcriptase inhibitors. Proc Natl Acad Sci USA 87: 8889–8893.
71. Batra JK, FitzGerald D, Gately M, Chaudhary VK and Pastan I (1990) Anti-Tac(Fv)-PE40, a single chain antibody Pseudomonas fusion protein directed at interleukin 2 receptor bearing cells. J Biol Chem 265: 15198–15202.
72. Berger EA, Clouse KA, Chaudhary VK, Chakrabarti S, FitzGerald DJ, Pastan I and Moss B (1989) CD4-Pseudomonas exotoxin hybrid protein blocks the spread of human immunodeficiency virus infection *in vitro* and is active against cells expressing the envelope glycoproteins from diverse primate immunodeficiency retroviruses. Proc Natl Acad Sci USA 86: 9539–9543.
73. Chaudhary VK, Gallo MG, FitzGerald DJ and Pastan I (1990) A recombinant single-chain immunotoxin composed of anti-Tac variable regions and a truncated diphtheria toxin. Proc Natl Acad Sci USA 87: 9491–9494.
74. Lorberboum GH, Garsia RJ, Gately M, Brown PS, Clark RE, Waldmann TA, Chaudhary VK, FitzGerald DJ and Pastan I (1990) IL2-PE664Glu, a new chimeric protein cytotoxic to human activated T lymphocytes. J Biol Chem 265: 16311–16317.
75. Nishioka Y, Sone S, Heike Y, Hamada H, Ariyoshi K, Tsuruo T, Ogura T (1992) Effector cell analysis of human multidrug resistant cell killing by mouse-human chimeric antibody against P-glycoprotein. Jpn J Cancer Res 83: 644–649.
76. Van Dijk J, Tsuruo T, Segal DM, Bolhuis RL, Colognola R, Van de Griend R, Fleuren GJ and Warnaar SO (1989) Bispecific antibodies reactive with the multidrug-resistance-related glycoprotein and CD3 induce lysis of multidrug-resistant tumor cells. Int J Cancer 44: 738–743.
77. Nitta T, Sato K, Yagita H, Okumura K and Ishii S (1990) Preliminary trial of specific targeting therapy against malignant glioma. Lancet 335: 368–371.
78. Gerlach JH, Bell DR, Karakousis C, Slocum HK, Kartner N, Rustum YM, Ling V and Baker RM (1987) P-glycoprotein in human sarcoma: evidence for multidrug resistance. J Clin Oncol 5: 1452–1460.
79. Goldstein LJ, Galski H, Fojo A, Willingham M, Lai SL, Gazdar A, Pirker R, Green A, Crist W, Brodeur GM, Lieber M, Crossman J, Gottesman MM and Pastan I (1989) Expression of a multidrug resistance gene in human cancers. J Natl Cancer Inst 81: 116–124.
80. Moscow JA, Fairchild CR, Madden MJ, Ransom DT, Wieand HS, O'Brien EE, Poplack DG, Cossman J, Myers CE and Cowan K (1989) Expression of anionic glutathione-S-transferase and P-glycoprotein genes in human tissues and tumors. Cancer Res 49: 1422–1428.
81. Dinota A, Tazzari PL, Michieli M, Visani G, Gobbi M, Bontadini A, Tassi C, Fanin R, Damiani D, Grandi M, Pileri S, Bolognesi A, Stirpe F, Baccarani M, Tsuruo T and Tura S (1990) *In vitro* bone marrow purging of multidrug-resistant cells with a mouse monoclonal antibody directed against Mr 170,000 glycoprotein and a saporin-conjugated antimouse antibody. Cancer Res 50: 4291–4294.
82. Aihara M, Aihara Y, Schmidt WG, Schmidt WI, Sikic BI, Blume KG and Chao NJ (1991) A combined approach for purging multidrug-resistant leukemic cell lines in bone marrow using a monoclonal antibody and chemotherapy. Blood 77: 2079–2084.
83. Tong AW, Lee J, Wang RM, Dalton WS, Tsuruo T, Fay JW and Stone MJ (1989) Elimination of chemoresistant multiple myeloma clonogenic colony-forming cells by combined treatment with a plasma cell-reactive monoclonal antibody and a P-glycoprotein reactive monoclonal antibody [published erratum appears in Cancer Res 50, 4551 (1990)] Cancer Res 49: 4829–4834.
84. Hamada H, Okochi E, Oh-hara T and Tsuruo T (1988) Purification of the Mr 22,000 calcium-binding protein (sorcin) associated with multidrug resistance and its detection with monoclonal antibodies. Cancer Res 48: 3173–3178.
85. Van der Bliek AM, Meyers MB, Biedler JL, Hes E and Borst P (1986) A 22-kd protein (sorcin/V19) encoded by an amplified gene in multidrug-resistant cells is homologous to the calcium-binding light chain of calpain. EMBO J 5: 3201–3208.
86. Beck WI, Mueller TJ and Tanzer LR (1979) Altered surface membrane glycoproteins in vinca alkaloid-resistant human leukemic lymphobalst. Cancer Res 39: 2070–2076.
87. Kartner N, Riordan JR and Ling V (1983) Cell surface P-glycoprotein associated with multidrug resistance in mammalian cell lines, Science 221: 1286–1288.
88. Richert N, Akijima DW, Shen DW, Gottesman MM and Pastan I (1985) Multiply drug-resistant human KB carcinoma cells have decreased amounts of a 75-kDa and 72-kDa glycoprotein. Proc Natl Acad Sci USA 82: 2330–2333.
89. Cianfriglia M, Cenciarelli C, Tombesi M, Barca S, Mariani M, Morrone S, Santoni A, Samoggia P, Alessio M and Malavasi F (1990) Murine monoclonal antibody recognizing a 90-kDa cell-surface determinant selectively lost by multidrug-resistant variants of CEM cells. Int J Cancer 45:

95–103.
90. Tsuruo T, Iida H, Tsukagoshi S and Sakurai Y (1981) Overcoming of vincristine resistance in P388 leukemia *in vivo* and *in vitro* through enhanced cytotoxicity of vincristine and vinblastine by verapamil. Cancer Res 41: 1967– 1972.
91. Tsuruo T, Iida H, Tsukagoshi S and Sakurai Y (1983) Potentiation of vincristine and adriamycin effects in human hemopoietic tumor cell lines by calcium antagonists and calmodulin inhibitors. Cancer Res 43: 2267–2272.
92. Rogan AM, Hamilton TC, Young RC, Klecker RW Jr and Ozols RF (1984) Reversal of adriamycin resistance by verapamil in human ovarian cancer. Science 224: 994–996.
93. Tsuruo T (1987) Treatment of tumors with calcium antagonists. ISI Atlas Sci 1: 325–327.
94. Tsuruo T (1991) Reversal of multidrug resistance by calcium channel blockers and other agents. In: Roninson IB (ed) Molecular and Cellular Biology of Multidrug Resistance in Tumor Cells, pp. 349-372, New York: Plenum Publishing Corp.
95. Ozols RF, Cunnian RE, Klecker RW, Hamilton TC, Ostchega Y, Parrillo JE and Young RC (1987) Verapamil and adriamycin in the treatment of drug-resistant ovarian cancer patients. J Clin Oncol 5: 641–647.
96. Cantwell B, Puamah P and Harris AL (1985) Phase I and phase II study of oral verapamil (VRP) and intravenous vindesine (VDN). Br J Cancer 52: 425.
97. Gottesman MM and Pastan I (1989) Clinical trials of agents that reverse multidrug-resistance. J Clin Oncol 7: 409–411.
98. Mickisch GH, Kossig J, Keilhauer G, Schlick E, Tschada RK and Alken PM (1990) Effects of calcium antagonists in multidrug resistant primary human renal cell carcinomas. Cancer Res 50: 3670–3674.
99. Thiebaut F, Tsuruo T, Hamada H, Gottesman MM, Pastan I and Willingham MC (1987) Cellular localization of the multidrug resistance gene product P-glycoprotein in normal human tissues. Proc Natl Acad Sci USA 84: 7735–7738.
100. Sugawara I, Nakahama M, Hamada H, Tsuruo T and Mori S (1988) Apparent stronger expression in the human adrenal cortex than in the human adrenal medulla of Mr 170,000-180,000 P-glycoprotein. Cancer Res 48: 4611–4614.
101. Cordon CC, O'Brien JP, Casals D, Rittman GL, Biedler JL, Melamed MR and Bertino JR (1989) Multidrug-resistance gene (P-glycoprotein) is expressed by endothelial cells at blood-brain barrier sites. Proc Natl Acad Sci USA 86: 695–698.
102. Thiebaut F, Tsuruo T, Hamada H, Gottesman MM, Pastan I and Willingham MC (1989) Immunohistochemical localization in normal tissues of different epitopes in the multidrug transport protein P170: evidence for localization in brain capillaries and crossreactivity of one antibody with a muscle protein. J Histochem Cytochem 37: 159–164.
103. Tatsuta T, Naito M, Oh-hara T, Sugawara I and Tsuruo T (1992) Functional involvement of P-glycoprotein in blood-brain barrier. J Biol Chem 267: 20383–20391.
104. Yang CP, DePinho SG, Greenberger LM, Arceci RJ and Horwitz SB (1989) Progesterone interacts with P-glycoprotein in multidrug-resistant cells and in the endometrium of gravid uterus. J Biol Chem 264: 782–788.

Address for correspondence: T. Tsuruo, The University of Tokyo, Institute of Molecular and Cellular Biosciences, Yayoi 1-1-1, Bunkyo-ku, Tokyo 113, Japan.

Cytotechnology **12**: 109–125, 1993.

Non-P-glycoprotein multidrug resistance in cell lines which are defective in the cellular accumulation of drug

Melvin S. Center
Division of Biology, Kansas State University, Manhattan, Kansas 66506, USA

Key words: drug transport, human tumor cell lines, membrane protein alterations, non-P-glycoprotein multidrug resistance, sequestration of drug, verapamil

Abstract

Non-Pgp mdr related to a defect in drug accumulation has now been documented in a number of different cell lines exposed to certain cytotoxic agents. In studies conducted thus far most isolates have been obtained after selection in either adriamycin or mitoxantrone. The work in this area is in its early stages and very little is known about the molecular events which contribute to this mode of drug resistance. At the present time no protein with drug binding properties comparable to Pgp has been identified in non-Pgp mdr isolates. Evidence based on the finding that all isolates do not respond in the same way to reversal agents such as verapamil suggests the possibility that more than one mechanism may exist for non-Pgp mdr. Future studies may thus reveal that cells contain a multiplicity of genes which upon transcriptional activation can function to alter drug transport processes and thus contribute to the development of mdr. Identifying and characterizing these genes will be important since they may function in transport systems of normal cells. The exact identity of proteins which contribute to non-Pgp mdr remains to be determined. One protein designated P190 has been found to be overexpressed in cell lines of human promyelocytic leukemia, lung and adenocarcinoma treated with adriamycin. The protein also is increased in some clinical samples from patients undergoing chemotherapy. P190 which has a minor sequence homology with Pgp can bind ATP and may thus contribute to the energy dependent drug efflux systems found in cells containing this protein. Transfection studies with a P190 cDNA should determine whether this protein actually contributes to drug resistance. Many other protein changes have been detected in non-Pgp mdr cells but the importance of these in resistance also remains to be determined. In some systems a particular protein change can be identified in multiple independent isolates suggesting a correlation between the development of resistance and the presence of this cellular alteration.

Experiments conducted thus far on the mechanism of non-Pgp mdr are intriguing. Studies utilizing fluorescence microscopy to follow the fate of daunomycin suggests that the drug passes to the interior of the cell and eventually localizes in the Golgi apparatus. Drug located at this site may move directly into an efflux pathway for rapid extrusion from the cell. Evidence also indicates that as drug leaves the Golgi some may be sequestered into other organelles such as lysosomes or mitochondria. Sequestration may thus be another means of protecting the cell from the cytotoxic action of the drug. Very little is known of the molecular details of these events and some new technological approaches may be required to gain insight into efflux and sequestration pathways. *In vitro* systems for drug transport would certainly be important in these studies.

A major question to be answered in the future is whether non-Pgp mdr actually contributes to clinical drug resistance. This will certainly be clarified as new probes which can selectively detect this type of resistance are developed. Some studies have shown that in experimental isolates a low level non-Pgp mdr can precede a Pgp mdr which appears after continuous treatment of cells with drug. Possibly these findings have clinical relevance.

Abbreviations: PM — plasma membranes; ER, endoplasmic reticulum; TPA — 12-0-tetradecanoylphorbol-13-acetate; DMSO — dimethyl sulfoxide; RA — retinoic acid; VP-16 — 4′-deimethylepipodophyllotoxin-9-(4,6-0-ethylidene)-β-D-glucopyranoside; MRP — Multidrug-resistance associated protein; IC_{50} — Amount of drug required to inhibit cellular growth by 50%; NBD — 7-chloro-4-nitrobenz-2-oxa 1,3-diazole; Pgp — P-glycoprotein; mdr — multidrug resistance; EGF — epidermal growth factor

Introduction

Treatment of experimental cell lines with a single cytotoxic agent such as an anthracycline or Vinca alkaloid often results in the development of a broad cross-resistance to a variety of agents which are distinct structurally and functionally (Biedler and Riehm, 1970; Ling and Thompson, 1974). This finding has proven to be of considerable interest since research in this area has provided insight into the basis of a similar type of multidrug resistance (mdr) which occurs in cancer patients undergoing chemotherapy. The molecular basis of mdr in many experimental isolates appears to be related to the ability of the cell to greatly decrease drug accumulation (Danø, 1973; Peterson *et al.* 1974; Skovsgaard, 1978). This occurs as a result of overexpression of a gene mdr which encodes a transmembrane protein, Pgp (Endicott and Ling, 1989). Pgp can apparently bind diverse cytotoxic agents and in an energy requiring reaction pump this material from the cell (Endicott and Ling, 1989). The mechanism by which Pgp accomplishes this feat remains unknown.

In the past few years evidence has been obtained that cell lines treated with a cytotoxic agent can develop mdr as a result of reduced accumulation of drug but these cells are devoid of detectable levels of Pgp. Understanding the molecular basis of this system should be of major importance in defining the diversity of drug transport mechanisms and cellular genes which contribute to this process. An analysis of this system may also provide new insight into the basis of clinical mdr. In this review, unless stated otherwise, non-Pgp mdr will refer only to those resistant cell lines which are devoid of Pgp but contain a defect in drug accumulation. Other types of non-Pgp mdr which may involve topoisomerases or glutathione transferase are reviewed in other chapters of this issue.

Non-Pgp mdr cell lines

A list of cell lines selected for multidrug resistance and which are defective in drug accumulation in the absence of Pgp is given in Table 1. A recent review has appeared which describes some properties of essentially all isolates developed since 1970 which are both Pgp positive and Pgp negative (Nielsen and Skovsgaard, 1992). Cells exhibiting a drug accumulation defect and subsequently shown to be Pgp negative were first described in 1985–86 with the selection of adriamycin resistant HL60 cells, HL60/AR (Bhalla *et al.*, 1985), HL60/ADR (Marsh *et al.*, 1985) and an adriamycin resistant human lung cell isolate COR-L23/R (Twentyman *et al.*, 1986).

Analysis of daunomycin accumulation in these cells reveals that initially the levels of drug in sensitive and resistant cells are the same. During longer incubation periods sensitive cells continue to accumulate drug whereas daunomycin levels in resistant cells reaches a plateau or declines (Bhalla *et al.*, 1985; Marsh *et al.*, 1986; Coley *et al.*, 1991). Steady state drug levels in the resistant

Table 1. Non-P-glycoprotein mdr cell lines which exhibit a defect in drug accumulation

Isolate	Drug used for selection[a]	Parental cell line	Verapamil reversal of resistance[b]	Reference
HL60/AR	ADR (100)	human leukemia, promyelocyte	+	Bhalla *et al.*, 1985
HL60/ADR	ADR (10)	human leukemia, promyelocyte	+	Marsh *et al.*, 1986
COR-L23/R	ADR (17)	human large cell lung	+	Twentyman *et al.*, 1986
GLC4/R	ADR (32)	human small cell lung	ND	Zijlstra *et al.*, 1987
SK-MEL-170/R	ADR (178)	human melanoma	ND	Panneerselvan *et al.*, 1987
H69/AR[c]	ADR (32)	human small cell lung	–	Mirski *et al.*, 1987
HT1080/R	ADR (129)	human fibrosarcoma	+	Slovak *et al.*, 1988
SW1573/120	ADR (10)	human squamous lung	–	Keizer *et al.*, 1989
PC4/40	ADR (40)	murine erythroleukemia	+	Slapak *et al.*, 1990
WiDr/R	Mitox (21)	human colon	–	Dalton *et al.*, 1988
MCF-7/Mitox	Mitox (120)	human breast	–	Taylor *et al.*, 1991
MCF-7/MX	Mitox (4,000)	human breast	–	Nakagawa *et al.*, 1992
KB/40a	VP16 (287)	human nasopharyngeal	+	Ferguson *et al.*, 1988
G5	MGBG (40)	Adenovirus transformed rat brain cells	–	Weber *et al.*, 1989
LALW-2	Therapy induced	human leukemic T-cells	ND	Haber *et al.*, 1989

[a]Drug designations: ADR — Adriamycin; Mitox — Mitoxantrone; MGBG — methylglyoxalbis-(quanylhydrazone). Numbers in parentheses refer to level of resistance to the selecting agent. [b]Reversal of resistance by verapamil is indicated by a + sign. The relative extent of resistance reversal varies, however, with different cell lines (see text). A + indicates that verapamil is capable of inducing an increase in drug accumulation and/or reducing drug IC_{50} values. A – sign indicates that verapamil has no effect on resistance levels. ND = effect of verapamil on resistance was not determined. [c]Recent studies indicate that H69/AR cells overexpress a protein (MRP) which has sequence homology with Pgp and other proteins which may be involved in transport processes (Cole *et al.*, 1992). Although H69/AR cells are not defective in drug accumulation (Cole *et al.*, 1991) the MRP protein may function to partition drug away from its cytotoxic target. This sequestration process may be operative in other non-Pgp mdr cell lines.

isolates are about 50% of that found in sensitive cells. Further analysis clearly demonstrates that the resistant cells contain an energy dependent drug efflux pump which presumably contributes to the observed reduction in steady state intracellular drug levels (Bhalla *et al.*, 1985; Marsh *et al.*, 1986; Hindenburg *et al.*, 1989; Coley *et al.*, 1991). The cells are cross-resistant to a variety of agents including daunomycin, vincristine, actinomycin D and colchicine. It is expected that the mechanism of resistance to these agents is also related to reduced drug accumulation. At the present time there is no information as to the basis of this crossresistance pattern. Despite the presence of a drug efflux system HL60/ADR, (McGrath and Center, 1987; Marsh and Center, 1988; McGrath and Center, 1988) HL60/AR (Cass *et al.*, 1989) and COR-L23/R (Reeve *et al.*, 1990) do not contain detectable levels of Pgp as determined with the anti-Pgp antibody C219 (Kartner *et al.*, 1985) in Western blot analysis. HL60/ADR cells exhibiting an 80-fold increase in resistance to adriamycin also do not contain Pgp as determined with Western blot analysis using a number of antisera prepared against synthetic peptides containing the deduced sequence of Pgp (Marquardt *et al.*, 1990). Furthermore, there is no detectable overexpression of mdr1 or mdr2 in 80-fold resistant HL60/ADR (McGrath *et al.*, 1989) or 100-fold resistant HL60/AR cells (Cass *et al.*, 1989). The absence of mdr1 expression in HL60/AR has also been documented in studies using the polymerase chain reaction (Gervasoni *et al.*, 1992). In contrast to HL60/ADR, HL60 cells isolated for resistance to vincristine (HL60/Vinc) contain an amplified mdr1 and high levels of Pgp (McGrath and Center, 1988). At the present time non-Pgp mdr has not been described in cell lines selected for resistance to Vinca alkaloids. It has also been observed that extensive treatment of HL60/ADR with vincristine or adriamycin still does not result in the presence of Pgp (M. Center, unpublished). As described below this is not the case for certain other cell lines in which non-Pgp mdr can be followed by the appearance of Pgp during continued exposure of cells to increasing levels of drug.

In the past few years several additional non-Pgp mdr isolates have been obtained after treating a variety of cell lines with increasing levels of adriamycin. The isolates GLC4/R (Zijlstra *et al.*, 1987) SK-MEL-170/R (Panneerselvam *et al.*, 1987) and HT1080/R (Slovak *et al.*, 1987) derived from human small cell lung, human melanoma and human fibrosarcoma cells respectively exhibit a major reduction in the cellular accumulation of drug and this seems to occur as a result of an enhanced drug efflux system. Development of mdr of SK-MEL-170 cells is also associated with reduced expression of GD3 ganglioside sites on the cell surface and a decreased binding of the complement component C3b (Panneerselvam *et al.*, 1987). The susceptibility of the resistant cells to killing by the R24 monoclonal anti-GD3 antibody and human complement is reduced about 60%. It was also observed that SK-MEL-170/R cells are resistant to the complement-enhancing activity of free and immobilized adriamycin. Thus exposure of SK-MEL-170 cells to adriamycin results in multiple phenotypical changes but the genetic relatedness of these alterations remains to be determined. A series of isolates selected for low and high level resistance to adriamycin have been derived from the human squamous lung cell line SW1573 (Keizer *et al.*, 1989; Versantvoort *et al.*, 1992). It was originally reported that all isolates contained Pgp (Keizer *et al.*, 1989) but later studies revealed that cells exhibiting about a 10-fold increase in resistance to adriamycin (SW1573/120) were actually negative for mdr1 expression (Versantvoort *et al.*, 1992). This early discrepancy seems to be related to a false positive detection of Pgp using the JSB-1 antibody (Scheper *et al.*, 1988). SW1573/120 cells are defective in the accumulation of drug and this seems to be reversed under conditions of ATP depletion (Versantvoort *et al.*, 1992). However, there is only a modest difference in the rate of efflux of [^{14}C]daunorubicin from sensitive and resistant cells. Thus although SW1573/120 cells seem to accumulate less drug than sensitive cells in an energy requiring reaction this particular property may not be related to an enhanced efflux system.

Resistance of GLC4/R and SW1573/R to agents such as anthracyclines and etoposides is also related to alterations in topoisomerase II (Zijlstra *et al.*, 1987; Eijdems *et al.*, 1992). Genetic transfer studies with the non-Pgp mdr SW1573/50 using somatic cell fusions demonstrate that distinct genetic lesions contribute to reduced drug accumulation and altered topoisomerase II (Eijdems *et al.*, 1992).

A non-Pgp mdr isolate has been selected after treatment of the human small lung cell line H69 with adriamycin (Mirski *et al.*, 1987). The cells are cross-resistant to etoposides and Vinca alkaloids, and contain reduced levels of topoisomerase II. Transport studies with radioactively labeled daunomycin, VP-16 or vinblastine suggest that the H69/AR isolate is not defective in cellular drug accumulation (Cole *et al.*, 1991). A cDNA has recently been cloned from mRNA of H69/AR cells which encodes a protein which has sequence homology with Pgp and other members of a family of proteins which function in a variety of transport processes (Cole *et al.*, 1992). The gene encoding this protein MRP is greatly overexpressed in the resistant isolate. A role for this gene in resistance of H69/AR is suggested by the finding that the level of expression of MRP is reduced as cells revert to drug sensitivity (Cole *et al.*, 1992). The function of MRP is not known, but it has been suggested that the encoded protein sequesters drug into certain organelles and thus away from potentially toxic targets such as the nucleus. It would seem indicated, however, that continued sequestration of drug would eventually have a deleterious effect on the cell. Since drug accumulation studies were conducted only during a period of 1 h (Cole *et al.*, 1991) it is possible that after a prolonged incubation a mechanism exists for removing drug from the cell. Also it is not known if sequestration occurs only with vincristine and resistance to anthracyclines and etoposides occurs as a result of altered topoisomerase levels.

Two studies have been carried out to analyze

mdr in a series of isolates selected for different levels of adriamycin resistance. Parental SW1573 cells (Table 1) contain a low level of mdr1 expression but this disappears as cells are selected for a 5—20-fold increase in resistance to adriamycin (Baas *et al.*, 1990). The low level resistant cells contain reduced topoisomerase II activity and are also defective in drug accumulation in the absence of Pgp (Baas *et al.*, 199; Eijdems *et al.*, 1992). Continued treatment of cells with adriamycin to a 50-fold increase in resistance results in the overexpression of mdr1 and the presence of Pgp (Baas *et al.*, 1990). Thus with SW1573 there is a sequential change in resistance properties as cells are exposed to increasing levels of adriamycin. Cells treated for prolonged periods with drug are thus resistant as a result of both non-Pgp and Pgp mechanisms for reducing drug levels and also as a result of a decrease in topoisomerase II. Slapak *et al.*, (1990) have conducted similar experiments with PC4 murine erythroleukemia cells and have analyzed sublines selected for resistance through a stepwise increase in the levels of adriamycin. Cells exhibiting a 44-fold increase in resistance have a non-Pgp mechanism for reducing drug accumulation. Raising the level of resistance to 98-fold results in the appearance of high levels of Pgp.

The result of these two studies have important implications relative to identifying mechanisms of resistance in experimental isolates or clinical samples. Thus based on these findings high level resistant cells which contain Pgp may also contain a non-Pgp mechanism which reduces cellular drug levels. This may be of particular importance in clarifying mechanisms of mdr in patients since a low level non-Pgp mode of resistance may precede a higher level of resistance in which cells contain Pgp.

Non-Pgp mdr cells have also been obtained after treatment of MCF-7 breast cancer cells or WiDr human colon carcinoma cells with the anthracenedione mitoxantrone. The isolates MCF7/Mitox (Taylor *et al.*, 1991), MCF7/MX (Nakagawa *et al.*, 1992) and WiDr/R (Dalton *et al.*, 1988) (Table 1) are all defective in drug accumulation, exhibit enhanced levels of energy dependent drug efflux in the absence of detectable Pgp. MCF7/Mx has an interesting cross-resistance pattern in that the cells are 4,000-fold resistant to mitoxantrone but exhibit only a 10-fold increase in resistance to adriamycin and VP-16. The cells are not cross-resistant to vinblastine (Nakagawa *et al.*, 1992). In contrast, MCF7/Mitox cells which are 1208-fold resistant to mitoxantrone have a substantial crossresistance to vinblastine (43-fold) and a slight resistance to adriamycin (8.3-fold) (Taylor *et al.*, 1991). It is also noteworthy that HL60 cells isolated for a 35-fold increase in resistance to mitoxantrone are also Pgp negative but the cells have no defect in drug accumulation (Harker *et al.*, 1989). The mechanism of resistance in these cells is not known. It seems indicated, however, that at least compared to anthracyclines and Vinca alkaloids mitoxantrone has a weak capability of inducing overexpression of mdr1.

Ferguson *et al.*, (1988) have established a series of non-Pgp mdr isolates of human KB cells selected for resistance to VP-16. Isolates of this series are cross resistant to agents such as adriamycin and vincristine and exhibit a major reduction in the cellular accumulation of drug. Despite reduced levels of drug the rate of efflux of [^{3}H]VP-16 from sensitive and resistant cells is essentially the same. These experiments are, however, complicated by the finding that within a 10-min efflux period close to 80% of the drug is lost from sensitive cells. It has also been observed that development of VP-16 resistance in KB cells is accompanied by reduced levels of topoisomerase II as determined both by Western blot analysis and measurement of enzyme activity (Ferguson *et al.*, 1988).

A multidrug resistant isolate which is Pgp negative and which is defective in drug accumulation has been obtained after treating adenovirus-transformed rat brain cells with methylglyoxal-bis(quanyl hydrazone) MGBG (Weber *et al.*, 1989). Cells isolated for a 30-fold resistance to MGBG are cross-resistant to a variety of other drugs including adriamycin (10-fold) colchicine (sevenfold) and vincristine (102-fold). The cells are highly defective in [^{3}H]colchicine accumulation but do not exhibit increased levels of drug efflux. These authors did not, however, find major differences in efflux of drug from sensitive AUXB1 and a highly resistant P-glycoprotein containing isolate CH^RC^5. Thus the basis of the accumulation defect in cells isolated for resistance to MGBG is not known.

Mechanisms of mdr have been examined in leukemic cells obtained from a patient who relapsed during treatment with a variety of drugs, including vincristine (Haber *et al.*, 1989). The cells LALW-2 were maintained by serial xenograft in nude mice and were not further exposed to drug treatment. The LALW-2 cells isolated in this way are highly resistant to Vinca alkaloids and actinomycin D but not to daunomycin, a drug not used in the treatment protocol. Resistance levels of these cells were obtained by using growth inhibition assays and comparing these results to those obtained in parallel experiments with the T-lymphoblast cell line CCRF-CEM. Also in comparison with CEM cells, LALW-2 is defective in accumulation of [^{3}H]colchine but this is not related to increased levels of drug efflux (Haber *et al.*, 1989). The results suggest that LALW-2 cells exhibit non-Pgp mdr but this is related to a membrane defect which reduces movement of drug into the cell.

Agents capable of reversing non-Pgp mdr

At the present time very little information has been obtained on agents capable of reversing drug accumulation defective non-Pgp mdr. The most extensive effort in this area has been to examine the resistance reversal activity of verapamil, an agent which is highly active in sensitizing Pgp containing mdr cells to cytotoxic drugs (Tsuruo *et al.*, 1981). As indicated in Table 1, verapamil can reverse resistance in some isolates but in others there is no response. This may indicate that there are multiple modes of non-Pgp mdr. The reversal effect of verapamil also appears to be related to the particular drug under study and the length of time cells are exposed to this agent (Cass *et al.*, 1989). The action of verapamil in reducing drug IC_{50} values appears to be highly selective for resistant isolates since this agent has little effect on the parental cell lines. Two isolates in which verapamil induces major resistance related phenotypic changes are HL60/AR and HL60/ADR cells. Treatment of either with this agent greatly increases drug accumulation and retention (Bhalla *et al.*, 1985; McGrath *et al.*, 1989) and studies with HL60/AR show that verapamil induces a 17-fold reduction of vincristine IC_{50} values (Cass *et al.*, 1989). Verapamil also reduces by 7-fold and 10-fold IC_{50} values for VP-16 and vincristine respectively in KB/40a (Ferguson *et al.*, 1988). Other non-Pgp mdr lines respond to verapamil but the effect seems to be somewhat less than that found for the HL60 adriamycin resistant cells and KB/40a. Thus in the presence of verapamil drug IC_{50} values are decreased about 3—4-fold in COR-L23/R (Coley *et al.*, 1991), HT1080/R (Slovak *et al.*, 1988) and PC40 (Slapak *et al.*, 1990). Cyclosporin A and its analog PSC-833 are also capable of reducing IC_{50} values in COR-L23/R but this effect is slightly less than that found with verapamil (Barrand *et al.*, 1992). With a number of other cell lines, particularly those selected for resistance to mitoxantrone, verapamil has essentially no effect in reducing drug IC_{50} values (Table 1).

Previous studies have shown that a photoaffinity analog of verapamil is capable of binding Pgp (Safa, 1988) and this probably occurs at sites on the protein which interacts with drug (Kessel and Wilberding, 1984). In similar types of experiments a photoactive derivative of verapamil did not bind to any resistance associated protein in membranes of HL60/ADR cells (McGrath *et al.*, 1989). Similar results were obtained with the photoactive calcium channel blocker azidopine (McGrath *et al.*, 1989), an agent also capable of binding to Pgp (Safa *et al.*, 1987). It thus seems possible that the action of verapamil in reversing non-Pgp mdr is not related to an interaction with a specific resistance associated drug transporter.

Recently, a number of additional agents have been tested for resistance reversal activity in HL60/ADR cells. Lysosomotropic agents such as chloroquine, methylamine or ammonium chloride have essentially no effect on levels of drug accumulation in sensitive or resistant cells (Marquardt and Center, 1992). Similar results were obtained with brefeldin A, an agent capable of disrupting Golgi (Fugiwara *et al.*, 1988). Although brefeldin A does not disrupt drug transport in HL60/ADR fluorescent microscopic studies indicates that daunomycin localizes in resistant cells in a perinuclear region and studies using organelle specific stains indicates that this site corresponds to the Golgi (Marquardt and Center, 1992; D. Marquardt, unpublished). Golgi may thus play a role in drug distribution and efflux in HL60/ADR cells, but brefeldin A does not appear to inactivate this function. A major increase in drug accumulation and retention does, however, occur in HL60/ADR cells incubated in the presence of the protonophores nigericin or monensin (Marquardt and Center, 1992). Of these two agents, nigericin is considerably more effective than monensin in increasing drug levels and inhibiting efflux. Nigericin and monensin have also been found to have a similar effect on drug accumulation and retention in mdr cell lines which contain Pgp (Klohs and Steinkampf, 1988a; Klohs and Steinkampf, 1988b; Sehested *et al.*, 1988). The function of these agents in reversing mdr is not known. Some evidence obtained by fluorescent microscopic analysis of daunomycin distribution in HL60/ADR, HL60/AR and COR-L23/R cells suggests that drug is localized in a perinuclear region which may be part of the Golgi apparatus (Hindenburg *et al.*, 1989; Marquardt and Center, 1992; Barrand, *et al.*, 1992). Possibly protonated drug becomes entrapped in acidic vesicles emanating from Golgi and is removed from the cell via an exocytotic pathway. Previous studies have shown that nigericin and monensin can disrupt Golgi and block transport of secretory vesicles to the cell surface (Tartakoff and Vassali, 1970; Tartakoff, 1983). Thus the ability of these protonophores to interfere with Golgi transport processes may contribute to their inhibitory effect on drug efflux systems. In this regard it has been shown that verapamil, monensin and chloroquine are capable of inhibiting secretion of very low density lipoprotein by a mechanism which may involve disruption of transmembrane proton gradients and a resulting swelling and distension of Golgi (Rustan *et al.*, 1987). Thus in certain non-Pgp mdr cells nigericin, monensin and verapamil may affect Golgi and thus block a vesicular drug transport pathway. If this is the case, the reason agents such as chloroquine do not affect drug transport in resistant cells is not known. Hindenburg *et al.*, (1987) have also suggested that reversal agents such as verapamil may alter solubility properties of drugs such as adriamycin and thereby change their cellular hydrophobic/hydrophilic interactions. Under these conditions the drugs may distribute in resistant cells similar to their sensitive counterparts and thus exhibit a cytotoxic effect. Some support for the concept that acidic vesicles are involved in drug transport is provided by the finding that bafilomycin A1, a selective inhibitor of vacuolar H^+-ATPase activity, at concentrations below 10 μM (Bowman *et al.*, 1988) induces a major increase in drug accumulation and retention in both HL60/ADR (Pgp negative) and HL60/Vinc (Pgp positive) cells (Marquardt and Center, 1991). The ability of bafilomycin A1 to enhance drug levels in HL60/Vinc does not appear to occur as a result of this agent binding to Pgp. This is based on the finding that bafilomycin A1 does not compete with azidopine for binding to Pgp, at concentrations where it is active in enhancing drug levels (Marquardt and Center 1991). NBD, an inhibitor of vacuolar H^+-ATPase, F_1F_0 mitochondrial ATPase and other ATPases as well (Mellman *et al.*, 1986) also induces an increase in drug accumulation and blocks efflux in the resistant isolates. However, selective inhibitors of F_1F_0 and E_1E_2 ATPase do not affect drug levels suggesting that the effect of NBD in reversing resistance may be related to its ability to inhibit vacuolar H^+-ATPases (Marquardt and Center, 1991). These studies have recently been extended with an analysis of vacuolar H^+-ATPase gene expression in sensitive and resistant HL60 cells. In eukaryotic cells vacuolar proton ATPase is a highly complex structure with the catalytic component consisting of five polypeptides (Nelson, 1989). Although this enzyme system has not been extensively studied in human cells, a cDNA of vacuolar H^+-ATPase subunit C from human brain tissue has been cloned and sequenced (Nelson *et al.*, 1990). Studies using this cDNA as a probe have shown that selection of HL60 cells for resistance to adriamycin or vincristine results in a major overexpression of the subunit C-H^+-ATPase mRNA (Ma and Center, 1992). In drug sensitive cells only very low levels of this mRNA can be detected. The levels of expression of the subunit C gene is greatly reduced as HL60/Vinc cells revert to drug sensitivity suggesting a correlation between the levels of the subunit C protein and the development of mdr. Although HL60/Vinc cells contain an amplification of mdr1, the subunit C gene is not amplified in these cells. In view of the multi-subunit structure of vacuolar H^+-ATPase it seems unlikely that selection of cells for drug resistance would result in an alteration in the levels of expression of all genes which encode proteins of the complex. Consistent with this is the finding that the gene encoding P32, an accessory subunit protein of vacuolar H^+-ATPase (Want *et al.*, 1988), is not overexpressed in resistant HL60/ADR cells (Ma and Center, 1992). These results taken together with the finding that bafilomycin can reverse resistance in HL60/ADR and HL60/Vinc suggest that vacuolar H^+-ATPase activity may contribute to drug transport processes in these resistant cells. This enzyme system has been shown to play a major role in regulating the internal acidic pH at a number of endomembrane sites including endoplasmic reticulum, Golgi, endosomes and lysosomes (Mellman *et al.*, 1986; Nelson 1989). Vacuolar H^+-ATPase may thus be important in maintaining an acidic environment in vesicles which have accumulated drug. The protonated form of the drug would be retained by the vesicle and eventually removed from the cell. Inhibitors of this enzyme activity could thus alter intravesicular pH and the transport of drug contained within vesicles would be disrupted.

Buthionine sulfoximine (BSO), an agent capable of reducing cellular glutathione levels, is capable of enhancing drug accumulation and retention in HL60/AR cells (Lutzky *et al.*, 1989). BSO does not affect drug accumulation in sensitive cells, although treatment with this agent reduces glutathione as in the resistant isolate. At the present time the mechanism by which BSO reverses resistance is not known.

Protein changes occurring in non-Pgp mdr

Although protein changes have been detected in certain non-Pgp mdr isolates it has been difficult to prove which of these actually contributes to resistance. Particularly lacking are non-Pgp mdr cells which have reverted to drug sensitivity, thus permitting a correlation between the presence of a protein and the levels of resistance. In the absence of revertants, attempts have been made to demonstrate that a particular protein change occurs in several independent non-Pgp mdr isolates. Detailed studies have been carried out to analyze cellular changes occurring in HL60 cells isolated for resistance to adriamycin. It was found early on that with surface membrane labeling techniques a resistance associated protein of 150–160 kd (P150) can be detected in plasma membranes of both HL60/AR and HL60/ADR cells (Bhalla *et al.*, 1985; Marsh *et al.*, 1986). Extensive analysis of P150 using HL60/ADR cells shows that it is phosphorylated *in vivo* and also *in vitro* in isolated plasma membranes or endoplasmic reticulum (Marsh *et al.*, 1986; McGrath and Center, 1988). P150 is also glycosylated since the *in vitro* phosphorylated protein binds tightly to BS-11 lectin (McGrath and Center, 1988). Analysis of [^{14}C]-glucosamine labeled plasma membrane proteins of sensitive and resistant cells reveals, however, that the levels of P150 proteins are essentially the same in the two cell types. Other studies have shown that [^{14}C]glucosamine labeled P150 from sensitive and resistant cells have identical glycopeptide maps after partial digestion with V8 protease and also have identical peptide maps after digestion of [^{125}I] labeled P150 with trypsin (Marsh and Center 1987). These results taken together suggest that the levels of P150 in sensitive and resistant cells are the same. However, as cells are selected for resistance, P150 is topologically altered in plasma membranes and possibly in endoplasmic reticulum as well. This topologically altered protein is possibly more accessible to the action of protein kinases and thus becomes phosphorylated in the resistant cell. HL60/AR cells contain twofold more protein kinase-C (PK-C) than the parental cell line (Aquino *et al.*, 1988) and this may be related to the presence of PK-Cγ in resistant but not sensitive cells (Aquino *et al.*, 1990a). Possibly PK-C contributes to the phosphorylation of P150. Using an *in vitro* phosphorylation system P150 can also be found in membranes of the Pgp containing isolate HL60/Vinc and in HL60/Vinc cells which have reverted to partial drug sensitivity (McGrath and Center, 1988). The revertant cells which exhibit about a 20-fold increase in resistance to vincristine contain extremely low levels of Pgp. It has also been observed that the P150 alteration occurs early during the selection process since the phosphorylated protein can be detected in HL60/ADR cells which exhibit a 10-fold increase in resistance to adriamycin (Marsh *et al.*, 1986). The results taken together suggest the possibility that phosphorylated P150 contributes to a low level of resistance and may function to reduce cellular drug levels. *In vitro* phosphorylation of P150 has also been examined in membranes isolated from HL60/ADR cells induced to differentiate in the presence of TPA, RA or DMSO (Marsh and Center 1986). Exposure of cells to these agents results in a major reduction in the phosphorylation level of this protein. Magnesium dependent protein kinase activity in membranes from cells treated with these agents is not altered relative to untreated membranes under conditions where there is a major decrease in P150 phosphorylation. Treatment of HL60/AR and the parental line with TPA does result, however, in an 80–90% decrease in PK-C activity (Aquino *et al.*, 1988). The basis of the finding that phosphorylation of P150 is reduced after inducer mediated differentiation of adriamycin resistant HL60 cells is not known. Possibly differentiation results in a reduction in the levels of P150 or the kinase system which phosphorylates this protein. It has also been found that HL60/AR cells contain a Ca^{2+} dependent protease which is absent in drug sensitive cells (Aquino *et al.*, 1990b). This protease could function to degrade P150 during the process of differentiation.

Gervasoni *et al.*, (1991) have analyzed membrane glycoproteins in HL60/AR cells and have found that high molecular weight proteins of 210–160 kd are hypoglycosylated. It was suggested that this alteration of surface membrane proteins may affect entry and efflux of drug in resistant cells.

An antibody (GSB1) has been obtained which recognizes two resistance associated proteins of 150 and 130 kd which are present in plasma membranes of HL60/ADR (Krishnamachary and Center, 1992). Since the P150 detected by GSB1 is not present in sensitive cells the protein would appear to be distinct from the P150 phosphoglycoprotein described above. The GSB1 antibody also recognizes a 180 kd protein which is present in plasma membranes of sensitive cells and is slightly increased in resistant membranes. These same protein changes detected by GSB1 in HL60/ADR have also been found in several other isolates including HL60/AR, HL60/DAUNO (Krishnamachary and Center, 1992), HL60/Vinc and in HL60/Vinc cells which have partially reverted to drug sensitivity (Krishnamachary, unpublished). The GSB1 antibody was prepared against a mixture of proteins of defined molecular weight obtained after separation of membrane proteins by polyacrylamide gel electrophoresis. Evidence indicates (see below) that P180, P150 and P130 recognized by this antibody are structurally distinct. Of interest is the finding that the levels of proteins P180 and P150 are greatly increased during inducer mediated differentiation of drug sensitive HL60 cells. Thus TPA induction of differentiation along the monocyte/macrophage pathway results in a major increase in the levels of P180 and this occurs both in plasma membranes and endoplasmic reticulum (Krishnamachary and Center, 1992). In contrast, differentiation of HL60 cells to granulocytes in the presence of DMSO results in the appearance of high levels of P150 in plasma membranes of the differentiated cells (Krishnamachary and Center, 1992). For reasons that are not known, induction of HL60/ADR cells to differentiate in the presence of either TPA or DMSO does not affect the levels of any protein detected by GSB1. Selection of HL60 cells for resistance to adriamycin thus results in the overexpression of two genes which are related to distinct differentiation pathways. In resistant cells the genes encoding P180 and P150 no longer respond to signals transduced by DMSO or TPA. Understanding the molecular basis of the deregulation of genes encoding P180 and P150 could provide further insight into mechanisms by which HL60 cells respond to inducers of differentiation.

The contribution of proteins recognized by GSB1 to drug resistance remains to be determined.

HL60/ADR cells also contain a major increase in the levels of mRNA which encodes the ribosomal protein S25 (Li *et al.*, 1991; Li and Center, 1992). Ribosomal protein genes S14 or S17 are not, however, overexpressed in this isolate (Li and Center, 1992). Although S25 mRNA is increased in HL60/ADR the levels of the S25 protein are highly regulated and the amount of this material in sensitive and resistant cells is essentially the same (Li and Center, 1992). In view of these results it seems unlikely that the S25 protein contributes to resistance in HL60/ADR cells.

Membranes isolated from HL60/ADR contain a protein of 190 kd (P190) which is capable of binding the photoactive agent 8-azidoATP^{32} (McGrath *et al.*, 1989; Marquardt *et al.*, 1990). This protein which is present primarily in the endoplasmic reticulum of the resistant isolate is essentially absent in drug sensitive cells. Unlabeled ATP blocks the interaction of P190 with 8-azidoATP^{32}, suggesting that the protein contains sequences involved in nucleotide binding (McGrath *et al.*, 1989). Although P190 and Pgp (Cornwall *et al.*, 1987) both contain nucleotide binding sites, the two proteins are distinct with regard to their interaction with other agents. Thus, unlike Pgp (Safa *et al.*, 1986; Safa *et al.*, 1987), P190 is not capable of binding photoactive derivatives of vinblastine, verapamil or azidopine (McGrath *et al.*, 1989). Certain lines of evidence suggest, however, that Pgp and P190 contain a minor sequence homology. Evidence for this was obtained during a study of antisera prepared against peptides synthesized according to the deduced sequence of Pgp (Marquardt *et al.*, 1990). Western blot analysis of 13 different antisera prepared against Pgp peptides revealed that one (ASP14) was capable of reacting both with Pgp and P190 (Marquardt *et al.*, 1990). The ASP14 antiserum was prepared against a synthetic peptide containing the sequence GTQLSGGQKQRIAIA. This sequence is highly conserved in the P-glycoproteins from rodents to humans (Gros *et al.*, 1986; Chen *et al.*, 1986) and is contained in a region which may be part of a nucleotide binding site. At the present time, however, it is not known if the exact sequence of the ASP14 peptide is contained in P190 or if the protein contains an homologous region which can be recognized by the antiserum. Recently we have found that the MRP gene associated with mdr in H69/AR cells (Table 2) (Cole *et al.*, 1992) is overexpressed in the HL60/ADR isolate (Ma, unpublished). Using an oligonucleotide probe based on the MRP sequence two mRNA species of about 8 and 6 kd can be detected. The basis of these two distinct mRNA forms is not known. Based on the deduced sequence of a full length MRP cDNA, this gene encodes a protein containing 1531 amino acids (Cole *et al.*, 1992; Cole and Deeley, 1993). A region of this protein is homologous to part of the sequence contained in the ASP14 synthetic peptide used to prepare antiserum which recognizes P190 (Marquardt *et al.*, 1990). Recent studies have also shown that P190 is glycosylated since treatment of HL60/ADR cells with tunicamycin results in a major reduction in the levels of P190 and the formation of a protein of about 165 kd which is reactive with the ASP14 antiserum (Marquardt, unpublished). Based on these findings it thus seems possible that the 6.0 kb MRP mRNA encodes the 165 kd form of protein P190. Indirect evidence that P190 may contribute to mdr is based on the finding that the protein has now been identified in non-Pgp mdr isolates derived from a number of different cell lines (Table 2). In addition to HL60/ADR the protein has been found to be overexpressed in HL60/AR (Marquardt and Center, 1990), HL60/DAUNO (Marquardt, unpublished), COR-L23/R (Barrand *et al.*, 1992a), GLC-4/R (Barrand *et al.*, 1992b) and in the adenocarcinoma line MOR selected for resistance to adriamycin (Barrand *et al.*, 1992b).

At the present time, mechanisms regulating genes which are overexpressed in resistant HL60 cells are not known. Of interest is the finding that HL60/AR cells contain a major increase in the levels of the transcription factor Spl (Borellini *et al.*, 1990). Spl could possibly contribute to the enhanced transcriptional activity of genes overexpressed in the HL60 resistant isolates.

The ASP14 serum has also been used to analyze membrane protein changes in MCF-7 cells selected for a 4,000-fold increase in resistance to mitoxan-

Table 2. Some cellular changes detected in non-Pgp mdr cell lines

Protein designation[a]	Comments	Reference
P150 (HL60/AR) (HL60/ADR)	P150 appears to be contained in membranes of sensitive and resistant cells. In resistant cells the protein is hyperphosphorylated.	Bhalla *et al.*, 1985; Marsh *et al.*, 1986, Marsh *et al.*, 1987; McGrath and Center, 1988.
P190 (HL60/ADR) (COR-L23/R) (GLC4/R) (MOR/R)	Protein overexpressed in many non-Pgp mdr lines. P190 is located in ER, is capable of binding ATP and contains a minor sequence homology with Pgp.	McGrath *et al.*, 1989; Marquardt *et al.*, 1990; Barrand *et al.*, 1992; Barrand et al., 1993.
P180 (HL60/ADR) P150 (HL60/ADR) P130 (HL60/ADR)	PM proteins reacting with GSB1 antiserum and which are overexpressed in resistant cells. Levels of P180 and P150 are elevated in sensitive cells treated with TPA or DMSO respectively.	Krishnamachry and Center, 1992.
Vacuolar H^+-ATPase subunit C (HL60/ADR)	Subunit C-H^+-ATPase mRNA levels are increased in resistant cells. Levels of the subunit C protein are not known at the present time.	Ma and Center, 1992.
Ribosomal S25 (HL60/ADR)	Ribosomal S25 mRNA is increased in the HL60/ADR isolate. Levels of S25 protein are the same in sensitive and resistant cells.	Li and Center, 1992.
Hypoglycosylated membrane proteins (HL60/AR)	Many surface membrane proteins of HL60/AR cells seem to be hypoglycosylated.	Gervasoni *et al.*, 1991.
Spl (HL60/AR)	The transcription factor Spl is greatly increased in HL60/AR cells.	Borellini *et al.*, 1990.
EGF receptor (COR-L23/R)	The level of the EGF receptors of COR-L23/R is reduced compared to the parental cell line.	Reeve *et al.*, 1990.
P85, P42 (MCF-7/MX)	Two membrane proteins overexpressed in MCF-7/MX cells. Proteins may have some sequence homology with Pgp.	Nakagawa *et al.*, 1992.
P110 (SW1573/2R120) (GLC4/R)	Protein recognized by LRP56 antiserum which is over-expressed and contained in cytoplasm of certain non-Pgp mdr cell lines.	Versantvoort *et al.*, 1992.
MRP (H69/AR) (J2/R)	MRP gene is associated with mdr in the indicated resistant isolates. The gene encodes a protein which has some sequence homology to Pgp and other proteins involved in transport processes.	Cole *et al.*, 1992.

[a]Non-Pgp mdr cell lines in parentheses indicate an isolate which contains the designated change either at the protein or mRNA level. Some properties of these cell lines are given in Table 1. Protein (P) designations are given as the molecular mass in kilodaltons as determined by Western blot analysis.

trone (MCF-7/MX, Table 1). The antibody reacts in Western blots with two proteins of 85 kd and 42 kd which are contained in plasma membranes of resistant but not sensitive cells (Nakagawa *et al.*, 1992). These proteins may thus contain some sequence homology with Pgp and may contribute to the enhanced drug efflux system of the MCF-7/MX isolate. Versantvoort *et al.* (1992) have developed an antibody LRP56 which stains a cytoplasmic protein in non-Pgp SW1573/2R120 and

GLC4/R fixed cells. Sensitive cells stain only weakly with this antiserum. The protein recognized by LRP56 appears to be contained exclusively in the cytoplasm of resistant cells and has a molecular mass of about 110 kd (Versantvoort *et al.*, 1992). In certain mdr cell lines changes have been found to occur in membrane proteins of defined function. Thus it has been observed that the level of epidermal grow factor receptors is greatly reduced in COR-L23 cells selected for resistance to adriamycin (Table 1) (Reeve *et al.*, 1990). How this membrane alteration may contribute to resistance is not known.

Cytogenetic analyses of non-Pgp mdr cells

A few reports have appeared which describe the results of a cytogenetic analysis of non-Pgp mdr cells. Three independently derived resistant sublines of SW-1573 (Table 1) were found to contain a heterozygous deletion of the short arm of chromosome 2 (p23-pter) (Nieuwint *et al.*, 1992). In contrast, a karyotype analysis of HL60/AR cells reveals a large homogeneously staining region (HSR) in the long arm of chromosome 7 (7q11-2) and translocation of the remainder of the long arm to another centromere (Gervasoni *et al.*, 1992). Other chromosomal changes were also noted in these cells. The HSR found in HL60/AR cells is near the region of mdr1 on chromosome 7 (Trent and Witlcowski, 1987). Cytogenetic analysis of WiDr/R cells also indicates the presence of an HSR on chromosome 7 (Dalton *et al.*, 1988). At the present time it is not known if the various cytogenetic changes observed actually contribute to the mdr of the cell isolates examined.

Mechanisms of non-Pgp mdr

Attempts to explore pathways of drug transport in non-Pgp mdr cells have utilized primarily fluorescence microscopy to examine the distribution of daunomycin in sensitive and resistant cells. The results obtained thus far suggest that in resistant isolates drug passes into the interior of the cell and is thereafter distributed into certain organelles which function in the drug transport process. Convincing evidence accrued through the use of a variety of fluorescent microscopic techniques suggest that in HL60/AR (Hindenburg *et al.*, 1989; Gervasoni *et al.*, 1991), HL60/ADR (Marquardt and Center, 1992; Marquardt, unpublished) and COR-L23/R (Barrand *et al.*, 1992) daunomycin after entering the cell eventually concentrates in a perinuclear region which appears to be that of the Golgi. This is based primarily on the finding that the fluorescence localization of daunomycin is very similar to that of agents such as NBD-ceramide (Hindenburg *et al.*, 1989; Center, unpublished) and fluorescent-labeled wheat germ agglutinin (Barrand *et al.*, 1992) which preferentially stain Golgi. In contrast, drug sensitive cells localize and retain drug in the nucleus. There may be two pathways by which drug reaches the Golgi in these isolates. Daunomycin which enters HL60/AR and COR-L23/R seems to be partitioned into Golgi and is prevented from reaching the nucleus (Gervasoni *et al.*, 1991; Barrand *et al.*, 1992). Incubation of HL60/AR cells with daunomycin during continuous metabolic inhibition by sodium azide/deoxyglucose does not prevent uptake of fluorescence into the Golgi apparatus. Drug contained in Golgi of HL-60/AR eventually localizes in lysosomes and mitochondria and this seems to be an energy requiring reaction (Hindenburg *et al.*, 1989). In contrast, incubation of sensitive and HL60/ADR cells with daunomycin results in the localization of drug in the nucleus in both cell types (Marquardt and Center, 1992). However, under efflux conditions daunomycin is retained in nuclei of sensitive cells but is rapidly removed from nuclei of HL60/-ADR. Some drug which leaves the nucleus of resistant cells relocates to a perinuclear region which has staining properties similar to Golgi (Marquardt *et al.*, 1992; Marquardt, unpublished). Two other points are of interest with regard to mechanisms of drug transport in non-Pgp mdr. Thus it has been found that in HL60/AR cells which exhibit a 100—200-fold increase in resistance to adriamycin the intracellular content of drug is 50—70% of that of sensitive cells (Gervasoni *et al.*, 1991). The finding that resistant cells can accumulate high levels of daunomycin suggests that the intracellular distribution of drug plays a major role in resistance. At the present time the fate of this accumulated drug during prolonged incubation periods is not known. Also the relative importance of partitioning drug into organelles and the rapid efflux of drug from the cell to total levels of resistance remains to be determined. It is also of interest to note that efflux of drug from HL60/AR and HL60/ADR is greatly reduced as the temperature is lowered from 37° to 18°C (Gervasoni *et al.*, 1991; Marquardt, unpublished). The reduced drug efflux occurring at 18°C may be related to a decrease in membrane fusion which takes place at this temperature (Pagano and Sleight, 1985). Based on these findings it may be speculated that intracellular drug is contained within membrane vesicles which undergo a series of fusion events which are involved in partitioning drug to certain organelles or removing drug via an efflux pathway.

Drug distribution patterns have also been analyzed in a series of SW1513 sublines (Table 1) which exhibit a 10—2,000-fold increase in resistance to adriamycin (Keizer *et al.*, 1989). Fluorescent microscopic analysis of daunomycin distribution indicates that both sensitive and non-Pgp mdr cells exhibiting a 10-fold increase in resistance accumulate drug primarily in nuclei. At intermediate levels of resistance drug is distributed equally in cytoplasm and the nucleus whereas with increasing resistance daunomycin is contained primarily in the cytoplasm. This latter pattern is representative of cells containing Pgp (Keizer *et al.*, 1989). Ratios of drug in nucleus and cytoplasm of low and high level resistant isolates have been quantified using laser scan fluorescence microscopy and image analysis (Schuurhuis *et al*, 1991; de Lange *et al.*, 1992). These studies have shown that the nuclear/cytoplasmic drug ratios decrease with increasing levels of resistance and seem to be characteristic of a non-Pgp or Pgp mode of mdr.

The results of these studies taken together suggest a highly complex mechanism for drug transport in non-Pgp mdr cells. It is indicated that some drug which enters the cell is sequestered into certain organelles. In this way cellular targets of the drug would be protected. It is also indicated that intracellular drug enters a pathway which leads directly to extrusion to the extracellular space. Although very little is known about the molecular basis of these events, mounting evidence indicates that Golgi plays a role in the drug transport system. Possibly Golgi is a site where drug concentrates and thereafter moves to other organelles for sequestration or leaves the cell in vesicles via exocytosis.

Clinical multidrug resistance

Tumor cells of patients with various types of cancer often contain increased levels of Pgp and evidence indicates that a good correlation exists between the presence of this protein and the development of clinical mdr (Goldstein *et al.*, 1989). The extent to which clinical mdr involves Pgp is not known, but there is also clear evidence that in many cases where this protein is absent, patients are still nonresponsive to chemotherapy (Shinn-Liang *et al.*, 1989; Gruber *et al.*, 1992; Volm *et al.*, 1992). At the present time the importance of drug accumulation defective non-Pgp mdr to resistance in patients undergoing chemotherapy remains to be determined. One study which addresses this question has, however, been conducted. In this study, samples from 152 patients with adult sarcoma were analyzed for the presence of Pgp and the non-Pgp mdr associated protein P190 (Table 2) using a variety of antisera, including ASP14 (Baker *et al.*, 1993). Although many Pgp positive samples were detected, P190 was also found to be overexpressed in some cases. This protein was present in six of 40 Pgp-negative patients and four of five positives were nonresponsive to chemotherapy as compared to 12 of 23 P190 negative cases. Although many more clinical samples need to be tested P190 may contribute to a non-Pgp mode of clinical drug resistance.

Acknowledgements

I wish to thank all colleagues who sent reprint material for use in this review. Work conducted in the authors' laboratory was supported by Public

Health Service grant CA-37585, National Institutes of Health and by a grant from Bristol-Myers Squibb.

References

Aquino A, Hartman KD, Kvode MC, Grant S, Huang K-P, Niu C-H and Glazer RI (1988) Role of protein kinase C in phosphorylation of vinculin in adriamycin-resistant HL60 leukemia cells. Cancer Res 48: 3324–3329.

Aquino A, Warren B, Omichinski J, Hartman KD and Glazer RI (1990a) Protein kinase Cγ is present in adriamycin-resistant HL60 leukemia cells. Biochem Biophys Res Commun 166: 723–728.

Aquino A, Johnson-Thompson M and Glazer RI (1990b) Enhanced Ca^{2+}-dependent proteolysis associated with adriamycin resistant HL60 cells. Oncology Res 2: 243–247.

Baas F, Jongsma APM, Broxterman HJ, Arceci RJ, Housman D, Scheffer GL, Riethorst A, van Groenigen M, Nieuwint AWM and Joenje H (1990) Non-P-glycoprotein mediated mechanism for multidrug resistance precedes P-glycoprotein expression during *in vitro* selection for doxorubicin resistance in a human lung cancer cell line. Cancer Res 50: 5392–5398.

Baker RM, Chen Y, Richard B, Fredericks WJ, Cason E, Karakousis C, Center M, Slocum HK and Rustum Y (1993) Relation of multidrug resistance markers assessed by immunoblotting to clinical course for disease in adult sarcoma. Proc Am Assoc Cancer Res 34: 232.

Barrand MA, Tsuruo T and Twentyman PR (1990) Differences between monoclonal antibodies in immunohistochemical detection of P-glycoprotein in human and mouse multidrug resistant cell lines. Br J Cancer 62: 510–514.

Barrand MA, Broxterman HJ, Wright KA, Rhodes T and Twentyman PR (1992) Immunodetection of a 190 kd protein in non-P-glycoprotein containing mdr cells derived from small cell and non-small cell lung tumours. Br J Cancer 65(suppl XVI): 21.

Barrand MA, Rhodes T, Center MS and Twentyman PR (1993) Chemosensitisation and drug accumulation effects of cyclosporin A, PSC-833 and verapamil in human mdr large cell lung cancer cells expressing a 190k membrane protein distinct from P-glycoprotein. Eur J Cancer 29: 408–415.

Bhalla K, Hindenburg A, Taub RN and Grant S (1985) Isolation and characterization of an anthracycline-resistant human leukemic cell line. Cancer Res 45: 3657–3662.

Biedler JL and Riehm H (1970) Cellular resistance to actinomycin D in Chinese hamster cells *in vitro*: crossresistance, radioautographic and cytogenetic studies. Cancer Res 30: 1174–1184.

Borellini F, Aquino A, Josephs SF and Glazer RI (1990) Increased expression and DNA binding activity of transcription factor Spl in doxorubicin resistant HL60 leukemia cells. Molec Cell Biol 10: 5541–5547.

Bowman EJ, Siebers A and Altendorf K (1988) Bafilomycins: A class of inhibitors of membrane ATPase from microorganisms, animal cells, and plant cells. Proc Natl Acad Sci USA 85: 7972–7976.

Callen DF, Baker E, Simmers RN, Seshadri R and Roninson IB (1987) Localization of the human multidrug resistance gene, Mdrl to 7q21.1. Human Genet 77: 142–144.

Cass CE, Janowska-Wieczorek A, Lynch MA, Sheinin H, Hindenburg AA and Beck WT (1989) Effect of duration of exposure to verapamil on vincristine activity against multidrug-resistant human leukemic cell lines. Cancer Res 49: 5798–5804.

Chen C-J, Chin JE, Ueda K, Clark DP, Pastan I, Gottesman MM and Roninson IB (1986) Internal duplication and homology with bacterial transport proteins in the mdrl (P-glycoprotein) gene from multidrug-resistant human cells. Cell 47: 381–389.

Cole SPC, Chanda ER, Dicke FP, Gerlach JH and Mirski SEL (1991) Non-P-glycoprotein-mediated multidrug resistance in a small cell lung cancer cell line: evidence for decreased susceptibility to drug-induced DNA damage and reduced levels of topoisomerase II. Cancer Res 51: 3345–3352.

Cole SPC, Bhardwaj G, Gerlach JH, Mackie JE, Grant CE, Almquist KC, Stewart AJ, Kurz EU, Duncan AMV and Deeley RG (1992) Overexpression of a transporter gene in a multidrug-resistant human lung cancer cell line. Science 258: 1650–1654.

Cole SPC and Deeley RG (1993) Multidrug resistance associated protein: sequence correction. Science 260: 879.

Coley HM, Workman P and Twentyman PR (1991) Retention of activity by selected anthracyclines in a multidrug resistant human large cell lung carcinoma line without P-glycoprotein hyperexpression. Br J Cancer 63: 351–357.

Cornwall MM, Tsuruo T, Gottesman MM and Pastan I (1987) ATP binding properties of P-glycoprotein from multi-drug resistant KB cells. FASEB J 1: 51–54.

Dalton WS, Cress AE, Alberts DS and Trent JM (1988) Cytogenetic and phenotypic analysis of a human colon carcinoma cell line resistant to mitoxantrone. Cancer Res 48: 1882–1888.

Danø K (1973) Active outward transport of daunomycin in resistant Ehrlich Ascite tumor cells. Biochim Biophys Acta 324: 466–483.

de Lange JHM, Schipper NW, Schuurhuis GJ, ten Kate TK, van Heijningen ThHM, Pinedo HM, Lankelma J and Baak JPA (1992) Quantification by laser scan microscopy of intracellular doxorubicin distribution. Cytometry 13: 571–576.

Eijdems EWHM, Borst P, Jongsma APM, de Jong S, de Vries EGE, van Groenigen M, Versantvoort CHM, Nieuwint AWH and Baas F (1992) Genetic transfer of non-P-glycoprotein-mediated multidrug resistance (MDR) in somatic cell fusion: dissection of a compound MDR phenotype. Proc Natl Acad Sci USA 89: 3498–3502.

Endicott JA and Ling V (1989) The biochemistry of P-glycoprotein-mediated multidrug resistance. Ann Rev Biochem 58: 137–171.

Ferguson RJ, Fisher MH, Stephenson J, Li D-h, Zhou B-S and

Cheng Y-C (1988) Combined modalities of resistance in etoposide-resistant human KB cell lines. Cancer Res 48: 5956–5964.

Fugiwara T, Oda K, Yokata S, Takatsuki A and Ikehara Y (1988) Brefeldin A causes disassembly of the Golgi complex and accumulation of secretory proteins in the endoplasmic reticulum. J Biol Chem 263: 18545–18552.

Gervasoni JE Jr, Taub RN, Rosado M, Krishna S, Stewart VJ, Knowles DM, Bhalla K, Ross DD, Baker MA, Lutzky J and Hindenburg AA (1991) Membrane glycoprotein changes associated with anthracycline resistance in HL60 cells. Cancer Chemother Pharmacol 28: 93–101.

Gervasoni JE Jr, Fields SZ, Krishna S, Baker MA, Rosado M, Thuraisamy K, Hindenburg AA and Taub RN. (1991) Subcellular distribution of daunorubicin in P-glycoprotein-positive and -negative drug-resistant cell lines using laser-assisted confocal microscopy. Cancer Res 51: 4955–4963.

Gervasoni JE Jr, Taub RN, Yu MT, Warburton D, Sabbath M, Gill S, Coppock DL, D'Alessandri J, Krishna S, Rosado M, Baker MA, Lutzky J, Chandra ER, Gerlach JH, Pinkoski MJ, Cole SPC and Hindenburg AA (1992) Homogeneous staining region in anthracycline-resistant HL60/AR cells not associated with mdrl application. Cancer Res 52: 1–6.

Goldstein LJ, Galski H, Fojo A, Willingham M, Lai S-L, Gazdar A, Pirker R, Green A, Crist W, Brodeur GM, Lieber M, Cossman J, Gottesman MM and Pastan I (1989) Expression of a multidrug resistance gene in human cancers. J Natl Cancer Inst 81: 116–124.

Gros P, Croop J and Housman D (1986) Mammalian multidrug resistance gene: complete cDNA sequence indicates strong homology to bacterial transport proteins. Cell 47: 371–380.

Gruber A, Vitols S, Norgren S, Areström I, Peterson C, Björkholm M, Reizenstein P and Luthman H (1992) Quantitative determination of mdrl gene expression in leukemic cells from patients with acute leukemia. Br J Cancer 66: 266–272.

Haber M, Norris MD, Kavallaris M, Bell DR, Davey PA, White L and Stewart BW (1989) Atypical multidrug resistance in a therapy-induced drug resistant human leukemia cell line (LALW-2): resistance to vinca alkaloids independent of P-glycoprotein. Cancer Res 49: 5281–5287.

Hamada H and Tsuruo T (1988) Purification of the 170- to 180-kilodalton membrane glycoprotein associated with multidrug resistance. J Biol Chem 263: 1454–1458.

Harker WG, Slade DL, Dalton WS, Meltzer PS and Trent JM (1989) Multidrug resistance in Mitoxantrone-selected HL60 leukemia cells in the absence of P-glycoprotein overexpression. Cancer Res 49: 4542–4549.

Hindenburg AA, Baker MA, Gleyzer E, Stewart VJ, Case N and Taub RN (1987) Effect of verapamil and other agents on the distribution of anthracyclines and on reversal of drug resistance. Cancer Res 47: 1421–1425.

Hindenburg AA, Gervasoni JE, Krishna S, Stewart VJ, Rosado M, Lutzky J, Bhalla K, Baker MA and Taub RN (1989) Intracellular distribution and pharmokinetics of daunorubicin in anthracycline-sensitive and resistant HL60 cells. Cancer Res 49: 4607–4614.

Kartner H, Evernden-Porelle D, Bradley G and Ling V (1985) Detection of P-glycoprotein in multidrug resistant cell lines by monoclonal antibodies. Nature (Lond) 316: 820–823.

Keizer HG, Schuurhuis GJ, Broxterman HJ, Lankelma J, Schoonen WGEJ, van Rijn J, Pinedo HM and Joenje H (1989) Correlation of multidrug resistance with decreased drug accumulation, altered subcellular drug distribution, and increased P-glycoprotein expression in cultured SW-1573 human lung tumor cells. Cancer Res 49: 2988–2993.

Kessel D and Wilberding C (1984) Mode of action of calcium antagonists which alter anthracycline resistance. Biochem Pharmacol 33: 1157–1160.

Klohs WD and Steinkampf RW (1988) Possible link between the intrinsic drug resistance of colon tumors and a detoxification mechanism of intestinal cells. Cancer Res 48: 3025–3030.

Klohs WD, and Steinkampf RW (1988) The effect of lysosomotropic agents and secretory inhibitors on anthracycline retention and activity in multiple drug-resistant cells. Mol Pharmacol 34: 180–185.

Krishnamachary N and Center MS (1992) Detection and characterization of membrane protein changes in multidrug resistant HL60 cells. Oncol Res 4: 23–28.

Li M, Latoud C and Center MS (1991) Cloning and sequencing a cDNA encoding human ribosomal protein S25. Gene 107: 329–333.

Li M and Center MS (1992) Regulation of ribosomal protein S25 in HL60 cells isolated for resistance to adriamycin. FEBS 298: 142–144.

Ling V and Thompson LH (1974) Reduced permeability in CHO cells as a mechanism of resistance to colchicine. J Cell Physiol 83: 103–116.

Lutzky J, Astor MB, Taub RN, Baker MA, Bhalla K, Gervasoni JE, Rosado M, Stewart V, Krishna S and Hindenburg AA (1989) Role of glutathione and dependent enzymes in anthracycline-resistant HL60/AR cells. Cancer Res 49: 4120–4125.

Ma L and Center MS (1992) The gene encoding vacuolar H^+-ATPase subunit C is overexpressed in multidrug resistant HL60 cells. Biochem Biophys Res Commun 182: 675–681.

Marquardt D, McCrone S and Center MS (1990) Mechanisms of multidrug resistance in HL60 cells: detection of resistance-associated proteins with antibodies against synthetic peptides that correspond to the deduced sequence of P-glycoprotein. Cancer Res 50: 1426–1430.

Marquardt D and Center MS (1991) Involvement of vacuolar H^+-Adenosine triphosphatase activity in multidrug resistance in HL60 cells. J Natl Cancer Inst 83: 1098–1102.

Marquardt D and Center MS (1992) Drug transport mechanisms in HL60 cells isolated for resistance to adriamycin: evidence for nuclear drug accumulation and redistribution in resistant cells. Cancer Res 52: 3157–3163.

Marsh W, Sicheri D and Center MS (1986) Isolation and characterization of adriamycin-resistant HL60 cells which are not defective in the initial intracellular accumulation of drug. Cancer Res 46: 4053–4057.

Marsh W and Center MS (1986) Dimethylsulfoxide, retinoic

acid and 12-0-tetradecanoylphorbol-13-acetate induce a selective decrease in the phosphorylation of P150, a surface membrane phosphoprotein of HL60 cells resistant to adriamycin. Biochem Biophys Res Commun 138: 9–16.

Marsh W and Center MS (1987) Adriamycin resistance in HL60 cells and accompanying modification of a surface membrane protein contained in drug-sensitive cells. Cancer Res 47: 5080–5086.

McGrath T and Center MS (1987) Adriamycin resistance in the absence of detectable P-glycoprotein. Biochem Biophys Res Commun 145: 1171–1176.

McGrath T and Center MS (1988) Mechanisms of multidrug resistance in HL60 cells: evidence that a surface membrane protein distinct from P-glycoprotein contributes to reduced cellular accumulation of drug. Cancer Res 48: 3959–3963.

McGrath T, Latoud C, Arnold ST, Safa AR, Felsted RL and Center MS (1989) Mechanisms of multidrug resistance in HL60 cells: analysis of resistance associated membrane proteins and levels of mdr gene expression. Biochem Pharmacol 38: 3611–3619.

Mirski SEL, Gerlach JH and Cole SPC (1987) Multidrug resistance in a human small cell lung cancer cell line selected in Adriamycin. Cancer Res 47: 2594–2598.

Nakagawa M, Schneider E, Dixon KH, Horton J, Kelley K, Morrow C and Cowan KH (1992) Reduced intracellular drug accumulation in the absence of P-glycoprotein (mdrl) overexpression in mitoxantrone-resistant human MCF-7 breast cancer cells. Cancer Res 52: 6175–6181.

Nelson N (1989) Structure, molecular genetics, and evolution of vacuolar H^+-ATPases. J Bioenerg Biomembr 21: 553–571.

Nelson H, Mandiyan S, Noumi T, Moriyama Y, Miedel MC and Nelson N (1990) Molecular cloning of cDNA encoding the C subunit of H^+-ATPase from bovine chromaffin granules. J Biol Chem 265: 20390–20393.

Nielsen D and Skovsgaard T (1992) P-glycoprotein as multidrug transporter: a critical review of current multidrug resistant cell lines (1992) Biochim Biophys Acta 1139: 169–183.

Nieuwint AWM, Baas F, Wiegant J and Joenje H (1992) Cytogenetic alterations associated with P-glycoprotein- and non-P-glycoprotein-mediated multidrug resistance in SW-1573 human lung tumor cell lines. Cancer Res 52: 4361–4371.

Pagano RE and Sleight R (1985) Defining lipid transport pathways in animal cells. Science (Washington DC) 229: 1051–1057.

Peterson RHF, O'Neil JA and Biedler JL. (1974) Some biochemical properties of Chinese hamster cells sensitive and resistant to actinomycin D. J Cell Biol 63: 223–779.

Reeve JG, Rabbits PH and Twentyman PR (1990) Non-P-glycoprotein-mediated multidrug resistance with reduced EGF receptor expression in a human large cell lung cancer cell line. Br J Cancer 61: 851–855.

Rustan AC, Nossen JØ, Tefre T and Drevon CA (1987) Inhibition of very-low-density lipoprotein secretion by chloroquine, verapamil and monensin takes place in the Golgi complex. Biochim Biophys Acta 930: 311–319.

Safa AR, Glover CJ, Meyers MB, Biedler JL and Felsted RL (1986) Vinblastine photoaffinity labeling of a high molecular weight surface membrane glycoprotein specific for multidrug-resistant cells. J Biol Chem 261: 6137–6140.

Safa AR, Glover CJ, Sewell JL, Meyers MB, Biedler JL and Felsted RL (1987) Identification of the multidrug resistance-related membrane glycoprotein as an acceptor for calcium channel blockers. J Biol Chem 262: 7884–7888.

Safa AR (1988) Photoaffinity labeling of multidrug-resistance-related P-glycoprotein with photoactive analysis of verapamil. Proc Natl Acad Sci USA 85: 7187–7191.

Scheper RJ, Bulte JWM, Brakkee JGP, Quak JJ, van der Schoot E, Balm AJM, Meijer CJLM, Broxterman HJ, Kuper CM, Lankelma J and Pinedo HM (1988) Monoclonal antibody JSB-1 detects a highly conserved epitope on the P-glycoprotein associated with multidrug resistance. Int J Cancer 42: 389–394.

Schuurhuis GJ, Broxterman HJ, de Lange JHM, Pinedo HH, van Heijningen THM, Kuiper CM, Scheffer GL, Scheper RJ, van Kalken CK, Baak JPA and Lankelma J (1991) Early multidrug resistance defined by changes in intracellular doxorubicin distribution, independent of P-glycoprotein. Br J Cancer 64: 857–861.

Sehested M, Skovsgaard T and Roed H (1988) The carboxylic ionophore monensin inhibits active drug efflux and modulates *in vitro* resistance in daunorubicin resistant Ehrlich ascites tumor cells. Biochem Pharmacol 37: 3305–3310.

Shen D-W, Lu Y-G, Chin K-V, Pastan I and Gottesman MM (1991) Human hepatocellular carcinoma cell lines exhibit multidrug resistance unrelated to mdrl gene expression. J Cell Sci 98: 317–322.

Shinn-Liang L, Goldstein LJ, Gottesman MM, Pastan I, Chun-Ming T, Johnson BE, Mulshine JL, Ihde DC, Kayser K and Gazdar AF (1989) Mdrl gene expression in lung cancer. J Natl Cancer Inst 81: 1144–1150.

Skovsgaard T (1978) Mechanism of cross-resistance between vincristine and daunorubicin in Ehrlich ascites tumor cells. Biochem Pharmacol 27: 1221–1227.

Slapak CA, Daniel JC and Levy SB (1990) Sequential emergence of distinct resistance phenotypes in murine erythroleukemic cells under adriamycin selection: decreased anthracycline uptake precedes increased P-glycoprotein expression. Cancer Res 50: 7895–7901.

Slovak ML, Hoeltge GA, Dalton WS and Trent JM (1988) Pharmacological and biological evidence for differing mechanisms of doxorubicin resistance in two human cell lines. Cancer Res 48: 2793–2797.

Tartakoff AM and Vassalli P (1978) Comparative studies of intercellular transport of secretory proteins. J Cell Biol 79: 694–707.

Tartakoff AM (1983) Perturbation of vesicular traffic with the carboxylic ionophore monensin. Cell 32: 1026–1028.

Taylor CW, Dalton WS, Parrish PR, Gleason MC, Bellamy WT, Thompson FH, Roe DJ and Trent JM (1991) Different mechanisms of decreased drug accumulation in doxorubicin and mitoxantrone resistant variants of the MCF-7 human breast cancer cell line. Br J Cancer 63: 923–929.

Trent JM and Witlcowski CM (1987) Clarification of the chromosomal assignment of the human P-glycoprotein/mdr1 gene possible coincidence with the cystic fibrosis and c-met oncogene. Cancer Genet Cytogenet 26: 187–190.

Tsuruo T, Lida H, Tsukagoshi S and Sakurai Y (1981) Overcoming of vincristine resistance in P388 leukemia *in vivo* and *in vitro* through enhanced cytotoxicity of vincristine and vinblastine by verapamil. Cancer Res 41: 1967–1972.

Twentyman PR, Fox NE, Wright KA and Bleehen NM (1986) Derivation and preliminary characterization of adriamycin resistant lines of human lung cancer cells. Br J Cancer 53: 529–537.

Versantvoort CHM, Broxterman HJ, Pinedo HM, de Vries EGE, Feller N, Kuiper CM and Lankelma J (1992) Energy-dependent processes involved in reduced drug accumulation in multidrug-resistant human lung cancer cell lines without P-glycoprotein expression. Cancer Res 52: 17–23.

Volm M, Mattern J, Efferth T and Pommerenke EW (1992) Expression of several resistance mechanisms in untreated human kidney and lung carcinomas. Anticancer Res 12: 1063–1068.

Wang S-Y, Moriyama Y, Mandel M, Hulmes JD, Pan Y-CE, Danho W, Nelson H and Nelson N (1988) Cloning of cDNA encoding a 32-kDa protein: an accessory polypeptide of the H^+-ATPase from chromaffin granules. J Biol Chem 263: 17638–17642.

Weber JB, Sircar S, Horwath J and Dion P (1989) Non-P-glycoprotein-mediated multidrug resistance in detransformed rat cells selected for resistance to methylglyoxal bis(quanylhydrazone). Cancer Res 49: 5779–5783.

Yoshimori T, Yamamoto A, Moriyama Y, Futai M and Tashiro Y (1991) Bafilomycin A1, a specific inhibitor of vacuolar-type H^+-ATPase, inhibits acidification and protein degradation in lysosomes of cultured cells. J Biol Chem 26: 17707–17712.

Zijlstra JG, de Vries EGE and Mulder NH (1987) Multifactorial drug resistance in an adriamycin-resistant human small cell lung carcinoma cell line. Cancer Res 47: 1780–1784.

Cytotechnology **12**: 127–135, 1993.

Topoisomerase I in multiple drug resistance

Augusto Pessina
Institute of Medical Microbiology, University of Milan, Via Pascal 36, 20133 Milan, Italy

Key words: camptothecin, mdr, topoisomerase I

Abstract

Topoisomerase I is a nuclear enzyme able to catalyse the relaxation of supercoiled DNA by introducing single-stranded breaks in DNA molecule. Its function seems important to prepare DNA for many processes such as recombination, DNA repair and RNA transcription. The most important drugs active as inhibitors of topoisomerase I are represented by camptothecin and its derivatives which were developed as promising anticancer drugs. Since selectivity of action is essential for an antitumor drug, many studies were performed to investigate the mechanisms by which cancer cells become resistant to drug treatment by developing a condition of multiple drug resistance (MDR). This article analyses the role of topoisomerase I in cell functions, considers the cellular effects of topo I poisons and discusses the ways by which tumoral cells may become resistant to these drugs with a special attention to MDR mechanisms.

Introduction

Anticancer chemotherapy is essentially based on the capacity to combine a cytotoxic mechanism with a selective action of a drug in order to distinguish tumor cells from host cells. However, during the tumor chemotherapy the cancer tissues, without modification of their anatomic and structural features, are able to change their functional characteristics becoming rapidly resistant to many cytotoxic drugs (multiple drug resistance). In this way the selectivity of the drug completely vanishes. For this reason, the knowledge of the mechanism generating this form of chemoresistance could permit to circumvent the phenomenon by making the therapy more efficient.

Among the cytotoxic drugs currently used there are chemicals that are active on specific nuclear enzymes called topoisomerases the function of which is to regulate the DNA topology, transposition, repartition, recombination, replication, transcription and other DNA processes (Liu, 1983; Kafian *et al.*, 1986; Zijlstra *et al.*, 1990). Two types of topoisomerases are known, one able to introduce single-stranded breaks (topo I) the other able to produce double-stranded breaks (topo II) in the DNA molecules (Wang, 1985; D'arpa, 1989). This paper will examine the role of topoisomerase I in cell function, consider the cellular effects of topoisomerase I poisons and will discuss the way by which tumoral cells may become resistant to these drugs. Particular attention will be given to the MDR phenomenon and to the *in vitro* culture systems developed for testing the resistance to chemotherapy by primary and established tumor cells.

Cellular function of topoisomerase I

Human DNA topoisomerase I is a monomeric

protein of 100 Kd (Gellert, 1981) encoded by a single copy gene located on chromosome 20 (Zhou *et al.*, 1989). Eukaryotic DNA topoisomerase I catalyzes the relaxation of supercoiled DNA acting as a swivel. It nicks one DNA strand and then closes the break after a rotation event. This process is ATP-independent mechanism and the reaction is stimulated by Mg^{++} ions and other polycations while it is inhibited by polyanions. DNA topoisomerase I is assumed to form two different complexes with DNA: a noncleavable complex and a cleavable complex in rapid equilibrium with each other. The cleavable complex represents a transient covalent intermediate formed by DNA + topoisomerase I and the treatment of this complex with a strong protein denaturant (such as SDS or alkali) produces a protein-linked single strand DNA breaks (See Fig. 1). The covalent linking of topo I to the 3′-phosphoryl end of the broken DNA strand is made possible by the presence of a tyrosine (Champoux, 1981). The consequence of this reaction is the rotation of the two broken ends of DNA leading to relaxation of the supercoiled DNA. Although the role of topoisomerase I has not been completely clarified (Liu, 1989) its function seems to be very important in the transcription processes (Zhang *et al.*, 1988) by promoting both the separation of the two DNA strands (for receiving the RNA-polymerase complex) and the rapid rotation of DNA (that allows to copy the base sequence by the new forming RNA strand) (Rose *et al.*, 1988). As recently reported by Wang *et al.* (1991) topoisomerase I appears to be essential also for the genomic insertion of viral DNA.

The most important observations concerning the function of DNA topoisomerase I in the DNA processing are summarized in Table 1.

Topoisomerase I assays

In order to check topoisomerases many methods were developed. A first group of techniques is based on the measurement of the DNA damage attributed to the direct action of topoisomerase I (or topo II). The nuclei of cells were incubated *in vitro* with drugs inhibiting topoisomerase and then the formation of a cleavable complex was evaluated by alkaline and neutral elution. These methods measure DNA strand breaks, DNA-DNA cross links and DNA-protein cross links (Khon *et al.*, 1981). These and similar techniques (as SDS sedimentation or filter binding assay), measure the DNA-bound protein assuming that the most part of the protein

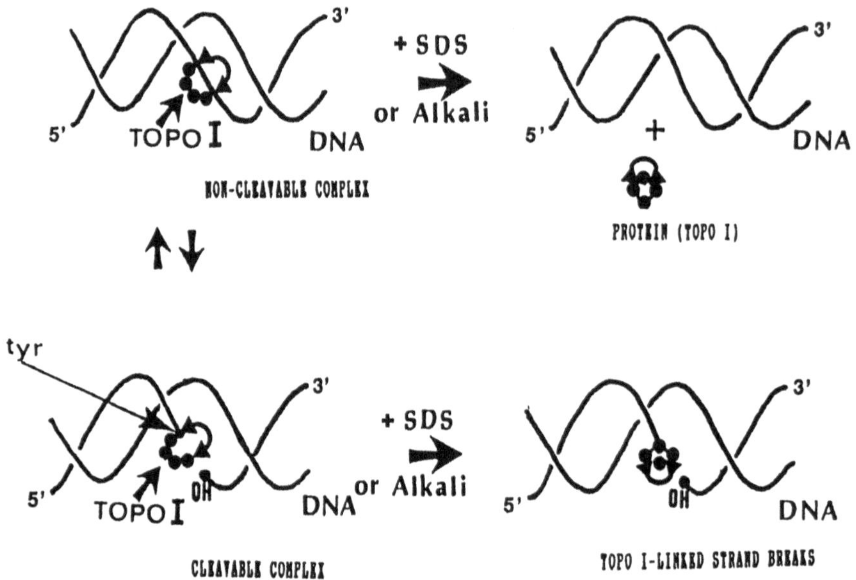

Fig. 1. Mechanism proposed for topoisomerase I. Exposure of the cleavable complex to a strong protein denaturant produces single strand DNA breaks linked to topoisomerase I. Topoisomerase I poisons as camptothecin convert the cleavable complex into a stable ternary complex. (See the text for explanation).

Table 1. Function described of DNA topoisomerase I

Function	Reference
RNA transcription	Javaherian *et al.*, Nucl Acids Res 11:461,1983 Egyhazi & Durban, Mol Cell Biol 12:4308,1987 Garg *et al.*, Proc Natl Acad Sci 84:3185,1987 Stewart & Schutz, Cell 50:1109,1987 Zhang *et al.*, Proc Natl Acad Sci 85:1060,1988 Camilloni *et al.*, Proc Natl Acad Sci 86:1989
SV40 DNA replication (*)	Brill *et al.*, Nature 326:414,1987 Yang *et al.*, Proc Natl Acad Sci 84:950,1987
Genomic insertion of viral DNA	Wang *et al.*, J Virol 65:2381-2392
Recombination and chromatin assembly	Bae *et al.*, Proc Natl Acad Sci 85:2076,1988 Christman *et al.*, Cell 55:413,1988 Sekiguki & Kmiec, Mol Cell Biol 83:195,1988
DNA repair	Musk & Steel, Br J Cancer 62:364,1990
Illegitimate recombination	Hsiang *et al.*, J Biol Chem 260:14873,1985 Nitiss *et al.*, Proc Natl Acad Sci 85:7501,1988 Eng *et al.*, Mol Pharmacol 34:755,1989

(*) Cell free systems. Function of DNA TOPO I in these systems can be substituted by DNA topo II.

linked was topoisomerase (Trask *et al.*, 1984; Liu *et al.*, 1983). For this reason, this type of measurement gives quite indirect and non-specific results. A more specific group of assays measures the direct activity of topoisomerase I by evaluating the relaxation of supercoiled plasmid DNA (Hsiang *et al.*, 1985). In this case the different topology of DNA is evaluated by gel electrophoresis.

A further group of techniques is based on the reaction of topoisomerase I with mono- or polyclonal antibodies both by using whole cells as well as blotting the cell extract and then testing it with specific antibody. These methods, although highly specific, are not predictive of the biological activity of the enzyme (Tsuruo *et al.*, 1988). Recently, Yang *et al.* (1993) developed a quantitative polymerase chain reaction (PCR) method to determine DNA topo isomerase I in cell lines and tumor samples. A review concerning the *in vitro* assays used to measure the activity of topoisomerases was published by Barrett *et al.*, 1990.

Drug interaction with topoisomerase I

As above described topoisomerase I acts by inducing transient breaks in a single DNA strand. In normal physiological conditions this process is reversible since the DNA-enzyme complex is resolved and DNA break ends are rapidly sealed to give the original form of the molecule. The two types of topoisomerases are sensitive to interaction with many drugs. The inhibitory action of coumarins, quinolones, acridines, anthracyclins, ellipcitines, epipodophyllotoxins on topoisomerase II is well-known.

Intercalator drugs of the acridines group (Pommier, 1985), actinomycin-D (Trask *et al.*, 1988), distamycin (Smith, 1990) and also heparin evidenced their capacity to inhibit *in vitro* the activity of topoisomerase I present in nuclear extracts (Ishii, 1987). Recently (Chen *et al.*, 1993) pointed out the capacity of Hoechst dye 33342 to interrupt the breakage/reunion reaction of mammalian DNA topoisomerase I by trapping reversible topo I cleavable complexes. So far, the most important

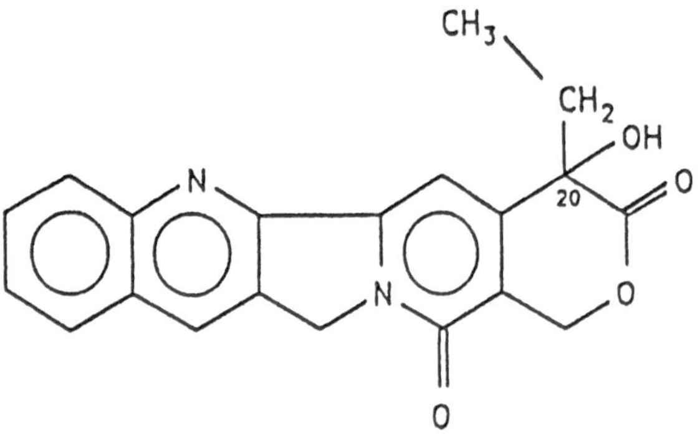

Fig. 2. Chemical structure of camptothecin.

drug active as inhibitor of topoisomerase I is represented by a plant alkaloid, camptothecin which was developed as an antitumor drug in 1966 (see the review of Horwitz, 1975). This molecule was isolated from Camptotheca acuminata (its chemical structure is reported in Fig. 2) and together with many of its derivatives shows a very high experimental activity as anticancer drug promising very interesting clinical results (Giovannella *et al.*, 1991). Recently, topoisomerase I inhibitors were also suggested as antiviral agents (Priel *et al.*, 1991). Camptothecin was recognized as a cytotoxic drug with a specific killing action on S-phase cells (Li *et al.*, 1972). Moreover, it inhibits both DNA and RNA synthesis by inducing cellular DNA degradation. Only in recent studies, however, has it been shown that these phenomena are related to the specific activity of camptothecin on topo I. In fact, by studying the products of topo I cleavable complexes it has been proved that the protein linked to broken DNA in detergent lysates of camptothecin treated cells was topo I (Hsiang 1985, Hsiang 1988).

The steps by which the formation of cleavable complex by camptothecin occurs is not fully clarified. It seems that camptothecin binds simultaneously DNA and topo I to give a stable ternary complex with the consequent block of the resealing step of strand-passing reaction (Hsiang 1985; Jaxel *et al.*, 1991). DNA cleavage pattern observed after denaturation shows wide variation in cleavage site intensity and moreover the cleavages occur at sites that were not detectably cleaved in the absence of the drug (Perez-Stable *et al.*, 1988). Since cleavages should occur at sequences with similar twist angle variation it has been proposed that camptothecin produces alterations of the topo I cleavage pattern as a consequence of alteration of the enzyme's normal recognition of twist angle variations (Shen, 1990).

Cellular effect of topoisomerase I poisons

The progressive knowledge that the mechanism of the action of camptothecin and its derivatives was directed against topo I allowing to better clarify its effect on cellular functions: inhibition of RNA and DNA synthesis as well as the cell death.

The inhibition of RNA synthesis is a reversible phenomenon probably due to the camptothecin altered cleavable complex which blocks the elongation process by preventing the progression of RNA polymerase molecules along the transcription unit (Zhang, 1988). If camptothecin is removed the phenomenon reverses probably because the topo I cleavable complex dissociates from the transcription unit and the RNA synthesis restarts.

The inhibition of DNA synthesis seems to be related to a blockage of the replication fork due to a progressive accumulation of cleavable complexes and after a remarkable level of their accumulation the phenomenon becomes irreversible (Horwitz, 1975). Studies performed in *in vitro* models with SV40 virus replicating in cell free systems suggested that the block of DNA synthesis may be due to the collision between the replication fork and the drug trapped cleavable complex (Hsiang *et al.*, 1989). The cell cycle stadium for the maximum expression of the cytotoxicity of camptothecin is the S-phase (Horwitz, 1975). It has been calculated that a cell in S-phase is about 1,000 times more sensitive to the effect of camptothecin than cells in G1 or G2 phase. This fact suggests that the interaction between the cleavage complex and the replication fork may be the cause of cell death. It also indicates that lethal effect of camptothecin may be due to the cleavable complexes directed at the genomic regions which are essential for cell viability (Hsiang *et al.*, 1989). The capacity of camptothecin and other topo I poisons to kill cells has been regarded as an interesting tool for the treatment of tumors in which topo I levels are very elevated. As reported by recent clinical studies, derivatives from camptothecin may have a particular role to play in the treatment of childhood solid tumors such as rhabdomyosarcoma and osteosarcoma (Houghton *et al.*, 1992).

Multiple drug resistance

Since for an antitumor drug, the selectivity of action is fundamental, many studies were directed to investigate the mechanisms by which cells become resistant to drug treatment (Clynes, 1993). For drugs acting on topoisomerases it has been reported that irreversible damage occurs with a higher frequency in cells containing high levels of these enzymes (that is the case of many tumoral cells) (van der Zee *et al.*, 1991). When a tumor cell population has a low content of topoisomerases it becomes refractory to the treatment and this type of resistance is called primary resistance. Different mechanisms of resistance have been characterized in many different experimental systems and currently they may be classified into three basic groups according to the mechanisms involved: drug transport (Endicott *et al.*, 1989), drug-target interaction (Beck, 1987) and drug detoxification (Deffie *et al.*, 1988).

The most important phenomenon of resistance made evident was the so called "multiple drug resistance" (MDR) that is caused by the drug transportation made by a "pump" system requiring energy and able to throw the drug out of the cell (Endicott *et al.*, 1989; Roninson, 1992).

The best example of MDR is represented by P-glycoprotein (P-gp), a member of a family of proteins acting as membrane carriers of different chemical compounds such as sugar, ions, hormones and proteins (this system is reviewed by Clynes *et al.*, 1990). The P-glycoprotein is present in normal tissues in which it protects the cells against toxic compounds providing an important factor of selectivity (Fojo *et al.*, 1987) that during chemotherapy protects normal cells whereas it does not defend neoplastic cells which do not contain P-gp. Unfortunately, this selectivity is lost when tumoral cells express P-glycoprotein by acquiring MDR phenotype and worst when the P-gp is associated with the expression of "ras" and "p53" oncogenes then an acceleration of tumor growth occurs (Chin *et al.*, 1992). P-glycoprotein is encoded by MDR-1 gene and may be the result both of a gene expression as well as of a gene amplification process (see the recent review of Chin *et al.*, 1993). The main part of the studies (performed on topoisomerase II poisons) pointed out that MDR cells are very resistant to Vinca alkaloids such as vincristine or vinblastine, moderately resistant to doxorubicin, etoposide and teniposide while maintaining a relative sensitivity to amsacrine (Qian *et al.*, 1990; Baguley *et al.*, 1990). However, it must be considered that some MDR mechanisms are not associated with increased expression of P-glycoprotein although modifications of the drug transport are involved. As observed by Scheper *et al.*, 1993, several p-glycoprotein-negative MDR cell lines overexpressed a Mr = 110,000 protein (p110) which may have a physiological function in several normal tissues, possibly involving a transport mechanism. Many *in vitro* cell lines with increased expression of P-glycoprotein and resistant to drug inhibition were obtained and also many spontaneous human tumors were described to become MDR.

For these reasons several *in vitro* methods have been developed in order to check the presence of MDR in tumor cells. Some methods are based on the direct detection of P-gp in hystological sections of tumoral tissues or cells by using specific monoclonal antibodies labelled with fluorescein (Thiebout *et al.*, 1987). Other ones check the MDR-1 gene expression by DNA-RNA hybridation techniques (Fojo *et al.*, 1987). Analytic methods based on the cytometry were developed by exploiting the capacity of MDR cells to throw out from the cytoplasm different fluorochromes (Nooter *et al.*, 1991; Ludescher *et al.*, 1991). Also the polymerase chain reaction (PCR) was recently used to analyze the expression of MDR-1 gene in tumor sample from breast cancer patients (Yang *et al.*, 1993).It is important to remember a last method which has the advantage of detecting other MDR mechanisms unrelated to the expression of P-gp. This aim is obtained by culturing tumor cells in the presence of two different drugs having similar structure but different susceptibility to MDR mediated by P-gp. Thus, the ratios between the IC_{50} values calculated for the two drugs may be considered as an index of MDR (Baguley *et al.*, 1990).

MDR and topoisomerase I inhibitors

As mentioned above the number of DNA strand breaks induced by topoisomerase inhibitors is dependent on the number of molecules of enzymes in the cells. Far this reason, the most simple mechanism of resistance to drugs acting against topoisomerases is consistent with the decrease of cell content of these enzymes. This is the case with the "cyclic" changes of resistance to topoisomerase-inhibitors which is directly related to the variation of the intracellular content of these enzymes along the cell cycle (Heck *et al.*, 1988; Barronco, 1984). During the G0-G1 phase, many enzymes utilized for DNA duplication process are degraded while their new synthesis starts just before the new DNA replication phase begins (S-phase) (Andrews, 1992). This phenomenon of resistance is called "cytokinetic resistance" and it seems to play an important role in the protection of normal tissues from toxics ingested with the food or administered during chemotherapy. This type of resistance may be the most probable mechanism which determines the insensitivity of many solid tumors to the chemotherapy. Thus, the quantitative reduction of topo I content seems to be the most frequently occurring event in the development of camptothecin resistance (Sugimoto *et al.*, 1990). A development of resistance to inhibitors of topo I occurring by down-regulation of the target enzyme, thus reducing the production of lethal enzyme-mediated DNA damages has also been suggested (Eng *et al.*, 1990). However, according to some authors,the resistance to topoisomerase inhibitors is not a cycle specific resistance but rather a combination between a decreased quantity of topoisomerases and a diminished direct sensitivity of the enzymes to the drug (Andoh *et al.*, 1987; Kjeldsen *et al.*, 1988).

This last mechanism producing resistance was well documented for topo II and was defined at-MDR (altered topo-MDR) (Beck *et al.*, 1987; Snow *et al.*, 1991).

For topo I the data is not so homogeneous although topo I alterations have been proposed to produce acquired resistance to camptothecin in mammalian cells.

Some cells have been characterized which exhibited altered topo I (Andoh *et al.*, 1987; Gupta *et al.*, 1988) and in some mutants it was even shown that the activity of topo I extracted from cells resistant to camptothecin was three times as active than topo I from wild type cells (Lefevre *et al.*, 1991). Studies with a mutated camptothecin resistant form of topoisomerase I (TOPO I-K5) indicated a higher cleavage/religation activity of topo I-K5 as a result of improved DNA binding and a concomitant shift in the equilibrium between cleavage and religation towards the religation step (Gromova *et al.*, 1993). Also Tanizawa *et al.* (1992) observed that resistance to camptothecin was associated with a marked reduction of drug-induced DNA strand breaks concluding that camptothecin resistance in DC3f/C10 cells is due to the qualitative alteration of DNA topo I. Ishimi *et al.* (1991) isolated a topo I mutant conferring camptothecin resistance but not able to affect its functions involved in cellular DNA replication. However, while topo II poisons are in general sensitive to MDR-1 mediated resistance, the topo I poisons such as camptothecin and its derivatives appear completely insensitive to MDR-1 overexpression (Liu, 1990). In fact, alterations of DNA topo I are unrelated to the acquisition of MDR characters (Oguro *et al.*, 1990) but neither does the P-glycoprotein associated with MDR phenotype modify the sensitivity to camptothecin and its derivatives (Giovanella *et al.*, 1991). Moreover, MDR-1 overexpression was shown to have little effect on camptothecin cytotoxicity and the capacity of this drug to overcame MDR-1 mediate resistance is most likely due to its unimpaired accumulation in MDR-1 cells (Chen *et al.*, 1991). Only very recent experimental evidence points out that MDR cells can be resistant also to camptothecin and its derivatives and that P-gp overexpression diminishes accumulation and toxicity of topo I poisons in hamster and human cells (Hendricks *et al.*, 1992).

Future perspectives

Data on topoisomerases are increasing every day in literature but much work is still necessary for a better understanding of the mechanisms regulating topoisomerases activity , the cytotoxic mechanism of topoisomerases poisons and the clinical implications concerning Multi Drug Resistance both in normal and in tumoral cells.

Moreover, it is important to underline that the improvement of the efficacy of this type of therapeutic approach will depend on three fundamental lines:

1. the development of new drugs able to circumvent MDR and at-MDR;
2. the refining of the drug combination strategy with particular attention to the association of chemotherapy with biological factors (e.g., hormones, TNF or other cytokines) able to increase topoisomerases activity and as a consequence the drug sensitivity of tumor cells;
3. the development of diagnostic tests able to check both the type and the level of resistance presented by tumor cells in a single patient (Marshall *et al.*, 1992; Hait *et al.*, 1992).

Acknowledgements

I wish to thank Dr G. Mancina for her support in the collection of references.

References

Andoh T, Ishii K, Suzuki Y, Ikegami Y, Kuzunoki Y, Takemoto Y and Okada K (1987) Characterization of a mammalian mutant with camptothecin resistant DNA topoisomerase I. Proc Natl Acad Sci USA 84: 5565–5569.

Andrews BJ (1992) Gene expression-dialogue with the cell cycle. Nature 355: 393–394.

Baguley BC, Holdway KM and Fray LM (1990) Design of DNA intercalators to overcome topoisomerase II-mediated multidrug resistance. J Natl Cancer Inst 82: 398–402.

Barrett JF, Sutcliffe YA and Gootz TD (1990) *In vitro* assay used to measure the activity of topoisomerases. Antimicrob Agents Chemother 34: 1–7.

Barronco SC (1984) Cellular and molecular effects of Adriamycin on dividing and nondividing cells. Pharmac Ther 24: 303–319.

Beck WT (1987) The cell biology of multiple drug resistance. Biochem Pharmacol 36: 2879–2888.

Beck WY, Cirtain MC, Danks MK, Felsted RL, Safa AR, Wolverton JS,Suttle DP and Trent JM (1987) Pharmacological molecular and cytogenetic analysis of "atypical" multidrug-resistant human leukemic lines. Cancer Res 47: 5455–5460.

Champoux JJ (1981) DNA is linked to the rat liver nicking-closing enzyme by a phosphodiester bond to tyrosine. J Biol Chem 256: 4805–4809.

Chen AY, Yu C, Potmesil M, Wall ME, Wani MC and Liu LF (1991) Camptothecin overcomes MDR1 mediated resistance in human KB carcinoma cells. Cancer Res 51: 6039–6044.

Chen AY, Yu C, Bodley A, Peng LF and Liu LF (1993) A new mammalian DNA topoisomerase I poison Hoechst 33342: cytotoxicity and drug resistance in human cell cultures. Cancer Res 53: 1332–1337.

Chin KV, Ueda K, Pastan I and Gottesman MM (1992) Modulation of activity of the promoter of the human MDR1 gene by ras and p53. Science 255: 459–462.

Chin KV, Pastan I and Gottesman MM (1993) Function and regulation of the human multidrug resistance gene. Adv Cancer Res 60: 157–180.

Clynes M, Redmond A and Heenan M (1990) Recent developments in research on multiple drug-resistance in cancer cells. Cancer J 3: 34–39.

Clynes M (1993) Cellular models for multiple drug resistance in cancer. *In Vitro* Cell Dev Biol 29A: 171–179.

D'arpa P and Liu FL (1989) Topoisomerase-targeting antitumor drugs. Biochim Biophys Acta 989: 163–177.

Deffie AM, Alam T, Seneviratne C, Beenken SW, Batra JK, Shea TC, Henner WD and Geldenberg GJ (1988) Multifactorial resistance to Adriamycin: relationship of DNA repair, glutathione transferase activity drug efflux, and P-glycoprotein in cloned cell lines of Adriamycin-sensitive and resistant P388 leukemia. Cancer Res 48: 3595–3602.

Endicott JA and Ling V (1989) The biochemistry of P-glycoprotein-mediated multridrug resistance. Annu Rev Biochem 58: 351–375.

Eng WK, McCabe FL, Tan KB, Mattern MR, Hofmann GA, Woessner RD, Hartzberg RP and Johnson RK (1990) Development of a stable camptothecin-resistant subline of P388 leukemia with reduced topoisomerase I content. Mol Pharmacol 38: 471–480.

Fojo AT, Ueda K, Slamon DJ, Poplank DG, Gottesman MM and Pastan I (1987) Expression of a multidrug-resistance gene in human tumors and tissues. Proc Natl Acad Sci USA 84: 265–269.

Gellert M (1981) DNA topoisomerases. Annu Rev Biochem 50. 879–910.

Giovanella BC, Hinz HR, Kozielski AJ, Stehlin JS, Silber R and Potmesil M (1991) Complete growth inhibition of human cancer xenografts in nude mice by treatment with 20-(S)-camptothecin. Cancer Res 51· 3052–3055.

Gromova II, Kjeldsen E, Svejstrup JQ, Alsner J, Christiansen K and Westergaard O (1993) Characterization of an altered DNA catalysis of a camptothecin resistant eukaryotic topoisomerase I. Nucleic Acid Res 21: 593–600.

Gupta RS, Gupta R, Eng B, Lock RB, Ross WE, Hertzberg RP, Caranfa MJ and Johnson RK (1988) Camptothecin resistant mutants of Chinese hamster ovary cells containing a resistant form of topoisomerase I. Cancer Res 48· 6404–6410.

Hait WN and Aftab DT (1992) Rational design and preclinical pharmacology of drugs for reversing multidrug resistance. Biochem Pharmacol 43: 103–107

Heck MMS, Hittelman WN and Earnshaw WC (1988) Differential expression of DNA topoisomerases I and II during the eukaryotic cell cycle. Proc Natl Acad Sci USA 85· 1086–1090.

Hendricks CB, Rowinsky EK, Grochow LB, Donehower RC and Kaufmann SH (1992) Effect of p-glycoprotein expression of the accumulation and cytotoxicity of topotecan (SK&f 104864) a new camptothecin analogue Cancer Res 52. 2268–2270.

Horwitz SB (1975) Antibiotics. In: Corcoran W and Hahn FE (eds) Mechanism of action of antimicrobial and antitumor agents. Vol III, (pp 48–57), New York: Springer Verlag.

Houghton PJ, Cheshire PJ, Myers I, Stewart CF, Synold TW and Oughton JA (1992) Evaluation of 9-dimethylaminomethyl-10-hydroxycamptothecin against xenografts derived from adult and childhood solid tumors Cancer Chemother Pharmacol 31: 229–259.

Hsiang YH, Herzberg R, Hecht S and Liu LF (1985) Camptothecin induces protein linked DNA breaks via mammalian DNA topoisomerases. J Biol Chem 260. 14873–14878.

Hsiang YH, Wu HY and Liu FL (1988) Proliferation-dependent regulation of DNA topoisomerase II in cultured human cells. Cancer Res 48: 3230–3235

Hsiang YH, Lihou MG and Liu LF (1989) Arrest of replication forks by drug-stabilized topoisomerase I-DNA cleavable complexes as a mechanism of cell killing by camptothecin Cancer Res 49 5077–5082

Ishii K, Futaki S, Uchiyama H, Nagasawa K and Andoh T (1987) Mechanism of inhibition of mammalian DNA topoisomerase I by heparin. Biochem J 241: 111–116.

Ishimi Y, Nishizawa M and Andoh T (1991) Characterization of a camptothecin-resistant human DNA topoisomerase I in an *in vitro* system for Simian virus 40 DNA replication Eur J Biochem 202: 835–839

Jaxel C, Capranico G, Wasserman K, Kerrigan D, Kohn KW and Pommier Y (1991) DNA sequence at sites of topoisomerase I cleavage induced by Camptothecin in SV40 DNA. In: Pommesil M, Kohn KW (eds) DNA Topoisomerases in Cancer, (pp 182–195) New York: Oxford University Press.

Kjeldsen E, Bonven BJ, Andoh T, Ishii E, Okada K, Bolund L and Westergaard O (1988) Characterization of a camptothecin resistant human DNA topoisomerase I. J Biol Chem 263· 3912–3916.

Khon KW, Ewig RAG,Erickson LC and Zwelling LA (1981) Measurement of strand breaks and cross links by alkaline elution. In. Friedberg EC and Hanawalt PC (eds) DNA Repair: A Laboratory Manual of Research Procedure, (pp 379–401) New York: Marcel Dekker, Inc.

Kafiani CA, Bronstein IB, Timofeen AV, Gromova II and Terskikh VV (1986) DNA topoisomerases and regulation of cell proliferation. Adv Enzyme Regul 25: 439–457.

Lefevre D, Riou JF, Ahomadegbe JC, Zhou DY, Benard J and Riou G (1991) Biochem Pharmacol 41: 1967–1979.

Li HL, Fraser TJ, Olin EJ and Bhuyan BK (1972) Action of camptothecin on mammalian cells in culture. Cancer Res 32: 2643–2650.

Liu LF (1983) DNA topoisomerase-enzymes that catalyse the breaking and rejoining of DNA. Crit Rev Biochem 15: 1–24.

Liu LF (1989) DNA topoisomerase poisons as antitumor drugs. Annu Rev Biochem 58. 351–375

Liu LF (1990) DNA topoisomerases in drug resistance and cancer chemotherapy. The Third Conference on DNA Topoisomerase in therapy Oct 15–18, New York, p. 25.

Liu LF, Rowe TC, Yang L, Tewey KM and Chen GL (1983) Cleavage of DNA by mammalian DNA topoisomerase II. J Biol Chem 258: 15365–15370.

Ludescher C, Gattringer C, Drach J, Hofmann J and Grunicke H (1991) Rapid functional assay for the detection of multi-drug-resistant cells using the fluorescent dye rhodamine-123 Blood 78· 1385–1387.

Marshall ES, Finlay GJ, Matthews JHL, Shaw JHF, Nixon J and Baguley BC (1992) Microculture-based chemosensitivity testing: a feasibility study camparting freshly explanted human melanoma cells with human melanoma cell lines J Natl Cancer Inst 84 341–344.

Nooter K and Herweijer H (1991) Multidrug resistance (mdr) genes in human cancer. Br J Cancer 63· 663–669.

Oguro M, Seki Y, Okada K and Andoh T (1990) Collateral drug sensitivity induced in CPT-11 (a novel derivative of camptothecin)-resistant cell lines Biomed Pharmacother 44 209–216

Perez-Stable C, Shen CC and Shen CKJ (1988) Enrichment and depletion of Hela topoisomerase I recognition sites among specific types of DNA elements. Nucl Acids Res 16: 7975–7993.

Pommier Y, Zwelling LA, Kao-Shan CS, Whang-Peng J and Bradley MO (1985) Correlation between intercalator induced DNA strand breaks and sister chromatid exchanges mutations and cytotoxicity in Chinese hamster cells. Cancer Res 45: 3143–3148

Priel E, Showalter SD and Blair DG (1991) Inhibition of human immunodeficiency virus (HIV-1) replication *in vitro* by noncytotoxic doses of camptothecin, a topoisomerase I inhibitor AIDS Res Hum Retrovirus 7: 65–72.

Qian X and Beck WT (1990) Binding of an optically pure photoaffinity analogue of verapamil LU-49888 to P-glycoprotein from multidrug resistant human leukemic cell lines. Cancer Res 50: 1132–1137.

Roninson IB (1992) The role of the MDR1 (P-glycoprotein) gene in multidrug resistance *in vitro* and *in vivo*. Biochem Pharmacol 43: 95–102.

Rose KM, Szopa J, Han FS, Cheng YC, Richter A and Scheer U (1988) Association of DNA topoisomerase and RNA polymerase I: a possible role for topoisomerase I in ribosomal gene transcripticn. Chromosoma 96: 411–416.

Scheper RJ, Broxterman HJ, Scheffer GL, Kaaijk P, Dalton WS, Heijningen van THM, van Kalken CK, Slovak ML, de Vries EGE, van der Valk P, Meijer CJLM and Pinedo HM (1993) Overexpression of a Mr 110,000 vesicular protein in non-P-glycoprotein-mediated multidrug resistance. Cancer Res 53: 1475–1479.

Shen CC and Shen CKJ (1990) Specificity and flexibility of the recognition of DNA helical structure by eukaryotic topoisomerase I. J Mol Biol 212: 67–78.

Smith PJ (1990) DNA topoisomerase dysfunction: a new goal for antitumor chemotherapy. Bioassay 12: 1657–1672.

Snow K and Judd W (1991) Characterization of adriamycin resistant and amsacrine-resistant human leukaemic T-cell lines. Br J Cancer 63: 17–28.

Sugimoto Y, Tsukahara S, Oh-Hara T, Isoe T and Tsuruo T (1990) Decreased expression of DNA topoisomerase I in camptothecin-resistant tumor cell lines as determined by a monoclonal antibody. Cancer Res 50: 6925–6930.

Tanizawa A and Pommier Y (1992) Topoisomerase I alteration in a camptothecin-resistant cell line derived from Chinese hamster DC3F cells in culture. Cancer Res 52: 1848–1854.

Thiebaut T, Tsuruo T, Hamada H, Gottsman MM, Pastan I and Willingham MC (1987) Cellular localization of the multidrug-resistance gene product P-glycoprotein in normal human tissues. Proc Natl Acad Sci USA 84: 7735–7738.

Trask DK, DiDonato JA and Muller MT (1984) Rapid detection and isolation of covalent DNA protein complexes: application to topoisomeraes I and II. EMBO J 3: 671–676.

Trask DK and Muller MT (1988) Stabilization of type I topoisomerse DNA covalent complexes by actinomycin D. Proc Natl Acad Sci USA 85: 1417–1425.

Tsuruo T, Matsuzaki T, Matsushita M, Saito H and Yokokura T (1988) Antitumor affect on CPT-11 a new derivative of camptothecin against pleiotropic drug resistant tumors *in vitro* and *in vivo*. Cancer Chemother Pharmac 21: 71–74.

Van der Zee AG, Hollema H, de Jong S, Boonstra H, Gouw A, Willemse PH, Zijlstra JG and de Vries EG (1991) P-glycoprotein expression and DNA topoisomerase I and II activity in benign tumors of the ovary and in malignant tumors of the ovary, before and after platinium/cyclophosphamide chemotherapy. Cancer Res 51: 5915–5920.

Wang JC (1985) DNA topoisomerases. Annu Rev Biochem 54: 665–697.

Wang HP and Rogler CE (1991) Topoisomerase I-mediated integration of hepadnavirus DNA *in vitro*. J Virol 65: 2381–2392.

Yang CH, Cowan KH and Schneider E (1993) Quantitative PCR for DNA topoisomerase I, II alpha, II beta and mdr1 in cell lines and tumor samples. Proc Annu Meet Am Assoc Cancer Res 34: A56.

Zhang H, Wang JC and Liu LF (1988) Involvement of DNA topoisomerase I in transcription of human ribosomal RNA genes. Proc Natl Acad Sci USA 85: 1060–1064.

Zijlstra JG, de Jong S, de Vries EGE and Mulder NH (1990) Topoisomerases, new targets in cancer chemotherapy. Med Oncol Tumor Pharmacother 7: 11–18.

Zhou BS, Bastow KF and Cheng YC (1989) Characterization of 3′region of the human DNA topoisomerase I gene. Cancer Res 49: 3922–3927.

Cytotechnology **12**: 137–154, 1993.

Topoisomerase II in multiple drug resistance

Glenn A. Hofmann and Michael R. Mattern
Department of Biomolecular Discovery, SmithKline Beecham Pharmaceuticals, P.O. Box 1539, King of Prussia, PA 19406, USA

Key words: adjuvant therapy, atypical multidrug resistance, P-glycoprotein, topoisomerase

Abstract

Topoisomerase II is a target of alkaloid, anthracycline and related antitumor agents. Two types of multiple drug resistance are associated with these enzymes. In classical (typical) multidrug resistance, inhibitors are actively effluxed from cells by P-glycoprotein. In atypical multidrug resistance, topoisomerase II is either reduced in cellular content or mutated to a form that does not interact with inhibitors. Because cytotoxicity of most antineoplastic topoisomerase II inhibitors is directly related to the number of active topoisomerase II molecules, a reduction in this number leads to resistance. In the topoisomerase II mechanism, through which the DNA linking number is altered, DNA double strands are cleaved, and the termini transiently bound covalently (5') or noncovalently (3') to the enzyme while a second double strand is passed through the break in the first. This transition state complex then decays to enzyme and DNA of altered linking number. Most cytotoxic topoisomerase II inhibitors stabilize these reaction intermediates as ternary complexes, which are converted to lethal lesions when cells attempt to utilize the damaged DNA as templates. Toxicity is related to topoisomerase II content as well as to drug concentration. Thus, multidrug resistance results from either 1) decreasing cellular content of the inhibitor by P-glycoprotein (typical) or 2) decreasing cellular content and/or activity of the target, topoisomerase II, as, for example, when its content or activity is modulated downward by decreased expression, deactivation, or by mutations to the TopII gene, producing an enzyme that reacts poorly with inhibitors (atypical). Mixed types, *i.e.*, both typical and atypical, are known. Attempts to abrogate or prevent both typical and atypical multidrug resistance to topoisomerase II inhibitors have been described.

Abbreviations: atMDR – atypical multidrug resistance; kDa – kilodaltons; MDR – multidrug resistance; Pgp – P-glycoprotein; TOPO II – topoisomerase II

Introduction

In the early 1980s, a novel target for conventional cancer chemotherapy was described – the nuclear enzyme DNA topoisomerase II, more commonly referred to simply as topoisomerase II (Liu *et al.*, 1980; Hsieh and Brutlag, 1980; Liu, 1983). A number of clinically useful antitumor agents, including DNA intercalators such as Adriamycin (doxorubicin), and ellipticine, and the epipodophyllotoxins etoposide (VP-16) and teniposide (VM-26), were shown to be inhibitors of this enzyme *in vitro* (Chen *et al.*, 1984) and in cells (Pommier *et al.*, 1984; Yang *et al.*, 1985). Thus, topoisomerase II became an object of intense study to understand more fully its enzymic mechanism and, using this knowledge, to design or discover inhibitors having improved therapeutic profiles.

Like other anticancer drugs, many topoisomerase II inhibitors are exported from cultured cells by the P-glycoprotein (Pgp) expressed at high levels in "classic" or "typical" MDR cell lines (Ling and Thompson, 1974; Baguley *et al.*, 1992; Ling, 1992). Thus, topoisomerase II inhibitors are among the several mechanistic classes of compound whose clinical usefulness may be compromised by classic MDR. In addition, any treatment that results in a drug's diminished cellular accumulation will confer resistance, irrespective of its mechanism of action. Resistance to topoisomerase II inhibitors might arise in cells that acquire the ability to metabolize the inhibitors to inactive forms, as well as in cells that actively export it.

Perhaps the most interesting finding regarding resistance to topoisomerase II inhibitors, however, has been a second mechanism of MDR, called "atypical" (Beck, 1989) owing to the novel mechanism of cell killing employed by most inhibitors of this enzyme. As described below, the inhibitor, the target (topoisomerase II), and the substrate (DNA) form a potentially lethal ternary complex in cells (Liu and D'Arpa, 1992; Takano *et al.*, 1992). The number of complexes per cell is limited by the cellular content of each component of the complex. Thus, resistance to complex-forming topoisomerase II inhibitors results from any decrease in the cellular content of functional topoisomerase II.

Evidence accumulated to date shows that tumors can become resistant to a spectrum of DNA topoisomerase II inhibitors (MDR) by two principal mechanisms. One depends upon topoisomerase II, while the other depends upon Pgp. To prevent or circumvent these two kinds of topoisomerase II-related MDR, appropriate strategies have been used. In the following sections, the mechanism of action of eukaryotic topoisomerase II will be reviewed to provide a basis to understand how cytotoxic topoisomerase II inhibitors 1) inhibit the enzyme reaction; and 2) kill cells. The principal mechanisms of MDR to these compounds, both in cultured cells and *in vivo*, will then be reviewed. Finally, mechanisms for overcoming the principal forms of MDR to topoisomerase II inhibitors will be discussed.

Topoisomerase II: mechanisms of catalysis, inhibition and resistance

Type II topoisomerases are Mg^{++}- and ATP-dependent enzymes that, as do all topoisomerases, catalyse interconversions among various topological forms of DNA, *i.e.*, forms that have the same primary structure but different linking numbers. The linking number is a topological parameter that, applied to biology, describes the number of topological revolutions that one strand of a constrained DNA molecule (that is, one whose strands are not free to swivel) makes about the other strand when the molecule is flattened into a plane (Bauer, 1978). The linking number is made up of two components: twist (or the number of helical turns in the molecule); plus a quantity called "writhe", which is the number of times the duplex axis crosses over itself. When the duplex axis crosses over, it occupies a different plane, and so writhe is directional, *i.e.*, it can be either positive or negative, and describes three-dimensional DNA structure, *e.g.*, supercoiling. Thus, linking number describes the relation between higher order structure of DNA and its twist; in cells, it cannot change except by the operation of DNA topoisomerases. The function of these enzymes is, in fact, to catalyse physiologically relevant changes in the DNA linking number within segments of constrained DNA in chromatin (called loops) (Cook and Brazell, 1975).

Topoisomerase action is perhaps better understood by less abstract terms, such as "relaxation of DNA supercoiling", which is often, but not always, the net effect of topoisomerase action in cells. "Type II" refers to the changing of the linking number of DNA by this class of topoisomerase in steps of two. In the enzyme reaction, one double-stranded DNA segment is passed through a transient break in a second segment of the DNA duplex. In contrast, type I DNA topoisomerases change the linking number in steps of one by producing single strand breaks in the DNA; they do not require ATP (Wang, 1985). Both of these critical enzymes are found in the cell nucleus and are involved in many aspects of cellular growth and division. Type II topoisomerases function in chromosome segregation, sister chromatid exchange, DNA replication and repair, maintenance of the nuclear matrix, chromosomal recombination, and transcription (Wang, 1985; Beck and Danks, 1991; Sullivan and Ross, 1991). Many of the functions of one class of topoisomerase can, in theory, be performed by the other class, although it is likely that the type I enzyme does not perform functions requiring double strand DNA scission (Wang, 1985).

Two isoforms of topoisomerase II, having subunit molecular weights of 170 kDa and 180 kDa [also referred to in the literature as p170 (or α) and p180 (or β), respectively] have been identified (Drake *et al.*, 1987; Chung *et al.*, 1989). They vary in their relative amounts in different cell lines as well as in different growth stages within a cell line (Drake *et al.*, 1989; Woessner *et al.*, 1991). The topoisomerase II holoenzyme is a homodimer (Liu *et al.*, 1980; Wang, 1985), the subunits being dissociated under the denaturing conditions of polyacrylamide gel electrophoresis. In the literature, the enzyme is usually described in terms of its subunit molecular mass. Both α and β isoforms catalyse ATP-dependent unknotting of P4 DNA and covalent binding to DNA stimulated by the topoisomerase II inhibitor amsacrine (*m*-AMSA); the inhibitors teniposide (VP-16) and merbarone, on the other hand, selectively inhibit p170, the more extensively studied of the two (Drake *et al.*, 1987).

The mechanism of action of eukaryotic topoisomerase II has been described as a five-stage process (Osheroff, 1986): 1) recognition and binding of topoisomerase II to DNA; 2) cleavage of the first double-stranded segment of DNA, resulting in a covalent bond between the 5'-terminus of each strand, and a topoisomerase II polypeptide tyrosyl subunit; 3) passage of a second double strand of DNA through the break, or "gate"; 4) religation of the strand-cleaved DNA; and 5) ATP-dependent enzyme turnover (Fig. 1).

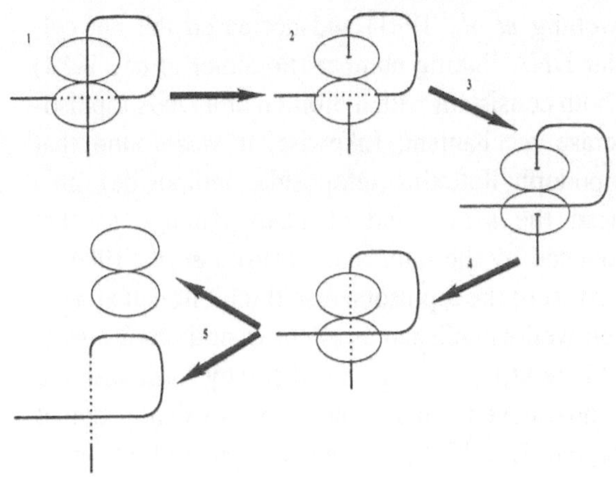

Fig. 1. Mechanism of DNA strand passage catalysed by topoisomerase II. In this illustration, the ovals represent the two subunits of the topoisomerase II homodimer – the active holoenzyme. They are associated with a continuous looped structure, representing a segment of double-stranded substrate DNA that binds near the active site of the enzyme. In the catalytic (strand passing) reaction, one DNA double-strand segment is passed through a break in a second double-strand segment at the point at which the two double-stranded regions intersect. (The dashed *horizontal* portion of the loop, depicted in 1 as lying behind the solid, *vertical* portion, is passed through the vertical portion, generating a DNA segment in which the dashed vertical portion now lies behind the horizontal [solid] portion). When the substrate DNA is part of a continuous molecule (for example, a covalent closed circle or a chromosomal DNA loop), this action of topoisomerase II changes the linking number of the DNA by 2.

In the enzyme reaction, topoisomerase II first binds to a recognized "consensus" sequence of DNA (1); second, a covalent bond is formed between a tyrosine of topoisomerase II and the 5'-phosphoryl terminus of each strand as the first double-stranded segment of DNA is cleaved (2); then a second double-strand of DNA is passed through the break (3); next, the break in the first DNA double-strand is religated (4); finally, active topoisomerase enzyme is regenerated in a step that requires ATP hydrolysis (5).

Soon after the discovery of topoisomerase II, two lines of experimental evidence established that it was the target of a number of antitumor agents. First, it was found that the intercalators Adriamycin, ellipticine, and amsacrine produced protein-associated DNA strand breaks and DNA-protein covalent cross-links in cells (Ross *et al.*, 1978;

Zwelling *et al.*, 1981) and decreased the net cellular DNA linking number (Pommier *et al.*, 1984) – both consistent with inhibition of a DNA topoisomerase mechanism. Likewise, it was found that epipodophyllotoxins (etoposide, teniposide) produced the same kind of DNA damage as that produced by the intercalators (Ross *et al.*, 1984).

Most of the topoisomerase II inhibitors that have been well-established as effective antitumor agents stabilize ternary complexes that they form with the strand-cleaved DNA and the covalently-bound enzyme (Liu, 1989; Liu and D'Arpa, 1992) (Fig. 1, step 2; Fig. 2, step 3), thus halting the catalytic reaction prior to the religation step (Fig. 1, step 4). The stabilized intermediate complexes present barriers to DNA fork progression, and so they become lethal lesions when cells attempt to utilize those regions of the genome as templates (Liu, 1989). Topoisomerase II inhibitors are generally classed as those which stabilize topoisomerase II-DNA complexes and those which inhibit by some other mechanism (Table 1).

It is clear from the above that 1) most known clinically useful topoisomerase II inhibitors kill cells by stabilizing enzyme-DNA complexes; and 2) the topoisomerase II inhibitors belonging to this class are associated with both typical and atypical MDR, while inhibitors belonging to the other class

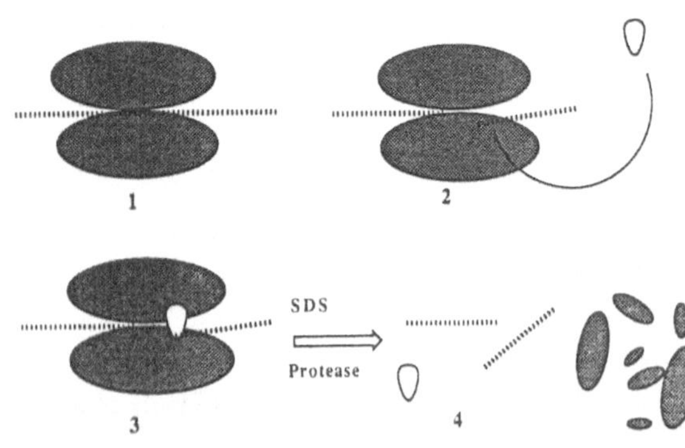

Fig. 2. Topoisomerase II inhibitor-induced DNA cleavage. The inhibitor (oval), strand-cleaved DNA, and topoisomerase II form a stable ternary complex, effectively inhibiting the catalytic reaction of topoisomerase II at the religation step (Fig. 1, step 4). In the presence of sodium dodecyl sulphate and proteinase, this stabilized complex yields DNA containing double-strand breaks.

are likely not associated with MDR (De Jong *et al.*, 1991).

Various topoisomerase II assays have been useful in evaluating MDR to the clinically relevant inhibitors of this enzyme. In the case of topoisomerase II inhibitors subject to classical MDR, it is often more convenient to assay for topoisomerase II activity that is detectable by the action of the inhibitor than to measure drug accumulation. In the case of atypical MDR, it is necessary to assay for some deficiency in topoisomerase II – either in its cellular content or its interaction with the inhibitor. Determining cellular content of topoisomerase II is a straightforward matter of immunoblot analysis or immunohistochemistry (Drake *et al.*, 1987; Efferth *et al.*, 1992). Two kinds of biochemical assay are also useful: assays of the DNA strand passing function by measuring P4 unknotting or DNA minicircle decatenation; and assays of inhibitor-induced DNA strand cleavage. DNA isolated from a mutant strain of P4 bacteriophage exists naturally in a topologically knotted conformation; to resolve these knots, it is necessary that the DNA undergo strand passing, which is catalysed by type II topoisomerases (Liu *et al.*, 1981). This method can be used to measure quantitatively the catalytic activity of topoisomerase II (Hofmann *et al.*, 1990). Likewise, catenated DNA can be resolved into individual double stranded DNA circles *via* strand passage (Marini *et al.*, 1980). An illustration of the P4 assay is given below (Fig. 3).

The cleavage function of the enzyme, which is directly related to the ability of inhibitors to kill cells by a topoisomerase II-associated mechanism, can be assayed by using intercalators such as amsacrine, which trap the reaction intermediate with the DNA in a strand-cleaved state; addition of proteinase and sodium dodecyl sulphate leads to DNA double-strand breakage (Fig. 2; also see below), which can be assayed in a variety of ways, including DNA filter elution (Kohn, 1979) and K^+-SDS precipitation (Trask *et al.*, 1983).

Given that each topoisomerase II-DNA interaction is a potential target for cytotoxic topoisomerase II inhibitors, it follows that the efficacy of an antineoplastic agent that works by the topoisomerase II cleavage mechanism is proportional to the

Table 1. Susceptibility to MDR of various toxic antineoplastic agents. Among the topoisomerase II inhibitors, both the DNA intercalating and the nonintercalating complex-forming inhibitors are affected by typical and atypical MDR, while fostriecin, novobiocin and merbarone, which do not stabilize ternary complex formation, have not been shown to be affected by either typical or atypical MDR. Tubulin binders are susceptible to typical but not to atypical MDR.

TOPO-II INHIBITION	MDR Type		COMPOUNDS AND STRUCTURES
	Typ	Atyp	
INTERCALATORS	+	+	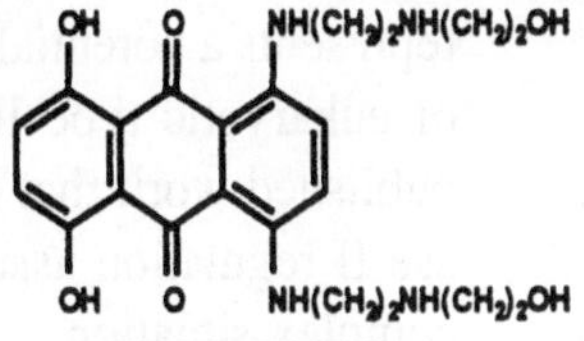 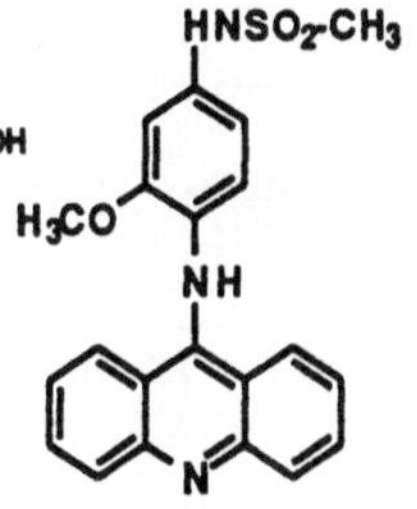 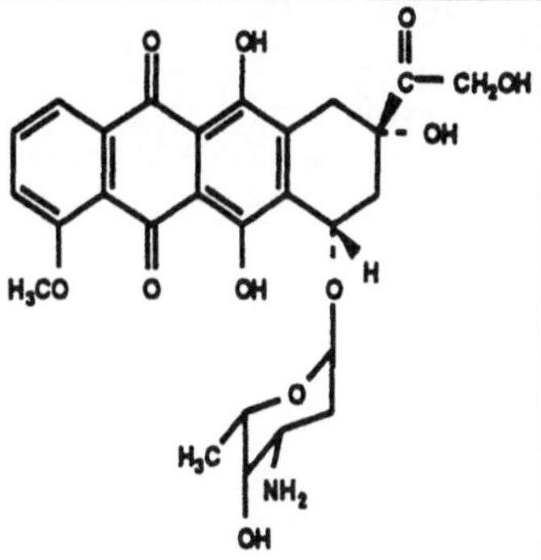 MITOXANTRONE, 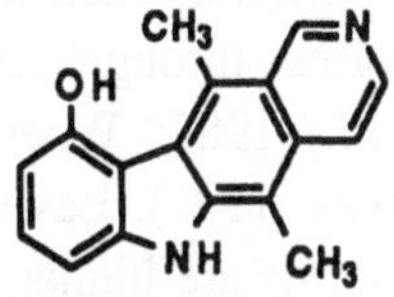 AMSACRINE, DOXORUBICIN, ELLIPTICINE
NON-INTERCALATORS	+	+	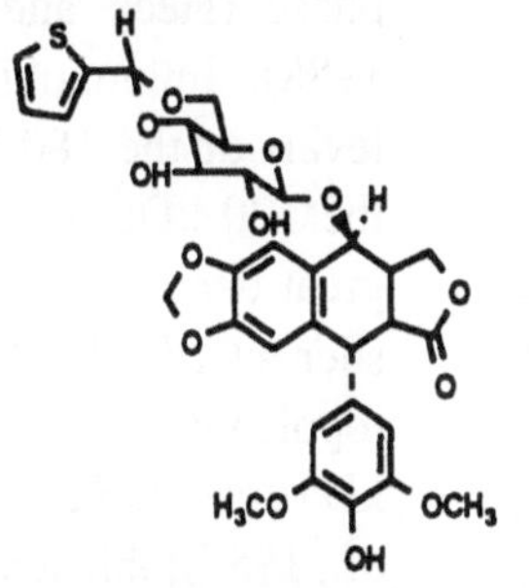 TENIPOSIDE, 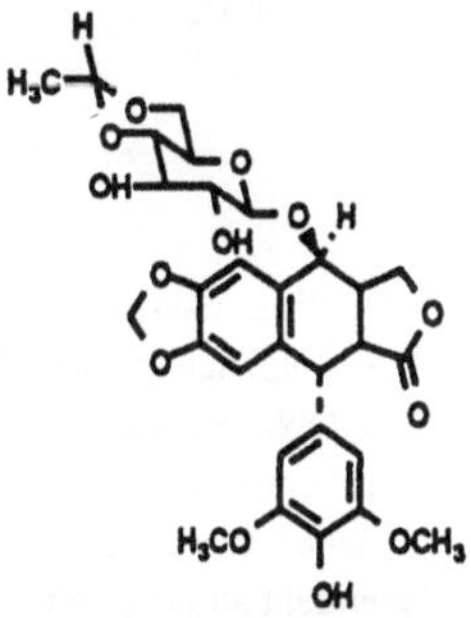 ETOPOSIDE
NON-INTERCALATORS	?	?	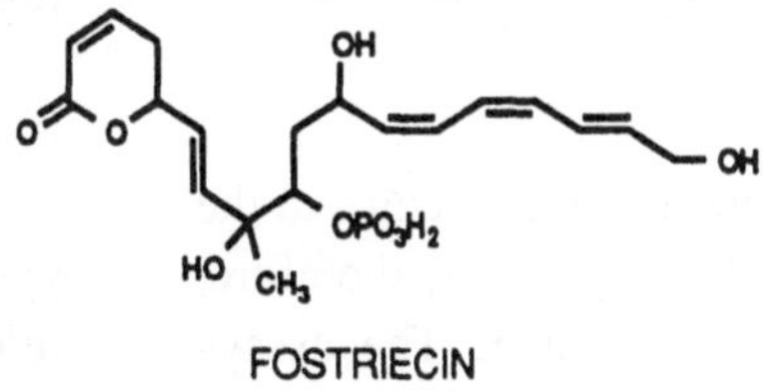 FOSTRIECIN,  MERBARONE, 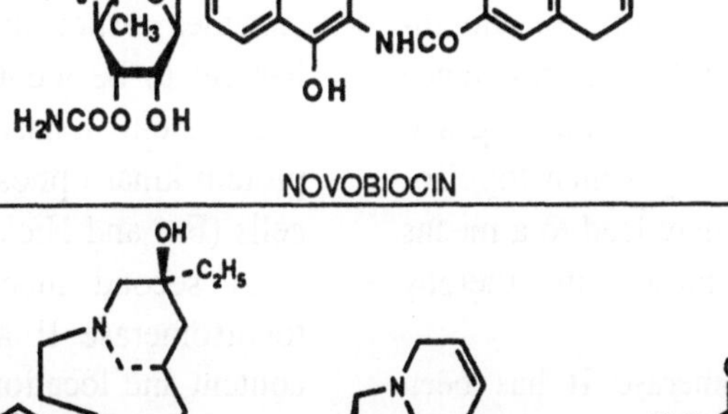 NOVOBIOCIN
TUBULIN BINDERS	+	-	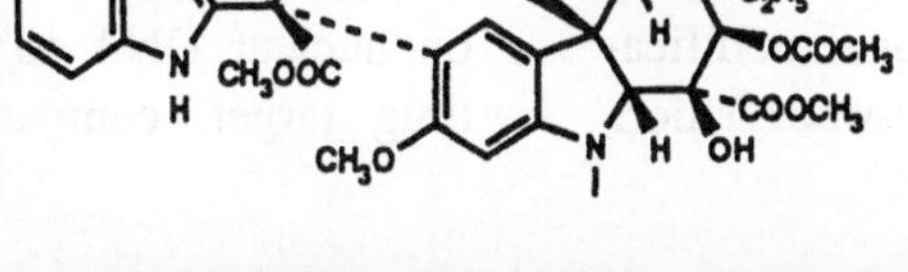 TAXOL 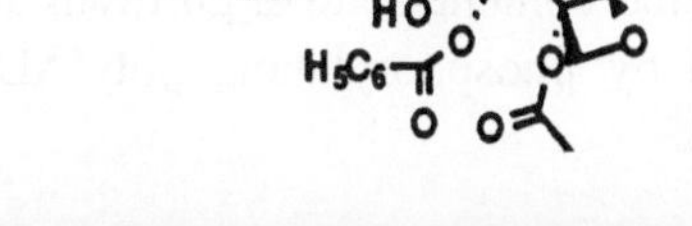

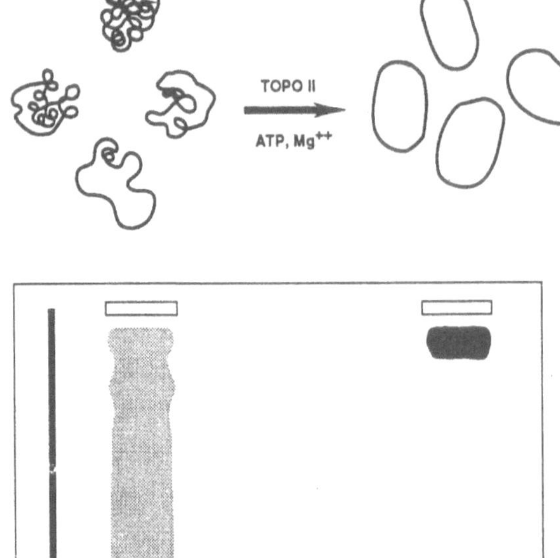

Fig. 3. P4 unknotting assay for topoisomerase II catalytic activity. P4 DNA is isolated in a topologically knotted form; the population of P4 DNA molecules is heterogeneous owing to differences in the numbers of knots among the individual molecules. Knotted DNA therefore migrates in agarose gels as a smear. ATP-dependent strand passing activity of topoisomerase II resolves the knots, producing a homogeneous population of circular DNA that migrates as a tight band in agarose gels.

number of active topoisomerase II molecules in the cell. Therapeutic advantage may result, therefore, from maximizing the target number. The assays described above (including immunoblotting with antibodies specific for topoisomerase II to determine cellular content) provide ways to estimate the relative numbers of available targets for the inhibitors, as well as to estimate degrees of resistance to inhibitors (Boege *et al.*, 1992). Identifying and understanding the mechanisms by which topoisomerase II is regulated in cells may lead to a means of upregulating cellular enzyme activity, thereby increasing efficacy.

The regulation of topoisomerase II has been studied extensively, and at many levels, including cellular content, state of posttranslational modification by phosphorylation, poly(ADP)-ribosylation, *etc.*, and subcellular distribution of the enzyme. All regulatory mechanisms apply, it must be kept in mind, to dimers made up of two 170 kDa or two 180 kDa subunits. In addition, topoisomerase II heterodimers may exist in cells. This possibility has not yet been addressed experimentally, but it represents a potential complication in the regulation of eukaryotic type II topoisomerases. Most of the published work that is concerned with topoisomerase II regulation assumes homodimers as the most complex situation.

Cells vary their topoisomerase II content as they progress through various steps of the cycle (Estey *et al.*, 1987; Woessner *et al.*, 1991; Beck and Danks, 1991). Levels of the 170 kDa enzyme (α isoform) are higher in actively growing cells than in stationary phase cells, and are highest during S-phase (Heck and Earnshaw, 1986; Heck *et al.*, 1988). Immunoblotting studies have shown that levels of the 180 kDa (β) isoform, unlike those of the 170 kDa (α) isoform, do not appear to vary a great deal as a function of cell cycle stage (Woessner *et al.*, 1991). With respect to inhibitors of topoisomerase II that are affected by MDR, the 170 kDa form appears to be the better target (Drake *et al.*, 1987); this may be why the complex-stabilizing inhibitors are somewhat proliferation-specific as regards toxicity (Sullivan *et al.*, 1986).

A principal way in which relative topoisomerase II activity may be regulated is by phosphorylation; *in vitro* and in cells, topoisomerase II is phosphorylated at specific serines, resulting in a 2– to 10-fold increase in activity (Ackerman *et al.*, 1985; Fry and Hickson, 1993). While protein kinase C and casein kinase II have been shown to phosphorylate purified topoisomerase II, the enzyme or enzymes responsible for phosphorylation in cells has yet to be identified. A recent report suggests that casein kinase and cyclic AMP-dependent protein kinase phosphorylate topoisomerase IIα in cells (Fry and Hickson, 1993).

A second mechanism of regulating cellular topoisomerase II activity is to vary the cellular content and location (compartmentalization) of the two (or more) isoforms. Since the enzyme must act on nuclear DNA to produce a potentially lethal drug target, compartmentalization could be an important way of controlling the cellular content of active topoisomerase II. It is not clear how compartmentalization is achieved, but it is known that the enzyme is a nuclear matrix protein and, thus, has a structural role in addition to its catalytic function. Changes in chromatin structure during the cell cycle may, *via* changes in the components of the nuclear matrix, make topoisomerase II more accessible or less accessible to DNA. The biochemical mechanism of such changes is not known, but it has been suggested that phosphorylation of topoisomerase II can alter its distribution within the nucleus (Beck and Danks, 1991), perhaps as part of a more general scheme of control over cell division.

It is appropriate to consider the following before typical and atypical MDR to topoisomerase II inhibitors are discussed in detail. It is well-known that tumors develop resistance to the complex-forming class of chemotherapeutic topoisomerase II inhibitor by overexpressing Pgp, altering topoisomerase II, overexpressing metabolizing enzymes such as glutathione-S-transferase, or any combination of these changes (Harris and Hochhauser, 1992; Lum *et al.*, 1993). A mutation in the TopII gene, for example, could produce a change in activity that results in altered binding of ATP to the enzyme; alternatively, it could change the cellular distribution of topoisomerase II. Phosphorylation or dephosphorylation could modulate the activity of the enzyme, thereby causing resistance or hypersensitivity to topoisomerase II inhibitors. Resistance could also be the result of changes in the isoform population; the p170 form, the more extensively studied of the two, was found to be diminished relative to the p180 form in amsacrine-resistant P388 murine leukemia cells (Drake *et al.*, 1987). It is thus possible that regulation of these forms is somehow involved in multidrug resistance.

Topoisomerase II in typical MDR

A number of cultured cell lines that overexpress Pgp (but have no deficiencies in topoisomerase II) and are resistant to vinblastine and vincristine (*i.e.*, are MDR in the classical sense) are cross-resistant to intercalating and nonintercalating inhibitors of topoisomerase II (Boiocchi *et al.*, 1992; Glisson and Altpeter, 1992). Their topoisomerase II is normal in amount and in its interaction with inhibitors. This apparent discrepancy is easily explained, since a) the ternary complex-stabilizing group of topoisomerase II inhibitors includes naturally occurring compounds like mitoxantrone, or semisynthetic analogues such as the epipodophyllotoxins and structurally related synthesized molecules, and b) such compounds are bound and exported preferentially by Pgp (Beck and Danks, 1991).

To overcome or prevent classical MDR to topoisomerase inhibitors, therapies that employ antagonists of Pgp have been used both experimentally and in clinical protocols to attempt reversal of the MDR phenotype by inhibiting Pgp or otherwise increasing net accumulation of the cytotoxic drug. Several structural and mechanistic classes of compound have been employed for this purpose. Among these are: calcium channel blockers (*e.g.*, verapamil), antimalarials, antibacterials, calmodulin inhibitors, steroids, and cyclosporins (*e.g.*, Cyclosporin A) (Lum *et al.*, 1993).

Verapamil, the first such compound widely used and, perhaps, still the standard adjuvant compound, has been reported to overcome MDR in combination with a number of clinically active MDR-sensitive topoisomerase II inhibitors (Darkin and Ralph, 1986; Slater *et al.*, 1986; Lum *et al.*, 1993). Cyclosporin A and other cyclosporins have also been employed for this purpose. Against many experimental tumors, for example, L1210 and P388 murine leukemias, verapamil and cyclosporin A have similar efficacies. Clinical trials have demonstrated that serum levels that are efficacious in the various *in vitro* studies can be achieved in patients (Slater *et al.*, 1993). A lipophilic compound from Elf-Sanofi (S 9788) has shown activity in clinical trials as an adjuvant; it is 300 times more potent than verapamil or Cyclosporin A. Pharmacokinetics determined using CCRF CEM wild-type and vinblastine-resistant (MDR) cells showed that this compound increases the toxicity of vinblastine, vincristine, and Adriamycin after 1 or 6 h of incubation in a concentration-dependent manner, and that it is active at achievable serum levels. Phase II trials are planned for S 9788 in combination with *Vinca* alkaloids and anthracyclines (Soudon *et al.*, 1993).

Topoisomerase II inhibitors have been modified chemically or, in some cases, in their physical milieux, to increase cellular retention. For example, Adriamycin has been incubated with MDR human KB cells as a transferrin conjugate (transferrin having cellular receptors to aid in active transport). Although the MDR cells are rendered more sensitive to killing by Adriamycin given this way, the wild-type cells are sensitized as well, raising questions about toxicity (Fritzer *et al.*, 1993). Adriamycin has also been encapsulated in liposomes to increase its cellular retention time. A second-generation liposomal formulation, TLED (Thermosensitive Liposome Encapsulated Doxorubicin), has been utilized to attempt to reverse MDR in the experimental system consisting of wild-type and MDR human breast cancer (MCF-7) cells. Liposomes were separated by exclusion chromatography after Adriamycin had been encapsulated in them by standard procedures. The temperature was shifted from 37° to 43°C to melt the liposomes and deliver the compound in an efficient manner. By employing liposomal delivery plus hyperthermia, it is possible to achieve the same effect as that obtained with Adriamycin plus verapamil. Unfortunately, potentiation in the case of TLED has been observed to be greater for the wild-type than for the MDR MCF-7 cells, raising some doubt about the clinical utility of this procedure.

A large number of additional compounds and other novel agents targeted against Pgp or some other molecule associated with decreased net drug accumulation have been studied, primarily using cultured cell lines, to find improved adjuvant treatments for clinical reversal or prevention of MDR. Among the compounds studied have been β-naphthoflavones (Scotto *et al.*, 1993); protein kinase C inhibitors (Fine *et al.*, 1988; Ganapathi *et al.*, 1991; Dickstein *et al.*, 1993; Scala *et al.*, 1993), and bacterial toxins (Mickisch *et al.*, 1993). MRK-16, a monoclonal antibody prepared against an epitope of Pgp, has been observed to lower by 50% the IC_{50} for toxicity toward the highly MDR human colon cell line HT-29 (Sela *et al.*, 1993). Perhaps one of the more novel adjuvants reported to date is a ribozyme designed to cleave specifically the mRNA for Pgp (Kobayashi *et al.*, 1993). Ribozymes and antisense oligonucleotides tend to have long half-lives in cells, and they may be useful anti-MDR agents in the future.

Potential limitations exist for the utilization of these, or for any adjuvant protocols, in cancer treatment. These include novel toxicities that may arise from the combination treatment, and alterations of the pharmacokinetics of one of the agents by the other agent so as to magnify its toxicity. An illustration of the latter problem can be seen in a study in which the chemosensitizer Cyclosporin A has been observed to produce changes in the tissue distribution and toxicity of Adriamycin given to mice or rats (Colombo *et al.*, 1993). No significant changes in serum levels were seen after 48 h, but concentrations increased in the liver, kidneys, adrenals, but not in the brain or heart. The results are consistent with an efflux block mediated by Cyclosporin A, leading to net retention of Adriamycin in the affected organs. The treated mice, however, experienced dramatic delayed toxicity (death) several days after treatment, with no deaths occurring among control animals or those treated with either agent alone. This result could have been produced by a toxicity peculiar to the combination treatment, or by a change in the pharmacokinetics of Adriamycin produced by Cyclosporin A.

Despite (or, perhaps, because of) the modest progress achieved in the clinic thus far against MDR, it is likely that in the next several years many of these novel treatments will undergo evaluation in combination with topoisomerase II inhibitors. Trials are now underway to find Pgp antagonists that have less toxicity and greater efficacy than those now in use.

Topoisomerase II in atypical MDR

The molecular basis of atypical multidrug resistance to the group of complex-stabilizing topoisomerase II inhibitors is the enzyme itself. Up to now, nearly all of the work reported has been performed using the 170 kDa subunit homodimeric form of topoisomerase II, which is the more abundant and the more extensively-studied, and which appears to be the more drug-sensitive of the two (Drake *et al.*, 1989). In one study, however, it was found that resistance to mitoxantrone in HL-60 cells was associated with reduced expression of topoisomerase IIβ (Harker *et al.*, 1991).

The identifying characteristic of atMDR cell lines is that they are not cross-resistant to the *Vinca* alkaloids (Roberts *et al.*, 1987; Beck, 1989), which are classical MDR drugs (tubulin binders), but, rather, appear to be cross-resistant *only* to other topoisomerase II inhibitors. The term "atypical" was proposed by Beck and colleagues (Danks *et al.*, 1987) for multidrug resistant cells having this property, and it has been in general use in recent years.

To understand atMDR, it is necessary to keep in mind the mechanism of cytotoxicity of the topoisomerase II inhibitors affected by this type of resistance (Isabella *et al.*, 1989; 1991). Toxicity depends not on inhibiting the catalytic activity of the enzyme in cells, but, rather, on forming and stabilizing ternary DNA complexes (Fig. 2). The earliest reports of what was subsequently identified as atypical MDR utilized the methods that were available to assay topoiosmerase II-related functions — *i.e.*, immunoblotting to determine cellular topoisomerase II content, biochemical assays to determine catalytic activity extractable from cell nuclei, and the formation of the stabilized ternary complexes, assayed as drug-induced protein-concealed DNA strand breakage, drug-induced DNA-protein covalent cross-linking, or extractable drug-dependent DNA cleavage activity. AtMDR cell lines were believed to be resistant because of either reduced cell content of topoisomerase II, or a mutation in the enzyme, which was present in normal amounts. The latter has been confirmed in more recent studies, some performed with the cell lines reported initially as atMDR; these studies took advantage of techniques such as DNA sequence amplification by polymerase chain reaction (PCR) and molecular hybridization to identify mutations in the TopII gene that likely were responsible for the observed resistance. We have listed several of these atMDR cell lines (Table 2) along with reported abnormalities involving topoisomerase II.

Two types of atMDR cell lines have been described, the type being defined according to whether there is a reduction in cellular content of topoisomerase II (*e.g.*, P388/AMSA, KB40a; see Table 2) or a qualitative change in the enzyme that renders it resistant to the antineoplastic topoisomerase II inhibitors (*e.g.*, CCRF-CEM-VM-1, HL-60/AMSA; see Table 2). Each type could have arisen by a genetic or by an epigenetic mechanism. Recently, a temperature-sensitive yeast topoisomerase II mutant has been described which has collateral temperature-dependent atMDR (Nitiss *et al.*, 1993). An examination of the more thoroughly characterized atMDR mammalian cell lines supports the notion that mutation in the TopII gene is a likely cause of most cases of atMDR, producing either decreased cell content or a mutant, noncleaving form of the enzyme.

Qualitative alterations to the normal-sized (170 kDa) topoisomerase II molecule, suggested by reduced capacity of the inhibitors to stabilize topoisomerase II-DNA complexes, directly suggests a simple point mutation in TopII. The cell lines Vpm^R-5, CCRF-CEM-VM-1, HL-60/AMSA, and KBM3/AMSA (all of which were characterized initially as being deficient, not in the cellular content of topoisomerase II, but in its drug-induced DNA cleavage activity) have been found to contain point mutations in a single, specific region of the TopII (actually, TopIIα) gene. All mutations detected have been G to A transitions (Bugg *et al.*, 1991; Hinds *et al.*, 1991; Chan *et al.*, 1993). Analyses such as these have been facilitated by gene amplification techniques (PCR), which permit the sequencing of the mutated regions in the genomic DNA, thereby providing confirmation of the mutation identified in the cDNA. The mutational hot spot in TopIIα is a region at the N-terminal third of the enzyme that codes for the putative ATP-binding site of topoisomerase IIα (amino acids 449–494). This region is highly conserved among topoisomerases from bacteriophage type II topoisomerases through bacterial gyrases to mammalian topoisomerase IIα (Chan *et al.*, 1993).

Table 2. Properties of some at-MDR mammalian cell lines

Cell line	Resistant to	Topoisomerase II-associated abnormalities	References
SW900 small cell lung carcinoma	etoposide, teniposide (natural)	Decreased etoposide-, teniposide-induced DNA cleavage	Long *et al.*, 1986
VpmR-5(CHO)	etoposide	Decreased Amsacrine, etoposide-induced DNA cleavage TopII point mutation G_{1478} to A Arg_{493} to Gln	Glisson *et al.*, 1986 Chan *et al.*, 1993
P388/AMSA	Amsacrine	Decreased cell content Decreased Amsacrine-induced DNA cleavage TopII rearrangement 1/2 alleles	Johnson and Howard, 1982 Per *et al.*, 1987 Tan *et al.*, 1989
CCRF-CEM-VM-1	teniposide	Decreased ATP binding Decreased activity Decreased teniposide-induced DNA-protein complexes TopII point mutation G_{1478} to A Arg_{449} to Gln	Danks *et al.*, 1988 Bugg *et al.*, 1991
KB/40a	etoposide	Decreased activity Decreased mRNA, cell content	Ferguson *et al.*, 1988
P388/ADR-3	doxorubicin (Adriamycin™)	Decreased activity Decreased cell content Truncated mRNA Mutant TopII allele?	Deffie *et al.*, 1989
GLC_4/ADR	doxorubicin (Adriamycin™)	Decreased activity Decreased Amsacrine, etoposide-induced DNA cleavage	De Jong *et al.*, 1990
HL-60/AMSA; KBM3/AMSA*	Amsacrine	Decreased Amsacrine-induced DNA cleavage TopII point mutation G_{1457} to A Arg_{486} to Lys	Hinds *et al.*, 1991 Lee *et al.*, 1992
CALc18/AMSA	Amsacrine	Decreased activity Decreased SSB Decreased DNA-protein cross-links Decreased Amsacrine-induced cleavage	Lefevre *et al.*, 1991
H69AR	doxorubicin (Adriamycin™)	Decreased etoposide, doxorubicin-induced SSB Decreased cell content	Cole *et al.*, 1991
CHO-SMR_5	9-(4,6-*O* -ethylidine-β-D-glucopyranosyl)-4'-demethyl epipodophyllotoxin	Decreased activity Decreased drug-induced DNA DSB Decreased DNA-protein cross-links Decreased mRNA	Webb *et al.*, 1991

* minimally cross-resistant to other topoisomerase II inhibitors.

Point mutations in this conserved region of TopIIα obviously can result in an enzyme that has only marginally impaired catalytic activity but diminished capacity to bind ATP and cleave DNA in the presence of inhibitor. This result has been obtained in studies with a number of cell lines (Table 2). Catalytic activity is not always unimpaired by the mutation, however; for example, CCRF-CEM human leukemia cells made resistant to teniposide are characterized by diminished topoisomerase II strand-passing activity as well as diminished drug-induced enzymic DNA cleavage (Danks *et al.*, 1988). Unless the amino acid substitution resulting from the point mutation selectively impairs stabilization of the topoisomerase II reaction intermediate by one particular drug, the mutation leads to multidrug resistance. In the case of the HL-60/AMSA and KBM3/AMSA cell lines, ratios of cross-resistance to topoisomerase II inhibitors other than amsacrine are not impressive, and it has been argued that these cells are, in fact, not multidrug resistant, but amsacrine-resistant (Lee *et al.*, 1992).

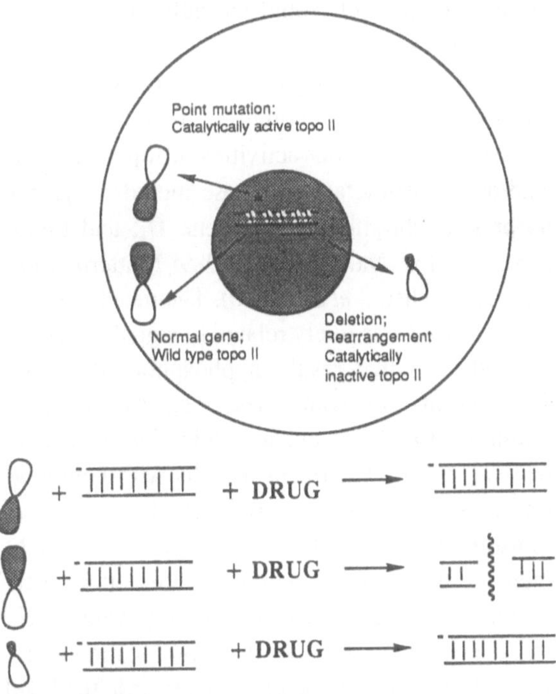

Fig. 4. Point mutations and deletion/rearrangement mutations in TopII can lead to resistance to cleavage of DNA by mutated topoisomerase II in the presence of an atMDR-type inhibitor, *e.g.*, amsacrine or etoposide (labelled "DRUG"). Point mutations may result in topoisomerase II enzymes with nearly normal catalytic activity; deletions and rearrangements would lead to catalytically inactive enzyme. Only the wild-type enzyme, however, cleaves in the presence of inhibitor.

Mutations other than point mutations could also lead to resistance to topoisomerase II inhibitors. The P388/AMSA subline, first described in 1982 (Johnson and Howard, 1982), has been demonstrated to have a reduced cellular content of normal topoisomerase II (170 kDa form), leading to an observed deficiency in protein-linked DNA strand breaks upon incubation of its *in vitro*-derived sublines with amsacrine and related topoisomerase II inhibitors (Per *et al.*, 1987). Subsequent studies with these cells revealed that the gene for the 170 kDa subunit form of topoisomerase II was rearranged in one of two alleles, and that it was hypermethylated (Tan *et al.*, 1989). Both of these actions on the gene would be expected to lead to a diminished cellular content of functional topoisomerase II. The doxorubicin-resistant P388 line described by Deffie *et al.* (1989) is probably similar; it contains a reduced quantity of normal-sized 170 kDa topoisomerase II as well as a truncated, presumably inactive, version of the enzyme. This is consistent with a gene deletion or rearrangement in one allele. It is possible that similar evaluations of most of the other atMDR cell lines identified will detect TopII genetic defects of these sorts.

Additional studies indicate that topoisomerase II may play a more important role than either detoxification or efflux mechanisms in the development of multidrug resistance by tumors. For example, in seven of a series of eight human lung cancer cell lines, topoisomerase II gene expression was positively correlated with sensitivity to epipodophyllotoxins (teniposide [VM-26] and etoposide [VP-16]), and doxorubicin, while no such correlation was observed between topoisomerase II expression and sensitivity to 5-fluorouracil, which does not inhibit topoisomerase II (Giaccone *et al.*, 1992). In another study of clinical tumor cell lines, involving six human xenografts, it was found that Adriamycin (doxorubicin)-responsive tumors had higher topoisomerase II mRNA levels than did unresponsive tumors. On the other hand, this study demonstrated no significant relationship between the amount of MDR1 or GST (Glutathione-S-transferase)π mRNA expressed and drug sensitivity (Kim *et al.*, 1992).

Of the large number of atMDR cell lines that have been analyzed, most have been produced by selection in the presence of an appropriate topoisomerase II inhibitor; some have, in addition, been exposed to mutagenizing treatment prior to this selection. Selection in the presence of topoisomerase II inhibitors that produce DNA lesions is itself potentially mutagenic (Wang, 1985; Liu and D'Arpa, 1992). It is, thus, not surprising that a number of atMDR lines contain mutations not only in their topoisomerase II alleles, but also in other genes. These cell lines are pleiotropic and not purely atMDR (McClean and Hill, 1992). An example of this type is the P388/doxorubicin line, generated *in vivo* by the same protocol used to produce P388/Amsacrine (Johnson and Howard, 1982); these cells overexpress the MDR-1 gene, are cross-resistant to *Vinca* alkaloids, and, thus, have both typical and atypical MDR characteristics. Other rodent and human cell lines with multifaceted MDR to topoisomerase II inhibitors have been described (Ferguson *et al.*, 1988; Spiridonidis *et al.*, 1989; Zwelling *et al.*, 1991; Deffie *et al.*, 1992; Kamath *et al.*, 1992; Volm *et al.*, 1992; Ishikawa *et al.*, 1993). A further complication is that most of the atMDR cell lines are not strictly diploid; thus, a simple pharmacologic-genetic model is often difficult to envision. This complexity may be closer to the actual situation faced by the clinical oncologist, however, as it is consistent with the known heterogeneity and hypermutability of malignant tumors. In a study of Pgp and topoisomerase I and II expression in a panel of malignant ovarian tumors, it was found that Pgp expression was highly variable, as were topoisomerase I and II activities (Van der Zee *et al.*, 1991). It may be that mutations occur in the topIIα and MDR-1 genes at a greater than random frequency. Thus, clinicians encountering the problem of multidrug resistance may have to consider several concurrent strategies for eliminating it.

Overcoming atMDR: strategies employing topoisomerase II

Adjuvant MDR-targeted therapy that employs Pgp antagonists or inhibitors of MDR-1 expression obviously is not indicated in the case of "pure" atMDR, where Pgp is not overexpressed in the multidrug resistant target tumor cells. Various alternate approaches to abrogate atMDR have been investigated using tissue culture and animal model systems; obviously, they are not as far along in development as are the MDR adjuvant studies. Studies undertaken to reverse atMDR are based upon two principles. The first is that the cytotoxicity of those topoisomerase II inhibitors affected by atMDR is related directly to the number of functioning topoisomerase II enzymes per cell (see Fig. 2). Resistance therefore is characterized by a decrease in this population, and so any treatment that increases the number of active topoisomerase II molecules per cell could partially abrogate this type of resistance.

Several kinds of agonist interactions at cell surfaces have been reported to induce transient increases in the cellular activities of topoisomerase I and/or topoisomerase II. These include: arginine vasopressin; thrombin; leukotriene D_4; and tumor necrosis factor (Nambi *et al.*, 1989; Mattern *et al.*, 1989, 1991; Utsugi *et al.*, 1990). The mechanism of these activations is likely related to signal transduction pathways that result in phosphorylation and activation of topoisomerases, *e.g.*, by activated protein kinase C or cyclic nucleotide-dependent kinases (Fig. 5). Treatment of cells with combinations consisting of these agonists and inhibitors of topoisomerases I or II has been demonstrated to potentiate the cytotoxic activity of the topoisomerase inhibitors in a synergistic fashion (Utsugi *et al.*, 1990). Thus, it is possible that combination treatments of this type can be used to decrease the level of atMDR in tumors.

The second principle, first enunciated as a result of studies of bacterial gyrase and type I topoisomerase enzymes (Gellert *et al.*, 1982), states that type I and type II cellular topoisomerase activities are regulated in a way that maintains cellular homeostasis with respect to higher order DNA

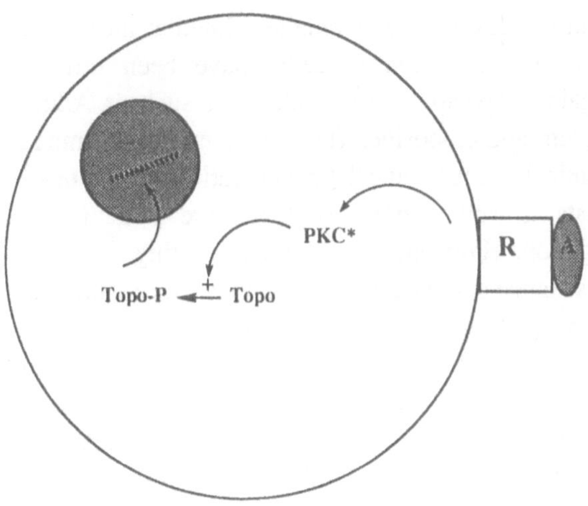

Fig. 5. Activation of cellular topoisomerase (I or II) by agonist interactions at the cell surface. An agonist (a hormone, cytokine, *etc.*) (A) binds to its cell surface receptor (R), which activates, *via* cellular signalling pathways, phosphorylating enzymes such as protein kinase C (PKC*, where the asterisk refers to the activated state). Activated PKC, in this example, phosphorylates a topoisomerase (Topo to Topo-P), thereby activating it. The proportion of topoisomerases interacting with nuclear DNA is thus increased, leading to increased sensitivity to topoisomerase inhibitors, or decreased resistance. If the "Topo" in this example is topoisomerase II, the model illustrates a possible way of decreasing atMDR, *i.e.*, by combination treatment with A plus topoisomerase II inhibitor.

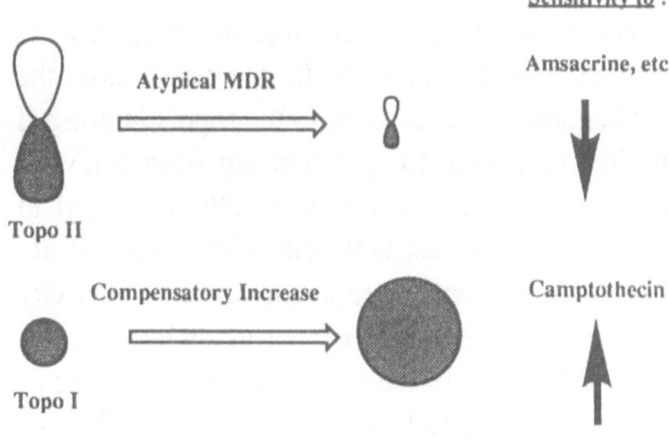

Fig. 6. Cellular compensation for loss of topoisomerase II content (and activity, thereby leading to atMDR) by increasing topoisomerase I content (or activity). This renders the cell hypersensitive to camptothecin, an inhibitor of topoisomerase I that kills cells by a mechanism similar to that used by topoisomerase II inhibitors such as amsacrine (see Fig. 1) Treatment of cells with camptothecin in combination with topoisomerase II inhibitors such as amsacrine or etoposide may help to prevent or overcome atMDR.

structure. According to a simplistic interpretation of this principle, the amount of change in DNA tertiary structure that is catalysed by type I topoisomerases in a given cell is balanced by the amount of change in DNA tertiary structure catalysed by type II topoisomerases in that cell. In mammalian cells, for example, a deficit in topoisomerase II activity might be partially offset by an increase in topoisomerase I activity (Fig. 6).

The situation as regards mammalian cells is very likely more complicated than that described above. Evidence that is consistent with this model, however, has been obtained in studies with *in vivo* or cultured cell lines; these studies correlate sensitivity to topoisomerase inhibitors with amount of topoisomerase I and II content and activity. For example, P388 cells that are resistant to the topoisomerase I inhibitor camptothecin, by virtue of greatly reduced cellular topoisomerase I content and activity, have been found to contain *increased* topoisomerase II enzyme and catalytic activity when compared to the wild-type, camptothecin-sensitive cells. This finding is consistent with a compensatory adjustment to topoisomerase I and II for the purposes of maintaining DNA topological balance. More important with respect to drug resistance, these topoisomerase I inhibitor-resistant cells have been found to be hypersensitive to topoisomerase II inhibitors amsacrine and etoposide (Eng *et al.*, 1989). What about atMDR cells that are resistant to topoisomerase II inhibitors? Hypersensitivity to camptothecin has been reported in one study of amsacrine-resistant human breast cancer cells (Lefevre *et al.*, 1991), but not in rodent cultured cell lines resistant to MDR topoisomerase II inhibitors in which it was assayed (Per *et al.*, 1987; Sullivan *et al.*, 1993). Additional evidence is also consistent with this notion. P388 cells that are made hyperresistant to camptothecin (1000-fold) by continuous exposure contain about 5% of normal topoisomerase I activity, but about twice the normal topoisomerase II activity. When these cells are grown in the absence of camptothecin, hyperresistance is lost, and they become merely resistant to camptothecin.

Moreover, loss of hyperresistance to camptothecin is accompanied by acquisition of resistance to topoisomerase II inhibitors. In this conversion, the cells become more resistant to the topoisomerase II inhibitor etoposide (IC_{50} increasing from 5 μM in the camptothecin-hyperresistant cells to 18 μM in the camptothecin-resistant cells) (Woessner *et al.*, 1993). Concurrently, topoisomerase I activity increases to about 60% of normal while the increase in topoisomerase II activity characteristic of the hyperresistant cells is reduced to about 125% of normal (Woessner *et al.*, 1993). These results suggest that combination treatments with topoisomerase I and topoisomerase II inhibitors have the potential to abrogate or prevent "atypical" multidrug resistance. Therapeutic synergy in animal models has recently prompted Phase I cancer trials of combination treatments using topotecan, a topoisomerase I inhibitor (related to camptothecin) that is now in Phase II/III clinical trial for solid tumors, and etoposide, an established cancer therapeutic agent (Eckardt *et al.*, 1993).

Conclusions

Inhibitors of two of the major targets of current front-line cancer chemotherapy – tubulin (formation) and topoisomerase II – share the common problem of MDR in the clinic, apparently in contrast to alkylating agents and other classes of antitumor drug. At least two mechanisms of MDR to topoisomerase II inhibitors have been described – classical, or typical efflux *via* Pgp, and a second mechanism termed atypical, which thus far has been characterized as arising from downregulation of the target enzyme, topoisomerase II. Because of the mutagenicity associated with the use of topoisomerase II inhibitors as well as the hypervariability of tumor cell populations, many malignant cell lines are resistant through both mechanisms.

In typical and atypical MDR, resistance is directly related to the intracellular concentration of both drug and target enzyme. To overcome classical MDR, strategies similar to those employed against MDR to tubulin binders such as vincristine and vinblastine (*i.e.*, strategies aimed at increasing net drug retention in cells) have been utilized against topoisomerase II inhibitors such as Adriamycin and etoposide. These have met with limited clinical success, and later generation Pgp antagonists as well as novel strategies are under investigation. Preventing or circumventing atypical resistance could be based upon either increasing the total cellular topoisomerase II activity through activation (for example, with receptor agonists such as tumor necrosis factor) or cotreatment with inhibitors of topoisomerase I, whose mechanism of cell killing is similar to that of complex-stabilizing topoisomerase II inhibitors, and whose activity may be increased in cells with downregulated topoisomerase II activity.

Acknowledgements

We thank our colleagues at SmithKline Beecham Pharmaceuticals for their many contributions to this work. We are particularly grateful to those alongside whom we have struggled during the past eight years to understand better those mysterious enzymes called DNA topoisomerases: Henry Bartus, Joan O'Leary Bartus, David Berges, David Berry, Rebecca Boyce, Ann Breen, Mary Jo Caranfa, Thomas Chung, Stanley Crooke, Fred Drake, Wai-Kwong Eng, Leo Faucette, Terry Francis, Lois Funk, Nabil Hanna, Robert Hertzberg, Ken Holden, Randall Johnson, William Kingsbury, Yan Leychkis, Francis McCabe, Christopher Mirabelli, Seymour Mong, Shau-Ming Mong, Rhonda Morrison, Ponnal Nambi, Donna Papsun, Israil Pendrak, Steven Per, Rodd Polsky, George Poste, David Rieman, Chiu-mei Sung, Jack Taggart, KB Tan, Richard Woessner, Charlene Wu, Susie Young, Young Yu, Teruhiro Utsugi and Joseph Zimmerman.

References

Ackerman P, Glover CVC and Osheroff N (1985) Phosphorylation of topoisomerase II by casein kinase II: modulation of eukaryotic topoisomerase II activity *in vitro*. Proc Natl Acad Sci USA 82: 3164–3168.

Baguley BC, Finlay GJ and Ching LM (1992) Resistance mechanisms to topoisomerase poisons: the application of cell culture methods. Oncol Res 4: 267–274.

Bauer WR (1978) Structure and reactions of closed duplex DNA. Annu Rev Biophys Bioeng 7: 287–313.

Beck WT (1989) Unknotting the complexities of multidrug resistance: The involvement of DNA topoisomerases in drug action and resistance. Journal of the NCI 81: 1683–1685.

Beck WT and Danks MK (1991) Mechanisms of resistance to drugs that inhibit DNA topoisomerases. Semin Cancer Biol 2: 235–244.

Boege F, Gieseler F, Biersack H and Meyer P (1992) The measurement of nuclear topoisomerase II inhibition *in vitro*: a possible tool for detecting resistance on a subcellular level in haematopoietic malignancies. Eur J Clin Chem Clin Biochem 30: 63–68.

Boiocchi M, Tumiotto L, Giannini F, Viel A, Biscontin G, Sartor F and Toffoli G (1992) P-glycoprotein but not topoisomerase II and glutathione-S-transferase-pi accounts for enhanced intracellular drug resistance in LoVo MDR human cell lines. Tumori 78: 159–166.

Bugg BY, Danks MK, Beck WT and Suttle DP (1991) Expression of a mutant DNA topoisomerase II in CCRF-CEM human leukemic cells selected for resistance to teniposide. Proc Natl Acad Sci USA 88: 7654–7658.

Chan VTW, Ng SW and Eder JP (1993) Molecular-cloning and identification of a point mutation in the topoisomerase II cDNA from an etoposide-resistant Chinese-hamster ovary cell line. J Biol Chem 268: 2160–2165.

Chen GL, Yang L, Rowe TC, Halligan BD, Tewey KM and Liu LF (1984) Nonintercalative antitumor drugs interfere with the breakage-reunion reaction of mammalian DNA topoisomerase II. J Biol Chem 259: 13560–13566.

Chung TDY, Drake FH, Tan KB, Per SR, Crooke ST, and Mirabelli CK (1989) Characterization and immunological identification of cDNA clones encoding two human topoisomerase II isozymes. Proc Natl Acad Sci USA 86: 9431–9435.

Cole SP, Chanda ER, Dicke FP, Gerlach JH and Mirski SE (1991) Non-P-glycoprotein-mediated multidrug resistance in a small cell lung cancer cell line: evidence for decreased susceptibility to drug-induced DNA damage and reduced levels of topoisomerase II. Cancer Res 51: 3345–3352.

Colombo T, Zucchetti M and D'Incalci M (1993) Cyclosporin A induces drastic modifications in Doxorubicin (Dx) distribution in mice. Proc Am Assoc Cancer Res 34: 322.

Cook PR and Brazell IA (1975) Supercoils in human DNA. J Cell Sci 19: 261–279.

Danks MK, Schmidt CA, Cirtain MC, Suttle DP and Beck WT (1988) Altered catalytic activity of and DNA cleavage by DNA topoisomerase II from human leukemic cells selected for resistance to VM-26. Biochemistry 27: 8861–8869.

Danks MK, Yalowich JC and Beck WT (1987) Atypical multiple drug resistance in a human leukemic cell line selected for resistance to teniposide (VM-26). Cancer Res 47: 1297–1301.

Darkin S and Ralph RK (1986) Potentiation of 4'-(9-acridinylamino)methane sulphon-*m*-anisidide action by verapamil. Cancer Lett 30: 25–33.

Deffie AM, Batra JK and Goldenberg GJ (1989) Direct correlation between topoisomerase II activity and cytotoxicity in Adriamycin-sensitive and resistant P388 leukemia cell lines. Cancer Res 49: 58–62.

Deffie AM, Bosman DJ and Goldenberg GJ (1989) Evidence for a mutant allele of the gene for DNA topoisomerase II in Adriamycin-resistant P388 murine leukemia cells. Cancer Res 49: 6879–6882.

Deffie AM, McPherson JP, Gupta RS, Hedley DW and Goldenberg GJ (1992) Multifactorial resistance to antineoplastic agents in drug-resistant P388 murine leukemia, Chinese hamster ovary, and human HeLa cells, with emphasis on the role of DNA topoisomerase II. Biochem Cell Biol 70: 354–364.

De Jong S, Zijlstra JG, De Vries EGE and Mulder NH (1990) Reduced topoisomerase II activity and drug-induced DNA cleavage activity in an Adriamycin-resistant human small cell lung carcinoma cell line. Cancer Res 50: 304–309.

De Jong S, Zijlstra JG, Mulder NH and De Vries EG (1991) Lack of cross-resistance to fostriecin in a human small-cell lung carcinoma cell line showing topoisomerase II-related drug resistance. Cancer Chemother Pharmacol 28: 461–464.

Dickstein B, Barnes K, Fojo T and Bates, SE (1993) Modulation of P-glycoprotein (Pgp)-mediated progesterone transport by protein kinase inhibition. Proc Am Assoc Cancer Res 34: 317.

Drake FH, Hofmann GA, Bartus HF, Mattern MR, Crooke ST and Mirabelli CK (1989) Biochemical and pharmacological properties of p170 and p180 forms of topoisomerase II. Biochemistry 28: 8154–8160.

Drake FH, Zimmerman JP, McCabe FL, Bartus HF, Per SR, Sullivan DM, Ross WE, Mattern MR, Johnson RK, Crooke ST and Mirabelli CK (1987) Purification of topoisomerase II from amsacrine-resistant P388 leukemia cells: Evidence for two forms of the enzyme. J Biol Chem 262: 16739–16747.

Eckardt JR, Burris HA, Rodriguez GA, Fields SM, Rothenberg ML, Moore TD, Smith SC, Ganapathi R, Weiss GR, Johnson RK, Kuhn JG and Von Hoff DD (1993) A phase I study of the topoisomerase I and II inhibitors topotecan and etoposide (E). Proc Am Soc Clin Oncol 12: 137.

Efferth T, Mattern J and Volm M (1992) Immunohistochemical detection of P-glycoprotein, glutathione-S-transferase, and DNA topoisomerase II in human tumors. Oncology 49: 368–375.

Eng W-K, McCabe FL, Tan KB, Mattern MR, Hofmann GA, Woessner RD, Hertzberg RP and Johnson RK (1990) Development of a stable camptothecin-resistant subline of P388 leukemia with reduced topoisomerase I content. Molec Pharmacol 38: 471–480.

Estey E, Adlakha RC, Hittelman WN and Zwelling LA (1987) Cell cycle stage dependent variations in drug-induced topoisomerase II mediated DNA cleavage and cytotoxicity. Biochemistry 26: 4338–4344.

Ferguson PJ, Fisher MH, Stephenson J, Li D-h, Zhou B-s and Cheng Y-c (1988) Combined modalities of resistance in etoposide-resistant human KB cell lines. Cancer Res 48: 5956–5964.

Fine RL, Patel J and Chabner BA (1988) Phorbol esters induce multidrug resistance in human breast cancer cells. Proc Natl Acad Sci USA 85: 582–586.

Fritzer M, Szekeres T, Berczi A and Goldenberg H (1993) Cytotoxic effects of a transferrin-Adriamycin conjugate in multidrug resistant KB cells. Proc Am Assoc Cancer Res 34: 319.

Fry AM and Hickson ID (1993) Phosphorylation modulates the enzymatic activity of human topoisomerase IIα protein. Proc Am Assoc Cancer Res 34: 421.

Ganapathi R, Kamath N, Constantinou A, Grabowski D, Ford J and Anderson A (1991) Effect of the calmodulin inhibitor trifluoperazine on phosphorylation of P-glycoprotein and topoisomerase II: relationship to modulation of subcellular distribution, DNA damage and cytotoxicity of doxorubicin in multidrug resistant L1210 mouse leukemia cells. Biochem Pharmacol 41: 21–26.

Gellert M, Menzel R, Mizuuchi K, O'Dea MH and Friedman DI (1982) Regulation of DNA supercoiling in *E. coli*. Cold Spring Harbor Symp Quant Biol 47: 763–767.

Giaccone G, Gazdar AF, Beck H, Zunino F and Capranico G (1992) Multidrug sensitivity phenotype of human lung cancer cells associated with topoisomerase II expression. Cancer Res 52: 1666–1674.

Glisson BS and Alpeter MD (1992) Multidrug resistance in a small cell lung cancer line: rapid selection with etoposide and differential chemosensitization with Cyclosporin A. Anti Cancer Drugs 3: 359–366.

Glisson BS, Gupta R, Smallwood-Kentro S and Ross WE (1986) Characterization of acquired epipodophyllotoxin resistance in a Chinese hamster ovary cell line: loss of drug-stimulated DNA cleavage activity. Cancer Res 46: 1934–1938.

Harker WG, Slade DL, Drake FH and Parr RL (1991) Mitoxantrone resistance in HL-60 leukemia cells: reduced nuclear topoisomerase II catalytic activity and drug-induced DNA cleavage in association with reduced expression of the topoisomerase II beta isoform. Biochemistry 30: 9953–9961.

Harris AL and Hochauser D (1992) Mechanisms of multidrug resistance in cancer treatment. Acta Oncol 31: 205–213.

Heck MMS and Earnshaw WC (1986) Topoisomerase II: a specific marker for cell proliferation. J Cell Biol 103: 2569–2581.

Heck MMS, Hittelman WN and Earnshaw WC (1988) Differential expression of DNA topoisomerases I and II during the eukaryotic cell cycle. Proc Natl Acad Sci USA 85: 1086–1090.

Hinds M, Deisseroth K, Mayes J, Altschuler E, Jansen R, Ledley FD and Zwelling LA (1991) Identification of a point mutation in the topoisomerase II gene from a human leukemia cell line containing an amsacrine-resistant form of topoisomerase II. Cancer Res 51: 4729–4731.

Hofmann GA, Mirabelli CK and Drake FH (1990) Quantitative adaptation of the bacteriophage P4 unknotting assay for use in the biochemical and pharmacological characterization of topoisomerase II. Anti-Cancer Drug Design 5: 273–282.

Hsieh T-s and Brutlag D (1980) ATP-dependent DNA topoisomerase from *D. melanogaster* reversibly catenates duplex DNA rings. Cell 21: 115–125.

Isabella P, Capranico G, Binaschi M, Tinelli S and Zunino F (1989) Evidence of topoisomerase II-dependent mechanisms of multidrug resistance in P388 leukemia cells. Molecular Pharmacology 37: 11–16.

Isabella P, Capranico G and Zunino F (1991) The role of topoisomerase II in drug resistance. Life Sci 48: 2195–2205.

Ishikawa H, Kawano MM, Okada K, Tanaka H, Tanabe O, Sakai A, Asaoku H, Iwato K, Nobuyoshi M and Kuramoto A (1993) Expressions of DNA topoisomerase I and topoisomerase II gene and the genes possibly related to drug resistance in human myeloma cells. Br J Haematol 83: 68–74.

Johnson RK and Howard WS (1982) Development and cross-resistance characteristics of a subline of P388 leukemia resistant to 4'-(9-acridinylamino)methanesulphon-*m*-anisidide. Eur J Cancer Clin Oncol 18: 479–484.

Kamath N, Grabowski D, Ford J, Kerrigan D, Pommier Y and Ganapathi R (1992) Overexpression of P-glycoprotein and alterations in topoisomerase II in P388 mouse leukemia cells selected *in vivo* for resistance to mitoxantrone. Biochem Pharmacol 44: 937–945.

Kim R, Hirabayashi N, Nishiyama M, Jinushi K, Toge T and Okada K (1992) Factors contributing to Adriamycin sensitivity in human xenograft tumors: the relationship between expression of the Mdr1, GST-pi, and topoisomerase II genes and tumor sensitivity to Adriamycin. Anticancer Res 12: 241–245.

Kobayashi H, Dorai T, Holland JF, and Ohnuma T (1993) Characteristics of the cleavage reaction of human multidrug resistance (MDR) mRNA by hammerhead ribozymes. Proc Am Assoc Cancer Res 34: 325.

Kohn KW, Ewig RAG, Erickson LC and Zwelling LA (1981) Measurements of strand breaks and cross-links by alkaline elution. In: EC Friedberg and PC Hanawalt (eds) DNA Repair: a Laboratory Manual of Research Techniques (pp. 397–401). New York: Marcel Dekker.

Lee MS, Wang JC and Beran M (1992) Two independent amsacrine-resistant myeloid leukemia cell lines share an identical point mutation in the 170 kDa form of human topoisomerase II. J Mol Biol 223: 837–843.

Lefevre D, Riou JF, Ahomadegbe JC, Zhou DY, Benard J and Riou G (1991) Study of molecular markers of resistance to *m*-AMSA in a human breast cancer cell line. Decrease of topoisomerase II and increase of both topoisomerase I and acidic glutathione S transferase. Biochem Pharmacol 41: 1967–1979.

Ling V and Thompson LH (1974) Reduced permeability in CHO cells as a mechanism of resistance to colchicine. J Cell Physiol 83: 103–116.

Ling V (1992) P-glycoprotein and resistance to anticancer drugs. Cancer 69: 2603–2609.

Liu LF (1983) DNA topoisomerases – enzymes that catalyse the breaking and rejoining of DNA. CRC Crit Rev Biochem

15: 1–24.

Liu LF (1989) DNA topoisomerase poisons as antitumor drugs. Annu Rev Biochemistry 58: 351–375.

Liu LF, Liu C-C and Alberts BM (1980) Type II DNA topoisomerases: enzymes that can unknot a topologically knotted DNA molecule *via* a reversible double-stranded break. Cell 19: 697–707.

Liu LF and D'Arpa P (1992) Topoisomerase II-targeting antitumor drugs: mechanisms of cytotoxicity and resistance. Important Adv Oncol 1992: 79–89.

Liu LF, Davis JL and Calendar R (1981) Novel topologically knotted DNA from bacteriophage P4 capsids: studies with DNA topoisomerases. Nucleic Acids Res 9: 3979–3989.

Long BH, Musial ST and Brattain MG (1986) DNA breakage in human lung carcinoma cells and nuclei that are naturally sensitive or resistant to etoposide and teniposide. Cancer Res 46: 3809–3816.

Lum BL, Gosland MP, Kaubisch S and Branimir BI (1993) Molecular targets in oncology: Implications of the multidrug resistance gene. Pharmacotherapy 13: 88–109.

Marini JC, Miller KG and Englund PT (1980) Decatenation of kinetoplast DNA by topoisomerases. J Biol Chem 25: 4976–4979.

Mattern MR, Mong S, Mong S-M, O'Leary Bartus J, Sarau HM, Clark MA, Foley JJ and Crooke ST (1990) Transient activation of topoisomerase I in leukotriene D_4 signal transduction in human cells. Biochem J 265: 101–107.

Mattern MR, Nambi P, O'Leary Bartus J, Mirabelli CK, Crooke ST and Johnson RK (1991) Regulation of topoisomerase I and II activities by cyclic nucleotide- and phospholipid-dependent protein kinases. Receptor 1: 181–190.

McClean S and Hill BT (1992) An overview of membrane, cytosolic and nuclear proteins associated with the expression of resistance to multiple drugs *in vitro*. Biochim Biophys Acta 1114: 107–127.

Mickisch GH, Pai LH, Gottesman MM, Pastan I and Schroeder FH (1993) MRK-16-Pseudomonas exotoxin conjugate reverses MDR in urogenital cancer and in MDR-transgenic mice. Proc Am Assoc Cancer Res 34: 317.

Nambi P, Mattern M, O'Leary Bartus J, Aiyar N and Crooke ST (1989) Stimulation of intracellular topoisomerase I activity by vasopressin and thrombin. Differential regulation by pertussis toxin. Biochem J 262: 486–489.

Nitiss JL, Liu YX and Hsiung Y (1993) A temperature-sensitive topoisomerase II allele confers temperature-dependent drug resistance on amsacrine and etoposide: a genetic system for determining the targets of topoisomerase II inhibitors. Cancer Res 53: 89–93.

Osheroff N (1986) Eukaryotic topoisomerase II. Characterization of enzyme turnover. J Biol Chem 261: 9944–9950.

Per SR, Mattern MR, Mirabelli CK, Drake FH, Johnson RK and Crooke ST (1987) Characterization of a subline of P388 leukemia resistant to amsacrine: evidence of altered topoisomerase II function. Molec Pharmacol 32: 17–25.

Pommier Y, Mattern MR, Schwartz RE, Zwelling LA and Kohn KW (1984) Changes in the deoxyribonucleic acid linking number due to treatment of mammalian cells with the intercalating agent 4'-(9-acridinylamino) methanesulfon-*m*-anisidide. Biochemistry 23: 2927–2932.

Roberts D, Lee T, Parganas E, Wiggins L, Yalowich J and Ashmun R (1987) Expressions of resistance and cross-resistance in teniposide resistant L1210 cells. Cancer Chemother Pharmacol 19: 123–130.

Ross W, Rowe T, Glisson B and Liu L (1984) Role of topoisomerase II in mediating epipodophyllotoxin-induced DNA cleavage. Cancer Res 44: 5857–5860.

Scala S, Dickstein B and Bates S (1993) Differential protein kinase modulation: Effect on P-glycoprotein (Pgp) transport. Proc Am Assoc Cancer Res 34: 317.

Scotto KW and Prochaska H-P (1993) Collateral sensitivity of multidrug resistant cells to a new family of compounds. Proc Am Assoc Cancer Res 34: 314.

Sela S, Hussain SR, Pearson J, Gottesman M, Longo D, Jacobson R and Rahman A (1993) Studies on modulation of multidrug resistance (MDR) in human colon HT-29 MDR-1 cancer cells. Proc Am Assoc Cancer Res 34: 324.

Slater LM, Murray SL, Wetzel MW, Sweet P and Stupecky M (1986) Verapamil potentiation of VP-16-213 in acute lymphatic leukemia and reversal of pleiotropic drug resistance. Cancer Chemother Pharmacol 16: 50–54.

Slater LM, Sweet P and Stupecky M (1993) Comparison of Cyclosporin A-, verapamil-, and cremophor EL-mediated enhancement of VP-16 cytotoxicity of L1210 and P388 leukemias *in vitro*. Proc Am Assoc Cancer Res 34: 318.

Soudon J, Berlion M, Lucas C, Haddad P, Bizzari JP and Calvo F (1993) *In vitro* time schedule modulation of multidrug resistance (MDR) with S 9788. Proc Am Assoc Cancer Res 34: 325.

Spiridonidis CA, Chatterjee S, Petzold SJ and Berger NA (1989) Topoisomerase II-dependent and -independent mechanisms of etoposide resistance in Chinese hamster cell lines. Cancer Res 49: 644–650.

Sullivan DM, Eskildsen LA, Groom KR, Webb CD, Latham MD, Martin AW, Wellhausen SR, Kroeger PE and Rowe TC (1993) Topoisomerase II activity involved in cleaving DNA into topological domains is altered in a multiple drug-resistant Chinese hamster ovary cell line. Molec Pharmacol 43: 207–216.

Sullivan DM, Glisson BS, Hodges PK, Smallwood-Kentro S and Ross WE (1986) Proliferation dependence of topoisomerase II-mediated drug action. Biochemistry 25: 2246–2256.

Sullivan DM and Ross WE (1991) Resistance to inhibitors of DNA topoisomerases. Cancer Treat Res 57: 57–99.

Takano H, Kohno K, Matsuo K, Matsuda T and Kuwano M (1992) DNA topoisomerase II-targeting antitumor agents and drug resistance. Anticancer Drugs 3: 323–330.

Tan KB, Mattern MR, Eng W-k, McCabe FL and Johnson RK (1989) Nonproductive rearrangement of DNA topoisomerase I and II genes: correlation with resistance to topoisomerase inhibitors. J Natl Cancer Inst 81: 1732–1735.

Trask DK, DiDonato JA and Muller MT (1984) Rapid detection and isolation of covalent DNA-protein complexes: application to DNA topoisomerase I and II. Nuc Acids Res 11: 2779–2800.

Utsugi T, Mattern MR, Mirabelli CK and Hanna N (1990) Potentiation of topoisomerase inhibitor-induced DNA strand breakage and cytotoxicity by tumor necrosis factor: enhancement of topoisomerase activity as a mechanism of potentiation. Cancer Res 50: 2636–2640.

Van der Zee AG, Hollema H, De Jong S, Boonstra H, Gouw A, Willemse PH, Zijlstra JG and De Vries EG (1991) P-glycoprotein expression and DNA topoisomerase I and II activity in benign tumors of the ovary and in malignant tumors of the ovary, before and after platinum/cyclophosphamide chemotherapy. Cancer Res 51: 5915–5920.

Volm M, Mattern J, Efferth T and Pommerenke EW (1992) Expression of several resistance mechanisms in untreated human kidney and lung carcinomas. Anticancer Res 12: 1063–1067.

Wang JC (1985) DNA topoisomerases: enzymes that control DNA conformation. Annu Rev Biochem 54: 665–697.

Webb CD, Latham MD, Lock RM and Sullivan DM (1991) Attenuated topoisomerase II content directly correlates with a low level of drug resistance in a Chinese hamster ovary cell line. Cancer Res 51: 6543–6549.

Woessner RD, Eng W-k, Hofmann GA, Rieman DJ, McCabe FL, Hertzberg RP, Mattern MR, Tan KB and Johnson RK (1993) Camptothecin hyperresistant P388 cells: drug-dependent reduction in topoisomerase I content. Oncology Res 4: 481–488.

Woessner RD, Mattern MR, Mirabelli CK, Johnson RK and Drake FH (1991) Proliferation- and cell cycle dependent differences in expression of the 170 kilodalton and 180 kilodalton forms of topoisomerase II in NIH-3T3 cells. Cell Growth and Differentiation 2: 209–214.

Yang L, Rowe TC and Liu LF (1985) Identification of topoisomerase II as the intracellular target of antitumor epipodophyllotoxins in simian virus 40-infected monkey cells. Cancer Res 45: 5872–5876.

Zwelling LA, Mayes J, Hinds M, Chan D, Altschuler E, Carroll B, Parker E, Deisseroth K, Radcliffe A, Seligman M, Li L and Farquhar D (1991) Cross-resistance of an amsacrine-resistant human leukemia line to topoisomerase II reactive DNA intercalating agents. Evidence for two topoisomerase II directed drug actions. Biochemistry 30: 4048–4055.

Address for correspondence: Glenn A. Hofmann, Department of Biomolecular Discovery, SmithKline Beecham Pharmaceuticals, P.O. Box 1539, King of Prussia, PA 19406, USA.

Cytotechnology **12**: 155–170, 1993.

Glutathione-related enzymes, glutathione and multidrug resistance

Jeffrey A. Moscow and Katharine H. Dixon
Medicine Branch, National Cancer Institute, Bethesda, MD 20892, USA

Key words: etoposide, doxorubucin, glutathione, glutathione peroxidase, glutathione reductase, glutathione S-transferase, multidrug resistance, P-glycoprotein, vincristine

Abstract

This review examines the hypothesis that glutathione and its associated enzymes contribute to the overall drug-resistance seen in multidrug resistant cell lines. Reports of 34 cell lines independently selected for resistance to MDR drugs are compared for evidence of consistent changes in activity of glutathione-related enzymes as well as for changes in glutathione content. The role of glutathione S-transferases in MDR is further analyzed by comparing changes in sensitivity to MDR drugs in cell lines selected for resistance to non-MDR drugs that have resulting increases in glutathione S-transferase activity. In addition, results of studies in which genes for glutathione S-transferase isozymes were transfected into drug-sensitive cells are reviewed. The role of the glutathione redox cycle is examined by comparing changes in elements of this cycle in MDR cell lines as well as by analyzing reports of the effects of glutathione depletion on MDR drug sensitivity. Overall, there is no consistent or compelling evidence that glutathione and its associated enzymes augment resistance in multidrug resistant cell lines.

Introduction

An appealing hypothesis emerged from the initial observation of increased glutathione S-transferase (GST) activity in a multidrug resistant cell line [1,2]: that glutathione-based detoxification systems, which were long known to guard normal tissues from chemical injuries, might also protect malignant tissues from chemotherapeutic agents. This idea had a certain historical logic, for it placed these detoxification systems, which were already known to act upon a broad range of substances, into the context of multidrug resistance (MDR). The idea possessed an attractive symmetry, for it appeared possible that glutathione-related metabolic pathways might cooperate with the drug efflux pump P-glycoprotein in drug excretion. And, as antibodies and cDNA probes for glutathione-dependent enzymes were available, the hypothesis was readily accessible to study. As a result, connections between MDR and glutathione-related enzymes have been extensively examined.

This article will review the literature regarding glutathione, glutathione-related enzymes and MDR. It will not address resistance to alkylating agents *per se*, a topic in which the interest in glutathione and its dependent enzymes is equal to that seen in the study of MDR. The review will first discuss GST isozymes, and will then examine the literature as it relates to the components of the glutathione redox cycle, including glutathione peroxidases, glutathione reductase, and glutathione itself.

Cytosolic GSTs and MDR

In the following discussion we will use the nomenclature system for human cytosolic GSTs that was described recently by Mannervik *et al.* [3]. In brief, alpha class GST subunits are identified as GSTA1-

1 and GSTA2-2; mu class GST subunits are identified as GSTM1a-1a, GSTM1b-1b, GSTM2-2 and GSTM3-3; and pi class GST as GSTP1-1.

It is a leap of logic to assume, since GSTs are detoxification enzymes with broad substrate specificity and MDR drugs have a variety of chemical structures, that GSTs must therefore play a role in detoxifying MDR drugs. In fact, though it is now widely assumed that cytosolic GSTs are capable of detoxifying MDR drugs, there is scant biochemical evidence to support these assumptions.

Although the ability to conjugate substrates with glutathione is the defining activity of GSTs, there is no evidence that any of the MDR drugs are conjugated with glutathione *en route* to detoxification and elimination. Specifically, glutathione conjugates of doxorubicin, vincristine, etoposide and actinomycin D have not been described. The single exception may be the participation of microsomal GST, a membrane-bound enzyme which is structurally dissimilar to the cytosolic GSTs, in the metabolism of mitoxantrone [4]. In contrast, cytosolic GST-mediated glutathione conjugation of the non-MDR antineoplastic agents chlorambucil [5,6] and melphalan [7], in addition to the cyclophosphamide metabolite acrolein [8], have been well-documented.

It is possible that GST isozymes could detoxify MDR drugs through pathways other than conjugation with glutathione. GSTs bind hydrophobic substances with high affinity and may participate in the intracellular transport and sequestration of these molecules [9], such as in the intracellular transport of bilirubin in the liver [10]. Thus, these isozymes may protect cells by binding toxins and preventing them from interacting with critical cellular targets. It has therefore been proposed that a possible role for GSTs in MDR may be to bind antineoplastic agents and present them to the P-glycoprotein pump for extrusion from the cell [11]. This hypothesis was supported by comparisons of P-glycoprotein with homologous bacterial transport proteins which require the activity of soluble periplasmic proteins to perform their transport functions [11]. However, Black *et al.* found no evidence that GST isozymes P1-1 and A1-1 could bind doxorubicin [12]. No biochemical evidence has thus far been published which demonstrates direct binding of GST isozymes to MDR drugs or their metabolites.

Since some GSTs, most notably certain alpha class isozymes, possess intrinsic peroxidase activity, it has been postulated that these isozymes could detoxify harmful organic peroxides. As we will discuss later, such toxic peroxide intermediates are created in the metabolism of doxorubicin [13]. However, the GST isozyme that is most frequently found to be elevated in MDR cell lines, GSTP1-1, possesses very low levels of peroxidase activity [14].

Is increased GST activity necessary for MDR?

At least 32 cell lines that were independently selected for resistance to MDR drugs have been described in which GST activity has been characterized [1,15–35]. These cell lines, whose characteristics are summarized in Table 1, were isolated by incubating the parental cell lines with the selecting drug, usually in serial passages of stepwise increasing drug concentrations. Most, but not all, express increased amounts of P-glycoprotein and increased levels of *mdr*-1 RNA.

Table 1 must be interpreted bearing in mind the following caveat. Most GSTs catalyze the conjugation of glutathione to chloro-2,4-dinitrobenzene (CDNB) at an easily measurable rate [9]. The GST activity cited in Table 1 reflects the ability of cell cytosols to catalyze this reaction. However, using this single assay for comparative purposes can be deceptive, as many tissues contain complex mixtures of GSTs and the rate of CDNB conjugation differs between isozymes. Complete characterization of the GST content of tissues requires identification of their subunits. In the studies outlined in Table 1, GST isozyme expression was not always determined. Studies which identified individual GST isozymes may have utilized immunological techniques or analysis of RNA expression.

Table 1 includes summaries of GST changes in 25 cell lines selected for resistance to doxorubicin in which bulk cytosolic GST activity is reported. Only 10 of these 25 resistant cell lines have demonstrated increased GST activity in comparison to

Table 1. Glutathione and glutathione-related enzymes in MDR cell lines

Parental cell line	Tumor origin	Fold resistance	GST fold change	Isozyme class increased	GSH fold change	GPx fold change	GRed fold change	Reference
Selected with doxorubicin								
V-79	Hamster fibroblast	3000	0.75	π	0.78	0.67	1.13	Medh *et al.* [15]
CHO	Hamster ovary	27	1.5	α				Hoban *et al.* [16]
H-134	Ovary	*	nc					Broxterman *et al.* [17]
A2780	Ovary	100	1.25		1.55	0.87		Hamilton *et al.* [18]
A2780	Ovary	1000	nc					Yusa *et al.* [19]
TR170	Ovary	2.3	nc	0.86	0.47	0.84		Hosking *et al.* [20]
Hattori	Breast	3.3	0.4				0.34	Yusa *et al.* [19]
MCF-7	Breast	192	44.7	π	0.9	12.9		Batist *et al.* [1]
MCF-7	Breast	900	0.5		1.07			Chen *et al.* [21]
MCF-7	Breast	2.6	2.6	π	1.11	1.15	1.95	Whelan *et al.* [22]
MCF-7	Breast	300	nc					Yeh *et al.* [23]
MCF-7	Breast	75			0.64			Taylor *et al.* (34)
H69	Lung-small cell ca.	73	10	π	0.1	1.24		Cole *et al.* [24]
SW1573	Lung-squamous	*	2					Cole *et al.* [24]
RPMI8226	Myeloma	40	nc		1.65	0.88		Bellamy *et al.* [25]
P388	Leukemia	37	0.33		1.95	3.33	1.14	Kramer *et al.* [26]
P388	Leukemia	5	1.4	π				Deffie *et al.* [27]
P388	Leukemia	10	1.5	π				Deffie *et al.* [27]
P388	Leukemia	103			0.97	1.36		Nair *et al.* [35]
HL60	Leukemia	111	nc		0.54	0.85	1.38	Lutzky *et al.* [28]
K562	Leukemia	128	nc					Yusa *et al.* [19]
K562	Leukemia	155	nc					Kato *et al.* [29]
FLC	Erythroleukemia	13	nc	α	0.74	1.44		Schisselbauer *et al.* [30]
SW620	Colon	108	0.7		nc			Lai *et al.* [31]
LS180	Colon	7.5	nc		nc			Lai *et al.* [31]
DLD-1	Colon	21	1.9	not π	1.3			Lai *et al.* [31]
SW620	Colon	75	2.5	π	1.65			Chao *et al.* [32]
Selected with etoposide								
MCF-7	Breast	5.1	6.7		1.24	1.45	0.58	Hosking *et al.* [20]
SuSAP	Testicular	5.1	nc		0.89	0.73	1.15	Hosking *et al.* [20]
HN-1P	Head and Neck	4	0.94		0.95	1.44	1.19	Hosking *et al.* [20]
Selected with vincristine								
MCF-7	Breast	13.7	1.3		1.15	5.25	0.58	Hosking *et al.* [20]
MCF-7	Breast	11.1	6.7		1.27	4.6	0.6	Whelan *et al.* [33]
Selected with colchicine								
MCF-7	Breast	24	70					Yusa *et al.* [19]
Selected with vinblastine								
CCRF-CEM	Leukemia	10	nc					Yusa *et al.* [19]

Fold resistance was calculated as the ratio of the IC_{50} of the resistant cell line to that of the parental cell line. GST activity in all cell lines was measured with CDNB as substrate. GPx – glutathione peroxidase activity, measured using either cumene hydroperoxide or hydrogen peroxide as substrate. GRed – glutathione reductase. * indicates MDR phenotype, fold-resistance not specified. nc – no change. Blank spaces indicate variable not determined.

the parental cell line, and only five out of 25 have increases of GST activity of 2-fold or greater. In contrast, 15 out of 25 resistant sublines had either no change of GST activity or decreased levels of GST activity. Thus, contrary to what is frequently stated, GST activity is most often unchanged in doxorubicin-selected cell lines. The isozyme pattern of four of the five cell lines with greater than 2-

fold increase in GST activity was characterized, and in each case demonstrated evidence of an increase in the GSTP1-1 isozyme, an isozyme with little intrinsic peroxidase activity.

There is no evidence, from Table 1, for a dose-response relating GST activity to the level of resistance to doxorubicin; in other words, GST activities were increased in some of the least resistant cell lines, and unchanged or decreased in some of the most resistant cell lines. There is also no evidence that GST changes may be cell line specific. For example, only two of four different MCF-7 human breast cancer cell lines selected for resistance to doxorubicin have shown increased GST activity, again with no direct relation between the degree of resistance and GST activity.

Despite the infrequent association of elevated GST activity and acquired drug resistance *in vitro*, several investigators have looked for associations between the variability of sensitivity towards MDR drugs of independent cell lines not selected for drug resistance and the cell lines' relative amounts of GST activity. These studies are difficult to interpret since different cell lines obviously possess a multitude of different characteristics. Nevertheless, a positive association between GST expression and doxorubicin resistance was reported for a series of colon cancer cell lines [36], lung cancer cell lines [37], but not with breast tumors in short-term culture [38].

Two studies have looked at cell lines established from the same patient before and after therapy, as a method of understanding mechanisms of drug resistance acquired *in vivo*. An increase in GSTP1-1 expression was seen in a neuroblastoma cell line established after chemotherapy in comparison to the pretreatment cell line [39]. GST activity was also increased in a series of small cell lung cancer cell lines established from a single patient whose disease evolved from drug sensitive to drug resistant [40]. Although these associations are provocative they provide neither evidence of a causal relationship nor a mechanism of action for GST expression in MDR.

Is increased GST activity sufficient for MDR?

It is likely that cells possess many alternate routes of detoxification. Although the data in Table 1 indicate that increased GST activity is not necessary for resistance, it does not determine whether increased GST activity does contribute to resistance in the cell lines where it is found to be elevated.

If increased GST activity imparts resistance to MDR drugs, then one would expect that cell lines with increased GST activity, regardless of the derivation of the cell line, would demonstrate cross-resistance to MDR drugs. To address this question, cell lines selected for resistance to non-MDR drugs, and which reported elevation of GST activity, were reviewed for reports of cross-resistance to MDR drugs. Of approximately two dozen reports of resistant cell lines to non-MDR drugs in which GST activity was characterized, half displayed increases in GST activity. Of these, half of the studies reported whether the resistant subline was cross-resistant to doxorubicin. Cell lines fulfilling these criteria are presented in Table 2 [41–44].

All of the seven cell lines in Table 2 have

Table 2. Doxorubicin cross-resistance in cell lines with increased GST activity after selection for resistance to alkylating agents

Parental cell line	Tumor	Selecting agent	Fold resistance	GST fold increase	Isozyme class	Doxorubicin cross-resistance	Reference
G6331	melanoma	cisplatinum	9	5.4	π	no	Wang *et al.* [41]
CHO	ovary	cisplatinum	10	6	π	no	Saburi *et al.* [42]
H69	lung	cisplatinum	11	6.7	π	no	Kasahara *et al.* [43]
SKOV-3	ovary	cisplatinum	23	1.3	π	3.3-fold	Shellard *et al.* [44]
G6331	melanoma	melphalan	4	4.9	π	no	Wang *et al.* [41]
G6331	melanoma	BCNU	4	3.4	π	no	Wang *et al.* [41]
G6331	melanoma	4-NC	11	3.1	π	no	Wang *et al.* [41]

increased GST activity relative to the parental cell line, and in all cases it is the GSTP1-1 isozyme that accounts for the increased activity. This is similar to the increase in GSTP1-1 isozyme expression seen in MDR cell lines with increased GST activity (Table 1). However, only one of the seven cell lines shown in Table 2 displayed cross-resistance to doxorubicin, and this was the cell line with the lowest fold-increase of GST activity. Thus, increased GSTP1-1 activity in non-MDR selected cell lines does not by itself appear to be capable of producing cross-resistance to doxorubicin.

Another clue to the role of increased GST activity in resistant cell lines comes from examination of MDR cell lines which revert to a drug-sensitive phenotype by serial passage in the absence of the selecting agent. For example, the largest increase in GST activity in the MDR cell lines in Table 1 was seen in an MCF-7 human breast cancer cell line selected for resistance to colchicine and which was cross-resistant to doxorubicin [19]. A revertant subline of this resistant cell line completely lost resistance to colchicine and doxorubicin, yet retained its elevated GST activity [19]. Thus, it is unlikely that the elevated GST activity alone could have accounted for the observed drug resistance. In contrast, however, Batist *et al.* did find decreased levels of GSTP1-1 in a drug sensitive cell line that had reverted from one of the drug resistant MCF-7 cells that expressed increased levels of GSTP1-1 [1].

Transfection of cells with GST expression vectors

Cell lines selected for drug resistance often display multiple phenotypic changes, making it difficult to ascribe a causal relationship between a single biochemical variable and the global transformation of the sensitive cell line into a resistant one. In order to examine the isolated effects of increased expression of individual GST isozymes, several laboratories have performed gene transfer experiments in which drug-resistance is determined in cells transfected with genes for the different GST isozymes. Each of the laboratories have used either different techniques (i.e., acute versus stable transfections) or different host models. These studies are summarized in Table 3.

The MCF-7 human breast cancer cell line has been stably transfected with expression vectors that contain the human pi-class GST (GSTP1-1) [14], a mu-class GST (GSTM1-1) [45], and the alpha-class GSTs (GSTA1-1 and GSTA2-2) [45,46]. MCF-7 cells are a good model to examine the effects of GST on drug resistance since the parental cell line has remarkably low levels of GST activity [14]. In each of the clones transfected with individual GST cDNAs, the increased levels of GST isozyme failed to confer significant levels of resistance to any of the antineoplastic drugs tested (Table 3). Although small differences in drug sensitivity were seen in some of the GST transfected MCF-7 cells, these differences were consistent with clonal variation of the parental cell line as has been seen in clonal analysis of other cell lines [47]. In particular, increased levels of all three major classes of GST isozymes did not confer resistance to doxorubicin in MCF-7 cells. Furthermore, when GSTP1-1 was transfected into cells previously transfected with the gene encoding P-glycoprotein there was no change in the sensitivity of the cells to doxorubicin or to other drugs associated with the MDR phenotype [11]. Thus, GSTP1-1 apparently does not interact with the P-glycoprotein drug efflux pump to enhance drug resistance in MCF-7 cells.

Three other studies have used stable transfection of expression vectors in order to examine the effect of increased GST expression on drug sensitivity (Table 3). Transfection of alpha-class GST expression vectors into T47D human breast cancer cells, while producing transfectants that were resistant to cumene hydroperoxide, did not result in resistance to daunorubicin [48]. In transformed mouse NIH3-T3 cells that were transfected with human GSTP1-1, two clones that expressed higher levels of this isozyme showed a 1.8 to 3-fold increase in the IC_{37} value for doxorubicin [49]. However, little or no significant difference was seen in the IC_{50} and IC_{90} values of any of the clones. It was concluded that the change in IC_{37} to doxorubicin in the GSTP1-1 transfected cells reflected an increased ability of these cells to repair sublethal damage in the cells. In another study, a GSTP1-1 expression vector was transfected into CHO cells; at the reported IC_{90}

Table 3. Relative anthracycline resistance of cell lines transfected with GST expression vectors

Cell line	Organism	Transfected subunit[a]	GST activity (CDNB)	Fold change in activity	Fold change resistance	Conclusion	Reference
MCF-7	Human	P1	28	4	0.4(IC_{50})	Not resistant	Moscow *et al.* [14]
			54	7	1.1(IC_{50})		
			91	11	1.3(IC_{50})		
		P1	40	10	1.1(IC_{50})	Not resistant	Fairchild *et al.* [11]
			78	20	1.4(IC_{50})		
			63	16	0.8(IC_{50})		
			50	17	0.6(IC_{50})		
			44	15	0.8(IC_{50})		
		A1	40	4	0.7(IC_{50})	Not resistant	Leyland-Jones *et al.* [46]
			55	6	1.0(IC_{50})		
			22	2	0.9(IC_{50})		
		A2	17	2	0.7(IC_{50})	Not resistant	Townsend *et al.* [45]
			28	3	0.7(IC_{50})		
			52	5	0.7(IC_{50})		
		M1a	56	6	0.8(IC_{50})	Not resistant	Townsend *et al.* [45]
			150	15	0.7(IC_{50})		
			340	35	1.0(IC_{50})		
T47D	Human	Yc	45	3	Nil	Not resistant	Lavoie *et al.* [48]
			32	2	Nil		
pT22-3	Mouse c-H-Ras	P1	360	3	1.8(IC_{37})	Resistant	Nakagawa *et al.* [49]
			594	4	3.0(IC_{37})		
COS	Monkey	P1	Not reported	Not reported	1.3(IC_{90})	Resistant	Puchalski *et al.* [51]
		Ya	Not reported	1.3	1.3(IC_{90})		
		Yb1	Not reported	Not reported	Nil		
CHO	Hamster	P1	Not reported	Not reported	1.1(IC_{90})	No conclusion	Miyazaki *et al.* [50]
			Not reported	Not reported	1.3(IC_{90})		

[a]Subunits are as follows: A1 — human alpha class; A2 — human alpha class; M1a — human mu class; P1 — human pi class; Ya — rat alpha class; Yc — rat alpha class; Yb1 — rat mu class.

value there was little change (1.1 to 1.3-fold) in resistance to doxorubicin in the transfected cells [50].

To avoid some of the problems associated with clonal variation that may occur during analysis of stably transfected cells, monkey kidney COS cells were transiently transfected with GST expression vectors and then selected by flow cytometry using the GST substrate, monochlorobimane [51]. The product of GST-mediated GSH conjugation with this compound is fluorescent and cells that expressed elevated levels of GST were sorted by fluorescence from those in the population that did not. The sensitivity to antineoplastic agents of the sorted cells containing increased GST levels was compared to cells containing lower levels of GST. When IC_{90} values were compared, COS cells expressing increased levels of GSTP1-1 were reported to be 1.3-fold more resistant to doxorubicin relative to controls. No difference was seen between transfected cells and control cells in sensitivity towards vinblastine.

In addition to mammalian cell lines, the effects of transfected GSTP1-1 and human alpha-class GSTA1-1 upon the sensitivities of the yeast *Saccharomyces cerevisiae* to doxorubicin have also been examined [12]. Since yeast can be transfected with great efficiency, the GST-transfected yeast contained very high levels of GST relative to the intracellular concentrations typically seen in human cell lines. Both transfected GST cDNAs conferred resistance to doxorubicin, but the reported level of resistance of the transfected yeast is difficult to compare to other studies. Usually, the level of resistance is calculated by the ratios of drug concentrations that produce a given level of growth inhibition. In this study, however, the level of resistance seen in the transfected yeast (3 to 10-fold to doxorubicin) was calculated by the ratio of the number of surviving yeast at a pre-selected drug concentration. Thus, this calculation cannot be compared to the levels of resistance described in other studies.

Given the vast differences in the model systems chosen for study, the data from these GST gene transfers into mammalian systems are actually not that dissimilar. Although the data from the transfection studies in Table 3 all show similar results, that GST transfectants with several-fold increases in total GST activity are less than 2-fold resistant to doxorubicin, the interpretation of these studies has widely varied. Some investigators believe that these small differences in resistance (2-fold or less) constitute meaningful drug resistance [49,51]. Others, including ourselves, have argued that such small differences in resistance may occur between subclones of the same cell line whether transfected or not, so that these minor differences may not necessarily be related to changes in GST expression [11,14,45,46]. In any event, the potency of transfected GST genes compares quite unfavorably to transfection of the *mdr*-1 gene, where levels of resistance conferred by gene transfer are considerably greater than control cells, and where the transfected gene confers predictable phenotypic changes [11].

GST and in vivo studies

Two major areas of investigation have dominated the study of the relation of GST expression to MDR in human tumors. The first area examines whether GST isozyme expression is associated with expression of the P-glycoprotein gene, *mdr*-1. The second area addresses the issue of whether altered GST expression in tumors has prognostic significance for patients receiving treatment with MDR drugs.

To address the question of whether GSTs are relevant in clinical MDR, it must first be determined whether these enzymes are expressed in human tumors. Studies from several laboratories have consistently shown that GST expression is often expressed at remarkably high levels in many types of tumor, and these high levels of expression often appear elevated when compared to GST levels in the normal tissues of origin [52–60]. This is particularly true of gastrointestinal malignancies [55–60]. The high levels of GST activity in human tumors are usually attributable to a single isozyme, GSTP1-1 [52–60]. Increased GST expression was not seen in a series of lung tumors, where normal lung parenchyma possesses intrinsically high levels of the GSTP1-1 isozyme [61].

A number of studies have examined human tumors for concomitant expression of GSTP1-1 and *mdr*-1. Direct correlations between *mdr*-1 and GSTP1-1 gene expression were reported for both lung tumors [62] and renal cell carcinoma cell lines [63]. A weak link between expression of the two genes has been suggested in chronic leukemias [64,65], but an association between GSTP1-1 expression and P-glycoprotein could not be demonstrated in acute leukemias [63–65]. A positive correlation between *mdr*-1 gene expression and GSTP1-1 expression in lymphomas was noted in one study [66], but not in another [53]. No association between GSTP1-1 and P-glycoprotein was seen in a study of gastrointestinal cancers [67] or multiple myeloma [68].

Studies relating GST expression to clinical outcome in leukemia have suggested that GST isozyme expression may be a prognostic marker in this heterogeneous group of diseases. Total GST activity [69] and specific GSTP1-1 expression [70] have shown a positive correlation with clinical response in two series of patients with leukemia. In contrast, GST activity and isozyme distribution did not predict response to chemotherapy in two series of patients with ovarian cancer [71,72], which in one study included the use of the MDR drug doxorubicin [72].

In breast cancer, the idea that GSTP1-1 expression might be a useful prognostic marker arose from the observation that GSTP1-1 expression is inversely correlated with estrogen receptor status [73,74]. Immunohistochemical examination of a series of 240 breast cancer specimens confirmed the inverse relationship between GSTP1-1 expression and estrogen receptor content, and furthermore suggested that GSTP1-1 expression was a strong predictor of relapse and death in the subgroup of women with axillary node-negative breast cancer, none of whom had received chemotherapy prior to relapse [75]. In contrast, women with axillary node-positive breast cancer who had received adjuvant chemotherapy, expression of GSTP1-1 was without predictive value [75,76]. These data suggest a possible relationship between GSTP1-1 expression and prognosis that is independent of any interaction between GSTP1-1 and chemotherapy.

GSTP1-1 expression was positively correlated with pathologically-defined biologically-aggressive features in human soft tissue sarcomas [77], again suggesting that GSTP1-1 expression may be a marker for a phenotype associated with poor clinical outcome. This phenotype may have nothing to do the GST-mediated drug-detoxification *per se* – it may be a phenotype more likely to express other drug resistance genes, to invade and metastasize to other sites or to resist chemotherapy through interaction with its local environment.

GST summary

After almost a decade of study, the hypothesis that GSTs participate in a mechanism of resistance that contributes to MDR has not been proven. In fact, the evidence against the hypothesis has become quite substantial. With the exception of mitoxantrone, enzyme-mediated glutathione conjugation of MDR drugs been described. There is no relation between the development of MDR drug resistance and increased GST expression (Table 1), nor is there any relation between the development of increased GST expression and MDR drug resistance (Table 2). Furthermore, in our opinion, GST transfection studies of drug sensitive cell lines have consistently demonstrated that marked increases in GST expression do not translate into meaningful levels of MDR drug resistance (Table 3).

At times, the study of the relationship of GSTs to MDR seems to have taken on a life of its own; studies supporting a positive association between GSTs and MDR have been frequently cited, while contradictory studies have often been ignored. However, the challenge in science is often to discard one hypothesis and embrace another. The literature regarding expression of GSTs in tumors, while not supporting a role for GSTs in MDR, has demonstrated a striking and consistent finding of the elevation of GSTP1-1 expression in many series of tumors relative to normal tissues. These clinical observations may be related to the laboratory observations that induction of pi-class GST expression is a marker of preneoplastic transformation in rat liver hyperplastic nodules, preneoplastic lesions which also express P-glycoprotein [78-80]. Although extrapolation from rat to human should be treated with caution [81], the mechanism by which GSTP1-1 expression increases during malignant transformation may reveal a step in oncogenesis. Increased GSTP1-1 expression has been related to both v-H-*ras* transformation of rat liver epithelial cells [82] and to the presence of Epstein-Barr virus in lymphoma [66]. So far, however, identification of factors which govern the specificity of GSTP1-1 gene expression has proven to be elusive [81,83,84].

Increased GST activity is observed in many models of drug resistance, but it is not known whether this reflects a nonspecific reaction to stress or whether GSTs have specific defensive roles. GSTs may be important factors in alkylating agent resistance, a topic which was not reviewed here. It is possible that GSTs do interact with MDR drugs, but that GST activity must be coupled with other, as yet unknown enzymes for effective MDR drug detoxification. GSTP1-1 expression may also be a marker for certain phenotypes arising from malignant transformation, such as a biologically aggressive phenotype in which GSTP1-1 acts as a prognostic factor, as suggested in nonmetastatic breast cancer. Clearly, a deeper understanding of the role of GSTP1-1 in malignancy must come through exploration of mechanisms that control its expression and of its role in tumor cell biology.

The glutathione redox cycle and MDR

The components of the glutathione redox cycle include the complementary enzymes glutathione peroxidase and glutathione reductase, and the substrate glutathione. Each component may play a significant role in the detoxification of hazardous organic peroxides. Several laboratories have demonstrated that at least one group of MDR drugs, the anthracyclines, produce these potentially toxic reactive oxygen intermediates: doxorubicin-mediated membrane lipid peroxidation has been demonstrated in liver and heart microsomes [85–87], in mitochondria [88] and in isolated nuclei [89]. These anthracycline-induced free-radicals may produce DNA strand breaks [90–91].

A biochemical role for glutathione peroxidase (GPx) activity has been suggested in the detoxification of anthracyclines. This activity has been associated with the selenium-dependent GPxs as opposed to the selenium-independent organic peroxidase activity of the GSTs [26]. There is substantial evidence that the glutathione peroxidases contribute to cellular defense against anthracycline-mediated cardiac toxicity [reviewed by Doroshow *et al.* in Ref. 93].

However, the evidence that glutathione and GPx play functional roles in tumor cell resistance is murky. On the one hand, there is ample evidence that anthracyclines form toxic free radical intermediates in tumor cells in tissue culture (94-96), and decreased anthracycline-induced free-radical formation has been documented in a multidrug resistant MCF-7 human breast cancer cell [13,97]. On the other hand, increased GPx activity has not been generally associated with MDR. Of 11 cell lines selected for doxorubicin resistance which report GPx activity (Table 1), only two resistant cell lines have 2-fold or greater increases in GPx activity [1,26]. In the case of intrinsic resistance, a comparison of five breast cancer cell lines showed no relation between resistance to doxorubicin and total GPx activity [98]. Finally, although Chinese hamster ovary cells selected for resistance to high oxygen concentrations developed a 2-fold increase in GPx activity and a 4-fold increase in glutathione content, this resistant cell line remained as sensitive as the parental cell line to equal intracellular concentrations of doxorubicin [99].

In the case of vincristine resistance, the story may be different. Both cell lines selected for vincristine resistance in Table 1 have elevated GPx activity [20,33]. These findings may be related to the demonstration that other peroxidases, horseradish peroxidase [100] and myeloperoxidase [101] catalyze the oxidative breakdown of vincristine. The intrinsic resistance of myeloid leukemias to vincristine may be related to the high levels of myeloperoxidase seen in this disease [101].

Analysis of GPx enzyme expression in cell lines and tumors has focused on the measurement of total peroxidase activity, using either cumene hydroperoxide or hydrogen peroxide as substrate. It has become apparent, however, that there are at least four different selenium-dependent glutathione peroxidase isozymes. The nucleotide sequences of these different GPx isozymes are distinctive in that they each utilize an internal UGA "stop" codon to encode the amino acid selenocysteine. The isozymes include the classic cytosolic GPx, hgpx1, for which cDNAs [102,103] and genomic sequences [104] have been isolated and characterized; a second cytosolic isozyme, GSHPx-GI, which is highly expressed in the gastrointestinal tract [105, 106]; plasma glutathione peroxidase, an isozyme originally isolated from human plasma [107] but also found in liver, kidney, heart, lung, breast and placenta tissues [108]; and a phospholipid hydroperoxide glutathione peroxidase [109,110]. In addition, there is evidence that other GPx isozymes may exist [111,112]. Reagents which could be used to distinguish between these isozymes, such as isozyme-specific substrates, molecular probes, and antibodies, are not yet widely available. Few of the studies of MDR cell lines cited in Table 1 examined expression of individual GPx isozymes, so it is possible that significant changes in particular glutathione peroxidase isozyme expression could be hidden within these data. For example, the change in GPx isozyme expression in a multidrug resistant MCF-7 cell line is complex: while total activity in this cell line is increased [1,26], it is associated with an increase in expression of both hgpx1 [98,104] and plasma glutathione peroxidase [108], while phospholipid hydroperoxide glutathione peroxidase activity is actually decreased in the resistant cell line in the presence of supplemental selenium [109].

In the MDR cell lines in which glutathione reductase has been measured, no trend was observed for altered enzyme activity in the resistant cell lines. Of the 11 MDR cell lines in which glutathione reductase activity was measured in Table 1, none had a greater than 2-fold increase in activity, and one cell line had a greater than 50% decrease in activity. Similarly, there was no trend in glutathione content in the resistant cell lines: none of the 22 MDR cell lines in which glutathione levels were measured in Table 1 had a greater than 2-fold increase in bulk glutathione content, and only one of the 21 cell lines had a greater than 50% decrease in glutathione content.

At present, there are no known specific inhibitors of GPxs, and the only known inhibitor of glutathione reductase, BCNU, is often toxic to cells at inhibitory concentrations. Thus, evaluation of the role of the entire glutathione redox cycle has frequently been made by depleting cells of glutathione, and assuming that decreased intracellular glutathione significantly decreases the capacity of the glutathione redox cycle.

Several studies have examined the effects of buthionine sulfoximine (BSO), aγ-glutamyl cysteine synthetase inhibitor which lowers intracellular glutathione (GSH), on anthracycline cytotoxicity. These studies, which are summarized in Table 4 [18,24–26,28,31,32,35,113–119], demonstrate wide variability in their results, which may reflect different methodologies or different intrinsic biological properties of the different cell lines. The effect of BSO on doxorubicin toxicity is summarized by the calculated Dose Modifying Factor (DMF), which is the ratio of doxorubicin IC_{50} values without and with BSO treatment. When the DMF values of BSO in drug resistant derivatives were compared to the parental cell lines they exhibited little or no difference (n = 3), an increase (n = 6), or a decrease (n = 2). In two xenograft models, growth delay was similar between the parental and resistant sublines after treatment with BSO. In addition, there does not appear to be any direct relationship between the degree of BSO-mediated glutathione depletion, relative to untreated cells, and the resulting DMF. This heterogeneity of findings makes it impossible to reach conclusions about the role of the glutathione concentrations in either intrinsic or acquired resistance. In cell lines where BSO appears to have a marked effect, it is not clear whether this is due to direct inhibition of the glutathione-redox cycle, to other effects of glutathione depletion, or to other toxic effects of BSO. For example, although it was suggested that BSO and verapamil acted synergistically in reversing MDR [26], another study demonstrated that BSO actually appeared to reverse MDR by increasing the cytotoxicity of verapamil [120]. Curiously, BSO treatment of both wild type and GST transfected yeast actually increased the resistance of both to doxorubicin [12].

Studies in which GPx levels have been selectively increased have suggested a role for GPx in MDR. Doroshow *et al.* increased intracellular GPx concentrations in drug sensitive cells by scrape loading, a procedure in which the cell membrane of tissue culture cells is disrupted with a plastic scraper in the presence of purified extracellular enzyme [121]. This technique resulted in a 10- to 20-fold increase in glutathione peroxidase activity in drug-sensitive MCF-7 human breast cancer cells, which was associated with an increase in the cloning efficiency of these cells after exposure to certain concentrations of doxorubicin [121]. In a gene transfer study, a cDNA for the hgpx1 GPx isozyme was transfected into T47D human breast cancer cells [122]. Although resistance to MDR drugs was not examined in this study, the transfected cells did become resistant to the quinone, menadione, which is chemically related to the anthracycline quinoid doxorubicin [122]. The determination of the role of GPxs in MDR awaits more definitive transfection studies with genes encoding all the different GPx isozymes.

Summary — glutathione redox cycle

Anthracyclines can exert toxic effects through the generation of dangerous peroxides and the glutathione redox cycle can detoxify these reactive species. Furthermore, there is significant evidence that the glutathione redox cycle is involved in protection from anthracycline-induced cardiac toxicity. Nevertheless, it is not clear from the current studies whether the glutathione redox cycle is actually utilized by malignant cells for protection

Table 4. Effect of BSO on MDR drug sensitivity

Cell line	Origin	Parental		Resistant		Reference
		Fold dec GSH	DMF*	Fold dec GSH	DMF*	
vs. doxorubicin						
P815	mastocytoma	nr	1.2	na	na	Arrick *et al.* [113]
HEp3	buccal ca.	.45	8	na	na	Lee *et al.* [114]
V79	hamster	.50	9	na	na	Russo & Mitchell [115]
A2780	ovary	.33	11.4	.24	1.5-5	Hamilton *et al.* [18]
SW620	colon	.11	1.5	.18	1.3	Lai *et al.* [31]
DLD-1	colon	.07	2.3	.06	1.1	Lai *et al.* [31]
LS180	colon	.19	1.1	.12	3.0	Lai *et al.* [31]
SW620	colon	nr	1.4	nr	1.8	Chao *et al.* [32]
H69	lung	nr	nr	.05	1	Cole *et al.* [24]
8226	myeloma	nr	1	.06	1	Bellamy *et al.* [25]
P388	leukemia	nr	1	nr	2	Kramer *et al.* [26]
P388	leukemia	.06	1.3	.05	5	Nair *et al.* [35]
MCF-7*	breast	nr	1	nr	4.4	Kramer *et al.* [26]
MCF-7*	breast	nr	nr	.23	5–7	Dusre *et al.* [116]
vs. daunorubicin						
HL60	leukemia	.08	.83	.07	17.5	Lutzky *et al.* [28]
vs. cyanomorpholino doxorubicin						
ES-2	ovary	.18	1	.15	3	Lau *et al.* [117]
Xenografts	**Origin**	**Sensitive**		**Resistant**		**Reference**
		Fold dec GSH	Growth delay	Fold dec GSH	Growth delay	
vs. doxorubicin						
16C	murine mammary adenocarcinoma	.16	1.3	.21	1.5	Lee *et al.* [114]
vs. vincristine						
TE-671	rhabdomyosarcoma	.15	2.1	.05	2.0	Rosenberg *et al.* [119]

DMF – dose modifying factor, the ratio of IC_{50}s for MDR drugs between cells exposed to BSO and cells exposed to MDR drug alone; nr – not reported; na – not applicable.

from anthracycline cytotoxicity. Certainly, overall levels of the components of this system do not consistently change in MDR cell lines, and the effects of BSO on resistant cell lines are highly variable. However, the reagents utilized in the analysis of the glutathione redox cycle have thus far been relatively nonspecific, and thus can not determine whether the relative expression of specific GPx isozymes is an important parameter in MDR. Comprehensive transfection studies with the various GPx isozyme genes should help to illuminate the role of GPxs in MDR.

Similarly, the measurement of overall glutathione levels may also be deceptive in assessment of the role of glutathione in drug detoxification. The ratio of reduced to oxidized glutathione (GSH to the disulfide GSSG) in cytosol and in organelles such as the endoplasmic reticulum may be important factors in its function [123]. Thus, future study of glutathione and its role in MDR may focus more upon qualitative aspects of its metabolism.

Acknowledgements

We thank Drs A.J. Townsend, B.K. Sinha and K.H. Cowan for their helpful reviews of the manuscript.

References

1. Batist G, Tulpule A, Sinha BK, Katki AG, Myers CE and Cowan KH (1986) Overexpression of a novel anionic glutathione transferase in multidrug-resistant human breast cancer cells. J Biol Chem 261. 15544–15549
2. Cowan KH, Batist G, Tulpule A, Sinha BK and Myers CE (1986) Similar biochemical changes associated with multidrug resistance in human breast cancer cells and carcinogen-induced resistance to xenobiotics in rats. Proc Natl Acad Sci USA 83: 9328–9332
3 Mannervik B, Awasthi YC, Board PG, Hayes JD, Di Ilio C, Ketterer B, Listowsky I, Morgenstern R, Muramatsu M, Pearson W, Pickett CB, Sato K, Widersten M and Wolf CR (1992) Nomenclature for glutathione transferases. Biochem J 282. 305–308
4 Wolf CR, Macpherson JS and Smyth JF (1986) Evidence for the metabolism of mitoxantrone by microsomal glutathione transferases and 3-methylcholanthrene-inducible glucuronosyl transferases Biochem Pharmacol 35 1577–1581.
5. Ciaccio PJ, Tew KD and LaCreta FP (1990) The spontaneous and glutathione S-transferase-mediated reaction of chlorambucil with glutathione Cancer Commun 2: 279–285.
6. Meyer PJ, Gilmore KS, Harris JM, Hartley JA and Ketterer B (1992) Chlorambucil/monoglutathionyl conjugate is sequestered by human alpha class glutathione S-transferases Br J Cancer 66: 433–438
7. Dulik DM, Fenselau C and Hilton J (1986) Characterization of melphalan-glutathione adducts whose formation is catalyzed by glutathione transferases. Biochem Pharmacol 35: 3405–3409.
8 Berhane K and Mannervik B (1990) Inactivation of the genotoxic aldehyde acrolein by human glutathione transferases of classes alpha, mu and pi. Mol Pharmac 37: 251–254.
9. Ketterer B, Meyer DJ and Clark AG (1988) Soluble glutathione transferase isozymes In Ketterer B and Seis H (Eds) Glintathione Conjugation Mechanisms and Biological Significance (pp 74–137), London, Academic Press.
10. Arias IM (1979) Ligandin. review and update of a multifunctional protein Medical Biology 57: 328–334.
11 Fairchild CR, Moscow JA, O'Brien EE and Cowan KH (1990) Multidrug resistance in cells transfected with human genes encoding a variant P-glycoprotein and glutathione S-transferase-pi. Mol Pharmacol 37. 801–809.
12. Black SM, Beggs JD, Hayes JD, Bartoszek A, Muramatsu M, Sakai M and Wolf CR (1990) Expression of human glutathione S-transferases in *Saccharomyces cerevisiae* confers resistance to the anticancer drugs adriamycin and chlorambucil Biochem J 268 309–315
13. Sinha BK, Mimnaugh EG, Rajagopalan S and Myers CE (1989) Adriamycin activation and oxygen free radical formation in human breast tumor cells· Protective role of glutathione peroxidase in adriamycin resistance. Cancer Res 49· 3844–3848.
14. Moscow JA, Townsend AJ and Cowan KH (1989) Elevation of pi class glutathione S-transferase activity in human breast cancer cells by transfection of the GST pi gene and its effect on sensitivity to toxins Mol Pharmacol 36 22–28
15. Medh RD, Gupta V, Zhang Y, Awasthi YC and Belli JA (1990) Glutathione S-transferase and P-glycoprotein in multidrug resistant Chinese hamster cells. Biochem Pharmacol 39 1641–1645.
16. Hoban PR, Robson CN, Davies SM, Hall AG, Cattan AR, Hickson ID and Harris AL (1992) Reduced topoisomerase II and elevated α-class glutathione S-transferase expression in a multidrug resistant CHO cell line highly cross-resistant to mitomycin C Biochem Pharmac 43 685–693.
17 Broxterman HJ, Pinedo HM, Kuiper CM, Schuurhuis GJ and Lankelma J (1989) Glycolysis in P-glycoprotein-overexpressing human tumor cell lines Effects of resistance-modifying agents Febs Lett. 247: 405–410
18. Hamilton TC, Winker MA, Louie KG, Batist G, Behrens BC, Tsuruo T, Grotzinger KR, McKoy WM, Young RC and Ozols RF (1985) Augmentation of adriamycin, melphalan, and cisplatin cytotoxicity in drug-resistant and -sensitive human ovarian carcinoma cell lines by buthionine sulfoximine mediated glutathione depletion Biochem Pharmacol 34· 2583–2586
19 Yusa H, Hamada H and Tsuruo T (1988) Comparison of glutathione S-transferase activity between drug-resistant and -sensitive human tumor cells· Is glutathione S-transferase associated with multidrug resistance? Cancer Chemother Pharmac 22 17–20
20 Hosking LK, Whelan RDH, Shellard SA, Bedford P and Hill BT (1990) An evaluation of the role of glutathione and its associated enzymes in the expression of differential sensitivities to antitumor agents shown by a range of human tumour cell lines. Biochem Pharmac 40: 1833–1842.
21. Chen Y-N, Mickley LA, Schwartz AM, Acton EM, Hwang J and Fojo AT (1990) Characterization of adriamycin-resistant human breast cancer cells which display overexpression of a novel resistance-related membrane protein J Biol Chem 265: 10073–10080
22 Whelan RDH, Hosking LK, Townsend AJ, Cowan KH and Hill BT (1989) Differential increases in glutathione S-transferase activities in a range of multidrug-resistant human tumor cell lines Cancer Commun 1: 359–365.
23 Yeh GC, Lopaczynska J, Poore CM and Phang JM (1992) A new functional role for P-glycoprotein: efflux pump for benzo(a)pyrene in human breast cancer MCF-7 cells

Cancer Res 52: 6692–6695.
24. Cole SP, Downes HF, Mirski SE and Clements DJ (1990) Alterations in glutathione and glutathione-related enzymes in a multidrug-resistant small cell lung cancer cell line. Mol Pharmac 37: 192–197.
25. Bellamy WT, Dalton WS, Meltzer P and Dorr RT (1989) Role of glutathione and its associated enzymes in multi-drug-resistant human myeloma cells. Biochem Pharmac 38: 787–793.
26. Kramer RA, Zakher J and Kim G (1988) Role of the glutathione redox cycle in acquired and de novo multidrug resistance. Science 241: 694–697.
27. Deffie AM, Alam T, Seneviratne C, Beenken SW, Batra JK, Shea TC, Henner WD and Goldenberg GJ (1988) Multifactorial resistance to adriamycin: relationship of DNA repair, glutathione transferase activity, drug efflux, and P-glycoprotein in cloned cell lines of adriamycin-sensitive and -resistant P388 leukemia. Cancer Res 48: 3595–3602.
28. Lutzky J, Astor MB, Taub RN, Baker MA, Bhalla K, Gervasoni JJ, Rosado M, Stewart V, Krishna S and Hindenburg AA (1989) Role of glutathione and glutathione-dependent enzymes in anthracycline-resistant HL60/AR cells. Cancer Res 49: 4120–4125.
29. Kato S, Ideguchi H, Muta K, Nishimura J and Nawata H (1990) Mechanisms involved in the development of Adriamycin resistance in human leukemic cells. Leuk Res 14: 567–573.
30. Schisselbauer JC, Crescimanno M, D'Alessandro N, Clapper M, Toulmond S, Tapiero H and Tew KD (1989) Glutathione, glutathione S-transferases, and related redox enzymes in adriamycin-resistant cell lines with a multidrug resistant phenotype. Cancer Commun 1: 133–139.
31. Lai G-M, Moscow JA, Alvarez MG, Fojo AT and Bates SE (1991) Contribution of glutathione and glutathione-dependent enzymes in thereversal of adriamycin resistance in colon carcinoma cell lines. Int J Cancer 49: 688–695.
32. Chao CCK, Huan Y-T, Ma CM, Chou W-Y and Lin-Chao S (1992) Overexpression of glutathione S-transferase and elevation of thiol pools in a multidrug-resistant human colon carcinoma cell line. Mol Pharmacol 41: 69–75.
33. Whelan RDH, Waring CJ, Wolfe CR, Hayes JD, Hosking LK and Hill BT (1992) Overexpression of P-glycoprotein and glutathione S-transferase pi in MCF-7 cells selected for vincristine resistance in vitro. Int J Cancer 52: 241–246.
34. Taylor CW, Dalton WS, Parrish PR, Gleason MC, Bellamy WT, Thompson FH, Roe DJ and Trent JM (1991) Different mechanisms of decreased drug accumulation in doxorubicin and mitoxantrone resistant variants of the MCF-7 human breast cancer cell line. Br J Cancer 63: 923–929.
35. Nair S, Singh SV, Samy TSA and Krishan A (1990) Anthracycline resistance in murine leukemic P388 cells. Biochem Pharmacol 39: 723–728.
36. Peters WHM and Roelofs HMJ (1992) Biochemical characterization of resistance to mitoxantrone and adriamycin in Caco-2 human colon adenocarcinoma cells: A possible role for glutathione S-transferases. Cancer Res 52: 1886–1890.
37. Carmichael J, Mitchell JB, Friedman N, Gazdar AF and Russo A (1988) Glutathione and related enzyme activity in human lung cancer cell lines. Br J Cancer 58: 437–440.
38. Keith WN, Stallard S and Brown R (1990) Expression of *mdr*-1 and GSTπ in human breast tumours: Comparison to in vitro chemosensitivity. Br J Cancer 61: 712–716.
39. Kuroda H, Sugimoto T, Ueda K, Tsuchida S, Hori Y, Inazawa J, Sato K and Sawada T (1991) Different drug sensitivity in two neuroblastoma cell lines established from the same patient before and after chemotherapy. Int J Cancer 47: 732–737.
40. de Vries EGE, Meijer C, Timmer-Bosscha H, Berendsen HH, de Leij L and Mulder NH (1989) Resistance mechanisms in three human small cell lung cancer cell lines established from one patient during clinical follow-up. Cancer Res 49: 4175–4178.
41. Wang YY, Teicher BA, Shea TC, Holden SA, Rosbe KW, al AA and Henner WD (1989) Cross-resistance and glutathione-S-transferase-pi levels among four human melanoma cell lines selected for alkylating agent resistance. Cancer Res 49: 6185–6192. PLS CHECK NAMES!
42. Saburi Y, Nakagawa M, Ono M, Sakai M, Muramatsu M, Kohno K and Kuwano M (1989) Increased expression of glutathione S-transferase gene in cis-diamminedichloroplatinum(II) resistant variants of a Chinese hamster ovary cell line. Cancer Res 49: 7020–7025.
43. Kasahara K, Fujiwara Y, Nishio K, Ohmori T, Sugimoto Y, Komiya K, Matsuda T and Saijo N (1991) Metallothionein content correlates with the sensitivity of human small cell lung cancer cell lines to cisplatin. Cancer Res 51: 3237–3242.
44. Shellard SA, Hosking LK and Hill BT (1991) Anomalous relationship between cisplatin sensitivity and the formation and removal of platinum-DNA adducts in two human ovarian carcinoma cell lines in vitro. Cancer Res. 51: 4557–4564.
45. Townsend AJ, Tu C-PD and Cowan KH (1991) Expression of human μ or α class glutathione S-transferases in stably transfected human MCF-7 breast cancer cells: effect on cellular sensitivity to cytotoxic agents. Mol Pharmacol 41: 230–236.
46. Leyland-Jones BR, Townsend AJ, Tu C-PD, Cowan KH and Goldsmith ME (1991) Antineoplastic drug sensitivity of human MCF-7 breast cancer cells stably transfected with a human alpha class glutathione S-transferase gene. Cancer Res 51: 587–594.
47. Ferguson PJ and Cheng YC (1989) Phenotypic instability of drug sensitivity in a human colon carcinoma cell line. Cancer Res 49: 1148–1153.
48. Lavoie L, Tremblay A and Mirault M-E (1992) Distinct oxido-resistance phenotype of human T47D cells transfected by rat glutathione S-transferase Yc expression vectors. J Biol Chem 267: 3632–3636.
49. Nakagawa K, Saijo N, Tsuchida S, Sakai M, Tsunokawa Y, Yokota J, Muramatsu M, Sato K, Terada M and Tew

KD (1990) Glutathione-S-transferase pi as a determinant of drug resistance in transfectant cell lines J Biol Chem 265 4296–301
50 Miyazaki M, Kohno K, Saburi Y, Matsuo K, Ono M, Kuwano M, Tsuchida S, Sato K, Sakai M and Muramatsu M (1990) Drug resistance to cis-diamminedichloroplatinum (II) in Chinese hamster ovary cell lines transfected with glutathione S-transferase pi gene Biochem Biophys Res Commun 166 1358–1364
51 Puchalski RB and Fahl WE (1990) Expression of recombinant glutathione S-transferase pi, Ya, or Yb1 confers resistance to alkylating agents Proc Natl Acad Sci USA 87 2443–2447
52 Shea TC, Kelley SL and Henner WD (1988) Identification of an anionic form of glutathione transferase present in many human tumors and human tumor cell lines Cancer Res 48 527–533
53 Moscow JA, Fairchild CR, Madden MJ, Ransom DT, Wieand HS, O'Brien EE, Poplack DG, Cossman J, Myers CE and Cowan KH (1989) Expression of anionic glutathione-S-transferase and P-glycoprotein genes in human tissues and tumors Cancer Res 49 1422–1428
54 di Ilio C, Sacchetta P, Del Boccio G, La Rovere G and Federici G (1985) Glutathione peroxidase, glutathione S-transferase and glutathione reductase activities in normal and neoplastic human breast tumors Cancer Lett 29 37–42
55 Peters WHM, Wormskamp NGM and Thies E (1990) Expression of glutathione S-transferases in normal gastric mucosa and in gastric tumors Carcinogenesis 11 1593–1596
56 Howie AF, Forrester LM, Glancey MJ, Sclager JJ, Powis G, Beckett GJ, Hayes JD and Wolf CR (1990) Glutathione S-transferase and glutathione peroxidase expression in normal and tumour human tissues Carcinogenesis 11 451–458
57 Moorghen M, Cairns J, Forrester LM, Hayes JD, Hall A, Cattan AR, Wolf CR and Harris AL (1991) Enhanced expression of glutathione S-transferases in colorectal carcinoma compared to nonneoplastic mucosa Carcinogenesis 12 13-17
58 Kodate C, Fukushi A, Narita T, Kudo K, Soma Y and Sato K (1986) Human placental form of glutathione S-transferase (GSTπ) as a new immunohistochemical marker for human colonic carcinoma Jpn J Cancer Res 77 226–229
59 Tsutsume M, Sugisake T, Makino T *et al* (1987) Oncofetal expression of glutathione S-transferase placental form in human stomach carcinomas Jpn J Cancer Res 78 631–633
60 Peters WHM, Boon CEW, Roelofs HMJ, Wobbes T, Nagengast FM and Kemers PG (1992) Expression of drug-metabolizing enzymes and P-170 glycoprotein in colorectal carcinoma and normal mucosa Gastroenterology 103 448–455
61 Carmichael J, Forrester LM, Lewis AD, Hayes JD and Wolf CR (1988) Glutathione S-transferase isozymes and glutathione peroxidase activity in normal and tumour samples from human lung Carcinogenesis 9 1617–1621
62 Volm M, Mattern J and Samsel B (1992) Relationship of inherent resistance to doxorubicin, proliferative activity and expression of P-glycoprotein 170, and glutathione S-transferase π in human lung tumors Cancer 70 764–769
63 Efferth T, Mattern J and Volm M (1992) Immunohistochemical detection of P-glycoprotein, glutathione S-transferase and DNA topoisomerase II in human tumors Oncology 49 368–375
64 Holmes J, Wareing C, Jacobs A, Hates JD, Padua RA and Wolf CR (1990) Glutathione-S-transferase pi expression in leukaemia A comparative analysis with *mdr*-1 data Br J Cancer 62 209–212
65 Gekeler V, Frese G, Noller A, Handgretinger R, Wilisch A, Schmidt H, Muller CP, Dopfer R, Klingebiel T, Diddens H, Probst H and Niethammer D (1992) *Mdr*-1/P-glycoprotein, topoisomerase and glutathione-S-transferase π gene expression in primary and relapse state adult and childhood leukaemias Br J Cancer 66 507–517
66 Cheng A-L, Su I-J, Chen Y-C, Lee T-C and Wang C-H (1993) Expression of P-glycoprotein and glutathione S-transferase in recurrent lymphoma the possible role of Epstein-Barr virus, and other predisposing factors J Clin Oncol 11 109–115
67 Satta T, Isobe K-i, Yamauchi M, Nakashima I and Takagi H (1991) Expression of MDR1 and glutathione S-transferase π genes and chemosensitivities in human gastrointestinal cancer Cancer 69 941–946
68 Linsenmeyer ME, Jefferson S, Wolf M, Mattthews JP, Board PG and Woodcock DM (1992) Levels of expression of the *mdr*-1 gene and glutathione S-transferase genes 2 and 3 and response to chemotherapy in multiple myeloma Br J Cancer 65 471–475
69 Koberda J and Hellman A (1991) Glutathione S-transferase activity of leukemic cells as a prognostic factor for response to chemotherapy in acute leukemias Med Oncol Tumor Pharmacother 8 35–38
70 Tidefelt U, Elmhorn-Rosenborg A, Paul C, Hao X-Y, Mannervik B and Eriksson LC (1992) Expression of glutathione S-transferase π as a predictor for treatment results at different stages of acute nonlymphoblastic leukemia Cancer Res 52 3281–3285
71 van der Zee AGJ, van Ommen B, Meijer C, Hollema H, van Bladeren PJ and de Vries EGE (1992) Glutathione S-transferase activity and isozyme composition in benign ovarian tumours, untreated malignant ovarian tumours, and malignant ovarian tumours after platinum/cyclophosphamide chemotherapy Br J Cancer 66 930–936
72 Murphy D, McGown AT, Hall A, Cattan A, Crowther D and Fox BW (1992) Glutathione S-transferase activity and isozyme distribution in ovarian tumour biopsies taken before or after cytotoxic chemotherapy Br J Cancer 66 937–942
73 Moscow JA, Townsend AJ, Goldsmith ME, Whang PJ, Vickers PJ, Poisson R, Legault PS, Myers CE and Cowan KH (1988) Isolation of the human anionic glutathione S-transferase cDNA and the relation of its gene expression to

estrogen-receptor content in primary breast cancer. Proc Natl Acad Sci USA 85: 6518–6522.

74. Howie AF, Miller WR, Hawkins RA, Hutchinson AR and Beckett GJ (1989) Expression of glutathione S-transferase B1, B2, Mu and Pi in breast cancers and their relationship to oestrogen receptor status. Br J Cancer 60: 834–837.
75. Gilbert L, Elwood L, Merino M, Masood S, Barnes R, Steinberg S, Lazarus D, Pierce L, d'Angelo T, Moscow JA, Townsend AJ and Cowan KH (1993) A pilot study of Pi-class glutathione S-transferase (GSTπ) in breast cancer: Correlation with estrogen receptor expression and prognosis in node-negative breast cancer. J Clin Oncol 11: 49–58.
76. Wright C, Cairns J, Cantwell BJ, Cattan AR, Hall AG, Harris AL and Horne CHW (1992) Response to mitoxantrone in advanced breast cancer: Correlation with expression of c-erbB-2 protein and glutathione S-transferases. Br J Cancer 65: 271–274.
77. Toffoli G, Frustaci S, Tumiotto L, Talamini R, Gherlinzoni F, Picci P and Boiocchi (1992) Expression of MDR1 and GSTπ in human soft tissue sarcomas: relation to drug resistance and biological aggressiveness. Ann Oncol 3: 63–69.
78. Cowan KH, Batist G, Tulpule A, Sinha BK and Myers CE (1986) Similar biochemical changes associated with multidrug resistance in human breast cancer cells and carcinogen-induced resistance to xenobiotics in rats. Proc Natl Acad Sci USA 83: 9328–9332.
79. Kitahara A, Satoh K, Nishimura K, Ishikawa T, Kazuo R, Sato K, Tsuda H and Itao N (1984) Purification, induction, and distribution of placental glutathione S-transferase: A new marker enzyme for preneoplastic cells in the rat chemical hepatocarcinogenesis. Proc Natl Acad Sci USA 82: 3964–3968.
80. Moore MA, Nakagawa K, Satoh K, Ishikawa T and Sato K (1987) Single GST-P positive liver cells; putative initiated hepatocytes. Carcinogenesis 8: 483–486.
81. Dixon KH, Cowell IG, Xia CL, Pemble SE, Ketterer B and Taylor JB (1989) Control of expression of the human glutathione S-transferase pi gene differs from its rat orthologue. Biochem Biophys Res Commun 163: 815–822.
82. Burt RK, Garfield S, Johnson K and Thorgeirsson SS (1988) Transformation of rat liver epithelial cells with v-H-*ras* or v-*raf* causes expression of MDR-1, glutathione S-transferase-P and increased resistance to cytotoxic chemicals. Carcinogenesis 9: 2329–2332.
83. Morrow CS, Goldsmith ME and Cowan KH (1990) Regulation of human glutathione S-transferase pi gene transcription: influence of 5′-flanking sequences and transactivating factors which recognize AP-1-binding sites. Gene 88: 215–225.
84. Morrow CS, Chiu J and Cowan KH (1992) Posttranscriptional control of glutathione S-transferase π gene expression in human breast cancer cells. J Biol Chem 267: 10544–10550.
85. Goodman J and Hochstein P (1977) Generation of free radicals and lipid peroxidation by redox cycling of adriamycin and daunomycin. Biochem Biophys Res Commun 77: 797–803.
86. Minmaugh EG, Trush MA, Ginsburg E and Gram TE (1982) Differential effects of anthracycline drugs on rat heart and liver microsomal reduced nicotinamide adenine diphosphate-dependent lipid peroxidation. Cancer Res 42: 3574–3582.
87. Minmaugh EG, Gram TE and Trush MA (1983) Stimulation of mouse heart and liver microsomal lipid peroxidation by anthracycline anticancer drugs: Characterization and effects of reactive oxygen scavengers. J Pharmacol Exp Ther 226: 806–816.
88. Julicher RH, Sterreberg L, Bast A, Riksen RO, Koomen JM and Noorhoek J (1986) The role of lipid peroxidation in acute doxorubicin-induced cardiotoxicity as studied in rat isolated heart. J Pharm Pharmacol 38: 277–282.
89. Minmaugh EG, Kennedy KA, Trush MA and Sinha BK (1985) Adriamycin-enhanced membrane lipid peroxidation in isolated rat nuclei. Cancer Res 45: 3296–3304.
90. Lown JW, Sim SK, Majumbar KC and Chang R (1977) Strand scission of DNA by bound adriamycin and daunorubicin in the presence of reducing agents. Biochem Biophys Res Commun 76: 705–710.
91. Lown JW, Chen HH, Plambeck JA and Acton EM (1982) Further studies on the generation of reactive oxygen species from activated anthracyclines and the relationship to cytotoxic action and cardiotoxic effects. Biochem Pharmacol 31: 575–581.
92. Eliot H, Gianni L and Myers C (1984) Oxidative destruction of DNA by the adriamycin – iron complex. Biochemistry 23: 928–936.
93. Doroshow JH, Akman S, Chu F-F and Esworthy S (1990) Role of glutathione-glutathione peroxidase cycle in the cytotoxicity of the anticancer quinones. Pharmacol Ther 47: 359–370.
94. Sinha BK, Katki AG, Batist G, Cowan KH and Myers CE (1987) Adriamycin-stimulated hydroxyl radical formation in human breast tumor cells. Biochem Pharmacol 36: 793–796.
95. Doroshow JH (1986) Prevention of doxorubicin-induced killing of MCF-7 human breast cancer cells by oxygen radical scavengers and iron chelating agents. Biochem Biophys Res Comm 135: 330–335.
96. Doroshow JH (1986) Role of hydrogen peroxide and hydroxyl radical formation in the killing of Ehrlich tumor cells by anticancer quinones. Proc Natl Acad Sci USA 83: 4514–4518.
97. Sinha BK, Katki AG, Batist G, Cowan KH and Myers CE (1987) Differential formation of hydroxyl radicals by Adriamycin in sensitive and resistant MCF-7 human breast tumor cells: Implications for the mechanism of action. Biochemistry 26: 3776–3781.
98. Townsend AJ, Morrow CS, Sinha BK and Cowan KH (1991) Selenium-dependent glutathione peroxidase expression is inversely related to estrogen receptor content in breast cancer. Cancer Commun 3: 265–270.
99. Keizer HG, van Rijn J, Pinedo HM and Joenje H (1988) Effect of endogenous glutathione, superoxide dismutases,

catalase, and glutathione peroxidase on Adriamycin tolerance of Chinese hamster ovary cells. Cancer Res 48: 4493–4497.

100. Rosazza JPN, Duffel MW, Elmarakby S and Ahm SH (1992) Metabolism of Caranthus alkaloids: From *Streptomyces griseus* to monoamine oxidase-B. J Nat Prod 55: 269.
101. Schlaifer D, Cooper MR, Attal M, Sartor AO, Trepel JB, Laurent G and Myers CE (1993) Myeloperoxidase: An enzyme involved in intrinsic vincristine resistance in human myeloblastic leukemia. Blood 81: 482–489.
102. Mullenbach GT, Tabrizi A, Irvine BD, Bell GI, Tainer JA and Hallewell RA (1988) Selenocysteine's mechanisms of incorporation and evolution revealed in cDNAs of three glutathione peroxidases. Protein Engineering 2: 239–246.
103. Sukenaga Y, Ishida K, Yakeda T and Tagaki K (1987) cDNA sequence coding for human glutathione peroxidase. Nucleic Acids Res 15: 7178.
104. Moscow JA, Morrow CS, He R, Mullenbach GT and Cowan KH (1992) Structure and function of the 5′ flanking sequence of the human cytosolic selenium-dependent glutathione peroxidase gene (hgpx1). J Biol Chem 267: 5949–5958.
105. Akasaka M, Mizoguchi J and Takahashi K (1990) A human cDNA sequence for a novel glutathione peroxidase related protein. Nucleic Acids Res 18: 4619.
106. Chu FF, Doroshow JH and Esworthy RS (1993) Expression, characterization, and tissue distribution of a new cellular selenium-dependent glutathione peroxidase, GSHPx-Gl. J Biol Chem 268: 2571–2576.
107. Takahashi K, Akasak M, Yamamoto Y, Kobayashi C, Mizoguchi J and Koyama J (1990) Primary structure of human plasma glutathione peroxidase deduced from cDNA sequences. J Biochem 108: 145–148.
108. Chu FF, Esworthy RS, Doroshow JH, Doan K and Liu X-F (1992) Expression of plasma glutathione peroxidase in human liver in addition to kidney, heart, lung and breast in humans and rodents. Blood 79: 3233–3238.
109. Maiorino M, Chu FF, Ursini F, Davies KJA, Doroshow JH and Esworthy RS (1991) Phospholipid hydroperoxide glutathione peroxidase is the 18-kDa selenoprotein expressed in human tumor cell lines. J Biol Chem 266: 7728–7732.
110. Schuckelt R, Bigelius-Flohe R, Maiorino M, Roveri A, Reumkens J, Strassburger W, Ursini F, Wolf B and Flohe L (1991) Phospholipid hydroperoxide glutathione peroxidase is a seleno-enzyme distinct from the classical glutathione peroxidase as evident from cDNA and amino acid sequencing. Free Rad Res Commun 14: 343–361.
111. Duan YJ, Komura S, Fiszer-Szafarz D and Yagi K (1988) Purification and characterization of a novel monomeric glutathione peroxidase from rat liver. J Biol Chem 263: 19003–19008.
112. Ghyselink NB and Dufaure J-P (1990) A mouse cDNA sequence for epididymal androgen-regulated proteins related to glutathione peroxidase. Nucleic Acids Res 18: 7144.
113. Arrick BA, Nathan CF and Cohn ZA (1983) Inhibition of glutathione synthesis augments lysis of murine tumors cells by sulfhydryl-reactive antineoplastics. J Clin Invest 71: 258–267.
114. Lee FYF, Vessey AR and Siemann DW (1988) Glutathione as a determinant of cellular response to doxorubicin. NCI Monographs 6: 211–215.
115. Russo A and Mitchell JB (1985) Potentiation and protection of doxorubicin cytotoxicity by cellular glutathione modulation. Cancer Treatment Rep 69: 1293–1296.
116. Dusre L, Mimnaugh EG, Myers CE and Sinha BK (1989) Potentiation of doxorubicin cytotoxicity by buthionine sulphoximine in multidrug-resistant human breast tumor cells. Cancer Res 49: 511–515.
117. Lau DHM, Lewis AD, Ehsan MN and Sikic BI (1991) Multifactorial mechanisms associated with broad cross-resistance of ovarian carcinoma cells selected by cyanomorpholino doxorubicin. Cancer Res 51: 5181–5187.
118. Lee FYF, Sciandra J and Siemann OW (1989) A study of the mechanism of resistance to adriamycin in vivo. Glutathione metabolism, P-glycoprotein expression and drug transport. Biochem Pharmac 38: 3697–3705.
119. Rosenburg MC, Colvin OM, Griffith OW, Bigner SH, Elion GB, Horton JK, Lilley E, Bigner DB and Friedman HS (1989) Establishment of a melphalan-resistant rhabdomyosarcoma xenograft with cross resistance to vincristine and enhanced sensitivity following buthionine sulphoximine-mediated glutathione depletion. Cancer Res 49: 6917–6922.
120. Ford JM, Yang JM and Hait WM (1991) Effect of buthionine sulphoximine on toxicity of verapamil and doxorubicin to multidrug resistant cells and to mice. Cancer Res 51: 67–72.
121. Doroshow JH, Akman S, Esworthy S, Chu F-F and Burke T (1991) Doxorubicin resistance conferred by selective enhancement of intracellular glutathione peroxidase or superoxide dismutase content in human breast cancer cells. Free Rad Res Commun 12–13: 779–781.
122. Mirault M-E, Tremblay A, Beudoin N and Tremblay M (1991) Overexpression of selenoglutathione peroxidase by gene transfer enhances the resistance of T47 human breast cancer cells to clastogenic oxidants. J Biol Chem 266: 20752–20760.
123. Hwang C, Sinskey AJ and Lodish HF (1992) Oxidized redox state of glutathione in the endoplasmic reticulum. Science 257: 1496–1502.

Address for offprints: J.A. Moscow, Building 10, Room 12N226, National Institutes of Health, Bethesda, MD 20892, USA.

Cytotechnology **12**: 171–212, 1993.

Pharmacologic circumvention of multidrug resistance

James M. Ford[1] and William N. Hait[2]
[1]*Division of Oncology, M-211, Stanford University Medical Center, Stanford, CA 94305, USA; and* [2]*Cancer Institute of New Jersey, University of Medicine and Dentistry of New Jersey, Robert Wood Johnson Medical School, 679 Hoes Lane, Piscataway, NJ 08854, USA*

Key words: cancer chemotherapy, chemosensitizer, modulation, multidrug resistance

Abstract

The ability of malignant cells to develop resistance to chemotherapeutic drugs is a major obstacle to the successful treatment of clinical tumors. The phenomenon multidrug resistance (MDR) in cancer cells results in cross-resistance to a broad range of structurally diverse antineoplastic agents, due to outward efflux of cytotoxic substrates by the *mdr*1 gene product, P-glycoprotein (P-gp). Numerous pharmacologic agents have been identified which inhibit the efflux pump and modulate MDR. The biochemical, cellular and clinical pharmacology of agents used to circumvent MDR is analyzed in terms of their mechanism of action and potential clinical utility. MDR antagonists, termed chemosensitizers, may be grouped into several classes, and include calcium channel blockers, calmodulin antagonists, anthracycline and *Vinca* alkaloid analogs, cyclosporines, dipyridamole, and other hydrophobic, cationic compounds. Structural features important for chemosensitizer activity have been identified, and a model for the interaction of these drugs with P-gp is proposed. Other possible cellular targets for the reversal of MDR are also discussed, such as protein kinase C. Strategies for the clinical modulation of MDR and trials combining chemosensitizers with chemotherapeutic drugs in humans are reviewed. Several novel approaches for the modulation of MDR are examined.

Abbreviations: ALL — acute lymphocytic leukemia; AML — acute myelogenous leukemia; CaM — calmodulin; CsA — cyclosporin A; MDR — multidrug resistance; P-gp — P-glycoprotein; PMA — phorbol 12-myristate 13-acetate; PKC — protein kinase C

Introduction

Clinical drug resistance to chemotherapeutic agents is a major obstacle for their curative potential in the treatment of cancer. Multidrug resistance (MDR) has been found to be an important and common cause of drug resistance in experimental systems, and is defined phenotypically by the ability of cells exposed to a single cytotoxic agent to develop resistance to a broad range of structurally and functionally unrelated drugs due to enhanced drug efflux. MDR is mediated by the increased expression of an energy-dependent drug transport pump, known as P-glycoprotein (P-gp) (Biedler and Riehm, 1970; Juliano and Ling, 1976). Cells selected for MDR with one drug display significant cross-resistance to other drugs affected by the MDR phenotype, which include the natural products and their semi-synthetic derivatives such as doxorubicin, mitoxantrone, vincristine, vinblastine, colchicine, actinomycin D, VP-16, taxol and topotecan, but not drugs such as bleomycin, methotrexate or alkylating agents. The most consistent alteration found in MDR cell lines is an increased expression of the 170–180 kD P-gp encoded for by the *mdr*1 gene (Gros *et al.*, 1986b), and the concomitant decrease in accumulation and retention of cytotoxic drugs due to their active efflux (Riordan and Ling, 1985).

A great deal has been learned regarding the biochemical and cellular pharmacology, and molecular biology of MDR, and is discussed in detail in other reviews within this volume. Briefly, structural and functional analyses of P-gp have revealed that it contains two highly hydrophobic membrane spanning domains and two nucleotide binding domains (Chen *et al.*, 1986; Gros *et al.*, 1986a), and belongs to a superfamily of proteins that behave as ion channels and/or transporters. For example, P-gp itself has been shown to function as a drug efflux pump and a volume-regulated chloride channel (Valverde *et al.*, 1992). Other proteins containing homologous transmembrane domains and ATP binding motifs include the bacterial periplasmic α-hemolysin protein export pump (Higgins *et al.*, 1982, 1986), an exporter of a hydrophobic, lipopeptide pheromone mating factor in yeast (McGrath and Varshavsky, 1989), the cystic fibrosis transmembrane conductance regulator in humans (Anderson *et al.*, 1991b; Bear *et al.*, 1992), proteins involved in peptide transport and presentation in MHC-restricted immune cells (Spies *et al.*, 1990; Trowsdale *et al.*, 1990), and additional related proteins potentially involved in mammalian drug resistance (Cole *et al.*, 1992; Scheper *et al.*, 1993). These and other studies suggested that P-gp may be capable of both transporter and ion channel activity, possibly explaining the complex pharmacologic profile displayed by P-gp. For example, we previously noted that P-gp shares properties with both active transport carrier molecules such as in gated-pore or membrane translocating models, and with ion channel proteins affecting hydrophobic, organic cations (Ford and Hait, 1990).

The drug efflux function and resulting cellular drug resistance conferred by P-gp has been found to be inhibited by a variety of pharmacologic agents, most often in a competitive manner (Ford and Hait, 1990). This has suggested the possible modulation of clinical drug resistance in human tumors by employing drugs to overcome MDR in combination with chemotherapeutic agents. Since P-gp has been found to be expressed in many human tumors, this strategy appears worth pursuing in clinical trials.

A great many questions remain regarding the pharmacologic inhibition of P-gp function, including potential mechanisms of action, the optimal agents for use against P-gp, and the clinical significance and sequelae of anti-MDR therapy. In this review, we will address these issues, and will discuss the biochemical, cellular and clinical pharmacology of drugs used for the circumvention of P-gp-mediated MDR in tissue culture and in humans.

Pharmacologic reversal of multidrug resistance *in vitro* by chemosensitizers

A primary goal in the investigation of P-gp-mediated MDR is to discover specific means by which to reverse or circumvent it. Therefore, a substantial number of studies have focused on the pharmacologic reversal of MDR. Through the identification of drugs that inhibit the MDR transporter, it is hoped that a better understanding of the biochemical mechanisms involved in this form of cellular drug resistance will emerge and that new agents for potential use in clinical trials will be discovered.

More than a decade ago, Tsuruo and coworkers first reported the pharmacologic reversal of MDR by showing that verapamil and trifluoperazine potentiated the antiproliferative activity of vincristine and produced an increased cellular accumulation of vincristine in an MDR murine leukemia cell line (Tsuruo *et al.*, 1981). Since this original observation, many compounds have been shown to antagonize MDR in a variety of *in vitro* tissue culture assays and *in vivo* tumor models when coadministered with chemotherapeutic agents to which the cells are resistant. In general, agents used to antagonize MDR, termed "chemosensitizers" or "MDR modulators", alter the drug accumulation defect present in MDR cells, and cause little or no changes in sensitive cells. Most investigators have determined the magnitude of the effects of chemosensitizers by comparing the concentrations of cytotoxic drug necessary to inhibit growth or kill cells by 50% in the absence and presence of a relatively nontoxic, fixed concentration of a chemosensitizer. This ratio reflects the fold-sensitization conferred by a particular chemosensitizer.

The chemosensitizers described to date may be grouped into seven broad categories: (*a*) calcium channel blockers, (*b*) calmodulin (CaM) antagonists, (*c*) noncytotoxic anthracycline and *Vinca* alkaloid analogs, (*d*) steroids and hormonal antagonists, (*e*) cyclosporines, (*f*) dipyridamole, and (*g*) miscellaneous hydrophobic, cationic compounds (summarized in Table 1). Although these compounds share only broad structural similarities, all are extremely lipophilic, and many are heterocyclic, positively charged substances. This suggests that there may be one or more specific receptor sites for anti-MDR drugs, which have unique structural requirements for efficient binding.

In the following sections we will review the effects of these various chemosensitizers *in vitro*, and will focus on the need to carefully define the structural requirements for antagonism of MDR by chemosensitizers and to elucidate their mechanism of action, in order to develop more specific and effective agents for clinical use.

Verapamil

In their initial report, Tsuruo's group analyzed the effects of the calcium channel blocker verapamil in a vincristine resistant P388 murine leukemia cell line (Tsuruo *et al.*, 1981). Since the alteration in drug transport in this cell line was attributable to the plasma membrane, verapamil was studied because of its known effects on membrane functions. Through cellular proliferation assays, it was shown that verapamil reversed the approximately 30-fold resistance to vincristine and 7-fold resistance to vinblastine displayed by these P388/VCR

Table 1. Pharmacologic agents with ability to reverse MDR *in vitro*[a]

Calcium channel blockers	Immunosuppressive drugs
Verapamil (6–10 μM)	Cyclosporin A (0.8–2 μM)
Nifedipine (35 μM)	11-Methyl-leucine cyclosporine (1 μM)
Nicardapine (3–10 μM)	SDZ PSC 833 (0.1 μM)
Niguldipine (10 μM)	SDZ 280-446 (0.1 μM)
Bepridil (4 μM)	FK506 (3 μM)
PAK-200 (5 μM)	Rapamycin (3 μM)
Calmodulin antagonists	Antibiotics
Trifluoperazine (3–5 μM)	Cefoperazone (1000 μM)
Prochlorperazine (4 μM)	Ceftriaxone (1000 μM)
Fluphenazine (3 μM)	Erythromycin (650 μM)
trans-Flupenthixol (3–5 μM)	Tetracycline (4000 μM)
Vinca alkaloid analogs	Miscellaneous compounds
Vindoline (20–50 μM)	Dipyridamole (5 – 10 μM)
	BIBW 22 (1 μM)
Steroidal agents	Quinidine (10 μM)
Progesterone (2 μM)	Chloroquine (10–50 μM)
Tamoxifen (3–10 μM)	Terfenadine (3–6 μM)
Toremifene (5–10 μM)	Reserpine (5 μM)
Megestrol acetate (5 μM)	Yohimbine (5 μM)
	Amiodarone (4 μM)
	Solutol HS 14 (4–14 μM)

[a]Concentrations in parenthesis are those shown to have effect in reversing MDR *in vitro*.

cells (Tsuruo *et al.*, 1981). Incubation of P388/VCR cells with verapamil for 5 h also caused a 10-fold increase in the accumulation of [^{3}H]-vincristine. Since verapamil was shown not to alter vincristine binding to tubulin, the apparent cytotoxic target of *Vinca* alkaloids (Owellen *et al.*, 1974), it was concluded that verapamil's pharmacologic chemosensitizing effect was due to alterations in drug accumulation (Tsuruo *et al.*, 1981).

In subsequent reports, Tsuruo's group demonstrated that verapamil fully reversed the 20-fold vincristine resistance and the 3-fold doxorubicin resistance in a MDR human acute myelogenous leukemia (AML) line K562/VCR (Tsuruo *et al.*, 1983b), and partially reversed the 40-fold doxorubicin resistance in P388/ADM cells (Tsuruo *et al.*, 1982). Again, this chemosensitizing effect was associated with increased accumulation of chemotherapeutic drug, which was shown to be associated with the inhibition of an energy-dependent efflux mechanism (Tsuruo *et al.*, 1982).

Since these early observations, many investigators have demonstrated the chemosensitizing activity of verapamil in various MDR cell lines (Ford and Hait, 1990). The sensitivity to cytotoxic drugs of most intrinsically sensitive cells from which these MDR lines were derived were not significantly affected by verapamil at noncytotoxic doses.

The effect of verapamil on cross-resistance to chemotherapeutic drugs other than those used for initial selection has also been investigated. In certain studies, verapamil produced as great an effect on cross-resistance to drugs as on primary resistance to the selecting agent. For example, Sikic's group showed that 6 μM verapamil caused a 7-fold sensitization to doxorubicin, associated with increased [^{14}C]-doxorubicin accumulation and retention, and a similar magnitude of reversal of cross-resistance to daunomycin, actinomycin D and mitoxantrone with a MES-SA human uterine sarcoma cell line selected for 100-fold resistance to doxorubicin (Harker *et al.*, 1986; Sikic *et al.*, 1989). Similarly, using an NIH 3T3 cell line transfected with and expressing an *mdr*1 gene derived from a colchicine resistant KB cell line, we found that 12 μM verapamil completely reversed primary resistance to colchicine (40-fold) and cross-resistance to doxorubicin (20-fold) (Ford *et al.*, 1990).

In other studies, verapamil was more effective in reversing resistance to the selecting agents than to the cross-resistant drugs. For example, we found that 5 μM verapamil caused a 13-fold reversal of 200-fold doxorubicin resistance in MCF-7 Adr^R cells, but resulted in only a 4-fold change in the 400-fold cross-resistance to colchicine (Ford *et al.*, 1990). In contrast, verapamil has also been shown to have a greater effect on cross-resistance than on primary resistance to selecting agents in some cell lines. Fojo *et al.* (1985) demonstrated that 20 μM verapamil completely reversed the 20- to 70-fold cross-resistance to doxorubicin, vinblastine and vincristine in colchicine selected KB human carcinoma cells, while causing only a 60-fold reduction in the 220-fold resistance to colchicine. Our data with vinblastine selected KB cells also shows a relatively greater reversal by verapamil of cross-resistance to doxorubicin and colchicine, than that of primary *Vinca* alkaloid resistance (Ford *et al.*, 1990).

These findings are of interest with respect to both the mechanism and pharmacology of the chemosensitizing activity of verapamil, and in defining the mechanisms by which P-gp transports structurally distinct molecules. Why MDR cell lines display different degrees of resistance and cross-resistance to drugs, and why some cross-resistance is refractory to modulation by verapamil remains unclear. One possibility is that mutations in the *mdr*1 gene (Choi *et al.*, 1988) or posttranslational modifications of P-glycoprotein (Hait and Aftab, 1992) may cause changes in the affinity of cytotoxic drugs or chemosensitizers for the putative drug-binding site(s). Alternatively, multiple drug binding sites within P-gp could explain these phenomena.

The effect of verapamil on drug accumulation in MDR cells has been carefully documented. Kessel and Wilberding (1985a) probed the effect of verapamil on anthracycline cellular kinetics in P388/ADR cells. Drug influx was not altered by verapamil nor by the metabolic inhibitor sodium azide, consistent with a diffusional model of anthracycline inward transport. Drug efflux from [^{3}H]-daunomycin loaded cells was also not affected by either modifying agent in P388 sensitive cells, while daunomycin efflux from MDR cells was inhibited by varying degrees in MDR cells by 2 to 20 μM verapamil.

Studies of the many other MDR chemosensitizers which have now been described demonstrate that most appear to function in a manner similar to that of verapamil, though with differing potencies, and perhaps at different sites on the P-gp molecule. Therefore, these critical studies of the effects of verapamil on the cellular pharmacology of MDR cell lines serve as a paradigm for the function of chemosensitizers in general. It has been clearly shown that verapamil inhibits the P-gp associated, energy-dependent drug efflux common to MDR cells and is a potent and effective antagonist of resistance to a number of drugs in most MDR cell lines *in vitro*. In addition, it is likely that the sensitization produced by verapamil is secondary to the resultant increase in intracellular accumulation of chemotherapeutic agents. A number of investigators have shown that photoactivated verapamil analogs bind to P-gp, and that verapamil inhibits the binding of many chemotherapeutic drugs as well as other chemosensitizers to P-gp (Akiyama *et al.*, 1988; Beck *et al.*, 1988; Cornwell *et al.*, 1987; Safa, 1988a, 1988b; Safa *et al.*, 1987), suggesting that the mechanism of action of verapamil is through blocking the binding of drug to P-gp.

In terms of the potential clinical utility for reversing MDR *in vivo*, verapamil is limited by its untoward cardiovascular effects in humans at plasma concentrations in the 2 to 6 μM range needed for antagonism of MDR *in vitro* (Candell *et al.*, 1979; DeFaire and Lundman, 1977). Therefore, more potent and less toxic chemosensitizing agents than verapamil have been investigated.

Verapamil analogs

Based on the success of verapamil in antagonizing MDR, a number of verapamil analogs with varying degrees of calcium channel blocking activity have now been investigated. In a study of 14 verapamil analogs, Kessel and Wilberding (1985b) found that the analog Ro 11-2933 was 10-fold more potent than verapamil in mediating this effect. The activity of the four most active verapamil analogs correlated with their effect on drug accumulation.

Yamaguchi and coworkers reported that SDB-ethylenediamine, a synthetic isoprenoid with a structurally similar carbon backbone to verapamil, produced a 2- to 5-fold increase in doxorubicin cytotoxicity in the P388/ADR cell line, as well as partial reversal of cross-resistance to vincristine, vinblastine and daunomycin (Yamaguchi *et al.*, 1986). This compound was also effective in P388/VCR cells and in colchicine selected MDR human epidermal carcinoma KB-Ch^R-24 (Nakagawa *et al.*, 1986; Yamaguchi *et al.*, 1986). SDB-ethylenediamine had far less activity as a calcium channel blocking agent than verapamil (Yamaguchi *et al.*, 1986). Furthermore, a radiolabeled, photoactivatable SDB analog photoaffinity labeled P-gp in MDR human KB-C2 cells, and a number of *Vinca* alkaloids, anthracyclines, and chemosensitizers, including verapamil, inhibited this binding (Akiyama *et al.*, 1989).

All of the previously discussed studies of the effect of verapamil on MDR have used a racemic mixture. Because the *S*-enantiomer of verapamil selectively binds to calcium channels (Weir *et al.*, 1992) recent studies have compared the effects of the *S*- and *R*-enantiomers of verapamil on MDR and found them to be equally active chemosensitizers (Gruber *et al.*, 1988; Pirker *et al.*, 1990; Plumb *et al.*, 1990; Qian and Beck, 1990). Similarly, Hollt *et al.* (1992) studied the effect of a variety of stereoisomers of phenylalkylamine and dihydropyridine structures for their ability to increase intracellular accumulation of [^{3}H]-vinblastine in a doxorubicin resistant, P-gp expressing Friend mouse leukemia cell line. The drugs (±)-verapamil, (±)-devopamil, (±)-emopamil, (±)-nimodepine, (±)-felodipine, (±)-nitrendipine and (±)-niguldipine all differed markedly in their calcium channel blocking activity, but were equally effective in increasing vinblastine levels in MDR cells. Therefore, the use of less cardiotoxic enantiomers of verapamil and its analogs may provide a means for reaching clinically effective anti-MDR levels in patients. Already this strategy has been incorporated into ongoing clinical trials (Wilson *et al.*, 1993a).

Thus, certain compounds structurally related to verapamil partially reverse MDR, but lack other physiological effects of verapamil on cell membranes. This suggests that through structure-activity relationships it may be possible to identify and exploit the structural features necessary for anti-MDR activity, while diminishing those important for blocking calcium channels.

Other calcium channel blockers

A number of calcium channel blockers structurally dissimilar to verapamil have been studied for chemosensitizing activity and have been found to be active (Helson, 1984; Tsuruo, 1989). For example, in Tsuruo's early investigations of chemosensitizers, he showed that the calcium channel blockers, prenylamine and caroverine, were as active, though 2- to 3-fold less potent, than verapamil for altering the sensitivity of P388/VCR and K562/VCR cells to vincristine and P388/ADM cells to doxorubicin (Tsuruo *et al.*, 1982, 1983).

Two other classes of structurally dissimilar calcium channel blockers, diltiazem and nifedipine and its analogs, had significant anti-MDR activity (Ramu, *et al.*, 1984; Tsuruo *et al.*, 1983a, 1985). While nifedipine is known to be a potent calcium channel blocker (Fleckenstein *et al.*, 1979), it was a poor antagonist of MDR (Ramu *et al.*, 1984, Tsuruo, *et al.*, 1983). However, the dihydropyridine analogs, niludipine, nimodipine, and nicardapine, were found to be potent antagonists of MDR, with 3.5—10 μM nicardapine fully reversing vincristine resistance in P388/VCR and K562/VCR cells (Tsuruo *et al.*, 1983a, 1985) and partially reversing doxorubicin resistance in P388/ADR cells (Ramu *et al.*, 1984; Tsuruo *et al.*, 1983a). Recently, the chemosensitizing activity of 200 newly synthesized dihydropyridine analogs in human MDR KB-C2 cells was reported (Niwa *et al.*, 1992). The lead compound, PAK-200, possessed the lowest calcium channel blocking activity, yet it reversed the >100-fold resistance to vincristine and low level resistance to doxorubicin in KB-C2 cells at a dose of 5 μM in several other human cell lines. These studies confirmed the lack of correlation between pharmacologic calcium channel antagonism and anti-MDR activity.

Thus, a number of calcium channel blockers from different structural classes have been identified which are active in the antagonism of MDR in cultured cells. Unfortunately, most of these compounds still harbor undesirable cardiovascular effects and will probably be unsuited for clinical use as chemosensitizers. However, the several dihydropyridine chemosensitizers which lack calcium channel blocking activity hold promise as relatively specific agents.

Calmodulin antagonists

The second class of MDR chemosensitizers to be identified include drugs previously known for their ability to inhibit CaM-mediated processes. The activity of these drugs for reversing MDR was first identified in Tsuruo's work on chemosensitizers (Tsuruo *et al.*, 1982), extending the observation of the anti-MDR effect of verapamil to several other drugs known to perturb intracellular calcium homeostasis. For example, nontoxic concentrations of the phenothiazine CaM antagonist, trifluoperazine, caused a 5- to 10-fold increase in vincristine and doxorubicin sensitivity in 20- and 40-fold resistant P388/VCR and P388/ADM cells, respectively (Tsuruo *et al.*, 1982), and fully reversed 17-fold K562/VCR resistance to vincristine (Tsuruo *et al.*, 1983). Trifluoperazine also caused a 4- to 5-fold increase in [^{3}H]-vincristine and [^{3}H]-doxorubicin accumulation, and did not significantly alter drug cytotoxicity or accumulation in sensitive cell lines (Tsuruo *et al.*, 1982, 1983). Equimolar concentrations of trifluoperazine and verapamil produced equivalent increases in drug accumulation in all three cell lines, though trifluoperazine possessed greater intrinsic cytotoxicity than verapamil (Tsuruo *et al.*, 1982, 1983). We found that verapamil caused a 2- to 3-fold greater reversal of primary and cross-resistance to anthracyclines, *Vinca* alkaloids and colchicine than equimolar concentrations of trifluoperazine in a variety of human and mouse MDR cell lines, and resulted in greater intracellular doxorubicin accumulation in MCF-7 Adr^R cells (Ford and Hait, 1990; Ford *et al.*, 1989).

A series of investigations by Ganapathi and colleagues have focused on the effect of trifluoperazine on cellular drug resistance, cross-resistance and drug accumulation. In P388/DOX cells which are 100-fold resistant to doxorubicin, 4–5 μM trifluoperazine caused a 5- to 10-fold increase in sensitivity to doxorubicin (Ganapathi and Grabowski, 1983; Ganapathi *et al.*, 1984). Trifluoperazine also had significant effects on the cytotoxicity of *Vinca* alkaloids in both MDR and sensitive P388 cells. It increased the sensitivity to vinblastine and vincristine by 2- to 10-fold in the 20- and 100-fold cross resistant P388/DOX cells, respectively, as well as in sensitive P388 cells (Ganapathi *et al.*, 1986). Conversely, the effect of trifluoperazine on the accumulation of [^{3}H]-vinblastine was different in the P388/DOX and P388 sensitive cells. It produced an 8-fold increase in resistant cells and only a 2-fold increase in sensitive cells (Ganapathi *et al.*, 1986a). Similarly, in a series of L1210 murine leukemia cell lines selected for progressive resistance to doxorubicin (5- to 40-fold) (Ganapathi and Grabowski, 1988), the magnitude of the effect of trifluoperazine was related to the degree of resistance to doxorubicin, but not to vincristine. Furthermore, trifluoperazine resulted in increased accumulation of vincristine in sensitive and resistant cells, but this did not correlate with the resultant cytotoxicity (Ganapathi *et al.*, 1991b). In these studies, the enhancement of *Vinca* alkaloid cytotoxicity by trifluoperazine did not clearly parallel its effect on cellular drug levels.

Thus, Ganapathi's group questioned the primary role of drug accumulation in modulating MDR by the phenothiazines. For example, they compared the cellular toxicity in P388 cells of similar intracellular concentrations of doxorubicin achieved by either exposing cells to 5 μM trifluoperazine or by increasing extracellular concentrations of doxorubicin, and found that greater cytotoxicity occurred with trifluoperazine treatment (Ganapathi *et al.*, 1986b). These results were confirmed in the previously described L1210 MDR cell lines, which were shown to display increasing levels of P-gp as well as decreased topoisomerase II activity with increasing levels of resistance (Ganapathi *et al.*, 1989). Higher nuclear concentrations of doxorubicin were required in resistant cells to achieve toxicity equal to that in sensitive cells. However, in the presence of trifluoperazine, equal doxorubicin cytotoxicity was achieved with lower nuclear doxorubicin concentrations than in sensitive cells (Ganapathi *et al.*, 1991a). Furthermore, studies of chromosomal damage and DNA single-strand breaks induced by doxorubicin suggest that trifluoperazine produces part of its anthracycline modulatory effect by potentiating DNA damage independent of the cellular concentrations of doxorubicin (Ganapathi *et al.*, 1990, 1991b).

The relationship of drug accumulation to cytotoxicity is difficult to assess since drug accumulation and retention may represent only one of many factors influencing the cytotoxicity of chemotherapeutic agents. Nevertheless, the disparity between modulation of accumulation and cytotoxicity in Ganapathi's studies suggests that trifluoperazine, a drug known to effect many cellular enzymes and receptors (Creese and Sibley, 1980; Hait and Lazo, 1986; Pang and Briggs, 1976; Ruben and Rasmussen, 1981) may alter cell sensitivity to drugs in additional ways unrelated to changes in accumulation. Indeed, several investigators have shown that CaM antagonists such as trifluoperazine modulate cell sensitivity to bleomycin through increased DNA damage or through inhibition of DNA repair mechanisms (Chafouleas *et al.*, 1984; Lazo *et al.*, 1985).

To identify CaM antagonists with greater activity and specificity for reversing MDR, we examined the structure-activity relationships for a series of 22 phenothiazine derivatives for potentiation of doxorubicin activity in 200-fold resistant MCF-7 Adr^R cells (Ford *et al.*, 1989). Hydrophobicity of the tricyclic ring, length of the alkyl bridge and the charge on the amino side chain were independently important for anti-MDR activity. Phenothiazines which fulfilled these criteria, such as prochlorperazine, fluphenazine and trifluoperazine, caused at most a 3-fold reversal of doxorubicin resistance in this study. However, by searching for related drugs which shared certain of these structural features, we identified the thioxanthenes as a novel class of chemosensitizers that possessed significantly greater activity. The presence of an exocyclic double bond confers stereoisomerism to these drugs, and the *trans*-isomer of each in a series of 16 thioxanthenes showed greater activity than the *cis*-isomer and its respective phenothiazine homolog for reversing doxorubicin resistance in MCF-7 Adr^R cells (Ford *et al.*, 1990). This effect was not related to differential uptake of the stereoisomers. Similar to the phenothiazines, the most potent thioxanthene chemosensitizers possessed halogenated tricyclic rings connected to piperazinyl or piperadinyl side groups. The lead compound, *trans*-flupenthixol, reversed MDR in a number of human and murine MDR cell lines and in sensitive cells transfected with the *mdr*1 gene, and increased doxorubicin accumulation in MCF-7 Adr^R cells to a greater extent than either its stereoisomer *cis*-flupenthixol, its phenothiazine structural homolog fluphenazine, or the calcium channel blocker verapamil, and inhibited [^{3}H]azidopine binding to P-gp (Ford *et al.*, 1990).

The clinical pharmacology and toxicology of *trans*-flupenthixol suggests it may be uniquely suited for *in vivo* use. Although tricyclic neuroleptics undergo extensive first-pass hepatic metabolism, *trans*-flupenthixol has a half-life of 34 h due to a large volume of distribution. Furthermore, tissue penetration by the parent drug and its metabolites is likely to be high because of its lipophilicity (Jorgensen *et al.*, 1982). Clinical trials of the antipsychotic effects of thioxanthenes in humans showed that *cis*-flupenthixol was far more effective than *trans*-flupenthixol, and that the latter drug was far less toxic (Johnstone *et al.*, 1978). In mice, *trans*-flupenthixol has been shown to be 100- to 1000-fold less potent than the active *cis*-isomer in a number of assays measuring the side effects and neuroleptic potential of antipsychotic drugs (Nielsen *et al.*, 1973; Nielson *et al.*, 1962). This observation may be explained by biochemical and crystallographic evidence that *cis*-flupenthixol is a potent antagonist of dopamine receptors (Huff and Molinoff, 1984; Post *et al.*, 1975), whereas *trans*-flupenthixol has virtually no activity as a dopamine antagonist, resulting in its apparent lack of extrapyramidal side effects (Nielsen *et al.*, 1973). Also, *trans*-flupenthixol has little or no α-adrenergic, β-adrenergic, or 5-HT_1-blocking activity (Hyttel *et al.*, 1984). Since extrapyramidal side effects have proven to be dose limiting in Phase I trials that combined trifluoperazine with bleomycin (Hait *et al.*, 1989a) or doxorubicin (Miller *et al.*, 1988), further preclinical development is warranted for this potentially useful clinical chemosensitizer.

In summary, certain CaM antagonists, such as the phenothiazines and thioxanthenes, are effective chemosensitizers, enhance cytotoxic drug accumulation and retention in MDR cells, and may have other effects on DNA damage and repair. Furthermore, specific structural features and spatial relationships appear to be critical for chemosensitizing activity for these classes of drugs. This suggests that application of these principles may enable the rational design of far more potent and specific chemosensitizers.

Anthracycline and vinca alkaloid analogs

Soon after the discovery that MDR cells accumulate less drug due to an active efflux mechanism (Skovsgaard, 1978), Skovsgaard tested the hypothesis that a specific drug efflux pump would be competitively inhibited by an excess of a nontoxic substrate (Skovsgaard, 1980). A useful anthracycline analog for this purpose was *N*-acetyl-daunorubicin, which lacks the free amino-group of daunomycin essential for DNA intercalation and thus has a lower affinity for DNA. As a result, there is less nuclear accumulation, cytotoxicity and higher cytoplasmic concentrations of this drug (Zunino *et al.*, 1972). Skovsgaard found that a 30-fold excess of *N*-acetyl-daunorubicin markedly inhibited active efflux of daunomycin in MDR, but not in sensitive, Ehrlich ascites cells, leading to increased net daunomycin accumulation (Skovsgaard, 1980). The effect of *N*-acetyl-daunorubicin on cellular resistance to daunomycin was unfortunately not examined *in vitro*, although a 1:20 combination of daunomycin and its *N*-acetyl analog increased the life span of mice inoculated with MDR Ehrlich ascites cells compared to no effect with either drug alone (Skovsgaard, 1980).

Inaba's group subsequently analyzed the chemosensitizing effects of three additional anthracycline analogs on MDR cells and found significant enhancement of vincristine toxicity in P388/VCR cells *in vitro* when the drugs were used in 100-fold excess of vincristine (Inaba *et al.*, 1984). One compound partially reversed the cross-resistance of these cells to daunomycin, implying that either a similar drug efflux mechanism effects both anthracyclines and *Vinca* alkaloids, or that the daunomycin analogs compete with both anthracyclines and *Vinca* alkaloids for separate binding sites on P-gp. Similarly, a number of relatively nontoxic *Vinca* alkaloid analogs, such as vindoline, effectively antagonized both the primary resistance and cross-resistance of MDR P388/ADM and P388/VCR cells (Inaba and Nagashima, 1986) when used in 1000-fold excess of the chemotherapeutic drug. A series of weakly cytotoxic stereoisomeric congeners of vinblastine, with poor antimicrotubule activity, were also shown to enhance the cytotoxicity of both anthracyclines and *Vinca* alkaloids in S180 MDR cell lines (Borman *et al.*, 1993).

Steroids and hormonal analogs

The expression of high levels of P-gp in human adrenal cortex, placenta and pregnant murine uterus (Arceci *et al.*, 1988), suggest a possible role for the pump in physiologic transport of steroid hormones. Therefore, several groups have studied whether steroids are natural substrates for P-gp and/or are chemosensitizers in combination with cytotoxic drugs. Horwitz's group found that progesterone, but not estradiol, increased labeled vincristine accumulation and inhibited azidopine photoaffinity labeling of endometrial P-gp. Progesterone also enhanced the cytotoxicity of vinblastine in murine J7.V1-1 macrophages, which overexpress the murine *mdr*1b gene (Yang *et al.*, 1989, 1990). Others have confirmed these findings, showing that progesterone inhibits the binding of vincristine or vinblastine to membranes from MDR cells (Naito *et al.*, 1989), and that progesterone directly labels human P-gp. Furthermore, the binding was inhibited by cold progesterone or verapamil to a greater extent than by other corticosteroids or by mineralocorticoids (Qian and Beck, 1990).

Work with steroid derivatives highlighted the observation that not all chemosensitizers are substrates for P-gp, as had been previously noted by Hait and Pierson (1990) for trifluoperazine. For example, Ueda *et al.* (1992) found that while mineralocorticoids were excellent substrates for P-gp transport, progesterone was not. Rather, progesterone inhibited transport and served as a potent chemosensitizer. In this study, human adrenal *mdr*1 cDNA was expressed at the apical surface of porcine kidney proximal tubule cells. Labeled vinblastine, cortisol, aldosterone and dexamethasone were efficiently transported across the epithelium, but progesterone was not. Nevertheless, progesterone inhibited steroid transport and azidopine photoaffinity labeling of P-gp, and increased sensitivity to vinblastine, with similar effectiveness to verapamil. These data suggest that the ability of a compound to act as a chemosensitizer and inhibit transport of other drugs by P-gp is not necessarily related to its own ability to function as a substrate.

A study by Safa's group adds to the complexity of understanding the interaction of steroids with P-gp. Megesterol acetate (Megace), an orally active congener of progesterone, functioned as a chemosensitizer in human MDR neuroblastoma and KB cells, and increased vincristine accumulation with 2- to 3-fold greater potency than progesterone (Fleming *et al.*, 1992). Paradoxically, Megace increased the binding of labeled azidopine to P-gp by up to 2-fold, while inhibiting labeled *Vinca* alkaloid binding by 50%. These results strongly suggest that separate drug binding sites exist on P-gp for certain chemosensitizing compounds such as the calcium channel blocker azidopine and drug substrates such as *Vinca* alkaloids. The ability of Megace to distinguish these binding sites more definitively than other chemosensitizers may be due to its structural similarity to a potential natural substrate for the MDR transporter.

In summary, certain natural steroids are excellent substrates for P-gp-mediated transport and suggest a role for this protein in their secretion. Others, such as progesterone, appear to bind P-gp and inhibit transport. Finally, Megace may be a potentially useful clinical chemosensitizer, since high oral doses producing plasma levels of up to 2 μM have been safely administered in trials of its anticachexia properties (Aisner *et al.*, 1988).

Synthetic steroid analogs have demonstrated particularly promise for clinical use. Ramu *et al.* (1984b) found the antiestrogen, tamoxifen, as well as other structurally related triparanol analogs, could partially overcome doxorubicin resistance in murine P388 MDR cells. Since that time, several other groups have demonstrated that tamoxifen and the related antiestrogen toremifene are active chemosensitizers at concentrations of 2—10 μM in a number of human P-gp expressing MDR cell lines, independent of their effect on estrogen receptors (Baker *et al.*, 1992; Berman *et al.*, 1991; Chatterjee and Harris, 1990a, 1990b; DeGregorio *et al.*, 1989; Foster *et al.*, 1988; Wiebe *et al.*, 1992).

The clinical pharmacology of antiestrogens also make them attractive for clinical use as chemosensitizers. Because of their relative lack of side effects, high serum concentrations have been achieved. For example, standard 20—40 mg/day doses of tamoxifen result in 0.8 μM serum levels (Furr and Jordan, 1984), and 200 mg/day result in serum levels of 5 μM, and is well tolerated (Jordan *et al.*, 1983). Toremifene administered at 400 mg/day results in steady state levels of up to 10 μM (DeGregorio *et al.*, 1989), with nausea and vertigo being dose-limiting. The major metabolite of toremifene, desmethyl-toremifene, possesses significant chemosensitizing activity itself. Ultrafiltrate plasma specimens containing unbound toremifene and its metabolites from patients treated with 20—400 mg/day toremifene demonstrated significant activity in enhancing doxorubicin cytotoxicity in an *in vitro* bioassay (DeGregorio *et al.*, 1989). Thus, the *in vitro* and clinical pharmacokinetics of these antiestrogens suggest they may be well-tolerated, effective chemosensitizers for use in combination with other chemotherapeutic agents in clinical drug resistance.

Cyclosporines and related immunosuppressive agents

Several immunosuppressive drugs, peptides with distinctly different pharmacologic and structural properties than other known chemosensitizing agents, have been found to possess unique and potent activities for modulating MDR. CsA, a hydrophobic cyclic peptide of 11 amino acids, was first investigated for potential anti-MDR activity in part due to reports of its CaM binding properties (Colombani *et al.*, 1985), though it was later shown that CsA does not specifically inhibit CaM-mediated processes (Hait *et al.*, 1986; Legrue *et al.*, 1986). The primary immunosuppressive activity of CsA is through specific inhibition of an early stage of T lymphocyte activation, by interaction with the cytosolic receptor protein, cyclophilin (Handschumacher *et al.*, 1984) and resultant inhibition of the protein phosphatase, calcineurin (Schreiber and Crabtree, 1992).

Slater *et al.* (1986) first studied the effect of CsA in MDR cells and found that it caused a 3- to 4-fold potentiation of daunomycin toxicity in Ehrlich ascites cells that had a very low level of resistance (2-fold) to daunomycin, as well as causing a 2-fold potentiation of daunomycin cytotoxicity in sensitive Ehrlich ascites cells. Since then, several groups have found CsA to reverse resistance and cross-resistance in MDR, but not sensitive cells (Hait *et al.*, 1989b; Twentyman, 1988a; Twentyman *et al.*, 1987), though cytotoxic drug potentiation by CsA has also been noted in certain sensitive cells (Chambers *et al.*, 1989; Gaveriaux *et al.*, 1989; Kloke and Osieka, 1985; Osieka *et al.*, 1986). For example, CsA caused an 11-fold enhancement of doxorubicin toxicity in sensitive Chinese hamster ovary Aux B1 cells, and a 62-fold enhancement in colchicine resistant MDR CH^RC5 cells (Chambers *et al.*, 1989), though CsA increased doxorubicin accumulation in the sensitive, but not MDR cell line (Chambers *et al.*, 1989). In contrast, CsA enhanced the sensitivity to doxorubicin in MDR P388/DOX cells, but did not increase the sensitivity of P388 cells, although a small increase in doxorubicin accumulation was seen in both lines (Hait *et al.*, 1989b). Nooter's group found that 0.5—3 μM CsA partially reversed the accumulation defect to daunomycin in another Chinese hamster ovary MDR cell line, CH^RB3, but did not affect daunomycin accumulation in sensitive AuxB1 cells, as determined by flow cytometry (Silbermann *et al.*, 1989). Similarly, Nooter *et al.* (1989) found CsA to increase daunomycin accumulation in P388/DAU cells. These findings suggest that the mechanism of action of CsA as a chemosensitizer may be more complex than other chemosensitizers, and may not be due solely to modification of P-gp-mediated drug transport.

Studies of the interaction of CsA with P-gp have been difficult to interpret. It has recently been reported that CsA itself is accumulated less in the P-gp associated CH^RC5 cells than the parent Aux B1 line (Goldberg *et al.*, 1988), and that CsA was transported in a saturable manner across the apical surface of proximal tubule cell epithelium expressing a *mdr*1 cDNA (Saeki *et al.*, 1993). However, excess verapamil was incapable of inhibiting CsA transport, indicating either that CsA does not specifically interact with P-gp, or that these chemosensitizers bind at different sites to P-gp. This suggests that CsA may serve as a P-gp substrate and competitively inhibit MDR in a manner similar to the nontoxic anthracycline analogs. Studies by Foxwell *et al.* (1989) demonstrated photolabeling of CH^RC5 P-gp by a radiolabeled CsA derivative, which was inhibited by 20-fold excess cold CsA, as well as verapamil and diltiazem. However, the extremely hydrophobic CsA molecules also binds to many other cellular proteins, and excess cold drug also inhibits those interactions.

Tamai and Safa (1991) reported detailed kinetic analyses of the effect of CsA on drug accumulation and *Vinca* alkaloid resistance in MDR DC-3F Chinese hamster cells, which display decreased accumulation of vincristine, vinblastine and CsA compared to wild-type DC-3F cells. Three µM CsA enhanced vincristine toxicity by 220-fold in the 2400-fold vincristine resistant MDR cells, but also produced a 34-fold potentiation of vincristine in sensitive cells. Ten µM CsA completely reversed the 22-fold decreased level of vinblastine accumulation displayed by the MDR cells, but caused only a 2-fold increase in vinblastine accumulation in sensitive cells. Furthermore, CsA was a potent, competitive inhibitor of vincristine uptake by MDR cell membranes. Also 0.5 µM CsA inhibited by 50% the photoaffinity labeling of P-gp by a derivative of vinblastine. This study suggests that CsA reverses MDR at least in part by interacting directly with P-gp and inhibiting outward transport of cytotoxic drugs, but also implies that CsA modulates the cytotoxicity of vincristine by other mechanisms. Thus, CsA appears to possess complex pharmacologic properties for modulating drug sensitivity, in accord with its known activity as an inhibitor of many important cellular enzymes.

Because of the profound pharmacologic effects of CsA, there has been great interest in exploring the anti-MDR activity of other, less immunosuppressive or nephrotoxic cyclosporine analogs. Indeed, initial studies of the nonimmunosuppressive analog, 11-methyl-leucine cyclosporin in P388/DOX cells (Hait *et al.*, 1987), and cyclosporin C (2-threonine), G (2-L-nor-valine) and H (11-D-methyl-valine) in a human small cell lung cancer line (Twentyman *et al.*, 1987) demonstrated modulation of MDR. Furthermore, the nonimmunosuppressive cyclosporines, 11-Me-Ile and O-acetyl-C_9 cyclosporin, were more effective chemosensitizers than identical concentrations of CsA (Twentyman, 1988a). Similarly, the nonimmunosuppressive 11-methyl-leucine analog was shown to be less potent but equally effective as CsA in sensitizing P388/-DOX cells to doxorubicin (Hait *et al.*, 1989b). 11-Methyl-leucine cyclosporin, as well as a 6-methyl-alanine analog, also increased the sensitivity of non-MDR Chinese hamster ovary sensitive as well as MDR cells to doxorubicin and epirubicin (Chambers *et al.*, 1989). In a study of the effect of 15 cyclosporine analogs on the sensitivity of MDR CH^RC5 and sensitive Aux B1 cells to colchicine, daunomycin and vincristine, no correlation was found between the immunosuppressive and anti-MDR properties of the derivatives (Gaveriaux *et al.*, 1989). Because nonimmunosuppressive analogs do not bind to cyclophilin (Handschumacher *et al.*, 1984), these results rule out the possibility that reversal of drug resistance is related to this cytosolic protein.

Subsequently, researchers at Sandoz screened nonimmunosuppressive cyclosporine analogs for anti-MDR activity, and discovered that the cyclosporin D analog ((3'-keto-Bmt^1)-[Val]^2-cyclosporin), PSC 833, was 10-fold more potent than CsA, causing a 60-fold reversal of 200-fold daunorubicin resistance in P388 MDR cells at a dose of 0.1 µM (Boesch *et al.*, 1991). One µM PSC 833 has also been shown to cause nearly complete reversion of drug resistance to taxol in 280-fold resistant P388 cells and 36-fold resistant CHO cells, while 30 μM verapamil was required for similar chemosensitizing activity (Jachez *et al.*, 1993). Preliminary results indicate that PSC 833 functions as a chemosensitizer by inhibiting P-gp function. For example, work from Sikic's laboratory shows that 2 μM PSC 833 can completely reverse the 1.5-, 3- and 5-fold drug accumulation defects of K562/R7 MDR human leukemia cells to VP-16, daunorubicin, and vinblastine, respectively (Duran *et al.*, 1993). This study also demonstrated that PSC 833 was 5- to 30-fold more potent than CsA and verapamil, respectively, in displacing vinblastine from MDR membrane preparations. Thus, this compound is the most potent described to date in standard *in vitro* assays by which chemosensitizers may be compared. Because of its lack of immunosuppressive side effects, and ability to be administered orally, results of current phase I trials of PSC 833 plus chemotherapeutic drugs are eagerly awaited.

Finally, two newly described immunosuppressive agents which are structurally distinct from the cyclosporines have been found to possess chemosensitizing activity. FK506 is a macrolide antibiotic recently shown to inhibit T-cell function with 10–100 times greater potency than CsA (Schreiber, 1991; Schreiber and Crabtree, 1992). Its structural analog, rapamycin, also potently inhibits T-cell proliferation, but by a mechanism distinct from either CsA or FK506 (Bierer *et al.*, 1990; Dumont *et al.*, 1990). FK506 also reverses vincristine and doxorubicin resistance in several murine and human MDR cell lines, inhibits vincristine efflux and displaces labeled azidopine binding to P-gp at concentrations of 3–10 μM (Naito *et al.*, 1992). Similar doses of rapamycin are also effective for chemosensitization of MDR cells (Arceci *et al.*, 1992). However, these concentrations necessary for anti-MDR activity are 1000-fold greater than required for their immunosuppressive effects.

In summary, the cyclosporines sensitize MDR cells to a variety of chemotherapeutic drugs, though it is uncertain whether the mechanism is through a direct effect on P-gp alone, or also through an indirect effect on cellular drug metabolism, through potentiation of chemotherapeutic drug toxicity, or a combination of these. CsA has been shown to potentiate a number of drugs *in vitro* and *in vivo* against both tumor cells and normal tissues (Twentyman, 1988b). This leads to the important possibility that CsA, in combination with other more specific chemosensitizing agents, such as verapamil or *trans*-flupenthixol, may act synergistically to antagonize MDR. The newly developed nonimmunosuppressive cyclosporine analogs, particularly PSC 833, appear to be excellent candidates for the clinical reversal of drug resistance, due to their increased potency and specificity for P-gp-mediated MDR.

Dipyridamole

The antithrombotic drug dipyridamole has been shown to be a unique and promising biochemical modulator of a variety of cytotoxic drugs, of both MDR and non-MDR classes. Dipyridamole is commonly used clinically for its ability to inhibit platelet function (FitzGerald, 1987). In addition to being an inhibitor of cyclic AMP phosphodiesterase, dipyridamole is a potent inhibitor of the salvage pathway for repletion of cellular nucleotide pools by nucleoside transport across cell membranes. This property has been exploited for the biochemical potentiation of antimetabolites such as methotrexate (Nelson and Drake, 1984), and 5-fluorouracil (Grem and Fischer, 1985). Dipyridamole also appears to modulate the cytotoxicity of cisplatin in various cell lines, although the mechanism remains obscure (Jekunen *et al.*, 1992). Recently, dipyridamole has been found to enhance the effects of several MDR related drugs against both sensitive and resistant cell lines, potentially through multiple mechanisms including inhibition of P-gp function. For example, Kusumoto *et al.* (1988) found that nontoxic concentrations of dipyridamole increased the cytotoxic effect of doxorubicin and augmented intracellular levels of doxorubicin in drug sensitive HeLa cells by approximately 2-fold. Similarly, Howell's group found that dipyridamole interacted in a synergistic manner with VP-16, doxorubicin and vinblastine in drug sensitive human ovarian carcinoma cells as measured by combination index plots (Howell *et al.*, 1989a, 1989b), though a poor correlation was found between the degree of enhancement of drug activity and the modest increase in intracellular drug accumulation. Although these studies suggest dipyridamole potentiates the activity of these cytotoxic drugs through mechanisms independent of P-gp, verapamil had similar activity, and the authors did not exclude the possibility that this relatively drug sensitive cell line actually expressed P-gp.

Further studies by this group suggest that dipyridamole may have multiple mechanisms for modulation of certain MDR related drugs. In human KB cells transfected with a *mdr*1 cDNA, dipyridamole enhanced the cytotoxicity of vinblastine with a high degree of synergy, but not VP-16 (Shalinsky *et al.*, 1990). Whereas dipyridamole did potentiate vinblastine toxicity and accumulation in wild-type, drug sensitive KB-3-1 cells, HT1080 cells, and a non-P-gp expressing doxorubicin resistant variant HT1080 line, these effects were additive or only slightly synergistic (Shalinsky *et al.*, 1990, 1991). Definitive studies of the interaction of dipyridamole with P-gp are lacking, though high concentrations of dipyridamole (100 μM) have been reported to interfere with azidopine photoaffinity labeling of P-gp (Asoh *et al.*, 1989). Therefore, dipyridamole appears to display complex pharmacologic effects in drug sensitive and resistant cells, reminiscent of CsA.

The multiple known mechanisms for modulation of chemotherapeutic drugs by dipyridamole make it a unique and interesting compound for clinical use. Phase I trials of dipyridamole in combination with 5-fluorouracil and folinic acid have demonstrated that doses of dipyridamole of up to 200 mg/m^2/day for 6 days are tolerable and result in mean peak plasma concentrations of 16 μM (Budd *et al.*, 1990). However, due to dipyridamole being more than 99.7% protein bound *in vivo*, less than 0.04 μM free dipyridamole existed in this trial.

Recently, Cheng and co-workers described an analog of dipyridamole, BIBW 22, which displays markedly increased potency for the reversal of MDR and inhibition of nucleoside transport (Chen *et al.*, 1993). BIBW 22 was 10-fold more potent than dipyridamole in reversing 30-fold vinblastine resistance in MDR KB cells and completely inhibited azidopine binding to P-gp at concentrations of 1 μM. Furthermore, BIBW 22 was a 7-fold more potent inhibitor of nucleoside transport than dipyridamole, and resulted in a 20-fold enhancement of 5-fluorouracil cytotoxicity at 1 μM concentrations. Thus, this compound may have unique clinical properties as a bifunctional modulator in trials of combination chemotherapy employing both antimetabolites and *Vinca* alkaloids.

Miscellaneous hydrophobic, cationic compounds

The search for agents to circumvent MDR has led to the identification of numerous compounds not belonging to the previously discussed groups of chemosensitizers and not otherwise pharmacologically related. Most of these compounds are lipophilic in nature, and share a broad structural similarity that includes a heterocyclic ring nucleus separated at a distance from a cationic, amino group. This diverse group of chemosensitizing agents includes: antiarrythmics such as amiodarone (Chauffert *et al.*, 1986; Van der Graaf *et al.*, 1991) and quinidine (Solary *et al.*, 1991; Tsuruo *et al.*, 1984); the alkaloid derivative, cepharanthine (Shiraishi *et al.*, 1987); the quinoline amines, chloroquine (Zamora and Beck, 1986; Zamora *et al.*, 1988) and a derivative, MS-073 (Sato *et al.*, 1991); the antimalarial, quinacrine (Inaba and Maruyama, 1988; Zamora *et al.*, 1988); the indole alkaloids, reserpine and yohimbine (Beck *et al.*, 1988; Kadam *et al.*, 1992; Pearce *et al.*, 1989); a triazinoaminopiperadine derivative, S 9788 (Leonce *et al.*, 1992; Pierre *et al.*, 1992); the nonsedating antihistamine, terfenadine (Seldane) (Hait *et al.*, 1993b); the antibiotics erythromycin (Hofsli and Nissen-Meyer, 1989a, 1989b), cefoperazone, ceftriaxone (Gosland *et al.*, 1989), tetracycline (Kavallaris *et al.*, 1993); and polyether compounds (Kawada *et al.*, 1992). Most have been reported to partially overcome resistance and cross-resistance to cytotoxic drugs, and to increase drug accumulation and retention in various MDR cell lines. Whether these many types of pharmacologic agents are acting through a common or independent mechanisms or targets to antagonize MDR is presently unclear.

It is important to note that not all hydrophobic cationic drugs reverse P-gp-mediated MDR. In fact, most of the chemotherapeutics affected by MDR share these features, suggesting that they may serve as partial or full agonists or antagonists of the process. For example, certain mitochondrial dyes such as rhodamine 123 or the bisquinaldinium, dequalinium, structurally resemble drugs that antagonize calmodulin as well as MDR, but share only the first property (Bodden *et al.*, 1986). However, MDR cells are cross-resistant to these types of drugs, and other calmodulin antagonists like trifluoperazine can completely reverse this phenomenon (Hait and Pierson, 1990). Therefore, it is possible that some of these compounds may serve as substrates for the P-gp multidrug transporter, but possess varying affinities resulting in either pure or mixed agonist versus antagonist activity, whereas others such as phenothiazines may have distinct or additional mechanisms of action.

Structure-activity relationships among chemosensitizers

The various classes of drugs with activity against MDR discussed produce a wide variety of cellular effects and have multiple cellular targets. While certain of these chemosensitizers may act at sites other than P-gp, or at different sites within the P-gp molecule, data suggests that many share a common target site for reversal of MDR. Thus, these compounds must share certain structural features important for their chemosensitizing activity. It may be possible to define cellular targets and structural features that enhance the interaction of chemosensitizers with these target(s) by studying the anti-MDR activity of drugs within a single pharmacological class. Using structural principles derived from such an analysis, it may be possible to rationally design agents with increased specificity and potency.

We studied the structure-activity relationships of a series of phenothiazines with individual molecular alterations for their ability to sensitize the human MDR breast cancer cell line, MCF-7 Adr^R, to doxorubicin (Ford *et al.*, 1989). Critical structure-activity relationships were identified for anti-MDR activity, including the hydrophobicity of the tricyclic ring, the charge on the terminal amino group and the length of the alkyl bridge connecting the two. For example, substitutions on the phenothiazine ring that increased hydrophobicity, such as a halogen at position 2, increased activity, while those that decreased hydrophobicity, such as hydroxyl groups, decreased activity. Side chains that contained positively charged tertiary amines were more potent chemosensitizers than primary or secondary amines, and incorporation of the amino moiety into cyclic structures were more effective than noncyclic, aliphatic amines. Finally, the distance between the amino group and the phenothiazine ring proved important, with a four carbon alkyl bridge being more effective than shorter chains.

The structural principles derived from this study allowed us to identify the thioxanthenes, a structurally similar class of chemosensitizers, which differ from the phenothiazines only by a carbon substitution for the nitrogen in the tricyclic ring, and by an exocyclic double bond to the side chain, creating stereoisomers for each thioxanthene analog. Similar to the phenothiazines, thioxanthenes with halogenated tricyclic rings, and piperazinyl amino side groups separated by a specific distance were particularly effective chemosensitizers (Ford *et al.*, 1990). We found that the *trans*-isomer of each thioxanthene pair was a more effective chemosensitizer than the *cis*-isomer, and that *trans*-flupenthixol was at least 5-fold more effective than any of the phenothiazines tested. Therefore, our study of structure-activity relationships with phenothiazines and thioxanthenes suggest that for these classes of drugs, compounds with tertiary, cationic amino groups incorporated into a cyclic ring structure in a particular spatial orientation and at a distance of at least three carbons from a hydrophobic conjugated ring were optimal for reversing MDR.

Ramu subsequently studied 232 phenothiazines and structurally related compounds, and came to similar conclusions regarding structural features important for activity against murine P388/ADR cells (Ramu and Ramu, 1992). Specifically, the presence of two hydrophobic phenyl rings with substitutions of carbonyl moieties at position 2, and a secondary or tertiary amine side group incorporated into a cyclic ring were found to enhance anti-MDR activity.

Studies of several other classes of chemosensitizers reached similar conclusions regarding the importance of a substituted amino group located at a certain distance from a ring structure. For example, the chemosensitizer perhexilene maleate possesses two cyclohexyl rings separated by a three-carbon chain from a piperadine amino group (Ramu *et al.*, 1984a). Replacement of the two rings with a noncyclic group resulted in a complete loss of anti-MDR activity (Ramu *et al.*, 1984), while replacement with phenyl rings enhanced activity (Ramu, 1989).

Structure-activity relationships of 43 dipyridamole analogs demonstrated that (*a*) the spatial relationships of the piperadine and diethanolamine substituents of dipyridamole were important to chemosensitizing activity, (*b*) tertiary amino groups were more effective than primary amino groups, and (*c*) replacement or loss of piperadine, but not diethanolamine groups resulted in a loss of anti-MDR activity (Ramu and Ramu, 1989).

Triparanol, an effective chemosensitizer analog of the triphenylethylene class of drugs, contains three substituted phenyl rings connected by a two carbon chain to a tertiary amine group. Related triphenylethylenes lacking the amino side chain lose activity for reversing MDR in P388/DOX cells (Ramu *et al.*, 1984b). Furthermore, incorporation of the tertiary amine into a cyclic pyridinyl structure enhanced activity.

A third major analysis of structural characteristics important for chemosensitizing activity comes from Beck's studies with indole alkaloids and other aromatic amines. Similar to the previously discussed structure-activity relationships, Beck and colleagues reported that quinoline derivatives such as primaquine and quinacrine, which possess conjugated ring structures attached to substituted amino side groups, display significant anti-MDR activity (Zamora *et al.*, 1988). Acridine, which completely lacks an amino side group, retains only partial activity at a 10-fold higher concentration. In a series of related alkaloid reserpine and yohimbine analogs, the presence of a pendant benzoyl moiety appeared to enhance activity (Pearce *et al.*, 1989). The addition of this group resulted in compounds such as reserpine, trimethoxybenzoylyohimbine and rescinnamine, with aromatic ring domains separated by a distance from a tertiary, cationic nitrogen. Compounds which lacked the former ring structure, such as reserpic acid and yohimbine, also lacked significant anti-MDR activity. Whereas these alkaloids are quite different in their basic structure from those compounds analyzed by our group (Ford *et al.*, 1989, 1990) and Ramu's group (Ramu, 1989; Ramu *et al.*, 1984, 1984a, 1984b; Ramu and Ramu, 1989, 1992), three-dimensional configurational analyses show similar spatial relationships between the hydrophobic, planar ring groups and the cationic, tertiary amines in these different classes of drugs (Pearce *et al.*, 1989).

Studies with several other groups of chemosensitizers agree with the principles emerging from these structure-activity relationships. For example, a number of dihydropyridine analogs were demonstrated to possess anti-MDR activity, and many of the more active of these possess one or more tertiary amine side groups attached to the conjugated dihydropyridine ring structure (Yoshinari *et al.*, 1989). Also, Sikic and colleagues identified several cephalosporin antibiotics with chemosensitizing activity (Gosland *et al.*, 1989). The most active of the five drugs studied, cefoperazone, was the only to contain an *N*-ethyl piperazine amino group, connected to the cephalosporin conjugated ring structure.

Lampidis and co-workers utilized a series of simple, aromatic (alkylpyridiniums) and nonaromatic (alkylguanidiums) organic cations to test the hypotheses derived from these studies of more structurally complex chemosensitizers (Dellinger *et al.*, 1992). By the stepwise addition of single alkyl carbons, compounds were synthesized with increasing lipophilicity. The cytotoxic activity and cellular accumulation of these compounds were studied in MDR and non-MDR cell lines, and it was found that aromaticity and the degree of lipophilicity determined the extent to which they were affected by the MDR mechanism.

Thus, systematic analyses of the structure-activity relationships of chemosensitizers such as the phenothiazines and thioxanthenes have defined particular structural features and spatial relationships important for anti-MDR activity. These imply a P-gp binding site with more stringent structural requirements for effective antagonism or partial agonist activity than previously imagined, and should help to more precisely define this protein domain. Based on our data with MDR and the structural requirements found important for the binding of phenothiazines to CaM (Prozialeck and Weiss, 1982), we have proposed a model incorporating the importance of hydrophobicity and molecular structure for the interaction of phenothiazines or thioxanthenes with P-gp (Hait and Aftab, 1992). In this schema, the P-gp drug binding site is contained within a transmembrane α-helical segment containing a hydrophobic region separated by one half turn of the helix from a hydrophilic region (Reid, 1983). The hydrophobic pocket is produced by two aromatic phenylalanines residues which are separated by two or three amino acids, and are oriented by the α-helix in such a way as to form a charge transfer complex with the aromatic, tricyclic groups in the phenothiazines or thioxanthenes. In fact, such phenylalanine repeat motifs exist in several transmembrane regions of human and murine P-gp (Chen *et al.*, 1986; Gros *et al.*, 1986a), which are known to be labeled by azidopine (Bruggemann *et al.*, 1989).

In CaM, the hydrophilic region is produced by acidic residues present in a proximal domain which interacts with the positively charged amino side chain of the drugs. Unlike the amino acid sequences found in the α-helical domains of CaM, there are no acidic amino acids within the α-helical transmembrane domains of P-gp. Since we found that a positively charged amino group appears important for the activity of chemosensitizers, this implies the existence of a hydrophilic region of P-gp which folds near the hydrophobic region and interacts with the basic side-chain of the drug.

Therefore, use of structure-activity relationships to better define determinants important for chemosensitizing activity may not only result in the discovery of more potent, less toxic agents for the reversal of MDR, but may also lead to defining the drug binding site important for this activity.

Mechanism of action of chemosensitizers in multidrug resistance

Our understanding of the mechanism of action of chemosensitizers for reversing MDR remains incomplete. The difficulty in studying their pharmacologic actions is due in part to incomplete understanding of P-gp's mechanism of efflux of cytotoxic drug substrates, but is also complicated by the multiple potential cellular targets for chemosensitizers. Furthermore, the broad and heterogeneous group of compounds that possess chemosensitizing activity suggest that a single, common target may not exist, and that a variety of interactions with different cellular targets may result in a similar effect on cytotoxic drug potentiation. In the following sections we will review several mechanisms that have been suggested to be relevant to the anti-MDR activity of chemosensitizers.

Interaction with P-glycoprotein

The preponderance of data from studies of chemosensitizers suggest they act by directly affecting the function of P-gp. However, conflicting evidence exists with regard to whether chemosensitizers of similar or different classes function as competitive or noncompetitive inhibitors of drug efflux, where and how many binding sites exist for chemosensitizers on P-gp, and whether chemosensitizers share binding sites with cytotoxic substrates.

One technique to analyze the effects of drugs on P-g has been to study cellular membranes enriched for the protein. For example, Cornwell *et al.* (1986) found that [^{3}H]-vinblastine associated in a specific, saturable, temperature-dependent manner with P-gp enriched MDR membrane vesicles from KB-C4 cells, but did not associate with vesicles from sensitive KB-3-1 or revertant KB-R1 cells. Drug accumulation was inhibited by excess unlabeled vinblastine, vincristine, and verapamil, less so by daunomycin, and poorly inhibited by colchicine

(Cornwell *et al.*, 1986).

Certain chemosensitizers bind directly to membranes enriched for P-gp, and this binding may be inhibited by other chemosensitizers and by chemotherapeutic drugs. For example, KB-V1 MDR vesicles bind 15- to 20-fold more verapamil and diltiazem than sensitive KB-3-1 vesicles, in a specific and saturable manner (Cornwell *et al.*, 1987). Experiments with photoactive, radiolabeled dihydropyridine and verapamil analogs, such as [^{3}H]-azidopine and [^{125}I]-NASAV, demonstrate irreversible binding to P-gp, which is effectively inhibited by nifedipine, nicardapine, verapamil, progesterone, *cis*- and *trans*-flupenthixol, and vinblastine, but poorly inhibited by trifluoperazine, chlorpromazine, fluphenazine, doxorubicin and colchicine (Ford *et al.*, 1990; Safa *et al.*, 1987; Safa, 1988a, 1988b; Yang *et al.*, 1988, 1989). Studies of the binding of chemotherapeutic agents to P-gp, and the competitive inhibition of this binding by chemosensitizers has demonstrated that [^{3}H]vincristine binds to MDR K562/ADM and Ehrlich ascites EHR2/DNR+ membrane preparations in a specific, temperature, pH and ATP/Mg^{2+}-dependent manner, with a K_d of 0.24 μM (Naito *et al.*, 1988; Sehested *et al.*, 1989). The chemosensitizers, verapamil, nicardapine, quinidine, CsA, progesterone, and to a lesser extent trifluoperazine, competitively inhibit this high affinity [^{3}H]-vincristine binding to K562/ADM membrane preparations (Naito *et al.*, 1988, 1989; Naito and Tsuruo, 1989). [^{125}I]-NASV photoaffinity labeling of KB-V1 and CEM/VLB$_{1K}$ P-gp was inhibited by 10 μM reserpine, cepharanthine, quinidine, the isoprenoid SDB-ethylenediamine, and active synthetic dihydropyridine chemosensitizers, but poorly inhibited by up to 100 μM concentrations of trifluoperazine, thioridazine and chlorpromazine (Akiyama *et al.*, 1988; Beck *et al.*, 1988; Nogai *et al.*, 1989). Labeled, photoactive colchicine analogs bind to P-gp from DC-3F/VCRd-5L cell membranes in a manner which is inhibited by *Vinca* alkaloids, doxorubicin and actinomycin D (Safa *et al.*, 1989). [^{125}I]-labeled daunomycin photoaffinity labeled P-gp from MDR CHO B30 cell membrane vesicles in a specific manner, which is inhibited by vinblastine, nortrendipine and verapamil (Busche *et al.*, 1989), although others have failed to demonstrate photoaffinity labeling of P-gp from EHR2/DNR plasma membrane vesicles by a radiolabeled doxorubicin derivative (Sehested *et al.*, 1989). These results suggest that *Vinca* alkaloids, colchicine and perhaps anthracyclines bind to P-gp, and that certain chemosensitizers compete for either a common drug binding site, for overlapping sites, or for sites which cause allosteric changes resulting in inhibition of cytotoxic drug binding.

Several studies have shown that chemosensitizers may also serve as substrates for the P-gp multidrug transporter, supporting their possible mechanism as competitive ligands for sites on P-gp. For example, Cano-Gauci and Riordan (1987) found MDR CHO cells accumulated less [^{3}H]verapamil than their sensitive counterparts. Similarly, Tsuruo's group showed a 3-fold decrease in the accumulation of [^{3}H]verapamil in MDR K562/ADM cells as compared with sensitive cells, which could be reversed by 5 μM nicardipine or vincristine (Yusa and Tsuruo, 1989). We have also demonstrated that MCF-7 Adr^R cells accumulate 2- to 4-fold less *trans*-flupenthixol than do sensitive MCF-7 cells after a 3 h incubation with 3–100 μM drug (Ford and Hait, unpublished observations). Tamai and Safa (1991) found that azidopine, CsA and vinblastine were effluxed from P-gp expressing cells in an active, energy dependent and saturable manner. Scatchard analysis demonstrated that azidopine bound P-gp at a single, saturable binding site with a dissociation constant of 1.2 μM, and that vinblastine and CsA inhibited this interaction in a noncompetitive fashion. However, Sehested *et al.* (1990) could not discern a difference in azidopine or verapamil accumulation or efflux in sensitive versus MDR cells, suggesting these chemosensitizers did not function as P-gp substrates. Similarly, we found that trifluoperazine accumulated identically in sensitive and resistant P388 cells (Hait and Pierson, 1990) and that the binding of CsA was also the same (Hait *et al.*, 1989). Thus, certain chemosensitizers may be transported by P-gp in a similar manner to cytotoxic drugs, thus functioning as competitive agonists, but others appear to be handled differently.

An independent line of evidence that chemosensitizers may interact with P-gp came from the recent purification and partial characterization of P-gp by Hamada and Tsuruo (1988a, 1988b). Enzymatically active P-gp was purified from human K562/ADM cells by immunoaffinity chromatography, and found to possess ATPase activity (Hamada and Tsuruo, 1988b). However, trifluoperazine and verapamil increased the ATPase activity associated with P-gp (Hamada and Tsuruo, 1988a). Since phenothiazines inhibit the activity of other ATPases (Pang and Briggs, 1976; Raess and Vincenzi, 1980), it was possible that trifluoperazine interferes with a P-gp-mediated drug efflux mechanism in a similar manner. In another study, partial purification and functional reconstitution of P-gp into phospholipid vesicles was performed to measure drug-stimulated ATP hydrolysis (Ambudkar *et al.*, 1992). Vinblastine, doxorubicin and verapamil were found to stimulate ATP hydrolysis, whereas non-MDR related drugs did not. Further studies of the ATPase function of P-gp may focus on its possible role as a target for chemosensitizers.

It is also possible that there are several sites on P-gp through which chemosensitizers interfere with the MDR process. Certain chemotherapeutic drugs encompassed by the MDR phenotype, such as doxorubicin, daunomycin and particularly colchicine, may not compete for the same binding sites on P-gp as the calcium channel blockers and *Vinca* alkaloids. A recent analysis by Bruggemann *et al.* (1992) of azidopine binding to P-gp has identified 2 distinct sites within the P-gp molecule, one between residues 198 and 440 of the amino half, and the other within the carboxy portion of the protein. By using mutant P-gp, however, the authors argue that a single functional binding site exists, formed by the homologous halves of the molecule folding together.

Chemosensitizers such as the phenothiazines may bind to alternative P-gp sites from other chemosensitizers and antagonize MDR. Alternatively, phenothiazines and other chemosensitizers may inhibit P-gp activity by interacting with physically separate sites on P-gp, such as the ATPase or the phosphorylation domains, or may act by altering the local membrane environment leading to structural or functional changes in P-gp. The final delineation of the P-gp drug binding sites will be necessary to elucidate these mechanisms. Hopefully, a more sophisticated understanding of these processes will also lead to more powerful methods to manipulate P-gp function.

Inhibition of protein kinase C

By understanding the factors that regulate the function of P-gp, it should be possible to identify additional ways to modify its actions. Posttranslational modification of P-gp through phosphorylation has been appreciated for many years (Carlsen *et al.*, 1977; Center, 1983; Fine *et al.*, 1988; Hamada *et al.*, 1987; Ma *et al.*, 1991; Mellado and Horwitz, 1987; Meyers, 1989), and a functional role has been suggested. Exposure of cells to substrates for P-gp including chemotherapeutic drugs and chemosensitizers produce hyper-phosphorylation of the molecule, i.e., phosphorylation above the basal levels (Center, 1985; Hamada *et al.*, 1987). In this section we will focus on our understanding of the phosphorylation of P-gp, and identify means by which interference with this reaction could sensitize cells to chemotherapy.

P-gp can be phosphorylated by a variety of serine/threonine kinases, an observation which is not surprising since the protein contains serine residues in the linker region that resemble the consensus phosphorylation sites of protein kinase C (PKC), cAMP-dependent protein kinase and calmodulin-dependent protein kinase II (Kemp and Pearson, 1990; Kennelly and Krebs, 1991; Mellado and Horwitz, 1987; Meyers, 1989). Whether one, all of these or additional kinases are the physiologic mediators of phosphorylation is under intense investigation. In this regard, PKC has received the most attention. For example, PKC can phosphorylate P-gp in cell membranes (Chambers *et al.*, 1990b) and in immunoprecipitates (Aftab and Hait, 1992), and translocation of PKC to cell membranes in response to phorbol 12-myristate 13-acetate (PMA) is temporally consistent with changes in drug accumulation (Chambers *et al.*, 1990a).

We had postulated that PKC might be a physiologic regulator of drug transport, and that its activity would be increased in selected MDR cell lines due to chronic enzyme inhibition by chemotherapeutic drugs (Palayoor *et al.*, 1987). In fact, we found that *Vinca* alkaloids and anthracyclines were competitive inhibitors of PKC, that doxorubicin displaced phorbol esters from MDR cell membranes, and that the activity of PKC was increased in certain, but not all MDR cell lines (Palayoor *et al.*, 1987). Since that time numerous reports have demonstrated that PKC activity is often increased in MDR cells compared to that of the parental lines (Aquino *et al.*, 1988; Fine *et al.*, 1988; O'Brian *et al.*, 1989; Posada *et al.*, 1989a, 1989b; Anderson *et al.*, 1991a).

Use of activators and inhibitors of PKC support the role of this enzyme in the MDR process. For example, treatment of sensitive cells with activators of PKC, such as phorbol esters, decreased the accumulation of chemotherapeutic drugs (Ido *et al.*, 1986; Ferguson and Cheng, 1987), and mimicked the MDR phenotype (Ferguson and Cheng, 1987). Treatment of MDR cells with PKC activators further increased drug resistance (Hamada *et al.*, 1987). Just as activators of PKC seemed to augment the MDR phenotype, inhibitors of PKC appear to diminish it (Dong *et al.*, 1991; O'Brian *et al.*, 1991a). In fact, many of the commonly used chemosensitizers, including phenothiazines, thioxanthenes, tamoxifen and CsA are inhibitors, albeit weak ones, of the enzyme (Aftab *et al.*, 1991; Ford *et al.*, 1990; Mori *et al.*, 1980; O'Brian *et al.*, 1985; Schatzman *et al.*, 1981; Walker *et al.*, 1989).

Phosphorylation of P-gp is associated with marked changes in drug transport. For example, using MCF-7 Adr^R cells we have recently shown that the steady state accumulation of vinblastine was reduced by 35% upon addition of PMA. This effect was only seen when the cells were co-incubated with verapamil, was not seen in sensitive cells, and was readily demonstrable in the absence of verapamil in the MCF-7/BC19 transfected line (Hait *et al.*, 1993c). We then embarked on a series of studies to elucidate the mechanism of action of the phorbol ester effect. These studies demonstrated that PMA had little effect on drug transport until a critical cellular-associated concentration of vinblastine was achieved. For example, when the concentration of vinblastine was less than 0.04 pmols/10^6 cells, incubation with PMA had no demonstrable effect on transport kinetics. However, at concentrations above this level, PMA markedly increased drug efflux and apparent uptake. This latter phenomenon was likely due to efflux from cell membranes, since the net change in total drug accumulation could be accounted for by the efflux component (Hait *et al.*, 1993c). These data suggested that the effect of phosphorylation of P-gp was to increase the velocity of transport and not the affinity of the transporter for its substrates. If this formulation proves true, then one might expect that inhibition of phosphorylation by inhibitors of PKC would have their most profound effects on velocity of transport, whereas drugs that compete for binding sites on P-gp would have their effects on the affinity of the transporter for its substrate. Therefore, the combination of these two classes of P-gp inhibitors might be expected to produce additive or even synergistic effects.

As discussed above, many modulators of MDR are also inhibitors of PKC. In addition, H-7 (Dong *et al.*, 1991; O'Brian *et al.*, 1991), staurosporine (Posada *et al.*, 1989b) and a cell permeant peptide inhibitor (O'Brian *et al.*, 1991b) are classic examples of PKC inhibitors which have been shown to sensitize MDR cells to chemotherapy.

Despite this compelling evidence for a critical role of PKC in MDR, several problems exist. For example, the function of PKC in cells that do not express P-gp is controversial. We studied the effect of PKC activity on drug transport and cytotoxicity in a mutant HL-60 line developed by Diamond and colleagues (Perrella *et al.*, 1986). These cells are resistant to the differentiation effects of phorbol esters and have down-regulated phorbol ester receptors and diminished activity of PKC. Upon removal of these cells from medium containing PMA, the phorbol ester receptor and the activity of PKC increased several fold, yet we found absolutely no difference in the sensitivity to anthracyclines or *Vinca* alkaloids, nor any difference in drug accumulation (Hait and DeRosa, 1991).

We have also studied the effect of PMA on sensitive MCF-7 cells and could not detect significant changes in drug accumulation or sensitivity (Aftab and Hait, 1992). Lastly, although many of the drugs cited above inhibit PKC, even staurosporine, a commonly used, relatively potent antagonist, has been shown to compete for azidopine binding sites on P-gp (Posada *et al.*, 1989b). Therefore, results using inhibitors of PKC to link the function of this enzyme to the MDR process must be cautiously interpreted.

A potentially exciting approach to the selective alteration of PKC activity in MDR cells may be possible. PKC exists as a family of at least eight isozymes which are differentially expressed in tissues (Nishizuka, 1988). The isozymes are divided into two subfamilies: α, β (I/II) and γ, which depend on calcium for activity (Ono *et al.*, 1988, 1989), and δ, ε, ζ and η/L (Bacher *et al.*, 1991; Osada *et al.*, 1990), which do not. The expression of individual isozymes in MDR cell lines has been recently investigated. Aquino *et al.* (1988) found PKC γ to be up-regulated in HL-60-AR cells, while Gollapudi *et al.* (1992) found overexpression of PKC α and β in P388/ADR cells. Recently, Blobe *et al.* (1993) have examined the previously identified up-regulation of PKC in MCF-7 MDR cells in considerable detail. These authors found that the 10-fold increase in enzymic activity was due to a selective increase in the expression of PKC α as determined by Western blot analysis and hydroxylapatite chromatography. To determine whether overexpression of a single PKC isoenzyme could affect the MDR phenotype, Glazer and colleagues transfected MCF-7 cells with PKC α. In these experiments, the overexpression of PKC α increased drug resistance (Yu *et al.*, 1991). It will be of enormous interest to see whether new, isoenzyme specific inhibitors of PKC might selectively modulate drug efflux in these MDR cell lines.

In conclusion, it appears that phosphorylation of P-gp may be a fundamental control point for the activity of the transporter. Drugs that inhibit the activity of this enzyme should be expected to sensitize cells to chemotherapy and act synergistically with modulators that displace drugs from p-gp.

Clinical relevance of multidrug resistance

The identification and analysis of the cellular and molecular biology of P-gp-mediated MDR *in vitro* have led to the investigation of whether experimental MDR has a clinical counterpart. Although clinical MDR is a common phenotypic occurrence, and P-gp expression is seen in clinically resistant tumors, a definitive link between the two awaits the results of on-going clinical research.

Expression of P-glycoprotein in normal human tissues

The distribution of P-gp in a variety of normal human tissues has been studied, and suggests a variety of possible physiologic functions for this protein. Immunohistochemical staining techniques with monoclonal antibodies against P-gp have shown high levels of P-gp in human adrenal cortex, kidney and placenta (Sugawara *et al.*, 1988a, 1988b). P-gp has also been localized at the cellular level to the apical biliary surface of hepatocytes, the columnar epithelial cells of colon and jejunum (Thiebaut *et al.*, 1987), and the apical brush border membrane surface of epithelial cells from renal proximal tubule cells (Lieberman *et al.*, 1989). The polarized expression of P-gp in these tissues suggests either a physiologic role in secreting substances from these tissues, such as bilirubin (Gosland *et al.*, 1993) or steroid hormones (Ueda *et al.*, 1992), or a protective role by excluding potentially toxic xenobiotics or mutagenic chemicals (Ferguson and Baguley, 1993).

Another possible function for P-gp was suggested by the observation that endothelial cells of human capillary blood vessels at the blood-brain barrier and blood-testis barrier also express P-gp (Cordon-Cardo *et al.*, 1989; Thiebaut *et al.*, 1989). In fact, in a murine model, primary cultures of cerebral capillary endothelium were shown to express P-gp and to actively exclude rhodamine 123, a substrate for P-gp (Hegmann *et al.*, 1992). This suggests that P-gp may serve to exclude various toxic compounds from the central nervous system and other pharmacologic sanctuaries.

Recently, analysis of P-gp expression in human hematopoietic cells has provided additional insight and raised additional questions regarding its potential physiologic role. Advances in flow cytometric techniques have allowed the isolation of highly purified, homogenous populations of peripheral blood cells and bone marrow stem cells. Rhodamine 123, a flourescent dye used for flow sorting studies to isolate primitive, hematopoietic progenitor cells from more differentiated bone marrow cells based on its preferential cell staining properties, is also a known substrate of P-gp. Chaudhary and Roninson (1991) first linked the observation that the decreased rhodamine staining of CD 34+ human stem cells was due to an active efflux process inhibited by P-gp chemosensitizers, and was associated with increased P-gp expression as measured by concurrent flow analysis of cells labeled with monoclonal antibodies to P-gp. The potential role for P-gp in the physiology of hematopoietic progenitor cells, possibly through secretion or presentation of specific peptide signals, remains obscure. However, the implication of this finding is profound with regard to the potential cytotoxic effect on stem cells and resultant myelosuppression in clinical trials of chemosensitizers in combination with cytotoxic drugs.

Similar studies using rhodamine exclusion of antibody binding to P-gp and of cell lineage specific membrane antigens have also found functional P-gp in significant proportions of peripheral blood lymphocytes (Chaudhary *et al.*, 1992; Drach *et al.*, 1992). While over half of heterogeneous B- or T-cell populations express P-gp, specific lymphocyte subsets demonstrate much higher proportions of P-gp positive cells. For instance, 80–90% of CD8+ T-suppressor cells and 90–95% of CD56+ natural killer cells efflux rhodamine and express P-gp (Chaudhary *et al.*, 1992), while less than 50% of CD4+ T-helper cells possess similar properties (Coon *et al.*, 1991). Furthermore, experiments with P-gp+ natural killer cells showed that treatment with chemosensitizing agents such as verapamil inhibited natural killer cell-mediated cytolytic activity against target cells, without effecting effector-to-target cell binding activity, in *in vitro* assays (Chong *et al.*, 1993). These results suggest that P-gp may play an active role in the function of peripheral blood lymphocytes, possibly by transporting cytokines or cytolysins involved in natural killer cell-mediated cytotoxicity.

Taken together, these data suggest that P-gp plays a critical role in a number of normal physiologic functions in human tissues, many of which may be related to secretion and/or protection of tissues from various naturally occurring toxins or commonly encountered xenobiotics. Further analysis of natural substrates for P-gp such as steroids and bilirubin, will increase our understanding of its normal function.

Expression of P-glycoprotein in human tumors

Detection of *mdr*1 expression in clinical tumor samples has been extensively studied to determine its potential role as a predictor of drug responsiveness. A clear relationship between the development of clinical drug resistance to MDR-related drugs and increased expression of *mdr*1 mRNA or P-gp would argue favorably for the use of chemosensitizers in treating such patients. However, most studies to date have not been able to separate the role of increased P-gp expression as a poor prognostic factor in general, from that of a specific determinate for chemoresistance to particular drugs.

A number of initial studies used *mdr*1 cDNA probes or anti-P-gp antibodies to measure the presence of the gene product in various tumors, and these results have been previously reviewed (Ford and Hait, 1990; Goldstein *et al.*, 1992; Roninson, 1992). The first large study analyzing clinical material, by Goldstein *et al.* (1989) demonstrated elevated *mdr*1 expression in untreated tumors derived from tissues known to normally overexpress *mdr*1, such as colon cancer, renal cell carcinoma, hepatoma, adrenocortical carcinoma and pheochromocytoma. Many other untreated tumors from tissues which do not normally overexpress P-gp had low or undetectable levels of *mdr*1 mRNA, such as breast cancer, non-small cell lung cancer, gastric cancer, carcinoma of the prostrate, and chronic phase chronic myelogenous leukemia (CML). These findings suggest that the *mdr*1 gene continues to be expressed in certain tissues after malignant transformation occurs. Kacinski *et al.* (1989a) showed significant expression of *mdr*1 mRNA in cells from nine of 16 untreated breast cancers using *in situ* hybridization and found a strong negative correlation between *mdr*1 cDNA hybrids and expression of progesterone receptors levels (Kacinski *et al.*, 1989b).

Two studies by Chan and colleagues in childhood malignancies have shown a strong association between P-gp expression and both lack of chemotherapeutic response and poor survival. They conducted a retrospective analysis of sequential tumor biopsies from 67 children with neuroblastoma using immunohistochemical staining for P-gp to demonstrate a highly significant difference in complete response rates, relapse-free survival and overall survival after treatment with Adriamycin containing regimens (Chan *et al.*, 1991). Analysis of samples from 30 children with rhabdomyosarcoma or undifferentiated sarcoma showed that while the presence or absence of P-gp expression did not predict for clinical responsiveness, since nearly all patients responded, it proved a highly significant predictor of relapse free and overall survival (Chan *et al.*, 1990). This highlights the fact that a clinical remission merely indicates the lack of detectable disease, which translates to a body burden of tumor of $\leq 1 \times 10^8 - 1 \times 10^9$ cells. Therefore, one interpretation of this data is that chemotherapy in P-gp positive patients reduced the tumor burden, but unlike what occurs in P-gp negative patients, failed to completely eradicate the disease. Therefore, the pretreatment expression of P-gp was prognostic of the success or failure of therapy for these two types of childhood malignancies. Nevertheless, whether P-gp was simply a marker for poor outcome, or a specific determinant of clinical drug resistance leading to treatment failure, cannot be directly concluded from these studies.

Recently, studies have begun to probe the role of P-gp in acquired drug resistance by measuring levels at presentation and at relapse after treatment with MDR-related drugs. For instance, in multiple myeloma, which initially responds to chemotherapy with vinblastine, Adriamycin and dexamethasone in 70% of patients, but nearly always relapses, quantitative immunohistochemical staining for P-gp in individual cells has shown that the myeloma cells of most patients express P-gp after relapse (Dalton *et al.*, 1989b; Salmon *et al.*, 1989), and that the percentage of cells expressing P-gp in treated patients is significantly greater than in untreated patients (Epstein and Barlogie, 1989; Epstein *et al.*, 1989).

Roninson and co-workers used a highly sensitive PCR technique to retrospectively analyse the correlation between low levels of *mdr*1 mRNA and response to chemotherapy in 100 ovarian and small cell lung cancer patients (Holzmayer *et al.*, 1992). A highly significant correlation was found between the absence of PCR detectable gene expression and responsiveness to chemotherapy. However, the clinical resistance was also seen to non-MDR drugs, suggesting coexistence of other drug resistance mechanisms within a heterogeneous tumor cell population.

Hematologic malignancies provide an excellent model to study the clinical significance of P-gp and its role in acquired drug resistance, due to their ease of sampling. Studies to date suggest that elevated *mdr*1 expression is an adverse prognostic factor in AML (Rischin and Ling, 1993). However, no prospective trials have been published which systematically measure P-gp expression in patients with initially drug sensitive leukemia at presentation and at drug resistant relapse, to determine if relative increases in P-gp expression correlate with the onset of drug resistance. A report from Stanford found elevated *mdr*1 mRNA levels in 19% of 31 patients with AML, acute lymphocytic leukemia (ALL) and CML in blast crisis at diagnosis, and in 50% of 10 different patients at time of relapse. There was a statistically significant decrease in the probability of achieving a complete response in the presence of elevated expression of *mdr*1 (Marie *et al.*, 1991). Of note, *in vitro* drug sensitivity to several MDR-related drugs was assayed in leukemic clones from 22 AML patients, and a statistically significant correlation between lack of *mdr*1 expression and daunorubicin sensitivity was found, though 40% of specimens expressing what was felt to be significant levels of *mdr*1 mRNA retained sensitivity to daunorubicin. Pirker *et al.* (1991) examined *mdr*1 mRNA levels by slot blot analysis in 63 patients with untreated de novo AML, and found that 53% of patients with detectable *mdr*1 gene expression achieved a complete response to an anthracycline containing induction regimen, versus 89% without detectable *mdr*1 mRNA (p = 0.008). Disease free and overall survival were also significantly better in the *mdr*1 negative group. A larger study of 150 AML patients measured P-gp expression at presentation by flow cytometry and found 47% with elevated P-gp in at least 20% of analyzed cells (Campos *et al.*, 1992). Again, there was a statistically significant difference in subsequent complete response rates between patients with or without elevated P-gp levels.

Expression of *mdr*1 has been less extensively studied in ALL than in AML, though several early reports suggested a lower prevalence of P-gp in the former (Ito *et al.*, 1989; Rothenberg *et al.*, 1989). Recently, Goasguen *et al.* (1993), reported a series of 59 patients presenting with de novo ALL (36 children, 23 adults), and also found expression of P-gp to be a negative prognostic factor. These investigators found a statistically significant difference in induction of remission for adults (56% versus 93%), and event-free survival, and a trend toward significance for second complete response (15% versus 38%) in patients expressing P-gp in ≥ 1% of leukemic cells (38% of patients) compared to no expression, respectively. Thus, while the prognostic significance of P-gp expression appears important in leukemia, the design of these studies cannot provide direct evidence for a functional role for P-gp in determining the presence or development of clinical drug resistance.

Whether P-gp expression plays a functional role in the failure of initial chemotherapy in intrinsically resistant tumors or in the development of acquired resistance is not known. It is important to remember that renal cell, adrenal and colon carcinoma are also clinically resistant to drugs that are not known to be affected by P-gp, such as alkylating agents, antimetabolites and bleomycin. Thus, mechanisms other than P-gp associated MDR must contribute to intrinsic clinical drug resistance. Though P-gp is often increased in acquired drug resistance, many of these patients are also resistant to non-MDR classes of drugs. This suggests several possibilities: (1) that *mdr*1 overexpression reflects a general response of cells to toxic stress, (2) that P-gp is one of many cellular defense mechanisms produced by exposure to toxins, or (3) that P-gp expression is a specific response to chemotherapy, and its antagonism could produce sensitivity to natural product chemotherapeutics.

These important studies prove that P-gp is expressed in both normal and neoplastic human tissues. They also demonstrate that immunological and molecular diagnostic techniques may be successfully used to detect P-gp and *mdr*1 gene expression in clinical samples, and suggests their use for identifying individual patients who may be appropriate candidates for trials designed to circumvent or modify MDR in humans.

At this time one can conclude that for certain tumors, increased P-gp appears to be a marker for the existence of multiple mechanisms of drug resistance. To prove an association between intrinsic or acquired drug resistance and P-glycoprotein expression, careful analyses of the possible correlations between tumor P-gp levels and treatment responses to specific drugs must be performed, and prospective sampling of pre- and post-treatment P-gp levels from individual patients must be determined. Finally, controlled trials designed to detect increased clinical response of tumors with intrinsic or acquired resistance to chemotherapeutic regimens containing inhibitors of P-gp function will be critical to understanding the therapeutic importance of clinical MDR.

Effects of chemosensitizers in human trials

The presence of detectable P-gp in many clinically drug resistant tumors suggests that the addition of a chemosensitizer at appropriate doses may sensitize human tumors *in vivo* to chemotherapy regimens employing MDR-related drugs. Furthermore, since the studies reviewed above show that P-gp expression is of prognostic relevance, but fail to clearly define a functional role for P-gp in clinical drug resistance, one way to implicate the functional relevance of the MDR transporter in intrinsic or acquired drug resistance will be through trials designed to determine whether inhibitors of P-gp can sensitize tumors to chemotherapy. Since P-gp is also expressed in many normal tissues, careful analysis of the toxicities encountered by combining P-gp inhibitors with cytotoxic drugs will be required to determine if a therapeutic advantage is achievable. While a number of phase I and phase II trials of various chemosensitizers have now been reported, a prospective, randomized study has not been published. Attempts to combine verapamil with single agent chemotherapy have been hindered by the intolerable cardiac effects encountered at levels of verapamil nearing the effective anti-MDR range (3 to 6 μM). In a phase I study of escalating doses of verapamil plus 1.5 mg/m^2 of vinblastine given for 5 days in 17 patients with advanced, unresponsive malignancies, no objective tumor responses were noted, and the majority of patients developed electrocardiographic changes (Benson *et al.*, 1985). Peak plasma verapamil concentrations achieved during the continuous infusion were 0.45 μM, well below the concentrations necessary to produce anti-MDR effects *in vitro*. Similarly, a pilot study of oral verapamil and doxorubicin (50 mg/m^2 every 3 weeks) in 13 patients refractory to chemotherapy (eight to doxorubicin) was limited by nausea, hypotension, and cardiac arrhythmias (Presant *et al.*, 1986). While one partial response and two minor responses were noted in patients who had received prior doxorubicin, plasma levels of doxorubicin or verapamil were not measured.

A pioneering trial of escalating doses of i.v. verapamil plus doxorubicin (50 mg/m^2) in eight patients with refractory ovarian cancer also resulted in significant toxicity, including severe hypotension, heart block and congestive heart failure, without objective responses to therapy (Ozols *et al.*, 1987). The verapamil infusion led to median and peak plasma verapamil levels of 2 and 5 μM, respectively. Although these concentrations are within the range of experimentally relevant *in vitro* anti-MDR concentrations, further verapamil dose-escalation is clearly limited by hemodynamic side effects. However, because none of the patients had received previous doxorubicin or other MDR-related drugs, but rather combinations of Cytoxan and cisplatin, the resistance of their tumors may not have involved P-gp expression.

Several well-designed trials of the use of verapamil with combination chemotherapy for lymphoma and myeloma have now been reported by the Arizona group. In the first trial, seven patients with multiple myeloma and one patient with non-Hodgkin's lymphoma, all refractory to standard treatment with 4 days infusional vincristine (0.4 mg/d) and doxorubicin (9 mg/m^2/d) and oral dexamethasone (40 mg/d) (VAD), were studied for tumor expression of P-gp, cellular accumulation of doxorubicin and vincristine, and response to treatment with the identical regimen plus verapamil (Dalton *et al.*, 1989a). Tumor cells from six out of eight patients were shown to express P-gp by immunoperoxidase staining of bone marrow biopsies or node biopsies. Analysis of tumor cells from two P-gp positive patients showed enhanced accumulation of doxorubicin or vincristine in the presence of verapamil *in vitro*. Finally, continuous infusion verapamil resulted in transient, objective remissions in three of six P-gp positive patients. The two P-gp negative patients did not respond, nor did three of six P-gp positive patients. Reversible cardiotoxicity was observed, but no increase in myelosuppression was noted with the addition of verapamil. A recent update of this study included 22 myeloma patients treated with VAD plus verapamil and showed four responses within a subset of 10 patients who had myeloma cells that expressed P-gp, whereas no responses were seen in five P-gp negative patients (Salmon *et al.*, 1991).

The Arizona group has also reported a trial combining infusional verapamil with a regimen containing cyclophosphamide (600 mg/m^2 on day one), continuous infusion vincristine (0.4 mg/day × 4 days) and doxorubicin (10 mg/m^2/day × 4 days) and oral dexamethasone (40 mg/day × 4 days) in 18 patients with relapsed, drug resistant lymphomas (Miller *et al.*, 1991). Overall, 13 patients responded and five achieved complete responses. Though an overall response rate of 72% was noteworthy in these heavily pretreated lymphoma patients, no correlation was seen between P-gp expression and response to the verapamil containing regimen, and peak verapamil serum levels were only 1–2 μm, making any conclusion regarding the chemosensitizing effect of verapamil in this situation difficult. In fact, this trial, together with the preliminary results reported form the NCI's trial of the EPOCH regimen (Wilson *et al.*, 1993b), suggest that the use of a continuous infusion delivery of MDR related drugs may, by itself, partially overcome clinical drug resistance in lymphomas. For instance, the EPOCH trial is studying the effect of adding *R*-verapamil (150 mg/m^2 po × 6 days) to infusional VP-16 (50 mg/m^2/day × 4 days), vincristine (0.4 mg/m^2/day × 4 days), and doxorubicin (10 mg/m^2/day × 4 days) with daily prednisone and a single bolus injection of cyclophosphamide (750 mg/m^2) in intermediate and low-grade lymphomas (Wilson *et al.*, 1993a). EPOCH, without verapamil, resulted in a 6% complete response and 94% partial response rate in 17 low-grade lymphomas, and an 83% overall response rate in 53 intermediate and high-grade lymphomas (Wilson *et al.*, 1993b). While the 77% complete response rate to EPOCH in patients with aggressive histology lymphomas which had relapsed from a complete response to primary chemotherapy was impressive, the complete response rate for all patients with intermediate or high-grade lymphomas to EPOCH was only 34%, equivalent to other salvage regimens for relapsed lymphoma, such as CEPP (29%) (Chao *et al.*, 1990) or CHAD (31%) (Valasquez *et al.*, 1988). Addition of *R*-verapamil to EPOCH resulted in a 23% response in 35 lymphoma patients having progressed on EPOCH. Interestingly, no response to *R*-verapamil plus EPOCH was seen in patients with very high or very low levels of *mdr*1 expression, the latter result suggesting the development of non-MDR mechanisms of drug resistance. A recent report of infusional cytoxan (187.5 mg/m^2/day × 4 days), doxorubicin (12.5 mg/m^2/day × 4 days) and VP-16 (60 mg/m^2/day × 4 days) in 58 patients with relapsed or drug resistant intermediate or high-grade non-Hodgkin's lymphoma, the majority of whom had received at least two of these same drugs in bolus form, showed a 17% complete response and 52% overall response rate (Sparano *et al.*, 1993). While these response rates are no better than other salvage regimens in this setting, a schedule-dependent effect in favor of infusional therapy may exist, since most patients had received prior bolus doxorubicin and cyclophosphamide. However, only 16% of patients had received prior VP-16. "Bolus" salvage regimens which include VP-16 for relapsed lymphoma also show overall response rates of 60% (Chao *et al.*, 1990).

Thus, these phase I/II trials suggest that clinical modulation by verapamil of an anthracycline/*Vinca* alkaloid based chemotherapeutic regimen in tumors known to express P-gp may be possible. Furthermore, the toxic effects of verapamil at relevant anti-MDR concentrations have been well documented, and include heart block, hypotension, bradycardia and junctional rhythms, congestive heart failure, peripheral edema and fluid retention (Pennock *et al.*, 1991). Clearly, the side effects of this first generation chemosensitizer are difficult to justify given the relatively poor efficacy seen to date.

The phenothiazine, trifluoperazine, was given with doxorubicin to 36 patients with either acquired or intrinsic tumor resistance (Miller *et al.*, 1988). Doxorubicin (60 mg/m^2) was given as a continuous infusion with an initial dose of 20 mg/day oral trifluoperazine, increasing up to 100 mg/day. One complete response and six partial responses were seen. All responses were in patients who had acquired, rather than intrinsic, resistance to doxorubicin. Dose-limiting side effects of this regimen were the extrapyramidal effects of trifluoperazine, including motor restlessness, akathesia, facial dystonia, and resting tremor. Trifluoperazine plasma levels ranged widely, but were all <0.3 μM, and thus at least 10-fold less than those found optimal *in vitro*. A phase I trial of i.v. prochlorperazine, another CaM antagonist in common clinical use, achieved serum concentrations of up to 2 μM of chemosensitizer in 13 patients with advanced, refractory solid tumors treated using 15–75 mg/m^2 prochlorperazine and 60 mg/m^2 doxorubicin (Sridhar *et al.*, 1993). Side effects included sedation, cramps, restlessness and hypotension.

The antiestrogens, tamoxifen and toremifene, have potential for clinical use in modulating MDR due to the expectation of minimal side effects at relevant doses (Ford and Hait, 1990). A phase I study of high dose tamoxifen combined with vinblastine in 53 patients with advanced, refractory malignancies has been reported (Trump *et al.*, 1992). Escalating dose tamoxifen was administered orally as a day one loading dose of 150–680 mg/m^2 followed by a maintenance dose of tamoxifen of

40–260 mg/m^2 twice a day for 12 days. Vinblastine (1.5 mg/m^2) was given as a continuous infusion for five days beginning on day nine of tamoxifen therapy. Reversible, dose-limiting toxicities were neurologic, and occurred mostly at high dose levels. Serum tamoxifen concentrations of 6 μM were achieved without toxicity for most patients. Of interest, two tumor samples were analyzed for tissue tamoxifen concentrations, which were found to be substantially higher than concurrent plasma concentrations. This is of great importance since most chemosensitizers have large volumes of distribution and may be expected to accumulate in certain tissues.

A second phase I trial treated 59 refractory solid tumor patients with 320 mg/day oral tamoxifen for 6 days with concurrent oral VP-16 at 300 mg/day for 3 days (Millward *et al.*, 1992). In approximately half the patients, mean plasma tamoxifen levels exceeded 3 μM, with significant levels of the anti-MDR active metabolite, *N*-desmethyl tamoxifen, also being achieved. Tamoxifen related toxicities included emesis and thromboembolic events in 10% of patients. Six objective responses were seen.

The cyclosporines are one of the most promising classes of chemosensitizers. Several studies have now been published using CsA in combination with MDR-related drugs. In 21 patients with myeloma progressing or resistant to VAD, Sonneveld *et al.* (1992) tested a continuous infusion of CsA at doses of 5–10 mg/kg/day in combination with VAD. A 48% response rate was seen, with 7 of 12 P-gp positive patients responding. The dose-limiting toxicity was musculoskeletal pain, and mild increases in serum bilirubin were seen at higher doses. However, peak serum concentrations were only 1 μM, at the lower limit of effective concentrations seen *in vitro.*

The Stanford trial of CsA plus VP-16 represents the most thorough clinical study published to date with regard to the clinical pharmacologic effects of CsA administration (Yahanda *et al.*, 1992). In this phase I study, 57 patients who failed to respond to VP-16 alone were treated with escalating doses of infusional CsA (5–21 mg/kg/day for three days) with VP16 (80–120 mg/m^2 × 3 days). Steady state levels of serum CsA of up to 4 μM were achieved with acceptable toxicity. Dose-limiting toxicities included nausea, vomiting, myelosuppression and hyperbilirubinemia. Nephrotoxicity and infectious complications were minimal. Pharmocokinetic studies demonstrated a marked increase of the area-under-the-curve for VP-16 at target levels of CsA, as a result of a decrease in both renal and nonrenal clearance mechanisms (Lum *et al.*, 1992). The 40% inhibition of renal clearance of VP-16 was consistent with P-gp-mediated drug transport in proximal renal tubules, since no significant decrease in overall renal function was seen. The 50% decrease in nonrenal VP-16 clearance may have been due to alterations in hepatic metabolism by competitive inhibition by CsA of the cytochrome P-450 enzyme system, or by inhibition of VP-16 transport by P-gp in the biliary tract. In addition, the increase in serum bilirubin was related to serum CsA concentrations >1μM, and felt to be due to inhibition of bilirubin transport by P-gp. This hypothesis was supported by *in vitro* studies demonstrating bilirubin to be a substrate for P-gp binding (Gosland *et al.*, 1991). It is of interest, therefore, that clinical studies of other chemosensitizers have not noted increases in bilirubin. This suggests that (*a*) hyperbilirubinemia may be a physiologic marker for *in vivo* MDR modulation, and trials lacking this laboratory finding have not achieved adequate serum concentrations of chemosensitizer, or (*b*) that CsA induced hyperbilirubinemia is not solely due to the effect of CsA on a P-gp-mediated process.

Few conclusions can be drawn from these published phase I and II trials regarding the potential efficacy of chemosensitizers for reversing clinical drug resistance mediated by P-gp. A number of problems exist in interpreting this data. First, few of the earlier studies have thoroughly sampled tumors for the presence of increased P-gp expression, and even fewer have prospectively selected patients proven to be *mdr*1 positive before entry into trials. Second, the concentration of chemosensitizer achieved in the serum without incurring significant toxicity has been lower than that required for maximum activity *in vitro.* Although the serum concentration may not accurately reflect the concentration of these lipophilic drugs at their target sites, this will remain a question until more accurate assessments of *in vivo* chemomodulation are available.

One approach that may circumvent this problem is the use of combinations of chemosensitizers with additive or synergistic anti-MDR activity but with nonoverlapping toxicities. For example, we demonstrated this possibility in cultured MDR cells by showing that the calcium channel blocker, verapamil, and the CaM antagonist, *trans*-flupenthixol, were additive in their effects in antagonizing MDR (Ford *et al.*, 1990). This suggested that clinical protocols may be designed using several chemosensitizing agents, each at less toxic doses, resulting in equivalent antagonism of P-gp-mediated drug resistance. In fact, a phase I/II trial of combined verapamil and tamoxifen with MDR related chemotherapy (vincristine, doxorubicin and VP-16) in 58 patients with small cell lung cancer has been reported (Figueredo *et al.*, 1990). Dose-limiting toxicities were peripheral neuropathy and myelosuppression, but no toxicity directly related to the chemosensitizers were seen.

The selection of appropriate tumors for studying MDR and its clinical modulation in phase II trials will be critical to testing the potential relevance of MDR. Criteria include selecting tumors which: (1) consistently overexpresses P-gp but do not harbor multiple mechanisms of resistance, (2) are either intrinsically resistant to drugs included in the MDR phenotype or acquire resistance to such drugs after initial treatment response, (3) are accessible for sequential biopsies, (4) result in reasonable life expectancy despite relatively early relapse, and (5) occur with relative frequency. As more specific and potent chemosensitizers with less clinical toxicity and side effects enter early clinical trials, it will be critical to carefully plan and coordinate prospective therapeutic trials and diagnostic procedures, by measuring MDR markers at enrollment and relapse, achieving adequate serum chemosensitizer levels, and analyzing both tumor and plasma chemosensitizer and cytotoxic drug pharmacokinetics.

A newly described P-gp transport substrate previously used as a radiopharmaceutical for cardiac perfusion imaging may provide a unique, noninvasive probe to measure human tissue expression of P-gp and *in vivo* reversal of P-gp function (Piwnica-Worms *et al.*, 1993). This lipophilic, cationic organotechnetium complex, [^{99m}Tc] SESTAMIBI, was shown to be actively transported by P-gp expressing cells, and this transport was inhibited by verapamil and CsA. Thus, this agent may be used scintigraphically to image P-gp expression in human tumors, which should take up less signal if P-gp positive, and chemosensitizer function *in vivo*, which if effective should result in increased SESTAMIBI signal in tumors.

New approaches for reversal of multidrug resistance

While the most promising chemosensitizers to emerge from *in vitro* studies have not yet been adequately tested in clinical trials, alternative approaches to the reversal of MDR are also of interest. The increased understanding of the molecular biology of MDR has revealed alternative ways in which to affect P-gp function, and the use of several novel pharmacologic, immunologic and molecular means to reverse MDR are reviewed below.

Combination chemosensitizers

The major limiting factor for achieving what are thought to be adequate human serum concentrations of chemosensitizers to reverse MDR are their intrinsic side effects. Therefore, a similar approach to that originally proposed for combination chemotherapy may prove effective for the pharmacologic use of chemosensitizers; that is, by combining several chemosensitizing agents with nonoverlapping toxicities, one may achieve an overall anti-MDR effect greater than that possible with individual agents at higher doses. We first explored this approach *in vitro* by using a rigorous statistical analysis of the combined effect of the CaM antagonist *trans*-flupenthixol plus verapamil for the reversal of MDR in a human MDR MCF-7 breast cancer cell line (Ford *et al.*, 1990). Using the isobologram method (Berenbaum, 1981), we found these two chemosensitizers to be additive in their ability to reverse doxorubicin resistance by a defined amount, although their individual potencies for this effect differed by more than 2-fold. Similarly, others have studied the effect of verapamil plus CsA *in vitro*. Hu *et al.* (1990) showed that these two chemosensitizers at doses of around 1 μM synergistically reversed vinblastine resistance in MDR human leukemic cells. This study was subsequently criticized with regard to the statistical method used to determine synergy (fractional product), and when the same dose-response curves were reanalyzed by the isobologram method, CsA and verapamil were found to be additive (Samuels and Ratain, 1991). However, Slater's group found CsA and verapamil to be highly synergistic in their effect on daunorubicin, and supra-additive for vincristine, in a human MDR ALL cell line (Osann *et al.*, 1992). An additional study reported a synergistic interaction between verapamil and quinine for potentiation of doxorubicin and vinblastine cytotoxicity in MDR human myeloma cells using clonogenic assays (Lehnert *et al.*, 1991). While this distinction between synergy and additivity is of interest with regard to the mechanism of action on P-gp function, only an antagonistic interaction between two chemosensitizers would prove disadvantageous for their potential combined clinical use. These findings clearly justify further *in vitro* and clinical testing of various combinations of chemosensitizers for the reversal of MDR.

Monoclonal antibodies

To devise ways to more directly target MDR reversing therapy to P-gp and to avoid the clinical side effects associated with pharmacologic chemosensitizers, monoclonal antibodies recognizing P-gp have been explored as potential inhibitors of P-gp. The utility of antibodies which recognize P-gp on intact cells for reversal of MDR was tested by Hamada and Tsuruo (1986). They found that the monoclonal antibody, MRK-16, which recognizes an external epitope of human P-gp, increased the intracellular accumulation and cytotoxicity of vincristine and actinomycin-D in MDR cells. In addition, MRK-16 alone induced regression of tumors (Tsuruo *et al.*, 1987), and enhanced the *in vivo* activity of vincristine (Pearson *et al.*, 1991) in MDR xenografts in nude mice. However, the effect of the MRK-16 monoclonal antibody, as well as the HYB-241 monoclonal antibody, was only effective for *Vinca* alkaloids, but not for anthracyclines (Rittmann-Grauer *et al.*, 1992). Subsequently, Roninson's group developed a murine monoclonal antibody, UIC2, that recognized a distinct external epitope of human P-gp. UIC2 strongly inhibited the efflux of the P-gp substrates rhodamine 123 and doxorubicin, and enhanced the cytotoxicity of a broad range of MDR related drugs in human MDR cells, including the *Vinca* alkaloids, doxorubicin, taxol and VP-16, but not the non-MDR-related drugs methotrexate, 5-fluorouracil or cisplatin (Mechetner and Roninson, 1992). Tsuruo and co-workers also studied the effect of the MRK-16 monoclonal antibody together with pharmacologic chemosensitizers for the reversal of MDR, and found that MRK-16 interacted synergistically with CsA but not verapamil or FK506, to enhance the cytotoxic effects of vincristine and doxorubicin in MDR K562 human leukemia cells (Naito *et al.*, 1993). Further investigation revealed that MRK-16 also caused an increased intracellular accumulation of CsA, but not verapamil, perhaps explaining their synergistic activity in reversing MDR. These findings with anti-P-gp antibodies suggest that their use together with MDR-related cytotoxic drugs with or without pharmacologic chemosensitizers, may be a potential therapeutic anti-MDR therapy.

Modification of monoclonal antibodies against P-gp have been made to enhance their antitumor activity. For example, Mickisch *et al.* coupled MRK-16 antibodies to *Pseudomonas* exotoxin, and found an increased activity for killing P-gp expressing bone marrow cells in transgenic mice compared to the native monoclonal antibody, alone (Mickisch *et al.*, 1992). Others have fused an anti-P-gp monoclonal antibody to doxorubicin containing liposomes, and found an enhanced cytotoxic effect against P-gp expressing cells compared to doxorubicin alone or liposomal doxorubicin without antibodies (Ahmad and Allen, 1992).

Thus, P-gp may serve as a target for potential immunologic therapies. The therapeutic use of antibodies in humans remains challenging. However, this approach to the treatment of MDR tumors may play a role alone, or in combination with pharmacologic drugs.

Differentiation therapy

A relationship between the expression of P-gp and the degree of cellular differentiation was suggested by Biedler who found that the most resistant P-gp expressing cell lines were the least tumorigenic in animals (Biedler *et al.*, 1991). This apparent paradox was later amplified upon by the observation by Roninson (1992) that in normal gastrointestinal epithelium, the most apical elements of the normal mucosa expressed P-gp whereas the most basal cellular crypt cells did not. In addition, when colon carcinoma cells were induced to differentiate with sodium butryate, the expression of P-gp was increased (Mickley *et al.*, 1989; Bates *et al.*, 1992). We have recently examined the role of P-gp expression in a transfected line, to more carefully delineate the unique role of P-gp expression in cellular differentiation. By infecting human myeloid blasts cells, K562 cells, with a retroviral vector containing a cDNA for the human *mdr*1 gene, we selected a series of clones that overexpressed P-gp but did not show numerous other mechanisms of drug resistance (Hait *et al.*, 1993a). This line, K562/Vbl, underwent erythroid differentiation in response to hemin or cytosine arabinoside as well as the noninfected parental line. In addition, we found that the induction of erythroid differentiation with either agent led to increased expression of MDR 1 transcription and translation. Therefore it appears that the expression of P-gp *per se* does not affect the process of cellular differentiation, but that differentiation leads to increased expression of this gene. This may be particular relevant to ongoing attempts at inserting the *mdr*1 gene into human bone marrow to protect the marrow against chemotherapy, since an alteration in the ability of these cells to under go normal differentiation would be a potentially lethal event. In addition, these results suggest that the MDR phenotype does not produce cross-resistance to certain differentiating agents, and offers a potential alternative to cytotoxic therapy in drug-resistant CML.

Regulation of mdr1 gene expression

Recent studies of the structure and function of the *mdr*1 promoter sequences have been reported (Goldsmith *et al.*, 1993; Madden *et al.*, 1993) which have begun to provide insight into the mechanism of regulating *mdr*1 gene expression in normal tissue and tumor cells. The presence of putative *cis*- and *trans*- activating transcriptional regulatory mechanisms suggest that modulation of P-gp expression may be possible at the genetic level.

Expression of the endogenous *mdr*1 gene has been shown to be modulated by heat shock and heavy metals through heat shock responsive elements in the *mdr*1 promoter (Chin *et al.*, 1993), estradiol (Jancis *et al.*, 1993), induction of PKC (Chaudhary and Roninson, 1992), various cytotoxic drugs, including non-MDR related chemotherapeutics (Chaudhary and Roninson, 1993), and other exogenous insults and toxins, such as carcinogens (Burt and Thorgeirssen, 1988), X-irradiation (Hill *et al.*, 1990) and serum starvation. Transient expression assays utilizing *mdr*1 promoter/reporter constructs also demonstrate regulatory activity by progesterone (Piekarz *et al.*, 1993), oncogenes and tumor suppressor genes (Chin *et al.*, 1993), and other DNA binding proteins (Ogura *et al.*, 1992; Goldsmith *et al.*, 1993).

Cho-Chang and colleagues have described a cAMP analog, 8-Cl-cAMP, which produces growth inhibiting activity in several MDR and non-MDR drug resistant tumor cell lines (Rohlff *et al.*, 1993b), possibly through interaction with specific cAMP-dependent protein kinases involved in regulating cell growth (Rohlff *et al.*, 1993a). Preliminary results using a *mdr*1-CAT reporter construct also showed that 8-Cl-cAMP inhibits *mdr*1 promoter activity and down regulates P-gp expression (Budillon *et al.*, 1993). The ability of this agent to specifically enhance cytotoxic drug activity or alter drug accumulation in P-gp expressing MDR cells has not been reported.

Therefore, a variety of *trans*- acting agents and *cis*- acting elements have been identified which suggest additional means for modulation of *mdr*1 gene expression. Further studies to determine the potential utility of this approach for reversal of MDR are indicated.

Concurrent reversal of multiple mechanisms of drug resistance

Alternative cellular mechanisms of resistance to cytotoxic drugs affected by P-gp-mediated MDR, such as alteration in the detoxifying enzymes glutathione and glutathione-*S*-transferases, or in topoisomerases have been extensively studied. Increasing evidence suggests that cellular drug resistance may be multifactorial, and that heterogeneity within tumors may occur with respect to drug resistance mechanisms. Several examples of this emerging principle have now been demonstrated *in vitro*. For example, a Chinese hamster lung cell line has been shown to have independent alterations in both topoisomerase II activity and P-gp expression (Larsen and Jacquemin-Sablon, 1989), and a series of L1210 cells selected for increasing resistance to doxorubicin, have been shown to display progressively increasing P-gp expression and decreasing topoisomerase II activity (Ganapathi *et al.*, 1989).

Indeed, clinical drug resistance may be so complex that the likelihood of a drug that affects only one mechanism of resistance to produce a meaningful effect on the clinical course of disease may be small. Future chemotherapeutic regimens, therefore, may need to include combinations of chemosensitizing strategies aimed at multiple drug resistance mechanisms. Several experimental studies have tested this hypothesis.

For example, Kramer reported that the combination of buthionine sulfoximine, an inhibitor of glutathione production, with the P-gp chemosensitizer verapamil could completely sensitize the MCF-7 Adr^R cell line to doxorubicin, which contains alterations in GSTπ activity and an over-expressed and amplified *mdr*1 gene (Batist *et al.*, 1986; Fairchild *et al.*, 1987), whereas each agent individually could not (Kramer *et al.*, 1988). Similarly, the combination of CsA and buthionine sulfoximine was found to completely reverse doxorubicin resistance in a MDR small cell lung cancer cell line more effectively than either chemosensitizer alone (Larsson *et al.*, 1991). A novel CaM antagonist, W-77, was found to possess bifunctional activity in inhibiting P-gp-mediated drug efflux and antagonizing glutathione-*S*-transferase activity (Maeda *et al.*, 1993). This agent was found to be less effective than verapamil for increasing doxorubicin accumulation in a MDR cell line with increased P-gp and glutathione-*S*-transferase activity, but more effective for reversal of doxorubicin resistance, suggesting a multifactorial basis for its mechanism of cellular sensitization.

This approach may not be simple. We found that the combination of buthionine sulfoximine with verapamil alone, in the absence of doxorubicin, proved extremely toxic to tumor cells, due to a specific enhancement of the toxicity of verapamil by buthionine sulfoximine (Ford *et al.*, 1991). Furthermore, otherwise nontoxic doses of verapamil were found to be lethal to animals receiving standard doses of buthionine sulfoximine in their drinking water. These findings emphasize the complexity involved in altering natural detoxification mechanisms utilized by both normal and neoplastic tissues. Clearly, the simultaneous clinical modulation of multiple mechanisms of drug resistance will present significant challenges.

Conclusions

The study of the cellular, biochemical and molecular biology and pharmacology of MDR has provided one of the most active and exciting areas within cancer research for translation into potential clinical benefit. While convincing evidence for the functional role of P-gp in mediating clinical drug resistance in humans remains scant, studies of the clinical expression of P-gp and trials of chemosensitizers with cancer chemotherapy suggest "resistance modification" strategies may be effective in some tumors with intrinsic or acquired drug resistance. However, even if P-gp associated MDR proves to be a relevant and reversible cause of clinical drug resistance, numerous problems remain to be solved before effective clinical chemosensitization may be achieved. Such factors as absorption, distribution, and metabolism, the effect of chemosensitizers on chemotherapeutic drug clearance, and toxicity to normal tissues expressing P-gp, and the most efficacious modulator regimens all remain to be studied *in vivo*. Clearly, the identification of more specific, potent and less clinically toxic chemosensitizers for clinical use remains critical to the possible success of this approach.

The finding that a number of pharmacological agents can antagonize a well-characterized form of experimental drug resistance provides promise for potential clinical applications. Further study of chemosensitizers in humans and the rational search or design of novel chemosensitizers with improved activity should define the importance of MDR to clinically resistant cancer.

References

Aftab DT, Ballas LM, Loomis CR and Hait WN (1991) Structure-activity relationships of phenothiazines and related drugs for inhibition of protein kinase C. Mol Pharmacol 40: 798–805.

Aftab DT and Hait WN (1992) Effects of phorbol 12-myristate 13-acetate on drug accumulation and P-glycoprotein phosphorylation in sensitive and multidrug resistant MCF-7 cells. Proc Am Assoc Cancer Res 33: 2821.

Ahmad I and Allen TM (1992) Antibody-mediated specific binding and cytotoxicity of liposome-entrapped doxorubicin to lung cancer cells *in vitro*. Cancer Res 52: 4817–4820.

Aisner J, Tchekmedyian NS, Tait N, Parnes H and Novak M (1988) Studies of high-dose megestrol acetate: potential applications in cachexia. Semin Oncol 15S: 68–75.

Akiyama S, Cornwell MM, Kuwano M, Pastan I and Gottesman MM (1988) Most drugs that reverse multidrug resistance inhibit photoaffinity labeling of P-glycoprotein by a vinblastine analog. Mol Pharmacol 33: 144–147.

Akiyama S, Yoshimura A, Kikuchi H, Sumizawa T, Kuwano M and Tahara Y (1989) Synthetic isoprenoid photoaffinity labeling of P-glycoprotein specific to multidrug-resistant cells. Mol Pharmacol 36: 730–735.

Ambudkar SV, Lelong IH, Zhang J, Cardarelli CO, Gottesman MM and Pastan I (1992) Partial purification and reconstitution of the human multidrug-resistance pump: characterization of the drug-stimulatable ATP hydrolysis. Proc Natl Acad Sci USA 89: 8472–8476.

Anderson L, Cummings J, Bradshaw T and Smyth JF (1991a) The role of protein kinase C and the phosphatidylinositol cycle in multidrug resistance in human ovarian cancer cells. Biochem Pharmacol 42: 1427–1432.

Anderson MP, Gregory RJ, Thompson S, Souza DW, Paul S, Mulligan RC, Smith AE and Welsh MJ (1991b) Demonstration that CFTR is a chloride channel by alteration of its anion selectivity. Science 253: 202–205.

Aquino A, Hartman KD, Knode MC, Grant S, Huang K-P, Niu C-H and Glazer RI (1988) Role of protein kinase C in phosphorylation of vinculin in Adriamycin-resistant HL-60 leukemia cells. Cancer Res 48: 3324–3329.

Arceci RJ, Croop JM, Horwitz SB and Housman D (1988) The gene encoding multidrug resistance is induced and expressed at high levels during pregnancy in the secretory epithelium of the uterus. Proc Natl Acad Sci USA 85: 4350–4354.

Arceci RJ, Stieglitz K and Bierer BE (1992) Immunosuppressants FK506 and rapamycin function as reversal agents of the multidrug resistance phenotype. Blood 80: 1528–1536.

Asoh K, Saburi Y, Sato S, Nogai I, Kohno K and Kuwano M (1989) Potentiation of some anticancer agents by dipyridamole against drug-sensitive and drug-resistant cancer cell lines. Jpn J Cancer Res 80: 475–481.

Bacher N, Zisman Y, Berent E and Livneh E (1991) Isolation and characterization of PKC-L, a new member of the protein kinase C-related gene family specifically expressed in lung, skin, and heart. Mol Cell Biol 11: 126–133.

Baker WJ, Wiebe VJ, Koester SK, Emshoff VD, Maenpaa JU, Wurz GT and DeGregorio MW (1992) Monitoring the chemosensitizing effects of toremifene with flow cytometry in estrogen receptor negative multidrug resistant human breast cancer cells. Breast Cancer Res Tr. 24: 43–49.

Bates SE, Currier SJ, Alvarez M and Fojo AT (1992) Modulation of P-glycoprotein phosphorylation and drug transport by sodium butyrate. Biochemistry 31: 6366–6372.

Batist G, Tulpule A, Sinha BK, Katkis AG, Myers CE and Cowan KH (1986) Overexpression of a novel anionic glutathione transferase in multidrug-resistant human breast cancer cells. J Biol Chem 261: 15544–15549.

Bear CE, Li C, Kartner N, Bridges RJ, Jensen TJ, Ramjeesingh M and Riordan JR (1992) Purification and functional reconstitution of the cystic fibrosis transmembrane conductance regulator (CFTR). Cell 68: 809–818.

Beck WT, Cirtain MC, Glover CJ, Felsted RL and Safa AR (1988) Effects of indole alkaloids on multidrug resistance and labeling of P-glycoprotein by a photoaffinity analog of vinblastine. Biochem Biophys Res Comm 153: 959–966.

Benson AB, Trump DL, Koeller JM, Egorin MI, Olman EA, Witte RS, Davis TE and Tormey DC (1985) Phase I study of vinblastine and verapamil given by concurrent iv infusion. Cancer Treat Rep 69: 795–799.

Berenbaum MC (1981) Criteria for analyzing interactions between biologically active agents. Adv Cancer Res 35: 269–335.

Berman E, Adams M, Duigou-Osterndorf R, Godfrey L, Clarkson B and Andreeff M (1991) Effect of tamoxifen on cell lines displaying the multidrug-resistant phenotype. Blood 77: 818–825.

Biedler JL, Casals D, Chang TD, Meyers MB, Spengler BA and Ross RA (1991) Multidrug resistant human neurobastoma cells are more differentiated than controls and retinoic acid further induces lineage-specific differentiation. In: Adv. Neuroblastoma Res. New York: Wiley-Liss, Inc.

Biedler JL and Riehm H (1970) Cellular resistance to actinomycin D in Chinese hamster cells in vitro: cross-resistance, radioautographic and cytogenetic studies. Cancer Res 30: 1174–1184.

Bierer BE, Mattila PS, Standaert RF, Herzenberg LA, Burakoff SJ, Crabtree B and Schreiber SL (1990) Two distinct signal transduction pathways in T lymphocytes are inhibited by alternative complexes formed between an immunophilin and either FK506 or rapamycin. Proc Natl Acad Sci USA 87: 9231.

Blobe GC, Sachs CW, Khan WA, Fabbro K, Stabel S, Wetsel W, Obeid LM, Fine RL and Hannun YA (1993) Selective regulation of expression of protein kinase C isoenzymes in multidrug-resistant MCF-7 cells: functional significance of enhanced expression of PKC α. J Biol Chem 268: 658–664.

Bodden WL, Palayoor ST and Hait WN (1986) Selective antimitochondrial agents inhibit calmodulin. Biochem Biophys Res Commun 135: 574–582.

Boesch D, Muller K, Pourtier-Manzanedo A and Loor F (1991) Restoration of daunomycin retention in multidrug-resistant P388 cells by submicromolar concentrations of SDZ PSC 833, a nonimmunosuppressive cyclosporin derivative. Exp Cell Res 196: 26–32.

Borman LS, Bornmann WG and Kuehne ME (1993) Modulation of drug cytotoxicity in wild-type and multidrug-resistant tumor cells by stereoisomeric series of C-20'-vinblastine congeners that lack antimicrotubule activity. Cancer Chemother Pharmacol 31: 343–349.

Bruggemann EP, Currier SJ, Gottesman MM and Pastan I (1992) Characterization of the azidopine and vinblastine binding site of P-glycoprotein. J Biol Chem 267: 21020–21026.

Bruggemann EP, Germann UA, Gottesman MM and Pastan I (1989) Two different regions of phosphoglycoprotein are photoaffinity-labeled by azidopine. J Biol Chem 264: 15483–15488.

Budd GT, Jayaraj A, Grabowski D, Adelstein D, Bauer L, Boyett J, Bukowski R, Murthy S and Weick J (1990) Phase I trial of dipyridamole with 5-fluorouracil and folinic acid. Cancer Res 50: 7206–7211.

Budillon A, Kelly K, Cowan K and Cho-Chung YS (1993) 8-Cl-cAMP, a site-selective cAMP analog as a novel agent that inhibits the promoter activity of multidrug-resistance gene. Proc Am Assoc Cancer Res 34: 1940.

Burt RK and Thorgeirssen SS (1988) Coinduction of MDR-1 multidrug resistance and cytochrome P-450 genes in rat liver by xenobiotics. J Natl Cancer Inst 80: 1383–1386.

Busche R, Tummler B, Riordan JR and Cano-Gauci DF (1989) Preparation and utility of a radioiodinated analogue of daunomycin in the study of multidrug resistance. Mol Pharmacol 35: 414–421.

Campos L, Guyout D, Archimbaud E, Calmard-Oriol P, Tsuruo T, Troncy J, Treille D and Fiere D (1992) Clinical significance of multidrug resistance P-glycoprotein expression on acute nonlymphoblastic leukemia cells at diagnosis. Blood 79: 473–476.

Candell J, Valle V and Soler M (1979) Acute intoxication with verapamil. Chest 75: 200–201.

Cano-Gauci DF and Riordan JR (1987) Action of calcium antagonists on multidrug resistant cells. Biochem Pharmacol 36: 2115–2123.

Carlsen SA, Till JE and Ling V (1977) Modulation of drug permeability in Chinese hamster ovary cells — possible role for phosphorylation of surface glycoproteins. Biochim Biophys Acta 467: 238–250.

Center MC (1985) Mechanisms regulating cell resistance to Adriamycin — evidence that drug accumulation in resistant cells is modulated by phosphorylation of a plasma membrane glycoprotein. Biochem Pharmacol 34: 1471–1476.

Center MS (1983) Evidence that Adriamycin resistance in Chinese hamster lung cells is regulated by phosphorylation of a plasma membrane glycoprotein. Biochem Biophys Res Commun 115: 159–166.

Chafouleas JG, Bolton WE and Means AR (1984) Potentiation of bleomycin lethality by anticalmodulin drugs: a role for calmodulin in DNA repair. Science 224: 1346–1348.

Chambers SK, Hait WN, Kacinski BM, Keyes SR and Handschumacher RE (1989) Enhancement of anthracycline growth inhibition in parent and multidrug-resistant Chinese hamster ovary cells by cyclosporin A and its analogues. Cancer Res 49: 6275–6279.

Chambers TC, Chalikonda I and Eilon G (1990a) Correlation of protein kinase C translocation, P-glycoprotein phosphorylation and reduced drug accumulation in multidrug resistant human KB cells. Biochem Biophys Res Commun 169: 253–259.

Chambers TC, McAvoy EM, Jacobs JW and Eilon G (1990b) Protein kinase C phosphorylates P-glycoprotein in multidrug resistant human KB carcinoma cells. J Biol Chem 265: 7679–7686.

Chan HSL, Haddad B, Thorner PS, DeBoer G, Lin YP, Oncrusek N, Yeger H and Ling V (1991) P-glycoprotein expression as a predictor of the outcome of therapy for neuroblastoma. N Engl J Med 325: 1608–1614.

Chan HSL, Thorner PS, Haddad G and Ling V (1990) Immunohistochemical detection of P-glycoprotein: prognostic correlation in soft tissue sarcoma of childhood. J Clin Oncol 8: 689–704.

Chatterjee M and Harris A (1990a) Enhancement of adriamycin cytotoxicity in a multidrug resistant Chinese hamster ovary (CHO) subline, CHO-Adr^r, by toremifene and its modulation by alpha 1 acid glycoprotein. Eur J Cancer 26: 432–436.

Chatterjee M and Harris AL (1990b) Reversal of acquired resistance to adriamycin in CHO cells by tamoxifen and 4-hydroxy tamoxifen: role of drug interaction with alpha 1 acid glycoprotein. Br J Cancer 62: 712–717.

Chao NJ, Rosenberg SA and Horning SJ (1990) CEPP(B): an effective and well-tolerated regimen in poor-risk, aggressive non-Hodgkin's lymphoma. Blood 76: 1293–1298.

Chaudhary PM, Mechetner EB and Roninson IB (1992) Expression and activity of the multidrug resistance P-glycoprotein in human peripheral blood lymphocytes. Blood 80: 2735–2739.

Chaudhary PM and Roninson IB (1991) Expression and activity of P-glycoprotein, a multidrug efflux pump, in human hematopoietic stem cells. Cell 66: 85–94.

Chaudhary PM and Roninson IB (1992) Activation of *MDR*-1 (P-glycoprotein) gene expression in human cells by protein kinase C agonists. Oncol Res 4: 281–290.

Chaudhary PM and Roninson IB (1993) Induction of multidrug resistance in human cells by transient exposure to different chemotherapeutic drugs. J Natl Cancer Inst 85: 632–639.

Chauffert B, Martin M, Hammann A, Michel MF and Martin F (1986) Amiodarone-induced enhancement of doxorubicin and 4'-deoxydoxorubicin cytotoxicity to rat colon cancer cells *in vitro* and *in vivo*. Cancer Res 46: 825–830.

Chen CJ, Chin E, Ueda K, Clark CP, Pastan I, Gottesman MM and Roninson IB (1986) Internal duplication and homology with bacterial transport proteins in the *mdr*l (P-glycoprotein) gene from multidrug resistant human cells. Cell 47: 381–389.

Chen HX, Bamberger U, Heckel A, Guo X and Cheng YC (1993) BIBW 22, a dipyridamole analogue, acts as a bifunctional modulator on tumor cells by influencing both P-glycoprotein and nucleoside transport. Cancer Res 53: 1974–1977.

Chin KV, Pastan I and Gottesman MM (1993) Function and regulation of the human multidrug resistance gene. Adv Cancer Res 60: 157–180.

Choi K, Chen C, Kriegler M and Roninson IB (1988) An altered pattern of cross-resistance in multidrug-resistant human cells results from spontaneous mutations in the mdr1 (P-glycoprotein) gene. Cell 53: 519–529.

Chong ASF, Markham PN, Gebel HM, Bines SD and Coon JS (1993) Diverse multidrug-resistance-modification agents inhibit cytolytic activity of natural killer cells. Cancer Immunol Immunother 36: 133–139.

Cole SPC, Bhardwaj G, Gerlach JH, Mackie JE, Grant CE, Almquist KC, Stewart AJ, Kurz EU, Duncan AMV and Deeley RG (1992) Overexpression of a transporter gene in a multidrug-resistant human lung cancer cell line. Science 258: 1650–1654.

Colombani PM, Robb A and Hess AD (1985) Cyclosporin A binding to calmodulin: a possible site of action on T lymphocytes. Science 228: 337–339.

Coon JS, Wang Y, Bines SD, Markham PN, Chong ASF and Gebel HM (1991) Multidrug resistance activity in human lymphocytes. Human Immunology 32: 134–140.

Cordon-Cardo C, O'Brien JP, Casals D, Rittman-Grauer L, Biedler JL, Melamed MR and Bertino JR (1989) Multidrug-resistance gene (P-glycoprotein) is expressed by endothelial cells at blood-brain barrier sites. Proc Natl Acad Sci USA 86: 695–698.

Cornwell MM, Gottesman MM and Pastan I (1986) Increased vinblastine binding to membrane vesicles from multidrug-resistant KB cells. J Biol Chem 261: 7921–7928.

Cornwell MM, Pastan I and Gottesman MM (1987) Certain calcium channel blockers bind specifically to multidrug resistant human KB carcinoma membrane vesicles and inhibit drug binding to P-glycoprotein. J Biol Chem 262: 2166–2170.

Creese I and Sibley DR (1980) Receptor adaptations to centrally acting drugs. Annu Rev Pharmacol Toxicol 21: 357–391.

Dalton WS, Grogan TM, Meltzer PS, Scheper RJ, Durie BGM, Taylor CW, Miller TP and Salmon SE (1989a) Drug-resistance in multiple myeloma and non-Hodgkin's lymphoma: detection of P-glycoprotein and potential circumvention by addition of verapamil to chemotherapy. J Clin Oncol 7: 415–424.

Dalton WS, Grogan TM, Rybski JA, Scheper RJ, Richter W, Kailey J, Broxterman HJ, Pinedo HM and Salmon SE (1989b) Immunohistochemical detection and quantitation of P-glycoprotein in multiple drug-resistant human myeloma cells: association with level of drug resistance and drug accumulation. Blood 73: 747–752.

DeFaire U and Lundman T (1977) Attempted suicide with verapamil. J Cardiol 6: 195–198.

DeGregorio MW, Ford JM, Benz CC and Wiebe VJ (1989) Toremifene: pharmacologic and pharmacokinetic basis of reversing multidrug resistance. J Clin Oncol 7: 1359–1364.

Dellinger M, Pressman BC, Calderon-Higginson C, Savaraj N, Tapiero H, Kolonias D and Lampidis TJ (1992) Structural requirements of simple organic cations for recognition by multidrug-resistant cells. Cancer Res 52: 6385–6389.

Dong Z, Ward NE, Fan D, Gupta KP and O'Brian CA (1991) *In vitro* model for intrinsic drug resitance: effects of protein kinase C activators on the chemosensitivity of cultured human colon cancer cells. Mol Pharmacol 39: 563–569.

Drach D, Zhao S, Drach J, Mahadevia R, Gattringer C, Huber H and Andreeff M (1992) Subpopulations of normal peripheral blood and bone marrow cells express a functional multidrug resistant phenotype. Blood 80: 2729–2734.

Dumont FJ, Melino MR, Staruch MJ, Koprak SL, Fischer PA and Sigal NH (1990) The immunosuppressive macrolides FK506 and rapamycin act as reciprocal antagonists in murine T cells. J Immunol 144: 1418.

Duran GE, Jaffrezou JP, Gosland MP, Ross KL and Sikic BI (1993) Inhibition of the multidrug transporter, P-glycoprotein, by the cyclosporin analogue PSC 833, cyclosporine and verapamil. Proc Am Assoc Cancer Res 34: 1922.

Epstein J and Barlogie B (1989) Tumor resistance to chemotherapy associated with expression of the multidrug resistance phenotype. Cancer Bull 41: 41–44.

Epstein J, Xiao H and Oba BK (1989) P-glycoprotein expression in plasma-cell myeloma is associated with resistance to VAD. Blood 74: 913–917.

Fairchild CR, Ivy SP, Kao-Shan C, Whang-Peng J, Rosen M, Israel MA, Melera PW, Cowan KH and Goldsmith ME (1987) Isolation of amplified and overexpressed DNA sequences from Adriamycin-resistant human breast cancer cells. Cancer Res 47: 5141–5148.

Ferguson LR and Baguley BC (1993) Multidrug resistance and mutagenesis. Mutation Res 285: 79–90.

Ferguson PF and Cheng Y (1987) Transient protection of cultured human cells against antitumor agents by 12-*O*-tetradecanoylphorbol-13-acetate. Cancer Res 47: 433–441.

Figueredo A, Arnold A, Goodyear M, Findlay B, Neville A, Normandeau R and Jones A (1990) Addition of verapamil and tamoxifen to the initial chemotherapy of small cell lung cancer. Cancer 65: 1895–1902.

Fine RL, Patel Ja and Chabner, BA (1988) Phorbol esters induce multidrug resistance in human breast cancer cells. Proc Natl Acad Sci USA 85: 582–586.

FitzGerald GA (1987) Dipyridamole. N Engl J Med 316: 1247–1256.

Fleckenstein A, Fleckenstein-Gun G, Byon YK, Haastert HP and Spah F (1979) Vergleichende untersuchungen uber die Ca^{++}-antagonistischen grundwirkungen von niludipin (Bay a 7168) und nifedipin (Bay a 1040) auf myokard, myomertrium und glatte gefassmuskulatur. Arzneim Forsch 29: 230–246.

Fleming GF, Amato JM, Agresti M and Safa AR (1992) Megestrol acetate reverses multidrug resistance and interacts with P-glycoprotein. Cancer Chemother Pharmacol 29: 445–449.

Fojo A, Akiyama S, Gottesman MM and Pastan I (1985) Reduced drug accumulation in multiply drug-resistant human KB carcinoma cell lines. Cancer Res 45: 3002–3007.

Ford JM, Bruggeman E, Pastan I, Gottesman MM and Hait WN (1990) Cellular and biochemical characterization of thioxanthenes for reversal of multidrug resistance in human and murine cell lines. Cancer Res 50: 1748–1756.

Ford JM and Hait WN (1990) Pharmacology of drugs that alter multidrug resistance in cancer. Pharmacol Rev 42: 156–199.

Ford JM, Prozialeck WC and Hait WN (1989) Structural features determining activity of phenothiazines and related drugs for inhibition of cell growth and reversal of multidrug resistance. Mol Pharmacol 35: 105–115.

Ford JM, Yang J and Hait WN (1991) Effect of buthionine sulfoximine on toxicity of verapamil and doxorubicin to multidrug resistant cells and to mice. Cancer Res 51: 67–72.

Foster BJ, Grotzinger KR, McKoy WM, Rubinstein LV and Hamilton TC (1988) Modulation of induced resistance to Adriamycin in two human breast cancer cell lines with tamoxifen or perhexiline maleate. Cancer Chemother Pharmacol 22: 147–152.

Foxwell BMJ, Mackie A, Ling V and Ryffel B (1989) Identification of the multidrug resistance-related P-glycoprotein as a cyclosporin binding protein. Mol Pharmacol 36: 543–546.

Furr BJA and Jordan, VC (1984) The pharmacology and clinical uses of tamoxifen. Pharmacol Ther 25: 127–205.

Ganapathi R and Grabowski D (1983) Enhancement of sensitivity to Adriamycin in resistant P388 leukemia by the calmodulin inhibitor trifluoperazine. Cancer Res 43: 3696–3699.

Ganapathi R and Grabowski D (1988) Differential effect of the calmodulin inhibitor trifluoperazine in modulating cellular accumulation, retention and cytotoxicity of doxorubicin in progressively doxorubicin-resistant L1210 mouse leukemia cells. Biochem Pharmacol 37: 185–193.

Ganapathi R, Grabowski D, Ford J, Heiss C, Kerrigan D and Pommier Y (1989) Progressive resistance to doxorubicin in mouse leukemia L1210 cells with multidrug resistance phenotype: reductions in drug-induced topoisomerase II-mediated DNA cleavage. Cancer Communication 1: 217–224.

Ganapathi R, Grabowski D, Hoeltge G and Neelon R (1990) Modulation of doxorubicin-incuced chromosomal damage by calmodulin inhibitors and its relationship to cytotoxicity in progressively doxorubicin-resistant tumor cells. Biochem Pharmacol 40: 1657–1662.

Ganapathi R, Grabowski D, Rouse W and Riegler F (1984) Differential effect of the calmodulin inhibitor trifluoperazine on cellular accumulation, retention, and cytotoxicity of anthracyclines in doxorubicin (Adriamycin)-resistant P388 mouse leukemia cells. Cancer Res 44: 5056–5061.

Ganapathi R, Grabowski D and Schmidt H (1986a) Factors governing the modulation of vinca-alkaloid resistance in doxorubicin-resistant cells by the calmodulin inhibitor trifluoperazine. Biochem Pharmacol 35: 673–678.

Ganapathi R, Kamath N, Constantinou A, Grabowski D, Ford J and Anderson A (1991a) Effect of the calmodulin inhibitor trifluoperazine on phosphorylation of P-glycoprotein and topoisomerase II: relationship to modulation of subcellular distribution, DNA damage and cytotoxicity of doxorubicin in multidrug resistant L1210 mouse leukemia cells. Biochem Pharmacol 41: R21–R26.

Ganapathi R, Kuo T, Teeter L, Grabowski D and Ford J (1991b) Relationship between expression of P-glycoprotein and efficacy of trifluoperazine in multidrug-resistant cells. Mol Pharmacol 39: 1–8.

Ganapathi R, Yen A, Grabowski D, Schmidt H, Turinic R and Valenzuela R (1986b) Role of the calmodulin inhibitor trifluoperazine on the induction and expression of cell cycle traverse perturbations and cytotoxicity of daunorubicin and doxorubicin (Adriamycin) in doxorubicin-resistant P388 mouse leukemia cells. Br J Cancer 53: 561–566.

Gaveriaux C, Boesch D, Boilsterli JJ, Bollinger P, Eberle MK, Hiestand P, Payne T, Traber R, Wenger R and Loor F (1989) Overcoming multidrug resistance in Chinese hamster ovary cells *in vitro* by cyclosporin A (Sandimmune) and non-immunosuppressive derivatives. Br J Cancer 60: 867–871.

Goasguen JE, Dossot JM, Fardel O, Le Mee F, Le Gall E, Leblay R, LePrise P, Chaperon J and Fauchet R (1993) Expression of the multidrug resistance-associated P-glycoprotein (P-170) in 59 cases of de novo acute lymphoblastic leukemia: prognostic implications. Blood 81: 2394–2398.

Goldberg H, Ling V, Wong PY and Skorecki K (1988) Reduced cyclosporin accumulation in multidrug-resistant cells. Biochem Biophys Res Commun 152: 552–558.

Goldsmith ME, Madden MJ, Morrow CS and Cowan KH (1993) A Y-box consensus sequence is required for basal expression of the human multidrug resistance (*mdr*1) gene. J Biol Chem 268: 5856–5860.

Goldstein LJ, Galski H, Fojo A, Willingham M, Lai S-L, Gazdar A, Pirker R, Green A, Crist W, Brodeur GM, Lieber M, Cossman J, Gottesman MM and Pastan I (1989) Expression of a multidrug resistance gene in human cancers. J Natl Cancer Inst 81: 116–124.

Goldstein LJ, Pastan I and Gottesman MM (1992) Multidrug resistance in human cancer. Crit Rev Oncol Hematol 12: 243–253.

Gollapudi S, Patel K, Jain V and Gupta S (1992) Protein kinase C isoforms in multidrug resistant P388/ADR cells: a possible role in daunorubicin transport. Cancer Lett 62: 69–75.

Gosland M, Brophy N, Duran G, Yahanda AM, Adler KM, Hardy RI and Sikic BI (1991) Bilirubin: a physiological substrate for the multidrug transporter. Proc Am Assoc Cancer Res 32: 426.

Gosland MP, Lum BL and Sikic BI (1989) Reversal by cefoperazone of resistance to etoposide, doxorubicin, and vinblastine in multidrug resistant human sarcoma cells. Cancer Res 49: 6901–6905.

Gosland MP, Vore M, Goodin S and Tsuboi C (1993) Estradiol-17β-(β-D-glucuronide): a cholestatic organic anion and substrate for the multidrug resistance transporter. Proc Am Assoc Cancer Res 34: 1842.

Grem JL and Fischer PH (1985) Augmentation of 5-flourouracil cytotoxicity in human colon cancer cells by dipyridamole. Cancer Res 45: 2967–2972.

Gros P, Croop J and Housman D (1986a) Mammalian multidrug resistance gene: complete cDNA sequence indicates strong homology to bacterial transport proteins. Cell 47: 371–380

Gros P, Neriah BY, Croop JM and Housman DE (1986b) Isolation and expression of a complementary DNA that confers multidrug resistance. Nature 323: 728–731.

Gruber A, Peterson C and Reizenstein P (1988) D-verapamil and L-verapamil are equally effective in increasing vincristine accumulation in leukemic cells *in vitro* Int J Cancer 41 224–226

Hait WN and Aftab DT (1992) Rational design and preclinical pharmacology of drugs for reversing multidrug resistance Biochem Pharmacol 43 103–107

Hait WN, Choudhury S, Srimatkandada S and Murren, JR (1993a) Sensitivity of K562 human chronic myelogenous leukemia blast cells transfected with a human multidrug resistance cDNA to cytotoxic drugs and differentiating agents. J Clin Invest 31: 2207–2215

Hait WN and DeRosa WT (1991) The role of the phorbol ester receptor/protein kinase C in the sensitivity of leukemic cells to Anthracyclines. Cancer Commun 3: 77–81.

Hait WN, Gesmonde JF, Murren JR, Yang JM, Chen HX and Reiss M (1993b) Terfenadine (Seldane) a new drug for restoring sensitivity to multidrug resistant cancer cells. Biochem Pharmacol 45: 401–406.

Hait WN, Harding MW and Handschumacher RE (1986) Calmodulin, cyclophilin and cyclosporin A Science 233 987–988

Hait WN and Lazo JS (1986) Calmodulin. a potential target for cancer chemotherapeutic agents J Clin Oncol 4: 994–1012

Hait WN, Morris S, Lazo JS, Figlin RJ, Durivage HJ, White K and Schwartz PE (1989a) Phase I trial of combined therapy with bleomycin and the calmodulin antagonist, trifluoperazine Cancer Chemother Pharmacol 23. 358–362.

Hait WN and Pierson NH (1990) Comparison of the efficacy of a phenothiazine and bisquinaldinium calmodulin antagonist against multidrug resistant cell lines Cancer Res 50 1165–1169

Hait WN, Stein JM, Koletsky AJ, Harding MW and Handschumacher RE (1989b) Activity of cyclosporin A and a non-immunosuppressive cyclosporin on multidrug resistant leukemic cell lines. Cancer Communications 1: 35–43.

Hait WN, Stein JM, Koletsky AJ, Slater LM, Harding MW and Handschumacher RE (1987) Modulation of doxorubicin (DOX) resistance by cyclosporin A (CsA) and a nonimmunosuppresive homolog. Proc Am Acad Cancer Res 28: 298.

Hait WN, Yang J and Aftab D (1993c) Effect of activation of protein kinase C on Vinca alkaloid transport in multidrug resistant cells. Proc Am Assoc Cancer Res 34: 1882.

Hamada H, Hagiwara K-I, Nakajima T and Tsuruo T (1987) Phosphorylation of the M_r 170,000 to 180,000 glycoprotein specific to multidrug-resistant tumor cells: effects of verapamil, trifluoperazine, and phorbol esters. Cancer Res 47: 2860–2865.

Hamada H and Tsuruo T (1986) Functional role for the 170- to 180-kDa glycoprotein specific to drug-resistant tumor cells as revealed by monoclonal antibodies. Proc Natl Acad Sci USA 83: 7785–7789.

Hamada H and Tsuruo T (1988a) Characterization of the ATPase activity of the M_r 170,000 to 180,000 membrane glycoprotein (P-glycoprotein) associated with multidrug resistance in K562/ADM cells. Cancer Res 48: 4926–4932.

Hamada H and Tsuruo T (1988b) Purification of the 170- to 180-kilodalton membrane glycoprotein associated with multidrug resistance: 170- to 180-kilodalton membrane glycoprotein is an ATPase. J Biol Chem 263: 1454–1458.

Handschumacher RE, Harding MW, Rice J, Drugge RJ and Speicher DW (1984) Cyclophilin: a specific cytosolic binding protein for cyclosporin A. Science 226. 544–547.

Harker WG, Bauer D, Etiz BB, Newman RA and Sikic BI (1986) Verapamil-mediated sensitization of doxorubicin-selected pleiotropic resistance in human sarcoma cells: selectivity for drugs which produce DNA scission. Cancer Res 46: 2369–2373.

Hegmann EJ, Bauer HC and Kerbel RS (1992) Expression and functional activity of P-glycoprotein in cultured cerebral capillary endothelial cells. Cancer Res 52: 6969–6975.

Helson L (1984) Calcium channel blocker enhancement of anticancer drug cytotoxicity – a review. Cancer Drug Delivery 1 353–361.

Higgins CF, Haag PD, Nikaido K, Ardeshir F, Garcia G and Ames GF (1982) Complete nucleotide sequence and identification of membrane components of the histidine transport operon of *S typhimurium*. Nature 298: 723–727.

Higgins CF, Hiles ID, Salmond GPC, Gill DR, Downie JA, Evans IJ, Holland IB, Gray L, Buckel SD, Bell AW and Hermodson MA (1986) A family of related ATP-binding subunits coupled to many distinct biological processes in bacteria. Nature 323: 448–452.

Hill BT, Deuchars K, Hosking LK, Ling V and Whelan KDH (1990) Overexpression of P-glycoprotein in mammalian tumor cell lines after fractionated X-irradiation in vitro. J Natl Cancer Inst 82 607–612.

Hofsli E and Nissen-Meyer J (1989a) Effect of erythromycin and tumour necrosis factor on the drug resistance of multidrug-resistant cells reversal of drug resistance by erythromycin. Int J Cancer 43: 520–525.

Hofsli E and Nissen-Meyer J (1989b) Reversal of drug resistance by erythromycin: erythromycin increases the accumulation of actinomycin D and doxorubicin in multidrug-resistant cells. Int J Cancer 44: 149–154.

Hollt V, Kouba M, Dietel M and Vogt G (1992) Stereoisomers of calcium antagonists which differ markedly in their potencies as calcium blockers are equally effective in modulating drug transport by P-glycoprotein. Biochem Pharmacol 43: 2601–2608.

Holzmayer TA, Hilsenbeck S, Von Hoff DD and Roninson IB (1992) Clinical correlates of MDR1 (P-glycoprotein) gene expression in ovarian and small-cell lung carcinomas. J Natl Cancer Inst 84: 1486–1491.

Howell SB, Hom D, Sanga R, Vick JS and Abramson IS (1989a) Comparison of the synergistic potentiation of etoposide, doxorubicin and vinblastine cytotoxicity by dipyridamole. Cancer Res 49: 3178–3183.

Howell SB, Hom DK, Sanga R, Vick JS and Chan TCK (1989b) Dipyridamole enhancement of etoposide sensitivity. Cancer Res 49: 4147–4153.

Hu XF, Martin TJ, Bell DR, Luise M and Zalcberg JR (1990) Combined use of cyclosporin A and verapamil in modulating multidrug resistance in human leukemia cell lines. Cancer Res 50: 2953–2957.

Huff RM and Molinoff B (1984) Assay of dopamine receptors with [α-3H] flupenthixol. J Pharmacol Exp Ther 232: 57–61.

Hyttel J, Arnt J and Bogeso KP (1984) Antipsychotic drugs: configurational stereoisomers. In: Smith DF (ed) CRC Handbook of Stereoisomers: Drugs in Psychopharmacology. (pp. 143–180) Boca Raton: CRC Press, Inc.

Ido M, Asao T, Sakurai M, Inagaki M, Saito M and Hidaka H (1986) An inhibitor of protein kinase C, 1-(5-isoquinolinylsulfonyl)-2-methylpiperazine (H-7) inhibits TPA-induced reduction of vincristine uptake from P388 murine leukemic cell. Leukemia Res 10: 1063–1069.

Inaba M and Maruyama E (1988) Reversal of resistance to vincristine in P388 leukemia by various polycyclic clinical drugs, with a special emphasis on quinacrine. Cancer Res 48: 2064–2067.

Inaba M and Nagashima K (1986) Non-antitumor vinca alkaloids reverse multidrug in P388 leukemia cells in vitro. Jpn J Cancer Res (Gann) 77: 197–204.

Inaba M, Nagashima K, Sakurai Y, Fukui M and Yanagi Y (1984) Reversal of multidrug resistance by non-antitumor anthracycline analogs. Gann 75: 1049–1052.

Ito Y, Tanimoto M, Kumazawa T, Okumura M, Morishima Y, Ohno R and Saito H (1989) Increased P-glycoprotein expression and multidrug-resistant gene (*mdr*1) amplification are infrequently found in fresh acute leukemia cells. Cancer 63: 1534–1538.

Jachez B, Nordmann R and Loor F (1993) Restoration of taxol sensitivity of multidrug-resistant cells by the cyclosporine SDZ PSC 833 and the cyclopeptolide SDZ 280–446. J Natl Cancer Inst 85: 478–483.

Jancis EM, Carbone R, Loechner KJ and Dannies PS (1993) Estradiol induction of rhodamine 123 efflux and the multidrug resistance pump in rat pituitary tumor cells. Mol Pharmacol 43: 51–56.

Jekunen A, Vick J, Sanga R, Chan TCK and Howell SB (1992) Synergism between dipyridamole and cisplatin in human ovarian carcinoma cells *in vitro*. Cancer Res 52: 3566–3571.

Johnstone EC, Crow TJ, Frith CD, Carney MWD and Price JS (1978) Mechanism of the antipsychotic effect in the treatment of acute schizophrenia. Lancet 1: 848–851.

Jordan VC, Bain RR, Brown RR, Gosden B and Santos MA (1983) Determination and pharmacology of a new hydroxylated metabolite of tamoxifen observed in patient sera during therapy for advanced breast cancer. Cancer Res 43: 1446–1450.

Jorgensen A, Andersen J, Bjorndal N, Dencker SJ and Lundin L (1982) Serum concentrations of cis(Z)-flupenthixol and prolactin in chronic schizophrenic patients treated with flupenthixol and cis(Z)-flupenthixol decanoate. Psychopharmacology 77: 58–65.

Juliano RL and Ling V (1976) A surface glycoprotein modulating drug permeability in Chinese hamster ovary cell mutants. Biochim Biophys Acta 455: 152–162.

Kacinski BM, Yee LD and Carter D (1989a) Quantitation of tumor cell expression of the P-glycoprotein (*mdr*1) gene in human breast carcinoma clinical specimens. Cancer Bull 41: 44–48.

Kacinski BM, Yee LD, Carter D, Li D and Kuo MT (1989b) Human breast carcinoma cell levels of MDR-1 (P-glycoprotein) correlate *in vivo* inversely and reciprocally with tumor progesterone receptor content. Cancer Communications 1: 1–6.

Kadam S, Maus M, Poddig J, Schmidt S, Tasmussen R, Novosad E, Plattner J and McAlpine J (1992) Reversal of multidrug resistance by two novel indole derivatives. Cancer Res 52: 4735–4740.

Kavallaris M, Madafiglio J, Norris MD and Haber M (1993) Resistance to tetracycline, a hydrophilic antibiotic, is mediated by P-glycoprotein in human multidrug-resistant cells. Biochem Biophys Res Comm 190: 79–85.

Kawada M, Sumi S and Umezawa K (1992) Circumvention of multidrug resistance in human carcinoma KB cells by polyether antibiotics. J Antibiotics 45: 556–562.

Kemp BE and Pearson RB (1990) Protein kinase recognition sequence motifs. Trends Biochem Sci 15: 342–346.

Kennelly PJ and Krebs EG (1991) Consensus sequences as substrate specificity determinants for protein kinases and protein phosphatases. J Biol Chem 266: 15555–15558.

Kessel D and Wilberding C (1985a) Anthracycline resistance in P388 murine leukemia and its circumvention by calcium antagonists. Cancer Res 45: 1687–1691.

Kessel D and Wilberding C (1985b) Promotion of daunorubicin uptake and toxicity by the calcium antagonist tiapamil and its analogs. Cancer Treatment Rep 69: 673–676.

Kloke O and Osieka R (1985) Interaction of cyclosporin A with antineoplastic agents. Klin Wochenschr 63: 1081–1082.

Kramer RA, Zakher J and Kim G (1988) Role of the glutathione redox cycle in acquired and de novo multidrug resistance. Science 241: 694–697.

Kusumoto H, Maehara Y, Anai H, Kusumoto T and Sugimachi K (1988) Potentiation of Adriamycin cytotoxicity by dipyridamole against HeLa cells *in vitro*. Cancer Res 48: 1208–1212.

Larsen AK and Jacquemin-Sablon A (1989) Multiple resistance mechanisms in Chinese hamster cells resistant to 9-hydroxyellipticine. Cancer Res 49: 7115–7119.

Larsson R, Bergh J and Nygren P (1991) Combination of cyclosporin A and buthionine sulfoximine as a pharmacological strategy for circumvention of multidrug resistance in small cell lung cancer cell lines selected for resistance to doxorubicin. Anticancer Res 11: 455–459.

Lazo JS, Hait WN, Kennedy KA, Braun D and Meandzija B (1985) Enhanced bleomycin-induced DNA damage and cytotoxicity with calmodulin antagonists. Mol Pharmacol 27: 387–393.

Legrue SJ, Turner R, Weisbrodt N and Dedman JR (1986) Does the binding of cyclosporine to calmodulin result in immunosuppression? Science 234: 68–71.

Lehnert M, Dalton WS, Roe D, Emerson S and Salmon SE (1991) Synergistic inhibition by verapamil and quinine of P-glycoprotein-mediated multidrug resistance in a human myeloma cell line model. Blood 77: 348–354.

Leonce S, Pierre A, Anstett M, Perez V, Genton A, Bizzari FP and Atassi G (1992) Effects of a new triazinoaminopiperidine derivative on Adriamycin accumulation and retention in cells displaying P-glycoprotein-mediated multidrug resistance. Biochem Pharmacol 44: 1707–1715.

Lieberman DM, Reithmeier RAF, Ling V, Charuk JHM, Goldberg H and Skorecki KL (1989) Identification of P-glycoprotein in renal brush border membranes. Bioch Biophys Res Commun 162: 244–252.

Lum BL, Kaubisch S, Yahanda AM, Adler KM, Jew L, Ehsan MN, Brophy NA, Halsey J, Gosland MP and Sikic BI (1992) Alteration of etoposide pharmacokinetics and pharmacodynamics by cyclosporine in a phase I trial to modulate multidrug resistance. J Clin Oncol 10: 1635–1642.

Ma L, Marquardt D, Takemoto L and Center MS (1991) Analysis of P-glycoprotein phosphorylation in HL-60 cells isolated for resistance to vincristine. J Biol Chem 266: 5593–5599.

Madden MJ, Morrow CS, Nakagawa M, Goldsmith ME, Fairchild CR and Cowan KH (1993) Identification of 5' and 3' sequences involved in the regulation of transcription of the human *mdr*1 gene *in vivo*. J Biol Chem 268: 8290–8297.

Maeda O, Terasawa M, Ishikawa T, Oguchi H, Mizuno K, Kawai M, Kikkawa F, Tomoda Y and Hidaka H (1993) A newly synthesized bifunctional inhibitor, W-77, enhances Adriamycin activity against human ovarian carcinoma cells. Cancer Res 53: 2051–2056.

Marie JP, Zittoun R and Sikic BI (1991) Multidrug resistance (*mdr*1) gene expression on adult acute leukemias: correlations with treatment outcome and in vitro drug sensitivity. Blood 78: 586–592.

McGrath JP and Varshavsky A (1989) The yeast *STE6* gene encodes a homologue of the mammalian multidrug resistance P-glycoprotein. Nature 340: 400–404.

Mechetner EB and Roninson IB (1992) Efficient inhibition of P-glycoprotein-mediated multidrug resistance with a monoclonal antibody. Proc Natl Acad Sci USA 89: 5824–5828.

Mellado W and Horwitz SB (1987) Phosphorylation of the multidrug resistance associated glycoprotein. Biochemistry 26: 6900–6904.

Meyers MB (1989) Protein phosphorylation in mulstidrug resistant Chinese hamster cells. Cancer Commun 1: 233–241.

Mickisch GH, Pai LH, Gottesman MM and Pastan I (1992) Monoclonal antibody MRK16 reverses the multidrug reistance of multidrug-resistant transgenic mice. Cancer Res 52: 4427–4432.

Mickley LA, Bates SE, Richert ND, Currier S, Tanaka S, Foss F, Rosen N and Fojo AT (1989) Modulation of the expression of a multidrug resistance gene by differentiating agents. J Biol Chem 264: 18031–18040.

Miller RL, Bukowski RM, Budd GT, Purvis J, Weick JK, Shepard K, Midha KK and Ganapathi R (1988) Clinical modulation of doxorubicin resistance by the calmodulin-inhibitor, trifluoperazine: a phase I/II trial. J Clin Oncol 6: 880–888.

Miller TP, Grogan TM, Dalton WS, Spier CM, Scheper RJ and Salmon SE (1991) P-glycoprotein expression in malignant lymphoma and reversal of clinical drug resistance with chemotherapy plus high-dose verapamil. J Clin Oncol 9: 17–24.

Millward MJ, Cantwell BMJ, Lien EA, Carmichael J and Harris AL (1992) Intermitent high-dose tamoxifen as a potential modifier of multidrug resistance. Eur J Cancer 28A: 805–-810.

Mori T, Takai Y, Minakuchi R, Yu B and Nishizuka Y (1980) Inhibitory action of chlorpromazine, dibucaine, and other phospholipid-interacting drugs on calcium-activated, phospholipid-dependent protein kinase. J Biol Chem 255: 8378–8380.

Naito M, Hamada H and Tsuruo T (1988) ATP/Mg^{2+}-dependent binding of vincristine to the plasma membrane of multidrug-resistant K562 cells. J Biol Chem 263: 11887–11891.

Naito M, Oh-hara T, Yamazaki A, Danki T and Tsuruo T (1992) Reversal of multidrug resistance by an immunosuppressive agent FK-506. Cancer Chemother Pharmacol 29: 195–200.

Naito M, Tsuge H, Kuroko C, Koyama T, Tomida A, Tatsuta T, Heike Y and Tsuruo T (1993) Enhancement of cellular accumulation of cyclosporine by anti-P-glycoprotein monoclonal antibody MRK-16 and synergistic modulation of multidrug resistance. J Natl Cancer Inst 85: 311–316.

Naito M and Tsuruo, T (1989) Competitive inhibition by verapamil of ATP-dependent high affinity vincristine binding to the plasma membrane of multidrug-resistant K562 cells without calcium ion involvement. Cancer Res 49: 1452–1455.

Naito M, Yusa K and Tsurou T (1989) Steroid hormones inhibit binding of *Vinca* alkaloid to multidrug reistance related P-glycoprotein. Biochem Biophys Res Commun 158: 1066–1071.

Nakagawa M, Akiyama S, Yamaguchi T, Shiraishi N, Ogata J and Kuwano M (1986) Reversal of multidrug resistance by synthetic isoprenoids in the KB human cancer cell line. Cancer Res 46: 4453–4457.

Nelson JA and Drake S (1984) Potentiation of methotrexate toxicity by dipyridamole. Cancer Res 44· 2493–2496

Nielsen IM, Pedersen V, Nymark M, Franch KF, Boeck V, Fjalland B and Christensen AV (1973) Comparative pharmacology of flupenthixol and some reference neuroleptics. Acta Pharmacol Toxicol (Copenh. 33: 353–362).

Nielson IM, Hougs W, Lassen N, Holm T and Petersen PV (1962) Central depressant activity of some thiaxanthene derivatives. Acta Pharmacol Toxicol 19. 87–100.

Nishizuka Y (1988) The molecular heterogeneity of protein kinase C and its implications for cellular regulation. Nature 334: 661–665.

Niwa K, Yamada K, Furukawa T, Shudo N, Seto K, Matsumoto T, Takao S, Akiyama S and Shimazu H (1992) Effect of a dihydropyridine analogue, 2-[benzyl(phenyl)amino]ethyl-1,4-dihydro-2,6-dimethyl-5-(5,5-dimethyl-2-oxo-1,3,2-dioxaphosphorinan-2-yl)-1-(2-morpholino-ethyl)-4-(3-nitrophenyl)-3-pyridinecarboxylate on reversing *in vivo* resistance of tumor cells to adriamycin Cancer Res 52. 3655–3660.

Nogai I, Kohno K, Kikuchi J, Kuwano M, Akiyama S, Kiue A, Suzuki K, Yoshida Y, Cornwell MM, Pastan I and Gottesman MM (1989) Analysis of structural features of dihydropyridine analogs needed to reverse multidrug resistance and to inhibit photoaffinity labeling of P-glycoprotein. Biochem Pharmacol 38· 519–527.

Nooter K, Oostrum R, Jonker R, Van Dekken H, Stokdijk W and Van Den Engh G (1989) Effect of cyclosporin A on daunorubicin accumulation in multidrug-resistant P388 leukemia cells measured by real-time flow cytometry. Cancer Chemother Pharmacol 23. 296–300.

O'Brian CA, Fan C, Ward NE, Dong Z, Iwamoto L, Gupta KP, Earnest LE and Fidler IJ (1991a) Transient enhancement of multidrug resistance by the bile acid deoxycholate in murine fibrosarcoma cells *in vitro*. Biochem Pharmacol 41. 797–806

O'Brian CA, Fan D, Ward NE, Seid C and Fidler IJ (1989) Level of protein kinase C activity correlates directly with resistance to Adriamycin in murine fibrosarcoma cells. FEBS Lett 246· 78–82

O'Brian CA, Liskamp RM, Solomon DH and Weinstein IB (1985) Inhibition of protein kinase C by tamoxifen. Cancer Res 45: 2462–2465.

O'Brian CA, Ward NE, Liskamp RM, De Bont DB, Earnest LE, Van Boom JH and Fan D (1991b) A novel N-myristylated synthetic octapeptide inhibits protein kinase C activity and partially reverses murine fibrosarcoma cell resistance to Adriamycin. Invest New Drugs 9 169–179.

Ogura M, Takatori T and Tsuruo T (1992) Purification and characterization of NF-R1 that regulates the expression of the human multidrug resistance (MDR1) gene. Nucleic Acids Res 20. 5811–5817.

Ono Y, Fujii T, Ogita K, Kikkawa U, Igarashi K and Nishizuka Y (1988) The stucture, expression, and properties of additional members of the protein kinase C family J Biol Chem 263: 6927–6932.

Ono Y, Fujii T, Ogita K, Kikkawa U, Igarashi K and Nishizuka Y (1989) Protein kinase C ζ subspecies from rat brain: its structure, expression and properties. Proc Natl Acad Sci 86: 3099–3103.

Osada S, Mizuno K, Saido TC, Akita Y, Suzuki K, Kuroki T and Ohno S (1990) A phorbol ester receptor/protein kinase, nPKCη, a new member of the protein kinase C family predominantly expressed in lung and skin. J Biol Chem 265. 22434–22440.

Osann K, Sweet P and Slater LM (1992) Synergistic interaction of cyclosporin A and verapamil on vincristine and daunorubicin resistance in multidrug-resistant human leukemia cells in vitro. Cancer Chemother Pharmacol 30: 152–154.

Osieka R, Seeber S, Pannenbacker R, Soll D, Glatte P and Schmidt CG (1986) Enhancement of etoposide-induced cytotoxicity by cyclosporin A. Cancer Chemother Pharmacol 18: 198–202.

Owellen RJ, Hartke CA, Dickerson RM and Hains FO (1974) Inhibition of tubulin-microtubule polymerization by drugs of the *Vinca* alkaloid class. Cancer Res 36: 1499–1502.

Ozols RF, Cunnion RE, Klecker RW, Hamilton TC, Ostchega Y, Parrillo JE and Young RC (1987) Verapamil and Adriamycin in the treatment of drug-resistant ovarian cancer patients. J Clin Oncol 5: 641–647.

Palayoor ST, Stein JM and Hait WN (1987) Inhibition of protein kinase C by antineoplastic agents: implications for drug resistance. Biochem Biophys Res Commun 148. 718–725.

Pang DC and Briggs FN (1976) Mechanism of quinidine and chlorpromazine inhibition of sarcotubular ATPase activity. Biochem Pharmacol 25 21–25

Pearce HL, Safa AR, Bach NJ, Winter MA, Cirtain MC and Beck WT (1989) Essential features of the P-glycoprotein pharmacophore as defined by a series of reserpine analogs that modulate multidrug resistance. Proc Natl Acad Sci USA 86 5128–5132

Pearson JW, Fogler WE, Volker K, Usui N, Goldenberg SK, Gruys E, Riggs CW, Domschlies D, Wiltrout RH, Tsuruo T, Pastan I, Gottesman MM and Longo DL (1991) Reversal of drug resistance in a human colon cancer xenograft expressing MDR1 complementary DNA by in vivo administration of MRK-16 monoclonal antibody. J Natl Cancer Inst 83· 1386–1391

Pennock GD, Dalton WS, Roeske WR, Appleton CP, Mosley K, Plezia P, Miller TP and Salmon SE (1991) Systemic toxic effects associated with high-dose verapamil infusion and chemotherapy administration. J Natl Cancer Inst 83· 105–110.

Perrella FW, Hellmig BD and Diamond L (1986) Up regulation of the phorbol ester receptor-protein kinase C in HL-60 variant cells Cancer Res 46 567–572

Piekarz RL, Cohen D and Horwitz SB (1993) Progesterone regulates the murine multidrug resistance *mdr*1b gene. J Biol Chem 268 7613–7616.

Pierre A, Dunn TA, Kraus-Berthier L, Leonce S, Saint-Dizier D, Regnier G, Dhainaut A, Berlion M, Bizzari JP and Atassi G (1992) *In vitro* and *in vivo* circumvention of multidrug resistance by Servier 9788, a novel triazinoaminopiperidine derivative. Investigational New Drugs 10: 137–148.

Pirker R, Keilhauer G, Raschack M, Lechner C and Ludwig H (1990) Reversal of multidrug resistance in human KB cell lines by structural analogs of verapamil. Int J Cancer 45: 916–919.

Pirker R, Wallner J, Geissler K, Linkesh W, Haas OA, Bettelheim P, Hopfner M and Scherrer R (1991) MDR1 gene expression and treatment outcome in acute myeloid leukemia. J Natl Cancer Inst 83: 708–712.

Piwnica-Worms D, Chiu ML, Budding M, Kronauge JF, Kramer RA and Croop JM (1993) Functional imaging of multidrug-resistant P-glycoprotein with an organotechnetium complex. Cancer Res 53: 977–984.

Plumb JA, Milroy R and Kaye SB (1990) The activity of verapamil as a resistance modifier in vitro in drug resistant human tumor cell lines is not stereospecific. Biochem Pharmacol 39: 787–792.

Posada J, Vichi P and Tritton TR (1989a) Protein kinase C in Adriamycin action and resistance in mouse sarcoma 180 cells. Cancer Res 49: 6634–6639.

Posada JA, McKeegan EM, Worthington KF, Morin MJ, Jaken S and Tritton TR (1989b) Human multidrug resistant KB cells overexpress protein kinase C: involvement in drug resistance. Cancer Commun 1: 285–292.

Post ML, Kennard U and Horn AS (1975) Stereoselective blockade of the dopamine receptor and the X-ray structures of alpha and beta-flupenthixol. Nature 256: 342–343.

Presant CA, Kennedy PS, Wiseman C, Gala K, Bouzaglou A, Wyres M and Naessig V (1986) Verapamil reversal of clinical doxorubicin resistance in human cancer. Am J Clin Oncol 9: 355–357.

Prozialeck WC and Weiss B (1982) Inhibition of calmodulin by phenothiazines and related drugs: structure-activity relationships. J Pharmacol Exp Ther 222: 509–516.

Qian X and Beck WT (1990) Binding of an optically pure photoaffinity analogue of verapamil, LU-49888, to P-glycoprotein from multidrug-resistant human leukemic cell lines. Cancer Res 50: 1132–1137.

Raess BU and Vincenzi FF (1980) Calmodulin activation of red blood cell (Ca^{2+} + Mg^{2+})-ATPase and its antagonism by phenothiazines. Mol Pharmacol 18: 253–258.

Ramu A (1989) Structure-activity relationship of compounds that restore sensitivity to doxorubicin in drug-resistant P388 cells. In: Kessel D (ed) Resistance to antineoplastic drugs. (pp. 63–80) Boca Raton: CRC Press, Inc.

Ramu A, Fuks Z, Gatt S and Glaubiger D (1984a) Reversal of acquired resistance to doxorubicin in P388 murine leukemia cells by perhexiline maleate. Cancer Res 44: 144–148.

Ramu A, Glaubiger D and Fuks Z (1984b) Reversal of acquired resistance to doxorubicin in P388 murine leukemia cells by tamoxifen and other triparanol analogues. Cancer Res 44: 4392–4395.

Ramu A and Ramu N (1992) Reversal of multidrug resistance by phenothiazines and structurally related compounds. Cancer Chemother Pharmacol 30: 165–173.

Ramu A, Spanier R, Rahamimoff H and Fuks Z (1984) Restoration of doxorubicin responsiveness in doxorubicin-resistant P388 murine leukaemia cells. Br J Cancer 50: 501–507.

Ramu N and Ramu A (1989) Circumvention of Adriamycin resistance by dipyridamole analogues: a structure-activity relationship study. Int J Cancer 43: 487–491.

Reid RE (1983) Drug interactions with calmodulin: the binding site. J Theor Biol 105: 63–76.

Riordan JR and Ling V (1985) Genetic and biochemical characterization of multidrug resistance. Pharmacol Ther 28: 51–75.

Rischin D and Ling V (1993) Multidrug resistance in leukemia. Cancer Treatment and Research 64: 269–293.

Rittmann-Grauer LS, Yong MA, Sanders V and Mackensen DG (1992) Reversal of *Vinca* alkaloid resistance by anti-P-glycoprotein monoclonal antibody HYB-241 in a human tumor xenograft. Cancer Res 52: 1810–1816.

Rohlff C, Clair T and Cho-Chung YS (1993a) 8-Cl-cAMP induces truncation and down-regulation of the RI-alpha subunit and up-regulation of the RII-beta subunit of cAMP-dependent protein kinase leading to type II holoenzyme-dependent growth inhibition and differentiation of HL-60 leukemia cells. J Biol Chem 268: 5774–5782.

Rohlff C, Safa B, Rahman A, Cho-Chung YS, Klecker RW and Glazer GI (1993b) Reversal of resistance to Adriamycin by 8-chloro-cyclic AMP in Adriamycin-resistant HL-60 leukemia cells is associated with reduction of type I cyclic AMP-dependent protein kinase and cyclic AMP response element-binding protein DNA-binding activities. Mol Pharmacol 43: 372–379.

Roninson IB (1992) The role of the *MDR*1 (P-glycoprotein) gene in multidrug resistance *in vitro* and *in vivo*. Biochem Pharmacol 43: 95–102.

Rothenberg ML, Mickley LA, Cole DE, Balis FM, Tsuruo T, Poplack DG and Fojo AT (1989) Expression of the *mdr*-1/P-170 gene in patients with acute lymphoblastic leukemia. Blood 74: 1388–1395.

Ruben L and Rasmussen H (1981) Phenothiazines and related compounds disrupt mitochondrial energy production by a calmodulin-independent reaction. Biochim Biophys Acta 637: 415–422.

Saeki T, Ueda K, Tanigawara Y, Hori R and Komano T (1993) Human P-glycoprotein transports cyclosporin A and FK506. J Biol Chem 268: 6077–6080.

Safa AR (1988a) Inhibition of azidopine binding to the multidrug resistance related gp 150–180 (P-glycoprotein) by modulators of multidrug resistance. Proc Am Assoc Cancer Res 29: 1160.

Safa AR (1988b) Photoaffinity labeling of the multidrug-resistance-related P-glycoprotein with photoactive analogs of verapamil. Proc Natl Acad Sci USA 85: 7187–7191.

Safa AR, Glover CJ, Sewell JL, Meyers MB, Biedler JL and Felsted RL (1987) Identification of the multidrug resistance-related membrane glycoprotein as an acceptor for calcium channel blockers. J Biol Chem 262: 7884–7888.

Safa AR, Mehta ND and Agresti M (1989) Photoaffinity labeling of P-glycoprotein in multidrug resistant cells with photoactive analogs of colchicine. Biochem Biophys Res

Comm 162: 1402–1408.

Salmon SE, Dalton WS, Grogan TM, Plezia P, Lehnert M, Roe DJ and Miller TP (1991) Multidrug-resistant myeloma: laboratory and clinical effects of verapamil as a chemosensitizer. Blood 78: 44–50.

Salmon SE, Grogan TM, Miller T, Scheper R and Dalton WS (1989) Prediction of doxorubicin resistance *in vitro* in myeloma, lymphoma, and breast cancer by P-glycoprotein staining. J Natl Cancer Inst 81: 696–701.

Samuels BL and Ratain MJ (1991) Letter to the editor. Cancer Res 51: 1749.

Sato W, Fukazawa N, Suzuki T, Yusa K and Tsuruo T (1991) Circumvention of multidrug resistance by a newly synthesized quinoline derivative, MS-073. Cancer Res 41: 2420–2424.

Schatzman RC, Wise BC and Kuo JF (1981) Phospholipid-sensitive calcium-dependent protein kinase: inhibition by antipsychotic drugs. Biochem Biophys Res Commun 98: 669–676.

Scheper RJ, Broxterman HJ, Scheffer GL, Kaajk P, Dalton WS, Van Heijningen THM, Van Kalken CK, Slovak ML, De Vries EGE, Van der Valk P, Meijer CJLM and Pinedo HM (1993) Overexpression of a M_r 110,000 vesicular protein in non-P-glycoprotein-mediated multidrug resistance. Cancer Res 53: 1475–1479.

Schreiber SL (1991) Chemistry and biology of the immunophilins and their immunosuppressive ligands. Science 251: 283–287.

Schreiber SL and Crabtree GR (1992) The mechanism of action of cyclosporin and FK506. Immunology Today 13: 136–142.

Sehested M, Bindslev N, Demant EJF, Slovsgaard T and Jensen PB (1989) Daunorubicin and vincristine vinding to plasma membrane vesicles from daunorubicin-resistant and wild type Ehrlich ascites tumor cells. Biochem Pharmacol 38: 3017–3027.

Sehested M, Skovsgaard T, Jensen PB, Demant EJF, Friche E and Bindslev N (1990) Transport of the multidrug resistance modulators verapamil and azidopine in wild type and daunorubicin resistant Ehrlich ascites tumour cells. Br J Cancer 62: 37–41.

Shalinsky DR, Andreeff M and Howell SB (1990) Modulation of drug sensitivity by dipyridamole in multidrug resistant tumor cells *in vitro*. Cancer Res 50: 7537–7543.

Shalinsky DR, Slovak ML and Howell SB (1991) Modulation of vinblastine sensitivity by dipyridamole in multidrug resistant fibrosarcoma cells lacking *mdr*1 expression. Br J Cancer 64: 705–709.

Shiraishi N, Akiyama S, Nakagawa M and Kobayashi M (1987) Effect of bisbenzylisoquinoline (biscoclaurine) alkaloids on multidrug resistance in KB human cancer cells. Cancer Res 47: 2413–2416.

Sikic BI, Scudder SA and Evans TL (1989) Multidrug (pleiotropic) resistance in the human sarcoma cell line MES-SA. In: Kessel D (ed) Resistance to Antineoplastic Drugs. (pp. 37–47) Boca Raton: CRC Press.

Silbermann MH, Boersma AWM, Janssen ALW, Scheper RJ, Herweijer H and Nooter K (1989) Effects of cyclosporin A and verapamil on the intracellular daunorubicin accumulation in Chinese hamster ovary cells with increasing levels of drug-resistance. Int J Cancer 44: 722–726.

Skovsgaard T (1978) Mechanisms of resistance to daunorubicin in Ehrlich ascites tumour cells. Cancer Res 38: 1785–1791.

Skovsgaard T (1980) Circumvention of resistance to daunorubicin by *N*-acetyldaunorubicin in Ehrlich ascites tumor. Cancer Res 40: 1077–1083.

Slater LM, Sweet P, Stupecky M, Wetzel MW and Gupta S (1986) Cyclosporin A corrects daunorubicin resistance in Ehrlich ascites carcinoma. Br J Cancer 54: 235–238.

Solary E, Velay I, Chauffert B, Bidan JM, Caillot D, Dumas M and Guy H (1991) Sufficient levels of quinine in the serum circumvent the multidrug resistance of the human leukemic cell line K562/ADM. Cancer 68: 1714–1719.

Sonneveld P, Durie BGM, Lokhurst HM, Marie JP, Solbu B, Suciu S, Zittoun R, Lowenberg B and Nooter K (1992) Modulation of multidrug-resistant multiple myeloma by cyclosporin. Lancet 3: 40255–40259.

Sparano JA, Wiernik PH, Leaf A and Dutcher JP (1993) Infusional cyclophosphamide, doxorubicin, and etoposide in relapsed or resistant non-Hodgkin's lymphoma: evidence for a schedule-dependent effect favoring infusional administration of chemotherapy. J Clin Oncol 11: 1071–1079.

Spies T, Bresnahan M, Bahram S, Arnold D, Blanck G, Mellins E, Pious D and DeMars R (1990) A gene in the human major histocompatibility complex class II regions controlling the class I antigen presentation pathway. Nature 348: 744–747.

Sridhar KS, Krishan A, Samy TSA, Sauerteig A, Wellham LL, McPhee G, Duncan RC, Anac SY, Ardalan B and Benedetto PW (1993) Prochlorperazine as a doxorubicin-efflux blocker; phase I clinical and pharmacokinetic studies. Cancer Chemother Pharmacol 31: 423–430.

Sugawara I, Kataoka I, Morishita Y, Hamada H, Tsuruo T, Itoyama S and Mori S (1988a) Tissue distribution of P-glycoprotein encoded by a multidrug-resistant gene as revealed by a monoclonal antibody, MRK 16. Cancer Res 48: 1926–1929.

Sugawara I, Nakahama M, Hamada H, Tsuruo T and Mori S (1988b) Apparent stronger expression in the human adrenal cortex than in the human adrenal medulla of Mr 170,000–180,000 P-glycoprotein. Cancer Res 48: 4611–4614.

Tamai I and Safa A (1991) Azidopine noncompetively interacts with vinblastine and cyclosporin A binding to P-glycoprotein in multidrug resistant cells. J Biol Chem 266: 16796–16800.

Thiebaut F, Tsuruo T, Hamada H, Gottesman MM, Pastan I and Willingham MC (1989) Immunohistochemical localization in normal tissues of different epitopes in the multidrug transport protein P170: evidence for localization in brain capillaries and crossreactivity of one antibody with a muscle protein. J Histochem Cytochem 37: 159–164.

Thiebaut F, Tsuruo T, Hamada H, Gottesman MM, Pastan I and Willingham MD (1987) Cellular localization of the multidrug-resistance gene product P-glycoprotein in normal human tissues. Proc Natl Acad Sci USA 84: 7735–7738.

Trowsdale J, Hanson I, Mockridge I, Beck S, Townsend A and Kelly A (1990) Sequences encoded in the class II region of the MHC related to the "ABC" superfamily of transporters. Nature 348: 741–744.

Trump DL, Smith DC, Ellis PG, Rogers MP, Schold SC, Winer EP, Panella TJ, Jordan VC and Fine RL (1992) High-dose oral tamoxifen, a potential multidrug-resistance-reversal agent: phase I trial in combination with vinblastine. J Natl Cancer Inst 84: 1811–1816.

Tsuruo T (1989) Circumvention of drug resistance with calcium channel blockers and monoclonal antibodies. In: Ozols RF (ed) Drug Resistance in Cancer Therapy. (pp. 73–95). Boston: Kluwer Publisher.

Tsuruo T, Hamada H, Sata S and Heike Y (1987) Inhibition of multidrug-resistant human tumor growth in athymic mice by anti-P-glycoprotein monoclonal antibodies. Jpn J Cancer Res 80: 627–631.

Tsuruo T, Iida H, Kitatani Y, Yokota K, Tsukagoshi S and Yakurai Y (1984) Effects of quinidine and related compounds on cytotoxicity and cellular accumulation of vincristine and Adriamycin in drug-resistant tumor cells. Cancer Res 44: 4303–4307.

Tsuruo T, Iida H, Nojiri M, Tsukagoshi S and Sakurai Y (1983a) Circumvention of vincristine and Adriamycin resistance *in vitro* and *in vivo* by calcium influx blockers. Cancer Res 43: 2905–2910.

Tsuruo T, Iida H, Tsukagoshi S and Sakurai Y (1981) Overcoming of vincristine resistance in P388 leukemia *in vivo* and *in vitro* through enhanced cytotoxicity of vincristine and vinblastine by verapamil. Cancer Res 41: 1967–1972.

Tsuruo T, Iida H, Tsukagoshi S and Sakurai Y (1982) Increased accumulation of vincristine and Adriamycin in drug-resistant P388 tumor cells following incubation with calcium antagonists and calmodulin inhibitors. Cancer Res 42: 4730 - 4733.

Tsuruo T, Iida H, Tsukagoshi S and Sakurai Y (1983b) Potentiation of vincristine and Adriamycin in human hematopoietic tumor cell lines by calcium antagonists and calmodulin inhibitors. Cancer Res 43: 2267–2272.

Tsuruo T, Kawabata H, Nagumo N, Iida H, Kitatani Y, Tsukagoshi S and Sakurai Y (1985) Potentiation of antitumor agents by calcium channel blockers with special reference to cross-resistance patterns. Cancer Chemother Pharmacol 15: 16–19.

Twentyman PR (1988a) Modification of cytotoxic drug resistance by non-immuno-suppressive cyclosporins. Br J Cancer 57: 254–258.

Twentyman PR (1988b) A possible role for cyclosporins in cancer chemotherapy. Anticancer Res 8: 985–994.

Twentyman PR, Fox NE and White DJG (1987) Cyclosporin A and its analogues as modifiers of Adriamycin and vincristine resistance in a multidrug resistant human lung cancer cell line. Br J Cancer 56: 55–57.

Ueda K, Okamura N, Hirai M, Tanigawara Y, Saeki T, Kioka N, Komano T and Hori R (1992) Human P-glycoprotein transports cortisol, aldosterone, and dexamethasone, but not progesterone. J Biol Chem 267: 24248–24252.

Valverde MA, Diaz M, Sepulveda FV, Gill DR, Hyde SC and Higgins CF (1992) Volume-regulated chloride channels associated with the human multidrug resistance P-glycoprotein. Nature 355: 830–833.

Van der Graaf WTA, De Vries EGE, Uges DRA, Nanninga AG, Meijer C, Vellenga E, Mulder POM and Mulder NH (1991) *In vitro* and *in vivo* modulation of multidrug resistance with amiodarone. Int J Cancer 48: 616–622.

Velasquez WS, Cabanillas F, Salvador P, McLaughlin P, Fridrik M, Tucker S, Jagannath S, Hagemeister FB, Redman JR and Swan F (1988) Effective salvage therapy for lymphoma with cisplatin in combination with high-dose Ara-C and dexamethasone (DHAP). Blood 71: 117–122.

Walker RJ, Lazzaro VA, Duggin GG, Horvath JS and Tiller DJ (1989) Cyclosporin A inhibits protein kinase C activity: a contributing mechanism in the development of nephrotoxicity? Bioch. Biophys Res Comm 160: 409–415.

Weir MR, Peppler R, Gomolka D and Handwerger BS (1992) Evidence that the antiproliferative effect of verapamil on afferent and efferent immune responses is independent of calcium channel inhibition. Transplantation 54: 681–685.

Wiebe V, Koester S, Lindberg M, Emshoff V, Baker J, Wurz G and DeGregorio M (1992) Toremifene and its metabolites enhance doxorubicin accumulation in estrogen receptor negative multidrug resistant human breast cancer cells. Investigational New Drugs 10: 63–71.

Wilson WH, Bates S, Kang YK, Fojo A, Bryant G, Wittes R, Stevenson MA, Steinberg S and Chabner BA (1993a) Reversal of multidrug resistance with *R*-verapamil and analysis of *mdr*-1 expression in patients with lymphoma refractory to EPOCH chemotherapy. Proc Am Assoc Cancer Res 34: 1266.

Wilson WH, Bryant G, Bates S, Fojo A, Wittes RE, Steinberg SM, Kohler DR, Jaffe ES, Herdt J, Cheson BD and Chabner BA (1993b) EPOCH chemotherapy: toxicity and efficacy in relapsed and refractory non-Hodgkin's lymphoma. J Clin Oncol 11: 1573–1582.

Yahanda AM, Adler KM, Fisher GM, Brophy NA, Halsy J, Hardy RI, Gosland MP, Lum BL and Sikic BI (1992) Phase I trial of etoposide with cyclosporine as a modulator of multidrug resistance. J Clin Oncol 10: 1624–1634.

Yamaguchi T, Nakagawa M, Shiraishi N, Yoshida T, Kiyosue T, Arita M, Akiyama S and Kuwano M (1986) Overcoming drug resistance in cancer cells with synthetic isoprenoids. J Natl Cancer Inst 76: 947–953.

Yang C-PH, DePinho SH, Greenberger LM, Arceci RJ and Horwitz SB (1989) Progesterone interacts with P-glycoprotein in multidrug-resistant cells and in the endometrium of gravid uterus. J Biol Chem 264: 782–788.

Yang CH, Mellado W and Horwitz SB (1988) Azidopine photoaffinity labeling of multidrug resistance-associated glycoproteins. Biochem Pharmacol 37: 1417–1421.

Yang CPH, Cohen D, Greenberger LM, Hsu SH and Horwitz SB (1990) Differential transport properties of two mdr gene products are distinguished by progesterone. J Biol Chem 265: 10282–10288.

Yoshinari T, Iwasawa Y, Miura K, Takahashi IS, Fukuroda T, Suzuki K and Okura A (1989) Reversal of multidrug resistance by new dihydropyridines with lower calcium antagonistic activity. Cancer Chemother Pharmacol 24: 367–370.

Yu G, Ahmad S, Aquino A, Fairchild CR, Trepel JB, Ohno S, Suzuki K, Tsuruo T, Cowan KH and Glazer RI (1991) Transfection with protein kinase C α confers increased multidrug resistance to MCF-7 cells expressing P-glycoprotein. Cancer Commun 3: 181–189.

Yusa K and Tsuruo T (1989) Reversal mechanism of multidrug resistance by verapamil: direct binding of verapamil to P-glycoprotein on specific sites and transport of verapamil outward across the plasma membrane of K562/ADM cells. Cancer Res 49: 5002–5006.

Zamora JM and Beck WT (1986) Chloroquine enhancement of anticancer drug cytotoxicity in multiple drug resistant human leukemic cells. Biochem Pharmacol 35: 4303–4310.

Zamora JM, Pearce HL and Beck WT (1988) Physical-chemical properties shared by compounds that modulate multidrug resistance in human leukemic cells. Mol Pharmacol 33: 454–462.

Zunino F, Gambetta R, Di Marco A and Zaccara A (1972) Interaction of daunomycin and its derivatives with DNA. Biochim Biophys Acta 277: 489–498.

Address for offprints: James M. Ford, Division of Oncology, M-211, Stanford University Medical Center, Stanford, CA 94305, USA.

Cytotechnology **12**: 213–230, 1993.

Multidrug resistance (MDR) genes in haematological malignancies

K. Nooter[1] and P. Sonneveld[2]
[1]*Department of Medical Oncology, Rotterdam Cancer Institute, Rotterdam; and* [2]*Department of Haematology, University Hospital Dijkzigt, Rotterdam, The Netherlands*

Key words: circumvention of MDR, clinical trials, haematological malignancies, MDR, multidrug resistance, P-glycoprotein

Abstract

The emergence of drug resistant cells is one of the main obstacles for successful chemotherapeutic treatment of haematological malignancies. Most patients initially respond to chemotherapy at the time of first clinical admission, but often relapse and become refractory to further treatment not only to the drugs used in the first treatment but also to a variety of other drugs. Laboratory investigations have now provided a cellular basis for this clinical observation of multidrug resistance (MDR). Expression of a glycoprotein (referred to as P-glycoprotein) in the membrane of cells made resistant *in vitro* to naturally occurring anticancer agents like anthracyclines, Vinca alkaloids and epipodophyllotoxins, has been shown to be responsible for the so-called classical MDR phenotype. P-glycoprotein functions as an ATP-dependent, unidirectional drug efflux pump with a broad substrate specificity, that effectively maintains the intracellular cytotoxic drug concentrations under a non-cytotoxic threshold value. Extensive clinical studies have shown that P-glycoprotein is expressed on virtually all types of haematological malignancies, including acute and chronic leukaemias, multiple myelomas and malignant lymphomas. Since in model systems for P-glycoprotein-mediated MDR, drug resistance may be circumvented by the addition of non-cytotoxic agents that can inhibit the outward drug pump, clinical trials have been initiated to determine if such an approach will be feasible in a clinical situation. Preliminary results suggest that some haematological malignancies, among which are acute myelocytic leukaemia, multiple myeloma and non-Hodgkin's lymphoma, might benefit from the simultaneous administration of cytotoxic drugs and P-glycoprotein inhibitors. However, randomised clinical trials are needed to evaluate the use of such resistance modifiers in the clinic.

Abbreviations: ALL – acute lymphocytic leukaemia; AML – acute myelocytic leukaemia; BM – bone marrow; CAT – chloramphenicol acetyltransferase; CLL – chronic lymphocytic leukaemia; CML – chronic myelocytic leukaemia; CR – complete remission; HCL – hairy cell leukaemia; MDR – multidrug resistance; MDS – myelodysplastic syndrome; MM – multiple myeloma; MoAb – monoclonal antibody; NHL – non-Hodgkin's lymphoma; PB – peripheral blood; PCR – polymerase chain reaction; PLL – prolymphocytic leukaemia; RMA – resistance modifying agent; VAD – vincristine, doxorubicin, dexamethasone

Introduction

Results of treatment of haematological malignancies with anticancer agents have steadily improved over the years following the introduction of more effective drugs and the establishment of better designed chemotherapy strategies. Still, chemotherapy failure due to cellular drug resistance remains a major problem in most patients suffering from leukaemias, lymphomas or multiple myelomas. A large variety of drug resistance mechanisms have been characterized, using *in vitro* cell lines made resistant against the different classes of anticancer agents. Alterations in target proteins, carrier mediated drug uptake, cellular drug metabolism, cellular repair mechanisms and cellular drug efflux, among others, can cause anticancer drug resistance *in vitro*. However, the relationship of these cellular biochemical alterations with inadequate therapeutic responses to chemotherapeutic treatment in patients with haematological malignancies remains to be established for most drug resistance mechanisms identified so far. An exception hereupon, certainly, is drug resistance caused by enhanced cellular drug efflux due to increased activity of a membrane-bound glycoprotein drug pump. This type of drug resistance is referred to as classical multidrug resistance (MDR) and represents a very intriguing development in drug resistance research (for reviews on MDR, see van der Bliek & Borst (1989) and Roninson (1991)). There is increasing evidence that MDR can play a crucial role in clinical drug resistance, especially of some haematological malignancies (for a review on clinical relevance of MDR, see Nooter and Herweijer (1991)).

In this Chapter we will first give a very brief summary of the molecular and cellular biology of the MDR phenotype and then will give an overview of the available data from the literature and from our own laboratories on the occurrence of the MDR phenotype in haematological malignancies and on the attempts to circumvent clinical drug resistance in acute nonlymphocytic leukaemias, multiple myelomas and lymphomas.

Molecular and cellular biology of MDR

In MDR cells, selection for resistance to "naturally occurring" drugs, *e.g.*, anthracyclines, vinca alkaloids, podophyllotoxins, and colchicine, results in the development of cross-resistance to other members of the MDR drug family. The MDR-related drugs are structurally dissimilar and have different intracellular targets. What these drugs have in common is that they are in general rather large (between 300–900 molecular weight), hydrophobic, enter the cell by passive diffusion, and have affinity for a glycoprotein (P-glycoprotein) that is overexpressed in the membranes of MDR cells. Classical MDR cells are not cross-resistant to alkylating agents (*e.g.*, chlorambucil and cyclophosphamide), antimetabolites (*e.g.*, cytarabine, methotrexate, and 5-fluorouracil), or cisplatin.

The classical MDR phenotype is characterized by a reduced ability to accumulate drugs, as compared to the parent cell lines, being most likely the main cause of multidrug resistance. The reduced drug accumulation in classical MDR is due to activity of an energy-dependent unidirectional drug efflux pump with broad substrate specificity.

P-glycoprotein

The MDR drug pump is composed of a transmembrane glycoprotein (P-glycoprotein) with a molecular weight of 170 kD (Chen *et al.*, 1986; Gerlach *et al.*, 1986; Gros *et al.*, 1986), and is encoded by the so-called *mdr* genes. The glycoprotein is generally called P-glycoprotein, whereby P stands for permeability, because it was originally thought that the glycoprotein regulated cellular permeability (Juliano and Ling, 1976). It uses energy in the form of ATP to transport drugs through a channel formed by the transmembrane segments (Hamada and Tsuruo, 1988; Horio *et al.*, 1988). Gottesman (1993) has generated the working model "hydrophobic vacuum cleaner" for the function of the P-glycoprotein drug pump, in which it is proposed that the major function of the multidrug transporter is to extrude drugs directly from the plasma membrane. In that way, drugs that enter the cell by passive diffusion, will be removed from the membrane before they can enter the cytoplasm.

P-glycoprotein is part of the ATP-binding cassette (ABC) superfamily of transport systems that now includes over thirty proteins that share extensive sequence similarity and domain organization (Higgens *et al.*, 1990). ABC transport proteins are found in bacteria, yeast, plants, unicellular eukaryotes like *Plasmodium*, in which pfMDR is also implicated in drug resistance, sponges, insects and mammals (Branko, 1992). Different P-glycoprotein isoforms have been identified, and these are encoded by a family of closely related genes. They are referred to as *pgp* genes in hamsters and rats and *mdr* genes in humans and mice (Ng *et al.*, 1989; Deuchars *et al.*, 1992). In humans, two P-glycoprotein isoforms (*mdr1* and *mdr3*) with 80% amino acid homology have been identified (Roninson *et al.*, 1986; van der Bliek *et al.*, 1987). Both the human *mdr1* and *mdr3* genes were found to be localized on the long arm of chromosome 7 and to be linked within 330 kilobases (Chin *et al.*, 1989). Direct proof for the role of *mdr1* in MDR was obtained by transfection experiments. Expression of a full length cDNA clone of the human *mdr1* gene in a drug-sensitive cell conferred a complete MDR phenotype (Ueda *et al.*, 1987). However, the human *mdr3* gene does not seem to be involved in drug resistance (Schinkel *et al.*, 1991) and no function of the gene product has yet been identified.

Due to the high degree of homology between the *mdr1* and *mdr3* gene products, it was initially speculated that the *mdr3* gene also encodes for an efflux pump with broad specificity (van der Bliek *et al.*, 1988a). However, there is no experimental evidence that the human *mdr3* gene and the homologous mouse *mdr2* gene are involved in MDR: transfection and expression of full length cDNA copies of these genes inserted into mammalian expression vectors have so far failed to induce resistance to drugs (Gros *et al.*, 1988; van der Bliek *et al.*, 1988a). The group of Borst in Amsterdam (The Netherlands Cancer Institute) has cloned a human *mdr3* cDNA, coding for a P-glycoprotein, into a mammalian expression vector and cotransfected it with a selectable marker into a drug-sensitive human cell line (Schinkel *et al.*, 1991). Stable *mdr3*-expressing clones were obtained. Although a significant fraction of the cells (5—10%) in one of the *mdr3*-expressing clones expressed as much P-glycoprotein in the cell membrane as a clearly drug-resistant *mdr1*-expressing clone, they found no resistance against a range of MDR-related drugs, including vincristine, colchicine, VP-16, daunorubicin, doxorubicin, actinomycin D, and gramicidin D.

P-glycoprotein inhibitors

Tsuruo *et al.* (1981) observed that noncytotoxic doses of the calcium channel blocker verapamil could restore the sensitivity to Vinca alkaloids in MDR cells. While originally it was thought that intracellular calcium fluxes were mediating this reversal of drug resistance, in the meantime it is known that verapamil does so by competitive inhibition of the P-glycoprotein drug pump (Akiyama *et al.*, 1988). As of now, a large number of such so-called resistance modifying agents (RMAs), which can serve as substrates for the P-glycoprotein drug pump, has been found including: other calcium antagonists, phenothiazines, alkaloids, analogs of triparanol, dipyridamole, dihydropyridine and cyclosporins. In most cases, the reversal of resistance by RMAs is accompanied by increased accumulation of cytotoxic agents by the resistant cells and the current hypothesis on the mode of action of RMAs is that they correct the defective cytotoxic drug accumulation by competing for outward transport directly by binding with P-glycoprotein (Akiyama *et al.*, 1988; Foxwell *et al.*, 1989; Nooter *et al.*, 1989; Twentyman, 1992).

In an attempt to explain the unusually broad substrate specificity of P-glycoprotein, Higgens and Gottesman (1992) postulated the "flippase" model. In that model, the primary determinant of specificity would be the ability of a substrate (drug) to intercalate into the lipid bilayer in an appropriate fashion; interactions with the drug binding site on the transport protein would be of secondary importance. In this way, the drug binding site on P-glycoprotein would be relatively non-selective, yet access to this binding site would be limited to those substrates that can intercalate appropriately into the lipid bilayer of the cell membrane.

Expression of the mdr1 gene in normal haemopoietic tissues

P-glycoprotein expression is not only found in drug resistant *in vitro* cell lines but also in normal specialized epithelial cells with secretory or excretory functions in organs like liver, pancreas, kidney, colon, skin and endothelial cells of capillary blood vessels (Cordon-Cardo *et al.*, 1990; Thiebaut *et al.*, 1987; van der Valk *et al.*, 1990). Although the natural substrates for the *mdr1* gene encoded P-glycoprotein are not yet known, it is likely that the P-glycoprotein drug pump plays a role in the normal physiology of the organism and in the process of detoxification of xenobiotic substances (Gottesman *et al.*, 1991).

The expression of P-glycoprotein has also been associated with a cell volume-regulated chloride channel (Valverde *et al.*, 1992). Drug transport requires ATP hydrolysis while, in contrast, ATP binding is sufficient to enable activation of the chloride channel. The chloride channel and drug transport activities of P-glycoprotein appear to reflect two distinct functional states of the protein that can be interconverted by changes in tonicity (Gill *et al.*, 1992). This raises the possibility that, especially in cells of the gastrointestinal tract, which must deal with a wide range of osmotic conditions, besides pumping out xenobiotic substances, P-glycoprotein is also involved in volume regulation.

Using *mdr1*-specific probes (monoclonal antibodies-MoAb- or nucleic acid probes) or functional drug-accumulation assays, many groups have looked at the expression of the *mdr1* gene in normal haemopoietic tissues. It appeared that, in the hands of most groups, immunocytochemistry with the available P-glycoprotein-specific MoAbs (C219, JSB1 and MRK16) is not sensitive enough for the detection of P-glycoprotein expression on normal haemopoietic cells (*e.g.* Weide *et al.*, 1990). However, with other, more appropriate techniques *mdr1* expression can be detected in normal bone marrow (BM) and peripheral blood (PB) cells. For example, low *mdr1* expression levels were found by dot blot assay, in total RNA isolated from heterogeneous cell populations obtained from total BM, spleen, or PB lymphocytes (Fojo *et al.*, 1987; Holmes *et al.*, 1990).

To detect cells expressing *mdr1* in normal and postchemotherapy BM, Marie *et al.* (1992a) used *in situ* RNA hybridization and RNA phenotyping by polymerase chain reaction (PCR) for *mdr1* mRNA detection.The presence of P-glycoprotein was evaluated by immunocytochemistry with MRK16 or C219. With *in situ* mRNA hybridization, a small subset of myeloid and lymphoid BM cells expressed *mdr1* mRNA in all cases tested. However, with MoAbs P-glycoprotein expression could not be detected in the same BM samples. Gruber *et al.* (1992) purified lymphocytic, monocytic and granulocytic subpopulations from PB from healthy blood donors. By using a quantitative RNA-RNA solution hybridization method, the average number of *mdr1* RNA transcripts per cell could be estimated. All lymphocyte samples and approximately 50% of the monocyte samples had detectable *mdr1* mRNA levels, while *mdr1* mRNA could not be detected in the granulocytes. In line with these results are the data provided by Coon *et al.* (1991) and Schluesener *et al.* (1992), which suggest that a P-glycoprotein-like transport system is present in normal lymphocytes. Using flow cytometric analysis, they observed that lymphocytes could efflux the fluorochromic probe rhodamine 123, a laser dye which is exported by P-glycoprotein. Verapamil and other known inhibitors of P-glycoprotein, could block the outward rhodamine 123 transport.

Another study (Chaudhary and Roninson, 1991) suggested that haemopoeitic stem cells may express detectable levels of *mdr1.* Attempts at purifying these cells have often made use of their poor staining with the dye rhodamine 123; a fact that was interpreted as indicating a low number or activity of mitochondria in the stem cell. Chaudhary and Roninson (1991) showed that treatment of human lymphoid BM cells with verapamil, led to increased staining, suggesting that *mdr1* is responsible for the active efflux of rhodamine from these cells. Subsequently, they stained human BM cells with CD34, a marker of human haemopoietic progenitor cells including the pluripotent haemopoietic stem cell, and analyzed rhodamine 123 efflux in the $CD34^+$ subpopulation. The majority of $CD34^+$ cells had low dye uptake. However, after addition of verapamil, all the cells stained bright. A large fraction of the $CD34^+$ cells also stained with MRK16, suggesting the presence of P-glycoprotein. Experimental and clinical studies indicate that the pluripotent haemopoietic stem cells are resistant to anticancer agents. This resistance has been attributed to the quiescent state of the stem cells (Visser and van Bekkum, 1990). However, the above mentioned results indicate that the resistance of stem cells to chemotherapeutic drugs may also be explained at least in part on the basis of P-glycoprotein expression. In a recent study (Drach *et al.*, 1992), PB and BM subpopulations were stained with lineage-specific MoAbs, sorted by fluorescence activated cell sorter (FACS), and subsequently assayed for *mdr1* expression by PCR. In PB, granulocytes ($CD15^+$) and monocytes ($CD14^+$) were negative for *mdr1* expression, while in major T-cell subsets ($CD4^+$ and $CD8^+$), B cells ($CD19^+$), and NK cells ($CD56^+$) a positive *mdr1* signal could be detected. In BM, erythroid precursors and monocytic cells ($CD33^{++}/CD34^-$) were negative for *mdr1* expression. The early progenitor cells ($CD34^+$), committed progenitor cells ($CD33^+/CD34^+$), myeloid precursor cells ($CD33^+/CD34^-$), and early ($CD10^+/CD19^+$) and mature ($CD10^-/CD19^+$) B cells were positive for *mdr1*.

Using Northern- and dot blot assays, expression of the human *mdr3* gene has been detected only in the liver (van der Bliek *et al.*, 1988b). However, Roninson and coworkers used a very sensitive and specific PCR assay for human *mdr1* and *mdr3* expression in normal human tissues (Chin *et al.*, 1989). In the colon, lung, stomach, oesophagus, breast, muscles, and bladder, only *mdr1* expression was detected (Chin *et al.*, 1989). In the liver, kidneys, adrenals and spleen, both *mdr1* and *mdr3* expression was observed. This distribution suggests that *mdr1* and *mdr3* gene products may be involved in some of the same processes or that coexpression of these mRNAs may reflect a common regulatory pathway.

Expression of the mdr1 gene in haematological malignancies

The emergence of drug resistant cells, either at initial presentation or at the time of relapse, is one of the main obstacles for successful chemotherapeutic treatment of haematological malignancies. For several reasons, it is an attractive hypothesis that the clinical observation of resistance to multidrug based chemotherapy of haematological malignancies is due to enhanced *mdr1* expression in the resistant tumour. In the first place MDR-related cytotoxic drugs (anthracyclines, vinca alkaloids, and podophyllotoxines) represent a substantial part of the chemotherapeutic arsenal for the treatment of haematological malignancies. Furthermore, in these cancers, initial periods of effective cytoreduction are often followed by a state of acquired drug resistance, making them particularly interesting to study with regard to acquired MDR. In addition to that, a great advantage of studying haematological malignancies, is the ease of obtaining biopsies from PB and BM. Therefore, the emergence of the MDR phenotype can be monitored throughout the development and progression of the disease. Consequently, many data are available on *mdr1* expression in untreated as well as in recurrent, chemotherapy treated haematological malignancies.

In almost all types of haematological malignancies, either untreated or treated, elevated *mdr1* levels have been reported. The *mdr1* expression levels can range from low to high and even in untreated tumours relatively high levels are sometimes observed. Most of the studies on the expression of the *mdr1* gene in human haemopoietic neoplasms have employed bulk techniques (Western-, Northern-, or slot blotting, and RNase protection) for the detection and quantification of P-glycoprotein or its mRNA.The disadvantage of such techniques is that the frequently observed contamination with nontumour cells in the biopsy as well as the heterogeneity within the tumour cell population with regard to the level of P-glycoprotein expression are ignored. But there are also studies that searched for expression of the gene in individual cells, by using either immunocytochemistry with specific antibodies or *in situ* hybridization with specific RNA probes. Although these *in situ* methods are more subjective in interpretation than are bulk methods, they provide specific information on, *e.g.*, the percentage of *mdr* positive cells, the expression levels in individual cells, and the morphology of the *mdr* expressing cells.

The search for *mdr1* gene expression in human haematological malignancies started with the finding of Ma *et al.* (1987). They detected P-glycoprotein expression in two adult patients with acute myelocytic leukaemia (AML) by immunocytochemical assay using the P-glycoprotein specific MoAb C219, developed by the group of Ling. In both patients, C219 staining could not be detected in the leukaemic cells at first clinical admission. However, as the patients relapsed after three- or four courses of combined chemotherapy containing daunorubicin, C219 positive leukaemic cells appeared in the PB.The proportion of positive staining cells and the intensity of staining increased as the disease progressed.

The group of Tsuruo had developed the P-glycoprotein specific MoAb MRK16, and used this antibody also on clinical specimens. They found that leukaemic samples of three of six chronic myelocytic leukaemia (CML) patients at blast crisis, that are inherently drug resistant, stained positive with MRK16 (Tsuruo *et al.*, 1987). Immunocytochemistry is a very attractive technique for the detection of P-glycoprotein, especially in a clinical setting. However, in the beginning most groups experienced difficulties in repeating the above mentioned immunocytochemical observations, and changed to RNA techniques on isolated total cellular RNA for the detection and quantification of *mdr1* expression. Goldstein *et al.* (1989), Holmes *et al.* (1989), Rothenberg *et al.* (1989), and Nooter *et al.* (1990b) reported elevated *mdr1* expression levels in AML, acute lymphocytic leukaemia (ALL), and myelodysplastic syndromes (MDS) by slot blot analysis, or RNase protection assay. A limitation of the RNA bulk techniques is that one cannot discern if a message is expressed equally in all cells or if there is heterogeneity of expression in the population. A leukaemia with a low percentage of cells expressing high levels of *mdr1* might give a low level of expression on the average. Yet, such a small clone of P-glycoprotein positive cells might be important clinically. In the ALL study of Rothenberg (Rothenberg *et al.*, 1989), leukaemic samples were also screened by *in situ* RNA hybridization with an *mdr1*-antisense probe and by immunocytochemistry with MRK16. With both techniques several distinct populations expressing high, moderate, or no *mdr1* could be identified.

Using slot blot or RNase protection, expression of *mdr1* could be detected in the majority of patients with chronic lymphocytic leukaemia (CLL) (Herweijer *et al.*, 1990; Holmes *et al.*, 1990; Shustik *et al.*, 1991). In initial studies with MRK-16, which recognizes an extracellular P-glycoprotein epitope, only a low percentage of lymphocyte samples from CLL patients stained positive for P-glycoprotein (Cumber *et al.*, 1990). However, the authors found that altered sialation of carbohydrate moieties on P-glycoprotein may mask the epitope recognized by the antibody. Aberrant sialation of lymphocytes in CLL has been well characterized (Brown *et al.*, 1985) and the cells have increased levels of sialyl transferase. Treatment of the CLL cells with neuraminidase to remove sialic acid residues, increased the proportion of P-glycoprotein-positive samples dramatically. This finding of epitope masking might explain, in part, the discordance in P-glycoprotein staining of clinical samples with MRK16, observed by many investigators. In multiple myeloma (MM) increased *mdr1* levels were detected by slot blot (Linsenmeyer *et al.*, 1992) as well as by immunocytochemistry with JSB1 (Dalton *et al.*, 1989), or C219 and MRK16 (Epstein *et al.*, 1989). non-Hodgkin's lymphomas (NHL) also expressed *mdr1* as determined by slot blot (Moscow *et al.*, 1989) or by immunohistochemistry (Pileri *et al.*, 1991; Miller *et al.*, 1991).

We reviewed here only a few of the many papers on *mdr1* expression in haematological malignancies. Many conflicting results have been published on the incidence of *mdr1* positive patients for a given disease, probably due to differences in the techniques used, and the patient populations studied. In one of our own studies (Herweijer *et al.*, 1990; Nooter, unpublished results) we

Table 1. mdr expression in haematological malignancies as determined by RNase protection assay

Malignancy		Expression	
		mdr1	*mdr3*
AML		19/25	0/25
CML	chronic	10/10	0/10
	blast-myeloid	2/3	0/3
	blast-lymphoid	1/1	1/1
ALL	T	9/12	0/12
	B	10/23	7/23
CLL	B	30/32	30/32
	T	2/2	0/2
PLL	B	0/9	9/9
	T	1/1	0/1
HCL	B	9/9	8/9
NHL	B	17/21	9/21
MM		13/17	0/17

screened a large variety of haematological malignancies for *mdr1* expression, using the same technique, including sample preparation. Such an approach allows for comparison of the frequency of elevated *mdr1* expression of the different neoplasms. Our data are summarized in Table 1, and show that in all neoplasms tested (AML, CML, ALL, CLL, hairy cell leukaemia -HCL-, NHL, and MM), elevated *mdr1* expression is frequently observed, except for B-cell prolymphocytic leukaemia (PLL).

Expression of the mdr3 gene in haematological malignancies

We have found that, besides the *mdr1* gene, also the *mdr3* gene is expressed at relatively high levels in certain types of human leukaemias (Herweijer *et al.*, 1990, 1991). The *mdr3* gene appeared to be expressed selectively in malignant cells of the B-cell lineage, specifically in B-cell ALL and CLL, B-cell PLL and HCL (Table 1). PLL cells from untreated patients appeared to express the *mdr3* gene without detectable levels of *mdr1* (Nooter *et al.*, 1990a). Further analysis of *mdr3* expression in B-cell malignancies revealed that the expression is associated with the differentiation (maturation) stage of the neoplasm. Undifferentiated null cell — and common ALL, representing the malignant counterpart of pro-B and pre-pre-B cells, respectively, had no *mdr3* expression, as determined by RNase protection assay. The same was found for the very mature B-cell tumours, like Waldenstrom and multiple myeloma: also no *mdr3* expression. However, B-cell PLL, CLL, and HCL had intermediate to high levels of *mdr3*. These malignancies represent the intermediate and mature B cells of normal B-cell development. With regard to the elucidation of any function of *mdr3* in B cells, it would be interesting to study *mdr3* expression in normal B-cell development. In another study (Sonneveld *et al.*, 1992a) we investigated the *mdr1* and *mdr3* expression in CLL patients in different disease stages. Expression of both genes could be found in almost all CLL patients. Coexpression of *mdr1* and *mdr3* was not interrelated, and prior treatment did not influence the level of *mdr1* or *mdr3* expression. In patients with advanced CLL (Rai stages 3 + 4) the *mdr3* expression was significantly higher than in early-stage CLL (Rai stages 0 to 2), while such a difference was not present for *mdr1*. Also in individual patients, *mdr3* expression was reduced on treatment-induced improvement of Rai stage, indicating that therapy with cytostatic agents does not upregulate *mdr3* expression. Therefore, *mdr3* seems an independent parameter of tumour load and of progressive disease in CLL.

Intrinsic and acquired MDR phenotype

Both from an academic as well as from a clinical point of view the question whether the MDR phenotype is an inherent characteristic of the malignant cell or whether it is induced during chemotherapeutic treatment is worthwhile studying. Both theories can be supported by clinical as well as by experimental data and need not to be mutually exclusive in the patients. In the majority of haematological neoplasms the clinical manifestation of MDR develops after repeated chemotherapeutic treatment. Therefore, it seems likely that the acquisition of clinical MDR in those cases occurs by selection of pre-existing *mdr1* expressing malignant cells.

Selection

Clinical evidence for selection of MDR clones from the malignant cell population is found in *e.g.* adult AML. Elevated *mdr1* expression levels are frequently observed in AML at diagnosis (Goldstein *et al.*, 1989; Herweijer *et al.*, 1990; Sato *et al.*, 1990a; Kuwazuru *et al.*, 1990; Marie *et al.*, 1991; Pirker *et al.*, 1991). The standard protocols for AML treatment contain MDR-related drugs, and it is very likely that such drug treatments select for *mdr1* expressing leukaemic cells. This assumption is in line with the observation that treated AML has generally higher *mdr1* expression levels than untreated ones (Herweijer *et al.*, 1990; Sato *et al.*, 1990a; Kuwazuru *et al.*, 1990; Marie *et al.*, 1991; Pirker *et al.*, 1991; Zhou *et al.*, 1992). Comparable evidence for selection has also been provided for MM: higher incidence and expression levels of *mdr1* in patients treated with the VAD (vincristine, doxorubicin, dexamethasone)-protocol than in patients at initial presentation (Grogan *et al.*, 1993; Sonneveld, unpublished results). In an impressive study from Tucson (Grogan *et al.*, 1993) about hundred MM patients were studied either before or after therapy and at the time of relapse. MM patients with no prior therapy had low incidence (6%) of *mdr1* expression, while those receiving chemotherapy with doxorubicin and/or vincristine had a significant higher incidence that was related to the cumulative drug dosages, and that finally became 100% at the highest drug levels. Again, these observations are in favour of selection rather than induction of the MDR phenotype.

With regard to the elevated *mdr1* expression levels in untreated haematological malignancies there are two options: either the neoplasm developed by malignant transformation of an haemopoietic progenitor cell that normally expresses *mdr1*, or the expression is a consequence of the malignant transformation which took place in the tumour cell. Both possibilities can be supported by laboratory observations. For example in AML, increased *mdr1* expression is associated with an immature phenotype as expressed by CD34 (Campos *et al.*, 1992; Zhou *et al.*, 1992). Such a correlation was also reported by List *et al.* (1991) for MDS and therapy-induced AML. Since normal CD34$^+$ haemopoietic progenitor cells have *mdr1* expression (Chaudhary and Roninson, 1991) CD34$^+$ leukaemias might be developed by malignant transformation of normal CD34$^+$ counterparts. Another example is the *mdr1* expression by the untreated lymphocytic leukaemias, which for the same reason might be determined by *mdr1* expression in the normal lymphocytic counterparts.

Consider the possibility that elevated *mdr* expression is a result of the process of malignant transformation, two molecules are of specific interest: the tumour suppressor gene *p53* and the *ras* oncogene, both of which are frequently associated with tumour progression.

Malignant transformation

Recently, it was found (Chin *et al.*, 1992) that the promoter of the human *mdr1* gene can be a target for the *c-Ha-ras-1* oncogene and the *p53* tumour suppressor gene products. The stimulatory effect of the *ras* gene product was not specific for the *mdr1* promoter alone, whereas a mutant *p53* specifically stimulated the *mdr1* promoter and wild-type *p53* exerted specific repression. These results imply that the *mdr1* gene could be activated during tumour progression associated with mutations in *ras* and *p53*. Mutations in *p53* and members of the *ras* oncogene family are among the most frequently found genetic aberrations in human neoplasms. Mutations in the *ras* gene family may also occur frequently in leukaemias (Shen *et al.*, 1987; Toksoz *et al.*, 1987), specifically in AML. About 25 to 50% of AML cases harbour a mutated *ras* oncogene (Needleman *et al.*, 1986; Bos *et al.*, 1987). Although, most findings suggest that mutational inactivation of the *p53* gene less often occur in haematological malignancies than in solid tumours, there is still evidence that it is involved in the tumorigenesis of several types of haematological neoplasms (Prokocimer *et al.*, 1986; Hu *et al.*, 1992; Mori *et al.*, 1992).

Induction by chemotherapy

Independent of the above mentioned considerations,

there is also increasing evidence that the *mdr1* promoter can be activated by chemical stress-inducing agents, including anticancer agents, like Vinca alkaloids and anthracyclines (Chin *et al.*, 1990; Kohno *et al.*, 1989; Tanimura *et al.*, 1992), suggesting that chemotherapeutic agents might themselves directly cause the activation of the *mdr1* gene. To evaluate conditions which increase *mdr1* gene expression, the group of Gottesman and Pastan (Chin *et al.*, 1990) investigated the induction of the *mdr1* gene by physical and chemical environmental insults in a renal adenocarcinoma cell line. They identified several heat shock consensus elements in the promoter region of the human *mdr1* gene. Exposure of the renal cells to heat shock, sodium arsenite, or cadmium chloride led to a 7- to 8-fold increase in *mdr1* mRNA levels. This increase in the level of P-glycoprotein in these renal cells correlated with a transient increase in resistance to vinblastine following heat shock and arsenite treatment. These results suggest that the *mdr1* gene is regulatable by environmental stress. However, the possibility that the *mdr1* gene plays an additional role in protection against such environmental insults as heat shock and metal toxicity remains a hypothesis to be tested. In two other studies (Kohno *et al.*, 1989; Tanimura *et al.*, 1992) human and rodent cell lines were used which stably expressed the chloramphenicol acetyltransferase (CAT) gene driven by the human *mdr1* promoter. Significant CAT activity was found after incubation with a variety of cytotoxic and cytostatic drugs including vincristine, etoposide, daunorubicin, doxorubicin, colchicine, and hydroxyurea, suggesting that the *mdr1* promoter could be activated directly, at the transcriptional level, by these agents.

Of special clinical interest is the finding that hydroxyurea could activate the *mdr1* promoter. Hydroxyurea is often used as a cytostatic agent in the palliative treatment of leukaemias with high peripheral cell counts. As an anecdote we mention here that in our own studies extremely high *mdr1* expression levels were observed in leukaemic blasts obtained from three end-stage AML patients that had been treated continuously for prolonged time periods with hydroxyurea.

Ex vivo studies with haematological malignancies

Characteristic features of MDR cells are their reduced drug accumulation and the resulting reduced drug sensitivity, which can both be restored by P-glycoprotein inhibitors. In view of the clinical attempts to overcome MDR with RMAs in combination with cytotoxic drugs, many groups have performed *ex vivo* studies on drug accumulation and cytotoxicity with PB or BM samples from leukaemias, lymphomas and myelomas.

In vitro drug accumulation

Most studies using fresh biopsies from clinically resistant leukaemia patients, that dealt with the question whether RMAs were able to stimulate net uptake of MDR-related drugs, were unable to show convincing results, probably due to inappropriate technology. Using a very sensitive, so-called on-line flow cytometric (FCM) method for the quantification of anthracycline accumulation kinetics in heterogeneous cell samples (Nooter *et al.*, 1990c), we were able to show that in leukaemic cells expressing *mdr1*, the steady-state accumulation of daunorubicin was impaired and could be increased significantly by cyclosporin A or verapamil (Herweijer *et al.*, 1990; Nooter *et al.*, 1990b). The daunorubicin content of individual cells can be measured by FCM (Nooter *et al.*, 1983). The method makes use of the fluorescent properties of the anthracycline drugs. Upon excitation with 488 nm laser light, anthracyclines fluoresce with an emission maximum around 600 nm. The extension of this technique in on-line FCM enables accurate measurement of complete drug accumulation curves over long periods (up to a few hours) in selected subpopulations in heterogeneous cell samples, like BM or PB. To establish the MDR phenotype of leukaemia cells expressing *mdr1*, we monitored samples of a large group of patients for drug accumulation *in vitro* and the effects of RMAs (Nooter *et al.*, 1990b; Herweijer *et al.*, 1990). In the data analysis, the leukaemic cells were selected from the total cell population on the basis of their light-scattering characteristics (Nooter *et al.*, 1983). In Fig. 1 a typical example is shown of an AML

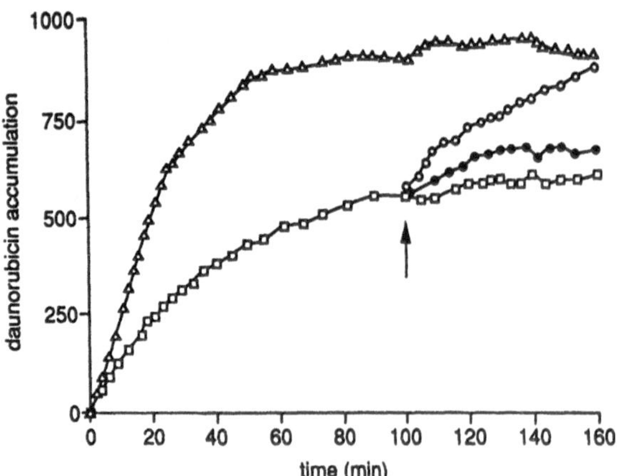

Fig. 1. Daunorubicin accumulation (expressed as fluorescence intensity) by leukaemia cells *ex vivo* from an AML patient. At time zero, daunorubicin (2 μM) was added to the cell suspension. Arrows indicate the time point of addition of cyclosporin (3 μM) (○), verapamil (10 μM) (□), or medium (●). The upper curve was made at the time of diagnosis, when the leukaemic blasts had no detectable *mdr1* expression. The other drug accumulation curves were made with leukaemic blasts, expressing relatively high levels of *mdr1*, obtained from the same AML patient after intensive chemotherapeutic treatment, at the time that the patient was refractory to further chemotherapeutic treatment.

patient with high mRNA *mdr1* expression (in this case 105 arbitrary units in an RNase protection assay). Addition of saline had no effect on the accumulation of daunorubicin by leukaemic blast cells, whereas that of cyclosporin was clearly increased. A much smaller increase resulted with verapamil. The upper curve in the figure represents the drug accumulation by AML cells from the same patient, at the time of diagnosis when the leukaemic blasts had no detectable *mdr1* expression. No effects were measured after addition of cyclosporin or verapamil. These results indicate that expression of *mdr1* mRNA in AML cells can result in an active outward drug pump that can be inhibited effectively by clinically-achievable concentrations of cyclosporin. Comparable experiments were performed by Cumber *et al.* (1991) who also measured the cellular accumulation of daunorubicin in clinical specimen. In PB samples from patients with CLL, the ability of cyclosporin A to increase steady-state drug accumulation *in vitro* correlated well with the levels of P-glycoprotein measured by immunofluorescent labelling of P-glycoprotein after treatment of the cells with neuraminidase to unmask the epitope recognized by MRK16.

Another aspect of the MDR phenotype that can be studied by on-line FCM very elegantly, is the energy-dependency of the MDR phenotype. In the next experiments we showed the ATP-dependency of the P-glycoprotein drug pump in an MDR cell line and in a clinical sample from an AML patient with abundant *mdr1* expression. In Fig. 2A the results with the MDR cell line (A2780/T100) are shown. The cells were preincubated for 30 min in glucose-free medium and at time zero daunorubicin was added and the intracellular drug accumulation was monitored. After about 45 min steady-state in drug accumulation was reached. In order to inhibit the oxidative phosphorylation, at time 60 min sodium azide was added to the incubation medium, resulting in a dramatic increase of the intracellular daunorubicin accumulation. The obvious explanation is that by incubation in glucose-free medium followed by the addition of sodium azide the cells are strongly depleted of ATP, resulting in a blockade of the efflux pump which finally leads to enhanced net drug uptake. Subsequently, the corollary experiment was performed. To the ATP-deprived cells, energy was added in the form of glucose (at time 90 min) and a dramatic decrease of drug accumulation is seen, with a steady-state level that is even lower than the one reached in glucose-free medium. Apparently, sufficient ATP can be generated by glycolysis in the presence of sodium azide to provide energy for the reactivation of the active efflux process. Comparable experiments were performed with *mdr1*-expressing leukaemia cells from a patient with refractory AML (Fig. 2B). The cells were preincubated for 30 min in glucose-free medium and at time zero daunorubicin was added and the intracellular accumulation was monitored. The addition of azide, at time 90 min resulted in an increased net-uptake of daunorubicin while the addition of glucose abolished this effect.

In vitro drug sensitivity

Two types of *in vitro* cytotoxicity studies have been performed: studies that tried to correlate *mdr1*

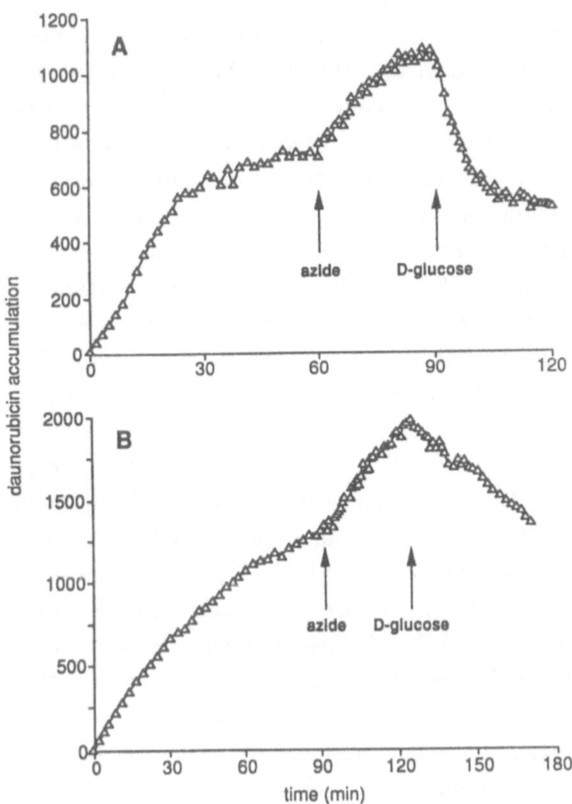

Fig. 2. Daunorubicin accumulation (expressed as fluorescence intensity in arbitrary units) by *mdr1*-transfected (A2780/T100) cells (A) and *mdr1*-expressing leukaemia cells *ex vivo* from a patient with refractory AML (B) in the absence and presence of energy. The cells were preincubated for 30 min in glucose-free medium. At time zero, daunorubicin (2 μM) was added to the cell supension. Arrows indicate time points of addition of sodium azide (10 mM) and glucose (10 mM).

expression in clinical samples with *in vitro* sensitivity/resistance to MDR-related cytotoxic drugs, and studies that tried to find evidence for circumvention of resistance *in vitro* with RMAs. The results of such studies must be regarded with caution because of a few obvious experimental pitfalls, among which are the sensitivity of the assay for *mdr1* expression, the heterogeneity of the clinical sample, and the technique of *in vitro* culturing of fresh biopsies. In two studies (Salmon *et al.*, 1989; Marie *et al.*, 1991) resistance to MDR-related drugs correlated with *mdr1* expression in tumour samples from patients with AML, MM, and lymphomas. Salmon *et al.* (1989) used immunocytochemistry to demonstrate P-glycoprotein staining on BM aspirates from myeloma patients and lymph node biopsies from patients with lymphoma. A correlation was then made with the sensitivity of the tumour cells to doxorubicin, assessed by clonogenic assay. The samples that stained positive for *mdr1* expression were all resistant *in vitro* to doxorubicin, while the negative *mdr1* samples were sensitive to this drug. In a series of AML patients Marie *et al.* (1991) demonstrated a correlation between overexpression of *mdr1* in the leukaemic cell sample and resistance of the clonogenic leukaemic cell (CFU-L) to MDR-related drugs, as estimated by clonogenic assay. However, in another AML study by the same authors (Marie *et al.*, 1992b), addition of the RMA cyclosporin A to the culture medium did not result in a change in CFU-L sensitivity to daunorubicin, despite the presence of a subset of P-glycoprotein positive cells. In a comparable study, Maruyama *et al.* (1989) had shown that the CFU-L sensitivity to daunorubicin in AML could be enhanced by verapamil. Unfortunately, the *mdr1* gene expression was not investigated in these patients.

Using a liquid culture assay, verapamil was demonstrated to increase the sensitivity of fresh leukaemic or myeloma cells to doxorubicin in 19 of 43 samples (Solary *et al.*, 1991). The capacity of verapamil to increase drug sensitivity *in vitro* did not correlate with the expression of P-glycoprotein, as estimated with immunocytochemistry using MRK16.

mdr1 expression as prognostic factor

Does the presence of *mdr1* expressing malignant cells limit successful chemotherapy? This of course, is the ultimate question. One of the strongest pieces of evidence that *mdr1* expression *in vivo* can induce acquired drug resistance, is provided by a transgenic mouse model (Galski *et al.*, 1989; Mickisch *et al.*, 1991). Transgenic mice expressing the human *mdr1* gene in the haematopoietic tissues, appeared to be resistant to leukopenia induced by the anticancer agent daunorubicin. Thus, it could be expected that leukaemic BM, expressing appropriate levels of *mdr1*, might also be resistant to chemotherapeutic treatment in the patient. The many clinical studies on *mdr1* expression indicate that the expression levels of *mdr1* in human leukaemias and myelomas can be as high as those of *in vitro* generated MDR cell lines. It has also been demonstrated that such *mdr1* expressing leukaemic cells have an impaired drug accumulation (Nooter *et al.*, 1990b; Herweijer *et al.*, 1990; Cumber *et al.*, 1991), and are resistant *in vitro* to MDR-related drugs (Salmon *et al.*, 1989; Sonneveld and Nooter, 1990; Marie *et al.*, 1991). Still, we do not know whether such levels of resistance (3- to 10-fold) can enable a malignant cell to survive the currently used chemotherapeutic treatment.

Recent clinical evidence suggests that in some specific haematological malignancies (*e.g.*, AML and MM) *mdr1* expression at initial presentation can indeed affect the outcome of subsequent chemotherapy: high levels of *mdr1* appeared to be associated with poor prognosis. In secondary AML, preceded by cytoxic exposure and/or a preleukemic phase, the prognosis is considerably worse than in *de novo* AML. In these patients a poor prognosis is associated with increased expression of *mdr1* (Holmes *et al.*, 1989; Sato *et al.*, 1990b; List *et al.*, 1991; Marie *et al.*, 1991; Campos *et al.*, 1992). A strong association of *mdr1* expression with other negative prognostic factors, including CD34 expression by AML blasts, a high blast cell proliferation rate and age has been identified (Campos *et al.*, 1992; Zhou *et al.*, 1992; Sonneveld, unpublished results). In these studies a negative prognostic weight on complete remission (CR) rate was attributed to *mdr1* expression, which was independent from other prognostic factors in most cases. In one of our own studies we found that *mdr1* and CD34 are expressed by the same leukemic blast cells, and that this association is frequently observed in secondary AML with monosomy 7 (Wittebol *et al.*, 1992).

Several groups have investigated the significance of *mdr1* overexpression as a prognostic factor in *de novo* AML (Table 2). Kuwazuru *et al.* (1990) used immunoblotting with C219 and in 9 of 17 untreated AML patients *mdr1* expression could be detected. Eighty percent of *mdr1*-positive patients were nonresponsive to chemotherapy, with only one CR and one partial response. In another study (Sato *et al.*, 1990a), AML patients whose leukaemic cells contained high *mdr1* transcript levels were difficult to induce into remission and, if CR was induced, that was of short duration. A Stanford study, using slot blot analysis, detected elevated *mdr1* mRNA levels in 45% of 19 cases of *de novo* AML at the time of initial diagnosis (Marie *et al.*, 1991). No known prognostic factors could be linked to *mdr1* expression, but the probability of achieving complete remission dropped from 70 to 40% with the presence of elevated *mdr1* expression. Another study, from Lyon (Campos *et al.*, 1992) reported measurements of P-glycoprotein expression by FACS analysis with MRK16 in 122 patients with *de novo* AML. Of patients expressing elevated levels of *mdr1*, 30% achieved CR, compared with 80% of those without detectable *mdr1* levels. In the same study it was also reported that elevated levels of *mdr1* in leukaemia cells correlated with the presence of CD34 antigen, although both markers independently conferred a negative prognostic value. By combining both markers, CD34 and *mdr1*, it was possible to define a subgroup with a very poor prognosis (both markers positive, CR rate: 20%) and a subgroup with a very good prognosis (both markers negative, CR rate: 100%). In a prospective study, Pirker *et al.* (1991) determined *mdr1* expression by slot blot analysis in *de novo* AML at diagnosis. Levels of *mdr1* were negative in 30% and positive in 70% of the patients. The CR rate in response to induction chemotherapy was 90% for *mdr1* negative patients and 55% for *mdr1* positive patients. Expression of the *mdr1* gene was observed in most patients who died early or had resistant disease. Kaplan-Meier curves revealed a decrease in disease-free survival of patients with detectable *mdr1* gene expression compared with the disease-free survival of *mdr1* negative patients. Zhou *et al.* (1992) used slot blot analysis and immunocytochemistry with C219, and found that CR rate was significantly lower (35%) in *mdr1* positive patients than in *mdr1* negative patients (75%). Most patients (70%) with resistant disease expressed *mdr1*. The overall conclusion

Table 2. Relevance of *mdr1* expression for prognosis in *de novo* AML

Author	Year	No. of patients	*mdr1* expression	
			Yes CR(%)	No CR(%)
Kuwazuru *et al.*	(1990)	17	10	80
Sato *et al.*	(1990a)	33	60	85
Marie *et al.*	(1991)	19	40	70
Pirker *et al.*	(1991)	63	55	90
Campos *et al.*	(1992)	122	30	80
Zhou *et al.*	(1992)	61	35	75

from these AML studies is that *mdr1* expression is clearly an unfavourable prognostic factor for lower CR rates, refractory disease, early death, and shorter disease-free survival.

In adult ALL, the frequency of *mdr1* expression has been less extensively investigated (Goldstein *et al.*, 1989; Rothenberg *et al.*, 1989; Herweijer *et al.*, 1990; Kuwazuru *et al.*, 1990; Musto *et al.*, 1991). In these studies the expression of *mdr1* in untreated ALL ranged from 10 to 70%, depending on the detection method, and no correlation could be made with clinical results. Other leukemias in which variable degrees of P-gp expression have been detected at diagnosis include chronic phase CML and CML blast crisis (Herweijer *et al.*, 1990; Weide *et al.*, 1990), and CLL (Herweijer *et al.*, 1990; Cumber *et al.*, 1990; Holmes *et al.*, 1990; Shustik *et al.*, 1991; Sonneveld *et al.*, 1992a). In none of these studies a prognostic significance of *mdr1* expression for response or survival was detected.

The results obtained so far for the lymphomas are not yet unequivocal. In one study (Niehans *et al.*, 1992) *mdr1* expression did not decrease the likelihood of response to induction chemotherapy. In another study (Miller *et al.*, 1991) newly diagnosed and untreated lymphoma patients had a very low (2%) incidence of *mdr1* expression, while previously treated and drug-resistant patients (64%) had detectable levels *mdr1*, suggesting that P-glycoprotein might contribute to drug resistance in lymphoma. Several groups have published increased expression of *mdr1* in MM: in untreated MM the number of *mdr1* positive patients is below 5%, and the incidence of *mdr1* positive patients increases dramatically after each course of chemotherapy with doxorubicin and/or vincristine (Grogan *et al.*, 1993; Linsenmeyer *et al.*, 1992; Sonneveld *et al.*, 1993). Increased *mdr1* expression is observed in 60 to 80% of VAD-refractory patients. It has been shown that *mdr1* has no impact on chemotherapy response if it is found before treatment with doxorubicin or vincristine, while in VAD-refractory patients it predicts for poor response (Epstein *et al.*, 1989; Grogan *et al.*, 1993).

Clinical trials with resistance modifying agents in leukaemias, lymphomas and multiple myelomas

Drug resistance of tumour cells which overexpress P-glycoprotein may be overcome by several approaches. In clinical practice one may use alternative drugs which cannot serve as a substrate for the P-glycoprotein drug pump. However, the number of active drugs can be very limited; *e.g.* in AML treatment the use of effective non-MDR drugs is limited to ara-C. A different approach can be taken by simple dose escalation of the drugs in order to achieve higher intracellular drug concentrations. Dose-escalation is however hampered by a concurrent increase of toxicity, and therefore it cannot be used without adequate stem cell support.

Recently the attention has focused on the RMAs, agents which inhibit the P-glycoprotein mediated efflux of cytostatic drugs. The finding that elevated *mdr1* expression can occur in haematological malignancies and that specific RMAs can circumvent MDR in model systems at concentrations that are clinically achievable, has stimulated the development of clinical protocols in which RMAs are used in conjunction with cytotoxic drugs. Promising results in pilot studies and phase I/II trials using different RMAs and MDR-related drugs in MM, lymphoma and AML patients have been reported. The clinical efficacy of the experimental protocols was assessed by the occurrence of otherwise unexpected tumour responses.

In MM verapamil has been added to VAD in VAD-refractory patients (Salmon *et al.*, 1991). In these heavily pretreated patients some responses were noted. Verapamil was also used in the treatment of drug-refractory lymphoma (Miller *et al.*, 1991) and in that study an unexpected high percentage of CRs was obtained. The dose-limiting toxicity of the verapamil infusion was cardiac dysfunction including hypotension, congestive heart failure, and cardiac arrhythmia. We reported treatment of a refractory AML patient with daunorubicin and cytarabine combined with cyclosporin A (Sonneveld and Nooter, 1990). In that case, the emergence of the MDR phenotype was monitored during clinical progression of the disease. At relapse, a decrease of daunorubicin accumulation by AML blasts was associated with elevated *mdr1* expression and a decreased *in vitro* sensitivity to daunorubicin. Intracellular daunorubicin accumulation and *in vitro* sensitivity could be completely restored by adding cyclosporin A to the cells. During progressive relapse, the patient was treated with reinduction therapy to which cyclosporin A was added and this resulted in a transient elimination of the *mdr1* positive AML clone. This study was followed by a larger one (Marie *et al.*, 1993) in which 16 patients with relapsed or refractory AML were treated with cyclosporin A as a modifier in combination with mitoxantrone and etoposide. This combination chemotherapy was effective in killing *mdr1* expressing leukaemia cells and even CRs were obtained. Other studies are in progress in Tucson (W.S. Dalton, personal communication) using cyclosporin A added to induction regimens in refractory AML.

Using a similar approach cyclosporin A was administered to VAD-refractory MM patients, leading to several unexpected long-lasting responses (Sonneveld *et al.*, 1992b). Before treatment *mdr1* expression could be detected in the myeloma cells in some of the patients. However, after the experimental treatment, in the majority of these patients no *mdr1* positive plasma cells were present, suggesting that cyclosporin plus VAD is effective against *mdr1* expressing myeloma cells. The dose-limiting toxicity of the cyclosporin infusion was musculoskeletal pain but also increased myelosuppression and hyperbilirubinaemia were noted. Hyperbilirubinaemia is a well-known side-effect of cyclosporin A in transplantation studies, and now appears to occur also in the *mdr1*-modulation studies (Yahanda *et al.*, 1992; Marie *et al.*, 1993; Sonneveld *et al.*, 1992b). The presence of P-glycoprotein on the luminal surfaces of the biliary tract (van der Valk *et al.*, 1990) makes it likely that the increase in blood bilirubin during cyclosporin A treatment is due to impairment of the P-glycoprotein pump in the epithelial cells of the bile duct. Such an inhibition of the endogenous P-glycoprotein pump not only can lead to reduced bile excretion, but also to reduced biliary cytotoxic drug elimination, as has been shown for colchicine (Speeg *et al.*, 1992), which might contribute to altered pharmacokinetics, *e.g.* increased area under the plasma concentration/time curve, of the cytotoxic drugs, and a subsequent alteration in the toxicity profile (Lum *et al.*, 1992). Another point of pharmacokinetic consideration is that the tissue distribution of the cytotoxic drugs might be changed by the haemodynamic effects of the RMAs (Bright and Buss, 1990; Fedeli *et al.*, 1989; Kerr *et al.*, 1986; Nooter *et al.*, 1987).

The overall conclusion is that modulation of drug resistance by non-cytotoxic RMAs seems promising and needs further clinical evaluation in prospective, randomised phase III trials.

Acknowledgements

Supported by grants from the Dutch Cancer Society.

References

Akiyama S, Cornwell MM, Kuwano M, Pastan I and Gottesman MM (1988) Most drugs that reverse multidrug resistance also inhibit photoaffinity labeling of P-glycoprotein by a vinblastine analog. Mol Pharmacol 33: 144–147.

Bos JL, Verlaan-de Vries M, van der Eb AL, Janssen JWG, Delwel R, Lowenberg B and Colly LP (1987) Mutations in N-*ras* predominate in acute myeloid leukemia. Blood 69: 1237–1241.

Branko K (1992) The multixenobiotic resistance mechanism in aquatic organisms. Crit Rev Toxicol 22: 23–43.

Bright JM and Buss DD (1990) Effects of verapamil on chronic doxorubicin-induced cardiotoxicity in dogs. J Natl Cancer Inst 82: 963–964.

Brown VA, Smith SK, Dewar E, Stockdill G and Maddy AH (1985) Surface glycoproteins as markers of the cellular status of B chronic lymphocytic leukaemia lymphocytes. Clinics Exp Immunol 62: 95–103.

Campos L, Guyotat D, Archimbaud E, Calmard-Oriol P, Tsuruo T, Troncy J, Treille D and Fiere D (1992) Clinical significance of multidrug resistance P-glycoprotein expression on acute nonlymphoblastic leukemia cells at diagnosis. Blood 79: 473–476.

Chaudhary PM and Roninson IB (1991) Expression and activity of P-glycoprotein, a multidrug efflux pump, in human hematopoietic stem cells. Cell 66: 85–94.

Chen C, Chin JE, Ueda K, Clark DP, Pastan I, Gottesman MM and Roninson IB (1986) Internal duplication and homology with bacterial transport proteins in the *mdr 1* (P-glycoprotein) gene from multidrug-resistant human cells. Cell 47: 381–389.

Chin JE, Soffir R, Noonan KE, Choi K and Roninson IB (1989) Structure and expression of the human MDR (P-glycoprotein) gene family. Mol Cell Biol 9: 3808–3820.

Chin KV, Tanaka S, Darlington G, Pastan I and Gottesman MM (1990) Heat shock and arsenite increase expression of the multidrug resistance (MDR1) gene in human renal carcinoma cells. J Biol Chem 265: 221–226.

Chin KV, Ueda K, Pastan I and Gottesman MM (1992) Modulation of activity of the promoter of the human *mdr1* gene by *ras* and *p53*. Science 255: 459–462.

Coon JS, Wang Y, Bines SD, Markham PN, Chong ASF and Gebel HM (1991) Multidrug resistance activity in human lymphocytes. Human Immunol 32: 134–140.

Cordon-Cardo C, O'Brien JP, Boccia J, Casals D, Bertino JR and Melamed MR (1990) Expression of the multidrug resistance gene product (P-glycoprotein) in human normal and tumor tissues. J Histochem Cytochem 38: 1277–1287.

Cumber PM, Jacobs A, Hoy T, Fischer J, Whittaker JA, Tsuruo T and Padua RA (1990) Expression of the multiple drug resistance gene (mdr-1) and epitope masking in chronic lymphatic leukaemia. Br J Haemat 76: 226–230.

Cumber PM, Jacobs A, Hoy T, Whittaker JA, Tsuruo T and Padua RA (1991) Increased drug accumulation *ex vivo* with cyclosporin in chronic lymphatic leukemia and its relationship to epitope masking of P-glycoprotein. Leukemia 5: 1050–1053.

Dalton WS, Grogan TM, Rybski JA, Scheper RJ, Richter L, Kailey J, Broxterman HJ, Pinedo HM and Salmon SE (1989) Immunohistochemical detection and quantitation of P-glycoprotein in multiple drug-resistant human myeloma cells: association with level of drug resistance and drug accumulation. Blood 73: 747–752.

Deuchars KL, Duthie M and Ling V (1992) Identification of distinct P-glycoprotein gene sequences in rat. Biochim Biophys Acta 1130: 157–165.

Drach D, Zhao S, Drach J, Mahadevia R, Gattringer C, Huber H and Andreeff M (1992) Subpopulations of normal peripheral blood and bone marrow cells express a functional multidrug resistant phenotype. Blood 80: 2729–2734.

Epstein J, Xiao H and Oba BK (1989) P-glycoprotein expression in plasma-cell myeloma is associated with resistance to VAD. Blood 74: 913–917.

Fedeli L, Colozza M, Boschetti E, Sabalich I, Aristei C, Guerciolini R, Del Favero A, Rossetti R, Tonato M, Rambotti P and Davis S (1989) Pharmacokinetics of vincristine in cancer patients treated with nifedipine. Cancer 64: 1805–1811.

Fojo AT, Ueda K, Slamon DJ, Poplack DG, Gottesman MM and Pastan I (1987) Expression of a multidrug-resistance gene in human tumors and tissues. Proc Natl Acad Sci USA 84: 265–269.

Foxwell BMJ, Mackie A, Ling V and Ryffel B (1989) Identification of the multidrug resistance-related P-glycoprotein as a cyclosporin binding protein. Mol Pharmacol 36: 543–546.

Galski H, Sullivan M, Willingham MC, Chin KV, Gottesman MM, Pastan I and Merlino GT (1989) Expression of a human multidrug resistance cDNA (*MDR*1) in the bone marrow of transgenic mice: resistance to daunomycin-induced leukopenia. Mol Cell Biol 9: 4357–4363.

Gerlach JH, Endicott JA, Juranka PF, Henderson G, Sarangi F, Deuchars KL and Ling V (1986) Homology between P-glycoprotein and a bacterial haemolysin transport protein suggests a model for multidrug-resistance. Nature 324: 485–489.

Gill DR, Hyde SC, Higgens CF, Valverde MA, Mintenig GM and Sepulveda FV (1992) Separation of drug transport and chloride channel functions of the human multidrug resistance P-glycoprotein. Cell 71: 23–32.

Goldstein LJ, Galski H, Fojo AT, Willingham M, Lai SL, Gazdar A, Pirker R, Green A, Crist W, Brodeur GM, Lieber M, Cossman J, Gottesman MM and Pastan I (1989) Expression of a multidrug resistance gene in human cancers. J Natl Cancer Inst 81: 116–124.

Gottesman MM, Willingham MC, Thiebaut F and Pastan I (1991) Expression of the *MDR*1 gene in normal human tissues. In: IB Roninson (ed) Molecular and Cellular Biology of Multidrug Resistance in Tumor Cells (pp. 279-289). New York: Plenum Press.

Gottesman MM (1993) How cancer cells evade chemotherapy: sixteenth Richard and Hinda Rosenthal Foundation award lecture. Cancer Res 53: 747–754.

Grogan TM, Spier CM, Salmon SE, Matzner M, Rybski J,

Weinstein RS, Scheper RJ and Dalton WS (1993) P-glycoprotein expression in human plasma cell myeloma correlation with prior chemotherapy Blood 81 490–495

Gros P, Croop J and Housman D (1986) Mammalian multidrug resistance gene complete cDNA sequence indicates strong homology to bacterial transport proteins Cell 47 371–380

Gros P, Raymond M, Bell J and Housman D (1988) Cloning and characterization of a second member of the mouse *mdr* gene family Mol Cell Biol 8 2770–2778

Gruber A, Vitols S, Norgren S, Arestrom I, Peterson C, Bjorkholm M, Reizenstein P and Luthman H (1992) Quantitative determination of *mdr1* gene expression in leukaemic cells from patients with acute leukaemia Br J Cancer 66 266–272

Hamada H and Tsuruo T (1988) Characterization of the ATPase activity of the Mr 170,000 to 180,000 membrane glycoprotein (P-glycoprotein) associated with multidrug resistance in K562/ADM cells Cancer Res 48 4926–4932

Herweijer H, Sonneveld P, Baas F and Nooter K (1990) Expression of *mdr1* and *mdr3* multidrug-resistance genes in human acute and chronic leukemias and association with stimulation of drug accumulation by cyclosporin J Natl Cancer Inst 82 1133–1140

Herweijer H, Nooter K, Beishuizen A, Sonneveld P, Oostrum R, Hesseling-Janssen ALW and van Dongen J (1991) Expression of *mdr1* and *mdr3* multidrug-resistance genes in hairy cell leukaemia Eur J Cancer 27 297–298

Higgens CF, Hyde SC, Mimmack MM, Gileadi U, Gill DR and Gallagher MP (1990) Binding protein-dependent transport systems J Bioenerg Biomemb 22 571–592

Higgens CF and Gottesman MM (1992) Is the multidrug transporter a flippase? Trends Biochem Sci 17 18–21

Holmes J, Jacobs A, Carter G, Janowska-Wieczorek A and Padua RA (1989) Multidrug resistance in haemopoietic cell lines, myelodysplastic syndromes and acute myeloblastic leukaemia Br J Haematol 72 40–44

Holmes JA, Jacobs A, Carter G, Whittaker JA, Bentley DP and Padua RA (1990) Is the *mdr1* gene relevant in chronic lymphocytic leukemia? Leukemia 4 216–218

Horio M, Gottesman MM and Pastan I (1988) ATP-dependent transport of vinblastine in vesicles from human multidrug-resistant cells Proc Natl Acad Sci USA 85 3580–3584

Hu G, Zhang W and Deisseroth AB (1992) P53 gene mutations in acute myelogenous leukaemia Br J Haematol 81 489–494

Juliano RL and Ling V (1976) A surface glycoprotein modulating drug permeability in Chinese hamster ovary cell mutants Biochim Biophys Acta 455 152–162

Kerr DJ, Graham J, Cummings J, Morrison JG, Thompson GG, Brodie MJ and Kaye SB (1986) The effect of verapamil on the pharmacokinetics of adriamycin Cancer Chemother Pharmacol 18 239–242

Kohno K, Sato S, Takano H, Matsuo K and Kuwano M (1989) The direct activation of human multidrug resistance gene (*mdr1*) by anticancer agents Biochem Biophys Res Commun 165 1415–1421

Kuwazuru Y, Yoshimura A, Hanada S, Utsunomiya A, Makino T, Ishibashi K, Kodama M, Iwahashi M, Arima T and Akiyama SI (1990) Expression of the multidrug transporter, P-glycoprotein, in acute leukemia cells and correlation to clinical drug resistance Cancer 66 868–873

Linsenmeyer ME, Jefferson S, Wolf M, Matthews JP, Board PG and Woodcock DM (1992) Levels of expression of the *mdr1* gene and glutathione S-transferase genes 2 and 3 and response to chemotherapy in multiple myeloma Br J Cancer 65 471–475

List AF, Spier CM, Cline A, Doll DC, Garewal H, Morgan R and Sandberg AA (1991) Expression of the multidrug resistance gene product (P-glycoprotein) in myelodysplasia is associated with a stem cell phenotype Br J Haematol 78 28–34

Lum BL, Kaubisch S, Yahanda AM, Adler KM, Jew L, Ehsan MN, Brophy NA, Halsey J, Gosland MP and Sikic BI (1992) Alteration of etoposide pharmacokinetics and pharmacodynamics by cyclosporin in a phase I trial to modulate multidrug resistance J Clin Oncol 10 1635–1642

Ma DDF, Davey RA, Harman DH, Isbister JP, Scurr RD, Mackertich SM, Dowden G and Bell DR (1987) Detection of a multidrug resistant phenotype in acute nonlymphoblastic leukaemia Lancet i 135–137

Marie JP, Zittoun R and Sikic BI (1991) Multidrug resistance (*mdr1*) gene expression in adult acute leukemias correlations with treatment outcome and *in vitro* drug sensitivity Blood 78 586–592

Marie JP, Brophy NA, Ehsan MN, Aihara Y, Mohamed NA, Cornbleet J, Chao NJ and Sikic BI (1992a) Expression of multidrug resistance gene *mdr1* mRNA in a subset of normal bone marrow cells Br J Haematol 81 145–152

Marie JP, Helou C, Thevenin D, Delmer A and Zittoun R (1992b) *In vitro* effect of P-glycoprotein (P-gp) modulators on drug sensitivity of leukemic progenitors (CFUL) in acute myelogenous leukemia (AML) Exp Hematol 20 565–568

Marie JP, Bastie JN, Coloma F, Suberville AM, Delmer A, Rio B, Delmas-Marsalet B, Leroux G, Casassus P, Baumelou E, Catalin J and Zittoun R (1993) Cyclosporin A as a modifier agent in the salvage treatment of acute leukemia (AL) Leukemia 7 821–824

Maruyama Y, Murohashi I, Nara N and Aoki N (1989) Effects of verapamil on the cellular accumulation of daunorubicin in blast cells and on the chemosensitivity of leukaemic blast progenitors in acute myelogenous leukaemia Br J Haematol 72 357–362

Mickisch GH, Licht T, Merlino GT, Gottesman MM and Pastan I (1991) Chemotherapy and chemosensitization of transgenic mice which express the human multidrug resistance gene in bone marrow efficacy, potency, and toxicity Cancer Res 51 5417–5424

Miller TP, Grogan TM, Dalton WS, Spier CM, Scheper RJ and Salmon SE (1991) P-glycoprotein expression in malignant lymphoma and reversal of clinical drug resistance with chemotherapy plus high-dose verapamil J Clin Oncol 9 17–24

Mori N, Wada M, Yokota J, Terada M, Okada M, Teramura M, Masuda M, Hoshino S, Motoji T, Oshimi K and Mizoguchi

H (1992) Mutations of the p53 tumour suppressor gene in haematologic neoplasms Br J Haematol 81 325–240

Moscow JA, Fairchild CR, Madden MJ, Ransom DT, Wieand HS, O'Brien EE, Poplack DG, Cossman J, Myers C and Cowan KH (1989) Expression of anionic glutathione-S-transferase and P-glycoprotein genes in human tissues and tumors Cancer Res 49 1422–1428

Musto P, Melillo L, Lombardi G, Matera R, Di Giorgio G and Carotenuto M (1991) High risk of early resistant relapse for leukaemia patients with presence of multidrug resistance associated P-glycoprotein positive cells in complete remission Br J Haematol 77 50–53

Needleman SW, Kraus MH, Srivastava SK, Levine PH and Aaronson S (1986) High frequency of N-*ras* activation in acute myelogenous leukemia Blood 67 753–757

Ng WF, Sarangi F, Zastawny RL, Veinot-Drebot L and Ling V (1989) Identification of members of the P-glycoprotein multigene family Mol Cell Biol 9 1224–1232

Niehans GA, Jaszsz W, Brunetto V, Perri RT, Gajl-Peczalska K, Wick MR, Tsuruo T and Bloomfield C (1992) Immunohistochemical identification of P-glycoprotein in previously untreated, diffuse large cell and immunoblastic lymphomas Cancer Res 52 3768–3775

Nooter K, van den Engh GJ and Sonneveld P (1983) Quantitative flow cytometric determination of anthracycline content of rat bone marrow cells Cancer Res 43 5126–5130

Nooter K, Oostrum R and Deurloo J (1987) Effects of verapamil on the pharmacokinetics of daunomycin in the rat Cancer Chemother Pharmacol 20 176–178

Nooter K, Oostrum R, Jonker R, van Dekken H, Stokdijk W and van den Engh G (1989) Effect of cyclosporin A on daunorubicin accumulation in multidrug-resistant P388 leukemia cells measured by real-time flow cytometry Cancer Chemother Pharmacol 23 296–300

Nooter K, Sonneveld P, Janssen A, Oostrum R, Boersma T, Herweijer H, Valerio D, Hagemeijer A and Baas F (1990a) Expression of the *mdr3* gene in prolymphocytic leukemia association with cyclosporin-A-induced increase in drug-accumulation Int J Cancer 45 626–631

Nooter K, Sonneveld P, Oostrum R, Herweijer H, Hagenbeek A and Valerio D (1990b) Overexpression of the *mdr1* gene in blast cells from patients with acute myelocytic leukemia is associated with decreased anthracycline accumulation that can be restored by cyclosporin-A Int J Cancer 45 263–268

Nooter K, Herweijer H, Jonker RR and van den Engh G (1990c) On-line flow cytometry – a versatile method for kinetic measurements In Z Darzynkiewicz and H Crissman (eds) Methods in Cell Biology (pp 631–645) Orlando Academic Press

Nooter K and Herweijer H (1991) Multidrug resistance (*mdr*) genes in human cancer Br J Cancer 63 663–669

Pileri SA, Sabattini E, Falini B, Tazzari PL, Gherlinzoni F, Michieli MG, Damiani D, Zucchini L, Gobbi M, Tsuruo T and Baccarani M (1991) Immunohistochemical detection of the multidrug transport protein P170 in human normal tissues and malignant lymphomas Histopathology 91 131–140

Pirker R, Wallner J, Geissler K, Linkesch W, Haas OA, Bettelheim P, Hopfner M, Scherrer R, Valent P, Havelec L, Ludwig H and Lechner K (1991) *mdr1* Gene expression and treatment outcome in acute myeloid leukaemia J Natl Cancer Inst 83 708–712

Prokocimer M, Shaklai M, Ben Bassat H, Wolf D, Goldfinger N and Rotter V (1986) Expression of p53 in human leukaemia and lymphoma Blood 68 113–118

Roninson IB, Chin JE, Choi K, Gros P, Housman DE, Fojo A, Shen DW, Gottesman MM and Pastan I (1986) Isolation of human *mdr* DNA sequences amplified in multidrug-resistant KB carcinoma cells Proc Natl Acad Sci USA 83 4538–4542

Roninson IB (1991) Molecular and Cellular Biology of Multidrug Resistance in Tumour Cells New York Plenum Press

Rothenberg ML, Mickley LA, Cole DE, Balis FM, Tsuruo T, Poplack DG and Fojo AT (1989) Expression of the *mdr*-1/P-170 gene in patients with acute lymphoblastic leukemia Blood 74 1388–1395

Salmon SE, Grogan TM, Miller T, Scheper R and Dalton WS (1989) Prediction of doxorubicin resistance *in vitro* in myeloma, lymphoma, and breast cancer by P-glycoprotein staining J Natl Cancer Inst 81 696–701

Salmon SE, Dalton WS, Grogan TM, Plezia P, Lehnert M, Roe DJ and Miller TP (1991) Multidrug-resistant myeloma laboratory and clinical effects of verapamil as a chemosensitizer Blood 78 44–50

Sato H, Preisler H, Day R, Raza A, Larson R, Browman G, Goldberg J, Vogler R, Grunwald H, Gottlieb A, Bennett J, Gottesman M and Pastan I (1990a) MDR1 transcript levels as an indication of resistant disease in acute myelogenous leukaemia Br J Haematol 75 340–345

Sato H, Gottesman MM, Goldstein LJ, Pastan I, Block AM, Sandberg AA and Preisler HD (1990b) Expression of the multidrug resistance gene in myeloid leukemias Leukemia Res 14 11–22

Schinkel AH, Roelofs MEM and Borst P (1991) Characterization of the human *mdr3* P-glycoprotein and its recognition by P-glycoprotein-specific monoclonal antibodies Cancer Res 51 2628–2635

Schluesener HJ, Koeppel C and Jung S (1992) Multidrug transport in human autoimmune T line cells and peripheral blood lymphocytes Immunopharmacology 23 37–48

Shen WP, Aldrich TH, Venta-Perez G, Franza BR and Furth ME (1987) Expression of normal and mutant *ras* proteins in human acute leukaemia Oncogene 1 157–165

Shustik C, Groulx N and Gros P (1991) Analysis of multidrug resistance (MDR1) gene expression in chronic lymphocytic leukaemia (CLL) Br J Haematol 79 50–56

Solary E, Bidan JM, Calvo F, Chauffert B, Caillot D, Mugneret F, Gauville C, Tsuruo T, Carli PM and Guy H (1991) P-glycoprotein expression and in vitro reversion of doxorubicin resistance by verapamil in clinical specimens from acute leukaemia and myeloma Leukemia 5 592–597

Sonneveld P and Nooter K (1990) Reversal of drug-resistance by cyclosporin-A in a patient with acute myelocytic leukaemia Br J Haematol 75 208–211

Sonneveld P, Nooter K, Burghouts JThM, Herweijer H,

Adriaansen HJ and van Dongen JJM (1992a) High expression of the *mdr3* multidrug resistance gene in advanced stage chronic lymphocytic leukemia. Blood 70: 1496–1500.

Sonneveld P, Durie BGM, Lokhorst HM, Marie JP, Solbu G, Suciu S, Zittoun R, Lowenberg B and Nooter K (1992b) Modulation of multidrug resistant multiple myeloma by cyclosporin. Lancet 340: 255–259.

Sonneveld P, Durie BGM, Lokhorst HM, Frutiger Y, Schoester M and Vela EE (1993) Analysis of multidrug-resistance (MDR-1) glycoprotein and CD56 expression to separate monoclonal gammopathy from multiple myeloma. Br J Haematol 83: 63–67.

Speeg KV, Maldonado AL, Liaci J and Muirhead D (1992) Effect of cyclosporin on colchicine secretion by a liver canalicular transporter studied *in vivo*. Hepatology 15: 899–903.

Tanimura H, Kohno K, Sato S, Uchiumi T, Miyazaki M, Kobayashi M and Kuwano M (1992) The human multidrug resistance 1 promoter has an element that responds to serum starvation. Biochem Biophys Res Commun 183: 917–924.

Thiebaut F, Tsuruo T, Hamada H, Gottesman MM, Pastan I and Willingham MC (1987) Cellular localization of the multidrug-resistance gene product P-glycoprotein in normal human tissues. Proc Natl Acad Sci USA 84: 7735–7738.

Toksoz D, Farr CJ and Marshall CJ (1987) *Ras* gene activation in a minor proportion of the blast population in acute myeloid leukaemia. Oncogene 1: 409–413.

Tsuruo T, Iida H, Tsukagoshi S and Sakurai Y (1981) Overcoming of vincristine resistance in P388 leukemia *in vivo* and *in vitro* through enhanced cytotoxicity of vincristine and vinblastine by verapamil. Cancer Res 41: 1967–1972.

Tsuruo T, Sugimoto Y, Hamada H, Roninson I, Okumura M, Adachi K, Morishima Y and Ohno R (1987) Detection of multidrug resistance markers, P-glycoprotein and *mdr1* mRNA, in human leukemia cells. Jpn J Cancer Res (Gann) 78: 1415–1419.

Twentyman PR (1992) Cyclosporins as drug resistance modifiers. Biochem Biopharmacol 43: 109–117.

Ueda K, Cardarelli C, Gottesman MM and Pastan I (1987) Expression of a full-length cDNA for the human "*MDR1*" gene confers resistance to colchicine, doxorubicin, and vinblastine. Proc Natl Acad Sci USA 84: 3004–3008.

Valverde MA, Diaz M, Sepulveda FV, Gill DR, Hyde SC and Higgens CF (1992) Volume-regulated chloride channels associated with the human multidrug-resistance P-glycoprotein. Nature 355: 830–833.

Van der Bliek AM, Baas F, ten Houte-de Lange T, Kooiman PM, van der Velde-Koerts T and Borst P (1987) The human *mdr3* gene encodes a novel P-glycoprotein homologue and gives rise to alternatively spliced mRNAs in liver. EMBO J 6: 3325–3331.

Van der Bliek AM, Kooiman PM, Schneider C and Borst P (1988a) Sequence of *mdr3* cDNA encoding a human P-glycoprotein. Gene 71: 401–411.

Van der Bliek AM, Baas F, van der Velde-Koerts T, Biedler JL, Meyers MB, Ozols RF, Hamilton TC, Joenje H and Borst P (1988b). Genes amplified and overexpressed in human multidrug-resistant cell lines. Cancer Res 48: 5927–5932.

Van der Bliek AM and Borst P (1989) Multidrug-resistance. In: GF van de Woude and G Klein (eds) Advances in Cancer Research 52 (pp. 165–203). New York: Acad. Press.

Van der Valk P, van Kalken CK, Ketelaars H, Broxterman HJ, Scheffer G, Kuiper CM, Tsuruo T, Lankelma J, Meijer CJLM, Pinedo HM and Scheper RJ (1990) Distribution of multidrug resistance-associated P-glycoprotein in normal and neoplastic human tissues. Ann Oncol 1: 56–64.

Visser JWM and van Bekkum DW (1990) Purification of pluripotent hematopoietic stem cells: past and present. Exp Hematol 18: 248–256.

Weide R, Dowding C, Paulsen W and Goldman J (1990) The role of the MDR-1/P-170 mechanism in the development of multidrug resistance in chronic myeloid leukemia. Leukemia 4: 695–699.

Wittebol S, te Boekhorst P, Hagemeijer A, van Dongen JJM, Schoester M and Sonneveld P (1992) Expression of the multidrug resistance (mdr-1) phenotype in acute myelocytic leukemia is associated with CD34 expression and monosomy 7 and predicts for poor survival. Blood 80: 202a.

Yahanda AM, Adler KM, Fischer GA, Brophy NA, Halsey R, Hardy RI, Gosland MP, Lum BL and Sikic BI (1992) Phase I trial of etoposide with cyclosporin as a modulator of multidrug resistance. J Clin Oncol 10: 1642–1634.

Zhou DC, Marie JP, Suberville AM and Zittoun R (1992) Relevance of *mdr1* gene expression in acute myeloid leukemia and comparison of different diagnostic methods. Leukemia 6: 879–885.

Address for correspondence: K. Nooter, Department of Medical Oncology, Rotterdam Cancer Institute, P.O. Box 5201, 3008 AE Rotterdam, The Netherlands.

Cytotechnology **12**: 231–256, 1993.

Human cell lines as models for multidrug resistance in solid tumours

Martin Clynes, Mary Heenan and Keara Hall
National Cell and Tissue Culture Centre/BioResearch Ireland, Dublin City University, Dublin 9, Ireland

Key words: antisense, cancer chemotherapy, carcinoma, clonal variation, cytokines, *mdr*-1 gene, multidrug resistance (MDR), P-glycoprotein, solid tumour

Abstract

In spite of our expanding knowledge on the molecular biology of cancer, relatively little progress has been made in improving therapy for the solid tumours which are major killers, e.g., lung, colon, breast. Significant advances over the past 10—15 years in chemotherapy of some tumours such as testicular cancer and some leukaemias indicates that, in spite of the undesirable side-effects, chemotherapy has the potential to effect cure in the majority of patients with certain types of cancer. Multidrug resistance, inherent or acquired, is one important limiting factor in extending this success to most solid tumours. *In vitro* studies described in this review are now uncovering a diversity of possible mechanisms of cross-resistance to different types of drug. Sensitive methods such as immunocytochemistry, RT-PCR or *in situ* RNA hybridisation may be necessary to identify corresponding changes in clinical material. Only by classifying individual tumours according to their specific resistance mechanisms will it be possible to define the multidrug resistance problem properly. Such rigorous definition is a prerequisite to design (and choice on an individual basis) of specific therapies suited to individual patients. Since a much larger proportion of cancer biopsies should be susceptible to accurate analysis by the immunochemical and molecular biological techniques described above than to direct assessment of drug response, it seems reasonable to hope that this approach will succeed in improving results for cancer chemotherapy of solid tumours where other approaches such as individualised *in vitro* chemosensitivity testing have essentially failed. Results from clinical trials using cyclosporin A or verapamil are encouraging, but these agents are far from ideal, and reverse resistance in only a subset of resistant tumours. Proper definition of the other mechanisms of MDR, and how to antagonize them, is an urgent research priority.

Abbreviations: MDR — multiple drug resistance; P-170 = pgp — P-glycoprotein = product of *mdr*-1 gene

Introduction

Multidrug resistance (MDR) is a phenomenon by which cancer cells *in vitro* and in patients' tumours may exhibit intrinsically, or develop, cross-resistance to a number of wide-ranging (but not universal) sets of structurally and mechanistically unrelated chemotherapeutic agents. The most frequently described set is the "classical MDR" pattern, involving overexpression of the plasma membrane efflux pump P-glycoprotein, conferring resistance to anthracyclines, epipodophyllotoxins, actinomycin D, vinca alkaloids and taxol. Different aspects of MDR have been discussed in other chapters of this special issue, and many reviews on the molecular biology and cellular biology, pharmacology and clinical aspects of this problem have been published (Bradley *et al.*, 1988; Clynes *et al.*, 1990; Clynes 1993; Croop *et al.*, 1988; Dietel 1993; Endicott and Ling 1989; Ferguson and Cheng 1989a; Germann *et al.*, 1993; Gottesman, 1993; Gottesman and Pastan 1993; Harris and Hochhauser 1992; Ford and Hait 1990; Hill 1991; Kaufman 1991; Nooter and Herweijer 1991; Roninson 1991).

This review focuses on a few specific questions, in particular:

- How can we identify (and quantitate the relative contributions of) the mechanisms causing resistance in a particular cell line?
- What mechanisms of resistance have been observed *in vitro* in human cell lines derived from different types of solid tumour, and how well do these mechanisms correlate with behaviour of patients' tumours *in vivo*?
- Do tissue-specific mechanisms of resistance exist?
- What new avenues of research are opening up which might allow development of better therapy of MDR in cancer patients?

1. What is the molecular basis of multidrug resistance?

1.1 Investigation of resistance mechanisms in MDR lines

Most of what we know about the mechanisms of multiple drug resistance has originated from demonstration of *correlations* between alteration in particular biochemical properties and resistance to a range of chemotherapeutic agents *in vitro*. The alteration may involve increased expression, e.g., P-glycoprotein (Gottesmann, 1993), reduced expression, e.g., topoisomerase II (Fry *et al.*, 1991; Giaccone *et al.*, 1992) or mutation, e.g., topoisomerase II (Bugg *et al.*, 1991; Nitiss *et al.*, 1993) of a specific protein.

Increased gene expression may be monitored at several levels. Altered expression of specific mRNAs can be evaluated using Northern blotting or more sensitive techniques such as RNase protection assays which are less affected than Northern blots by partial mRNA degradation (Zaman *et al.*, 1993) and reverse transcriptase – PCR (O'Driscoll *et al.*, this issue). Altered gene expression may also be detected by differential or subtractive hybridisation strategies and such approaches are proving useful in identifying changes associated with MDR and also with phenomena such as senescence and carcinogenesis (Calvet, 1991; Cole *et al.*, 1992b; Cole and Deeley 1993; Murano *et al.*, 1991; Schalken 1990). At the level of specific protein, relevant techniques include immunocytochemistry/immunofluorescence (Van der Valk *et al.*, 1990; Verelle *et al.*, 1991; Volm *et al.*, 1991a); Western blotting (Georges *et al.*, 1990) and where high levels of protein are involved, SDS-PAGE and staining for proteins (e.g., using sensitive silver staining techniques). In some instances, assessment of functional activity is possible, for example for topoisomerase II activity (Hofmann *et al.*, 1990) glutathione-S-transferase activity (See Moscow and Dixon, this issue) or drug uptake (Spoelstra, *et al.*, 1992).

One of the problems with correlative evidence, however, is that so many different properties in addition to those just mentioned have been associated or correlated with resistance in at least some cell lines. Such properties include:

Elevated levels of non-P-glycoprotein membrane-bound proteins including an 85 kD protein (Sugimoto *et al.*, 1993) and a 110,000 kD protein whose pattern of expression seems to relate inversely to that of P-glycoprotein (Scheper *et al.*, 1993; Miguel Izquierdo, personal communication) and overexpression of C subunit of a vacuolar H^+-ATPase (Ma and Center, 1992). Some MDR lines are also radioresistant whereas others are not (as discussed below, Section 3.5); some MDR lines show altered susceptibility to immune effector cells (Yanovich *et al.*, 1986) but this is not a universal property of MDR lines (Harker *et al.*, 1990) and may not be related to P-glycoprotein expression (Rivoltini *et al.*, 1990). Some MDR lines show altered protein kinase C levels (Lee *et al.*, 1992;

Dolfini *et al.*, 1993) but again this may not be a universal correlate of resistance (Epand and Stafford, 1993). Overexpression of the cytosolic calcium-binding protein sorcin has been reported in some P-glycoprotein-mediated MDR cells, but its functional role, if any, in MDR is unclear (see Reeve, *et al.*, 1990, for references). EGF receptor levels are reduced in some MDR lines and increased or unchanged in others (Reeve *et al.*, 1990; Slovak *et al.*, 1991). Other changes which have been associated with MDR include altered translocation of a nucleolar phosphoprotein (B23) (Sweet *et al.*, 1989), elevated cytoplasmic calcium levels (Nygren *et al.*, 1991), increased resistance to mechanical disruption (Redmond *et al.*, 1990) and altered membrane lipid composition (Alon *et al.*, 1991). It should also be remembered that environmental conditions *in vitro*, (and probably also *in vivo*) alter drug sensitivity, as for example in reversible hypoxia-induced drug resistance (Sakata *et al.*, 1991) and confluence-dependent resistance (Dimanche-Boitrel *et al.*, 1992).

When we try to determine the relevance of particular biochemical alterations to resistance in human cancer patients, we generally use probes (genetic or immunochemical) developed using cell culture models. In general, we can obtain only correlative evidence from patient studies, as we attempt to define which of the mechanisms operating *in vitro* have any relevance to clinical resistance. Of course, in limited instances where circumvention therapy is being used, and the cellular targets are at least partly understood, as in the case of verapamil and cyclosporin A (Ford and Hait, 1990; Pennock *et al.*, 1991; Sonneveld *et al.*, 1992), we can obtain some direct evidence for operation of a specific resistance mechanism in cancer patients.

In cell culture systems, certain experimental approaches are available which permit evaluation of the direct role of a particular mechanism in determination of drug resistance. Some of these approaches also permit some sort of quantitative evaluation of the contribution of a particular mechanism to the resistance level observed. In view of the very large number of biochemical changes which have been found to be associated or correlated with MDR in cell lines *in vitro*, it is most important that the field should be narrowed down to those changes which are not just associated with, but actually directly involved in determining the drug resistance phenotype. It is also important that researchers should try to assess quantitatively to what extent each mechanism contributes to resistance; it is not sufficient, for example, to demonstrate P-glycoprotein overexpression in a cell line and then to assume that no other mechanisms of resistance may be relevant in the cell line or tissue being examined. Nielsen and Skovsgaard (1992) have recently published a critical evaluation of published results on cell lines with increased expression of P-glycoprotein. While in general a good correlation between P-glycoprotein overexpression, levels of resistance, and reduced drug accumulation and/or increased efflux was found, exceptions existed, e.g., cell lines with increased resistance and increased P-glycoprotein expression, but without increased drug efflux.

The approaches to demonstrating direct involvement of a gene product in determination of resistance, and to evaluating quantitatively the contribution of each mechanism to resistance include:

a. cDNA transfection
b. Use of antisense oligonucleotides
c. Use of specific antagonists or inhibitors to reverse resistance or to reverse a resistance-related function (e.g., drug uptake or afflux).

1.2 cDNA transfection

Transfection with cDNA has, for example, demonstrated that the *mdr*-1 gene product can confer multidrug resistance on sensitive murine (Schurr *et al.*, 1989) and human (Ueda *et al.*, 1987) cell lines. Of equal importance, cDNA transfection has been used to demonstrate that certain gene products which had been associated with multidrug resistance could not on their own confer drug resistance on sensitive cell lines. Examples of such negative results include transfection of human BRO melanoma cells with cDNA for human *mdr*-3 (*alias* mdr-2), (Schinkel *et al.*, 1991) and transfection with a cDNA for an 85 kD resistance associated protein (Sugimoto *et al.*, 1993); in both cases, transfectants expressing the relevant protein did not exhibit altered drug-resistance patterns. Even apparently clearcut results such as these have to be interpreted with some caution, however. Even single mutations in a cDNA could render the protein product inactive, or alter its substrate specificity (Choi *et al.*, 1988; Kioka *et al.*, 1989; Devine *et al.*, 1992). Exact patterns of cross-resistance differ even when functional *mdr*-1 cDNA is used to transfect different cell lines, or when different transfectants of the same line are examined (Fairchild *et al.*, 1990; Lincke *et al.*, 1990; Schurr *et al.*, 1989; Sugimoto and Tsuruo, 1987). The contribution of cellular background to expression of different MDR phenotypes is still poorly understood, but the possibility exists that different cells may contain different patterns of accessory molecules (e.g., proteins) which modify functional expression of P-glycoprotein or other resistance-determining gene products. Point mutations in the cDNA (Choi *et al.*, 1988; Kioka *et al.*, 1989; Devine *et al.*, 1992) as well as posttranslational modifications such as glycosylation and phosphorylation (Kramer *et al.*, 1993; Bates *et al.*, 1992) may also determine specificity for different drugs. The precise importance of glycosylation and phosphorylation in P-glycoprotein function is still somewhat unclear, and the most recent developments are discussed by Germann elsewhere in this special issue.

1.3 Transfection with antisense oligonucleotides

If a particular cellular function can be blocked by treatment with specific antisense oligonucleotides (but not by the corresponding sense oligonucleotide) it provides strong evidence for direct involvement of the relevant gene product in supporting that function (Uhlmann and Peyman, 1990). Successful reversal of drug resistance by treatment with different *mdr*-1-specific antisense oligonucleotides has been reported by Rivoltini *et al.*, (1990), Clynes *et al.* (1992), and Efferth and Volm (1993) although other workers have been unable to demonstrate such an effect (various personal communications). It is of interest that Clynes *et al.* (1992) found a *human mdr*-1-specific antisense oligonucleotide to be capable of reversing resistance in both human and hamster MDR lines.

A related approach is the use of hammerhead ribozymes. This approach has been used, for example, to demonstrate the role of c-*fos* gene expression in determination of *cis*-platin resistance (Scanlon *et al.*, 1991; Funato *et al.*, 1992).

1.4 Use of specific antagonists

Specific antagonists of, for example, P-glycoprotein function have proved useful in evaluation of the role of specific proteins in resistance. Experimental approaches include examining the effect of such inhibitors on resistance (Ford and Hait, 1990) or on some aspect of drug transport such as drug accumulation or efflux (Spoelstra *et al.*, 1992). Verapamil and Cyclosporin A have been used as antagonists of P-glycoprotein function (Ford and Hait, 1990) but even here there is a need for caution. Nygren *et al.* (1991), for example, have demonstrated verapamil reversal of resistance in a human lung small cell carcinoma MDR cell line which does not over-express P-glycoprotein. It is clear that definite mechanistic conclusions should not be drawn based on use of a single antagonist.

Buthionine sulfoximine (BSO) is an inhibitor of gamma-glutamyl-cysteine synthetase, and therefore depletes cellular glutathione (GSH) levels; it may be used to probe for the involvement of GSH-dependent systems in resistance (Dusre *et al.*, 1989; Lau *et al.*, 1991; Richardson and Sieman, 1992).

1.5 Clonal variation in multidrug-resistant cell lines

The possible importance of clonal variation in multidrug-resistant cell lines has not received much attention. The majority of MDR cell lines described in the literature are not cloned. Where multiple mechanisms of resistance are shown to exist, it is not always clear whether the different mechanisms coexist within the same cells, or are segregated in different cellular subpopulations. Clonal variation with respect to drug resistance has been observed in human tumour cell lines (Ferguson and Cheng, 1989b). It is unclear, during selection of MDR lines, whether subpopulations within the parental line are the main targets for selection, or whether random genetic change within any cells in the population is the basis for selection. Glisson and Alpeter (1992) suggested that the pulse treatment method which they used to select VP-16-resistant MDR variants might tend to select for pre-existing subpopulations. Yang and Trujillo (1990) and Yang *et al.* (1992, 1993) have studied adriamycin-resistant sublines, as well as two *cis*-platin-resistant subclones (Yang *et al.*, 1993) of the human colon carcinoma line LoVo; Dolfini *et al.*, (1993) have also studied MDR subclones of this line. Yang *et al.*, found quite different patterns of cross-resistance and also different mechanisms of resistance between the *cis*-platin resistant clones and between the adriamycin-resistant sublines. By plating cells at clonal densities in different concentrations of adriamycin, we have demonstrated substantial clonal variation in resistance in human MDR lung (Clynes *et al.*, 1992) and ovarian (Redmond *et al.*, 1993) cell lines. Since the short half-life of adriamycin in culture medium (approx. 29 h — Hildebrand-Zanki and Kern, 1986) could result in survival and subsequent growth of quiescent cells in the population, leading to apparent resistance in this assay, the experiments were repeated using changes of fresh drug media, with identical results (M. Heenan and M. Clynes, unpublished). Subsequently, 10 independent clones were isolated from the human lung squamous cell carcinoma MDR line, DLKP-A; these showed different patterns of cross-resistance to adriamycin, vincristine and VP-16, and spanned a 10-fold difference in level of adriamycin resistance (M. Heenan and M. Clynes, manuscript in preparation). It is difficult to understand how the more sensitive cells can survive in drug concentrations which should be lethal for them, and we have suggested that they survive by metabolic cooperation in co-culture (Chanson and Meda, 1993) with the more resistant clones, which are very actively extruding drug (Redmond *et al.*, 1993; I. Cleary and M. Clynes, manuscript in preparation). Gap junctions are known to function, although sometimes in an altered fashion, in cancer cells (Holder *et al.*, 1993). Flow cytometry, based on adriamycin fluorescence, can be used to sort more resistant cells in mixed populations, and the results lend further support to the idea of clonal variation with respect to drug resistance in unselected tumour cell populations (Watson, 1991; Zou *et al.*, 1992). Tumor cell heterogeneity *in vivo* is being studied actively (Arvan, 1992; Fey *et al.*, 1993; Zangwill *et al.*, 1993; Nguyen *et al.*, 1993) and includes variation in DNA repair (Powell and McMillan, 1991) and in response to chemotherapy (Asselin, 1992). If significant heterogeneity with respect to drug resistance levels and mechanisms exists within individual tumours, it could be important in determining response to treatment with drugs and circumvention agents, and in the general biological behaviour of the tumour.

2. MDR in cell lines derived from different solid tumour types

2.1 Are mechanisms of MDR tissue- or cell type-specific?

It is not as yet completely clear whether different mechanisms of MDR apply in different tissues or cell types, or whether any of the mechanisms may operate, through aberrant/ectopic gene expression, in any type of tumour. P-glycoprotein expression, for example, has been reported to occur in a wide range of different tumour types, even those derived from tissues not normally expressing P-glycoprotein (van der Valk *et al.*, 1990).

We and others (e.g., Toffoli *et al.*, 1992) have found that some cell lines quite readily generate MDR variants by adaptation in the presence of adriamycin or other drugs, but other lines do not. We have also found some lines which readily generate variants resistant to *cis*-platin but not to adriamycin or vincristine. What determines the diversity in ability to generate variants resistant to different drugs is not known. Within a tumour or cell line which has developed an MDR pattern, it is difficult to ascertain with certainty whether a preexisting subpopulation (expressing, for example, P-glycoprotein) has been selected, or whether the tumour cells have generated new variants. *In vitro* studies indicate that the latter can occur during prolonged exposure to increasing levels of chemotherapeutic drugs. On the other hand, *mdr*-1 gene expression is regulated by a number of external stimuli. Chaudhary and Roninson (1993) have recently demonstrated that transient exposure of cells to chemotherapeutic agents can induce expression of *mdr*-1 mRNA (and that this induction can be blocked by protein kinase C inhibitors). Exposure to P-glycoprotein antagonists may also increase *mdr*-1 gene expression (Herzog *et al.*, 1993). Expression of *mdr*-1 mRNA can also be altered by environmental stimuli such as heat shock (Chin *et al.*, 1990) and oncogene products such as *ras* and p53 may interact with the *mdr*-1 promotor, causing positive or negative regulation (Chin *et al.*, 1992).

In this section we will review what is known about mechanisms of MDR in cell lines derived from different tumour sites, and how well these mechanisms correlate with studies on biopsies from tumours of the same sites. The discussion will be mainly, but not exclusively, confined to cell lines of human origin.

It is worth emphasising that cell line cross-contamination is a surprisingly frequent event even in the most experienced of cell culture laboratories (Gold, 1986). It is very important to bear this in mind if several drug-resistant lines are being used in a laboratory. Appropriate precautions should be observed, e.g., separate media, reagents, pipettes, etc., for different cell lines; stock cultures of different lines should never be handled at the same time; use of a "clean-out" interval in lamina flow cabinet between handling of different cell lines. Positive authentication of the lines should also occur periodically using the now readily accessible technique of DNA fingerprinting and/or other established characterisation techniques (Hay *et al.*, 1992; MacLeod *et al.*, 1992).

2.2 Ovarian carcinoma

Ovarian cancer is the fourth most frequent cause of cancer deaths in women. Ovarian carcinomas are frequently responsive to chemotherapy, combinations of *cis*-platinum, cyclophosphamide and adriamycin being most commonly used (ten Tije and Wils, 1992; van der Zee *et al.*, 1992). Unfortunately, after an initial response, many tumours reappear in a drug un-responsive form.

Van der Zee *et al.* (1991) have found large variability in expression of P-glycoprotein, topoisomerase I and topoisomerase II in ovarian tumours. Holzmayer *et al.* (1992) found, using a sensitive RT-PCR technique, that expression of *mdr*-1 mRNA (even at a low level, possibly representing higher level expression in a small proportion of the tumour cells) correlated well with resistance in ovarian cancer patients to chemotherapy not only with adriamycin, but also with drugs like *cis*-platin and cyclophosphamide which are not believed to be substrates for P-glycoprotein. These authors suggested that a subpopulation of the tumour cells, overexpressing *mdr*-1, may also possess other mechanisms of resistance, relevant to cyclophosphamide and to *cis*-platin. Redmond *et al.*, (1993) selected (by adriamycin exposure) an ovarian MDR line (OAW-42-A) resistant to adriamycin (and to a lesser extent to epipodophyllotoxins and vinca alkaloids) but also to *cis*-platin. This line may therefore be a good model for studying the multiple mechanisms of resistance postulated by Holzmayer *et al.* Redmond *et al.*, also showed that verapamil or *mdr*-1-specific antisense oligonucleotide treatment caused only partial reversal of resistance to adriamycin, again indicating the additional involvement of non-P-glycoprotein-mediated mechanisms.

A number of *in vivo* studies have indicated that glutathione-related systems may not be important determinants of *cis*-platin resistance in ovarian tumours (Murphy *et al.*, 1992; Van Der Zee *et al.*, 1992) even though an ovarian MDR line ES-2R selected with an adriamycin derivative and cross-resistant to VP-16, bleomycin, adriamycin and ionizing radiation, (but not to vincristine) had sensitivity restored by BSO treatment and showed elevated levels of GST-π and other GST-related enzymes (Lau *et al.*, 1991). This line, however, also shows reduced topoisomerase II content, increased generation time and possibly enhanced DNA repair. There is evidence (from BSO treatment) that intracellular GSH levels may be important in radiosensitivity in human ovarian carcinoma cells, at least *in vitro* (Louie *et al.*, 1985).

Bradley *et al.* (1989) derived, by exposure to vinca alkaloids, a series of MDR variants of the human ovarian carcinoma line SKOV3; these spanned a range of resistance from 2-fold to 10,000-fold to vinca alkaloids and 2-fold to 500-fold to adriamycin. Resistance to adriamycin increased (generally but not exactly) in parallel with resistance to the vinca alkaloids. Resistance to vincristine was generally higher than to vinblastine (except at low levels of resistance) and these in turn were greater than fold-resistances to colchicine. At lower levels of resistance, increases in mdr-1 mRNA and protein were observed; at higher levels, *mdr*-1 gene amplification was observed also.

Bénard *et al.* (1989) also described vincristine-selected MDR variants of a human ovarian carcinoma line (IGROVI), which were cross-resistant to vinca alkaloids, adriamycin, actinomycin D and VP-16. Sensitivity was enhanced by verapamil in the resistant cells, but although *mdr*-1 transcript levels were elevated (as determined by Northern blot), there appeared to be no *mdr*-1 gene amplification as evaluated by Southern blot analysis. Broxterman *et al.*, (1990) found that both verapamil and cyclosporin A enhanced daunorubicin accumulation in the human ovarian carcinoma MDR line, 2780AD; however, cyclosporin A was more potent in this effect, and unlike verapamil, it did not cause a rapid increase in ATP consumption and lactate production. These authors therefore suggested that verapamil and cyclosporin A antagonise P-glycoprotein activity by different mechanisms. Reduced drug accumulation may be an important mechanism of cisplatin resistance in human ovarian carcinoma lines, but the precise molecular mechanisms involved are still obscure, and it is unlikely that P-glycoprotein is involved (Loh *et al.*, 1992). In a recent study of eight human ovarian carcinoma lines, Sekiya *et al.* (1992) found *mdr*-1 mRNA overexpression in only the most vincristine- and adriamycin-resistant cell lines, and gene amplification in none of the lines. In the other lines, differences in resistance did not correlate clearly with altered P-glycoprotein expression.

The literature to date, therefore, indicates that P-glycoprotein may play some role in drug resistance in human ovarian cancer cells, and that expression of *mdr*-1 mRNA may be a useful marker of resistance (even to non-MDR drugs such as *cis*-platin and cyclophosphamide). Different patterns of cross-resistance may be found in different ovarian tumours *in vitro* and *in vivo* (Wilson *et al.*, 1987). It is clear however that other, hitherto undiscovered, mechanisms are also involved and it is important to identify these if ovarian carcinoma is to be effectively subclassified and treated.

2.3 Colorectal carcinoma

Colorectal carcinoma is generally considered to be inherently resistant to chemotherapy, although drugs are sometimes used in an attempt to limit progress of the disease. Various xenobiotic metabolising enzymes may play a part in drug resistance in human colon cells (Masaad *et al.*, 1992). Since P-glycoprotein is expressed in normal colon, and has also been detected in some colon tumours (see Kramer *et al.*, 1993 for references) it has been suggested that P-glycoprotein may be involved in intrinsic resistance of colorectal cancers to chemotherapy. The level or pattern (cytoplasmic, luminal, paranuclear) of expression of P-glycoprotein (unlike PCNA and EGF receptor) as determined by immunohistochemistry, however, was found not to predict poor prognosis in colorectal cancer (Mayer *et al.*, 1993). Overexpression of P-glycoprotein has been reported in a number of MDR colon carcinoma lines (Toffoli *et al.*, 1992) including LS-180-derived sublines (Herzog *et al.*, 1993) in which P-glycoprotein shows increased expression following treatment with antagonists such as verapamil, nifedipine and cyclosporin A, probably via post-transcriptional mechanisms; MDR variants of LoVo (Danova *et al.*, 1992); and MDR variants of SW620 which (in addition to increased *mdr*-1 overexpression) have increased Glutathione-S-Transferase levels and thiol pools (Chao *et al.*, 1992). Kramer *et al.* (1993) noted *mdr*-1 mRNA expression in liver metastases of 7 out of 11 patients with colon carcinoma. These authors also surveyed a range of human colorectal cancer cell lines for *mdr*-1 mRNA expression, P-170 biosynthesis and surface expression, daunomycin accumulation and cross-resistance to different drugs. Their results emphasise the importance of phosphorylation status and surface localization of the mature

170 kD protein in determination of resistance. They found that etoposide resistance did not clearly correlate with P-glycoprotein levels, and that the pattern of cross-resistance to different drugs differed significantly in different lines,leading them to conclude that non-P-glycoprotein mechanisms of resistance must also be important in colorectal tumour cell lines. Bates *et al.* (1992) showed that sodium butyrate (a differentiation promoting agent) decreases phosphorylation of P-glycoprotein making it essentially nonfunctional in export of adriamycin, actinomycin D and vinblastine, but still functional for colchine export; it was noted that P-170 contains many potential protein kinase C phosphorylation sites, although other protein kinases may be important also.

Dalton *et al.*, (1988) have described a mitoxantrone- and adriamycin-resistant variant of WiDr which does not overexpress P-glycoprotein. Yang and Trijillo (1990) and Yang *et al.* (1992 and 1993) have found quite different patterns of cross-resistance in different adriamycin-resistant sublines and *cis*-platin resistant clones isolated from LoVo, again indicating multiple mechanisms (possibly other membrane proteins in this case) co-existing not only within individual cells but in the cell population. These postulated additional mechanisms remain obscure, but there is evidence from work on human colorectal cell lines of involvement of elevated protein kinase C (Dolfini *et al.*, 1993), DNA repair enzymes as well as glutathione-related enzymes (Redmond *et al.*, 1991) altered drug uptake, distribution or sequestration mechanisms (Toffoli *et al.*, 1991) and altered expression of cytoskeletal proteins (E. Dolfini, personal communication). Confluent cultures of human colon cancer cells are more resistant to drugs than are their proliferating counterparts (Dimanche-Boitrel *et al.*, 1992).

2.4 Lung carcinoma

Small cell carcinoma is frequently responsive to chemotherapy, while nonsmall cell carcinoma (NSCLC) has generally been considered as an intrinsically resistant tumour type. There is increasing interest in the mechanism of this inherent resistance, and in fact recent studies have indicated that at least some NSCLCs may show response to chemotherapy. Clinical trials (Millward *et al.*, 1993) showing improved median survival in advanced NSCLC patients treated with verapamil in addition to vindesine and ifosfamide may indicate a role for P-glycoprotein in inherent resistance *in vivo*.

Recent *in vitro* work also supports the idea that a range of different drug sensitivities may exist within the NSCLC tumour range (Volm *et al.*, 1991a; 1992; Wilbur *et al.*, 1992). Low levels of *mdr*-1 mRNA are found in a proportion of both normal and malignant lung biopsies (Lai *et al.*, 1989). RT-PCR analysis of human lung and ovarian tumour tissue indicates that even low levels of *mdr*-1 expression may be a marker (although not necessarily sole determinant) of multidrug resistance *in vivo* (Holzmayer *et al.*, 1992). Recent work by Volm *et al.*, (1991a) using immunocytochemistry (which, in contrast to Western blotting allows assessment of protein expression in a small subpopulation of cells) established a correlation between expression of P-glycoprotein, GST-π, smoking history and adriamycin sensitivity for NSCLC; no correlation with topoisomerase II activity was found in this study. Milroy *et al.* (1992) did not detect P-glycoprotein expression (by C219-immunocytochemistry) in cell lines derived from four untreated and three postchemotherapy relapsed small cell lung carcinomas.

Reduced levels or alterations in catalytic activity of topoisomerase II have been correlated with increased resistance in a number of small cell and nonsmall cell lung carcinoma lines *in vitro* (de Jong *et al.*, 1990; Long *et al.*, 1991; Kasahara *et al.*, 1992; Giaccone *et al.*, 1992); generally, reduced activity of topoisomerase II increases resistance to anthracyclines and epipodophyllotoxins (but not to vinca alkaloids) and is sometimes associated with hypersensitivity to *cis*-platinum (Long *et al.*, 1991).

Overexpression of P-glycoprotein has been reported in many human MDR lung cell lines, both small cell (Reeve *et al.*, 1989; Jensen *et al.*, 1993) and nonsmall cell (Baas *et al.*, 1990; Clynes *et al.*, 1992). Tsai *et al.* (1993) have reported a close correlation between chemosensitivity in human

NSCLC lines, and overexpression of HER-2/neu mRNA; no correlation with *ras* mutations was observed. The biochemical basis for the observed correlation is not known.

Coley *et al.* (1991) have described an adriamycin-resistant variant of the human large-cell carcinoma line CORL-3 which is cross-resistant to vinca alkaloids, VP-16 and certain anthracyclines, but does not over-express P-glycoprotein, has reduced levels of EGF receptors (Reeve *et al.*, 1990) and does not show altered resistance on verapamil treatment; a non-P-glycoprotein ATP-dependent efflux mechanism may be involved.

MDR variants with different resistance mechanisms can be selected from the same parental cell line by exposure to the same drug. Baas *et al.* (1990) described variants of the human squamous lung carcinoma line SW1573 selected with different levels of adriamycin; the lower-level resistant variants did not overexpress *mdr*-1 mRNA or P-glycoprotein (in fact the levels in these resistant lines appeared to be reduced compared to the parental cell line) whereas the higher level variants expressed both. During longer-term selection with higher adriamycin levels, mdr-1 gene amplification also occurred. Both types of MDR variant were cross-resistant to anthracyclines, VP-16, gramicidin D and vincristine, but the lower-level variants showed preferential resistance towards VP-16 (Kuiper *et al.*, 1990). With increasing resistance, decreased drug accumulation, increased efflux and decreased localisation of adriamycin in the nucleus relative to the cytoplasm occurred (Keizer *et al.*, 1989). The altered nuclear: cytoplasmic ratio was clear for both P-170-positive and -negative lines (Schuurhuis *et al.*, 1991). Cytosolic pH also increased with increasing resistance but was restored to normal by verapamil treatment (Keizer and Joenje, 1989) but the relevance of intracellular pH changes to resistance determination is not clearly established (Boscoboinik *et al.*, 1990). Different cytogenetic alterations have been reported in non-P-170-(deletion or translocation in short arm of chromosome 2) and P-170 (rearrangement of long arm of chromosome 7) expressing variants of SW-1573 (Nieuwint *et al.*, 1992).The non-P-170 variants also have lower levels of topoisomerase II than the parental line, but in cell fusion experiments between lethally γ-irradiated resistant cells and drug-sensitive cells, the hybrids express an MDR phenotype, reduced drug accumulation (and the unexpected *reduced mdr*-1 mRNA levels seen in the low level resistant variant) but the topoisomerase II level reduction is not seen (Eijdems *et al.*, 1992).

A monoclonal antibody raised against one of the low-level resistant SW1573 variants detects a 110 kD vesicular protein present in some normal epithelial cells and overexpressed in a variety of MDR lines; this protein may have a role in resistance determination (Scheper *et al.*, 1993; M. Izquierdo and R. Scheper, Personal Communication).

Clynes *et al.* (1992) have described an adriamycin-selected MDR variant (DLKP-A) of a poorly differentiated human lung squamous cell carcinoma line which overexpresses P-glycoprotein as determined by Western blot and immunocytochemistry (and also overexpresses *mdr*-1 mRNA (O'Driscoll and Clynes, unpublished). Treatment of this line with verapamil or with *mdr*-1-specific antisense oligonucleotides results in only partial reversal of adriamycin resistance, indicating the co-existence of other resistance mechanisms. Western blotting and RT-PCR experiments suggest that glutathione-S-transferase and topoisomerase II levels are unaltered in this line (G. Grant, M. Heenan and M. Clynes, manuscript in preparation). Adriamycin uptake experiments showing only partial reversal of the drug accumulation deficit in the resistant variant by verapamil or cyclosporin A (I. Cleary and M. Clynes, manuscript in preparation) suggest that a non-P-glycoprotein-mediated drug transport mechanism may exist in this cell line.

A camptothecin-11-resistant variant of an NSCLC line with low-level cross-resistance to adriamycin and 5-fluorouracil has been described, and has a reduced level of DNA topoisomerase I (Kanzawa *et al.*, 1990).

Long *et al.* (1991) have described variants of the human adenocarcinoma line A549, selected with VP-16 or VM-26, which are cross-resistant to adriamycin (but not to actinomycin D, colchicine or mitomycin C) and which are hypersensitive to *cis*-

platin. These lines do not overexpress P-glycoprotein (at protein or mRNA levels), and have reduced topoisomerase II mRNA and protein levels.

A non-P-glycoprotein-expressing MDR variant (H69AR) of the human small cell lung cancer line NCI-H69 was selected by exposure to adriamycin; it is cross-resistant to vinca alkaloids, VP-16, and other anthracyclines, but not to bleomycin, 5-fluorouracil or carboplatin (Mirski *et al.*, 1987). Differences in drug accumulation are apparently not involved in resistance, and verapamil does not reverse resistance. Monoclonal antibodies raised to the resistant line detected specific antigenic differences between the parental and resistant cell line (Mirski and Cole, 1989, 1991); one of the differences relates to overexpression of a 36 kD calcium- and phospholipid-binding protein, Annexin II/Lipocortin, in the resistant line (Cole *et al.*, 1992a). The resistant line also has reduced levels of topoisomerase II and decreased susceptibility to drug-induced DNA damage (Cole *et al.*, 1991). Using differential hybridisation of cDNA libraries from parental and resistant cell lines, Cole *et al.* (1992b) identified a novel gene (for sequence correction see Cole and Deeley, 1993), termed MRP, which is overexpressed in the resistant variant, and whose levels are again reduced in a revertant line. The MRP gene is amplified in H69AR cells, and also in an adriamycin-resistant variant of the fibrosarcoma line HT1080; the chromosomal location of the MRP gene has been investigated in these cell lines (Slovak *et al.*, 1993). Zaman *et al.*, (1993) used an RNase protection assay to estimate expression of MRP mRNA in ten SW-1573-derived non-P-170 MDR lines (in which they found no alteration in expression) and in a resistant subline of the small cell lung carcinoma line GLC4 (in which they found a 25-fold overexpression). We have found (O'Driscoll and Clynes, unpublished) no obvious alteration (by RT-PCR) in MRP mRNA levels in the lung squamous cell carcinoma MDR line DLKP-A (Clynes *et al.*, 1992) or in the ovarian carcinoma MDR line OAW-42-A (Redmond *et al.*, 1993) relative to their sensitive parental lines. It has been suggested that the MRP gene product may be a 190 kD protein which has been associated with MDR in several cell lines (see Center, elsewhere in this issue). The results to date indicate that MRP may have a role in some cases of resistance, at least in small cell lung carcinoma. Using antisera to peptides from the deduced MRP sequence, Krishnamachary and Center (1993) have identified a 190 kD protein, found predominantly in endoplasmic reticulum in adriamycin-resistant HL60 cells, which may be the product of the MRP gene.

It is interesting to note that adriamycin selection starting with the same parental cell line can lead to MDR lines with completely different mechanisms. Reeve *et al.*, (1989) found, in contrast to H69AR, amplification and overexpression of the *mdr*-1 gene in another MDR variant (termed LX4) of NC1-H69.

Glisson and Alpeter (1992) selected an MDR variant of H69 cells, by exposure to pulses of VP-16; *mdr*-1 mRNA was overexpressed in the variant, but drug accumulation was only slightly reduced and topoisomerase II levels were unchanged. Cyclosporin A (but not verapamil) enhanced VP-16 cytostasis in the variant, although both agents had the same effect on drug accumulation. Once again, a hitherto unknown mechanism of resistance may be involved; the authors suggest that the selection method may have led to isolation of a pre-existing subpopulation from the H69 cell line. The cell line was cross resistant to VP-16 (20×), vincristine (20×), and daunomycin (5×), but not to X-rays or *cis*-platin. Ohta *et al.* (1993) have established a vindesine resistant variant of H69 which is cross-resistant to vincristine, colchicine and taxol, but not to VP-16, adriamycin and *cis*-platin; *mdr*-1 mRNA is not overexpressed, and alterations in cellular tubulin may be involved. This line illustrates the fact that not all selections with "MDR" drugs result in multidrug-resistant cell lines. Other cell lines resistant to microtubule-directed drugs have been described (Cabral and Barlow, 1989). Twentyman *et al.* (1992) have established a panel of human NSCLC and small-cell lung carcinoma lines which are resistant to *cis*-platin and melphalan; drug uptake is altered in some of the lines, as are levels of glutathione-S-transferases, but not of glutathione. Kasahara *et al.* (1992) have established a correlation between *cis*-platin resistance and metallothionein content in *cis*-platin-resistant human small cell lung cancer lines derived from H69. Jensen *et al.* (1993) have examined the cytotoxicity of 19 anticancer agents to a range of VP-16-resistant MDR small cell lung cancer lines; different combinations of sensitivity and resistance to classically "MDR" and "non-MDR" drugs were observed. It seems clear that a great diversity of mechanism of MDR exists in small cell lung carcinoma lines.

Takigawa *et al.* (1992) have described an MDR variant of the human small cell carcinoma line SBC-3, by continuous exposure to increasing concentrations of VP-16; this line was cross-resistant to adriamycin, VM-26, vinca alkaloids, mitomycin C and 4-hydroxy cyclophosphamide, and slightly resistant to *cis*-platin and showed collateral sensitivity to bleomycin. (^{3}H)-VP-16 accumulation was reduced, *mdr*-1 mRNA and P-glycoprotein were overexpressed, and topoisomerases I and II showed reduced activity in the resistant line.

U-1285/100 and U-1285/250 are variants of the small cell lung carcinoma line U-1285, which have been adapted to grow in the continuous presence of 100 ng/ml and 250 ng/ml adriamycin respectively. Both are cross-resistant to *cis*-platin and vincristine. U1285/100 was also found to be resistant to VP-16, while the resistance level of U-1285/250 to this drug was not determined (Nygren *et al.*, 1991). U-1690/40 and U-1690/150 are MDR variants of the SCLC cell line U1690 which were selected by growing in the continuous presence of 40 ng/ml and 150 ng/ml Adriamycin, respectively (Nygren *et al.*, 1991). Both of the MDR lines were found to be resistant to adriamycin and vincristine. U-1690/40 was also shown to be resistant to VP-16 and to *cis*-platin. The level of resistance of U-1690/150 to these drugs was not determined. There was no evidence in either of the resistant lines of P-glycoprotein or *mdr*-1 mRNA over-expression (Western blotting using the antibody C219). However, adriamycin resistance was completely reversed by verapamil in the U1690 sublines but only partially in the U1285 sublines. Slightly more adriamycin was taken up by U1690/ 150 compared to U-1690 and this effect was only slightly intensified by the presence of verapamil. Conversely, slightly less adriamycin was taken up by the resistant sublines of U1285 compared to the parental line and this effect was not altered by the presence of verapamil. The concentration of cytoplasmic free Ca^{2+} was significantly elevated in the resistant cell lines; a partial revertant line showed lower levels of elevation. These cell lines may exhibit novel resistance mechanisms.

An adriamycin-selected MDR variant of the human small cell carcinoma line GLC_4, called GLC_4/ADR does not overexpress P-glycoprotein, has reduced levels and activity of topoisomerase II and shows reduced drug uptake; This cell line is also radioresistant (Zijlstra *et al.*, 1987; Meijer *et al.*, 1987; de Jong *et al.*, 1990, 1993).

The three cell lines GLC-14, GLC-16 and GLC-19 were established from one patient with SCLC during the course of disease. GLC-14 was derived from a biopsy taken before any therapy was initiated. GLC-16 was established from a biopsy taken after treatment consisting of cyclophosphamide, adriamycin and etoposide (CDE) had begun. GLC-19 was derived after the patient had received radiotherapy (Berendsen *et al.*, 1988). The cell line GLC-14 was found to be the most sensitive to the drugs which were tested. GLC-16 was found to be resistant to vincristine, vinblastine, colchicine, VP-16, melphalan and to X-irradiation. GLC-19 was resistant to vincristine, vinblastine, VP-16, and melphalan, and to X-irradiation. GLC-16 was found to be most resistant to doxorubicin and etoposide, but GLC-19 showed greater resistance to melphalan than GLC-16 (Berendsen *et al.*, 1988). The changes in clinical behaviour of the tumour are mirrored in the changing resistance status of the derived cell lines. Clinically, the tumour changed from being sensitive (GLC14) to chemotherapy to being completely resistant (GLC19) during the course of the disease. The level of the P-glycoprotein, as detected by western blotting (using JSB-1), was most pronounced in GLC16 and absent in GLC19. Glutathione levels of 12.9, 15.5 and 16.6 µg/mg protein were found in GLC14, GLC16 and GLC19, respectively. GST activity, catalase activity and total sulphydryl groups also increased from GLC14 to GLC19 (de Vries *et al.*, 1989). Compared to GLC14, intracellular adriamycin levels in GLC16 and GLC19 were reduced. After 1 h exposure to adriamycin, there was an increase in the number of DNA strand breaks in GLC19 compared to GLC14 and GLC16. None of the cell lines could repair adriamycin-induced DNA strand breaks within 1 h. However, comparable levels of X-ray induced strand breaks could be repaired by all cell lines.

A recent investigation by Campline *et al.* (1993) found no correlation between drug sensitivity of a panel of 20 human small cell lung carcinoma lines, and levels of glutathione or glutathione-related enzymes.

It is clear, in summary, that a range of different molecular events can confer multidrug resistance in lung tumour cell lines. When the changes most relevant to small-cell and nonsmall cell lung carcinomas *in vivo* are identified, it should be possible to improve treatment of this serious disease.

2.5 Melanoma

Some human cell line model systems for multidrug resistance in melanoma have been described. Lincke *et al.* (1990) showed that transfection of the *mdr*-1 gene into BRO human melanoma cells resulted in expression of a "classical MDR" phenotype. Rittman-Grauer *et al.* (1992) reported reversal of resistance in one of these transfectants grown as a xenograft in nude mice by a monoclonal antibody which binds to an extracellular epitope of P-glycoprotein. Lemontt *et al.* (1988) selected MDR variants of a human melanoma cell line with low levels (4 to 24-fold) of resistance to vinca alkaloids, anthracyclines and actinomycin D. The variants showed increased levels of mdr mRNA, but not gene amplification. The cell lines showed decreased accumulation and enhanced efflux of ^{3}H-Vinblastine; the efflux alteration was reduced by treatment with rotenone or deoxyglucose. In contrast, however, Ramachandran *et al.* (1993) have described five human melanoma lines with different levels (up to 6-fold) of resistance to doxorubicin; although expression of *mdr*-1 and GST-π genes was detected, and immunoreactive P-glycoprotein was present, verapamil did not significantly alter drug retention or cytotoxicity. It appears, therefore, that non-P-170 mechanisms of resistance are predominant in these cell lines.

Panneerselvam *et al.* (1987) described human melanoma cells resistant to both adriamycin and to adriamycin-enhanced killing by complement plus anti-ganglioside antibody. The mechanisms of resistance, while co-expressed, may be different, since immobilized adriamycin may be capable of exerting the latter effect, whereas adriamycin uptake may be necessary for direct cytotoxicity. It is worth mentioning that some experiments (Vichi and Tritton, 1992) have been interpreted as indicating that extracellular drug action is important also in direct adriamycin cytotoxicity.

Fuchs *et al.* (1991) reported no significant expression of *mdr*-1 mRNA (by Northern blotting) or P-glycoprotein (by immunocytochemistry) in the vast majority of chemoresistant primary and metastatic melanomas examined, although low but significant levels of expression could have been missed by the methodology used.

2.6 Breast cancer

Development of chemoresistance is a serious problem in treatment of breast cancer. P-glycoprotein expression has been detected in a subset of human breast carcinomas by methods including immunocytochemistry (Wishart *et al.*, 1990), slot blot analysis (Wallner *et al.*, 1991) and RNA *in situ* hybridisation (Kacinski *et al.*, 1989). While the precise relationship of this expression to chemosensitivity is still unclear, there is some indication of a relationship between high level of P-glycoprotein expression and *in vitro* chemosensitivity (Keith *et al.*, 1990) and clinical behaviour (Schneider *et al.*, 1989). MDR variants of the human breast carcinoma line MCF-7 overexpress P-glycoprotein, and also GST-π (with associated changes in other glutathione-related enzymes), (Fairchild *et al.*, 1987; Mimnaugh *et al.*, 1991; Whelan *et al.*, 1992). Pretreatment levels of GST alpha, mu and pi (by Western blot) do not correlate with clinical outcome in human breast carcinoma (Peters *et al.*, 1993) although these results do not preclude possible significance of expression in a small subset of cells. Transfection of MCF-7 cells with cDNA for GST-α did not alter sensitivity to chlorambucil, melphalan, adriamycin or *cis*-platin (Leyland-Jones *et al.*, 1991).

Fairchild *et al.* (1990) found that transfection of GST-π cDNA into parental MCF-7 cells, or into cells previously rendered adriamycin-resistant (by transfection with a variant *mdr*-1 cDNA cloned from an adriamycin-resistant MCF-7 line) did not change resistance levels to adriamycin, actinomycin D or vinblastine. P-glycoprotein may act as an efflux pump for benzo(a)pyrene in MCF-7 cells (Yeh *et al.*, 1992). The ability of dioxin and other inducers to induce transcription of cytochrome P450IAI mRNA in MCF-7 cells is lost in adriamycin-resistant MCF-7 cells (Ivy *et al.*, 1988) resulting in resistance to compounds such as benzo(a)pyrene.

Mitroxantrone-resistant variants of MCF-7 have been described. The variants described by Nakagawa *et al.* (1992) are cross-resistant to adriamycin and VP-16, but not to vinca alkaloids; neither topoisomerase II nor P-glycoprotein appear to be altered in expression levels, but enhanced drug efflux appears to be associated with resistance. Taylor *et al.* (1991) described a variant of MCF-7 also selected for resistance to mitoxantrone but in this case cross-resistant to vinca alkaloids, *cis*-platin, VP-16 and adriamycin (and some other drugs); neither *mdr*-1 mRNA nor P-glycoprotein are overexpressed, and resistance is not modulated by verapamil, although drug accumulation is reduced. As in other cellular models, non-P-glycoprotein pumps seem to be part of the overall picture.

2.7 Renal cancer

Renal tumours are generally poorly responsive to chemotherapy, and are classified as inherently resistant. P-glycoprotein is expressed in some cells of normal kidney (Van der Valk *et al.*, 1990). Immunohistochemical studies and circumvention experiments with verapamil, trifluoperazine and buthionine sulfoximine in a panel of human renal carcinoma cell lines were consistent with a role for P-170 in resistance to vinca alkaloids and adriamycin, and with a role for glutathione in resistance to adriamycin and *cis*-platin (Mickisch *et al.*, 1990; Volm *et al.*, 1991b; Efferth *et al.*, 1992). Levels of *mdr*-1 gene expression may be varied, in the renal adenocarcinoma line HTB-46, by environmental changes, e.g., heat shock, arsenite treatment (Chin *et al.*, 1990). In the survey by Efferth *et al.*, (1992) reduced topoisomerase II levels, as well as elevated P-170 and GST-π correlated with chemoresistance in human renal carcinoma cells *in vitro*. Nishiyama *et al.*, (1993) detected (by immunoblotting with C219) P-glycoprotein expression in 33 of 38 patients with renal cell carcinoma and in 3 of 17 patients with transitional cell carcinoma; the protein was 5—10 kD smaller than that derived from normal renal tissues.

2.8 Cancer of brain and nervous system

Gliomas and neuroblastomas are often treated with chemotherapy, but many of these tumours prove to be chemoresistant. The possible role of P-glycoprotein in resistance is obscure, however, because in biopsies from gliomas and related tumours, and from neuroblastomas, P-170 expression is frequently observed in normal cells, as well as in some tumour cells (Becker *et al.*, 1991; Favrot *et al.*, 1991). Since P-glycoprotein is expressed by endothelial capillary cells in the brain (Cordon-Cardo *et al.*, 1989; Sugawara *et al.*, 1990), it is possible that P-glycoprotein may play a role in the so-called blood-brain barrier, and in this way, expressed on normal cells, help to confer phenotypic chemoresistance on brain tumours (Nabors *et al.*, 1991). Verapamil can enhance toxicity of several antitumour drugs in glioma and neuroblastoma cell lines *in vitro* and as xenografts in nude mice (Kaba *et al.*, 1985; Merry *et al.*, 1986; Ikeda *et al.*, 1987) and *mdr*-1 mRNA is certainly overexpressed in some neuroblastoma cell lines (van der Bliek *et al.*, 1988). In a rat glioblastoma MDR cell line model verapamil was found to completely reverse the drug transport defect, but to only partially restore wild-type sensitivity, pointing to the co-existence of nonverapamil-sensitive resistance mechanisms (Huet *et al.*, 1992).

2.9 Other solid tumours

Scattered information is available in the literature

on mechanisms of MDR in cells derived from other types of solid tumour. *Mdr*-1 mRNA shows increased expression in human musculo-skeletal tumours which are chemoresistant or metastatic, but not in benign tumours (Nagata *et al.*, 1992). Verapamil enhances adriamycin toxicity in the human bladder carcinoma line T24 (Simpson *et al.*, 1984). Katagiri *et al.* (1993) found higher level expression, by immunohistochemistry, of P-170 and GST-π in non-seminomatous germ cell tumours in comparison with the more chemosensitive seminomas. Tawa *et al.* (1990) found progressively increasing levels of *mdr*-1 mRNA in a patient with undifferentiated sarcoma of the liver, in samples taken at initial diagnosis, and at various stages during post-chemotherapy relapse. Shen *et al.* (1991) have described a panel of human hepatocellular carcinoma lines expressing different patterns of drug resistance, which appear not to be P-glycoprotein mediated. Immunocytochemistry with the monoclonal antibody U1C2, which recognizes an external epitope on the *mdr*-1 gene product, indicated expression of the protein in 60% of head and neck squamous cell carcinomas, and in the small sample of patients examined, expression appeared to correlate well with response to chemotherapy (Kelley *et al.*, 1993). An MDR variant of the human fibrosarcoma line HT1080 does not overexpress P-glycoprotein, and neither does resistance appear to be due to altered activity of topoisomerase II (Zwelling *et al.*, 1990); overexpression of the recently-discovered MRP gene may be involved in resistance in this cell line (Slovak *et al.*, 1993).

2.10 Summary

The balance of experimental evidence to date indicates that common rather than tissue-specific mechanisms of MDR are most important. P-glycoprotein overexpression appears to be widespread in human tumours, and may confer resistance in at least a subset of various human solid tumour types. Topoisomerase II-related resistance seems also to occur in many tumour types. Alterations in glutathione levels and related enzyme systems is widespread also in MDR lines, but the contribution of such systems to MDR in patients remains controversial. It appears that there are several other mechanisms of MDR, some of them involving drug transport systems, which remain to be fully elucidated. Overall, investigation of mechanisms in MDR cell line models is proving fruitful in gaining an understanding of what determines MDR in human solid tumours.

3. Possible approaches to treatment of MDR

As we come to understand what molecular changes underlie MDR in human tumours, and what pattern of cross-resistance each mechanism implies, we need to look for new ways to treat such tumours, whether by antagonising the specific mechanism(s) involved, or by exploiting sensitivities remaining in resistant cancer cells.

3.1 Circumvention of resistance

Use of chemicals which enhance toxicity of chemotherapeutic drugs to MDR cells, in particular when they antagonize a specific mechanism of resistance, is a promising approach to overcoming MDR. This area has been extensively reviewed by Ford and Hait (1990 and elsewhere in this special issue), and clinical trials of verapamil and cyclosporin A have yielded promising results (Pennock *et al.*, 1991; Sonnenveld *et al.*, 1992; Millward *et al.*, 1993). The bulk of circumvention work has looked at effects on resistance to a single drug only (often adriamycin) and in some systems verapamil, for example, does not reverse VP-16 resistance to the same extent as adriamycin resistance (Sehested *et al.*, 1992). This could be due to coexistence of several resistance mechanisms.

In evaluating effects of circumvention agents, it is important to note that many MDR lines are more sensitive to verapamil (and to other circumvention agents) than are the sensitive parental cells (Biedler *et al.*, 1993).

3.2 Inhibition of invasion and metastasis

Novel therapies aimed at inhibiting tumour cell invasion or metastasis should in principle be unaffected by the MDR status of the tumour cells. Some novel aspects of action of proteases in tumour invasion are discussed by McDonnell and Fingleton elsewhere in this issue, and anti-angiogenic agents have been shown in animal models to potentiate the activity of chemotherapeutic drugs (Teicher *et al.*, 1992).

3.3 Use of non-MDR drugs

Most MDR lines are not cross-resistant to certain classes of antitumour agent such as nucleotide analogues, alkylating agents, hormones and *cis*-platin-related compounds. Some examples of co-resistance and even hypersensitivity to such drugs in MDR lines have been reported, however. Arkin *et al.* (1989) for example, have described a variant of a human leukemic cell line adapted for resistance to the antifolate Trimetrexate which also expresses P-170 and is cross-resistant to classical MDR-type drugs. Some other cell lines selected for resistance to methotrexate are cross-resistant to vinca alkaloids and actinomycin D (Haber *et al.*, 1989). In principle, it should be possible to use probes for different types of MDR and other drug resistance patterns, to determine the best panel of drugs to be used for particular patients. In general, however, studies on resistance to "classical MDR" drugs and "other" drugs have tended to proceed in isolation. A considerable amount is known about resistance to alkylating agents (Waxman, 1990), analogues of nucleic acid precursors and anti-metabolites (Fox *et al.*, 1991; Saunders, 1989) and hormones (Maddox, 1989). Resistance to *cis*-platin is found in some MDR lines, and collateral sensitivity in others; some mechanisms of *cis*-platin resistance (Rothenberg *et al.*, 1989) may also confer resistance to anthracyclines and to epipodophyllotoxins. Resistance to *cis*-platin and related drugs may involve altered drug uptake (Loh *et al.*, 1992; Mistry *et al.*, 1992), altered patterns of DNA damage and repair (Fram *et al.*, 1990; Schmidt and Chaney, 1993) and alterations in glutathione-related and metallothionein-related enzyme systems (Fujiwara *et al.*, 1990; Yang *et al.*, 1993; Niimi *et al.*, 1992; Mansouri *et al.*, 1989). Expression of a mutant C-Ha-*ras* oncogene in mouse NIH 3T3 cells induces *cis*-platin resistance associated with a decrease in drug accumulation and an increase in metallothionein content (Isonishi *et al.*, 1991), and an anti-*fos* ribozyme can reverse *cis*-platin resistance in some cell lines (Funato *et al.*, 1992).

It must be remembered, however, that many factors including side effects as they relate to particular patients taking into account age, nutritional and immunological status, etc., limit the flexibility in choice of drugs and drug combinations (Fischer and Knobf, 1989; Pratt and Ruddon, 1979).

3.4 Cytokines, immunotherapy and MDR

Cytokines are gradually finding a role in cancer therapy; the usefulness of α-interferon in treatment of Hairy Cell Leukemia is the best current example (Vedantham *et al.*, 1992). Although rescue of hemopoietic cells and stimulation of immune effector cell function are certainly relevant to the activity of some cytokines (e.g., GM-CSF, IL-2) in cancer patients, high levels of some cytokines seem to have an intrinsic inhibitory or cytotoxic effect on tumour cells. Jones *et al.* (1990) reported enhancement of carmustine (BCNU) toxicity by addition of human recombinant tumour necrosis factor alpha (TNF-α) in murine B16 melanoma cells *in vitro* and *in vivo*. Safrit and Bonavida (1992) also found the combination of TNF-α and adriamycin to result in enhanced toxicity, in a range of human tumour cell lines; a similar synergistic activity of TNF-α and VP-16 was reported by Valenti *et al.* (1993). Morimoto *et al.*, (1993) have found that treatment with a monoclonal antibody against Fas antigen (a 36 kD cell surface polypeptide possibly involved in mediation of some actions of TNF-α) can restore sensitivity in some adriamycin-resistant and *cis*-platinum-resistant tumour lines. These authors use carefully-defined criteria for definition of synergistic effects. Scala *et al.* (1991) found that human recombinant α 2b interferon enhanced adriamycin cytotoxicity in a multidrug resistant variant of the human colon carcinoma line LoVo, but not in the parental line. Some human solid tumour-derived cell lines, for example from colorectal or small cell lung carcinomas (Langdon *et al.*, 1991) are sensitive to growth inhibition by interferons; it is interesting to note that interferon-resistant variants of human carcinoma lines have been described, and that they show a more aggressively malignant phenotype when implanted in nude mice (Toth and Thomas, 1992). Reeve *et al.* (1990) reported decreased EGF Receptor levels in a non-P-glycoprotein expressing human large cell lung carcinoma MDR line, although other papers report increased or unchanged levels in MDR lines (Slovak *et al.*, 1991). Reddy *et al.* (1991) found that the human glioblastoma multiform cell line GBM-18, made multidrug resistant by transfection with the expression vector PHaMDR1/A (containing the complete *mdr*-1 cDNA), retained sensitivity to interferon-β and interferon-γ; it is clearly of potential therapeutic interest to know what substances retain cytostatic or cytotoxic activity in multidrug-resistant cells. Logothetis *et al.* (1991) reported synergistic action of 5-fluorouracil and recombinant human interferon α-2a in treatment of patients with metastatic urothelial tumors for whom primary chemotherapy had failed; the mechanism of this effect is not yet clearly established. Weisenthal *et al.* (1991) have proposed that prior chemotherapy may lead to release of substantial amounts of tumour antigens which prime the immune system resulting in enhanced antitumour activity during subsequent treatment with certain of the cytokines (e.g., interferon γ, TNF-α).

There is evidence that several MDR cell lines are more susceptible to immune effector cells such as Natural Killer (NK) and Lymphokine-Activated Killer (LAK) cells (Yanovich *et al.*, 1986; Kimmig *et al.*, 1990; Miura *et al.*, 1993), although this effect is not observed in all MDR lines (Harker *et al.*, 1990). Antagonism of P-glycoprotein activity by verapamil, or of mdr-1 mRNA translation capacity by antisense oligonucleotide treatment, did not alter the high susceptibility of a human colon carcinoma (LoVo)-derived MDR line to LAK cells (Rivoltini *et al.*, 1990). Interestingly, by analogy with derivation of MDR lines, Tachibana *et al.* (1992) have isolated variants of a small cell lung carcinoma line which are resistant to LAK cells, by prolonged culture of the sensitive parental line in the presence of LAK cells. Unfortunately, cancer cell populations seem to possess the ability to generate variants to every known type of therapeutic agent. Cianfriglia *et al.* (1991) reported loss of expression of Lymphocyte Homing Receptor mRNA in MDR variants of a human T-lymphoblastoid line.

In summary, while this field of research is just beginning to develop, it appears likely that many MDR lines may be sensitive to certain cytokines and to immune effector cells, and that such sensitivity may point to one avenue for improving chemotherapy of resistant tumours.

3.5 Radiation sensitivity in MDR cells

Radiation sensitivity in tumours *in vivo* is believed to be related in many cases to nongenetic factors such as proliferation status and oxygen levels (Brown and Koong, 1991; Hall, 1991). Inherent cellular differences in radiosensitivity can also occur, and in some cases cells are both chemo- and radio-resistant (Schimm *et al.*, 1988; Louie *et al.*, 1985; Miller *et al.*, 1992; de Jong *et al.*, 1993). There are, however, many instances of MDR lines which are not cross-resistant to irradiation (Belli, 1989; Mitchell *et al.*, 1988; Oshita *et al.*, 1992; Wallner and Li, 1987) and we have found no difference in radiation sensitivity between parental lines and MDR variants of three human squamous cell carcinoma lines (DLKP-A, SKMES-1A, SKLU-1A) and one human ovarian carcinoma line (OAW42-A) (K. Kavanagh, M. Heenan, M. Maher and M. Clynes, unpublished). Dempke *et al.* (1992) found that X-irradiation induced stable etoposide and vincristine resistance in ovarian cell lines (for further details see review by Hill in this issue), and this may imply a danger of emergence of drug-resistance in cancer cells surviving after radiation therapy. In contrast, we have failed to observe altered drug-resistance in human lung squamous cell carcinoma lines taken through several cycles of irradiation (M. Heenan and M. Clynes, unpublished). Some caution needs to be observed when assessing the relevance of *in vitro* radiosensitivity to *in vivo* effects; immortalisation of human fibroblasts by transfection with the plasmid pSV2 *neo*, for example, causes increased radioresistance (Parris *et al.*, 1990) and also as mentioned earlier environmental factors and cell proliferation status can play an important role *in vivo*.

3.6 Novel molecular biological therapies

Novel molecular biological approaches using antisense oligodeoxynucleotides, triple helix-forming oligodeoxynucleotides and site-specific splice-inducing ribozymes can be used, in specific experimental situations, to downregulate expression of specific oncogenes, or of genes involved in determination of chemo-and radio-resistance (Scanlon *et al.*, 1991; Carter and Lemoine, 1993). In spite of the formidable problems to be solved in relation to efficient delivery of adequate amounts, and the unknown safety implications, these approaches could allow very specific therapies directed against cells expressing undesirable properties such as drug- or radio-resistance.

Conclusion

Human cell lines established from different sites are useful models for study of mechanisms of multidrug resistance (MDR) in human solid tumours. Many different biochemical alterations have been correlated with MDR in cell lines, including altered levels of P-glycoprotein, of other transport proteins, of topoisomerase II, and of glutathione and related enzyme systems, and also alterations in intracellular drug distribution. Many other molecular changes have been found in MDR cells; ideally therefore, direct involvement of a gene product in (and assessment of its quantitative contribution to) resistance determination should be assessed by methods such as use of specific antagonists or antisense oligonucleotides, or by cDNA transfection. Mechanisms of MDR have been investigated in cell lines derived from carcinomas of lung, ovary, colon, rectum, breast, brain, nervous system, skin, kidney and liver; in general, MDR mechanisms do not appear to be tumour-type specific. Sensitive methods such as immunocytochemistry and RT-PCR may be necessary to detect expression of MDR-related genes in subpopulations of tumour cells *in vivo*. Classification of MDR human tumours according to resistance mechanism(s) is an important research goal. This classification must lead on to development of novel mechanism-specific treatment methods.

References

Alon N, Busche R, Tummler B and Riordan J (1991) Membrane lipids of MDR cells; Chemical composition and physical state. In: Roninson I (ed) Molecular and Cellular Biology of MDR in tumour cells. Plenum Press, N.Y. pp. 263–278.

Arkin H, Ohnuma T, Kamen BA, Holland JF and Vallabhajosula S (1989) Multidrug resistance in a human leukemic cell line selected for resistance to trimetrexate. Cancer Res 49: 6556–6561.

Arvan DA (1992) Tumor cell heterogeneity: an overview. Clinica Chimica Acta 206: 3–7.

Asselin BL (1992) Evaluation of tumor cell heterogeneity based on response to chemotherapy. Clinica Chimica Acta 206: 33–44.

Baas F, Jongsma A, Broxterman H, Arceci R, Housman D, Scheffer G, Riethorst A, Van Groenigen M, Nieuwint A and Joenje H (1990) Non-P-glycoprotein-mediated mechanism for multidrug resistance precedes P-glycoprotein expression during *in vitro* selection for doxorubicin resistance in a human lung cancer cell line. Cancer Res 50: 5392–5398.

Bates SE, Currier SJ, Alvarez M and Fojo AT (1992) Modulation of P-glycoprotein phosphorylation and drug transport by sodium butyrate. Biochemistry 31: 6366–6372.

Becker I, Becker KF, Meyermann R and Hollt V (1991) The multidrug-resistance gene MDR-1 is expressed in human glial tumors. Acta Neuropathol 82: 516–519.

Belli JA (1989) Interaction between radiation and drug damage in mammalian cells. IV radiation response of adriamycin-resistant V79 cells. Radiat Res 119: 88–100.

Bénard J, De Silva J, Teyssier JR and Riou G (1989) Overexpression of mdr-1 gene with no DNA amplification in an MDR human ovarian carcinoma cell line. Int J Cancer 43: 471–477.

Berendsen HH, De Leij L, de Vries E, Mesander G, Mulder N, de Jong B, Buys C, Postmus P, Poppema S, Sluiter H and The H (1988) Characterization of three small cell lung cancer cell lines established from one patient during longitudinal follow-up. Cancer Res 48: 6891–6899.

Bhushan A, Abramson R, Chiu J-F and Tritton TR (1992) Expression of c-*fos* in human and murine multidrug resistant cells. Mol Pharmacol 42: 69–74.

Biedler JL, Chang T, Druskin H, Meyers MB and Spengler BA (1993) Modulation of multidrug resistance gene expression by chemosensitizing agents. Proc Am Assoc Cancer Res 34: 578.

Boscoboinik D, Gupta RS and Epand RM (1990) Investigation of the relationship between altered intracellular pH and multidrug resistance in mammalian cells. Br J Cancer 61: 568–572.

Bradley G, Juranka PF and Ling V (1988) Mechanisms of multidrug resistance. Biochim Biophys Acta 948: 87–128.

Bradley G, Naik M and Ling V (1989) P-glycoprotein expression in MDR human ovarian carcinoma cell lines. Cancer Res 49: 2790–2796.

Brown JM and Koong A (1991) Therapeutic advantage of hypoxic cells in tumors: a theoretical study. J Natl Cancer Inst 83: 178–185.

Broxterman HJ, Pinedo HM, Schuurhuis GJ and Lankelma J (1990) Cyclosporin A and Verapamil have different effects on energy metabolism in MDR tumour cells. Br J Cancer 62: 85–88.

Bugg BY, Danks MK, Beck WT and Suttle DP (1991) Expression of a mutant DNA topoisomerase II in CCRF-CEM human leukemic cells selected for resistance to teniposide. Proc Natl Acad Sci 88: 7654–7658.

Cabral F and Barlow SB (1989) Mechanisms by which mammalian cells acquire resistance to drugs that affect microtubule assemply. FASEB J 3: 1593–1599.

Calvet JP (1991) Molecular approaches for analyzing differential gene expression: differential cDNA library construction and screening. Pediatr Nephrol 5: 751–757.

Campling BG, Baer K, Baker HM, Lam YM and Cole SPC (1993) Do glutathione and related enzymes play a role in drug resistance in small cell lung cancer cell lines? Br J Cancer 68: 327–335.

Carter G and Lemoine NR (1993) Antisense technology for cancer therapy; does it make sense? Br J Cancer 67: 869–876.

Chanson M and Meda P (1993) Rat pancreatic acinar cell coupling: Comparison of extent and modulation *in vitro* and *in vivo*. In: Hall, JE, Zampighi, GA and Davis, RM (eds.) Progress in Cell Research, Vol. 3, Elsevier. Ch. 28, pp. 199–205.

Chao CC, Huang YT, Ma CM, Chou WY and Lin-Chao S (1992) Overexpression of glutathione S-transferase and elevation of thiol pools in a multidrug resistant human colon cancer cell line. Mol Pharmacol 41: 69–75.

Chaudhary PM and Roninson IB (1993) Induction of multidrug resistance in human cells by transient exposure to different chemotherapeutic drugs. J Natl Cancer Inst 85: 632–639.

Chin K, Tanaka S, Darlington G, Pastan I and Gottesman MM (1990) Heat shock and arsenite increase expression of the multidrug resistance (*MDR*-1) gene in human renal carcinoma cells. J Biol Chem 265: 221–226.

Chin K, Ueda K, Pastan I and Gottesman MM (1992) Modulation of activity of the promoter of the human MDR-1 gene by *ras* and P53. Science 255: 459–462.

Choi K, Chen CJ, Kriegler M and Roninson IB (1988) An altered pattern of cross-resistance in MDR human cells results from spontaneous mutation in the mdr-1 (P-glycoprotein) gene). Cell 53: 519–529.

Cianfriglia M, Yassen A, Tombesi M, Samoggia P, Barca S and Caserta M (1991) Expression of lymphocyte homing receptor gene is lost in MDR variants of human T-lymphoblastoid CCRF-CEM cells. Int J Cancer 49: 394–397.

Clynes M, Redmond A and Heenan M (1990) Recent developments in research on multiple drug-resistance in cancer cells. The Cancer Journal 3: 34–39.

Clynes M (1993) Cellular models for multiple drug resistance in cancer. *In Vitro* Cell. Dev Biol 29A: 171–179.

Clynes M, Redmond A, Moran E and Gilvarry U (1992) Multiple drug-resistance in variant of a human nonsmall cell lung carcinoma cell line DLKP-A. Cytotechnology 10: 75–89.

Cole SP, Pinkoski MJ, Bhardwaj G and Deeley RG (1992a) Elevated expression of annexin II (lipocortin II, p 36) in a multidrug resistant small cell lung cancer cell line. Br J Cancer 65: 498–502.

Cole SP, Bhardwaj G, Gerlach JH, Mackie JE, Grant CE, Almquist KC, Stewart AJ, Kurz EU, Duncan AM and Deeley RG (1992a) Overexpression of a transporter gene in a multidrug-resistant human lung cancer line. Science 258: 1650–1654.

Cole SPC, Bhardwaj G, Gerlach JH, Almquist KC and Deeley RG (1993) A novel ATP-binding cassette transporter gene overexpressed in multidrug-resistant human lung tumour cells. Proc Am Assoc Cancer Res 34: 579.

Cole SP and Deeley RG (1993) "Multidrug resistance – associated protein: sequence correction". Science 260: 879.

Cole SP, Chanda ER, Dicke FP, Gerlach JH and Mirski SE (1991) Non-P-glycoprotein-mediated MDR in a small cell lung cancer cell line: evidence for decreased susceptibility to drug-induced DNA damage and reduced levels of topoisomerase II. Cancer Res 51: 3345–3352.

Coley HM, Workman P and Twentyman P (1991) Retention of activity by selected anthracyclines in a multidrug resistant human large cell lung carcinoma line without P-glycoprotein hyperexpression. Br. J. Cancer 63: 351–357.

Cordon-Cardo C, O'Brien JP, Casals D, Rittman-Grauer L, Biedler JL, Melamed MR and Bertino JR (1989) Multidrug-resistance gene (P-glycoprotein) is expressed by endothelial cells at blood-brain barrier sites. Proc Natl Acad Sci USA 86: 695–698.

Croop JM, Gros P and Housman DE (1988) "Genetics of multidrug resistance". J. Clin. Invest. 81: 1303–1309.

Dalton WS, Cress AE, Alberts DS and Trent JM (1988) Cytogenetic and phenotypic analysis of a human colon carcinoma cell line resistant to mitoxantrone. Cancer Res 48: 1882–1888.

Danova M, Giordano M, Erba E, Palmer S, Candiloro V, Riccardi A, Ucci G, Mazzini G, D'Incalci M and Ascari E (1992) Flow cytometric analysis of MDR-associated antigen (p-glycoprotein) and DNA ploidy in human colon cancer. J Cancer Res Clin Oncol 118: 575–580.

De Jong C, Zijlstra J, De Vries E and Mulder N (1990) Reduced DNA topoisomerase II activity and drug-induced DNA cleavage activity in an adriamycin-resistant human small cell lung carcinoma cell line. Cancer Res 50: 304–309.

De Jong S, Kooistra AJ, De Vries EG, Mulder NH and Zijlstra JG (1993) topoisomerase II as a target of VM-26 and 4'-(9-Acridinylamino) methane sulfonyl-m-aniside in atypical MDR human small cell lung carcinoma cells. Cancer Res 53: 1064–1071.

Dempke W, Whelan R and Hill BT (1992) Expression of resistance to etoposide and vincristine *in vitro* and *in vivo* after x-irradiation of ovarian tumor cells. Anti-cancer Drugs 3: 395–399.

Devine SE, Ling V and Melera PW (1992) Amino acid substitutions in the 6th transmembrane domain of P-glycoprotein alter MDR. Proc Natl Acad Sci USA 89: 4564– 4568.

De Vries EG, Meijer C, Timmer-Bosscha H, Berendsen HH, Deleij L, Scheper RJ and Mulder NH (1989) Resistance mechanisms in three human small cell lung cancer cell lines established from one patient during clinical follow-up. Cancer Res 49: 4175–4178.

Dietel M (1993) Second international symposium on cytostatic drug resistance. Cancer Res 53: 2683–2688.

Dimanche-Boitrel M, Pelletier H, Genne P, Petit J, Legrimellec C, Canal P, Ardiet C, Bastian G and Chauffert B (1992) Confluence-dependent resistance in human colon cancer cells: role of reduced drug accumulation and low intrinsic chemosensitivity of resting cells. Int J Cancer 50: 677–682.

Dolfini E, Dasdia T, Perletti G, Romagnoni M and Piccinini F (1993) Analysis of calcium-dependent protein kinase C isoenzymes in intrinsically resistant cloned lines of LoVo cells; reversal of resistance by kinase inhibitor 1-(5-isoqinolinylsulfonyl)-2-methylpiperazine. Anticancer Research, (in press).

Dusre L, Mimnaugh EG, Myers CE and Sinha BK (1989) Potentiation of doxorubicin cytotoxicity by buthionine sulfoximine in multidrug-resistant human breast tumor cells. Cancer Res 49: 511–515.

Efferth T, Mattern J and Volm M (1992) Immunohistochemical detection of P-glycoprotein, GST and DNA topoisomerase II in human tumors. Oncology 49: 368–375.

Efferth T and Volm M (1993) Modulation of P-glycoprotein-mediated MDR by monoclonal antibodies, immunotoxins or antisense oligodeoxynucleotides in kidney carcinoma and normal kidney cells. Oncology 50: 303–308.

Eijdems EW, Borst P, Jongsma AP, De Jong S, De Vries EG, Van Groenigen M, Versantvoort CH, Nieuwint AW and Baas F (1992) Genetic transfer of non-P-glycoprotein-mediated multidrug resistance (MDR) in somatic cell fusion; dissection of a compound MDR phenotype. Proc Natl Acad Sci USA 89: 3498–3502.

Endicott JA and Ling V (1989) The biochemistry of P-glycoprotein-mediated multidrug resistance. Ann Rev Biochem 58: 137–171.

Epand RM and Stafford AR (1993) Protein kinases and multidrug resistance. The Cancer Journal 6: 154–158.

Fairchild CR, Ivy SP, Kao-Shan CS, Whang-Peng J, Rosen N, Israel MA, Melera PW, Cowan KH and Goldsmith ME (1987) Isolation of amplified and overexpressed DNA sequences from adriamycin-resistant human breast cancer cells. Cancer Res 47: 5141–5148.

Fairchild CR, Moscow JA, O'Brien EE and Cowan KH (1990) Multidrug resistance in cells transfected with human genes encoding a variant P-glycoprotein and glutathione-S-transferase π. Mol Pharmacol 37: 801–809.

Favrot M, Combaret V, Goillot E, Wagner JP, Bouffet E, Mazingue F, Thyss A, Bordigoni P, Delsol G, Bailly C, Fontainiere B and Philip T (1991) Expression of P-glycoprotein restricted to normal cells in neuroblastoma biopsies. Br J Cancer 64: 233–238.

Ferguson PJ and Cheng YC (1989a) Critical issues relating to clinical drug resistance. Cancer Bull 41: 7–13.

Ferguson PJ and Cheng YC (1989b) Phenotypic instability of drug sensitivity in a human colon carcinoma cell line. Cancer Res 49: 1148–1153.

Fey MF, Zimmermann A, Borisch B and Tobler A (1993) Studying clonal heterogeneity in human cancers. Cancer Res 53: 921.

Fischer DS and Knobf MT (1989) The cancer chemotherapy handbook year book medical publishers, Chicago.

Ford JM and Hait WN (1990) Pharmacology of drugs that alter mdr in cancer. Pharmacol Rev 42: 155–199.

Fox M, Boyle JM and Kinsella AR (1991) Nucleoside salvage and resistance to antimetabolite anticancer agents. Br J Cancer 64: 428–436.

Fram RJ, Woda BA, Wilson JM and Robichaud N (1990) Characterization of acquired resistance to CIS-diamminedichloroplatinum (II) in BE human colan carcinoma cells. Cancer Res 50: 72–77.

Fry AM, Chresta CM, Davies SM, Walker MC, Harris AL, Hartley JA, Masters JR and Hickson ID (1991) Relationship between topoisomerase II levels and chemosensitivity in human tumor cell lines. Cancer Res 51: 6592–6595.

Fuchs B, Ostmeier H and Suter L (1991) P-glycoprotein expression in malignant melanoma. J Cancer Res Clin Oncol 117: 168–171.

Fujiwara Y, Sugimoto Y, Kasahara K, Bungo M, Yamakido M, Tew KD and Saijo J (1990) Determinants of drug response in a cisplatin-resistant human lung cancer cell line. Jpn J Cancer Res 81: 527–535.

Funato T, Yoshida E, Jiao L, Tone T, Kashani-Sabet M and Scanlon KJ (1992) The utility of an anti-*fos* ribozyme in reversing *cis*-platin resistance in human carcinomas. Advan Enzyme Regul 32: 195–209.

Georges E, Bradley G, Gariepy J and Ling V (1990) Detection of P-glycoprotein isoforms by gene-specific monoclonal antibodies. Proc Natl Acad Sci USA 87: 152–156.

Germann UA, Pastan I and Gottesman MM (1993) P-glycoproteins: mediators of multidrug resistance. Semin Cell Biol 4: 63–76.

Giaccone G, Gazdar AF, Beck H, Zunino F and Capricano G (1992) Multidrug sensitivity phenotype of human lung cancer cells associated with topoisomerase II expression. Cancer Res 52: 1666–1674.

Glisson BS and Alpeter MD (1992) Multidrug resistance in a small cell lung cancer cell line: Rapid selection with etoposide and differential chemosensitization with cyclosporin A. Anti-Cancer Drugs 3: 359–366.

Gold M (1986) A conspiracy of cells. State Univ. of New York Press, Albany, N.Y.

Gottesman MM (1993) How cancer cells evade chemotherapy.

Cancer Res 53: 747–754.

Gottesmann MM and Pastan I (1993) Biochemistry of multidrug resistance mediated by the multidrug transporter. Ann Rev Biochem (in press).

Haber M, Reed C, Kavallaris M, Norris M and Stewart B (1989) Resistance to drugs associated with the MDR phenotype following selection with high-concentration methotrexate. J Natl Cancer Inst 81: 1250–1254.

Hall EJ (1991) Hypoxia revisited. J Natl Cancer Inst 83: 156.

Harker WG, Corrynne T, Mcgregor JR, Slade L and Samlowski WE (1990) Human tumor cell line resistance to chemotherapeutic agents does not predict resistance to natural killer or lymphokine-activated killer cell-mediated cytolysis. Cancer Res 50: 5931–5936.

Harris AL and Hochhauser D (1992) Mechanisms of multidrug resistance in cancer treatment. Acta Oncol 31: 205–213.

Hay RJ, Caputo J and Macy ML (1992) ATCC quality control methods for cell lines. (2nd edition) American Type Culture Collection, 12301 Parklawn Drive, Rockville, Maryland 20852–1776, USA.

Herzog CE, Tsokos M, Bates SE and Fojo AT (1993) Increased mdr-1/P-glycoprotein expression after treatment of human colon carcinoma cells with P-glycoprotein antagonists. J Biol Chem 268: 2946–2952.

Hildebrand-Zanki SU and Kern DH (1986) A rapid bioassay to determine stabilities of anticancer agents under conditions of the clonogenic assay. *In Vitro* Cell Dev Biol 22: 247–252.

Hill BT (1991) Interactions between antitumour agents and radiation and the expression of resistance. Cancer Treat Rev 18: 149–190.

Holder JW, Elmore E and Barrett JC (1993) Gap junction function and cancer. Cancer Res 53: 3475–3485.

Holzmayer TA, Hilsenbeck S, Von Hoff DD and Roninson IB (1992) Clinical correlates of MDR-1 (P-glycoprotein) gene expression in ovarian and small-cell lung carcinomas. J Natl Cancer Inst 84: 1486–1491.

Hofmann GA, Mirabelli CK and Drake FH (1990) Quantitative adaptation of the bacteriophage P4 unknotting assay for use in the biochemical and pharmacological characterization of topoisomerase II. Anti-Cancer Drug Design 5: 273–282.

Hong WS, Saijo N, Sasaki Y, Minato K, Nakano H, Nakagawa K, Fujiwara Y, Nomura K and Twentyman PR (1988) Establishment and characterisation of *cis*-platin resistant sublines of human lung cancer cell lines. Int J Cancer 41: 462–467.

Huet S, Schott B and Robert J (1992) P-glycoprotein overexpression cannot explain the complete doxorubicin-resistance phenotype in rat glioblastoma cell lines. Br J Cancer 65: 538–544.

Ikeda H, Nakano G, Nagashima K, Sakamoto K, Harasawa N, Kitamura T, Nakamura T and Nagamachi Y (1987) Verapamil enhancement of antitumor effect of *cis*-Diamminedichloroplatinum (II) on nude mouse-grown human neuroblastoma. Cancer Res 47: 231–234.

Isonishi S, Hom DK, Thiebaut FB, Mann SC, Andrews PA, Basu A, Lazo JS, Eastman A and Howell SB (1991) Expression of the C-Ha-*ras* oncogeny in mouse NIH 3T3 cells induces resistance to cisplatin. Cancer Res *51* 5903–5909.

Ivy SP, Tulpule A, Fairchild CR, Auerbuch SD, Myers CE, Nebert DW, Baird WM and Cowan KH (1988) Altered regulation of P-450IAI expression in an MDR MCF-7 human breast cancer cell line. J Biol Chem 263: 19119–19125.

Jensen PB, Vindelov L, Roed H, Demant EJF, Sehested M, Skovsgaard T and Hansen HH (1989) *In vitro* evaluation of the potential of aclarubicin in the treatment of small cell carcinoma of the lung (SCCL). Br J Cancer 60: 838–844.

Jensen PB, Christensen IJ, Sehested M, Hansen HH and Vindelov, L (1993) Differential cytotoxicity of 19 anticancer agents in wild-type and and etoposide resistant small cell lung cancer cell lines. Br J Cancer 67: 311–320.

Jones AL, Millar JL, Miller BC, Powell B, Selby P, Winkley A, Lakhani S, Gore ME and McElwain TJ (1990) Enhanced anti-tumour activity of carmustine (bcnu) with tumour necrosis factor *in vitro* and *in vivo.* Br J Cancer 62: 776–780.

Kaba K, Tani E, Morimura T and Matsumoto T (1985) Potentiation of vincristine effect in human and murine gliomas by calcium channel blockers or calmodulin inhibitors. J Neurosurg 63: 905–911.

Kacinski BM, Yee LD and Carter D (1989) Quantitation of tumor cell expression of the P-glycoprotein (*mdr*-1) gene in human breast carcinoma clinical specimens. Cancer Bull 41: 44–48.

Kanzawa F, Sugimoto Y, Minato K, Kasahara K, Bungo M, Nakagawa K, Fujiwara Y, Liu LF and Saijo N (1990) Establishment of a camptothecin analogue (CPT-11) – resistant cell line of Human nonsmall cell lung cancer: characterization and mechanism of resistance. Cancer Res 50: 5919–5924.

Kasahara K, Fujiwara Y, Nishio K, Ohmori T, Sugimoto Y, Komiya K, Matsuda and Saijo N (1991) Metallothionein content correlates with the sensitivity of human small cell lung cancer cell lines to cisplatin. Cancer Res 51: 3237–3242.

Kasahara K, Fujiwara Y, Sugimoto Y, Nishio K, Tamura T, Matsuda T and Saijo N (1992) Determinants of response to the DNA topoisomerase II inhibitors doxorubicin and etoposide in human lung cancer cell lines. J Natl Cancer Inst 84: 113–118.

Katagiri A, Tomita Y, Nishiyama T, Kimura M and Sato S (1993) Immunohistochemical detection of P-glycoprotein and GST-π-1 in testis cancer. Br J Cancer 68: 125–129.

Kaufmann S (1991) DNA topoisomerases in chemotherapy. Cancer Cells 3: 24–27.

Keith WN, Stallard S and Brown R (1990) Expression of *mdr*-1 and GST-π in human breast tumours; comparison to *in vitro* chemosensitivity. Br J Cancer 61: 712–716.

Keizer HG and Joenje H (1989) Increased cytosolic pH in MDR human lung tumor cells; effect of verapamil. J Natl Cancer Inst 81: 706–709.

Keizer HM, Schuurhuis GJ, Broxterman HJ, Lankelma J, Schoonen WE, van Rijn J, Pinedo HM and Joenje H (1989) Correlation of multidrug resistance with decreased drug accumulation, altered subcellular drug distribution and increased P-glycoprotein expression in cultured SW-1573 human lung tumour cells. Cancer Res 49: 2988–2993.

Kelley DJ, Pavelic ZP, Gapany M, Stambrook P, Pavelic L, Gapany S and Gluckman JL (1993) Detection of P-glycoprotein in squamous cell carcinomas of the head and neck. Arch Otolaryngol Head Neck Surg 119: 411–414.

Kimmig A, Gekeler V, Neumann M, Frese G, Handgretinger R, Kardos G, Diddens H and Niethammer D (1990) Susceptibility of MDR human leukemia cell lines to human IL-2-activated killer cells. Cancer Res 50: 6793–6799.

Kioka N, Tsubota J, Kakehi Y, Komano T, Gottesman MM, Pastan I and Ueda K (1989) P-glycoproetin gene (*mdr*-1) cDNA from human adrenal: Normal P-glycoprotein carries Gly-185 with an altered pattern of multidrug resistance. Biochem Biophys Res Commun 162: 224–231.

Kohno K, Kikuchi J, Sato S, Takano H, Saburi Y, Asoh K and Kuwano M (1988) Vincristine-resistant human cancer KB cell line and increased expression of multidrug-resistance gene. Jpn J Cancer Res (Gann) 79: 1238–1246.

Kramer R, Weber TK, Morse B, Arceci R, Staniunas R, Steele G and Summerhayes IC (1993) Constitutive expression of multidrug resistance in human colorectal tumours and cell lines. Br J Cancer 67: 959–968.

Krishnamachary N and Center MS (1993) The MRP gene associated with a non-P-glycoprotein multidrug resistance encodes a 190 kDa membrane bound glycoprotein. Cancer Res 53: 3658–3661.

Kuiper CM, Broxterman HJ, Baas F, Schuurhuis GJ, Haisma HJ, Scheffer GL, Lankelma J and Pinedo HM (1990) Drug transport variants without P-glycoprotein overexpression from a human squamous lung cancer cell line after selection with doxorubicin. J Cell Pharmacol 1: 35–41.

Lai SL, Goldstein LJ, Gottesman MM, Pastan I, Tsai CH, Johnson BE, Mulshine JL, Ihde DC, Kayser K and Gazdar AF (1989) MDR-1 gene expression in lung cancer. J Natl Cancer Inst 81: 1144–1150.

Langdon SP, Rabiasz GJ, Anderson L, Ritchie AA, Fergusson RS, Hay FG, Miller EP, Mullen P, Plumb J, Miller WR and Smyth JF (1991) Characterisation and properties of a small cell lung cancer cell line and xenograft WX 322 with marked sensitivity to alpha-interferon. Br J Cancer 63: 909–915.

Lau DH, Lewis AD, Ehsan MN and Sikic BI (1991) Multifactorial mechanisms associated with broad cross-resistance of ovarian carcinoma cells selected by cyanomorpholino-doxorubicin. Cancer Res 51: 5181–5187.

Lee SA, Karaszkiewicz JW and Anderson WB (1992) Elevated levels of nuclear protein kinase C in multidrug-resistant MCF-7 human breast carcinoma cells. Cancer Res 52: 3750–3759.

Lemontt JF, Azzaria M and Gros P (1988) Increased MDR gene expression and decreased drug accumulation in MDR human melanoma cells. Cancer Res 48: 6348–6353.

Leyland-Jones BR, Townsend AJ, Tu CD, Cowan KH and Goldsmith ME (1991) Antineoplastic drug sensitivity of human MCF-7 breast cancer cells stably transfected with a human α class GST gene. Cancer Res 51: 587–594.

Lincke CR, Van der Bliek A, Schuurhuis GJ, Van der Velde-Koerts T, Smit JJ and Borst P (1990) Multidrug resistance phenotype of human BRO melanoma cells transfected with a wild-type human MDR-1 cDNA. Cancer Res 50: 1779–1785.

Logothetis CJ, Hossan E, Selca A, Dexeus FH and Amato RJ (1991) Fluorouracil and recombinant human interferon alfa-2a in the treatment of metastatic chemotherapy-refractory urothelial tumors. J Natl Cancer Inst 83: 285–288.

Loh SY, Mistry P, Kelland LR, Abel G and Harrap KR (1992) Reduced drug accumulation as a major mechanism of acquired resistance to cisplatin in a human ovarian carcinoma line: circumvention studies using novel platinum (II) and IV) ammino/amine complexes. Br J Cancer 66: 1109–1115.

Long SH, Wang L, Lorico A, Wang RCC, Brattain MG and Casazza AM (1991) Mechanisms of resistance to etoposide and tenopiside in acquired resistant colon and lung carcinoma cell lines. Cancer Res 51: 5275–5284.

Louie KG, Behrens BC, Kinsella TJ, Hamilton TC, Grotzinger KR, Mckoy WM, Winkler MA and Ozols RF (1985) Radiation survival parameters of antineoplastic drug-sensitive and -resistant human ovarian cancer cell lines and their modification by buthionine sulfoximine. Cancer Res 45: 2110–2115.

Ma L and Center MS (1992) The gene encoding vacular H^+-ATPase subunit C is overexpressed in multidrug-resistant HL60 cells. Biochem Biophys Res Commun 182: 675– 681.

MacLeod RA, Haene B and Drexler HG (1992) Cells, lines and DNA fingerprinting. *In Vitro* Cell Dev Biol 28A: 591–594.

Maddox AM (1989) Mechanism of resistance to hormone therapy. Cancer Bull 41: 52–55.

Mansouri A, Henle KJ, Benson AM, Moss AJ and Nagle WA (1989) Characterization of a cisplatin-resistant subline of murine RIF-1 cells and reversal of drug resistance by hyperthermia. Cancer Res 49: 2674–2678.

Massaad L, Dewaziers I, Ribrag V, Janot F, Beaune PH, Morizet J, Gouyette A and Chabot GG (1992) Comparison of mouse and human colon tumors with regard to phase I and phase I drug-metabalizing enzyme systems. Cancer Res 52: 6567–6575.

Mayer A, Takimoto M, Fritz E, Schellander G, Kofler K and Ludwig H (1993) The prognostic significance of proliferating cell nuclear antigen, EGF receptor and *mdr* gene expression in colorectal cancer. Cancer 71: 2454–2460.

Meijer C, Mulder NH, Timmer-Boscha H, Zijlstra JG and De Vries EGE (1987) Role of free radicals in an adriamycin-resistant human small cell lung cancer cell line. Cancer Res 47: 4613–4617.

Merry S, Fetherston CA, Kaye SB, Freshney RI and Plumb JA (1986) Resistance of human glioma to adriamycin *in vitro*: the role of membrane transport and its circumvention with verapamil. Br J Cancer 53: 129–135.

Mickisch GH, Roehrich K, Koessig J, Forster S, Tschada RK and Alken PM (1990) Mechanisms and modulation of MDR in primary human renal cell carcinoma. J Urol 144: 755–759.

Miller PR, Hill AB, Slovak ML and Shimm DS (1992) Radiation resistance in a doxorubicin-resistant human fibrosarcoma cell line. Am J Clin Oncol 15: 216–221.

Milroy R, Plumb JA, Bastone P, Maclay A, Wishart GC, Hay FB, Candlish W, Adamson R, Khan MZ, Banham S and Kaye SB (1992) Lack of expression of P-glycoprotein in 7 small cell lung cancer cell lines established both from untreated and from treated patients. Anti-Cancer Res 12: 193–200.

Millward MJ, Cantwell BM, Munro NC, Robinson A, Corris PA and Harris AL (1993) Oral verapamil with chemotherapy for advanced NSCLC: A randomized study. Br J Cancer 67: 1031–1035.

Mimnaugh EG, Fairchild CR, Freuhauf JP and Sinha BK (1991) Biochemical and pharmacological characterisation of MCF-7 drug-sensitive and Adr^R MDR human breast tumor xenografts in athymic nude mice. Biochem Pharmacol 42: 391–402.

Mirski SE, Gerlach JH and Cole SP (1987) Multidrug resistance in a human small cell lung cancer line selected in adriamycin. Cancer Res 47: 2594–2598.

Mirski SE and Cole SP (1989) Antigens associated with MDR in H69AR, a small cell lung cancer cell line. Cancer Res 49: 5719–5724.

Mirski SE and Cole SP (1991) MDR-associated antigens on drug-sensitive and -resistant human tumour cell lines. Br J Cancer 64: 15–22.

Mistry P, Kelland LR, Loh SY, Abel G, Murrer BA and Harrap KR (1992) Comparison of cellular accumulation and cytotoxicity of cisplatin with that of tetraplatin and amminedibutyratodichloro (cyclohexylamine) platinum (IV) (JM221) in human ovarian carcinoma cell lines. Cancer Res 52: 6188–6193.

Mitchell JB, Gamson J, Russo A, Friedman N, Degraff W, Carmichael J and Glatstein E (1988) Chinese hamster pleiotropic MDR cells are not radioresistant. NCI Monogr 6: 187–191.

Miura K, Nishio K, Heike Y and Saijo N (1993) Cytolytic activity of natural killer and lymphokine-activated killer cells against human tumor cell lines resistant to chemotherapeutic agents. The Cancer Journal 6: 91–96.

Miyamoto H (1986) Establishment and characterization of an adriamycin-resistant subline of human small cell lung cancer cells. Acta Med Okayama 40: 65–73.

Morimoto H, Yonehara S and Bonavida B (1993) Overcoming TNF- and drug resistance of human tumor cell lines by combination treatment with anti-fas antibody and drugs or toxins. Cancer Res 53: 2591–2596.

Murano S, Thweatt R, Shmookler-Reis RJ, Jones RA, Moerman EJ and Goldstein S (1991) Diverse gene sequences are overexpressed in Werner syndrome fibroblasts undergoing premature replicative senescence. Mol Cell Biol 11: 3905–3814.

Murphy D, McGown AT, Hall A, Cattan A, Crowther D and Fox BW (1992) Glutathione S-transferase activity and isoenzyme distribution in ovarian tumour biopsies taken before or after cytotoxic chemotherapy. Br J Cancer 66: 937–942.

Nabors MW, Griffin CA, Zehnbauer BA, Hruban RH, Phillips PC, Grossman SA, Brem H and Colvin MC (1991) Multidrug resistance gene (MDR-1) expression in human brain tumors. J Neurosurg 75: 941–946.

Nagata Y, Yamasaki H, Fukuda K, Tanaka S, Tsurou T, Shuntoh H and Kuno T (1992) Expression of the MDR gene in human musculoskeletal tumors. J Env Pathol Tox Oncol 11: 131–137.

Nakagawa M, Schneider E, Dixon KH, Horton J, Kelley K, Morrow C and Cowan KH (1992) Reduced intracellular drug accumulation in the absence of P-glycoprotein (*mdr*-1) overexpression in mitoxantrone-resistant human MCF-7 breast cancer cells. Cancer Res 52: 6175–6181.

Nieuwint AW, Baas F, Wiegant J and Joenje H (1992) Cytogenetic alterations associated with P-glycoprotein- and non-P-glycoprotein mediated MDR in SW-1573 human lung tumor cell lines. Cancer Res 52: 4361–4371.

Nielsen D and Skovsgaard T (1992) P-glycoproetin as multidrug transporter: a critical review of current multidrug resistant cell lines. Biochim Biophys Acta 1139: 169–183.

Nguyen HN, Sevin BU, Averette HE, Ramos R, Genjei P and Perras J (1993) Evidence of tumor heterogeneity in cervical cancers and lymph node metastases as determined by flow cytometry. Cancer 71: 2543–2550.

Niimi S, Nakagawa K, Sugimoto Y, Nishio K, Fujiwara Y, Yokoyama S, Terashima Y and Saijo N (1992) Mechanism of cross-resistance to a camptothecin analogue (CPT-11) in a human ovarian cancer cell line selected by cisplatin. Cancer Res 52: 328–333.

Nishiyama K, Shirahama T, Yoshimura A, Sumizawa T, Furukawa T, Ichikawa-Haraguchi M, Akiyama SI and Ohi Y (1993) Expression of the multidrug transporter, P-glycoprotein in renal and transitional cell carcinomas. Cancer 71: 3611–3619.

Nishio M, Ohata M, Uetani K, Suruda T, Kobayashi H, Funasako M, Nishio K and Saijo N (1993) Calcitriol increases adriamycin sensitivity in human nonsmall cell lung cancer cells. The Cancer Journal 6: 97–101.

Nitiss JL, Liu YX and Hsiung Y (1993) A temperature-sensitive topoisomerase II allele confers temperature-dependent drug resistance on amsacrione and etopoiside: a genetic system for determining the targets of topoisomerase II inhibitors. Cancer Res 53: 89–93.

Nooter K and Herweijer H (1991) Multidrug resistance (*mdr*) genes in human cancer. Br J Cancer 63: 663–669.

Nygren P, Larsson R, Gruber A, Peterson C and Bergh J (1991) Doxorubicin-selected MDR small call lung cancer cell lines characterised by elevated cytoplasmic Ca^{2+} and resistance modulation by Verapamil in the absence of P-glycoprotein overexpression. Br J Cancer 64: 1011–1018.

Ohta S, Nishio K, Kubo S, Nishio M, Ohmori T, Takahashi T and Saijo N (1993) Characterisation of a vindesine-resistant human small-cell lung cancer cell line. Br J Cancer 68: 74–79.

Oshita F, Fujiwara Y and Saijo N (1992) Radiation sensitivities in various anticancer-drug-resistant human lung cancer cell lines and mechanism of radiation cross-resistance in a cisplatin-resistant cell line. J Cancer Res Clin Oncol 119: 28–34.

Panneerselvam M, Bredehorst R and Vogel C (1987) Resistance of human melanoma cells against the cytotoxic and complement-enhancing activities of doxorubicin. Cancer Res 47: 4601–4607.

Parris CN, Masters JR and Green MH (1990) PSV3neo transfection and radiosensitivity of human cancer cell lines. Cancer Letts 55: 171–175.

Pennock GD, Dalton WS, Roeske WR, Appleton CP, Mosley K, Plezia P, Miller JP and Salmon SE (1991) Systemic toxic effects associated with high dose verapamil infusion and chemotherapy administration. J Natl Cancer Inst 83: 105–110.

Peters WH, Roelofs HM, Van Putten WL, Jansen JB, Klijn JG and Foekens JA (1993) Response to adjuvant chemotherapy in primary breast cancer; no correlation with expression of GSTs. Br J Cancer 68: 86–92.

Powell S and McMillan TJ (1991) Clonal variation of DNA repair in a human glioma cell line. Radiother Oncol 21: 225–232.

Pratt WB and Rudden RW (1979) The anticancer drugs. Oxford University Press, N.Y.

Ramachandran C, Yuan ZK, Huang XL and Krishan A (1993) Doxorubicin resistance in human melanoma cells: mdr-1 and gst-π gene expression. Biochem Pharmacol 45: 743–751.

Reddy PG, Graham GM, Datta S, Guarini L, Moulton TA, Jiang H, Gottesman MM, Ferrone S and Fisher PB (1991) Effect of recombinant fibroblast interferon and recombinant immune interferon on growth and antigenic phenotype of mdr human glioblastoma multiform cells. J Natl Cancer Inst 83: 1307–1315.

Redmond A, Law E, Gilvarry U and Clynes M (1990) Establishment of two MDR variants of the human tumor cell line Hep-2. Cell Biol Toxicol 6: 293–300.

Redmond A, Moran E and Clynes M (1993) Multiple drug resistance in the human ovarian carcinoma cell line OAW42-A. Eur J Cancer 29A: 1078–1081.

Redmond SM, Joncourt F, Buser K, Ziemieck A, Altermatt HJ, Fey M, Margison G and Cerny T (1991) Assessment of P-glycoprotein, glutathione-based detoxifying enzymes and O^6-alkylguanine-DNA alkyltransferase as potential indicators of constitutive drug resistance in human colorectal tumors. Cancer Res 51: 2092–2097.

Reeve JG, Rabbitts PH and Twentyman PR (1989) Amplification and expression of mdr-1 gene in a multidrug resistant variant of small cell lung cancer cell line NCI-H69. Br J Cancer 60: 339–342.

Reeve JG, Rabbitts PH and Twentyman PR (1990) Non-P-glycoprotein-mediated MDR with reduced EGF receptor expression in a human large cell lung cancer cell line. Br J Cancer 61: 851–855.

Richardson ME and Siemann DW (1992) Thiol manipulation as a means of overcoming drug resistance in a novel cyclophosphamide-induced resistant cell line. Int J Radiat Oncol Biol Phys 22: 781–784.

Rittmann-Grauer LS, Yong MA, Sanders V and Mackensen DG (1992) Reversal of vinca alkaloid resistance by anti-P-glycoprotein monoclonal antibody HYB-241 in a human tumor xenograft. Cancer Res 52: 1810–1816.

Rivoltini L, Colombo M, Supino R, Ballainari D, Tsuruo T and Perianti G (1990) Modulation of MDR by verapamil or *MDR*-1 antisense oligonucleotides does not change the high susceptibility to lymphokine-activated killers in MDR-activated carcinoma (LoVo) line. Int J Cancer 46: 727–732.

Roninson IB (ed) (1991) molecular and cellular biology of multidrug resistance in tumor cells. Plenum Publishing Corp.

Rothenberg ML, Ozols RF and Hamilton TC (1989) Cisplatin drug resistance. Cancer Bull 41: 48–51.

Safrit JT and Bonavida B (1992) Sensitivity of resistant human tumor cell lines to tumor necrosis factor and adriamycin used in combination: correlation between down-regulation of TNF-mRNA induction and overcoming resistance. Cancer Res 52: 6630–6637.

Sakata K, Kwok TT, Murphy BJ, Laderoute KR, Gordon GR and Sutherland RM (1991) Hypoxia-induced drug resistance-comparison to P-glycoprotein-associated drug resistance. Br J Cancer 64: 809–814.

Saunders PP (1989) Acquired resistance to antimetabolites: application of antimetabolite-resistant cell lines in studies of drug metabolism. The Cancer Bull 41: 20–25.

Scala S, Pacelli R, Iaffaioli RV, Normanno N, Pepe S, Frasci G, Genua G, Tsuruo T, Tagliaferri P and Bianco AR (1991) Reversal of adriamycin resistance by recombinant α-interferon in MDR human colon carcinoma LoVo-doxorubicin cells. Cancer Res 51: 4898–4902.

Scanlon KJ, Jiao L, Funato T, Wang W, Tone T, Rossi JJ and Kashani-Sabet M (1991) Ribozyme-mediated cleavage of c-fos mRNA reduces gene expression of DNA synthesis enzymes and metallothionein. Proc Natl Acad Sci USA 88: 10591–10595.

Schalken JA (1990) The use of differential hybridisation analysis to identify markers for diagnosis of urinogenital cancer. Eur Urol 18: (Suppl. 2): 52–57.

Scheper RJ, Broxterman HJ, Scheffer GL, Kaaijk P, Dalton WS, Van Heijningen TH, Van Kalken C, Slovak ML, De Vries EG, Van der Valk P, Meijer CJ and Pinedo HM (1993) Overexpression of a Mr 110,000 vesicular protein in Non-P-glycoprotein-mediated multidrug resistance. Cancer Res 53: 1475–1479.

Schimm DS, Olson S and Hill AB (1988) Radiation resistance in a MDR human T-cell leukemia line. Int J Radiat Oncol Biol Phys 15: 931–936.

Schinkel AH, Roelofs ME and Borst P (1991) Characterization of the human MDR-3 P-glycoprotein and its recognition by P-glycoprotein-specific monoclonal antibodies. Cancer Res 51: 2628–2635.

Schmidt W and Chaney SG (1993) Role of carrier ligand in platinum resistance of human carcinoma cell lines. Cancer Res 53: 799–805.

Schneider J, Bak M, Efferth T, Kaufmann M, Mattern J and Volm M (1989) P-glycoprotein expression in treated and untreated human breast cancer. Br J Cancer 60: 815–818.

Schurr E, Raymond M, Bell JC and Gros P (1989) Characterisation of the MDR protein expressed in cell clones stably transfected with mouse mdr-1 cDNA. Cancer Res 49:

2729–2734.

Schuurhuis GJ, Broxterman HJ, de Lange JH, Pinedo HM, van Heijningen TH, Kuiper CM, Scheffer GL, Scheper RJ, van Kalken CK, Baak JP and Lankelma J (1991) Early MDR defined by changes in intracellular doxorubicin distribution, independent of P-glycoprotein. Br J Cancer 64: 857–861.

Sehested M, Friche E, Jensen PB and Demant EF (1992) Relationship of VP-16 to the classical MDR phenotype. Cancer Res 52: 2874–2879.

Sekiya S, Nunoyama T, Shirasawa H, Kimura H, Kawata M, Iijima N, Sugimoto Y, Tsuruo T and Takamizawa H (1992) Expression of a human multidrug resistance gene in human ovarian carcinoma cell lines. Arch Gynacol Obstet 251: 79–86.

Shen D, Lu Y, Chin K, Pastan I and Gottesman MM (1991) Human hepatocellular carcinoma cell lines exhibit MDR unrelated to MDR-1 gene expression. J Cell Sci 98: 317–322.

Simpson WG, Tseng MT, Anderson KC and Hart JI (1984) Verapamil enhancement of chemotherapeutic efficacy in human bladder cancer cells. J Urol 132: 574–576.

Slovak ML, Mirski SE, Cole SP, Gerlach JH, Yohem KH and Trent JM (1991) Tumorigenic MDR HT1080 cells do not overexpress receptors for EGF. Br J Cancer 64: 296–298.

Slovak ML, Ho JP, Bhardwaj G, Kurz EU, Deeley RG and Cole SPC (1993) Localization of a novel MDR-associated gene in the HT1080/DR4 and H69AR human tumor cell lines. Cancer Res 53: 3221–3225.

Sonneveld P, Durie BG, Lokhorst HM, Marie JP, Solbu G, Suciu S, Zittoun R, Lowenberg B and Nooter K (1992) Modulation of multidrug-resistant multiple myeloma by cyclosporin. The Lancet 340: 255–259.

Spoelstra EC, Westerhoff HV, Dekker H and Lankelma J (1992) Kinetics of daunorubicin transport by P-glycoprotein of intact cancer cells. Eur J Biochem 207: 567–579.

Sugawara I, Hamada H, Tsuruo T and Mori S (1990) Specialised localization of P-glycoprotein recognised by MRK-16 monoclonal antibody in endothelial cells of the brain and the spinal cord. Jpn J Cancer Res 81: 727–730.

Sugimoto Y and Tsuruo T (1987) DNA-mediated transfer and cloning of a Human MDR gene of Adriamycin-resistant myelogenous leukemia K562. Cancer Res 47: 2620–2625.

Sugimoto Y, Tsukahara S, Oh-hara T, Toshiyuki I and Tsuruo T (1990) Decreased expression of DNA topoisomerase I in camptothecin-resistant tumour cell lines as determined by a monoclonal antibody. Cancer Res 50: 6925–6930.

Sugimoto Y, Tsukahara S, Oh-hara T, Liu LF and Tsuruo T (1990) Elevated expression of DNA topoisomerase II in camptothecin-resistant human tumor cell lines. Cancer Res 50: 7962–7965.

Sugimoto Y, Hamada H, Tsukahara S, Noguchi K, Yamaguchi K, Sato M and Tsuruo T (1993) Molecular cloning and characterisation of the complementary DNA for the Mr 85,000 protein overexpressed in adriamycin-resistant human tumor cells. Cancer Res 53: 2538–2543.

Sweet P, Chan PK and Slater LM (1989) Cyclosporin A and verapamil enhancement of daunorubicin-produced nucleolar protein B23 translocation in daunorubicin-resistant and - sensitive murine tumor cells. Cancer Res 49: 677–680.

Tachibana I, Watanabe M, Tanio Y, Hayashi S, Hosoe S, Saito S, Matsunashi M, Osaki T, Shigedo Y, Masuno T and Kawase I (1992) Generation of a small cell lung cancer variant resistant to LAK cells; association with resistance to a LAK cell-derived, cytostatic factor. Cancer Res 52: 3310–3316.

Takigawa N, Ohnoshi T, Ueoka H, Kiura K and Kimura T (1992) Establishment and characterisation of an etoposide-resistant human small cell lung cancer cell line. Acta Med Okayama 46: 203–212.

Tawa A, Inoue M, Ishihara S, Hara J, Yamura-yagi K, Okumura K, Okada A, Nihei A, Taguchi J, Kanai N, Tsuruo T and Kawa-ha K (1990) Increased expression of the MDR Gene in undifferentiated sarcoma. Cancer 66: 1980–1983.

Taylor CW, Dalton WS, Parrish PR, Gleason MC, Bellamy WT, Thompson FH, Roe DJ and Trent JM (1991) Different mechanisms of decreased drug accumulation in doxorubicin and mitoxantrone-resistant variants of the MCF-7 human breast cancer cell line. Br J Cancer 63: 923–929.

Teicher BA, Sotomayor EA and Huang ZD (1992) Antiangiogenic agents potentiate cytotoxic cancer therapies against primary and metastatic disease. Cancer Res 52: 6702–6704.

Ten Tije BJ and Wils J (1992) Intraperitoneal cisplatin in the treatment of refractory or recurrent advanced ovarian carcinoma. Oncology 49: 442–444.

Toffoli G, Viel A, Tumiotto L, Biscontin G, Rossi C and Boiocchi M (1991) Pleiotropic-resistant phenotype is a multifactorial phenomenon in human colon carcinoma cell lines. Br J Cancer 63: 51–56.

Toffoli G, Viel A, Tumiotto L, Maestro R, Biscontin G and Boiocchi M (1992) Expression of the *mdr*-1 gene in human colorectal carcinomas-relationship with MDR inferred from analysis of human colorectal carcinoma cell lines. Cancer Chemother Pharmacol 29: 283–289.

Toth CA and Thomas P (1992) Type I interferon resistance in a colorectal cancer cell line is associated with a more agressive phenotype *in vito*. Br J Cancer 65: 365–368.

Tsai C, Chang K, Perng R, Mitsudomi T, Chen M, Kadoyama C and Gazdar AF (1993) Correlation of intrinsic chemoresistance of nonsmall cell lung cancer cell lines with HEP-2/*neu* gene expression but not with *ras* gene mutations. J Nat Cancer Inst 85: 897–901.

Twentyman PR, Fox NE, Wright KA and Bleehen NM (1986) Derivation and preliminary characterisation of adiramycin resistant lines of human lung cancer cells. Br J Cancer 53: 529–537.

Twentyman PR, Wright KA, Mistry P, Kelland LR and Murrer BA (1992) Sensitivity to novel platinum compounds of panels of human lung cancer cell lines with acquired and inherent resistance to *cis*-platin. Cancer Res 52: 5674–5680.

Ueda K, Cardarelli C, Gottesman MM and Pastan I (1987) Expression of a full-length cDNA for the human MDR-1 gene confers resistance to colchicine, doxorubicin and Vinblastine. Proc Natl Acad Sci USA 84: 3004–3008.

Uhlmann E and Peyman A (1990) Antisense Oligonucleotides:

A new therapeutic principle. Chem Rev 90: 543–584.

Valenti M, Cimoli G, Parodi S, Mariani GL, Venturini M, Conte P and Russo P (1993) Potentiation of TNF-mediated cell killing by VP-16 on human ovarian cancer cell lines. *in vitro* results and clinical implications. Eur J Cancer 29A: 1157–1161.

Van der Bliek AM, Baas F, Van der Velde-Koerts T, Biedler JL, Meyers MB, Ozols RF, Hamilton TC, Joenje H and Borst P (1988) Genes amplified and overexpressed in human MDR cell lines. Cancer Res 48: 5927–5932.

Van der Valk P, Van Kalken CK, Ketelaars H, Broxterman HJ, Scheffer G, Kuiper CM, Tsuruo T, Lankelma J, Meijer CJ, Pinedo MM and Scheper RJ (1990) Distribution of MDR-associated P-glycoprotein in normal and neoplastic human tissues; analysis with 3 monoclonal antibodies recognising different epitopes of the P-glycoprotein molecule. Ann Oncol 1: 56–64.

Van der Zee AG, Hollema H, de Jong S, Boonstra H, Gouw A, Willemse PH, Zijlstra JG and De Vries EG (1991) P-glycoprotein expression and DNA topoisomerase I and II activity in benign tumors of the ovary and in malignant tumors of the ovary, before and after platinum/cyclophosphamide chemotherapy. Cancer Res 51: 5915–5920.

Van der Zee AG, Van Ommen B, Meijer C, Hollema H, Van Bladeren J and de Vries G (1992) Glutathione S-transferase activity and isoenzyme composition in benign ovarian tumours, untreated malignant ovarian tumours and malignant ovarian tumours after platinum/cyclophosphamide chemotherapy. Br J Cancer 66: 930–936.

Vedantham S, Gamliel H and Golomb HM (1992) Mechanism of interferon action in hairy cell leukemia: a model of effective cancer biotherapy. Cancer Res 52: 1056–1066.

Verelle P, Meissonnier F, Fonck Y, Feillel V, Dionet C, Kwiatkowski F, Plagne R and Chassagne J (1991) Clinical relevance of immunohisto-chemical detection of MDR P-glycoprotein in breast carcinoma. J Natl Cancer Inst 83: 111–116.

Vichi P and Tritton TR (1992) Adriamycin: Protection from cell death by removal of extracellular drug. Cancer Res 52: 4135–4138.

Volm M, Mattern J and Samsel B (1991a) Overexpression of P-glycoprotein and GST-π in resistant nonsmall cell lung carcinomas of smokers. Br J Cancer 64: 700–704.

Volm M, Pommerenke EW, Efferth T, Lohrke H and Mattern J (1991b) Circumvention of MDR in human kidney and kidney carcinoma *in vitro*. Cancer 67: 2484–2489.

Volm M, Mattern J and Samsel B (1992) Relationship of inherent resistance to doxorubicin, proliferative activity and expression of P-glycoprotein 170 GST-π in human lung tumors. Cancer 70: 764–769.

Wallner J, Depisch D, Hopfner M, Haider K, Spona J, Ludwig H and Pirker R (1991) MDR-1 gene expression and prognostic factors in primary breast carcinomas. Eur J Cancer 27: 1352–1355.

Wallner KE and Li GC (1987) Effect of cisplatin resistance on cellular radiation response. Int J Radiat Oncol Biol Phys 13: 587–591.

Watson JV (1991) Introduction to flow cytometry. Cambridge University Press.

Waxman DJ (1990) Glutathione S-transferases: Role in alkylating agent resistance and possible target for modulation chemotherapy – A Review. Cancer Res 50: 6449–6454.

Weisenthal LM, Dill PL and Pearson FC (1991) Effect of prior cancer chemotherapy on human tumor-specific cytotoxicity *in vitro* in response to immunopotentiating biologic response modifiers. J Natl Cancer Inst 83: 37–42.

West IC (1990) What determines the substrate specificity of the multidrug resistance pump? Trends Biochem Sci 15: 42–46.

Whelan RD, Waring CJ, Wolf CR, Hayes JD, Hosking LK and Hill BT (1992) Overexpression of P-glycoprotein and GST-π in MCF-7 cells selected for vincristine resistance *in vitro*. Int J Cancer 52: 241–246.

Wilbur DW, Camacho ES, Hillard DA, Dill PL and Weisenthal LM (1992) Chemotherapy of nonsmall cell lung carcinoma guided by an *in vitro* drug resistance assay measures total tumor cell kill. Br J Cancer 65: 27–32.

Wilson AP, Ford CHJ, Newman CE and Howell A (1987) *Cis*-platinum and ovarian carcinoma. *In vitro* chemosensitivity of cultured tumour cells from patients receiving high dose *cis*-platinum as first line treatment. Br J Cancer 56: 763–773.

Wishart GC, Plumb JA, Going JJ, McNicol AM, McArdle CS, Tsuruo T and Kaye SB (1990) P-glycoprotein expression in primary breast cancer detected by immunocytochemistry with two monoclonal antibodies. Br J Cancer 62: 758–761.

Yang LY and Trujillo HM (1990) Biological characterization of multidrug-resistant human colon carcinoma sublines induced/selected by two methods. Cancer Res 50: 3218–3225.

Yang LY, Trujillo JM and Su YZ (1992) Further characterization of two distinct adriamycin-resistant sublines from LoVo human colon carcinoma cells. Anti-Cancer Res 12: 473–480.

Yang LY, Trujillo JM, Siciliano MJ, Kido Y, Siddik ZH and Su YZ (1993) Distinct P-glycoprotein expression in two subclones simultaneously selected from a human colon carcinoma cell line by *cis*-diamminedichloroplatinum (II). Int J Cancer 53: 478–485.

Yanovich S, Hall RE and Weinert C (1986) Resistance to Natural killer cell-mediated cytolysis by a pleiotropic drug-resistant human erythroleukemia (K562-R) cell line. Cancer Res 46: 4511–4515.

Yeh GC, Lopaczynska J, Poore CM and Phang JM (1992) A new functional role for P-glycoprotein efflux pump for benzo(a)pyrene in human breast cancer MCF-7 cells. Cancer Res 52: 6692–6695.

Zaman GJ, Versantvoort CH, Smit JJ, Eijdems EW, De Haas M, Smith AJ, Broxterman NJ, Mulder NH, De Vries EG, Baas F and Borst P (1993) Analysis of the expression of MRP, the gene for a new putative transmembrane drug transporter in human MDR cancer cell lines. Cancer Res 53: 1747–1750.

Zangwill BC, Balsara G, Dunton C, Varello M, Rebane BA, Hernandez E and Atkinson BF (1993) Ovarian carcinoma heterogeneity as demonstrated by DNA ploidy. Cancer 71: 2261–2267.

Zijlstra JG, DeVries EGE and Mulder NH (1987) Multifactorial

drug resistance in an adriamycin-resistant human small cell lung carcinoma cell line. Cancer Res 47: 1780–1784.
Zou CP, Van NT and Kuo MT (1992) Isolation and Characterisation of putative intrinsic multidrug resistant CHO cells by FACS. Anti-Cancer Res 12: 427–432.
Zwelling LA, Slovak ML, Doroshow JH, Hinds M, Chan D, Parker E, Mayes J, Sie KL, Meltzer PS and Trent JM (1990) HT1080/DR4: A P-glycoprotein negative human fibrosarcoma cell line exhibiting resistance in topoisomerase II – reactive drugs despite the presence of a drug-sensitive topoisomerase II. J Natl Canc Inst 82: 1553–1561.

Address for offprints: Martin Clynes, National Cell and Tissue Culture Centre, Dublin City University, Dublin 9, Ireland.

Cytotechnology **12**: 257–264, 1993.

Studies on low-level MDR cells

G. Belvedere and E. Dolfini
Istituto di Ricerche Farmacologiche Mario Negri, Via Eritrea 62, 20157 Milan, Italy

Key words: low resistance, MDR, P-glycoprotein

Abstract

Acquired or spontaneous resistance is a major clinical problem in the treatment of cancer. Low levels of MDR gene expression or P-glycoprotein have been correlated with a high level of drug resistance in vitro and a poor response to chemotherapy in some tumors. A strong correlation between MDR mRNA, P-glycoprotein levels and degree of drug resistance has not been found in several resistant model tumor cell lines. In some cell lines at low and high level of resistance different mechanisms seem to be involved.

MDR resistance as a clinical problem

The majority of human metastatic cancers are either intrinsically resistant to chemotherapy or acquire resistance to multiple drugs after an initial promising response to treatment.

A particularly important type of tumor drug resistance, termed multidrug resistance (MDR), is manifested by cross-resistance to a number of structurally and functionally unrelated drugs (doxorubicin, VP16, m-AMSA, vinblastine, vincristine). This type of resistance has been shown to be the result of decreased drug accumulation in resistant cells due to energy-dependent drug efflux. Acquisition of the MDR phenotype in human cells is the result of the expression of the MDR1 gene (Ueda *et al.*, 1987; Choi *et al.*, 1988), which encodes a 170 kDa membrane glycoprotein, called P-glycoprotein (P-gp), believed to function as an efflux pump for various hydrophobic compounds (Chen *et al.*, 1986; Gottesman and Pastan, 1988).

To study the phenomenon of MDR several cell lines showing even 1000-fold resistance in respect to the parental line have been selected after continuous exposure to increasing levels of doxorubicin (Dx) or vinblastine (VBL) (Chen *et al.*, 1990; Bradley *et al.*, 1989). Most of these cells show amplification of the MDR gene and a high expression of P-gp that can precede gene amplification (Riordan *et al.*, 1985; Shen *et al.*, 1986). However, analysis of MDR sublines of the human epidermoid carcinoma cell line KB has shown that even very low levels of MDR1 gene expression can confer a several-fold increase in the level of drug resistance, which may be clinically relevant (Shen *et al.*, 1986; Chin *et al.*, 1989).

MDR1 gene expression is frequently observed in different human tumors, both untreated and treated with chemotherapeutic drugs, as well as in some normal tissues (Goldstein *et al.*, 1989; Thiebaut *et al.*, 1987).

A study of MDR1 RNA levels in over 600 human cancers has shown that cancers derived from tissues that normally express the MDR1 gene such as cancers of the colon, kidney, pancreas, adrenal and liver usually express high levels of MDR1 RNA (Goldstein *et al.*, 1989). Using a semiquantitative slot blot assay, MDR1 RNA levels in these tumors are comparable to levels seen in selected tissue culture lines with relatively low drug resistance (4- to 6-fold) (Akiyama *et al.*, 1985; Shen *et al.*, 1986). Thus, it seems likely that a component of the intrinsic resistance of these tumors is due to the expression of P-gp and, in some cases, direct evidence suggests that this is the case (Kakehi *et al.*, 1988).

High expression of MDR1 is sometimes found in untreated cancers, such as leukemias, lymphomas, nonsmall cell lung cancers with neuroendocrine properties and chronic myelogenous leukemia in blast crisis in which the cell of origin of the tumor does not show high MDR1 RNA levels (Goldstein *et al.*, 1989).

A recent study by Chan *et al.* (1991) suggests that any appearance of P-gp, as detected histochemically, predicts poor response to treatment in childhood sarcoma and neuroblastoma.

Elevated MDR1 RNA levels are seen in many tumors that have acquired resistance to chemotherapy and have recurred after successful initial treatment with anticancer drugs. Such cancers include breast and ovarian cancer, neuroblastoma, leukemias and lymphomas.

There are many cancers, including many drug-sensitive tumors, in which MDR1 RNA levels are low or undetectable by standard techniques. Examples include untreated ovarian and breast cancer, Wilms tumor and small cell and nonsmall cell lung cancer (Lai *et al.*, 1989).

P-glycoprotein and MDR gene expression in tumors and cell lines

As regards the expression of classical MDR resistance, it is necessary that the reduced drug uptake responsible for the reduced sensitivity to the antitumor drugs is correlated to the increased expression of the 170 kDa P-gp that is the product of the MDR gene(s) whose mRNA level is elevated in resistant cells because of the amplification of the gene or alterations in its posttranscriptional control.

Several studies, reported below, have shown that these requirements do not coexist in different cell lines and also in sublines at low and high levels of resistance originated from the same line indicating that the regulation of the MDR is under a complex control.

In the human ovarian carcinoma cell lines SKOV3 resistant to VBL and vincristine (VCR) with a degree of resistance spanning from 4- to 2000-fold the increase in P-gp level (detected by Western blot) paralleled the increase in multidrug resistance. However, at a more detailed analysis the expression of the P-gp appears under a complex control. At low level of resistance, only an increase in P-gp m-RNA and protein was observed. At intermediate to high levels of resistance P-gp gene amplification becomes evident. At high level of resistance an example was observed where only the amount of P-gp was increased without a concomitant increase in mRNA or gene copy (Bradley *et al.*, 1989).

That amplification of the mdr gene is not a prerequisite for high levels of resistance is shown also by a study with V79/MEN and P388/MEN cells resistant to menogaril (Badiner *et al.*, 1987). It was observed that V79/MEN cells were 3.2-fold resistant while P388/MEN were 25-fold resistant to menogaril. However, mdr amplification was 2-fold for P388/MEN compared to the parent line, but a 30-fold amplification was observed for V79/MEN. Increased levels of mdr RNA without amplification of the mdr gene have been also observed in five clones of human melanoma isolated by single-step selection in culture medium containing VCR, VBL or colchicine and expressing classical MDR resistance at relatively low levels (4- to 24-fold) (Lemontt *et al.*, 1988).

In Dx-selected sublines of the human squamous lung cancer cell line SW1537, P-gp-mediated resistance is preceded by another mechanism that confers resistance to anthracyclines, VP16, Vinca alkaloids, colchicine and gramicidin D (Baas *et al.*, 1990). The emergence of non-P-gp mediated MDR is accompanied by a reproducible loss of MDR1 mRNA, which is present at low level in the parental SW1573 cell line. Upon selection for high level of Dx resistance (greater than 50-fold) the MDR1 gene becomes transcriptionally activated. A detailed analysis of the cross-resistance patterns suggests that both mechanisms for non-P-gp and P-gp-mediated MDR can be operational at the same time.

Not only the levels, but also the localization of the P-gp may be different in cells showing different levels of drug resistance. A series of human squamous lung cancer cell lines with increasing levels of resistance to Dx were developed from a SW1573 lung tumor cell line and stained for P-gp using the JSB-IM antibody (Broxterman *et al.*, 1989). A subline with a 4- to 6-fold resistance to daunorubicin and VCR showed rather uniform positive staining for P-gp apparently at cytoplasmatic sites, only in cells with higher degrees of resistance (greater than 10-fold) plasma membrane associated P-gp was visible. This different localization of P-gp may have a relevance in the different sensitization to daunorubicin produced by verapamil. In fact a 30% reduction of the resistance index was observed in the low resistance cell variant and an 84-fold sensitization was observed in a 70-fold resistant cell line.

A different correlation between the level of resistance and the degree of sensitization was observed in a panel of mouse tumor cell lines (Twentyman *et al.*, 1990). At low dose of the sensitizers verapamil and cyclosporin A the sensitization was greater in lines expressing low level of P-gp than in lines showing high levels although a generally good correlation between the degree of resistance and the amount of P-gp was found.

These differences in sensitization may be due to the selection of different mutants in the cell population. In low level VCR resistant chinese hamster ovary cell lines, two main groups of mutants have been identified, membrane mutants cross-resistant to taxol and whose VCR resistance is reversible by verapamil and those that appear tubuline mutants which have increased sensitivity to taxol and resistance to VCR in the presence of verapamil (Brewer and Warr, 1987).

MDR lines possess multiple phenotypic changes, suggesting that P-gp function could be complemented by some additional mechanisms associated with cytotoxic selection. Low level of drug resistance have been observed in unselected NIH3T3 mouse cells after retrovirus mediated transfer of mouse *mdr*-1 cDNA. To determine whether cytotoxic selection contributes to the MDR phenotype of *mdr*-1 expressing cells, NIH3T3 cells infected with a recombinant retrovirus carrying the human MDR1 gene were selected by two different procedures: noncytotoxic selection for increased P-gp expression on the cell surface by multiple rounds of immunofluorescence labelling and flow sorting or one or more steps of selection with a cytotoxic drug (Choi *et al.*, 1991). The level of MDR in both type of infectants showed an excellent correlation with the P-gp density in the plasma membrane expressed as immunoreactivity with a P-gp specific antibody normalized by reactivity with an antibody against an unrelated antigen. Cytotoxic selection conferred no additional increase in resistance relative to P-gp density.

The presence of low levels of drug resistance may also be relevant in the response to differentiating agents. In erythroleukemia cells (MELC) accelerated hexamethylenebisacetamide-induced commitment to differentiation is characteristic of MELC cells with a low level (2 to 5-fold) of VCR resistance in four independently derived cell lines (Richon *et al.*, 1991). Low level VCR-resistant cell lines do not display increased levels of P-gp or *mdr*-1, *mdr*-2 or *mdr*-3 mRNA, nor do they exhibit cross-resistance to colchicine or Dx. These cells show increased level of protein kinase C activity, reduced accumulation of VCR and restoration of VCR sensitivity in the presence of verapamil that does not alter accelerated response to hexamethylenebisacetamide.

In low-level resistant cells MDR has also been observed to be modulated by the levels of glutathione (GSH) and GSH transferase (GST). Five human melanoma cell lines with 1- to 6-fold Dx resistance were isolated and tested for drug retention, MDR1 expression and GST π gene expression (Ramachandran *et al.*, 1992). All the cell lines showed high Dx retention and efflux blockers such as trifluoperazine and verapamil did not enhance drug retention and cytotoxicity. A significant correlation between Dx chemosensitivity and GST π mRNA but not MDR1 mRNA expression was noted. Although MDR1 and P-gp are expressed in these cells, drug resistance seems to depend on GST π but not P-gp expression.

In two human melanoma cell lines FCCM-2 and FCCM-9 exhibiting low level Dx resistance (2- to 6-fold) efflux blockers did not enhance Dx cytotoxicity. GSH content was 52% higher in FCCM-2 and 30% lower in FCCM-9 cell lines compared to the sensitive cells. GST activity was 2.3- and 7.2-fold higher in FCCM-2 and FCCM-9 cells, respectively, compared to the parental cells. GSH peroxidase and reductase activities were also significantly higher in the resistant cell lines. GST π content was significantly higher in the resistant cells. Depletion of >90% GSH by butyl sulfoxime increased Dx sensitivity 2.1- and 1.6-fold in FCCM-2 and FCCM-9 cells, respectively (Nair *et al.*, 1992).

In vitro selection of low and high resistant cell sublines

To study the mechanisms responsible for MDR highly resistant tumor cell lines were developed in which the resistance was induced by continuous exposure to a drug that selected for that subpopulation (Chen *et al.*, 1986; Gerlach *et al.*, 1986; Cornwell *et al.*, 1986), or by combination of a drug treatment following an initial exposure to either a chemical mutagen (Ahluwalia *et al.*, 1986) or radiation (Hill *et al.*, 1984).

In the clinic, however, patients may develop drug resistance at relapse after only several courses of chemotherapy. In order to simulate the resistance obtained in the clinic it has been pulse selected an MDR subline from the human colon tumor line LoVo. The mechanism of resistance of this cell line was compared to that of a LoVo line obtained after exposure to increasing amount of the selecting drug (Yang and Trujillo, 1990).

The pulse selected LoVo subline (SRA1.2) was obtained by a procedure consisting of nine 1-h pulse treatments with 1.2 μg/ml of adriamycin (Adr). The resistant subline obtained after continuous treatment was produced by exposing LoVo cells to Adr in stepwise increments of concentrations ranging from 0.01 to 1.2 μg/ml over 16 months. After phenotypic drug resistance developed, the Adr1.2-resistant subclone was maintained by continuous exposure to 1.2 μg/ml of Adr.

The Adr resistant SRA1.2 LoVo subline was 2.5-fold resistant to Adr, while the Adr1.2 subline was 97-fold resistant, respectively. To verify the MDR phenotype of the resistant LoVo sublines it was assessed the cross-resistance to different antitumor agents (VCR, VBL, vindesine, VP16, mitomycin C and amsacrine). The degrees of resistance to each agent differed between the sublines: Adr1.2 was most resistant to Adr, whereas SRA1.2 was more resistant to the Vinca alkaloids than to the other agents including Adr.

The stability of the resistance was different in the two sublines. The level of resistance in Adr1.2 grown in drug free medium decreased from 95 to 3 times within 10 months. The resistance in the SRA1.2 subline, however, was rather stable through the entire 10-month testing period.

Adr1.2 had an higher doubling time (55.5 h) compared to SRA1.2 and LoVo, 30.2 and 31.5 h, respectively. This subline showed also an altered cell cycle distribution with an higher G_1 (65.8%) and a lower S phase (19%) subpopulation compared to LoVo (G_1 54.1 and S 31.9%) and SRA1.2 (G_1 55.4 and S 31.2%), respectively. The higher proportion of nonproliferating cells and the decreased cell cycle progression rate may explain the reduced cytotoxicity of many anticancer drugs to Adr1.2.

Immunoblotting with the ^{125}I-labeled monoclonal antibody C219 against p170 detected a 150 kDa plasma membrane protein in Adr1.2, but not in LoVo or SRA1.2 cells. A lower amount of protein was present in an Adr1.2 partial revertant that showed a 3-fold resistance, a level comparable to that of SRA1.2. Enhanced drug efflux was also observed in Adr1.2 (26%), but not in SRA1.2 when compared to LoVo.

It is conceivable that the prevailing mechanisms for MDR acquisition, uncovered by observations made in the laboratory may apply to only a specific type of MDR developed by mutations after continuous treatment with antitumor drugs. However, other types of MDR may develop in tumors not because of drug induced mutations but rather as a result of spontaneous mutations. Although a large difference was observed in the mechanisms of resistance of these two sublines of LoVo cells, showing high and low levels of resistance to Adr, it is not known whether these differences are due to the difference in the resistance levels or the methods used to induce the resistance.

In another study directed to clinical application it was investigated whether particular resistance phenotypes characterize all cells progressing to higher resistance levels and what factors (e.g. drug dosage, exposure time) influence the appearance of these phenotypes (Slapak *et al.*, 1990).

Two murine erythroleukemia cell lines PC4 and C7D were grown in suspension cultures containing Adr (5 ng/ml). Once adapted to 5 ng/ml the culture was split with one sample maintained with 5 ng/ml and another with 10 ng/ml. The drug concentration was increased by successive doublings up to 160 ng/ml.

After five passages with 5 ng/ml PC4 cells displayed about 2.5-fold resistance, the resistance level of this subline (PC4-5) increased after passaging with this concentration and reached 4.9-fold after 70 passages. A parallel increase of resistance to VP16 was observed, but the cells never acquired detectable VCR resistance.

The exposure of C7D cells to 5 ng/ml of Adr resulted in the acquisition of 1.5-fold resistance after five passages. After 15 passages the resistance of the C7D-5 subline increased to 3-fold and remained at this level after 70 passages, at this time VP16 resistance increased to 11-fold, but no VCR resistance was observed.

Passage of PC4-5 into 10 ng/ml of Adr increased the resistance to 9.5-fold for Adr and to 25-fold for VP16, but PC4-10 cells were still not resistant to VCR. After selection into 20 ng/ml of Adr the PC4-20 subline was 17- and 62-fold resistant to Adr and VP16, respectively, PC4-20 cells were also 2-fold resistant to VCR. The Adr concentration was doubled up to 160 ng/ml, at this concentration the PC4-160 subline was 180-fold resistant to Adr and 133- and 21-fold resistant to VP16 and VCR, respectively.

Verapamil pretreatment had little effect on Adr cytotoxicity on the PC4-5, PC4-10 and PC4-20 mutants; however, with the PC4-40 and PC4-80 mutants verapamil pretreatment restored the drug sensitivity to an intermediate level between the PC4-10 and PC4-20 mutants.

The expression of the mdr gene started to increase only after exposure of PC4 cells to 80 ng/ml of Adr, but the uptake of daunorubicin started to decrease after exposure to 40 ng/ml of Adr.

The resistance of the C7D-5 mutant increased 10-fold after exposure to 10 ng/ml of Adr, the resultant C7D-10 subline was also 15-fold resistant to VP16 and 3.1-fold to VCR. Parental C7D cells were also exposed directly to 10 ng/ml of Adr and the subline that was obtained, C7D-10A, had a 5.1-fold resistance.

After exposure of C7D-10 mutant to 20 ng/ml of Adr, the resistance increased to 12-fold for Adr and 22- and 5-fold to VP16 and VCR, respectively. At an Adr concentration of 160 ng/ml the resistance level was 54-fold for Adr and 50- and 21-fold to VP16 and VCR, respectively.

Verapamil pretreatment had essentially no effect on Adr toxicity on C7D-5 or C7D-10A, but increased the susceptibility of C7D-10 cells to the same level as C7D-10A. With C7D-20 cells, verapamil treatment decreased Adr resistance to a level similar to that of C7D-5.

The mdr expression did not increase until exposure of C7D cells to 20 ng/ml of Adr. The uptake of Adr was significantly increased in C7D-5 mutants in respect to the parental line, it was unchanged in the C7D-10A mutants and decreased after exposure of the C7D-5 cells to 10 ng/ml of Adr.

These findings seem to indicate that the time of overexpression of the mdr gene appears to be somewhat dependent on the individual cell line. Although the murine genome contains a family of three mdr genes (*mdr*-1, *mdr*-2 and *mdr*-3) expressed differently in a tissue specific manner (Croop *et al.*, 1989), the difference in mdr expression with the concentration of Adr does not depend on the different expression of these genes because the probe used hybridized all the three genes (Croop *et al.*, 1989).

It appears that the increase in P-gp is, initially at least, contributing minimally to the resistance phenotypes and that other mechanisms of resistance, besides MDR, are involved.

Spontaneous resistance

To investigate and characterize the probably different mechanisms involved in acquired and spontaneous resistance to Dx, two monoclonal cell sublines (LoVo/C17 and LoVo/C15) were isolated from human colon adenocarcinoma cells (LoVo) without previous Dx treatment. The sensitivity of these sublines to Dx, the expression of the MDR1 gene, DNA topoisomerase II and GSH-related enzymes was compared in LoVo/C17, the parental cell line (LoVo) and in a resistant line (LoVo/Dx) selected by continuous exposure to Dx (Monti *et al.*, 1993; Dolfini *et al.*, 1993a).

LoVo/Dx and LoVo/C17 were 44- and 3-fold resistant respectively to Dx. Expression of MDR1 was observed in LoVo/Dx but not in LoVo/C17, accordingly drug efflux was significantly higher in LoVo/Dx but not in the cloned line. No differences were detected in either topoisomerase II expression or GSH-related enzyme activity among the three cell lines. The presence of keratin 18 was demonstrated in all cell lines by immunochemistry, and at the mRNA level. A progressive increase in mRNA level for vimentin was seen in LoVo/C15, LoVo/C17 and LoVo/Dx, this effect was also observed by immunoperoxidase reaction (Dolfini *et al.*, 1993a).

The enzyme protein kinase C (PKC) has been observed to be involved in the expression of drug resistance. In MCF7 cells it was shown that phorbol ester treatment induced MDR and that drug resistance was associated with activation of PKC, increased phosphorylation of the 20 kDa region of the P-gp and decreased intracellular accumulation of anticancer drugs; in addition resistance was reverted by inhibitors of PKC (Fine *et al.*, 1988). Overexpression of PKC was also observed in multidrug resistant KB cells (Posada *et al.*, 1989), and the level of PKC was also found to be directly related to resistance to Dx in murine fibrosarcoma cells (O'Brian *et al.*, 1989).

Two isoforms, immunologically identified as β and αPKC were isolated from the cytosolic fraction of LoVo, LoVo/C15 and LoVo/C17 and one single peak of αPKC was obtained from the particulate fraction. Resistant LoVo/C17 cells showed a significant increase of PKC activity; preincubation with an inhibitor of PKC (H7) induced PKC inhibition and reversal of drug resistance (Dolfini *et al.*, 1993b).

The usefulness of MDR1, GST-π or topoisomerase II mRNA expression detected by dot blot analysis as an indicator of spontaneous resistance to Dx was investigated in 15 fresh human tumor specimens. MDR1 and GSH-π expression, which is known to be a marker for Dx resistance, was detected in six (66.7%) and seven (77.8%) of the nine clinically resistant tumors, respectively. However, in 67% of the responsive tumors MDR1 and GST-π were detected. Thus these two markers were not indicators of clinical response to Dx. In contrast topoisomerase II mRNA expression was significantly correlated with clinical response. The expression of topoisomerase II was detected at high level in five (83.3%) of the six clinically responsive tumors, and the other nine tumors resistant to Dx exhibited undetectable or low levels of topoisomerase II (Kim *et al.*, 1991).

In another study it was shown that in human colorectal tumors the constitutive levels of MDR1 mRNA/P-gp expression may not necessarily result in the functional expression of the MDR phenotype. While low levels of MDR1 mRNA/P-gp were detected in 5/8 well differentiated colon cell lines only 2/8 were functionally MDR. In contrast, 10/11 moderate and poorly differentiated lines expressed MDR1 mRNA/P-gp and of these 9/11 were functionally MDR. The phosphorylation status of the mature 170 kDa P-gp and the surface localization of this glycoprotein showed the strongest correlation with functionality (Kramer *et al.*, 1993).

References

Ahluwalia GS, Cohen MB, Kang GJ, Arnold ST, McMahon JB, Dalal M, Wilson YA, Cooney DA, Balzarini J and John DG (1986) Arabinosyl-5-azacytosine: mechanisms of native and acquired resistance. Cancer Res 46: 4479–4485.

Akiyama SI, Fojo ATT, Hanover JA, Pastan I and Gottesman MM (1985) Isolation and genetic characterization of human KB cell lines resistant to multiple drugs. Somatic Cell Mol Genet 11: 117–126.

Baas F, van Groeningen M, Nieuwint AW and Joenje HAD (1990) Two different mechanisms for multidrug resistance in a human lung cancer. Proc Ann Meet Am Assoc Cancer Res

31: A2186.

Badiner GJ, Bichay TJ and Bhuyan BK (1987) Menogaril resistance results in amplification of MDR gene in chinese hamster V79 cells but not in mouse leukemia P388 cells. Proc Ann Meet Am Assoc Cancer Res 28: A281.

Bradley G, Naik M and Ling V (1989) P-glycoprotein expression in multidrug-resistant human ovarian carcinoma cell lines. Cancer Res 49: 2790–2796.

Brewer F and Warr JR (1987) Verapamil reversal of vincristine-resistance and cross-resistance patterns of vincristine-resistant chinese hamster ovary cells. Cancer Treat Rep 71: 353–359.

Broxterman HJ, Pinedo HM, Kuiper CM, van-der Hoeven JJ, de Lange P, Quak JJ, Scheper RJ, Keizer HF, Schuurhuis GJ and Lankelma J (1989) Immunohistochemical detection of P-glycoprotein in human tumor cells with a low degree of drug resistance. Int J Cancer 43: 340–343.

Chan HS, Haddad G, Thorner PS, DeBoer G, Liu GP, Ondrusek N Yeger H, and Ling V (1991) P-glycoprotein expression as a predictor of the outcome of therapy for neuroblastoma. N Engl J Med 325: 1608–1614.

Chen CJ, Chin JE, Ueda K, Clark DP, Pastan I, Gottesman MM and Roninson IB (1986) Internal duplication and analogy with bacterial transport proteins in the *mdr*-1 (P-glycoprotein) gene from multidrug-resistant tumor cells. Cell 47: 381–389.

Chen YN, Mickley LA, Schwartz AM, Acton EM, Hwang J and Fojo AT (1990) Characterization of adriamycin-resistant human breast cancer cells which display overexpression of a novel resistance-related membrane protein. J Biol Chem 265: 10073–10080.

Chin JE, Soffir R, Nooan KE, Choi K and Roninson IB (1989) Structure and expression of the human MDR (P-glycoprotein) gene family. Mol Cell Biol 9: 3808–3820.

Choi K, Chen CJ, Kriegler M and Roninson IB (1988) An altred pattern of cross-resistance in multidrug resistant human cells results from spontaneous mutations in the *mdr*-1 (P-glycoprotein) gene. Cell 53: 519–529.

Choi K, Frommel TO, Stern RK, Perez CF, Kriegler M, Tsuruo T and Roninson IB (1991) Multidrug resistance after retroviral transfer of the human MDR1 gene correlates with P-glycoprotein density in the plasma membrane and is not affected by cytotoxic selection. Proc Nat Acad Sci USA 88: 7386–7390.

Cornwell M, Gottesman MM and Pastan I (1986) Increased vinblastine binding to membrane vesicles from multidrug-resistant KB cells. J Biol Chem 261: 7921–7928.

Croop JM, Raymond M, Haber D, De Vault A, Arceci RJ, Cross P and Houman DE (1989) The three mouse multidrug resistance (mdr) genes are expressed in a tissue-specific manner in normal mouse tumors. Mol Cell Biol 9: 1346–1350.

Dolfini E, Broggini M, Ceratto N, Barbieri C, Scanziani E and Dasdia T (1993a) Acquisition of vimentin expression and loss of epithelial markers in doxorubicin intrinsically resistant LoVo cell clones. Proc Ann Meet Am Assoc Cancer Res 34: A141.

Dolfini E, Dasdia T, Perletti G, Romagnoni M and Piccinini F (1993b) Analysis of calcium-dependent protein kinase-c isoenzymes in intrinsically resistant cloned lines of LoVo cells: reversal of resistance by kinase inhibitor 1-(5-isoquinolinylsulfonyl)-2-methylpiperazine. Anticancer Res (in press).

Fine RL, Patel J and Chabner BA (1988) Phorbol esters induce multidrug resistance in human breast cancer cells. Proc Natl Acad Sci USA 85:582–586.

Gerlach JH, Endicott JA, Juranka PF, Henderson G Sarangi F, Deuchars KL and Ling V (1986) Homology between P-glycoprotein and a bacterial haemolysin transport protein suggests a model for multidrug resistance. Nature 324: 485–489.

Goldstein LJ, Galski H, Fojo A, Willingham M, Lai SL, Gazdar A, Pirker R, Green A, Crist W, Brodeur GM, Lieber M, Cossman J, Gottesman MM and Pastan I (1989) Expression of a multidrug resistance gene in human cancers. J Nat Cancer Inst 81: 116–124.

Gottesman MM and Pastan I (1988) The multidrug transporter: a double edge sword. J Biol Chem 263: 12163–12166.

Hill BT, Whelan RDH and Bellamy AS (1984) Identification and differential drug responses to mechanisms of resistance in vincristine-resistant cell lines developed either by exposure to the drug or to fractional radiation. Cancer Treat Rev 11 (Suppl) A 73–79.

Kakehi Y, Kanamura H, Yoshida O, Ohkubo H, Nakamishi S, Gottesman MM and Pastan I (1988) Measurement of multidrug-resistance messenger RNA in urogenital cancers; elevated expression in renal cell carcinoma is associated with intrinsic drug resistance. J Urol 139: 862–865.

Kim R, Hirabayashi N, Nishiyama M, Saeki S, Toge T and Okada AD (1991) Expression of MDR1, GST-π and topoisomerase II as an indicator of clinical response to adriamycin. Anticancer Res 11: 429–431.

Kramer R, Weber TK, Morse B, Arceci R, Staniunas R, Steele Jr G and Summerhayes IC (1993) Constitutive expression of multidrug resistance in human colorectal tumors and cell lines. Br J Cancer 67: 959–968.

Lai SL, Goldstein LJ, Gottesman MM, Pastan I, Tsai CM, Johnson BE, Mulshine JL, Ihde DC, Kayser K and Gazdar AF (1989) MDR1 gene expression in lung cancer. J Nat Cancer Inst 81:1144–1150.

Lemontt JF, Azzaria M and Gros P (1988) Increased MDR gene expression and decreased drug accumulation in multidrug-resistant human melanoma cells. Cancer Res 48: 6348–6353.

Monti E, Perletti G, Broggini M, Dasdia T and Dolfini E (1993) Characterization of acquired and spontaneous resistance to doxorubicin in LoVo cells. Proc Ann Meet Am Assoc Cancer Res 34: A138.

Nair S, Yuan ZK, Singh SV and Krishan A (1992) Potentiation of doxorubicin cytotoxicity following glutathione depletion in human melanoma cell lines. Proc Ann Meet Am Assoc Cancer Res 33: A2720.

O'Brian CA, Fan D, Ward NE, Seid C and Fidler IJ (1989) Level of protein kinase C activity correlates directly with resistance to adriamycin in murine fibrosarcoma cells. FEBS Lett 246: 78–82.

Posada JA, McKeegan EM, Worthington KF, Marin MJ, Jaken S and Tritton TR (1989) Human multidrug resistant KB cells overexpress protein kinase C: involvement in drug resistance. Cancer Commun 1: 285–292.

Ramachandran C, Yuan ZK, Huang XL and Krishan A (1992) MDR1 and glutathione S-transferase π gene expression in human melanoma cells with low level of doxorubicin resistance. Proc Ann Meet Am Assoc Cancer Res 33: A2719.

Richon VM, Weich N Leng L, Kiyokawa H, Ngo L, Rifkind RA and Marks PA (1991) Characteristics of erythroleukemia cells selected for vincristine resistance that have accelerated inducer-mediated differentiation. Proc Nat Acad Sci USA 88: 1666–1670.

Riordan JR, Deuchars K, Kartner N, Alan N, Trent J and Ling V (1985) Amplification of P-glycoprotein genes in multidrug resistant mammalian cell lines. Nature 316: 817–819.

Shen DW, Fojo A, Chin JE, Roninson IB, Richert N, Pastan I and Gottesman MM (1986) Human multidrug-resistant cell lines: increased *mdr*-1 expression can precede gene amplification. Science 232: 643–645.

Slapak CA, Daniel JC and Levy SB (1990) Sequential emergence of distinct resistance phenotypes in murine erythroleukemic cells under adriamycin selection: decreased anthracycline uptake precedes increased P-glycoprotein expression. Cancer Res 50: 7895–7901.

Thiebaut F, Tsuruo T, Hamada H, Gottesman MM, Pastan I and Willingham M (1987) Cellular localization of the multidrug-resistance gene product P-glycoprotein in normal human tissues. Proc Nat Acad Sci USA 84: 7735–7738.

Twentyman PR, Reeve JG, Koch G and Wright KA (1990) Chemosensitization by verapamil and cyclosporin A in mouse tumor cells expressing different levels of P-glycoprotein and CP22 (sorcin). Br J Cancer 62: 89–95.

Ueda K, Cardarelli C, Gottesman MM and Pastan I (1987) Expression of full-length cDNA for the human "MDR1" gene confers resistance to colchicine, doxorubicin and vinblastine. Proc Nat Acad Sci USA 84: 3004–3008.

Yang LY and Trujillo JM (1990) Biological characterization of multidrug-resistant human colon carcinoma sublines induced/selected by two methods. Cancer Res 50: 3218–3225.

Address for offprints: G. Belvedere, Istituto di Ricerche Farmacologiche Mario Negri, Via Eritrea 62, 20157 Milan, Italy.

Cytotechnology **12**: 265–288, 1993.

Differing patterns of cross-resistance resulting from exposures to specific antitumour drugs or to radiation *in vitro*

Bridget T. Hill
Laboratory of Cellular Chemotherapy, Imperial Cancer Research Fund, 44 Lincoln's Inn Fields, London WC2A 3PX, UK

Key words: cross-resistance patterns, drug resistance, fractionated X-irradiation, non-P-glycoprotein-mediated multidrug resistance, P-glycoprotein-mediated multidrug resistance, P-glycoprotein regulation, topoisomerase II

Abstract

This article reviews the patterns of cross-resistance identified in various P-glycoprotein-mediated and non-P-glycoprotein-mediated drug resistant mammalian tumour cell lines. The differing patterns of cross-resistance and the variable levels of resistance expressed are summarised and discussed. Although the mechanism by which P-glycoprotein can recognise and transport a large group of structurally-unrelated substrates remains to be defined, the recent evidence indicating that membrane associated domains participate in substrate recognition and binding is summarised, and other possible explanations for these variable cross-resistance patterns are considered. Amongst the non-P-glycoprotein-overexpressing multidrug resistant cell lines, two subsets are clearly identifiable, one lacking and the other expressing cross-resistance to the Vinca alkaloids. Resistance mechanisms implicated in these various sublines and possible explanations for their differing levels and patterns of cross-resistance are summarised.

Clinical resistance is identified in patients following treatment not only with antitumour drugs, but also after radiotherapy. Experimental data providing a biological basis for this observation are summarised. A distinctive multiple drug resistance phenotype has been identified in tumour cells following exposure *in vitro* to fractionated X-irradiation characterised by: the expression of resistance to the Vinca alkaloids and the epipodophyllotoxins but not the anthracyclines and overexpression of P-glycoprotein which is post-translationally regulated, but without any concomitant overexpression of P-glycoprotein mRNA.

Finally, the possible clinical relevance of these variable patterns of cross-resistance to the antitumour drugs commonly used in the clinic is considered.

Introduction

The initial enthusiasm of clinicians in the 1950s associated with the finding that certain human cancers, mainly the leukaemias, could be treated effectively with individual antitumour drugs was tempered when response durations were short and relapses almost inevitably occurred (Farber *et al.*, 1948). However, the subsequent use of combinations of drugs lead to spectacularly improved therapies, even in certain solid tumours, including for example the pioneering work in breast cancer by Greenspan in the 1960s (Greenspan *et al.*, 1963). Indeed, considerable progress has been made in the last 25 years resulting in a significant number of cures in several types of cancers including Hodgkin's disease, large cell lymphoma, acute lymphocytic leukaemia, testicular cancer and early stage breast cancer (*cf.* reviews by Chabner, 1986; DeVita Jr, 1989). Unfortunately, however, most other malignancies, whilst initially responsive to chemotherapy are not curable and others remain refractory to therapy with any currently available antitumour agents. The development of drug resistance appears to represent a significant factor in these chemotherapy failures (Chabner, 1986; DeVita Jr, 1989). This has provided the impetus for detailed laboratory investigations aimed at elucidating the mechanisms associated with this phenomenon, so as to devise ways of overcoming or circumventing this major clinical problem. To those of us involved in these endeavours, on occasions it appears surprising, not that tumours fail to respond to chemotherapy, but that any such tumour cell populations can be eradicated with our present armamentarium of antitumour agents. It has become increasingly clear that once a tumour cell "acquires" the resistance phenotype, it becomes "equipped" with a seemingly endless array of mechanisms for evading the cell killing effects of currently available antitumour drugs. Clearly, therefore, the chance of cure rests on "early" treatment and effective destruction of the tumour cell population whilst it expresses drug sensitivity and before resistance develops (Goldie and Coldman, 1979; Goldie and Coldman, 1984; Skipper *et al.*, 1964).

As our laboratory knowledge of the plethora of drug resistance phenotypes increases, the full extent of the clinical problem becomes harder to grasp. Twenty five years ago Kessel and his colleagues (*cf.* review by Kessel, 1986) first noted that murine leukaemias selected for resistance to Vinca alkaloids were found to be cross-resistant to actinomycin D and daunomycin, drugs of unrelated structure. This pleiotropic drug pattern was subsequently identified in a number of rodent cell lines (Bech-Hansen *et al.*, 1976; Biedler and Riehm, 1970; Dano, 1972; Riehm and Biedler, 1971). Further research established the essential features of this phenotype: broad cross-resistance to a variety of apparently dissimilar "natural product" drugs; decreased drug accumulation and retention; and, overexpression of a novel plasma membrane glycoprotein called P-glycoprotein (Pgp) (Juliano and Ling, 1976; Kartner *et al.*, 1983). Beck and coworkers extended these observations to human leukaemic cell lines (*cf.* Beck, 1984; Beck *et al.*, 1979) and in the succeeding years numerous laboratories have developed and described cell lines from a range of different human tumour types which express a defined phenotype termed, by many, as multidrug resistance (see Table 1). It is, however, salutary to remember that this is only one specific form of multidrug resistance or MDR originally termed "classical" or "typical" MDR, caused by an increased level of Pgp, a fact reinforced by the seemingly exponentially expanding scientific literature in this field. In this review it will be referred to as Pgp-mediated MDR. In 1987 the existence of an "atypical" form of MDR was defined (Beck *et al.*, 1987; Danks *et al.*, 1987). The essential features of this specific phenotype included: cross-resistance to a variety of natural product antitumour drugs, but not including the Vinca alkaloids and colchicine; unaltered drug uptake, accumulation and efflux of drugs; a lack of overexpression of Pgp or its mRNA; and, an alteration in topoisomerase II. More recent advances, however, have provided evidence that other forms of non-Pgp-mediated or nonclassical MDR also exist. For example, Cole and her colleagues (1992) have identified overexpression of a new transporter gene termed MRP (multidrug resistance – associated protein), in an Adriamycin-selected lung cancer cell line showing the "classic" cross-resistance pattern to anthracyclines and epipodophyllotoxins and also to the Vinca alkaloids, but in the absence of Pgp overexpression or any detectable alteration in drug accumulation (see Table 2). In addition, certain tumour cell lines selected for anthracycline resistance appear to manifest modifications in several of the defined resistance associated proteins, including both Pgp and the topoisomerases and possibly other novel proteins (*cf.* for example Baas *et al.*, 1990; Deffie *et al.*, 1992; Fairchild *et al.*, 1987; Sinha *et al.*, 1988; Zijlstra *et al.*, 1987). Finally, my laboratory has reported that

Table 1. Summary of resistance and cross resistance patterns, biochemistry and molecular biology of a range of Pgp-mediated MDR cell lines

Cell line designation	Selecting agent (origin)	Levels of resistance and cross resistance									Altered drug up-take / efflux	P-glycoprotein:			References (see below)
		Act D	ADR	COLC	DAUNO	VCR	VBL	VP-16	TAXOL	MITOX		Overexpression		Amplification	
												Protein	mRNA	DNA	
Hamster:															
AuxB1	(ovary)														
CH^{R}C5	Colchicine	nr	25	180	76	120	30	33	9	36	+/nr	+	+	+	1–4
DNR^{R}51	Daunomycin	nr	~30	25	41	nr	22	nr	nr	nr	+/nr	+	+	+	1,2
DC-3F	(lung fibroblasts)														
-VCRd-5L	Vincristine	1000	1640	470	178	2740	nr	nr	nr	nr	+/+	+	+	+	7–9,11
-DMXX	Daunomycin	267	8360	441	1530	126	nr	nr	nr	nr	+/nr	+	+	+	6,7,9–11
-ADIV	Actinomycin D	1000	110	71	29	89	239	nr	nr	nr	+/nr	+	+	+	5,7,9
-ADX	Actinomycin D	10000	761	180	54	593	860	nr	1670	nr	+/nr	+	+	+	6,7,9,10
-ADXC	Actinomycin D	70000	2240	412	130	1260	nr	nr	nr	nr	+/nr	+	+	+	9,11
Mouse:															
P388	(leukaemia)														
-ADR3	Adriamycin	18	5	nr	11	nr	nr	4	nr	13	+/+	+	+	–	12,13
-ADR7	Adriamycin	18	10	12	16	nr	nr	6	nr	11	+/+	+	+	–	12,13
L1210	(leukaemia)														
-ADM	Adriamycin	12	45	nr	24	nr	nr	nr	nr	nr	+/nr	+	+	+	14,15
-DNR	Daunomycin	20	51	nr	30	nr	nr	nr	nr	nr	+/nr	+	+	+	14,15
J774.2	(macrophage-like)														
-V1-1	Vinblastine	8	234	382	nr	nr	1083	nr	643	nr	+/nr	+	+	+	16
-V3-1	Vinblastine	21	166	268	nr	nr	1017	nr	163	nr	+/nr	+	+	–	16
-C1-20	Colchicine	61	81	550	nr	nr	17	nr	65	nr	+/nr	+	+	+	16
-T1-50	Taxol	88	128	58	nr	nr	43	nr	833	nr	+/nr	+	+	+	17
Human:															
A2780	(ovarian ca[a])														
2780AD	Adriamycin	nr	170	nr	nr	15	nr	nr	nr	nr	+/nr	+	+	+	18
CCRF/CEM	(lymphoblastoid)														
-VLB100	Vinblastine	nr	120	103	111	500	420	20	nr	21	+/+	+	+	+	19
-VCR100	Vincristine	nr	556	399	528	923	912	588	nr	nr	+/+	+	+	+	20
DLKP	([b]scc lung)														
/A	Adriamycin	nr	322	4	nr	80	nr	36	nr	nr	nr	+	nr	nr	21

(continued)

Table 1. Summary of resistance and cross resistance patterns, biochemistry and molecular biology of a range of Pgp-mediated MDR cell lines (continued)

Cell line designation	Selecting agent (origin)	Levels of resistance and cross resistance: Act D	ADR	COLC	DAUNO	VCR	VBL	VP-16	TAXOL	MITOX	Altered drug uptake / efflux	P-glycoprotein: Over-expression: Protein	mRNA	Amplification: DNA	References (see below)
HL60	(AML)														
/VCR	Vincristine	nr	15	nr	nr	140	nr	nr	nr	nr	+/nr	+	+	nr	22
NCI-H69	(small cell lung ca[a])														
/LX4	Adriamycin	nr	100	26	nr	430	nr	88	nr	nr	+/nr	+	+	+	23–25
K562	(leukaemia)														
/ADM	Adriamycin	108	135	nr	41	606	122	11	>3800	nr	+/+	+	+	+	26
/VCR	Vincristine	nr	10	nr	nr	88	nr	nr	nr	nr	+/nr	+	+	+	27,28
KB-3-1	(epidermoid ca[a])														
-8	Colchicine	1.5	1.1	2.1	nr	1.0	1.2	nr	nr	nr	+/+	+	+	–	29–32
-8–5–11	Colchicine	58	23	40	nr	36	51	nr	nr	nr	+/+	+	+	+	29–32
-8–5–11–24	Colchicine	42	26	128	nr	59	20	nr	nr	n r	+/+	+	+	+	29–32
-A1	Adriamycin	123	97	19	nr	nr	43	nr	nr	nr	+/+	+	+	+	29–32
-V1	Vinblastine	100	420	170	nr	nr	213	nr	nr	nr	+/nr	+	+	+	29–32
-C1	Colchicine	22	21	323	106	450	nr	13	nr	nr	+/nr	+	+	–	33
-VJ300	Vincristine	92	10	21	18	394	nr	5	nr	nr	+/+	+	+	–	30–32
MCF-7	(breast ca[a])														
/ADR	Adriamycin	357	192	nr	nr	>250	375	100	nr	nr	+/+	+	+	+	35,35
/D40	Adriamycin	nr	75	nr	nr	190	93	11	nr	153	+/+	+	+	+	36
/VCR6E	Vincristine	nr	1.8	nr	nr	13.7	3.0	3.0	nr	6.2	+/nr	+	+	–	37,38
/VCREMS	EMS[c] + vincristine	nr	2.7	nr	nr	11.1	nr	3.2	nr	nr	+/nr	+	–	–	39
MES-SA	(sarcoma)														
Dx1	Adriamycin	90	25	nr	25	100	5	15	nr	nr	nr/nr	nr	nr	nr	40
Dx5	Adriamycin	1200	100	nr	160	240	105	30	nr	nr	+/+	+	+	+	41
SH-SY5Y	(neuroblastoma)														
/VCR	Vincristine	270	290	510	nr	1420	nr	nr	nr	nr	nr	+	+	+	9,11,42,43
BE(2)-C	(neuroblastoma)														
/CHCb(0.2)	Colchicine	65	40	54	nr	87	nr	nr	nr	nr	nr	+	+	+	44,45
/ADR(5)	Adriamycin	65	104	38	nr	100	nr	nr	nr	nr	nr	+	+	+	44,45
/ACT(0.2)	Actinomycin D	124	57	24	nr	64	nr	nr	nr	nr	nr	+	+	+	44,45
/VCR(10)	Vincristine	324	126	98	nr	567	nr	nr	nr	nr	nr	+	+	+	44,45

(continued)

Table 1. Summary of resistance and cross resistance patterns, biochemistry and molecular biology of a range of Pgp-mediated MDR cell lines (continued)

Cell line designation	Selecting agent (origin)	Levels of resistance and cross resistance									Altered drug uptake / efflux	P-glycoprotein: Overexpression		P-glycoprotein: Amplification	References (see below)
		Act D	ADR	COLC	DAUNO	VCR	VBL	VP-16	TAXOL	MITOX		Protein	mRNA	DNA	
OV1	(ovarian ca[a])														
/VCR	Vincristine	110	30	nr	nr	705	nr	20	nr	nr	nr	+	+	–	46
SKOV3	(ovarian ca[a])														
/VCR0.015	Vincristine	nr	16	4	nr	8	4	nr	nr	nr	nr	–	+	nr	47
/VCR0.1	Vincristine	nr	32	8	nr	64	260	nr	nr	nr	nr	+	+	–	47
/VCR0.25	Vincristine	nr	260	64	nr	510	1000	nr	nr	nr	nr	+	+	+	47
/VCR2	Vincristine	nr	510	260	nr	1000	4100	nr	nr	nr	nr	+	+	+	47
/VLB0.01	Vinblastine	nr	4	2	nr	3	4	nr	nr	nr	nr	+	–	–	47
/VLB0.03	Vinblastine	nr	8	5	nr	16	64	nr	nr	nr	nr	+	+	–	47
/VLB0.06	Vinblastine	nr	87	130	nr	2600	490	nr	nr	nr	nr	+	+	+	47
/VLB1	Vinblastine	nr	260	510	nr	10000	2000	nr	nr	nr	nr	+	+	+	47
SW1573	(scc lung)														
-IR500	Adriamycin	nr	250	62	70	540	nr	270	nr	nr	+/+	+	+	–	48–51
-IR10000	Adriamycin	nr	2000	220	nr	nr	nr	750	nr	nr	+/+	+	+	+	48–51
-2R160	Adriamycin	nr	94	nr	35	480	nr	120	nr	nr	+/+	+	+	–	48–51
8226	(myeloma)														
-Dox6	Adriamycin	nr	17	nr	14.5	27	9	7.5	nr	6.2	–/+	+	+	–	52,53
-Dox40	Adriamycin	nr	138	nr	nr	269	85	48	nr	95	–/+	+	+	+	52,53

[a]ca — carcinoma; [b]scc — squamous cell carcinoma; [c]EMS — ethylmethane sulphonate.

References:

1. Bech-Hansen *et al.*, 1976
2. Ling *et al.*, 1983
3. Hill *et al.*, 1990.
4. McClean *et al.*, 1993c
5. Biedler and Riehm, 1970
6. Riehm and Biedler, 1971
7. Peterson *et al.*, 1983
8. Sirotnak *et al.*, 1986
9. Scotto *et al.*, 1986
10. De Bruijn *et al.*, 1986
11. Biedler *et al.*, 1988
12. Deffie *et al.*, 1988
13. Deffie *et al.*, 1992
14. Volm *et al.*, 1989
15. Volm *et al.*, 1991
16. Kirschner *et al.*, 1992
17. Horwitz *et al.*, 1989
18. Rogan *et al.*, 1984
19. Beck *et al.*, 1979
20. Conter and Beck, 1984
21. Clynes *et al.*, 1992
22. McGrath and Center, 1988
23. Reeve *et al.*, 1990
24. Coley *et al.*, 1989
25. Coley *et al.*, 1991
26. Sugimoto *et al.*, 1987
27. Tsuruo *et al.*, 1983
28. Tsuruo *et al.*, 1986
29. Akiyama *et al.*, 1985
30. Shen *et al.*, 1986
31. Fojo *et al.*, 1985
32. Fojo *et al.*, 1987
33. Kohno *et al.*, 1988
34. Fairchild *et al.*, 1987
35. Kramer *et al.*, 1988
36. Taylor *et al.*, 1991
37. Hill *et al.*, 1989
38. Whelan and Hill, 1993
39. Whelan *et al.*, 1992
40. Harker and Sikic, 1985
41. Sikic *et al.*, 1989
42. Meyers *et al.*, 1985
43. Van der Bliek *et al.*, 1988
44. Biedler *et al.*, 1991
45. Biedler, unpublished data
46. Benard *et al.*, 1989
47. Bradley *et al.*, 1989
48. Keizer *et al.*, 989
49. Baas *et al.*, 1990
50. Kuiper *et al.*, 1990
51. Versantvoort *et al.*, 1992
52. Dalton *et al.*, 1986
53. Dalton *et al.*, 1989

Table 2. Summary of resistance and cross resistance patterns, biochemistry and molecular biology of a range of non-Pgp-mediated MDR cell lines

Cell line designation	Selecting agent (origin)	Levels of resistance and cross resistance										Altered drug uptake/ efflux	Pgp overexpression (protein and mRNA)	Altered Topo II level activity	Other membrane proteins overexpressed:		References (see below)
		Act D	ADR	COLC	DAUNO	VCR	VBL	VP-16	VM-26	TAXOL	MITOX				MRP	Others	
CCRF/CEM	(lymphoblastoid)																
VM-5	Teniposide	2.4	84	0.9	6.6	0.3	3	41	47	nr	16	–/–	–	+	nr	nr	1–3
VM-1	Etoposide	11.4	116	nr	nr	7.2	0.9	129	141	nr	nr	–/–	–	+	nr	nr	2
VP-1	Teniposide	nr	7.6	nr	nr	1	1.9	15	17.6	nr	nr	nr	?/[a]	+	nr	nr	4
COR-L23	(large cell lung ca[b])																
/R	Adriamycin	nr	17	12	nr	34	nr	50	nr	nr	nr	+/+	–	nr	+	nr	5–7
GLC4	(small cell lung ca[b])																
/ADR	Adriamycin	1	123	≤0.5	nr	≥6	nr	67	72	nr	10	+/+	–	+	+	110kD	7–11
K562	(leukaemia)																
/VP-16	Etoposide	1	8	1.4	nr	nr	nr	17	nr	nr	nr	+/–	–	+	nr	85kD	12
/VP-5	Etoposide	nr	5	nr	nr	1	1	11	11	nr	nr	–/–	–	+	nr	nr	13
H69	(small cell lung ca[b])																
/AR	Adriamycin	nr	54	51	360	35	7	40	100	10	42	–/–	–	+	+	nr	14–18
H209/V6	Etoposide	nr	4	nr	nr	nr	1	22	nr	1	nr	–/–	–	+	–	nr	19
HL60	([c]AML)																
/ADR	Adriamycin	nr	80	nr	nr	nr	nr	nr	nr	nr	nr	+/+	–	nr	+	nr	20–22
/MX2	Mitoxantrone	2	4	nr	4	1	1	15	24	nr	35	–/–	–	+	nr	nr	23
HT1080	(fibrosarcoma)																
/DR4	Adriamycin	17	222	nr	nr	25	2	837	nr	nr	nr	nr/+	–	+	+	110kD	11,24–27
SW1573	([d]scc lung)																
2R50	Adriamycin	nr	5	2	3	6	nr	14	nr	nr	nr	+/–	–	+	–	110kD	7,10,11,
2R120	Adriamycin	nr	20	nr	4	17	nr	45	nr	2	7	+/–	–	+	–	110kD	28–30
EPG85	(gastric ca[b])																
/275NOV	Mitoxantrone	nr	11	1.5	4	1	1	nr	nr	nr	186	nr/+	–	+	–	nr	31,32

(continued)

Table 2. Summary of resistance and cross resistance patterns, biochemistry and molecular biology of a range of non-Pgp-mediated MDR cell lines (continued)

Cell line designation	Selecting agent (origin)	Levels of resistance and cross resistance										Altered drug uptake/ efflux	Pgp over-expression (protein and mRNA)	Altered Topo II level/ activity	Other membrane proteins overexpressed:		References (see below)
		Act D	ADR	COLC	DAUNO	VCR	VBL	VP-16	VM-26	TAXOL	MITOX				MRP	Others	
MCF-7	(breast ca[b])	nr	8.3	nr	nr	22	43	26	nr	nr	1208	nr/+	–	–	–	110kD	10,11,33
/MITOX	Mitoxantrone	nr	900	nr	326	nr	5	nr	26	nr	nr	–/nr	–	+	nr	95kD	34
/AdrVp	Adriamycin + Verapamil																
8226	(myeloma)	nr	3	nr	nr	1	nr	8	nr	nr	37	+/–	–	nr	–	160kD	33,35
/MR20	Mitoxantrone	nr	12	nr	nr	1	1	18	nr	nr	5	–/–	–	+	–	nr	36,37
/Dox$_1$V	Adriamycin + Verapamil																
KB	(epidermoid ca[b])	nr	8	nr	nr	13	nr	29	nr	nr	2.0	+/–	–	+	nr	nr	38
/11c	Etoposide	nr	23	nr	nr	13	nr	145	nr	nr	2.4	+/–	–	+	nr	nr	38
/17c	Etoposide	nr	40	nr	nr	14	nr	186	nr	nr	2.7	+/–	–	+	nr	nr	38
/120a	Etoposide	nr	34	nr	nr	15	nr	287	nr	nr	6.2	+/–	–	+	nr	nr	38
/140a	Etoposide																

[a]No reversal of VCR or VP-16 resistance by verapamil; [b]carcinoma; [c]acute myelogenous leukaemia; [d]squamous cell carcinoma.

References:

1. Danks *et al.*, 1987
2. Danks *et al.*, 1988
3. Danks *et al.*, 1989
4. Patel and Fisher, 1993
5. Twentyman *et al.*, 1986
6. Barrand *et al.*, 1993
7. Versantvoort *et al.*, 1992
8. De Jong *et al.*, 1990
9. De Jong *et al.*, 1993
10. Scheper *et al.*, 1993
11. Zaman *et al.*, 1993
12. Sugawara *et al.*, 1991
13. *cf.* Beck and Danks, 1991
14. Mirski *et al.*, 1987
15. Cole *et al.*, 1991
16. Cole, 1992
17. Cole *et al.*, 1992
18. Gervasoni *et al.*, 1992
19. Mirski *et al.*, 1993
20. McGrath and Center, 1987
21. Marquardt *et al.*, 1990
22. Marquardt and Center, 1992
23. Harker *et al.*, 1991
24. Slovak *et al.*, 1988
25. Zwelling *et al.*, 1990
26. Slovak *et al.*, 1991
27. Slovak *et al.*, 1993
28. Kuiper *et al.*, 1990
29. Baas *et al.*, 1990
30. Nieuwint *et al.*, 1992
31. Dietel *et al.*, 1990
32. Kellner *et al.*, 1993
33. Taylor *et al.*, 1991
34. Chen *et al.*, 1990
35. Dalton *et al.*, 1993
36. Dalton, 1990
37. Dalton, unpublished data
38. Ferguson *et al.*, 1988.

exposure of tumour cell *in vitro* to X-irradiation appears to result in a further novel form of Pgp-mediated MDR (Hill, 1991; Hill *et al.*, 1990). These irradiated cells clearly express a distinctive phenotype characterised by cross-resistance to Vinca alkaloids and epipodophyllotoxins, but not to the anthracyclines and overexpression of Pgp, but without any concomitant elevation in Pgp message, associated with increased protein stability (Hill, 1991; Hill *et al.*, 1990, 1993; McClean and Hill, 1993; McClean *et al.*, 1993c).

This article reviews briefly the patterns of cross-resistance identified in various Pgp-mediated and non-Pgp-mediated drug resistant cell lines and then details the distinctive MDR phenotype associated with exposure of tumour cells to X-irradiation. It is selective and not comprehensive, but further examples are provided in two excellent recent reviews (Beck and Danks, 1991; Nielsen and Skovsgaard, 1992). It is appreciated that different methods of evaluating drug sensitivities have been used by various investigators, so that valid direct comparisons cannot be made reliably between different cell lines. However, generally within each group methodologies have been consistent and so attempts to compare and contrast relative levels of resistance and cross-resistance have been made in this review.

Patterns of cross-resistance in drug-selected MDR cell lines

Pgp-overexpressing MDR sublines

Table 1 summarises data on various "classical" MDR sublines derived from hamster, mouse or human tumours in which Pgp overexpression has been identified. Several general points can be drawn from an analysis of these published data. In the original studies using rodent cell lines, high levels of resistance were selected for, ranging from 30–50,000 and more generally being in excess of 100-fold, to the drugs available at that time, predominantly colchicine, daunomycin, actinomycin D and the Vinca alkaloids. These lines invariably proved most resistant to the selecting agents used and Pgp overexpression was accompanied by altered drug accumulation and some degree of gene amplification. In the more recent studies involving human tumours, these MDR sublines appear to have been selected by exposure either to the Vinca alkaloids or anthracyclines, perhaps reflecting the clinical usage of these particular drugs, but I have not been able to find any examples in the literature of Pgp-overexpressing MDR sublines resulting from exposure to the epipodophyllotoxins. This raises the possibility that the expression of cross-resistance to these agents, identified as characteristic of "classical" MDR sublines may be coincidental. However, unpublished data from my laboratory indicates that etoposide can be used to select for a Pgp-overexpression MDR subline of a testicular teratoma (SuSa) cell line, although it should be stressed that this resulted only after further selection, using a higher etoposide concentration, of a low level resistant subline which expressed the "atypical" MDR phenotype associated with an alteration in topoisomerase II. Baas *et al.* (1990) also reported that non-Pgp-mediated mechanisms for MDR preceded Pgp expression during *in vitro* selection for Adriamycin resistance in the human squamous lung cancer cell line SW-1573, although this sequence of events was not noted in a series of highly resistant etoposide-selected KB cell lines (Ferguson *et al.*, 1988). This concept that differential resistance mechanisms may operate or predominate depending on the level of resistance expressed is one that I have discussed in earlier reviews (Hill, 1986; Hill, 1990) and will be considered again below. However, it is important to note from Table 1 that Pgp overexpression has been identified not only in highly resistant sublines, but also in those expressing more modest levels of resistance, namely those most likely to be encountered clinically (*cf.* for example the detailed study by Bradley *et al.*, 1989). In these latter cases, however, Pgp overexpression was not associated with gene amplification. In the human tumour Pgp-overexpressing MDR sublines listed in Table 1, the highest level of resistance expressed is frequently *not* to the selection agent used. This appears to be the case, especially with Adriamycin, since for example the following cell lines all proved appreciably more resistant to vincristine: NCI-H69/LX4,

K562/ADM, MCF-7/ADR, MCF-7/D40, MES-SA/Dx1 and Dx5, SW1573/1R500 and 2R160, as well as 8226/Dox40. On the other hand, the Vinca alkaloid-selected resistant human tumour lines generally appear to reflect the situation noted first in the rodent lines, proving more resistant to vinblastine and vincristine than to the anthracyclines. Although, interestingly, in the SKOV3 series selected for resistance to either vincristine or to vinblastine (Bradley *et al.*, 1989), it is only at levels of resistance less than 10-fold that the cells proved most resistant to the actual selecting drug.

Examination of the different patterns of cross-resistance and the variable levels of resistance expressed by these various Pgp-overexpressing sublines, reveals no real consistency. Whilst it has been shown generally that increasing the concentration of the selection agent results in a more highly resistant subline, the associated responses to other drugs appear either to increase concordantly (for example, in the SKOV3/VCR and VLB sublines and the MES-SA and 8226 series), remain the same or even to decrease (for example, in the KB colchicine series). Such variations clearly cannot be ascribed to the tumour cell type or assay conditions used since these were constant in these examples and they do not appear to be specific for individual drugs. Although the mechanism by which Pgp can recognise and transport a large group of structurally unrelated substrates remains to be defined, recent evidence indicates that membrane-associated domains participate in substrate recognition and binding. Functional analyses of chimeric (Dhir and Gros, 1992) and mutant Pgps highlight transmembrane (TM) domains as important determinants of substrate interactions. Human, murine and hamster Pgps bearing mutations within or near TM domains express drug resistance phenotypes distinct from that of their wild-type counterparts: a single Ser to Phe substitution within TM11 strongly modulated the degree of resistance to colchicine and Adriamycin, but not to vinblastine (Gros *et al.*, 1991); two adjacent amino acid substitutions within the predicted TM6 domain of Pgp favoured maximal resistance to actinomycin D rather than to colchicine (Devine *et al.*, 1992); substitution of a Gly to Val within the interval separating the predicted TM2 and TM3 domains of human *mdr1* caused opposite effects on the degree of drug resistance and transport of colchicine (increase) and vinblastine (decrease) (Choi *et al.*, 1988; Safa *et al.*, 1990); expression of the mutant Pgp also resulted in significantly increased resistance to etoposide and decreased resistance to vincristine and actinomycin D (Safa *et al.*, 1990). Clearly such analyses are still at an early stage. Other possible explanations of these variable patterns of cross-resistance may involve: (i) alterations in N-linked glycosylation of Pgp or differences in Pgp precursors (Greenberger *et al.*, 1988, 1987); (ii) expression of different members of the *mdr* gene family (Riordan *et al.*, 1985; Scotto *et al.*, 1986); (iii) alternate expression or splicing of the *mdr* gene (Hsu *et al.*, 1989; Van der Bliek *et al.*, 1987); (iv) altered post-translational regulation of Pgp (Bradley *et al.*, 1989; Hill *et al.*, 1990, 1993; Whelan *et al.*, 1992); (v) modulation of the specificity of Pgp transport by phosphorylation (Bates *et al.*, 1992); (vi) stochastic processes with the selection of one cross-resistance pattern over another being randomly generated. Several studies designed to evaluate these possibilities further are now under way. Few groups have determined whether similar resistance patterns emerge from independently-selected sublines using identical drug exposure conditions. Similar cross resistance patterns were noted in independently-isolated *mdr1* transfectants of human BRO melanoma cells (Lincke *et al.*, 1990) and in two *mdr1* transfected and one nontransfected resistant K562 clones (Hait *et al.*, 1993). Furthermore, it remains to be established whether the differences observed in cross-resistant patterns of human MDR cell lines has any relevance to clinical MDR.

Non-Pgp-overexpressing sublines

A list of the main characteristics of certain non-Pgp-expressing cell lines has been compiled in Table 2. The selection agents used include the epipodophyllotoxins, Adriamycin and mitoxantrone. Clearly this is a much more heterogeneous group of cell lines. Generally, overall levels of resistance are lower in these non-Pgp expressing lines than those cited as examples of Pgp-mediated MDR in Table 1. One possible explanation for this, however, may relate to the fact that in these more recent studies emphasis has been placed on developing model systems expressing more clinically-relevant levels of drug resistance.

Examination of the data summarised in Table 2 reveals that whilst sublines selected for resistance to mitoxantrone appear most resistant to this selecting drug, this is not the case for the majority of the lines selected by exposure to Adriamycin, teniposide or to etoposide, which have variously proved most highly resistant to colchicine, etoposide or to Adriamycin. Indeed, Baas *et al.* (1990) reported that in their Adriamycin-selected non-Pgp-mediated SW-1573 sublines, preferential resistance to etoposide was noted. The original definition of "atypical"-MDR specified a lack of cross-resistance to the Vinca alkaloid, exemplified by the CEM/VM-1 subline and also shown by the following sublines listed in Table 2: CEM/VP1, K562/VP-5, H209/V6, HL60/MX2, EPG85/275NOV, 8226/MR20 and Dox_1V. However, another subset of lines can now be identified which clearly express cross-resistance to vincristine, including COR-L23/R, GLC4/ADR, H69/AR, HT1080/DR4, sublines SW1573-2R50 and 2R120 and MCF-7/MITOX and the four etoposide-selected KB sublines. Interestingly though, I am not aware of any example of these latter type lines being derived following Vinca alkaloid selection. No consistent pattern of decreased drug accumulation and retention by these non-Pgp-expressing cell lines is apparent and these factors appears unrelated to their Vinca alkaloid sensitivities. Some recent studies, however, suggest that altered subcellular drug distribution may constitute an important mechanism in certain of these sublines, since sequestration of drugs in these resistant cells may be modified so that the drugs are less able to reach their intracellular sites of action (Cole *et al.*, 1992; Marquardt and Center, 1992). It is probable that an alteration in topoisomerase II or in a modifying activity for this enzyme is implicated in all these non-Pgp-expressing MDR cell lines. However, with the recent identification of MRP, the new member of the ATP-binding cassette transmembrane transporter superfamily in the Adriamycin-selected lung cancer cell line H69AR, other resistance associated membrane and or cytoplasmic proteins (*cf.* recent review by McClean and Hill, 1992) are likely to be implicated in various forms of non-Pgp-expressing MDR. Indeed, overexpression of a 110 kd vesicular protein has now been detected in the GLC4/ADR, SW157/2R120, HT1080DR4, 8226MR40 and MCF-7/MITOX cell lines (Scheper *et al.*, 1993) and an 85 kd membrane protein was detected in the K562/VP-16 cells, although this latter protein has also been identified in two Pgp-overexpressing human tumour sublines, K562/ADM and 2780/AD (Hamada *et al.*, 1988; Sugimoto *et al.*, 1993). Dalton *et al.* (1993) also recently reported a new 160 kD protein in their 8226/MR20 cells. A novel resistance-related 95 kD membrane protein was also identified in a non-Pgp-expressing Adriamycin-resistant MCF-7 subline selected with incremental increases of Adriamycin in the presence of verapamil (Chen *et al.*, 1990). It is also pertinent to note that whilst *MRP* has been detected in lines COR-L23/R, GLC4/ADR, H69/AR and HT1080/DR4, its overexpression cannot account for all forms of non-Pgp-mediated MDR since, as Zaman *et al.* (1993) recently reported, ten SW-1573 non-Pgp MDR sublines expressed *MRP* mRNA levels equal to the parental line. It is, therefore, essential to stress that functions for these more recently described proteins remain to be defined. Results of gene transfer studies have firmly established *mdr1* expression as the major and perhaps the only determinant of the Pgp-related MDR phenotype (*cf.* review by Schinkel and Borst, 1991). Genetic transfer of non-Pgp-mediated MDR in somatic cell fusion has permitted a dissection of this phenotype and provided clear evidence of its multifacted nature: an MDR phenotype with reduced drug accumulation and loss of *mdr1* Pgp mRNA was cotransferred to the acceptor cells, but the decrease in topoisomerase II gene expression was not (Eijdems *et al.*, 1992). Furthermore, whilst altered topoisomerase II-mediated drug resistance is a recessive phenotype (*cf.* Beck, 1989) recent exciting advances involving genetic suppressor elements (GSEs) derived from recessive drug resistant genes have established that GSE-mediated inhibition of topoiomerase IIα induces a resistance phenotype specific to topoisomerase II poisons (Gudkov *et al.*, 1993) In the meantime, experiments are in progress to determine whether overexpression of the *MRP* gene results in non-Pgp-related MDR in transfected cells, or whether suppression of *MRP* expression reverses this resistance (Zaman *et al.*, 1993).

There are few reports of studies aimed at providing an explanation for the differential levels of cross resistance identified in these non-Pgp-overexpressing sublines. In terms of topoisomerase II related mechanisms, since it is known that the two isoforms differ in their drug sensitivities, with the α-form proving, for example, more sensitive to teniposide (Drake *et al.*, 1989), downregulation of the α or β isoform could lead to changes in resistance patterns. Indeed a specific deletion in topoisomerase II α has been described in an amsacrine resistant P388 subline (Drake *et al.*, 1987), whilst a depletion of topoiosmerase II β was associated with mitoxantrone resistance in an HL60 subline (Harker *et al.*, 1991). This latter study though emphasised that multiple alteration may be associated with resistance including reduced enzyme activity, altered cellular distribution and reduced sensitivity to drug-induced DNA damage. Other factors which may be implicated, as reviewed recently (Hochhauser and Harris, 1993) include point mutations of the gene, altered phosphorylation or upregulation of topoisomerase I to compensate for the depletion of topoisomerase II.

Patterns of cross-resistance in X-irradiated MDR sublines

Drug sensitivity evaluations

Clinical drug resistance has been identified in certain subsets of patients following their treatment not only with chemotherapy, but also after radiotherapy (*cf.* review by Hill, 1991). This has been manifest not only by reduced response rates, but also by significantly shorter response and survival durations. These observations lead to our original proposal that exposure to radiation could "induce" or "select" for drug resistance and thus subsequent chemotherapy would provide a positive selection pressure for these resistant tumour cells resulting in the outgrowth of a drug resistant cell population (Hill, 1990). To investigate this hypothesis, therefore, we established an *in vitro* model system to determine whether exposure of cultured tumour cells to fractionated X-irradiation resulted in any expression of drug resistance in the surviving population. Using L5178Y murine lymphoblasts we demonstrated the expression of resistance to vincristine and to etoposide under these conditions (Bellamy and Hill, 1984; Hill and Whelan, 1982). This led us to extend these investigations to include the well characterised Chinese hamster ovary (CHO) model and a wide range of human tumour cell types (Bedford *et al.*, 1987; Dempke *et al.*, 1992; Hill, 1990, 1991; Hill *et al.*, 1990, 1988; McClean *et al.*, 1993b, 1993c; Whelan and Hill, 1993). Characteristically all these sublines established after *in vitro* exposure to X-irradiation proved resistance to multiple drugs.

The derivation and fundamental growth characteristics of these radiation-pretreated sublines have been reviewed recently (Hill, 1991). It must be emphasised that none of these sublines expressed any significant resistance to a subsequent challenge with acute X-rays, *i.e.,* this exposure to fraction X-irradiation *in vitro* does *not* select for radiation resistance, as discussed earlier (Hill, 1990; Hill, 1991) In contrast, however, characteristically *in vitro* exposure of all these various tumour cell lines to fractionated X-irradiation resulted in the expression of resistance to the Vinca alkaloids and to the epipodophyllotoxins, but *not* to the anthracyclines (see Table 3). It is striking that the same pattern of responses to the MDR-associated drugs was identified in all the lines tested of both human and rodent origin, although the precise levels of resistance has tended to vary. Furthermore, essentially identical results were obtained when a second line was developed independently from the same parental cells, as exemplified by the two CHO/DXR-10 and the two SuSa/DXR sublines (*cf.* Table 3). It should be noted, however, that the responses of these irradiated sublines to other drugs not implicated in MDR, namely cisplatin, methotrexate and

Table 3. Summary of origins, derivation and *in vitro* drug responses of X-radiation-pretreated sublines

Cell line	Tumour origin	Fractionation protocol	Resultant subline designation	Patterns of resistance			References (see below)
				Vincristine	Etoposide	Adriamycin	
Rodent							
L5178Y	Murine lymphoma	10 × 2 Gy	L5178Y/DXR	R+++	R++++	NC	1
CHO-AuxB1	Chinese hamster ovary	10 × 9 Gy	CHO/DXR-10I	R+++	R++	NC	2
Human		10 × 9 Gy	CHO/DXR-10II	R+++	R+	NC	2
HN-1	[a]scc of the tongue	11 × 4–5 Gy	HN-1/DXR-11	R++	R++	NC	3,4
MCF-7	breast carcinoma	10 × 5–6 Gy	MCF-7/DXR-10	R+++	R+++	NC	5
RT112	[a]tcc bladder	8 × 6 Gy	RT112/DXR-8	R+	R++	NC	4,6
SuSa	testicular teratoma	13 × 1 5 Gy	SuSa/DXR-13	R++	R++	NC	4
		10 × 3 Gy	SuSa/DXR-10	R++	R+++	NC	4
SKOV-3	ovarian carcinoma	10 × 5 Gy	SKOV-3/DXR-10	R+++	R++	NC	7
JA-T	ovarian carcinoma	10 × 5 Gy	JA-T/DXR-10	R+++	R+++	NC	7
LoVo	colon carcinoma	8 × 4 Gy	LoVo/DXR-8	R+	R+++	NC	8
ZR-75-1	breast carcinoma	10 × 4 Gy	ZR-75–1/DXR-10	R+++	R++	NC	5

Data are presented as the ratio of IC_{50} values for a 24 h drug exposure of each subline compared with the parental line;
R++++ = resistance with a ratio > 10, R+++ = resistance with a ratio of 3–10; R++ = resistance with a ratio of 1.8–2.9;
R+ = resistance with a ratio of 1.4–1.7;
[a]scc — squamous cell carcinoma; tcc — transitional cell carcinoma.

References:
1 Bellamy and Hill, 1984
2 Hill *et al.*, 1990
3. Hill and Bellamy, 1984
4. Hill *et al.*, 1988
5. Whelan and Hill, 1993
6. Bedford *et al.*, 1987
7. Dempke *et al.*, 1992
8. Hill, 1991

5-fluorouracil, lacked consistency with resistance, unchanged responses and collateral sensitivity being variably expressed (Hill, 1991). The levels of drug resistance expressed to the Vinca alkaloids and epipodophyllotoxins in these radiation-pretreated sublines are relatively modest, *i.e.*, 2- to 10-fold, but they were derived by clonogenic assays, are statistically significant and probably represent the approximate orders of resistance most likely to be encountered clinically (Hill, 1991). The highest levels of resistance are generally expressed to vincristine, with the exception of the L5178Y/-DXR-10, LoVo/DXR-8 and SuSa/DXR-13 sublines which proved more resistant to etoposide, strikingly so in the case of the former two sublines.

The observed lack of cross-resistance to Adriamycin was somewhat unexpected. In order to confirm this finding we have recently evaluated responses to daunomycin in a number of these sublines (see Table 4). In addition, comparative drug sensitivity evaluations have been made between a number of these DXR-irradiated sublines and drug-selected resistant sublines derived from the same parental cells and expressing comparable modest levels of resistance to either vincristine or to etoposide. These data, summarised in Table 4, provide confirmatory evidence of a definite lack of cross-resistance to Adriamycin and, where tested, also to daunomycin and form the basis for our proposal that *in vitro* exposure of tumour cells to fractionated X-irradiation results in the expression of a distinctive resistance phenotype characterised by the expression of resistance to the Vinca alkaloids and epipodophyllotoxins, but *not* to the anthracyclines. In addition, this specific resistance phenotype is still expressed after xenografting JA-T/DXR-10 cells and subsequent drug treatment of these tumour-bearing mice (Dempke *et al.*, 1992).

Identification of mechanisms associated with the differential expression of resistance to the Vinca alkaloids and epipodophyllotoxins, but not to the anthracyclines in tumour sublines established following exposure to fractionated X-irradiation

The consistent lack of resistance to Adriamycin appeared to argue against concluding that these X-ray-pretreated sublines might express the "classic" MDR phenotype, involving overexpression of Pgp. Indeed, initial investigations aimed at detecting Pgp by Western blotting using the C219 monoclonal antibody failed to identify any reactivity in the MCF-7/DXR-10 and HN-1/DXR-11 sublines (Hill, 1991; Whelan and Hill, 1993). However, in more recent detailed investigations we have obtained clear evidence of significant Pgp overexpression in both the two independently-derived CHO/DXR-10 sublines (Hill *et al.*, 1990) and in two ovarian carcinoma sublines JA-T/DXR-10 and SKOV-3/DXR-10 (McClean *et al.*, 1992), using the C219 antibody. This protein was also cross-reactive with the C494 antibody, indicating that the gene product which was overexpressed following X-irradiation was the *pgp1* or *mdr1* gene product (Hill *et al.*, 1993). Results with the irradiated human ovarian tumour sublines are particularly interesting since they demonstrate the ability to detect Pgp overexpression in these cell lines exhibiting only 2- to 3-fold levels of resistance to vincristine and etoposide. These observations therefore suggest, that if these findings are reflected clinically, Pgp overexpression should be identifiable in certain human tumours following radiotherapy. Our current laboratory investigations are centred on developing sensitive immunocytochemical techniques for Pgp expression, such as those described recently (Chan *et al.*, 1990; Grogan *et al.*, 1990; Toth *et al.*, 1993) which will then be used to screen all the irradiated sublines derived from the different tumour types, as well as human tumour biopsy material. At present definite Pgp overexpression has only been identified in the irradiated *ovarian* sublines, although the significance, if any, of this remains to be established. Indeed, Mattern *et al.* (1991) reported Vinca alkaloid resistance and overexpression of Pgp in human lung carcinoma xenografts after their exposure to fractionated X-irradiation *in vivo*.

To investigate whether this distinctive resistance phenotype is dominantly or recessively expressed, intraspecific hybrids were established by fusion of X-ray-pretreated, drug resistant CHO cells (DXR-10I and DXR-10II) with drug sensitive cells (E29-pro+). These hybrids proved resistant to colchicine

Table 4. Comparative *in vitro* drug responses of resistant sublines selected following exposure either to specific antitumour drugs or to X-irradiation

Cell line	Selection agent	Levels of drug resistance:				References (see below)
		Adriamycin	Daunomycin	Vincristine	Etoposide	
CHO-DXR-10I	X-irradiation	1.1	1.2	6.7	2.3	1,2
CHO-DXR-10II	X-irradiation	1.2	0.9	8.0	2.2	1,2
CHO-CHRA3	colchicine	2.5	nr	4–6	4.4	3,4
HN-1/DXR-11	X-irradiation	1.1	1.0	2.8	2.9	5,6
HN-1/VP2	etoposide	2.7	1.8	2.6	3.9	6,7
MCF-7/DXR-10	X-irradiation	1.0	nr	3.7	2.0	8
MCF-7/VCR3E	vincrirstine	2.6	nr	5.0	1.2	8
MCF-7/VCR6E	vincrirstine	1.8	nr	13.7	3.0	8
MCF-7/VCREMS	vincrirstine	2.7	nr	11.1	3.2	9
SuSa/DXR-10	X-irradiation	0.9	nr	2.0	2.8	10,11
SuSa/VP-10	etoposide	1.6	nr	1.8	4.7	10,11
SuSa/VPC3	etoposide	2.3	nr	4.1	21.0	11
SKOV-3/DXR-10	X-irradiation	1.1	nr	2.2	2.1	12
SKOV-3/VCR0.01	vincrirstine	6.0	nr	2.0	nr	13

Data are presented as the ratio of IC_{50} values for a 24 h drug exposure of each subline compared with the parental line.

References:

1. Hill *et al.*, 1990
2. McClean and Hill, 1992
3. Bech-Hansen *et al.*, 1976
4. McClean and Hill, unpublished data
5. Hill and Bellamy, 1984
6. Lock and Hill, unpublished data
7. Lock and Hill, 19888.Whelan and Hill, 1993
9. Whelan *et al.*, 1992
10. Hill *et al.*, 1988
11. Hosking and Hill, unpublished data
12. Dempke *et al.*, 1992
13. Bradley *et al.*, 1989.

(2-fold) and to vincristine (5- to 7-fold), but not to Adriamycin (McClean *et al.*, 1993a). The hybrid lines exhibited levels of Pgp comparable to those of the unfused X-ray-pretreated parent cell line, suggesting that Pgp overexpression is a dominant trait in these hybrid lines. Overall these data are consistent with the hypothesis that this distinctive resistance phenotype is a consequence of the dominant genetic alteration resulting from exposure to X-irradiation.

In the "classical" MDR Pgp-overexpressing sublines, an increase in Pgp protein has almost always been associated with an increase in Pgp mRNA, sometimes accompanied at the higher levels of resistance by gene amplification (see Table 1). However, in these irradiated CHO/DXR-10 sublines, Pgp overexpression occurred, despite a lack of Pgp gene amplification or of any significant alteration in Pgp mRNA levels. Recent Northern blot analysis using the SKOV-3/DXR-10 subline appears to confirm this finding (Hill *et al.*, 1993). These observations, therefore, suggest that overexpression of Pgp resulting from exposure to X-irradiation may be under different regulation from that operating in the "classical" Pgp-overexpressing MDR tumour cells.

Expression of Pgp has been found to be regulated by various mechanisms including gene amplification (Riordan *et al.*, 1985; Scotto *et al.*, 1986), transcriptional activation (Shen *et al.*, 1986), as well as at the level of translational or posttranslationally (Bradley *et al.*, 1989; Whelan *et al.*, 1992). CHO cells selected clonally with increasing concentrations of colchicine exhibited an increase in Pgp gene copy number, Pgp mRNA and protein, at each step in the selection procedure (Kartner *et al.*, 1985; Riordan *et al.*, 1985). In contrast, Shen *et al.* (1986) observed that in colchicine-selected human tumour KB sublines, elevation in Pgp mRNA only occurred prior to *mdr1* gene amplification. In agreement with these findings, Bradley *et al.* (1989) reported that in their series of SKOV-3 human ovarian carcinoma cell lines selected with increasing concentrations of vincristine or vinblastine, initially Pgp mRNA and Pgp levels increased without any amplification of the *mdr1* gene and, at higher levels of selection, amplification of the DNA followed. Interestingly, at a later stage in the selection process, Pgp continued to increase although no further increase in Pgp mRNA was observed. This apparent lack of coordinate increase in Pgp protein and mRNA expression was also reported in a EMS mutagenised vincristine-selected MCF-7 resistant subline (Whelan *et al.*, 1992). Adding to the complexity of Pgp regulation, it has been reported recently that the *mdr1* gene promoter contains negative elements within the AP-1 site (Ikeguchi *et al.*, 1991). In addition, wildtype p53 can repress *pgp1* promoter activity in CHO cells, while mutant p53 appears to enhance activity of the *pgp1* promoter (Zastawny *et al.*, 1992).

Our observation that Pgp overexpression in these irradiated sublines was not matched by a concomitant elevation of Pgp mRNA implies that either translational or posttranslational modifications of Pgp are involved in this overexpression of the protein following X-ray treatment. Initially the turnover of Pgp in the CHO/DXR-10II cells was examined to investigate whether any differences relative to the CHRC5 cell line, which does express concomitant elevations in Pgp mRNA (Riordan *et al.*, 1985), could be identified. Since CHRC5 cells are a highly resistant "classical" MDR cell line (145-fold resistant colchicine), Pgp turnover was also examined in a subline of comparable resistance to these DXR-10 cells, namely the 7-fold colchicine-resistant cell line CHRA3, selected in a single step from AuxB1 parental cells with colchicine (Ling and Thompson, 1974). CHRA3 cells overexpressed Pgp and amplified pCHP1-homologous sequences 10- to 20-fold on Southern blots (Kartner *et al.*, 1985; Riordan *et al.*, 1985). Stability of Pgp in these resistant CHO cells was estimated by following the loss of ^{35}S-labelled-Pgp from cells with time, by quantitative immunoprecipitation and subsequent analysis on SDS-PAGE gels. Using either the C219 or C494 monoclonal antibodies, we were able to demonstrate that while the half-life of Pgp was of the order of 10-17 h in drug-selected CHO cells, irrespective of their level of drug resistance, Pgp in the X-ray-pretreated CHO/DXR-10II cells had a half-life of >40 h (McClean and Hill, 1993). Comparable studies carried out with the SKOV-3/DXR-10 cells have also revealed a prolonged Pgp half-life of 36 h (Hill *et al.*, 1993), contrasting with the shorter 16 h period identified in the "classical" Pgp-overexpressing MDR subline SKVCR0.25. Our results with drug-selected sublines are comparable with those reported by other groups (see Table 5). It is, however, of interest that Muller and Ling (1992) reporting preliminary data showed that the half-life of Pgp was very dependent on the growth conditions of the cells and could increase up to 72 h when cells were deprived of serum. In our experiments, therefore, care was taken to ensure that all the irradiated cells were in logarithmic growth throughout the 40 h period of the study. Nevertheless, since exposure to a stress such as serum deprivation resulted in a reduced rate of Pgp turnover, it is conceivable that the increased stability of Pgp in these cells pretreated with X-irradiation could be stress associated. Arguing against this, however, is our recent demonstration that a series of X-ray exposures is not necessary for expression of this distinctive MDR phenotype since it has also been identified following exposure of CHO cells to a single supralethal X-ray dose of 30 Gy (McClean *et al.*, 1993b). Indeed these data implicate radiation exposure itself, with both single-dose and fractionated doses selecting for similar genetic mutants.

Overall, our results provide one of the first detailed examples of Pgp being regulated by a post-translational increased stability. It is, of course, also possible that the rate of synthesis of Pgp may be altered in the DXR-10 cells, relative to the parental cells or their "classical" MDR counterpart. However, existing techniques are not sufficiently sensitive to permit an examination of this possibility in these cells with their low level of Pgp expression. Yoshimura *et al.* (1989) reported that altered glycosylation or phosphorylation of Pgp did not affect its half-life, however, this possibility will be investigated in these irradiated sublines. Although the means by which X-irradiation causes the increased expression of Pgp remains to be established, the finding that irradiation of cells can alter stability of a protein has a precedent. It has been reported that normal p53, the tumour suppressor gene product (Levine, 1992), accumulated when nontransformed 3T3 mouse cells were irradiated with either UV or with gamma irradiation (Kastan *et al.*, 1991; Maltzman and Czyzyk, 1984). Following UV irradiation the accumulation was found to be mediated by a posttranslational stabilisation, with the half-life increasing from 35 to >150 min.

This demonstration of Pgp regulation at the level of posttranslational stability may, therefore, provide a basis for the overexpression of Pgp in irradiated tumour cells in the absence of any elevation in Pgp mRNA. Interestingly, recent data suggest that increased protein stabilisation may also be an important mechanism for the regulation of N-*myc* in aggressive MDR neuroblastoma (Domenech *et al.*, 1993). Clearly this area of research is likely to attract further interest in the future.

Investigations of mechanisms associated with etoposide resistance in three of the irradiation-pretreated sublines, derived from HN-1, MCF-7 and SuSa parental cells have shown that they consistently sustained less etoposide-induced DNA single-strand breakage compared with their parental cells (Hosking *et al.*, 1993; Lock and Hill, 1988; Whelan and Hill, 1993): data consistent with modifications of topoisomerase II activity. However, no significant differences were detectable in the amount of immunoreactive topoisomerase II in the HN-1/DXR-11 or SuSa/DXR-10 sublines and neither was altered topoisomerase II catalytic activity identified in the former subline (Hill *et al.*, 1991). In interpreting these results it is important to realise that in view of the modest levels of etoposide resistance expressed by these irradiated sublines, it must be questionable as to whether the sensitivity of the techniques used in these investigations are adequate reliably to detect differences of only two- to three-fold.

Future studies will be directed towards determining whether the *MRP* protein is also involved in this distinctive resistance phenotype expressed by tumour cells following their exposure to X-irradiation or indeed whether other novel resistance-associated proteins (*cf.* recent review by McClean and Hill, 1992) are implicated. In addition, we plan to determine whether the expression of any other, possibly stress-related or general "housekeeping" proteins can be "switched-on" by *in vitro* exposure to fractionated X-irradiation and,

Table 5. Summary of data relating to turnover of Pgp in MDR sublines

Cell line	Selecting agent	P-glycoprotein half-life (h)	References (see below)
Hamster:			
CHO/CHRC5	Colchicine	17 ± 3	1
CHO/CHRC5	Colchicine	12–20	2
CHO/CHRA3	Colchicine	10±0.5	1
CHO/DXR-10II	X-irradiation	>40	1
Mouse:			
J7774.2/V1-1	Vinblastine	16.8 ± 0.5	3
J7774.2/V3-1	Vinblastine	17.4 ± 0.5	3
Human:		16 ± 0.5	4
SKVCR0.25	Vincristine		
SKVCR0.25	Vincristine	12–20	2
SKVCR2.0	Vincristine	12–20	2
SKOV-3/DXR-10	X-irradiation	39 ± 3	4

References
1. McClean and Hill, 1993
2. Muller and Ling, 1992
3. Cohen *et al.*, 1990
4. Hill *et al.*, 1993

if so, whether this results from regulation at the transcriptional or posttranslational level. In this way we aim to establish whether this posttranslational regulation of Pgp is specific for this drug resistance associated protein or a more general effect of exposure to X-irradiation.

Concluding comments

The fundamental basis for the expression of different patterns of cross-resistance in these various drug- or X-ray-selected tumour cell lines remains to be defined. In the last 10 years major advances have been made in terms of identifying drug resistance-associated genes and their products. Unfortunately though, even in our relatively straightforward laboratory model systems, resistance is clearly multifactorial: many different biochemical changes have been associated with MDR cells, although any cause or role generally remains to be established. Whilst Pgp appears a sufficient determinant of the cellular level of "classical" MDR (Roninson, 1992), the differential patterns of cross-resistance identified could result from some modification having occurred in the primary resistance mechanism. Indeed, there is already some evidence, as discussed earlier, that different resistance mechanisms may operate or predominate depending on the level of resistance expressed and the selection agent used. In the former case, relatively few studies appear to have examined the possible sequential emergence of distinct resistance phenotypes. For example: Slapak *et al.* (1990) reported that decreased anthracycline uptake preceded increased Pgp expression under Adriamycin selection, Baas *et al.* (1990) provided evidence that non-Pgp mechanisms for MDR preceded Pgp overexpression, whilst data from Eijdems *et al.* (1992) indicated that non-Pgp-mediated MDR is the major mechanism operating in low-level resistant Adriamycin selected in nonsmall cell lung cancer cell lines. Evidence of differential patterns of cross resistance following selection with X-rays as opposed to antitumour drugs has been presented above, but it has not been possible readily to discriminate between various drugs and cells so as to identify any truly drug-specific characteristic pattern of sensitivities.

With our increasing knowledge showing that Pgp expression can be differentially regulated, various groups have now established that the human *mdr1* gene may be susceptible to stress induction. As discussed recently (Chaudhary and

Roninson, 1993; Kohno *et al.*, 1988, 1992), *mdr1* can be induced by treatment with heat shock, certain differentiating agents as well as agents that activate protein kinase C. Furthermore, it has been shown that *mdr1* mRNA and Pgp can be induced in response to cellular damage by cytotoxic drugs, irrespective of whether the drugs are transported by Pgp. The fact that this induction can be prevented by protein kinase C inhibitors provided a potentially useful pharmacological approach towards preventing the emergence of MDR during cancer chemotherapy.

Current clinical studies aimed at overcoming or circumventing Pgp-associated MDR are mainly concentrating on the use of resistance modulators or chemosensitisers, including verapamil or cyclosporin A (Pennock *et al.*, 1991; Sonneveld *et al.*, 1992). Laboratory studies investigating this approach have provided some interesting data suggesting that the MDR phenotype expressed can be modified by coselecting tumour cells in drug plus verapamil. For example, two groups have demonstrated that the presence of verapamil together with Adriamycin negated the selection advantage of Pgp, with the surviving cells expressing other mechanisms of Adriamycin resistance (Chen *et al.*, 1990; Dalton, 1990). Interestingly, the non-Pgp over-expressing Adriamycin resistant (900-fold) MCF-7 subline (Chen *et al.*, 1990) exhibited a unique cross-resistance pattern too, proving highly resistant to daunomycin (326-fold), less resistant to teniposide (26-fold) and yet only 5-fold resistant to vinblastine. Furthermore, Abbaszadegan *et al.* (1993) recently described a human myeloma cell line ($MDR_{10}V$) selected from a Pgp positive cell line 8226/Dox40 by continuous exposure to Adriamycin and verapamil, which expressed reduced Pgp at both protein and mRNA levels coupled with less gene amplification.

Development of improved strategies to overcome topoisomerase II-mediated drug resistance are still at the drawing board stage. Possibilities include modulation of phosphorylation which may change drug sensitivities and attempts to identify selective inhibitors of the two isoforms which may perhaps have differential roles in normal tissues, as opposed to tumours. Furthermore, recent evidence for distinct drug interaction domains on topoisomerase II for the two broad groups of topoisomerase II-targeted drugs may shed light on the varied resistance patterns identified (Roninson, 1992).

However, whilst our knowledge of the molecular and genetic events associated with the expression of MDR steadily extends, the fact remains that when one sets out to select for resistance using a specific antitumour drug *in vitro*, no accurate prediction can yet be made as to the cross-resistance pattern that will emerge. The clinical relevance, if any, of these variable patterns of cross-resistance identified in human MDR tumour cell lines is still unclear. This statement serves to highlight the need to develop and use appropriate experimental model systems. Obviously it would be valuable if it were possible accurately to identify alternative drugs likely to be useful clinically after resistance to initially used drugs developed. At present, any such assumptions should only be made with extreme caution.

Acknowledgements

I am very grateful to the following for providing up-to-date information on their own cell lines for inclusion in Tables 1 and 2: Drs. W.T. Beck; W.T. Bellamy; J. Benard; J.L. Biedler; H.J. Broxterman; M.S. Center; S. Cole; W.J. Dalton; E.G.E. de Vries; M. Dietel; P.J. Ferguson; L.M. Fisher; G.J. Goldenberg; W.G. Harker; S.B. Horwitz; S. Mirski; M.L. Slovak; T. Tsuruo; P.R. Twentyman and M. Volm. The secretarial assistance provided by Mary Wallace was exemplary and she remained cheerful in spite of my numerous redrafts of this manuscript.

References

Abbaszadegan MR, Futscher BW and Dalton WS (1993) Evaluation of verapamil resistance in a multi-drug resistant myeloma cell line. Proc Am Assoc Cancer Res 34:324.

Akiyama SI, Fojo A, Hanover JA, Pastan I and Gottesman MM (1985) Isolation and genetic characterisation of human KB cell lines resistant to multiple drugs. Somatic Cell Mol Genetic 11:117–126.

Baas F, Jongsma A, Broxterman H, Arceci R, Housman D, Scheffer G, Riethorst A, Van Groeningen M, Nieuwint A and Joense H (1990) Non P-glycoprotein mediated mechanism for MDR precedes P-glycoprotein expression during *in vitro* selection for doxorubicin resistance in human cancer cell line. Cancer Res 50:5392– 5398.

Barrand MA, Rhodes T, Center MS and Twentyman PR (1993) Chemosensitisation and drug accumulation effects of cyclosporin A, PSC-833 and verapamil in human MDR large cell lung cancer cells expressing a 190k membrane protein distinct from P-glycoprotein. Eur J Cancer 29A: 408–415.

Bates SE, Currier SJ, Alvarez M and Fojo AT (1992) Modulation of P-glycoprotein phosphorylation and drug transport by sodium butyrate. Biochem 31: 6366–6372.

Bech-Hansen NTG, Till TE and Ling V (1976) Pleitropic phenotype of colchicine resistant CHO cells: Cross-resistance and collateral sensitivity. J Cell Physiol 88: 23–32.

Beck WT (1984) Cellular pharmacology of Vinca alkaloid resistance and its circumvention. Adv Enzyme Regul 22: 207–227.

Beck WT (1989) Unknotting the complexities of multidrug resistance: The involvement of DNA topoisomerases in drug action and resistance. J Natl Cancer Inst 81: 1683–1685.

Beck WT, Cirtain MC, Danks MK, Felsted RL, Safa AR, Wolverton JS, Suttle DP and Trent JM (1987) Pharmacological, molecular, and cytogenetic analysis of "atypical" multidrug-resistant human leukemic cells. Cancer Res 47: 5455–5460.

Beck WT and Danks MK (1991) Characteristics of multidrug resistance in human tumor cells. In: IB Roninson (ed) Molecular and Cellular Biology of Multidrug Resistance in Tumor Cells (pp 3–55). Plenum Press, New York.

Beck WT, Muller TJ and Tanzer LR (1979) Altered surface membrane glycoproteins in Vinca alkaloid resistant human leukemic lymphoblasts. Cancer Res 39: 2070–2076.

Bedford P, Shellard SA, Walker C, Whelan RDH, Masters JRW and Hill BT (1987) Differential expression of collateral sensitivity or resistance to cisplatin in human bladder carcinoma cell lines pre-exposed *in vitro* to either X-irradiation or cisplatin. Int J Cancer 40: 681–686.

Bellamy AS and Hill BT (1984) Murine L5178Y cells resulting in altered drug sensitivities from fractionated radiation exposure *in vitro*. J Natl Cancer Inst 72: 411–417.

Benard J, Da Silva J, Teyssier J-R and Riou G (1989) Overexpression of *MDR1* gene with no DNA amplification in a multiple drug resistant human ovarian carcinoma cell line. Int J Cancer 43: 471–477.

Biedler JL, Casals D, Chang T-D, Meyers MB, Spengler BA and Ross RA (1991) Multidrug-resistant human neuroblastoma cells are more differentiated than controls and retinoic acid further induces lineage-specific differentiation. In: AE Evans, GJ D'Anglio, AG Knudson Jr, RC Seeger (eds) Advances in Neuroblastoma Research 3 (pp 181–191). Wiley-Liss, New York.

Biedler JL, Meyers MB and Spengler BA (1988) Cellular concomitants of multidrug resistance. In: PV Woolley III, KD Tew (eds) Mechanisms of Drug Resistance in Neoplastic Cells (pp 41–68). Academic Press, New York.

Biedler JL and Riehm H (1970) Cellular resistance to actinomycin D in Chinese hamster cells *in vitro*: Cross-resistance, radioautographic and cytogenetic studies. Cancer Res 30: 1174–1184.

Bradley G, Naik M and Ling V (1989) P-glycoprotein expression in multidrug resistant human ovarian carcinoma cell lines. Cancer Res 49: 2790–2796.

Chabner BA (1986) The oncologic end game. J Clin Oncol 4: 625–638.

Chan HSL, Thorner PS, Haddad G and Ling V (1990) Immunohistochemical detection of P-glycoprotein: Prognostic correlation in soft tissue sarcoma of childhood. J Clin Oncol 8: 689–704.

Chaudhary PM and Roninson IB (1993) Induction of multidrug resistance in human cells by transient exposure to different chemotherapeutic drugs. J Natl Cancer Inst 85: 632–639.

Chen Y-N, Mickley LA, Schwartz AM, Acton EM, Hwang J and Fojo AT (1990) Characterization of Adriamycin-resistant human breast cancer cells which display overexpression of a novel resistance-related membrane protein. J Biol Chem 265: 10073–10080.

Choi K, Chen C-J, Kriegler M and Roninson IB (1988) An altered pattern of cross-resistance in multidrug-resistant human cells results from spontaneous mutations in the *mdr*1 (P-glycoprotein) gene. Cell 53: 519–529.

Clynes M, Redmond A, Moran E and Gilvarry U (1992) Multiple drug-resistance in variant of a human nonsmall cell lung carcinoma cell line, DLKP-A. Cytotechnol 10: 75–89.

Cohen D, Yang C-HP and Horwitz SB (1990) The products of the mdr1a and mdr1b genes from multidrug resistant murine cells have similar degradation rates. Life Sci 46: 489–495.

Cole SPC (1992) The MERCK Frosst Award. Multidrug resistance in small cell lung cancer. Cancer J Physiol Pharmacol 70: 313–329.

Cole SPC, Bhardwaj G, Gerlach JH, MacKie JE, Grant CE, Almquist KC, Stewart AJ, Kurz EU, Duncan AMV and Deeley RG (1992) Overexpression of a transporter gene in a multidrug-resistant human lung cancer cell line. Science 258: 1650–1654.

Cole SPC, Chanda ER, Dicke FP, Gerlach JH and Mirski SEL (1991) Non-P-glycoprotein-mediated multidrug resistance in a small cell lung cancer cell line: Evidence for decreased susceptibility to drug-induced DNA damage and reduced levels of topoisomerase II. Cancer Res 51: 3345–3352.

Coley HM, Twentyman PR and Workman P (1989) Identification of anthracyclines and related agents that retain preferential activity over Adriamycin in multidrug-resistant cell lines and further resistance modification by verapamil and cyclosporin A. Biochem Pharmacol 38: 4467–4475.

Coley HM, Workman P and Twentyman PR (1991) Retention of activity by selected anthracyclines in a multidrug resistant human large cell lung carcinoma line without P-glycoprotein hyperexpression. Br J Cancer 63: 351–357.

Conter V and Beck WT (1984) Acquisition of multiple drug resistance by CCRF-CEM cells selected for different degrees of resistance to vincristine. Cancer Treat Rep 68: 831–839.

Dalton WS (1990) Reversing multidrug resistance in the laboratory and the clinic. Proc Am Assoc Cancer Res 31: 520–521.

Dalton WS, Durie BGM, Alberts DS, Gerlach JH and Cress AE (1986) Characterization of a new drug-resistant human myeloma cell line that expresses P-glycoprotein. Cancer Res 46: 5125.

Dalton WS, Gleason-Guzman MC and Foley NE (1993) Mitoxantrone selects for non-P-glycoprotein resistance in a human myeloma cell line. Proc Am Assoc Cancer Res 34: 305.

Dalton WS, Grogan TM, Rybski JA, Scheper RJ, Richter L, Kailey J, Broxterman HJ, Pinedo HM and Salmon SE (1989) Immunohistochemical detection and quantitation of P-glycoprotein in multiple drug-resistant human myeloma cells: association with level of drug resistance and drug accumulation. Blood 73: 747–752.

Danks MK, Schmidt CA, Cirtain MC, Suttle DP and Beck WT (1988) Altered catalytic activity of and DNA cleavage by DNA topoisomerase II from human leukemic cells selected for resistance to VM-26. Biochemistry 27: 8861–8869.

Danks MK, Schmidt CA, Deneka DA and Beck WT (1989) Increased ATP requirement for activity of a complex formation by DNA topoisomerase II from human leukemic CCRF-CEM cells selected for resistance to teniposide. Cancer Commun 1: 101–109.

Danks MK, Yalowich JC and Beck WT (1987) Atypical multiple drug resistance in a human leukemic cell line selected for resistance to teniposide. Cancer Res 47: 1297–1301.

Dano K (1972) Cross resistance between vinca alkaloids and anthracyclines in Ehrlich ascites tumour *in vivo*. Cancer Chemother Rep 56: 701–708.

De Bruijn MHL, Van der Bliek AM, Biedler JL and Borst P (1986) Differential amplification and disproportionate expression of five genes in three multidrug-resistant Chinese hamster lung cell lines. Mol Cell Biol 6: 4717–4722.

Deffie AM, Alam T, Seneviratne C, Beenken SW, Batra JK, Shea TC, Henner WD and Goldenberg GJ (1988) Multifactorial resistance to Adriamycin: Relationship of DNA repair, glutathione transferase activity, drug efflux, and P-glycoprotein in cloned cell lines of Adriamycin-sensitive and -resistant P388 leukemia. Cancer Res 48: 3595–3602.

Deffie AM, McPherson JP, Gupta RS, Hedley DW and Goldenberg GJ (1992) Multifactorial resistance to antineoplastic agents in drug-resistant P388 murine leukemia, Chinese hamster ovary and human HeLa cells, with emphasis on the role of DNA topoisomerase II. Biochem Cell Biol 70: 354–364.

De Jong S, Kooistra AJ, De Vries EGE, Mulder NH and Zijlstra JG (1993) Topoisomerase II as a target of VM-26 and 4'-(9-acridinylamino) methanesulfon-*m*-aniside in atypical multidrug resistant human small cell lung carcinoma cells. Cancer Res 53: 1064–1071.

De Jong S, Zijstra JG, De Vries EGE and Mulder NH (1990) Reduced DNA topoisomerase II activity and drug-induced cleavage activity in an Adriamycin-resistant human small cell lung carcinoma cell line. Cancer Res 50: 304–309.

Dempke WCM, Whelan RDH and Hill BT (1992) Expression of resistance to etoposide and vincristine *in vitro* and *in vivo* after X-irradiation of ovarian tumour cells. Anti-Cancer Drugs 3: 395–399.

Devine SE, Ling V and Melera PW (1992) Amino acid substitutions in the sixth transmembrane domain of P-glycoprotein alter multidrug resistance. Proc Natl Acad Sci USA 89: 4564–4568.

DeVita Jr VT (1989) The problem of resistance: Keynote address. In: E Mihich (ed) Pezcoller Foundation Symposia Trento – Italy 1: Drug Resistance – Mechanisms and Reversal (pp 7–27) John Libbey, Rome.

Dhir R and Gros P (1992) Functional analysis of chimeric proteins constructed by exchanging homologous domains of two P-glycoproteins conferring distinct drug resistance profiles. Biochem 31: 6103–6110.

Dietel M, Arps H, Lage J and Niendorf A (1990) Membrane vesicle formation due to acquired mitoxantrone resistance in human gastric carcinoma cell line EPG 85-257. Cancer Res 50: 6100–6106.

Domenech C, Spengler BA, Ross RA and Biedler JL (1993) Prolonged half-life in a multidrug-resistant N-*myc*-amplified human neuroblastoma cell line. Proc Am Assoc Cancer Res 34: 16.

Drake FH, Hofmann GA, Bartus HF, Mattern MR, Crooke ST and Mirabelli CK (1989) Biochemical and pharmacological properties of p170 and p180 forms of topoisomerase II. Biochemistry 28: 8154–8160.

Drake FH, Zimmerman JP, McCabe FL, Bartus HF, Per SR, Sullivan DM, Ross WE, Mattern MR, Johnson RK, Crooke ST and Mirabelli CK (1987) Purification of topoisomerase II from Amsacrine-resistant P388 leukemia cells. J Biol Chem 262: 16739–16747.

Eijdems EWHM, Borst P, Jongsma APM, de Jong S, De Vries EGE, Van Groeningen M, Versantvoort CHM, Nieuwint AWM and Baas F (1992) Genetic transfer of non-P-glycoprotein-mediated multidrug resistance (MDR) in somatic cell fusion: Dissection of a compound MDR phenotype. Proc Natl Acad Sci USA 89: 3498–3502.

Fairchild CR, Ivy SP, Kao-Shan C-S, Whang-Peng J, Rosen N, Israel MA, Melera PW, Cowan KH and Goldsmith ME (1987) Isolation of amplified and overexpressed DNA sequences from Adriamycin-resistant human breast cancer cells. Cancer Res 47: 5141–5148.

Farber S, Diamond LK, Mercer RD, Sylvester RF and Wolff JA (1948) Temporary remissions in acute leukemia in children produced by folic acid antagonist, 4-aminopteroyl-glutamic acid (aminopterin). New Engl J Med 238: 787–793.

Ferguson PJ, Fisher MH, Stephenson J, Li D-H, Zhou B-S and Cheng Y-C (1988) Combined modalities of resistance in etoposide-resistant human KB cell lines. Cancer Res 48: 5956–5964.

Fojo AT, Ueda K, Slamon DJ, Poplack DG, Gottesman MM and Pastan I (1987) Expression of a multidrug-resistance gene in human tumours and tissues. Proc Natl Acad Sci USA 84: 265–269.

Fojo AT, Whang-Peng J, Gottesman MM and Pastan I (1985) Amplification of DNA sequences in human multidrug-resistant KB carcinoma cells. Proc Natl Acad Sci USA 82: 7661—7665.

Gervasoni JE, Taub RN, Yu MT, Warburton D, Sabbath M, Gilleran S, Koppock DL, D'Alessanderi J, Kirshna S, Rosado M, Baker MA, Lutsky J, Chanda ER, Gerlach JH, Pinoski M, Cole SPC and Hindenburg AA (1992) Homogeneously staining region in anthracycline-resistant HL60/AR cells not associated with MDR1 amplification. Cancer Res 52: 5244—5249.

Goldie JH and Coldman AJ (1979) A mathematical model for relating the drug sensitivity of tumours to their spontaneous mutation rate. Cancer Treat Rep 63: 1727—1733.

Goldie JH and Coldman AJ (1984) The genetic origin of drug resistance in neoplasms: implications for systemic therapy. Cancer Res 44: 3643—3653.

Greenberger LM, Lothstein L, Williams SS and Horwitz SB (1988) Distinct P-glycoprotein precursers are overproduced in independently isolated drug-resistant cell lines. Proc Natl Acad Sci USA 85: 3762—3766.

Greenberger LM, Williams SS and Horwitz SB (1987) Biosynthesis of heterogeneous forms of multidrug resistance-associated glycoproteins. J Biol Chem 262: 13685—13789.

Greenspan EM, Feiber M, Lesnick G and Edelman S (1963) Response of advanced breast cancer to the combination of the antimetabolite methotrexate and the alkylating agent thiotepa. J Mount Sinai Hosp 30: 246—267.

Grogan T, Dalton W, Rybski J, Spier C, Meltzer P, Richter M, Gleason M, Pindur J, Cline A, Scheper R, Tsuruo T and Salmon S (1990) Optimization of immunocytochemical P-glycoprotein assessment in multidrug-resistant plasma cell myeloma using three antibodies. Lab Invest 63: 815—824.

Gros P, Dhir R, Croop J and Talbot F (1991) A single amino acid substitution strongly modulates the activity and substrate specificity of the mouse mdrl and mdr3 drug efflux pumps. Proc Natl Acad Sci USA 88: 7289—7293.

Gudkov AW, Zelnick C, Kazarov AR, Thimmapaya R, Suttle DP, Beck WT and Roninson IB (1993) Isolation of genetic suppressor elements, inducing resistance to topoisomerase II-interactive cytotoxic drugs, from human topoisomerase II cDNA. Proc Natl Acad Sci USA 90: 3231—3235.

Hait WN, Choudhury S, Srimatkandada S and Murren JR (1993) Sensitivity of K562 human chronic myelogenous leukemia blast cells transfected with a human multidrug resistance cDNA to cytotoxic drugs and differentiating agents. J Clin Invest 91: 2207—2215.

Hamada H, Akochi E, Watanabe M, Oh-Hara T, Sugimoto Y, Kawabata H and Tsuruo T (1988) M_r85,000 membrane protein specifically expressed in Adriamycin-resistant human tumor cells. Cancer Res 48: 7082—7087.

Harker WG and Sikic BI (1985) Multidrug (pleiotropic) resistance in doxorubicin-selected variants of the human sarcoma cell line MES-SA. Cancer Res 45: 4091—4096.

Harker WG, Slade DL, Drake FH and Parr RL (1991) Mitoxantrone resistance in HL-60 leukemia cells: Reduced nuclear topoisomerase II catalytic activity and drug-induced DNA cleavage in association with reduced expression of the topoisomerase II β isoform. Biochemistry 30: 9953—9961.

Hill BT (1986) *In vitro* human tumour model systems for investigating drug resistance. Cancer Surveys 5: 129—149.

Hill BT (1990) *In vitro* drug-radiation interactions using fractionated X-irradiation regimens. In: BT Hill, AS Bellamy (eds) Antitumor Drug-Radiation Interactions (pp 207—222). CRC Press Inc., Boca Raton, Florida.

Hill BT (1991) Interactions between antitumour agents and radiation and the expression of resistance. Cancer Treat Rev 18: 149—190.

Hill BT and Bellamy AS (1984) Establishment of an etoposide-resistant human epithelial tumour cell line *in vitro*: Characterisation of patterns of cross-resistance and drug sensitivities. Int J Cancer 33: 599—608.

Hill BT, Deuchars K, Hosking LK, Ling V and Whelan RDH (1990) Overexpression of P-glycoprotein in mammalian tumor cell lines after fractionated X-irradiation *in vitro*. J Natl Cancer Inst 82: 607—612.

Hill BT, Hosking LK, Shellard SA and Whelan RDH (1989) Comparative effectiveness of mitoxantrone and doxorubicin in overcoming experimentally induced drug resistance in murine and human tumour cell lines *in vitro*. Cancer Chemother Pharmacol 23: 140—144.

Hill BT and Whelan RDH (1982) Establishment of vincristine-resistant and vindesine-resistant lines of murine lymphoblasts *in vitro* and characterisation of their patterns of cross-resistance and drug sensitivities. Cancer Chemother Pharmacol 8: 163—169.

Hill BT, Whelan RDH, Hosking LK, Hinds MD, Mayes J and Zwelling LA (1991) A lack of detectable modification of topoisomerase II activity in a series of human tumour cell lines expressing only low levels of etoposide resistance. Int J Cancer 47: 899—902.

Hill BT, Whelan RDH, Hosking LK, Shellard SA, Bedford P and Lock RB (1988) Interactions between antitumor drugs and radiation in mammalian tumor cell lines: Differential drug responses and mechanisms of resistance following fractionated X-irradiation or continuous drug exposure *in vitro*. Natl Cancer Inst Monogr 6: 177—181.

Hill BT, Whelan RDH and McClean S (1993) Overexpression and posttranslational regulation of P-glycoprotein resulting from exposure of human ovarian tumour cells to fractionated X—irradiation. Proc Am Soc Clin Oncol 12: 118.

Hochhauser D and Harris AL (1993) The role of topoisomerase II α and β in drug resistance. Cancer Treat Rev 19: 181—194.

Horwitz SB, Liao L-L, Greenberger L and Lothstein L (1989) Mode of action of taxol and characterization of a multidrug-resistant cell line selected with taxol. In: D Kessel (ed) Resistance to Antineoplastic Drugs (pp 109—125). CRC Press, Boca Raton, Florida.

Hosking L, Shellard S, McClean S and Hill B (1993) Modified topoisomerase II expression appears to predominate in drug- and X-ray-selected etoposide resistant human testicular teratoma sublines. Br J Cancer 67 (Suppl. XX): 76.

Hsu SI-H, Lothstein L and Horwitz SB (1989) Differential overexpression of three *mdr* gene family members in multidrug-resistant J774.2 mouse cells. J Biol Chem 264: 12053–12062.

Ikeguchi M, Teeter LD, Eckersberg T, Ganapathi R and Kun MT (1991) Structural and functional analyses of the promoter of the murine multidrug resistance gene mdr3/mdr1a reveal a negative element containing the AP-1 binding site. DNA and Cell Biol 10: 639–647.

Juliano RL and Ling V (1976) Surface glycoprotein modulating drug permeability in Chinese hamster ovary cell mutants. Biochim Biophys Acta 455: 152–162.

Kartner N, Everden-Porelle D, Bradley G and Ling V (1985) Detection of P-glycoprotein in multidrug-resistant cell lines by monoclonal antibodies. Nature 316: 820–823.

Kartner N, Riordan JR and Ling V (1983) Cell surface P-glycoprotein associated with multidrug resistance in mammalian cells. Science 221: 1285–1288.

Kastan MB, Onyekwere O, Sidaransky D, Vogelstein B and Craig RW (1991) Participation of p53 protein in the cellular response to DNA damage. Cancer Res 51: 6304–6311.

Keizer HG, Schuurhuis GJ, Broxterman HJ, Lankelma J, Schoonen WGEJ, Van Rijn J, Pinedo HM and Joenje H (1989) Correlation of multidrug resistance with decreased drug accumulation, altered subcellular drug distribution, and increased P-glycoprotein expression in cultured SW-1573 human lung tumor cells. Cancer Res 49: 2988–2993.

Kellner V, Boege F, Gieseler F and Dietel M (1993) Alterated DNA topoisomease II in mitoxantrone resistant human Gastric carcinoma celline. Proc Am Assoc Cancer Res 34: 330.

Kessel D (1986) Circumvention of resistance to anthracyclines by calcium antagonists and other membrane-perturbing agents. Cancer Surveys 5: 109–127.

Kirschner LS, Greenberger LM, Hsu SI-H, Yang C-PH, Cohen D, Piekarz RL, Castillo G, Han EK-H, Yu L and Horwitz SB (1992) Biochemical and genetic characterization of the multidrug resistance phenotype in murine macrophage-like J774.2 cells. Biochem Pharmacol 43: 77–87.

Kohno K, Kikuchi J, Sato S-I, Takano H, Saburi Y, Asoh K-I and Kuwano M (1988) Vincristine-resistant human cancer KB cell line and increased expression of multidrug-resistance gene. Jpn J Cancer Res 79: 1238–1246.

Kohno K, Sato S-I, Uchiumi T, Takano H, Tanimura H, Miyazaki M, Matsuo K-I, Hidaka K and Kuwano M (1992) Activation of the human multidrug resistance 1 (MDR1) gene promoter in response to inhibitors of DNA topoisomerases. Int J Oncol 1: 73–77.

Kramer RA, Zakher J and Kim G (1988) Role of the glutathione redox cycle in acquired and de novo multidrug resistance. Science 241: 694–697.

Kuiper CM, Broxterman HJ, Baas F, Schuurhuis GJ, Haisma HJ, Scheffer DL, Lankelma J and Pinedo HM (1990) Drug transport variants without P-glycoprotein overexpression from a human squamous lung cancer cell line after selection with doxorubicin. J Cell Pharmacol 1: 35–41.

Levine AJ (1992) The p53 tumour supressor gene and product. In: AJ Levine (ed) Cancer Surveys 12: Tumour Supressor Genes, the Cell Cycle and Cancer (pp 59–78). Cold Spring Harbour Press, New York.

Lincke CR, Van der Bliek AM, Schuurhuis GJ, Van der Velde-Koerts T, Smit JJM and Borst P (1990) Multidrug resistance phenotype of human BRO melanoma cells transfected with a wild-type human *mdr*1 complementary DNA. Cancer Res 50: 1779–1785.

Ling V, Kartner N, Sudo T, Siminovitch L and Riordan JR (1983) Multidrug-resistance phenotype in Chinese hamster ovary cells. Cancer Treat Rep 67: 869–874.

Ling V and Thompson LH (1974) Reduced permeability in CHO cells as a mechanism of resistance to colchicine. J Cell Physiol 83: 103–116.

Lock RB and Hill BT (1988) Differential patterns of antitumour drug responses and mechanisms of resistance in a series of independently-derived VP-16-resistant human tumour cell lines. Int J Cancer 42: 373–381.

Maltzman W and Czyzyk L (1984) UV irradiation stimulates levels of p53 tumor antigen in nontransformed mouse cells. Mol Cell Biol 4: 1689–1694.

Marquardt D and Center MS (1992) Drug transport mechanisms in the HL60 cells isolated for resistance to Adriamycin: Evidence for nuclear drug accumulation and redistribution in resistant cells. Cancer Res 52: 3157–3163.

Marquardt D, McCrone S and Center MS (1990) Mechanisms of multidrug resistance in HL-60 cells: Detection of resistance associated proteins with antibodies against synthetic peptides that correspond to the deduced sequence of P-glycoprotein. Cancer Res 50: 1426–1430.

Mattern J, Efferth T and Volm M (1991) Overexpression of P-glycoprotein in human lung carcinoma xenografts after fractionated irradiation *in vivo*. Radiat Res 127: 335–338.

McClean S, Dempke WCM, Whelan RDH, Hosking LK and Hill BT (1992) Overexpression of P-glycoprotein in human ovarian carcinoma cells following exposure to fractionated X-irradiation *in vitro*. Proc Am Assoc Cancer Res 33: 470.

McClean S and Hill BT (1992) An overview of membrane, cytosolic and nuclear proteins associated with the expression of resistance to multiple drugs *in vitro*. Biochim Biophys Acta 1114: 107–127.

McClean S and Hill BT (1993) Posttranslational regulation of P-glycoprotein in mammalian tumour cells expressing a distinctive multiple drug resistance phenotype after exposure to fractionated X-irradiation. Proc Am Assoc Cancer Res 34: 313.

McClean S, Hosking LK and Hill BT (1993a) Dominant expression of multiple drug resistance after *in vitro* X-irradiation exposure in intraspecific Chinese hamster ovary hybrid cells. J Natl Cancer Inst 85: 48–53.

McClean S, Hosking LK and Hill BT (1993b) Expression of P-glycoprotein-mediated drug resistance in Chinese hamster ovary cells surviving *in vitro* exposure to a single lethal X-ray dose of 30Gy. Int J Radiat Biol 63: 765–773.

McClean S, Whelan RDH, Hosking LK, Hodges GM, Thompson FH, Meyers MB, Schuurhuis GJ and Hill BT (1993c) Characterisation of the P-glycoprotein overexpressing drug resistance phenotype exhibited by Chinese hamster ovary cells following their *in vitro* exposure to fractionated X-irradiation. Biochim Biophys Acta 1177: 117–126.

McGrath T and Center MS (1987) Adriamycin resistance in HL-60 cells in the absence of detectable P-glycoprotein. Biochem Biophys Res Commun 145: 1171–1176.

McGrath T and Center MS (1988) Mechanisms of multidrug resistance in HL60 cells: Evidence that a surface membrane protein distinct from P-glycoprotein contributes to reduced cellular accumulation of drug. Cancer Res 48: 3959–3963.

Meyers MB, Spengler BA, Chang T-D, Melera PW and Biedler JL (1985) Gene amplification-associated cytogenetic aberrations and protein changes in vincristine-resistant Chinese hamster, mouse, and human cells. J Cell Biol 100: 588–597.

Mirski SEL, Evans CD, Almquist KC, Slovak ML and Cole SPC (1993) An altered topoisomerase II α in a drug resistant small cell lung cancer cell line selected in VP-16. Cancer Res (in press).

Mirski SEL, Gerlach JH and Cole SPC (1987) Multidrug resistance in a human small cell lung cancer cell line selected in adriamycin. Cancer Res 47: 2594–2598.

Muller C and Ling V (1992) P-glycoprotein is affected by serum deprivation and high cell density in multidrug resistant cells. Proc Am Assoc Cancer Res 33: 452.

Nielsen D and Skovsgaard T (1992) P-glycoprotein as a multidrug transporter: A critical review of current multidrug resistant cell lines. Biochim Biophys Acta 1139: 169–183.

Nieuwint AWM, Baas F, Wiegant J and Joenje H (1992) Cytogenic alterations associated with P-glycoprotein- and non-P-glycoprotein-mediated multidrug resistance in SW-1573 human lung tumor cell lines. Cancer Res 52: 4361–4371.

Patel S and Fisher LM (1993) Novel selection and genetic characterisation of an etoposide-resistant human leukaemic CCRF-CEM cell line. Br J Cancer 67: 456–463.

Pennock GD, Dalton WS, Roeske WR, Appleton CP, Mosley K, Plezia P, Miller TP and Salmon SE (1991) Systemic toxic effects associated with high-dose verapamil infusion and chemotherapy administration. J Natl Cancer Inst 83: 105–110.

Peterson RHF, Meyers MB, Spengler BA and Biedler JL (1983) Alteration of plasma membrane glycopeptides and gangliosides of Chinese hamster cells accompanying development of resistance to daunorubicin and vincristine. Cancer Res 43: 222–228.

Reeve JG, Rabbits PH and Twentyman PR (1990) Non-P-glycoprotein mediated multidrug resistance with reduced EGFR expression in a human large cell lung cancer cell line. Br J Cancer 61: 851–855.

Riehm H and Biedler JL (1971) Cellular resistance to daunomycin in Chinese hamster cells *in vitro*. Cancer Res 31: 409–412.

Riordan JR, Deuchars K, Kartner N, Alon N, Trent J and Ling V (1985) Amplification of P-glycoprotein genes in multi-drug resistant mammalian cell lines. Nature 316: 817–819.

Rogan AM, Hamilton TC, Young RC, Klecker Jr RW and Ozols RF (1984) Reversal of Adriamycin resistance by verapamil in human ovarian cancer. Science 224: 994–996.

Roninson IB (1992) The role of the *MDR*1 (P-glycoprotein) gene in multidrug resistance *in vitro* and *in vivo*. Biochem Pharmacol 43: 95–102.

Safa AR, Stern RK, Choi K, Agresti M, Tamai I, Mehta ND and Roninson IB (1990) Molecular basis of preferential resistance to colchicine in multidrug-resistant human cells conferred by Gyly-185 $\rightarrow$ to Val-185 substitution in P-glycoprotein. Proc Natl Acad Sci USA 87: 7225–7229.

Scheper RJ, Broxterman HJ, Scheffer GL, Kaaijk P, Dalton WS, Van Heijningen THM, Van Kalken CK, Slovak ML, De Vries E, Van der Valk P, Meijer CJLM and Pinedo HM (1993) Overexpression of a M_r 110,000 vesicular protein in non-P-glycoprotein-mediated multidrug resistance. Cancer Res 53: 1475–1479.

Schinkel AH and Borst P (1991) Multidrug resistance mediated by P-glycoproteins. Semin Cancer Biol 2: 213–226.

Scotto KW, Biedler JL and Melera PW (1986) Amplification and expression of genes associated with multidrug resistance in mammalian cells. Science 232: 751–755.

Shen DW, Fojo A, Chin JE, Roninson IB, Richert N, Pastan I and Gottesman MM (1986) Human multidrug resistant cell lines: increased mdr1 expression can precede gene amplification. Science 232: 643–645.

Sikic BI, Skudder SA and Evans TL (1989) Multidrug (pleiotropic) resistance in the human sarcoma cell line MES-SA. In: D Kessel (ed) Resistance to Antineoplastic Drugs (pp 37–47). CRC Press, Boca Raton, Florida.

Sinha VK, Haim N, Dusre L, Kerrigan D and Pommier Y (1988) DNA strand breaks produced by etoposide (VP-16, 213) in sensitive and resistant human breast tumor cells: Implications for the mechanism of action. Cancer Res 48: 5096–5100.

Sirotnak FM, Yang C-H, Mines LS, Oribe E and Biedler JL (1986) Markedly altered membrane transport and intracellular binding of vincristine in multidrug-resistant Chinese hamster cells selected for resistance to Vinca alkaloids. J Cell Physiol 126: 266–274.

Skipper HE, Schabel Jr FM and Wilcox WS (1964) Experimental evaluation of potential anticancer agents. XII, on the criteria and kinetics associated with "curability" of experimental leukemia. Cancer Chemother Rep. 35: 1–111.

Slapak CA, Daniel JC and Levy SB (1990) Sequential emergence of distinct resistance phenotypes in murine erythroleukemia cells under Adriamycin selection: Decreased anthracycline uptake precedes increased P-glycoprotein expression. Cancer Res 50: 7895–7901.

Slovak ML, Coccia M, Meltzer PS and Trent JM (1991) Molecular analysis of two human doxorubicin-resistant cell lines: Evidence for differing multidrug resistance mechanisms. Anticancer Res 11: 423–425.

Slovak ML, Ho J, Bhardwag G, Kurz EU, Deeley RG and Cole SPC (1993) Localisation of novel multidrug resistance-associated gene in the HT1080/TR4 and H69AR human tumour cell lines. Cancer Res 53: 3221–3225.

Slovak ML, Hoeltge GA, Dalton WS and Trent JM (1988) Pharmacological and biological evidence for differing mechanisms of doxorubicin resistance in two human tumor cell lines. Cancer Res 48: 2793–2797.

Sonneveld P, Durie BGM, Lokhorst HM, Marie J-P, Solbu G, Suciu S, Zittoun R, Lowenberg B and Nooter K (1992) Modulation of multidrug-resistant multiple myeloma by cyclosporin. Lancet 340: 255–259.

Sugawara I, Iwahashi T, Okamoto K, Sugimoto Y, Ekimoto H, Tsuruo T, Ikeuchi T and Mori S (1991) Characterization of an etoposide-resistant human K562 cell line, K/eto. Jpn J Cancer Res 82: 1035–1043.

Sugimoto Y, Hamada H, Tsukahara S, Noguchi K, Yamaguchi K, Sato M and Tsuruo T (1993) Molecular cloning and chatacterisation of the cDNA for the M_r85,000 protein overexpressed in Adriamycin-resistant human tumor cells. Cancer Res 53: 2538–2543.

Sugimoto Y, Roninson IB and Tsuruo T (1987) Decreased expression of the amplified *mdr*1 gene in revertants of multidrug-resistant human myelogenous leukemia K562 occurs without loss of amplified DNA. Mol Cell Biol 7: 4549–4552.

Taylor CW, Dalton WS, Parrish PR, Gleason MC, Bellamy WT, Thompson FH, Rowe DJ and Trent JM (1991) Different mechanisms of decreased drug accumulation in doxorubicin and mitoxantrone resistant variants of the MCF7 human breast cancer cell lines. Br J Cancer 63: 923–929.

Toth K, Voughan MM, Slocum HK, Arredondo MA, Takita H, Baker RM and Rustum YM (1993) New immunoperoxidase "sandwich" staining method for mdr1 P-glycoprotein detection with JSB-1 monoclonal antibody in formal and fixed, paraffin embedded tissues. Proc Am Assoc Cancer Res 34: 312.

Tsuruo T, Iida H, Ohkochi E, Tsukagoshi S and Sakurai Y (1983) Establishment and properties of vincristine-resistant human myelogenous leukemia K562. Gann 74: 751–758.

Tsuruo T, Iida-Saito H, Kawabata H, Oh-Hara T, Hamada H and Utakoji T (1986) Characteristics of resistance to Adriamycin in human myelogenous leukemia K562 resistant to Adriamycin and in isolated clones. Jpn J Cancer Res 77: 682–692.

Twentyman PR, Fox NE, Bright KA and Bleehen NM (1986) Derivation and preliminary characterisation of adriamycin resistant lines of human lung cancer cell lines. Br J Cancer 53: 529–537.

Van der Bliek AM, Baas F, Houte de Lange TT, Kooiman PM, Van der Velde-Koerts T and Borst P (1987) The human *mdr*3 gene encodes a novel P-glycoprotein homologue and gives rise to alternatively spliced mRNAs in liver. EMBO J 6: 3325–3331.

Van der Bliek AM, Baas F, Van der Velde-Koerts T, Biedler JL, Meyers MB, Ozols RF, Hamilton TC, Joenje H and Borst P (1988) Genes amplified and overexpressed in human multidrug-resistant cell lines. Cancer Res 48: 5927–5932.

Versantvoort CHM, Broxterman HJ, Pinedo HM, De Vries EGE, Feller N, Kuiper CM and Lankelma J (1992) Energy dependent processes involved in reduced drug accumulation in multidrug-resistant human lung cancer cell lines without P-glycoprotein expression. Cancer Res 52: 17–23.

Volm M, Bak Jr M, Efferth T and Mattern J (1989) Induced multidrug resistance in murine leukemia L1210 and associated changes in a surface membrane glycoprotein. J Cancer Res Clin Oncol 115: 17–24.

Volm M, Mattern J and Pommerenke EW (1991) Timecourse of MDR gene amplification during *in vivo* selection for doxorubicin-resistance and during reversal in murine leukaemia L1210. Anticancer Res 11: 579–586.

Whelan RDH and Hill BT (1993) Differential expression of steroid receptors, HSP27 and pS2 in a series of drug resistant breast tumor cell lines derived following exposure to antitumour drugs or to fractionated X-irradiation. Breast Cancer Res Treat 26: 23–39.

Whelan RDH, Waring CJ, Wolf CR, Hayes JD, Hosking LK and Hill BT (1992) Over-expression of P-glycoprotein and glutathione S-transferase pi in MCF-7 cells selected for vincristine resistance *in vitro*. Int J Cancer 52: 241–246.

Yoshimura A, Kuwazuru Y, Sumizuwa T, Ikeda S-I, Ichikawa M, Usagawa T and Akiyama S-I (1989) Biosynthesis, processing and half-life of P-glycoprotein in a human multidrug resistant KB cell. Biochim Biophys Acta 992: 307–314.

Zaman GJR, Versantvoort CHM, Smit JJM, Eijdems EWHM, De Haas M, Smith AJ, Broxterman HJ, Mulder NH, De Vries EGE, Baas F and Borst P (1993) Analysis of the expression of *MRP*, the gene for a new putative transmembrane drug transporter, in human multidrug resistant lung cancer cell lines. Cancer Res 53: 1747–1750.

Zastawny RL, Benchimol S and Ling V (1992) Modulation of P-glycoprotein promoter activity by p53. Proc Am Assoc Cancer Res 33: 452.

Zijlstra JG, De Vries EGE and Mulder NH (1987) Multifactorial drug resistance in an Adriamycin-resistant human small cell lung carcinoma cell line. Cancer Res 47: 1780–1784.

Zwelling LA, Slovak ML, Doroshow JH, Hinds M, Chan D, Parker E, Mayes J, Sie KL, Meltzer PS and Trent JM (1990) A P-glycoprotein negative human fibrosarcoma cell line exhibiting resistance to topoisomease II-reactive drugs despite the presence of a drug-sensitive toposiomerase II. J Natl Cancer Inst 82: 1553–1561.

Address for correspondence: Bridget T. Hill, Cellular Chemotherapy Laboratory, Imperial Cancer Research Fund, P.O. Box 123, Lincoln's Inn Fields, London WC2A 3PX, UK.

Cytotechnology **12**: 289–314, 1993.

The use of reverse transcriptase-polymerase chain reaction (RT-PCR) to investigate specific gene expression in multidrug-resistant cells

Lorraine O'Driscoll, Carmel Daly, Mohamad Saleh* and Martin Clynes
National Cell and Tissue Culture Centre/Bioresearch Ireland, Dublin City University, Glasnevin, Dublin 9, Ireland

Key words: cancer, multidrug resistance, quantitative PCR, reverse transcriptase polymerase chain reaction (RT-PCR)

Abstract

Expression of specific genes at the level of mRNA can be studied using techniques such as Northern blot, slot/dot blot, RNase protection assay, *in situ* hybridisation and RT-PCR. In this article these methods of analysis are compared; RT-PCR offers higher levels of specificity and sensitivity than traditional methods of RNA analysis and as such has become the method of choice for the study of gene expression. The RT-PCR technique is described in detail with sections dealing with RNA extraction, choice of primers (including the use of cDNA sequence data bases), PCR and RT-PCR protocols in addition to the limitations of the method. The study of one particular mRNA transcript (*MDR*1) using RT-PCR is discussed in detail. Recently described methods for quantitation of PCR products are discussed. Quantitative PCR would appear to offer a method of studying gene expression in a more extensive way than has been possible to date.

Introduction

Control of transcription plays a critical role in the multistep process that regulates gene expression. Gene transcript levels within a cell change in response to a wide variety of signals that occur during cell development, differentiation and normal physiological function as well as in response to disease. Changes in transcription levels cause variations in the steady state levels of individual mRNAs. Thus, analysis of mRNA levels of a gene is vital in a broad range of research areas.

Traditionally, levels of individual mRNAs have been analysed by procedures such as Northern blot (Alwine *et al.*, 1977; Thomas, 1980), RNA slot/dot blot (Kafatos *et al.*, 1979), RNase protection assay (Reyes and Wallace, 1987) and *in situ* hybridisation (for details of method see Ausubel *et al.*, 1991b). Application of the PCR technique provides another method of RNA analysis (for review see Wright and Wynford-Thomas, 1990; Larrick, 1992). This PCR based technique has been variously termed RNA-PCR (Kawasaki, 1991), RT-PCR (Rappolee *et al.*, 1988a), RNA phenotyping (Rappolee *et al.*, 1988b) or message amplification phenotyping (MAPPing) (Brenner *et al.*, 1989). RT-PCR, because of its specificity, sensitivity, speed and accuracy, has become increasingly popular for the

*Visiting Postdoctoral Research Fellow from Scientific Studies & Research Centre, P.O. Box 4470, Damascus, Syria.

study of short-lived low copy number mRNA transcripts (Rappolee *et al.*, 1988a; Murphy *et al.*, 1990; Hoof *et al.*, 1991; Horikoshi *et al.*, 1992) and for the detection of unique mRNA transcripts from a background of normal cells (Chelly *et al.*, 1988; Kawasaki *et al.*, 1988; Lee *et al.*, 1987). In addition, recent advances allowing quantitation of PCR products has meant that gene expression can be studied in a more extensive way than has been possible to date.

In the past, expression of eukaryotic mRNAs has been studied using the Northern blot technique (Alwine *et al.*, 1977; Thomas, 1980). However, this method is insensitive, (5–10 µg quantities of purified polyadenylated RNA (poly $(A)^+$ RNA) is required to produce a signal) and cumbersome, necessitating the use of specific ^{32}P-labelled probes for detection following electrophoresis and transfer of mRNA to filters. However, Northern blotting is semi-quantitative and does permit mRNA to be sized enabling, for example, the study of splicing patterns of mRNA. Another method used in the analysis of RNA is the slot/dot blot technique described by Kafatos *et al.* (1979). This method provides a semi-quantitative estimate of amounts of particular mRNAs present in a sample but is limited by its lack of sensitivity (1–10 µg of purified poly $(A)^+$ RNA is necessary to produce a signal). In addition, the method is not dependable for detecting extremely rare sequences because of background problems.

RNase protection assays are more sensitive than Northern blotting, (100 ng to 1 µg of poly $(A)^+$ RNA is required for each assay). The main disadvantages of this method is that it requires hybridisation for each mRNA, as well as the use of radioactive detection methods and sequencing gels.

A more sensitive technique for the analysis of mRNA is *in situ* hybridisation (see Ausubel *et al.* (1991b) for details of method) in which the temporal and spatial expression patterns of mRNA can be determined within complex cell populations and tissues. Using this method 10 to 100 molecules of mRNA can be detected in a single cell and 3-D information about transcript distribution and cellular localization can be obtained. However, *in situ* hybridisation can be technically difficult and does not lend itself well to the processing of a large number of samples. In addition, it is not possible to quantify mRNA amounts using this method.

These methods are not always sensitive enough to detect mRNA in samples limited by either low cell number or low copy number per cell. In addition, they permit only crude quantitation of mRNA.

Reverse transcriptase-polymerase chain reaction (RT-PCR)

Polymerase chain reaction (PCR) is an *in vitro* technique for the selective amplification of defined target DNA sequences. This technique was developed by Mullis and co-workers at the Cetus Corporation (Mullis and Faloona, 1987). In order to amplify a specific target sequence using PCR, some prior DNA sequence information about the target DNA locus is normally required. This information is needed to design two oligonucleotide primers which, when added to denatured DNA, will specifically bind to their complementary sequences on opposite strands, immediately flanking the desired target i.e., the region to be amplified. The primers anneal to opposite strands in such a way that the extension reaction directs the synthesis of DNA towards each other, both occurring in the 5′ to 3′ direction (see Fig. 1). In the presence of a suitable thermostable DNA polymerase enzyme and DNA precursors (the four deoxynucleotide triphosphates – dATP, dCTP, dGTP, dTTP), the synthesis of new strands of DNA, complementary to the individual primed DNA strands, is initiated.

DNA generally exists as a double stranded molecule. This must be denatured to single strands for the primers to recognise and bind to their complementary sequences. Denaturation occurs at 90–95°C and the temperature is then lowered to allow the primers to anneal. The optimum annealing temperature may range from 50°C to 75°C; increasing the temperature improves the primer stringency. (The optimum temperature will depend on the A + T and G + C make-up of the primers, as will be discussed later). Extension of the annealed primers by the incorporation of free deoxy-

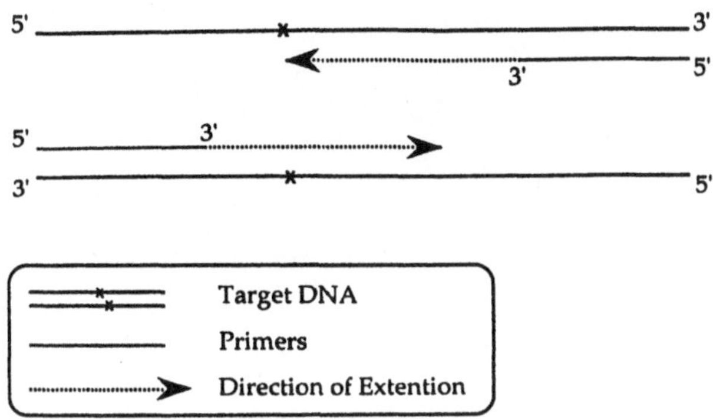

Fig. 1. Primers anneal to their specific sites of recognition on opposite strands, flanking the target DNA region to be amplified. The DNA polymerase enzyme catalyses the incorporation of the free deoxynucleotide triphosphates, resulting in extension of the primers and so the synthesis of new complementary strands of DNA.

nucleotide triphosphates is catalysed by a DNA polymerase enzyme at approximately 72°C.

This completes one cycle of the polymerase chain reaction. PCR is a *chain* reaction because newly synthesized strands from one cycle act as templates for further DNA synthesis in subsequent cycles. This cycle is repeated many times resulting in an exponential increase in the amount of DNA synthesized, until eventually it reaches a plateau (see Fig. 2 and Protocol 3: Appendix 1).

RT-PCR, an adaptation of the basic PCR method enables mRNA to be studied in a similar way to DNA. More than twenty years ago reverse transcriptase (also known as RNA-directed DNA polymerase), an animal retro-viral enzyme capable of synthesizing DNA on an RNA template was discovered by Termin (1972). Reverse transcriptase enzymes have been shown to have three enzymatic activities; they can:

1. copy an RNA molecule to yield double-stranded DNA-RNA, using a primer and joining deoxynucleotide triphosphates in a 3'–5' linkage;
2. degrade RNA in a DNA-RNA hybrid; and
3. copy a primed single strand of DNA to form double-stranded DNA (Freifelder, 1983a).

The RT-PCR method is based on these properties of reverse transcriptase enzymes.

mRNA is extracted from cells, either in isolation or as part of a total RNA extraction. For RT-PCR we have found total RNA to be adequate. However, the extraction of undegraded RNA from cells can often prove difficult. This is due to both the labile nature of RNA itself and the presence in the cells of active ribonuclease (RNase) enzymes.

The life-time of mRNA is short compared to other types of RNA molecules. This has an important regulatory function in the cell. Degradation of mRNA (to control the formation of a particular protein) proceeds primarily from the 5'-P terminus by RNase enzyme activity (Freifelder, 1983b). Ribonuclease enzymes are formed on ribosomes on the cytoplasmic surface of the endoplasmic reticulum (ER). They then pass into the internal compartments of the ER where they are packaged in secretory granules. This causes the synthetic pathways occurring in the cytoplasm to be separated from the degradative activity of the RNase enzymes. However, the disruption of the cell which occurs when attempting to extract RNA liberates RNases and subjects the RNA to degradation by these enzymes (Jackson *et al.*, 1991). To alleviate this problem, cells are generally lysed in a chemical environment containing guanidinium thiocyanate.

Guanidinium thiocyanate is one of the most effective protein denaturants known. The use of guanidinium to lyse cells was originally developed to allow purification of RNA from cells high in endogenous ribonucleases (Cox, 1968; Ullrich *et al.*, 1977; Chirgwin *et al.*, 1979). Whereas RNases can recover activity after many forms of treatment (such as boiling), they are inactivated by 4M guanidinium thiocyanate and reducing agents such as β-mercaptoethanol (Sela *et al.*, 1957).

After disruption of the cells using guanidinium thiocyanate, β-mercaptoethanol and a mild detergent, RNA can be recovered by sedimentation through a cesium chloride gradient (or by organic extraction and ethanol precipitation). The protocol outlined below takes advantage of the fact that RNA can be separated from DNA and protein by virtue of its density. To achieve this, the cell lysate is layered on a cushion of a dense solution of

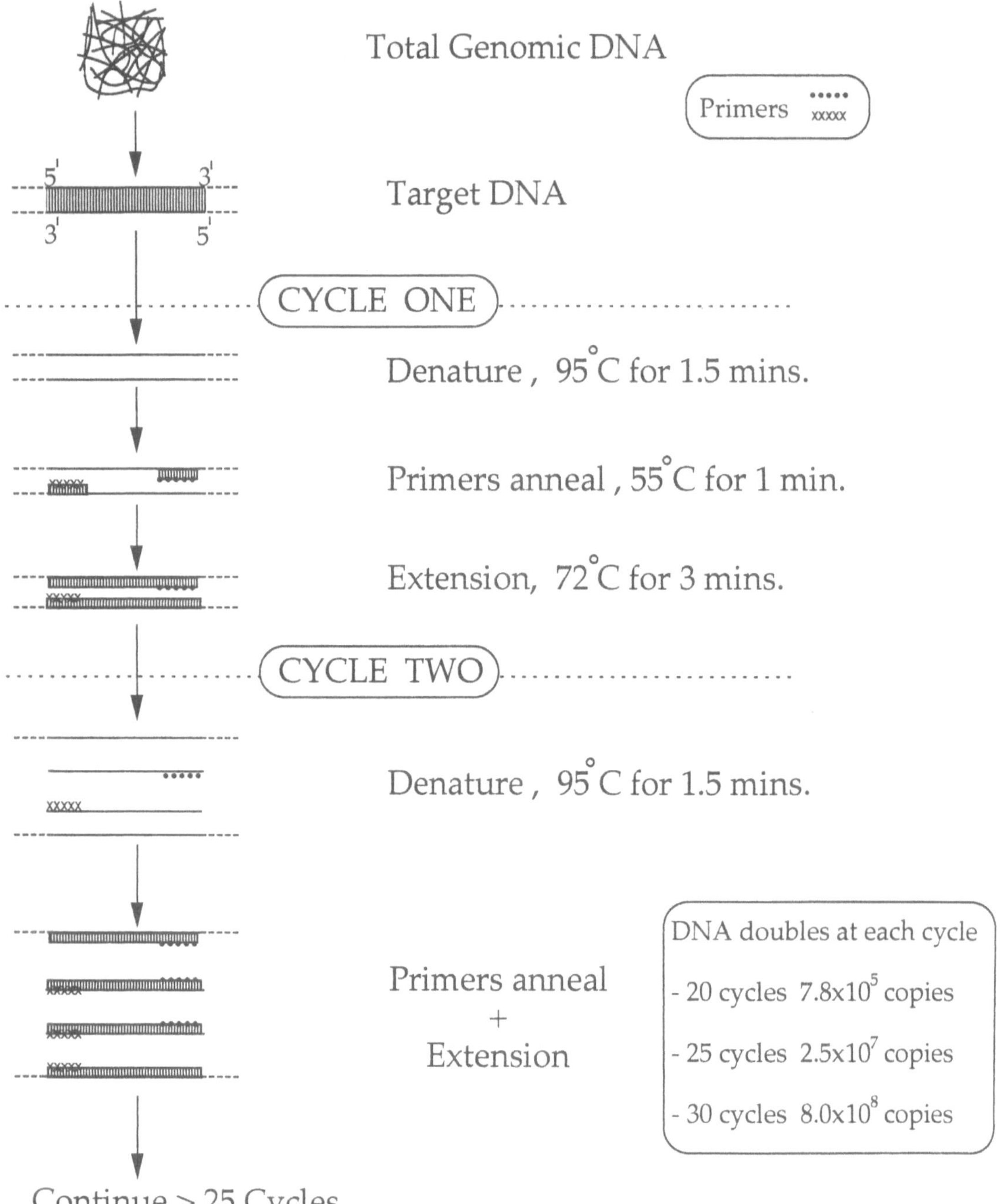

Fig. 2. Each of the primers is chosen so that it recognises one specific DNA sequence and does not bind to any other DNA sequence present in the DNA specimen. Each cycle of the polymerase chain reaction consists of a denaturation step where double-stranded DNA is melted to single strands; an annealing step, where the primers bind to their specific recognition sites on the template; and an extension step, where new strands of DNA are formed, complementary to the primed single strands, by DNA polymerase enzyme extension of the primers. The primers become incorporated into the newly synthesised strands of DNA and so become the boundaries of the amplified region. New strands of DNA formed at each cycle act as template for the next cycle, and so for a finite time DNA strands are synthesised exponentially.

cesium chloride. The buoyant density of RNA in cesium chloride (> 1.8 g/ml) is much greater than that of other cellular components (Glisin *et al.*, 1974). During centrifugation, RNA forms a pellet on the bottom of the tube, DNA is suspended at the interface and protein floats in the supernatant solution (see Protocol 1: Appendix 1).

mRNA must be selectively primed for it to act as a template on which double-stranded DNA, known as complementary DNA (cDNA), can be formed by a reverse transcriptase enzyme (See Protocol 2: Appendix 1). The cDNA area of interest, which is a direct copy of the mRNA, can then be amplified and studied as in a typical PCR (see Protocol 3: Appendix 1).

Oligo (dT) primers (for eukaryotic RNA), random hexanucleotide primers, or the 3′ antisense gene-specific primer used in the PCR reaction, may be used for this purpose. In choosing the primers for the RT reaction the following should be considered:

Oligo (dT) primers recognise and bind to the poly $(A)^+$ tail (3′) of mRNA allowing it to be copied to cDNA, selectively (see Fig. 3). (A polyadenylated tail occurs on the 3' end of most, but not all, eukaryotic mRNAs (Ausubel *et al.*, 1991c). The possibility of having long stretches of cDNA or even full-length cDNA made from mRNA have been found to be more likely using oligo (dT) primers.

Random primers, as their name suggests, bind randomly to the mRNA template and, again, the reverse transcriptase enzyme catalyses the formation of cDNA strands. These primers, rather than oligo (dT), are claimed to minimise the effects of mRNA secondary structure and the distance of the amplified sequences from the poly $(A)^+$ tail. This is suggested to be of particular importance for the analysis of partially degraded samples of cellular RNA, such as are commonly isolated from tumours and other clinical specimens (Noonan and Roninson, 1991). Because of their random binding, a mixture of short and long cDNA stretches may result.

The 3′ gene-specific complementary primer binds selectively to the 3′ region flanking the mRNA locus of interest. Some advantages of using the downstream gene-specific primer include the fact that cDNA including the specific area of interest should result. It is also convenient as an aliquot of this primer will be used in both the RT and the PCR reactions.

Alternatively, the advantage of using oligo (dT) primers or hexanucleotide primers over gene-specific primers for cDNA synthesis is that the products of a single RT reaction can be used for amplification of multiple mRNA sequences, either in the same PCR reaction, or separately. In this way, the cDNA of interest can be amplified simultaneously with an internal standard (see Protocol 2: Appendix 1).

Applications of RT-PCR

Adaptation of PCR to the detection of RNA has increased the sensitivity of detection of a particular mRNA species by several orders of magnitude (with the exception of *in situ* hybridisation). RT-PCR is 1,000 to 10,000 times more sensitive than the traditional RNA blot techniques (Byrne *et al.*, 1988; Wang *et al.*, 1989; Rappolee *et al.*, 1989; Mocharla *et al.*, 1990; Singer-Sam *et al.*, 1990) e.g., Mocharla *et al.* (1990) have reported that RT-PCR amplified cDNA was obtained from as little as 5 pg of human pancreatic or parotid total RNA, yet transcripts were not detected on Northern blots unless at least 15—30 ng of pancreatic total RNA was used. As a result of this sensitivity, RT-PCR can be used to analyse RNA of extremely low abundance (Chelly *et al.*, 1988), mRNAs in small numbers of cells (1 to 1,000 cells (Kawasaki *et al.*, 1988; Rappolee *et al.*, 1989)) or in small amounts of RNA (as little as 6 pg of total RNA (Rappolee *et al.*, 1989; Wang *et al.*, 1989; Mocharla *et al.*, 1990; Kawasaki, 1991)). Expression of multiple mRNAs can be determined simultaneously from a single sample of RNA, a difficult task by traditional methods. Purification of poly $(A)^+$ mRNA is rarely necessary for RT-PCR analysis. RT-PCR should be of particular benefit in clinical studies, where biopsy samples may not provide enough material for accurate analysis by traditional immunological and biochemical methods.

The sensitivity of PCR can be one of the disadvantages of the system. As a result of the exponential nature of PCR, even trace contamination of reagents, samples and apparatus, with cloned or PCR derived sequences, may lead to false positives.

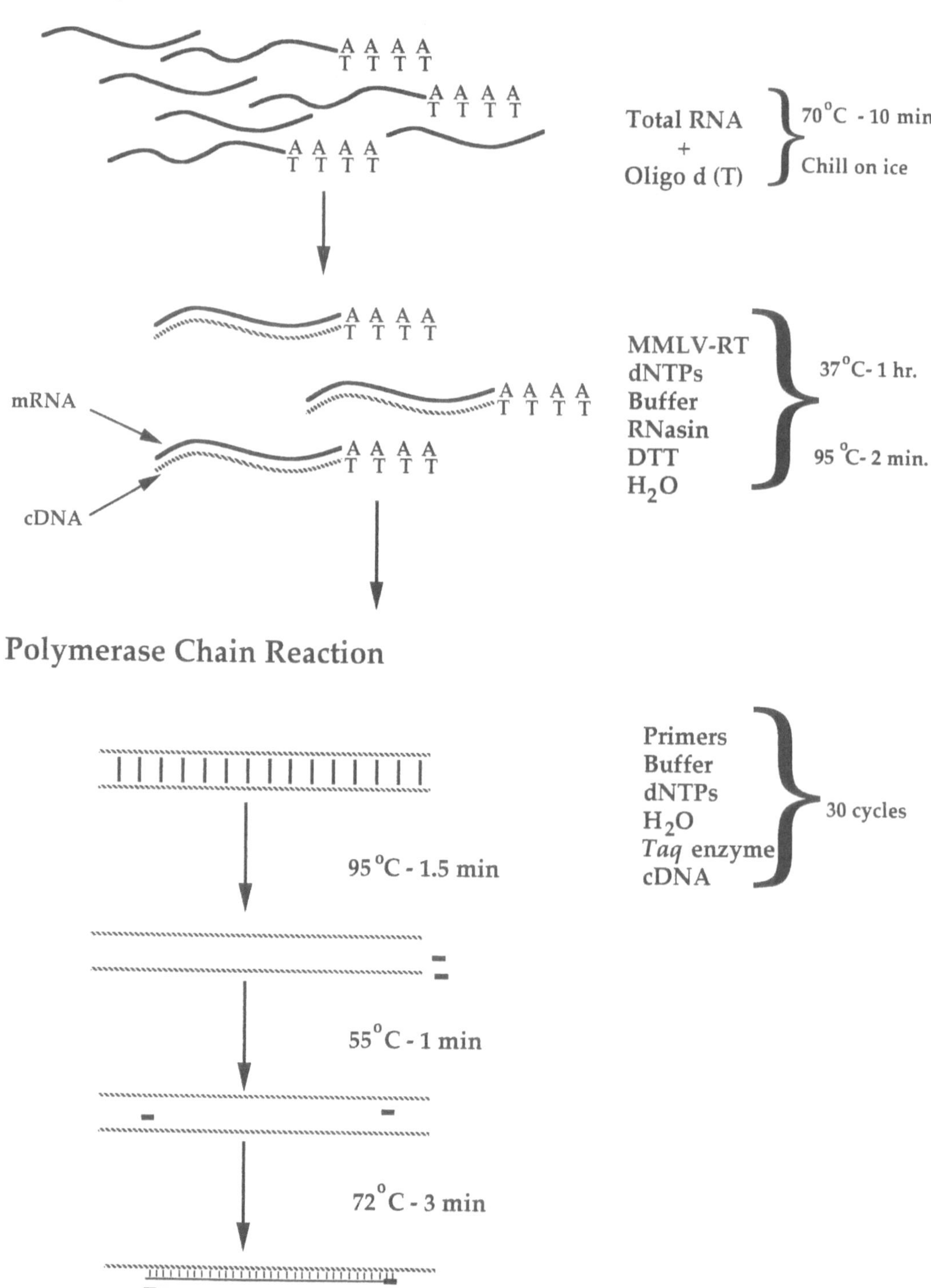

Fig. 3. Oligo (dT) may be used to prime the poly $(A)^{+}$ tail of mRNA. cDNA is formed on this template by reverse transcription. Specific cDNA regions of interest can then be amplified by PCR for their analysis.

Unlike the results obtained using Northern blotting, RT-PCR does not provide information on the size of the transcripts of interest. Although RT-PCR provides an ideal method to survey the expression of multiple genes simultaneously in diverse populations of cells, subpopulations of cells may provide unacceptable background signals. Expression of genes in cells should be treated with caution as it may not indicate physiological significance e.g., using RT-PCR Chelly *et al.* (1988) detect expression of several tissue specific genes in cells not expected to express these genes; this phenomenon has been termed "illegitimate" or "leaky" transcription. This leads to the question as to whether levels detected by PCR are clinically or physiologically relevant and whether in all cases functional protein expression results. For example, in the case of *MDR1* (the multidrug resistance gene) low level expression may be as a result of leaky transcription or possibly may represent subpopulations of P-glycoprotein expressing cells within a tumour. A positive result may be obtained from an RNA preparation from a tumour which may only represent the expression of a gene in a subpopulation of tumour cells. These cells would have a growth advantage in the course of chemotherapy. Low level expression may be significant for some tumour types. Clinical correlative studies are necessary to establish whether low levels of *MDR1* expression would be prognostically significant for different tumours; recent results by Holzmayer *et al.* (1992) suggest that such correlations may indeed be clinically relevant.

RT-PCR can be used for the detection of transcripts that display variable expression patterns in tissues and during development (Chelly *et al.*, 1988; Rappolee *et al.*, 1988a; Chin *et al.*, 1989) e.g., the dystrophin gene, defective in patients with muscular dystrophy is expressed at very low levels (representing only 0.01–0.001% of total muscle mRNA) making it difficult to study by conventional methods. RT-PCR was successfully used to study levels of dystrophin mRNA in clinical samples (Chelly *et al.*, 1988).

RT-PCR in the study of multidrug-resistance

The multidrug resistance gene *MDR1* has also been studied by RT-PCR, as conventional methods were often unsuccessful at detecting expression of this gene (Roninson *et al.*, 1986; Gekeler *et al.*, 1990; Fuqua *et al.*, 1990; Murphy *et al.*, 1990). The multidrug resistance phenotype is characterised by cross resistance to a large group of lipophilic cytotoxic compounds including plant alkaloids and antitumour antibiotics. Multidrug resistance is associated with decreased intracellular drug accumulation and correlated with the increased expression of MDR genes, which encode membrane glycoproteins (P-glycoproteins) of approximately 170 kDa (Juliano and Ling, 1976; Roninson *et al.*, 1986; Ueda *et al.*, 1986; Roninson, 1987; Gottesman and Pastan, 1988). Expression of a single human gene, designated *MDR1*, is sufficient to confer the multidrug resistance phenotype to drug sensitive cells (Gros *et al.*, 1986; Shen *et al.*, 1986a; Ueda *et al.*, 1987; Choi *et al.*, 1988). Even very low levels of *MDR1* gene expression can confer a severalfold increase in the level of drug resistance which may be clinically significant (Shen *et al.*, 1986a; Chin *et al.*, 1989). *MDR1* expression has been analysed by conventional RNA slot/dot blotting techniques (Goldstein *et al.*, 1989; Fojo *et al.*, 1987; Bourhis *et al.*, 1989) and Northern blots (Shen *et al.*, 1986b; Roninson *et al.*, 1986; Fojo *et al.*, 1987; Kramer *et al.*, 1993). However, in cells with a several fold increase in cellular drug resistance *MDR1* expression levels are close to the limits of sensitivity of conventional methods even when large amounts of RNA are used for the assay (Shen *et al.*, 1986b; Fojo *et al.*, 1987; Noonan *et al.*, 1990). RNA blotting techniques may not be specific or sensitive enough to detect low levels of *MDR1* gene expression which may be sufficient for clinically relevant multidrug resistance (Shen *et al.*, 1986b; Noonan *et al.*, 1990; Noonan and Roninson 1991; Holzmayer *et al.*, 1992).

There are a number of additional problems in the study of gene expression in multidrug resistant cells under clinical conditions. The amount of available tissue is often limited; analysis of RNA from these samples is often problematic because

RNA may have suffered significant degradation. Detection of *MDR1* is often complicated by the existence of a homologous gene, *MDR3*, which is apparently not associated with resistance to chemotherapeutic drugs; *MDR3* gene products often show cross reactivity with some *MDR1* reactive antibodies and nucleic acid probes. When using *MDR1* and *MDR3* DNA probes high stringency hybridisation conditions are required to avoid cross hybridisation of *MDR3* probes with *MDR1* RNA. This limits the sensitivity of the assay and as a result cells with a low degree of resistance may escape notice. MDR expression in tumours is often heterogeneous with only a subpopulation of tumour cells expressing the gene (Weinstein *et al.*, 1990). Even if some cells in the tumour express the gene at a relatively high level, heterogeneity may lead to a very low signal when total RNA extracted from the tumour is analysed. Subpopulations of P-glycoprotein expressing cells within the tumour may have a growth advantage in the course of chemotherapy (Noonan *et al.*, 1990). The advancement of PCR based technology has provided a sensitive, specific and quantitative protocol for measuring the levels of *MDR1* mRNA in clinical samples which overcomes many of these drawbacks. PCR methods require much less tissue than traditional methods and so are applicable to the detection of *MDR1* in clinical samples, even from a heterogeneous population of cells. In addition, by careful choice of sequence-specific primers, it is possible to amplify selectively even such highly related sequences as *MDR1* and *MDR3*.

Noonan *et al.* (1990) used a quantitative PCR assay to study *MDR1* mRNA expression in a large number of both normal and tumour tissues. They report *MDR1* expression which was rarely observed using standard (slot/blot) assays. Similarly, Chin *et al.* (1989) used PCR to detect the presence of *MDR1* and *MDR3* mRNA in a number of multidrug resistant cell lines which had not been detected by Northern hybridisation using *MDR1* and *MDR3* genomic clones (Roninson *et al.*, 1986). Both of these studies, while agreeing with the results obtained using standard blot techniques on tissue-specific distribution of *MDR1* RNA (Fojo *et al.*, 1987), report that the PCR assay also detects *MDR1* expression in some tissue which had not been evident using Northern or slot blots. Fuqua *et al.* (1990), Gekeler *et al.* (1990), Murphy *et al.* (1990), Hoof *et al.* (1991) and Hegewisch-Becker *et al.* (1993) report the detection of *MDR1* mRNA (using RT-PCR) in cells and tissues in which other workers (Fojo *et al.*, 1987), using conventional RNA blot techniques, have either not detected expression or found expression to be extremely low. PCR would appear to provide a clinical tool comparable to Northern analysis in accuracy but greatly exceeding its sensitivity.

In conclusion, RT-PCR is superior in ease, speed, sensitivity and resolution to mRNA analysis by *in situ* hybridisation, RNA (Northern) blot and RNase protection assays and as such provides the only method of accurate mRNA analysis if quantities of RNA are limited.

Precautions/artifacts/limitations

As with all PCR studies, general precautions must be taken and good laboratory procedures exercised. These include prealiquoting all solutions in small amounts and discarding the remains after each use; physically separating the reaction preparation from the amplified products; the use of separate pipettes etc. for setting-up the reaction and analysing the product; and optimising the conditions for each set of templates and primers. Primers must be chosen carefully, avoiding the possibility of primer-dimer formation etc. (see "Criteria to follow when choosing primers"). This is of particular importance where more than one primer pair is included in a reaction which would be the case if primers were also included to amplify an internal control, such as β-actin. Extra care must be taken in this case to ensure that no two of the four primers included will form "primer-dimers". More obvious precautions include autoclaving all necessary solutions, eppendorfs, pipette tips etc. and wearing gloves, as previously discussed. Reaction components should be added on ice and the accuracy of the pipettes monitored regularly. This last point, although somewhat obvious, is extremely important for these procedures as the volumes involved are quite small.

Due to the sensitivity of this technique a very small inaccuracy in setting up the reaction components could result in a very large inaccuracy in the final product.

Good experimental design is very important. This involves including negative controls to detect contamination and positive controls to ensure the reaction is conducted successfully (discussed later). All reactions should be set up at least in duplicate. To standardise the procedure, it is generally considered better to prepare a master-mix of all the ingredients including the enzyme and aliquoting (on ice) just prior to the template addition. The minimum number of PCR cycles required should be carried out in order to minimize the chances of a rare contaminating template being amplified.

Extra precaution (as previously discussed) must be taken when studying RNA e.g., baking glassware, DEPC-treating solutions. Furthermore, it must be possible to discriminate between the RNA product and that of contaminating DNA. Although great care is taken when extracting RNA after sedimenting through the cesium chloride cushion, the possibility of DNA contamination (from the cesium chloride: guanidinium thiocyanate interface) should not be ruled out. To overcome this, primers from different exons should be used whenever possible. Different product sizes will then result if mRNA or DNA is used as the template.

To confirm amplification of the expected molecules, cleavage with a restriction endonuclease enzyme whose site of recognition in the amplified fragment is known, should be included. Alternatively, hybridisation with an internal oligonucleotide can be done. This is claimed to increase the sensitivity by allowing detection of amplification products not visualised by ethidium bromide staining (Ausubel *et al.*, 1991c).

Quantitative PCR

Although RT-PCR has many advantages over RNA blot methods it had been difficult to obtain quantitative data on the expression of individual genes using PCR based methods. RT-PCR requires two enzymatic steps, synthesis of the cDNA template and PCR amplification. Both steps must be considered in any attempt to quantitate mRNA. Efficiency of cDNA synthesis can be monitored by the incorporation of ^{32}P labelled dNTPs into TCA precipitable material. As PCR amplification is an exponential process, small variations in amplification efficiency or in any of the variables that control the reaction (e.g., concentration of polymerase, dNTPs, magnesium, DNA or primers) can drastically affect the yield of products and in so doing obscure differences in the initial amounts of target sequences (Gilliland *et al.*, 1990a; Bloch, 1991). In addition, the efficiency of PCR decreases at the later stages of amplification due to depletion of reaction components, diminished enzymatic activity and accumulation of products (see "RT-PCR. Methods and Applications", (Clontech Labs) 1991). Therefore, any attempt to quantitate mRNA levels by PCR must be limited to the analysis of products generated only during the exponential phase of the amplification. Under these conditions RT-PCR can yield reasonably precise information about relative changes in mRNA levels.

PCR product concentration is proportional to the starting target DNA as long as product accumulation remains exponential (Chelly *et al.*, 1988; Singer-Sam *et al.*, 1990). The point at which exponential accumulation plateaus can be roughly estimated by noting the point at which continued cycles do not produce significantly increased product yield. RT-PCR can detect changes in mRNA levels of greater than two-fold without the use of any internal standard. Singer-Sam *et al.* (1990) used RT-PCR quantitatively to measure accumulation of RNA transcripts in total mouse RNAs derived from male germ cells at various spermatogenic stages. By limiting the amount of input template and the number of cycles of PCR, they obtained a signal which was linearly related to specific RNA levels over at least a thousand fold range, with reproducibility adequate for any study where changes in RNA levels were greater than two-fold.

For more accurate quantitation of mRNA levels corrections for reaction-to-reaction variation in amplification efficiency are necessary. Amplification of replicate samples of plasmid results in product yield which can vary by as much as six-fold (Gilliland *et al.*, 1990b). RT-PCR, although renowned for its high sensitivity, specificity, ease and speed, can also be a technique of variable efficiency between preparations and between RNA species. This variability may be due to a number of reasons including the fact that the procedure consists of a number of individual techniques – extracting RNA; quantifying the RNA spectrophotometrically; forming cDNA on the mRNA template; and amplifying the region of interest. Taking into account e.g., the extraction efficiency, the number of very small volume additions etc., it is understandable that some amount of variability may occur even when extreme care is exercised.

For this reason, it is very important to include an internal control in each RT-PCR. Such a control may be the product of an ubiquitously expressed gene (e.g., a "house-keeping" gene product). As this is extracted at the same time, by the same procedure, using the same solutions, from the same population of cells (subjected to the same contaminants/inhibitors); quantified also as part of the total RNA and included in the same RT and PCR reactions as the mRNA to be analysed, the products of the two should be directly comparable.

The uses of such a control are two-fold. If no signal from the sequence of interest is obtained in a given sample, the internal control will verify whether this is a true (if internal control is present) or false (if both are absent) negative. It is not uncommon for amplifications to fail, especially since some experimental samples contain contaminants that interfere with DNA replication (Ausubel *et al.*, 1991c) or inhibitors of *Taq* DNA polymerase e.g., porphyrin compounds derived from heme, that may be a problem if contaminating the DNA extracted from blood cells (Higuchi, 1989). Secondly, the internal control allows for quantitation since it normalises for several factors including variation in the amount of sample RNA; efficiency of the RT reaction; efficiency of amplification and the amount of product loaded on the gel.

The internal control PCR product should differ enough in size to be resolved from the product of interest, but close enough so as to minimise the probability of differences in amplification efficiency. It should also be distinguishable in size from artifactual "primer-dimers" that may be produced.

When such an internal control is used for this purpose, either β-actin or β_2-microglobulin are generally chosen. Esterase D has also been favoured by some researchers (Cole *et al.*, 1991).

β-Actin

Actin is the most abundant structural protein in eukaryotic cells (Nakajima-Iijima *et al.*, 1985; Pollack, 1980; Firtel, 1981; Fulton, 1981). Its amino acid sequence has been highly conserved during evolution (ranging from 1.1 to 6.2% difference in amino acids). At least six isoforms exist in vertebrates. Four actin types are involved in muscle contraction including skeletal, cardiac, aortic type smooth muscle and stomach-type smooth muscle forms. The muscle-type actins are tissue-specific.

The cytoplasmic actins (nonmuscle) – β and γ – coexist in many cell types and are involved in a variety of functions. Multiple gene copies related to the cytoplasmic actins have been found to exist, most of which seem to be pseudogenes (Engel *et al.*, 1981; Moos and Gallwitz, 1982; Moos and Gallwitz, 1983; Ponte *et al.*, 1983). Abnormal expression of cytoplasmic actin has been found to be associated with neoplastic transformation of human fibroblasts (Leavitt *et al.*, 1982; Nakajima-Iijima *et al.*, 1985).

From the primer sequences published for amplification of β-actin it seems to be very difficult to choose sequences which are specific for β-actin and do not recognise β-actin pseudogene, mutated β-actin, α-actin or γ-actin. It is important to be aware of this, particularly if a range of tissue types are to be studied, as the levels of the various actin forms may differ greatly from tissue to tissue and so may not act as a good standard. It may be important to choose different isoforms of actin as internal standards, depending on the tissue type being studied. β-actin is generally favoured as a standard by many researchers including Horikoshi *et al.* (1992), Avraham *et al.* (1992) and Lönn *et al.* (1992).

β_2-microglobulin

β_2-microglobulin has also been used as an internal

control (Noonan *et al.*, 1990). β_2-microglobulin is a small (12 kDa) polypeptide found in serum and associated with the major histocompatibility complex class 1 heavy chain on the surface of nearly all cells (Güssow *et al.*, 1987). It has been claimed by others as a gene that is ubiquitously expressed and its expression correlates with the cell surface area in most cell types (Noonan *et al.*, 1990). Noonan *et al.*, (1990) suggested that as the amount of β_2-microglobulin is not constant in different cell types, but is proportional to the cell surface area, it may not be the best control for genes encoding for intracellular proteins.

In tumour types, including small cell lung cancer, the amount of β_2-microglobulin and another HLA-related gene may be decreased or eliminated (Doyle *et al.*, 1985). Horikoshi *et al.* (1992) studied the ratios of β_2-microglobulin: β-actin and found a considerable variation among the tissues tested. This suggested that either one or both are not expressed at similar levels among different tissues. β-actin was found to be more constantly expressed and so was favoured as the internal standard.

Esterase D

Esterase D has also been chosen as an internal control by Cole *et al.*, (1991) who claimed it to be less dependent on cell cycle than other genes e.g., β and γ actins. Because of the nature of their study they favoured esterase D over β_2-microglobulin as they believed β_2-microglobulin to be often undetected in small cell lung cancer. However, the polymorphic enzyme esterase D (13q14) has been used as a marker for loss of heterozygosity in linkage analysis studies of the retinoblastoma gene (RB1). Mutations of the RB1 gene have been found in many tumour types and in particular, in small cell lung cancer. In 77% of small cell lung cancers studied, absence or trace levels of RB1 mRNA have been reported (Macdonald and Ford, 1991). It is possible therefore that esterase D may also be affected in this way. This would make it an unreliable control if its presence in all tumour types studied was uncertain.

It therefore would seem that no one endogenous sequence will be suitable as an internal control for all studies. As discussed in "Criteria to follow when choosing Primers", good experimental design will also involve choosing the best possible internal control – whether this be β-actin, β_2-microglobulin, esterase D or some other endogenous sequence.

For more reliable and precise quantitation of PCR amplification of specific mRNA species, a process termed competitive PCR has been developed. This method involves the co-amplification of a competitive template that uses the same primers as those of the target cDNA but can be distinguished from the target cDNA after amplification. Gilliland *et al.* (1990b) have compared the co-amplification of target cDNA to cloned genomic sequences. Aliquots of the PCR mixture containing cDNA copies of the RNA to be assayed are added to serial dilutions of a competitive DNA fragment differing from the cDNA of interest by the presence a small intron i.e., primers are chosen in separate exons and flank the intron. Amplification of the competitor yields a PCR product larger than the target mRNA region. The dilution at which the control and cDNA products are equivalent indicates that the starting concentration of cDNA prior to PCR is equal to the known starting concentration of the competing control plasmid. Using this method, the reference templates share with the target sequence the same primer sites and near totality of the amplified sequence so that the two templates compete for the same primer set and subsequently amplify at the same rate (competitive PCR), as any variable affecting amplification has the same effect on both. However, there may be differences in the amplification efficiency of genomic and complementary DNAs i.e., differences in the size of the target sequences may affect amplification. Therefore, the presence of an intron in genomic sequence may have an affect on quantitation.

In an attempt to minimize sequence variation a number of studies have been carried out which use a competitive template that contains a single base pair change (from the target) which results in the creation of a restriction site (Becker-Andre and Hahlbrock, 1989; Gilliland *et al.*, 1990b). This method has been termed PCR aided titration assay (PATTY). Competitive templates containing changes in a single base pair can easily be synthesized by using PCR for site directed mutagenesis (Higuchi *et al.*, 1988). After PCR amplification, the competitor can be identified by restriction digestion. In their studies Gilliland *et al.* (1990b) compared the efficiency of using competitive templates containing small introns and those containing a single base change to create a restriction site. They report that the amplification efficiencies for each were identical.

However, there are a number of potential problems when using competitors containing small introns or altered restriction sites. They do not provide a control for the efficiency of the reverse transcription step of RT-PCR. In addition, quantitation may be complicated by the possible formation of heterodimers between native and mutated amplified products during later amplification cycles (Becker-Andre and Hahlbrock, 1989; Gilliland *et al.*, 1990b). Since heterodimers are not recognised as substrates for restriction enzymes, quantitation can be artificially skewed towards uncleaved products. In addition, a competitive template must be constructed for each gene to be analysed which is often tedious and involves mutagenesis and cloning.

In order to provide a control for the reverse transcriptase step of the process, cRNA templates have been utilised (Becker-Andre and Hahlbrock, 1989; Wang *et al.*, 1989; Wang and Mark, 1990; Gaudette and Crain, 1991). RNA competitors are derived from the *in vitro* transcription of the mutated DNA fragment containing the recognition sequence of T7 RNA polymerase at one extremity (Wang *et al.*, 1989). By reverse transcription and amplification of the target mRNA and cRNA strand in the same tube, variable effects due to differences in sample preparation, conditions of reverse transcription or the PCR amplification are internally controlled and will affect the yield of PCR product equally for the target mRNA and the standard cRNA.

In an attempt to overcome both the problem of heterodimer formation and the necessity to create different competitors for each gene Wang *et al.* (1989) have used a synthetic RNA as an internal standard. This artificial gene consists of twelve target gene upstream primers connected in sequence followed by the complementary sequences of their downstream primers in the same order. The intervening DNA bears no sequence relation to the target RNA. It has been shown that amplification efficiency is primarily determined by the primer sequences unless there are significant differences in denaturing or polymerase extension characteristics due to high G/C content or secondary structure (Wang *et al.*, 1989; Siebert and Larrick, 1992). Differences in sequence content between target and competitor RNAs is believed to have a minimal effect on amplification efficiency. Using this cRNA in competitive PCR, differences in the reverse transcriptase step can be analysed and, in addition, the same standard can be used to quantitate a number of different genes. The fact that target and competitor sequences are different eliminates the problem posed by heterodimer formation.

Using competitive RT-PCR it is possible to obtain quantitative information on mRNA distribution comparable to RNA blot analysis. mRNA species can be accurately quantitated from 1 ng of total RNA or as few as 10 cells (Becker-Andre and Hahlbrock, 1989; Gilliland *et al.*, 1990a; Bej *et al.*, 1991; Gaudette and Crain, 1991). Wang *et al.* (1989) report that they have used quantitative PCR to reliably quantitate the amount of a specific mRNA in a sample of < 0.1 ng of total RNA. They state that they can measure 10^4 molecules which is 1,000 times more sensitive than the dot blot assay.

Due to its sensitivity, speed and accuracy quantitative PCR can be used for the study of gene expression in a more extensive way than has been possible to date, allowing analysis of changes in the levels of expression of specific RNA molecules. This is of benefit in the diagnosis and analysis of cancer, metabolic disorders and autoimmune diseases. A number of studies have been carried out using competitive PCR to analyse the expression of *MDR*1 in normal tissues, tumour derived cell lines and clinical specimens of untreated tumours (Noonan and Roninson, 1988; Murphy *et al.*, 1990; Noonan *et al.*, 1990). These studies have resulted in the detection and quantitation of *MDR1* gene expression which was rarely possible by standard assays.

Criteria to follow when choosing primers

The following is a list of guidelines for the design of oligonucleotide primers. Due to the uniqueness of each situation's requirements, it may not be possible, or indeed necessary, to conform to all of the guidelines listed.

Complementarity to template

The fundamental requirement for a primer is that it should hybridise efficiently to the sequence of interest with negligible hybridisation to other sequences present in the sample. (If sufficient amounts of template are available, this can be tested by performing oligonucleotide hybridisation, Ausubel *et al.*, 1991c).

Target length

The distance between the primers for which optimum amplification can be achieved is generally considered to between 180 and 500 bp. However, much longer targets may be amplified efficiently. Amplifications of sequences up to 10 kbp in length are sometimes attempted successfully (Ausubel *et al.*, 1991c) but long sequences may be difficult to amplify consistently. A considerable drop-off in synthesis efficiency has been found with distances > 3 kbp (Jeffreys *et al.*, 1988).

The sequence length amplified will obviously be determined by the requirements of the study. For the purpose of RFLP analysis, for example, it is desirable to have a reasonably short distance between primers with the restriction site approximately central (Ivinson and Taylor, 1991). Alternatively, small distances between primers lessens the ability to obtain much sequence information or to reamplify with nested internal oligonucleotides, should that be necessary (Ausubel *et al.*, 1991c). The fidelity of the DNA polymerase enzyme used in the PCR reaction (i.e., the number of errors produced per nucleotide synthesized) should be taken into consideration when deciding on the target length. *Taq* DNA polymerase does not have a 3′–5′ proofreading exonuclease activity to remove nucleotides that have been misinserted during polymerisation. The longer the sequence being amplified with *Taq*, the more mismatches will be present. Other DNA polymerase enzymes e.g., the T4 and native T7 DNA polymerases are proofreading-proficient and are very accurate with regards to base substitution and one base frameshift mutations. The Klenow fragment of *E. coli.* DNA polymerase I also contains 3′→5′ exonuclease, but the proofreading activity is much weaker than that of T4 or native T7 DNA polymerase. However, the T4, native T7 and Klenow DNA polymerase are heat labile and operate optimally at 37°C. Using such low temperatures increases the probability of nonspecific annealing of primers to other regions of the DNA and also the likelihood of secondary structure formation being a problem (Eckert and Kunkel, 1991).

For many applications, primers are designed to be exactly complementary to the regions flanking the target sequence. For other purposes, such as engineering of mutations, introducing new restriction sites or for efforts to detect or clone gene homologs when sequence information is lacking, base-pair mismatches will be intentionally or unavoidably created. As far as is possible, it is advisable to include these mismatches as a "tag" near the 5′ end of the primer. This should have no effect on the amplification reaction. When incorporating a modification at an internal point in the primer, the fewer the number of modifications, the better i.e., 1—2 changes (Clarkson *et al.*, 1991).

Overall, it is best to include any changes as far from the 3′ end of the primer as possible; at least 10—12 bases from the 3′ end of the primer (Clarkson *et al.*, 1991). A mismatch in the 3′ area could prevent extension. When only the protein sequence is available, mixed oligonucleotide primers, known as degenerate oligonucleotide primers, derived from an amino acid sequence may be used. This has had some, but in many cases, limited success (Compton, 1990). Again, the less degenerate the oligonucleotides, especially at the 3′ end, the better (Ausubel *et al.*, 1991c). This can be achieved by choosing primers with conserved amino acids which have nondegenerate codons (e.g., Trp and Met) at the 3′ end.

Primer length

The length of a primer contributes to its specificity. It is generally considered ideal to choose oligonucleotide primers between 18 and 30 bases in length; however, shorter and longer primers will work. Low complexity DNA e.g., plasmids or previously amplified DNA, can be successfully amplified with shorter primers. This is because the template itself is more "selective" than, for example, a total genomic DNA sample, with a large proportion of the sample consisting of the template of interest. For this reason, the chances of the primers annealing to and amplifying an unwanted template are minimised. Longer primers are often favoured to improve specificity. (However, it is unlikely that primers longer than 30 bases will help improve specificity significantly – Ausubel *et al.*, 1991c). The length of a pair of primers should be similar, if not equal.

Annealing temperature

The annealing temperature of primers is determined by their length and base composition. Increasing this temperature enhances discrimination against incorrectly annealing primers and so reduces the extension of incorrect nucleotides at the 3′ end of the primer. It is considered advisable to choose primers whose melting temperature (*Tm*) i.e., the temperature at which half of the duplex is dissociated (Kimmel, 1987) are between 55°C and 75°C (Rappolee, 1990). The appropriate PCR annealing temperature may be equal to the melting temperature or 5°C below this temperature (Innis and Gelfand, 1990). The *Tm* for a given primer can be estimated using the following equation (Bej *et al.*, 1991):
Tm = 2°(no. of A+T residues) + 4°(no. of G+C residues)

Base composition

Whenever possible, primers should have a balanced G/C and A/T concentration. Primers with a random base distribution should be selected, avoiding stretches of polypurines or polypyrimidines. This, again, is to try to prevent nonspecific binding e.g., stretches of T's binding to poly $(A)^+$ tails, and to inhibit secondary structure formation. A pair of primers should not be complementary to each other, especially at their 3′ end, to avoid the formation of "primer-dimers". (This results when the 3′ end of one primer anneals to the 3′ end of the other primer, forming an amplified product which will compete with the target of interest in the PCR; as well as being, basically, a waste of primers). Individual primers should not be self-complementary e.g., they should not possess palidromes. This is to avoid secondary structure formation.

Location of primers on template

- Primers should frame a sequence as far 3′ as possible so that cDNA produced by oligo (dT) priming need not be full-length to act as template in the PCR (Rappolee, 1990). However, when the genomic structure is not known, primers separated by 300 to 400 bases in the 5′ portion of the coding region should be chosen. This is because exons larger than 300 bases in this area are fairly rare in vertebrates. Avoiding areas which may possibly be degraded in mRNA should be a main consideration (Hawkins, 1988).
- Where possible, areas that are likely to present problems of secondary structure (as detected by a number of computer programs e.g., OLIGO (Rychlik and Rhoads, 1989) should be avoided.
- Primers should flank a sequence that crosses an intron so that DNA contamination can be diagnosed.
- Primers should span introns so that primer-DNA annealing will not occur.
- Primers should frame a sequence with a diagnostic restriction site for validation purposes.
- Primers should frame a sequence that covers a cDNA insert possessed by the laboratory, whenever possible, so that the cDNA can be used as a positive control (discussed above) and/or for validation by Southern blot analysis.
- Primers can be chosen specifically for a single member of a gene family by choosing an area

that is unique for that member. Alternatively, a region conserved by a particular gene family may be selected for a more general study of the family. Primers can be chosen for regions of interspecies homology so that the same set of primers can be used for each species or so that primers are sufficiently complementary to hybridise with cDNA from a species with unknown sequences of the gene of interest. Alternatively, primers unique to a particular species may be chosen.

Sequence banks

Primers should be selected in such a way as to minimise the possibility of their binding to other templates present in the specimen. If one primer binds to an extra template this will result in a waste of primers, etc. However, if both primers recognise and bind to the extra template, not only will reaction components be wasted, but an extra region may be amplified unintentionally.

To prevent this happening, the primer sequences can be sent to a DNA data-base e.g., EMBL data bank. (See Appendix 2). At EMBL, the submitted sequence is compared to all the stored sequences, and the first 50 sequences for which it shows strongest homology are listed. For the first 30 of these, the sequences are aligned, their percentage homology calculated and the position on the template shown.

It must be remembered, however, that this is not a comprehensive list, but *only* the first 50 sequences of common homology. Other templates not listed could possibly share enough sequence homology to allow the primers anneal.

Examples of primers chosen in our laboratory

A number of primers have been chosen in this laboratory to enable mRNA of multidrug resistance (MDR) related factors to be detected. All primers do not fit all suggested criteria; primers chosen represent our best attempts to compromise between the conflicting requirements of the different criteria. See Table 1 for details.

MDR1

The primers published by Noonan *et al.* (1990) are both 100% homologous to the *MDR1* mRNA template, resulting in a region of 157 bp being amplified. A diagnostic restriction enzyme recognition site for this *MDR1* region was chosen in this laboratory. Digestion with Mae 1 results in products of 84 bp and 73 bp. The primers are equal in length and the annealing temperature for both fit into the recommended range of 55°C to 75°C. Neither primer contains palindromes or stretches of polypurines or polypyrimidines. The region amplified by these primers is from position 2,596 to 2,752 bp and the poly $(A)^+$ tail commences at position 4,223 bp. The primers flank a sequence that crosses an intron so that DNA contamination can be diagnosed. If the relevant area of DNA was amplified, a band of 1,257 bp would result in comparison to a band of 157 bp from cDNA. The primers were selected to amplify only *MDR1* (avoiding *MDR3*). Both primers share homology with a number of other human nucleotide sequences. However, their only common homologies are *MDR1* mRNA (for which they were selected) and *MDR1* gene. By virtue of the fact that the selected region has an intron present in the DNA, if contaminating DNA is amplified, it should be possible to identify it by its longer size in comparison to the amplified region of cDNA.

Topoisomerase II, topoisomerase IIα and topoisomerase IIβ

Two isozymes of Topo II have been identified in mammalian cells – Topo IIα (170 kDa) and Topo IIβ (180 kDa). These enzymes differ not only in size but in other biochemical respects, such as the sequence specificity of binding to DNA (Drake *et al.*, 1989; Jenkins *et al.*, 1992). Primers were chosen which allow amplification of Topo II in general by selecting a region that is common to both isozymes. Specific primers were also chosen for Topo IIα and Topo IIβ so that their levels can be studied independently. This allows the levels of Topo IIα and Topo IIβ to be analysed individually, the overall levels of Topo II evaluated, and the results compared.

Table 1. Primers to amplify cDNA formed by reverse transcription on mRNA templates of MDR related factors

		Primer length (bases)	A+T: G+C	*Tm*	Amplified RNA sequence length (bases)	Diag-nostic R.E.	Restriction product length (bases)	Corresponding DNA length (bases)	Location on template
MDR1	(a)*	20	10:10	60	157	Mae 1	84 + 73	1,257	2596 – 2615
	(b)*	20	11:9	56				*(Chen et al., 1990)*	2733 – 2752
									(Chen et al., 1986 ; 1990)
MDR 3	(a)	24	16:8	64	321	Ssp 1	194 + 127	N.K.	2580 – 2603
	(b)	24	15:9	66					2877 – 2900
									(van der Bliek et al., 1988)
MRP	(a)	21	13:8	58	203	Bvu 1	129 + 74	N.K.	1317 – 1337
	(b)	21	11:10	62		Ban II	129 + 74		1499 – 1519
						Apa 1	129 + 74		*(EMBL no. L05628)*
Topo I	(a)	30	17:13	86	180	EcoN 1	115 + 65	2,280 bp	1123 – 1152
	(b)	27	16:11	82				*(D'Arpa et al., 1988;*	1276 – 1302
								Kunze et al., 1991)	*(D'Arpa et al., 1988)*
Topo II	(a)	18	10:8	52	216	Hind III	129 + 87	N.K.	1395 – 1412
	(b)	26	16:10	72					1585 – 1610
									(Tsai-Pflugfelder et al., 1988)
Topo IIα	(a)	21	11:10	62	139	Alu 1^	87 + 52	N.K.	4052 – 4072
	(b)	26	12:14	80					4165 – 4190
									(Tsai-Pflugfelder et al., 1988)
Topo IIβ	(a)	30	19:11	82	118			N.K.	4335 – 4364
	(b)	28	18:10	76		Hph 1^	69 + 49		4425 – 4452
									(Jenkins et al., 1992)
									(EMBL no. X68060)
GST π	(a)	18	8:10	56	270	Sty 1	171 + 99	749	58 – 75
	(b)	24	14:10	68				*(EMBL no. X08058)*	304 – 327
									(Moscow et al., 1989)
GST α	(a)$	30	18:12	84	330	Hind III	206 + 124	4,630	235 – 264
	(b)	30	14:16	92		Hsu 1	206 + 124	*(Klöne et al., 1992;*	535 – 564
						Msp 1	228 + 102	*Tu and Qian, 1986;*	*(Tu and Qian, 1986)*
								EMBL no.s X65726, X65734)	
CYP1A1	(a)	21	12:9	60	327	Hinc II	246 + 81	755	1021 – 1041
	(b)	21	14:7	56				*(Jaiswal et al., 1985a)*	1327 – 1347
									(Jaiswal et al., 1985b)
β-Actin	(a)	29	16:13	84	383	Hin f 1	193 + 190	590	619 – 647
	(b)	22	12:10	64		Hph 1	266 + 117	*(EMBL no. M10277)*	980 – 1001
									(EMBL no. X00351, M10277)
β_2.Micro-	(a)	20	11:9	58	114	Mse 1	68 + 46	1,974	271 – 290
globulin*	(b)	20	12:8	56				*(Güssow et al., 1987)*	365 – 384
									(Güssow et al., 1987)

Topo II

When choosing primers for Topo II in general, the C-terminus and the extreme N-terminal regions were avoided as these are the main areas where IIα and IIβ diverge in structure. The chosen primers are both 100% homologous to the Topo II mRNA template, resulting in a region of 216 bp being amplified. Digesting with Hind III restriction enzyme produces bands of 129 bp and 87 bp. The primers are not equal in length. However, this seemed unavoidable when attempting to maximise their specificity. Many sequences were chosen as potential primers, but the primer pair 1395–1412 and 1585–1610 (Tsai-Pflugfelder *et al.*, 1988) were favoured by comparison to the others as they are independently homologous to fewer other human sequences and they do not share common homologies other than Topo II. The annealing temperature for Topo II(a) primer is slightly below the recommended range; the Topo II(b) primer is within this range. Neither primer contains palindromes or stretches of polypurines or polypyrimidines. Due to the unavailability of the Topo II genomic sequence, it is uncertain whether or not these primers span or cross introns.

Topo IIα (IIα)

The longest region of divergence between IIα and IIβ was considered to be the best area to choose specific primers for IIα and IIβ respectively. This is the C-terminal region (Jenkins *et al.*, 1992). As the region of divergence between IIα and IIβ is relatively small, primers that span the longest region possible were favoured. This is so that the product and the restriction enzyme digestion product can be easily detected on a gel. A number of potential 3′ primers were not chosen for IIα as they contained stretches of polypurines and polypyrimidines. To avoid these and other problem areas, primers 4052–4072 and 4165–4190 (Tsai-Pflugfelder *et al.*, 1988) were chosen. Although the 139 bp region amplified is homologous with that of IIβ to a small extent (Jenkins *et al.*, 1992) the primers are specific for IIα. Digestion with Alu 1 restriction enzyme results in stretches of 87 bp and 52 bp being produced. The annealing temperature for the 5′ primer is within the recommended range, but that for the 3′ primer is slightly high. Both primers are independently homologous to a number of other human sequences, but do not seem to share common homologies.

IIβ (IIβ)

The IIβ primers (4335–4364 and 4425–4452, Jenkins *et al.*, 1992; EMBL accession no. X68060) amplify a 118 bp region unique to IIβ. Digestion with Hph 1 restriction enzyme results in 69 bp and 49 bp products. The primers are quite long – 30 bp and 28 bp, respectively, to increase their specificity. They have an annealing temperature slightly above the recommended range. Neither primer contains palindromes or lengths of polypurines or polypyrimidines. Both these primers, too, are independently homologous to a number of human sequences, but they apparently do not share common human homologies (with the exception of IIβ).

β-Actin

As discussed in "Quantitative PCR", it is very difficult to find primers that are specific for β-actin

Notes to Table 1:

* This primer was not selected in our laboratory, but was published by Noonan *et al.* (1990).
When numbering all primers "A" of the start codon was considered as position "1".
^ β_2-microglobulin may be used as an internal control when amplifying this sequence if diagnostic digestion with a restriction enzyme is to be included. This is because the region amplified with the primers chosen for β-actin has also a site present for this restriction enzyme, whereas the β_2-microglobulin region does not.
$ GST-α primers were chosen on the human liver cDNA templates (clone pGTH1 and pGTH2) and the human kidney cDNA template (GSTα 12 k), according to the sequences published by Klöne and co-workers (1992). GSTα (a) primer is located on exon 4 and primer (b) spans from exon 6 to exon 7 (Klöne *et al.*, 1992).
N.K. = Not known.

and do not recognise other actin forms. The β-actin primer pair published here are also homologous to mutant β-actin (EMBL accession no. X63432). This was unavoidable when the other actin forms were avoided and the guidelines for choosing primers were considered, as the normal (EMBL accession no. X00351) and the mutant forms differ only by one nucleotide in the amplified region.

Glutathione-S-transferase π (GST π)

The cDNA sequence of GST π (Moscow *et al.*, 1989) apparently differs from that of human myocardial fatty acid ethyl esters-III (FAEE-III) (Bora *et al.*, 1991) by only six nucleotides. As primers specific for GST π could not be found, an effort was made to find a restriction enzyme recognition site that is present in the GST π amplified region but not in FAEE-III, or *vice versa* (due to the four nucleotide sequence difference in this region). However, such a site was not identified. According to Bora and co-workers (1991) the sequence differences between GST π and FAEE-III are not due to sequence errors, but reflect inherent functional differences. This should be considered when studying certain tissue types as expression of FAEE-III mRNA has been reported in normal human liver, placenta and heart (Bora *et al.*, 1991).

From the above examples, it is obvious that choosing specific primers representing all the suggested guidelines is not possible in all situations.

Appendix 1

Preparation for RNA extraction

RNases are ubiquitous in the environment and so precautions must be taken to avoid their introduction while extracting RNA.

- General laboratory glassware and plasticware are often contaminated by RNases. Glassware should be treated by baking at 180°C for 8 h, or more. Plasticware should be rinsed with chloroform. All spatulas which may come in contact with any of the solution components should be baked (as for glassware), the chemicals weighed out onto baked tin-foil and a stock of chemicals kept separate, for "RNA extraction only", from general stocks.
- Sterile, disposable plasticware is essentially free of RNases and can be used for the preparation and storage of RNA without pretreatment. However, polyallomer ultracentrifuge tubes, eppendorf tubes, pipette tips, etc. for the RT reaction (and any other glassware not baked) can be filled with DEPC* (0.1% (v/v) in water) and stored at 37°C for 2 h (or more), rinsed several times with sterile ultra-pure water and then autoclaved (Sambrook *et al.*, 1989). We have not found it necessary to DEPC-treat eppendorfs and tips in our laboratory.
- Solutions that come in contact with the RNA, after lysing the cells in guanidinium solution, should all be prepared in baked glassware, made from sterile ultra-pure water and treated with 0.1% DEPC before autoclaving, with the exception of Tris containing solutions (DEPC reacts with amines and so is inactivated by Tris).
- Gloves should be worn at all times to protect both the operator and the experiment. This, again, prevents the introduction of RNases and foreign RNA/DNA into the RT and PCR reactions. Gloves should be changed frequently.

Protocol 1: total RNA isolation

The following procedure is conducted in a laminar flow to maintain sterile conditions and prevent the introduction of contaminants.

Cultured cells

Cells are grown in 135 mm diameter petri-dishes, until approximately 80% confluent. The medium is removed. The following procedure is then followed:

1. Rinse twice in phosphate-buffered saline (PBS).
2. Lyse in guanidinium thiocyanate (GnSCn) solution; (It is generally recommended that the cell lysate be homogenised for 1—2 mins. This serves to shear the nuclear DNA and so prevent the formation of an impenetrable layer at the top of the cushion of CsCl which might block sedimentation of the RNA to the bottom of the centrifuge tube).
3. Layer onto 5.7 M cesium chloride (CsCl) cushion in ultracentrifuge tube.
4. Centrifuge at 26,000 rpm at 15°C for 21—24 h; (Centrifuging in a swinging-bucket rotor causes the RNA to be deposited at the bottom of the centrifuge tube and so prevents it coming in contact with the cell lysate).
5. Remove GnSCn and the "jelly-like" layer just below the GnSCn/CsCl interface by aspiration; change pasteur pipette and remove all but approximately 1 ml of the CsCl layer (to prevent disruption of the RNA pellet).
6. Quickly invert the tubes to drain.
7. Cut off the bottom of the tube containing the RNA pellet using a heated scalpel blade.

---

*DEPC (diethyl procarbonate) is a strong, but not absolute, inhibitor of RNases. *Warning*: DEPC is suspected to be carcinogenic – due care should be exercised.

8. Rinse the pellet with 95% ethanol at room temperature.
9. Resuspend the RNA pellet in 200 µl of DEPC-treated water by gently pipetting it up-and-down, whilst keeping tube on ice.
10. Rinse suspension remains into an eppendorf tube using a further 200 µl of DEPC-treated water.
11. Precipitate RNA out of solution by the addition of 3M sodium acetate (to result in a final concentration of 0.3 M) and two volumes of ice-cold absolute ethanol (overnight at –20°C or 30 min at –80°C).
12. Pellet RNA by centrifuging at 4°C; remove supernatant and dry pellet briefly. (Ensure that the pellet does not dry completely as this greatly decreases its solubility. The solubility of RNA can be improved by heating to 55–60°C with intermittent vortexing or by passing the RNA solution through a pipette tip, if necessary).
13. Resuspend pellet in 50 µl DEPC-treated water. (Susan Mc Donnell, pers. comm).

Whole tissue

In the case of whole tissue, the tissue is stored in liquid nitrogen immediately upon its removal from the body, until ready to be analysed. At this stage it is pulverised very finely using a mortar and pestle and homogenised in denaturing (GnSCn) solution, without thawing. The above procedure (for cultured cells) is then followed from (3).

RNA quantitation

RNA is quantified spectrophotometrically at 260 nm and 280 nm. The A_{260}/A_{280} ratio of RNA is approximately 2. Partially solubilised RNA has a ratio < 1.6 (Ausubel *et al.*, 1991a).

The yield of RNA from most lines of cultured cells is 100–200 µg per 90 mm plate (Sambrook *et al.*, 1989).

(For further information see: Sambrook *et al.*, (1989) Ch. 7 "Extraction, Purification and Analysis of Messenger RNA from Eukaryotic Cells" In: Molecular Cloning: A Laboratory Manual. (second addition) pp. 7.2–7.83 and Ausubel *et al.* (1991a) Ch. 4. "Preparation and Analysis of RNA". In: Current Protocols in Molecular Biology, Vol. 1 pp. 4.0.1–4.10.9).

Protocol 2: typical RT reaction

Component:	Volume:
Oligo (dT) 12–18 primers (1 µg/µl)	1.0 µl
Total RNA (1 µg/µl)	1.0 µl
H_2O	3.0 µl

70°C for 10 min; chill on ice.
(This should get rid of any RNA secondary structure formation and allow the oligo (dT) primers to bind to the poly $(A)^+$ region of the mRNA).

Then add:	Volume:
5X Buffer	4.0 µl
DTT (100 mM)	2.0 µl
RNasin (40 U/µl)	1.0 µl
dNTPs (10 mM each)	1.0 µl
H_2O	6.0 µl
MMLV-RT (200 U/µl)	1.0 µl

37°C for 1 h, followed by 95°C for 2 min (Susan Mc Donnell, pers. comm).

5X Buffer

250 mM Tris-HCl, pH 8.3; 375 mM KCl; 15 mM $MgCl_2$.

The successful use of the PCR buffer instead of this 5X buffer in the reverse transcriptase reaction has been reported by Kawasaki (1989). Better results were achieved using the PCR buffer, at least when reverse transcribing short RNA sequences. This, it is suggested, simplifies the protocol by eliminating the possibility of differences in optimal monovalent and divalent ionic strengths between the two buffers.

RNasin

An inhibitor of ribonuclease enzymes is included in the RT reaction. This is to prevent degradation of RNA by ribonucleases. RNasin is a broad spectrum ribonuclease inhibitor which acts by binding noncovalently to ribonucleases enzymes. RNasin is unstable at temperatures of 50°C or above. As ribonucleases are capable of retaining their degradative properties under denaturing conditions, such temperatures should be avoided for the RT reaction to prevent release of active ribonuclease enzymes.

Dithiothreitol (DTT)

The concentration of DTT is critical as RNasin requires 1mM DTT to prevent its dissociation from the RNases (Susan Mc Donnell, pers. comm).

Deoxynucleotide triphosphates (dNTPs)

The dNTPs (dATP, dCTP, dGTP and dTTP) are aliquoted as a mix at a concentration of 10 mM each. These are the precursors of the cDNA strand.

Moloney murine leukemia virus reverse transcriptase (MMLV-RT)

MMLV-RT is the reverse transcriptase enzyme used to form cDNA on the primed mRNA template. (MMLV H^- RT, a recombinant enzyme has been engineered to have no RNase H activity. This is claimed to result in greater yields of cDNA including full-length sequences).

RT-PCR can be simplified by the fact that *Taq* DNA polymerase also exhibits reverse transcriptase activity at 68°C

(Jones and Foulkes, 1989). RT-PCR using one enzyme was conducted by Shaffer *et al.* (1990) in a study of interleukin-2 mRNA in gibbon T cells.

The use of *Taq* DNA polymerase as the enzyme catalysing the RT reaction may have some advantages over MMLV-RT. The RT reaction in the presence of reverse transcriptase enzyme generally takes approximately an hour. Using *Taq* DNA polymerase, cDNA can be formed in a number of minutes (Shaffer *et al.*, 1990; Singer-Sam *et al.*, 1990). *Taq* DNA polymerase enzyme is much more heat-stable than MMLV-RT, which allows the reaction to be conducted at higher temperatures. This should increase the primer stringency and help eliminate RNA secondary structure. However, optimum reverse transcription using *Taq* DNA polymerase may require the addition of magnesium (Bej *et al.*, 1991).

Trouble-shooting

Rappolee (1990) reported first-strand cDNA synthesis by reverse transcription to be a technique of variable efficiency between preparations and between RNA species. To overcome secondary structure problems in RNA, a number of recommendations are made. These include: increasing the RT reaction temperature from 37 to 55°C (*Taq* DNA polymerase seems to be more stable at high temperatures than reverse transcriptase enzymes); increasing the reverse transcription enzyme concentration from 1X to 3X or more and using random hexanucleotide primers and/or 3′ antisense oligonucleotide priming. To improve sensitivity, the first strand synthesis may be repeated after heat denaturation. Reverse transcriptases have some amount of heat stability, but it is often advisable to add a fresh aliquot of enzyme for greater efficiency if a second cycle of reverse transcriptase is included. (This further addition of enzyme may not be necessary if *Taq* DNA polymerase reverse transcriptase activity is taken advantage of).

Polymerase chain reaction

Once the cDNA copy has been created using the mRNA template, a typical PCR reaction is conducted as below. Concentrations, temperatures, volumes and times may need to be varied to optimise reaction conditions to suit the need of a particular template and primer pair. (For further details see: Innis and Gelfand, 1990; Taylor, 1991).

Protocol 3: PCR reaction

Component:	Volume:
H_2O	24.5 µl
10X Buffer ($MgCl_2$ free)	5 µl
$MgCl_2$ (25 mM)	3 µl
dNTP (1.25 mM)	8 µl
Target primer (1) (250 ng/µl)	1 µl
Target primer (2) (250 ng/µl)	1 µl
Internal control primer (1) (250 ng/µl)	1 µl
Internal control primer (2) (250 ng/µl)	1 µl
Taq DNA polymerase (5 U/µl)	0.5 µl

Template:	Volume:
cDNA (from RT reaction)*	5 µl

*Heat cDNA to 95°C for 3 min, followed by cooling on ice, before adding to the PCR reaction components.

Add a drop of mineral oil;

Amplify using the following procedure:

- 95°C for 1.5 min (denature ds DNA).
- 30 cycles: 95°C for 1.5 min (denature);
  55°C for 1 min (anneal);
  72°C for 3 min (extend).
- 72°C for 7 min (extend).
- Hold at 4°C.

(Susan Mc Donnell, pers. comm).

The result of the RT-PCR can be analysed by agarose gel electrophoresis. Ethidium bromide intercalates with the cDNA forming a product that is readily visible under ultraviolet trans-illumination.

10X buffer

100 mM Tris-HCl (pH 9.0 at 25°C); 50 mM KCl; 1% Triton X-100.

The 10X buffer is generally supplied with the *Taq* DNA polymerase enzyme.

$MgCl_2$: (25 mM)

Magnesium chloride is usually supplied with the *Taq* DNA polymerase enzyme, but as a separate entity to the 10X buffer. The magnesium chloride concentration in the buffer has been found to greatly influence the PCR result. It is generally advised to vary the concentration of magnesium chloride (usual range 1–8 mM) to determine the optimum concentration of this component.

Deoxynucleotide triphosphates (dNTPs)

The dNTPs for the PCR are aliquoted as a mix (dATP, dCTP, dGTP and dTTP) at a final concentration of 1.25 mM each. These are the required precursors for the synthesis of new cDNA strands.

Primers (see "criteria to follow when choosing primers")

A primer pair is included in the PCR which is specific for the target cDNA region of interest. It is also advisable to include a primer pair which will result in amplification of an ubiquitously occurring cellular component. This will serve as an internal control (see "Quantitative PCR").

Taq DNA polymerase enzyme

(Previously discussed).

Mineral oil

The purpose of the mineral oil is to prevent evaporation of the reaction components when subjected to the high temperatures required for melting the double-stranded DNA to single-strands.

Solutions for total RNA extraction

4M guanidinium thiocyanate:

Guanidinium thiocyanate salt (Sigma G-6639)	50 g
N lauroyl sarcosine (Sigma L-5125)	0.5 g
1 M Na citrate, pH 7	5 ml

Bring to 100 ml with H_2O; stir until dissolved (can heat gently); check pH is approximately 7; filter through 0.45 μ filter; store at room temperature in dark bottle.

Before use add:

β-mercaptoethanol (Sigma M6250)	700 μl/100 ml
Antifoam A (30%) (Sigma A-5758)	330 μl/100 ml

5.7M CsCl

CsCl (C-3032)	95.8 g
1M Na citrate, pH 7	2.5 ml

Bring to 100 ml with H_2O; filter sterilize; DEPC-treat and autoclave; store at room temperature.

1M sodium citrate

Na citrate (RDH 32320)	29.4 g

Bring to 80 ml with H_2O; pH to 7 with HCL; bring to 100 mls; filter sterilise; store at room temperature.

Note: H_2O used in all solutions should be sterile ultra-pure H_2O. (Susan McDonnell, pers. comm.)

Appendix 2

Using cDNA sequence data-bases to check uniqueness of primers

The most critical element in choice of primers for PCR is obviously that the primers are unique i.e., that they do not cross-react with sequences likely to be present in the mRNA of the cells/tissues being studied. This involves choice of primers using the criteria described above, and checking via a DNA data-base (e.g., EMBL, GenBank) what similar sequences exist. This approach is essential if no references are available on RT-PCR of the mRNA in question. Even where references do exist, there is need for checking because:

- the cDNA data-base is expanding rapidly and new cross-reactions may have been discovered since the choice of primers was published;
- occasionally, published primers have not been well chosen and cross-reactions have been overlooked; also, occasionally, probably due to typographical errors, incorrect primer sequences are published.

Because of the excellent search facilities available, checking sequence uniqueness of primer sequences is a straightforward procedure. Such searching is routine in many laboratories, but for readers without experience in this area we wish to emphasize here how simple the procedures are, and that no great expertise in either molecular biology or computing is required.

For example, EMBL DNA data bases can be accessed easily via E-Mail by linking to large mainframe computers (e.g., VAX) which are accessible to most universities or hospital laboratories. In our laboratory, we have installed software (KERMIT) to allow our IBM-PCs to be used as VAX terminals for E-Mail. Once E-Mail has been accessed on the PC (through use of KERMIT commands and the appropriate user number and password given to us by the Central Computer Office in the University) we can send a primer sequence to be checked (e.g., to EMBL) by the following procedure:

Note:
Italics = prompts received;
Bold = to be typed in;
[R] = "return" key;
CTRL/Z = press "control" key and "Z", simultaneously.

Username: **xxxxxxxx** [R]
Password: **yyyyyyyy** [R]

$ **mail**
Mail> **send**
To: **In%"Fasta@embl-heidelberg.de"** [R]
Subj: (not necessary to choose subject) [R]
Enter your message below. Press CTRL/Z when complete or CTRL/C to quit:
Lib email [R] (can choose to check a particular library in this way)
Align 20 [R] (can choose how many alignments required. Default is 30)
Seq [R]
CCCATCATTGCAATAGCAGG[R] (The example given here is the *MDR1* primer (a) sequence – see Table 1).
end [R]
CTRL/Z [R]

If you have an accession number e.g., for a cDNA sequence, and wish to obtain the cDNA sequence, the address and procedure is slightly different:

Mail> **send** [R]
To: **In%"Netserv@embl-heidelberg.de"** [R]
Subj: [R]

Enter your message below. Press CTRL/Z when complete or CTRL/C to quit:
GET NUC:X00351 [R] (The example shown here is the EMBL accession number of human mRNA for β-actin).
CTRL/Z

To access returned mail:

Mail> **dir** [R] (will give a list of your directory of returned messages)
*Mail>***3** [R] (choose the number of any of the listed return messages e.g., no. 3)

By pressing [R] you can read through the message or

Mail> **print** [R]
CTRL/Z (if want message printed).

After use, when have exited by using CTRL/Z command

$ **lo** (to exit system).

Depending on the pressure on the EMBL facility at the time, a list of related sequences will be sent back to us in minutes or hours. (Note: The location and hence E-Mail address for EMBL is scheduled to change in mid-1995).

A more detailed account of facilities available and relevant software is beyond the scope of this review. The reader is referred to recent summaries of services provided via EMBL (Rice *et al.*, 1993) and Genbank (Benson *et al.*, 1993) to an introductory book on the subject (Bishop and Rawlings, 1987) and to supplement *19*, "Current Protocols in Molecular Biology" (1992) Unit 7.7 which provides a good introduction to databases and to some software useful for sequence analysis and comparison.

References

Alwine JC, Kemp DJ and Stark GR (1977) Method for detection of specific RNAs in agarose gels by transfer to diazobenzyloxymethyl-paper and hybridisation with DNA probes. Proc. Nat. Acad. Sci. USA 74(12): 5350–5354.

Ausubel MA, Brent R, Kinston RE, Moore DD, Seidman JG, Smith JA and Struhl K (eds) (1991a) Current protocols in molecular biology (pp. 4.0.1–4.1.0.9). Greene Publishing Associates and Wiley Interscience, New York.

Ausubel MA, Brent R, Kinston RE, Moore DD, Seidman JG, Smith JA and Struhl K (eds) (1991b) Current protocols in molecular biology (pp. 14.0.1–14.6.13). Greene Publishing Associates and Wiley Interscience, New York.

Ausubel MA, Brent R, Kinston RE, Moore DD, Seidman JG, Smith JA and Struhl K (eds) (1991c) Current protocols in molecular biology (pp. 15.0.1–15.5.16). Greene Publishing Associates and Wiley Interscience, New York.

Avraham H, Vannier E, Chi SY, Dinarello CA and Groopman JE (1992) Cytokine gene expression and synthesis by human megakaryocytic cells. Int. J. Cell Cloning 10: 70–79.

Becker-Andre M and Hahlbrock K (1989) Absolute mRNA quantification using the polymerase chain reaction (PCR). A novel approach by a PCR aided transcript titration assay (PATTY). Nuc. Acids Res. 1722: 9437–9446.

Bej AK, Mahbubani MH and Atlas RM (1991) Amplification of nucleic acids by polymerase chain reaction (PCR) and other methods and their applications. Crit. Rev. Biochem. Mol. Biol. 26(3/4): 301–334.

Benson D, Lipman DJ and Ostell J (1993) GenBank. Nuc. Acids Res. 21: 2963–2965.

Bishop MJ and Rawlings CJ (eds) (1987) Nucleic acids and protein sequence analysis: a practical approach. IRL Press, Oxford.

Bloch W (1991) A biochemical perspective of the PCR reaction. Biochemistry 30(11): 2735–2747.

Bora PS, Bora NS, Wu X and Lange LG (1991) Molecular cloning, sequencing, and expression of human myocardial fatty acid ethyl ester synthase-III cDNA. J. Biol. Chem. 266: 16774–16777.

Bourhis J, Goldstein LJ, Riou G, Pastan I, Gottesman MM and Benard J (1989) Expression of a human multidrug resistance gene in ovarian carcinomas. Cancer Res. 49: 5062–5065.

Brenner CA, Tam AW, Nelson PA, Engleman EG, Suzuki N, Fry KE and Larrick JW (1989) Message amplification phenotyping (MAPPing) a technique to simultaneously measure multiple mRNAs from small numbers of cells. Biotechniques 7: 1096–1103.

Byrne BC, Li JJ, Shinsky J and Poiesz BJ (1988) Detection of HIV-1 RNA sequences by *in vitro* DNA amplification. Nuc. Acids Res. 16(9): 4165.

Chelly J, Kaplan J-C, Maire P, Gautron S and Kahn A (1988) Transcription of the dystrophin gene in human muscle and nonmuscle tissue. Nature 333: 858–860.

Chen C-J, Chin JE, Ueda K, Clark DP, Pastan I, Gottesman MM and Roninson IB (1986) Internal duplication and homology with bacterial transport proteins in the *mdr1* (p-glycoprotein) gene from multidrug-resistant human cells. Cell 47: 381–389.

Chen C-J, Clark D, Ueda K, Pastan I, Gottesman MM and Roninson IB (1990) Genomic organization of the human multidrug resistance (*MDR1*) gene and origin of P-glycoproteins. J. Biol. Chem. 265: 506–514.

Chin JE, Soffir R, Noonan KE, Choi K and Roninson I (1989) Structure and expression of the human MDR (P-glycoprotein) gene family. Mol. Cell Biol. 9(9): 3808–3820.

Chirgwin JJ, Przybyla AE, MacDonald RJ and Rutter WJ (1979) Isolation of biologically active ribonucleic acid from sources enriched in ribonuclease. Biochemistry 18: 5294–5299.

Choi K, Chen C-J, Kriegler M and Roninson IB (1988) An altered pattern of cross resistance in multidrug resistant human cells results from spontaneous mutations in the mdr1 (P-glycoprotein) gene. Cell 53: 519–529.

Clarkson T, Güssow D and Jones PT (1991) General application of PCR to gene cloning and manipulation. In: McPherson MJ, Quirke P and Taylor GR (eds) PCR. A Practical

Approach (pp. 187–214). IRL Press, Oxford University Press.

Cole SPC, Chanda ER, Dicke FP, Gerlach JH and Mirski SEL (1991) Non-P-glycoprotein-mediated multidrug resistance in a small cell lung cancer cell line: Evidence for decreased susceptibility to drug-induced DNA damage and reduced levels of topoisomerase II. Cancer Res. 51: 3345–3352.

Compton T (1990) Degenerate primers for DNA amplification. In: Innis MA, Gelfand DH, Sninsky JJ and White TJ (eds) PCR Protocols: A Guide to Methods and Applications (pp. 39–45). Academic Press Inc., San Diego, CA.

Cox RA (1968) The use of guanidinium chloride in the isolation of nucleic acids. Methods in Enzymol. 12b: 120–129.

D'Arpa P, Machlin PS, Ratrie III H, Rothfield NF, Cleveland DW and Earnshaw WC (1988) cDNA cloning of human DNA topoisomerase I: Catalytic activity of a 67.7-kDa carboxyl-terminal fragment. Proc. Natl. Acad. Sci. USA 85: 2543–2547.

Doyle A, Martin WJ, Funa K, Gazdar A, Carney D, Martin SE, Linnoila I, Cuttitta F, Mulshine J, Bunn P and Minna J (1985) Markedly decreased expression of class I histocompatibility antigens, proteins, and mRNA in human small-cell lung cancer. J. Exp. Med. 161: 1135–1151.

Drake FH, Hofmann GA, Bartus HF, Mattern MR, Crooke ST and Mirabelli CK (1989) Biochemical and pharmacological properties of p170 and p180 forms of topoisomerase II. Biochemistry 28: 8154–8160.

Eckert KA and Kunkel TA (1991) The fidelity of DNA polymerases used in the polymerase chain reactions. In: McPherson MJ, Quirke P and Taylor GR (eds) PCR. A Practical Approach (pp. 225–244). IRL Press, Oxford University Press.

Engel JN, Gunning PW and Kedes L (1981) Isolation and characterization of human actin genes. Proc. Natl. Acad. Sci. USA 78: 4674–4678.

Firtel RA (1981) Multigene families encoding actin and tubulin. Cell 24: 6–7.

Fojo AJ, Ueda K, Slamon DL, Poplack DG, Gottesman MM and Pastan I (1987) Expression of a multidrug resistance gene in human tumours and tissues. Proc. Nat. Acad. Sci. USA 84: 265–269.

Freifelder D (ed) (1983a) Eukaryotic viruses. In: Molecular Biology. A comprehensive introduction to prokaryotes and eukaryotes (pp. 839–921). Jones and Bartlett Publishers, Inc., Boston, Portola Valley.

Freifelder D (ed) (1983b) Transcription In: Molecular Biology. A comprehensive introduction to prokaryotes and eukaryotes (pp. 369–425). Jones and Bartlett Publishers, Inc., Boston, Portola Valley.

Fulton AB (1981) How do eucaryotic cells construct their cytoarchitecture? Cell 24: 4–5.

Fuqua SAW, Fitzgerald SD and McGuire WL (1990) A simple polymerase chain reaction method for detection and cloning of low abundance transcripts. Biotechniques 9: 206–211.

Gaudette MF and Crain WR (1991) A simple method for quantifying specific mRNAs in small numbers of early mouse embryos. Nuc. Acids Res. 19(8): 1879–1884.

Gekeler V, Weger S and Probst H (1990) MDR1/P-glycoprotein gene segments analysed from various human leukemic cell lines exhibiting different multidrug resistance profiles. Biochem. Biophys. Res. Commun. 169: 796–802.

Gilliland G, Perrin S and Bunn HF (1990a) Competitive PCR for quantitation of mRNA. In: Innis MA, Gelfand DH, Sninsky JJ and White TJ (eds) PCR Protocols: A Guide to Methods and Applications (pp. 60–69). Academic Press Inc., San Diego, CA.

Gilliland G, Perrin S, Blanchard K and Bunn HF (1990b) Analysis of cytokine mRNA and DNA: Detection and quantitation by competitive polymerase chain reaction. Proc. Nat. Acad. Sci. USA 87: 2725–2729.

Glisin V, Crkvenjakov R and Byus C (1974) Ribonucleic acid isolation by cesium chloride centrifugation. Biochemistry 13: 2633–2637.

Goldstein LJ, Galshi H, Fojo A, Willingham M, Lai SL, Gazdar A, Pirker R, Green A, Crist W, Brodeur GM, Lieber M, Cossman J, Gottesman MM and Pastan I (1989) Expression of a multidrug resistance gene in human cancers. J. Natl. Cancer Inst. 81: 116–124.

Gottesman MM and Pastan I (1988) The multidrug transporter, a double edged sword. J. Biol. Chem. 263: 12163–12166.

Gros P, Ben Neriah Y, Croop JM and Housman DE (1986) Isolation and expression of a cDNA (*mdr*) that confers multidrug resistance. Nature 323: 728–731.

Güssow D, Rein R, Ginjaar I, Hochstenbach F, Seemann G, Kottman A and Ploegh HL (1987) The human β_2-microglobulin gene: Primary structure and definition of the transcriptional unit. J. Immunol. 139: 3132–3138.

Hawkins JD (1988) A survey on intron and exon length. Nuc. Acids. Res. 16: 9983–9908.

Hegewisch-Becker S, Fliegner M, Tsuruo T, Zander A, Zeller W and Hossfeld DK (1993) P-glycoprotein expression in normal and reactive bone marrows. Br. J. Cancer 67: 430–435.

Higuchi R (1989) Simple and Rapid Preparation of Samples for PCR. In: Erlich HA (ed) PCR Technology. Principles and Applications for DNA Amplification (pp. 31–43). Stockton Press.

Higuchi R, Krummel B and Saiki RK (1988) A general method of *in vitro* preparation and specific mutagenesis of DNA fragments: study of protein and DNA interactions. Nuc. Acids Res. 16: 7351–7367.

Holzmayer TA, Hilsenbeck S, von Hoff DD and Roninson IB (1992) Clinical correlates of MDR1 (P-glycoprotein) gene expression in ovarian and small-cell lung carcinomas. J. Natl. Cancer Inst. 84(19): 1486–1491.

Hoof T, Riordan JR and Tummler B (1991) Quantitation of mRNA by the kinetic polymerase chain reaction assay: A tool for monitoring P-glycoprotein gene expression. Analytical Biochemistry 196: 161–169.

Horikoshi T, Danenberg KD, Stadlbauer THW, Volkenandt M, Shea LCc, Aigner K, Gustavsson B, Leichmas L, Frosing R, Ray M, Gibson NW, Spears CP and Dandenberg PV (1992) Quantitation of thymidylate synthetase, dihydrofolate reductase and DT-diaphorase: Gene expression in human tumors using the polymerase chain reaction. Cancer Res. 52:

10—116.

Innis MA and Gelfand DH (1990) Optimization of PCRs. In: Innis MA, Gelfand DH, Sninsky JJ and White TJ (eds) PCR Protocols: A Guide to Methods and Applications (pp. 3—27). Academic Press Inc., San Diego, CA.

Ivinson AJ and Taylor GR (1991) PCR in genetic diagnosis. In: McPherson MJ, Quirke P and Taylor GR (eds) PCR. A Practical Approach (pp. 15—27). IRL Press, Oxford University Press.

Jackson DP, Hayden JD and Quirke P (1991) Extraction of nucleic acid from fresh and archival material. In: McPherson MJ, Quirke P and Taylor GR (eds) PCR. A Practical Approach (pp. 29—50). IRL Press, Oxford University Press.

Jaiswal AK, Gonzalez FJ and Nebert DW (1985a) Human P_1—450 gene sequence and correlation of mRNA with genetic differences in benzo[a]pyrene metabolism. Nuc Acids Res. 13: 4503—4520.

Jaiswal AK, Gonzalez FJ and Nebert DW (1985b) Human dioxin-inducible cytochrome P_1—450: Complementary DNA and amino acid sequence. Science 228: 80—83.

Jeffreys AJ, Wilson V, Neumann R and Keyte J (1988) Amplification of human minisatellites by the polymerase chain reaction. Towards DNA fingerprinting of single cells. Nuc. Acids Res. 16: 10953—10971.

Jenkins JR, Ayton P, Jones T, Davies SL, Simmons DL, Harris AL, Sheer D and Hickson ID (1992) Isolation of cDNA clones encoding for the β isozyme of human DNA topoisomerase II and localisation of the gene to chromosome 3p24. Nuc. Acids Res. 20: 5587—5592.

Jones MD and Foulkes NS (1989) Reverse transcription of mRNA by *Thermus aquaticus* DNA polymerase. Nuc. Acids Res. 17: 8387—8388.

Juliano RL and Ling V (1976) A surface glycoprotein modulating drug permeability in chinese hamster ovary cell mutants. Biochim Biophys Acta 455: 152—162.

Kafatos FC, Weldon Jones C and Efstratiadis A (1979) Determination of nucleic acid sequence homologies and relative concentrations by a dot hybridisation procedure. Nuc. Acids Res. 7(6): 1541—1552.

Kawasaki ES, Clark SS, Coyne MY, Smith SI, Champlin R, Wilte ON and McCormack FP (1988) Diagnosis of chronic myeloid and acute lymphocyte leukemias by detection of leukemia-specific mRNA sequences amplified *in vitro*. Proc. Nat. Acad. Sci. USA 85: 5698—5702.

Kawasaki E (1989) Amplification of RNA sequences via complementary DNA (cDNA). In: Amplifications. A forum for PCR Users (Technical Bulletin) 3: 4—6.

Kawasaki ES (1991) Amplification of RNA In: Innis MA, Gelfand DH, Sninsky JJ and White TJ (eds) PCR Protocols. A Guide to Methods and Applications (pp. 21—27). Academic Press Inc., San Diego, CA.

Kimmel AR (1987) Selection of clones from libraries: Overview. In: Berger SL and Kimmel AR (eds) Methods in Enzymol. Guide to Molecular Cloning Techniques 152: 393—504.

Klöne A, Hussnatter R and Sies H (1992) Cloning, sequencing and characterization of the human alpha glutathione S-transferase gene corresponding to the cDNA clone pGTH2. Biochem. J. 285: 925—928.

Kramer R, Weber TK, Morse B, AreciR, Staniunas R, Steele JG and Summer Layes IC (1993) Constitutive expression of multidrug resistance in human colorectal tumours and cell lines. Br. J. Cancer 67: 959—968.

Kunze N, Yang G, Dolberg M, Sundarp R, Knippers R and Richter A (1991) Structure of the human type I DNA topoisomerase gene. J. Biol. Chem. 266: 9610—9616.

Larrick JW (1992) Message amplification phenotyping (MAPPing) – principles, practice and potential. Trends BioTechnology 10: 146—152.

Leavitt J, Bushar G, Kakunaga T, Hamada H, Hirakawa T, Goldman D and Merril C (1982) Variations in expression of mutant β actin accompanying incremental increases in human fibroblast tumorigenicity. Cell 28: 259—268.

Lee MS, Chang KS, Cabanillas F, Freireich EJ, Trijillt JM and Stass SA (1987) Detection of minimal residual cells carrying the +(14;18) by DNA sequence amplification. Science 237: 175—178.

Lönn U, Lönn S, Nylen U and Stenkvist B (1992) Appearance and detection of multiple copies of the *mdr*-1 gene in clinical samples of mammary carcinoma. Int. J. Cancer 51: 682—686.

Macdonald F and Ford CHJ (1991) Diagnostic and prognostic applications of tumor suppressor genes. In: Read AP and Brown T (eds) Oncogenes and Tumor Suppressor Genes (pp. 71–79). βIOS Scientific Publishers, Oxford, U.K.

Mocharla H, Mocharla R and Hodes ME (1990) Coupled reverse transcription-polymerase chain reaction (RT-PCR) as a sensitive and rapid method for isozyme genotyping. Gene 93: 271—275.

Moos M and Gallwitz D (1982) Structure of a human beta-actin-related pseudogene which lacks intervening sequences. Nuc. Acids Res. 10: 7843—7849.

Moos M and Gallwitz D (1983) Structure of two human beta-actin-related processed genes one of which is located next to a simple repetitive sequence. EMBO J. 2: 757—761.

Moscow JA, Fairchild CR, Madden MJ, Ransom DT, Wieand HS, O'Brien EE, Poplack DG, Cossman J, Myers CE and Cowan KH (1989) Expression of anionic glutathione-S-transferase and P-glycoprotein genes in human tissues and tumors. Cancer Res. 49: 1422—1428.

Mullis K and Faloona F (1987) Specific synthesis of DNA *in vitro* via a polymerase catalysed chain reaction. Methods in Enzymol. 155: 335—350.

Murphy LD, Herzoy CE, Rudich JB, Fojo AT and Bates SE (1990) Use of the polymerase chain reaction in the quantitation of MDR-1 gene expression. Biochemistry 29: 10351—10356.

Nakajima-Iijima S, Hamada H, Reddy P and Kakunaga T (1985) Molecular structure of the human cytoplasmic β-actin gene: Interspecies homology of sequences in the introns. Proc. Natl. Acad. Sci. USA 82: 613—6137.

Noonan KE and Roninson IB (1988) mRNA phenotyping by enzymatic amplification of randomly primed cDNA. Nuc. Acids Res. 16(21): 10366.

Noonan KE, Beck C, Holzmayer TA, Chin JE, Wunder JS,

Andrulis IL, Gazdar AF, Willman CL, Griffith B, Von Hoff DD and Roninson IB (1990) Quantitative analysis of MDR1 (multidrug resistance) gene expression in human tumors by polymerase chain reaction. Proc. Nat. Acad. Sci. USA 87: 7160–7164.

Noonan KE and Roninson IB (1991) Quantitative estimation of *MDR1* mRNA levels by polymerase chain reaction. In: Roninson IB (ed) Molecular and Cellular Biology of Multidrug Resistance in Tumor Cells (pp. 319–333). Plenum Press, New York.

Pollack R (1980) Hormones, anchorage and oncogenic cell growth. In: Burchenal JH and Oettgen HF (eds) Cancer: Achievements, Challenges, and Prospects for the 1980s (pp. 501–515). New York: Grune and Stratton.

Ponte P, Gunning P, Blau H and Kedes L (1983) Human actin genes are single copy for alpha-skeletal and alpha-cardiac actin but multicopy for beta- and gamma-cytoskeletal genes: 3′ untranslated regions are isotype specific but are conserved in evolution. Mol. Cell Biol. 3: 1783–1791.

Rappolee DA (1990) Optimizing the sensitivity of RT-PCR. In: Amplifications. A Forum for PCR Users (Technical Bulletin) 4: 5–7.

Rappolee DA, Mark D, Banda MJ and Werb Z (1988a) Wound macrophages express TGF-α and other growth factors *in vivo*: Analysis by mRNA phenotyping. Science 241: 708–712.

Rappolee DA, Brenner CA, Schultz R, Mark D and Werb Z (1988b) Developmental expression of PDGF, TGF-α and TGF-β genes in pre-implantation mouse embryos. Science 241: 1823–1825.

Rappolee DA, Wang A, Mark D and Werb Z (1989) Novel methods for studying mRNA phenotypes in single or small numbers of cells. J. Cell. Biochem. 39: 1–11.

Reyes AA and Wallace RB (1987) Mapping of RNA using S_1 nuclease and synthetic oligonucleotides. Methods in Enzymol. 154: 87–101.

Rice CM, Fuchs R, Higgins DG, Stoehr PJ and Cameron GN (1993) The EMBL data library. Nuc. Acids Res. 21: 2967–2971.

Roninson IB, Chin JE, Choi K, Gros P, Housman DE, Fojo A, Shen DW, Gottesman MM and Pastan I (1986) Isolation of human mdr DNA sequences amplified in multidrug resistant KB carcinoma cells. Proc. Nat. Acad. Sci. USA. 83: 4538–4542.

Roninson IB (1987) Molecular mechanism of multidrug resistance in tumour cells. Clin. Physiol. Biochem. 5: 140–151.

RT-PCR. Methods and Applications. Book 1 (1991) Clontech Laboratories, Inc., Paulo Alto, CA.

Rychlik W and Rhoads RE (1989) A computer program for choosing optimal oligonucleotides for filter hybridization, sequencing and *in vitro* amplification of DNA. Nuc. Acids Res. 17: 8543–8551.

Sambrook J, Fritsch EF and Maniatis (1989) Extraction, purification and analysis of messenger RNA from eukaryotic cells. In: Ford N, Nolan C and Ferguson M (eds) Molecular Cloning. A Laboratory Manual (second edition) (pp. 7.2–7.83). Cold Spring Harbor Laboratory Press.

Sela M, Anfinsen CB and Harrington WF (1957) The correlation of ribonuclease activity with specific aspects of tertiary structure. Biochim. Biophys. Acta 26: 502–512.

Shaffer AL, Wojnar W and Nelson W (1990) Amplification, detection, and automated sequencing of gibbon interleukin-2 mRNA by *Thermus aquaticus* DNA polymerase reverse transcription and polymerase chain reaction. Anal. Biochem. 190: 292–296.

Shen DW, Fojo A, Roninson IB, Chin JE, Soffir R, Pastan I and Gottesman MM (1986a) Multidrug resistance in DNA mediated transformants is linked to transfer of the human mdr1 gene. Mol. Cell. Biol. 6: 4039–4045.

Shen DW, Fojo A, Chin JE, Roninson IB, Richert N, Pastan I and Gottesman MM (1986b) Human multidrug resistant cell lines: Increased mdr1 expression can precede gene amplification. Science 232: 643–645.

Siebert PD and Larrick JW (1992) Competitive PCR. Nature 359: 557–558.

Singer-Sam J, Robinson MO, Bellve AR,Simon MI and Riggs AD (1990) Measurement by quantitative PCR of changes in HPRT, PGK-1, PGK-2, APRT, MTase and Zfy gene transcripts during mouse spermatogenesis. Nuc. Acids Res. 18(5): 1255–1259.

Taylor GR (1991) Polymerase chain reaction: basic principles and automation. In: McPherson MJ, Quirke P and Taylor GR (eds) PCR. A Practical Approach (pp. 1–14). IRL Press, Oxford University Press.

Termin HM (1972) RNA-directed DNA synthesis. Scientif. Am. (Jan) pp. 24–33.

Thomas PS (1980) Hybridisation of denatured RNA and small DNA fragments transferred to nitrocellulose. Proc. Natl. Acad. Sci. USA. 77(9): 5201–5205.

Tsai-Pflugfelder M, Liu LF, Liu AA, Tewey KM, Whang-Peng J, Knutsen T, Huebner K and Croce CM (1988) Cloning and sequencing of cDNA encoding human DNA topoisomerase II and localization of the gene to chromosome region 17q21-22. Proc. Natl. Acad. Sci. USA 85: 7177–7181.

Tu C-P D and Qian B (1986) Human liver glutathione-S-transferases: Complete primary sequence of an H_a subunit cDNA. Biochem. Biophys. Res. Commun. 141: 229–237.

Ueda K, Cornwell MM, Gottesman MM, Pastan I, Roninson IB, Ling V and Riordan JR (1986) The MDR1 gene responsible for multidrug resistance codes for P-gycoprotein. Biochem. Biophys. Res. Commun. 141: 956–962.

Ueda K, Cardrelli C, Gottesman MM and Pastan I (1987) Expression of a full length cDNA for the human MDR1 gene confers resistance to colchine, doxorubicin and vinblastine. Proc. Natl. Acad. Sci. USA 84: 300–3008.

Ullrich A, Shine J, Chirgwin J, Pictet R, Tischer E, Rutter WJ and Goodman HM (1977) Rat insulin genes: Construction of plasmids containing the coding sequences. Science 196: 1313–1319.

van der Bliek AM, Kooiman PM, Schneider C and Borst P. (1988) Sequence of *mdr3* cDNA encoding a human P-glycoprotein. Gene 71: 401–411.

Wang AM, Doyle MV and Mark DF (1989) Quantitation of

mRNA by the polymerase chain reaction. Proc. Natl. Acad. Sci. USA 86: 9717–9721.

Wang AM and Mark DF (1990) Quantitative PCR. In: Innis MA, Gelfand DH, Sninsky JJ and White TJ (eds) PCR Protocols: A Guide to Methods and Applications (pp. 70–75). Academic Press Inc., San Diego, CA.

Weinstein RS, Kuszak JR, Klushens LF and Coon JS (1990) P-glycoproteins in pathology: the multidrug resistance gene family in humans. Human Pathol. 21: 34–48.

Wright PA and Wynford-Thomas D (1990) The polymerase chain reaction, miracle or mirage? A critical review of its uses and limitations in diagnosis and research. J. Pathol. 162: 99–117.

Address for offprints: Martin Clynes, National Cell and Tissue Culture Centre, Dublin City University, Glasnevin, Dublin 9, Ireland.

Cytotechnology **12**: 315–324, 1993.

Mathematical models for multidrug resistance and its reversal

Seth Michelson
Department of Biomathematics, Syntex Discovery Research, MS S3–1B, 3401 Hillview Avenue, Palo Alto, CA 94303, USA

Key words: facilitated diffusion, mathematical models, MDR

Abstract

Mathematical models describing drug resistance are briefly reviewed. One model which describes the molecular function of the P-glycoprotein pump in multidrug resistant (MDR) cell lines has been developed and is presented in detail. The pump is modeled as an energy dependent facilitated diffusion process. A partial differential equation linked to a pair of ordinary differential equations forms the core of the model. To describe MDR reversal, the model is extended to add an inhibitor. Equations for competitive, one-site noncompetitive, and two-site noncompetitive inhibition are derived. Numerical simulations have been run to describe P-glycoprotein dynamics both in the presence and absence of these kinds of inhibition. These results are briefly reviewed. The character of the pump and its response to inhibition are discussed within the context of the models. All discussions, descriptions, and conclusions are presented in nonmathematical terms. The paper is aimed at a scientifically sophisticated but mathematically innocent audience.

I. Introduction

Experiments *in vitro* with cultured tumor cell lines have revealed a series of very complicated mechanisms, both at the genetic and biochemical levels, that can account for drug resistance (Warr and Atkinson, 1988; Marx, 1986; Curt *et al.*, 1984; Schimke, 1984; Harris, 1984). Many are reviewed in this special issue. The most common forms of resistance are: 1) decreased drug uptake, 2) increased drug efflux, 3) increased degradation/metabolism of drug, 4) increased drug-target concentration, and 5) alteration in drug-target properties. Cells can become resistant to only a single selective agent (single drug resistance), or, in some cases, to agents which are structurally and mechanistically diverse (multiple or pleiotropic drug resistance) (Pastan and Gottesman, 1987; Moscow and Cowan, 1988; Gottesman and Pastan, 1988; Croop *et al.*, 1988).

Classical multidrug resistance (MDR) falls into the second category outlined above. The mechanism underlying MDR involves the overexpression of an energy-dependent efflux pump in the plasma membrane (Ling and Thompson, 1974; Juliano and Ling, 1976; Kartner *et al.*, 1983; Inaba *et al.*, 1979; Tsuruo *et al.*, 1982). The pump is a transmembrane glycoprotein referred to as p170 or P-glycoprotein. Often, one finds that stepwise selection of cells with one agent, say doxorubicin, leads to the generation of a cell line that is also resistant to other natural product agents including anthracyclines, vinca alkaloids, podophyllotoxins, and colchicine (reviewed by Pastan and Gottesman, 1987; Gottesman and Pastan, 1988).

Coincident with the development of these experimental models has been the evolution of a mathematical theory to describe it. And while the aim of this paper is to describe in detail a particular model of the P-glycoprotein pump (see Sections III.C and IV below), I recognize that mathematical models are no different from the experiments they mimic in that they must be viewed in the appropriate context. Therefore, I have decided to "zoom in" on the appropriate models of MDR by moving from the general case of models for the emergence of resistance (Section II.A) to models of tumor-wide resistance (Section II.B) to models of cellular resistance (Section II.C) to models of MDR (Section III) and its reversal (Section IV).

Recognizing that the audience for this special issue probably spans a spectrum of mathematical sophistication, I have decided to present the mathematics in as straightforward a manner as possible. Where necessary, mathematical equations are presented. However, the thrust of this paper is to describe in plain English the assumptions underlying the theory, the results of those theoretical speculations, and the limitations one must expect of the theory.

II. General models of drug resistance

A. Stochastic models of emergence

By stochastic, the mathematician means random. In these models, the emergence of a drug resistant subclone is considered a fortuitous result of random mutations expressed during tumor evolution.

Given the complexities in both definition and characterization of drug resistance, it has been difficult to develop manageable mathematical models to describe its emergence. Goldie and Coldman have proposed some of the most notable examples (Goldie and Coldman, 1979, 1984; Goldie *et al.*, 1982; Coldman and Goldie, 1983; Coldman *et al.*, 1985). Their models are phenomenological, predicting that single agent resistance depends upon an unspecified cellular mutation. They predict that the frequency of cross-resistance to several agents should be the product of the individual underlying single mutation frequencies, and account for multiple resistance on the *tumor level* in this manner. Furthermore, in their earlier work, they assumed that "resistant" meant completely resistant, i.e., that no clinically acceptable level of drug could kill the resistant subclone. This assumption has been relaxed in subsequent models.

Simple pencil and paper analyses point out the limitations of this model. Even if the viable mutation rate is small (i.e. one mutation in 10^6 mitoses), tumors presenting at an early clinical stage (approximately 1 cm^3) have negligible chance of being homogeneously chemosensitive. In fact, there are probably multiple subpopulations already present in a tumor of that size. If one also assumes, as Goldie and Coldman have, that resistant cells are *totally* resistant, one must conclude that upon presentation, the chances of curing any moderately sized tumor are negligible. Even if caught early, massive levels of a vast array of drugs would be needed to cure the cancer.

Goldie and Coldman address multiple drug resistance on the cellular level incidentally. They assume that individual subpopulations emerge independently, and that these subclones are resistant to different drugs. MDR, as we understand it, must then be quite rare, the product of two mutations, yielding a cell that is doubly mutated. Therefore, the odds of achieving resistance on the cellular level to three or more drugs are astronomical. Thus resistance to an array of drugs is a tumor-wide, not cellular, phenomenon. Day (1986) has extended their work to include asymmetry in growth, mutation and death rates to show that, in multiple resistant tumors, optimal treatment is always achieved with multiple drug regimens, but that the sequencing strategy employed may depend upon the underlying transition rates established for each subpopulation independently. However, Day states quite clearly that this type of model may be insufficient when considering MDR as expressed by the P-glycoprotein pump.

B. Deterministic models of therapy

Deterministic models are mathematical models, usually described by differential equations, which are fairly robust to small population variances. The biological effects being modeled act on total populations, and are strictly defined by dynamic equations which predict the mean behavior of the population. Fluctuations about the mean (i.e., what any single individual might add to the effect) are considered negligible in these models.

Like the stochastic models outlined above, deterministic models have been developed which portray drug resistance as a classifier. These models usually define a resistant subpopulation in generic terms; e.g., varying transition rates for death, growth, etc. Overall tumor growth is modeled as a process in which two populations vie for survival in a hostile (drug treated) environment. Invariably, the loss rates for the resistant subpopulation are constants, fixed at smaller magnitudes than those defined for sensitive population. None of these models directly describes resistance mechanistically.

For example, in three related papers (Gregory *et al.*, 1988; Hokanson *et al.*, 1986; Birkhead *et al.*, 1986) tumor growth-rate and overall tumor volume at the time of presentation are used as predictors of response in the presence of single drug therapy. However, the authors state explicitly that they only meant to describe resistance as a phenomenon, and that no attempt was made to model it mechanistically. As a case in point, they assumed that sensitivity and resistance are nonacquirable traits, fixed in an altering environment. This kind of assumption would totally ignore the effects of gene amplification and DHFR upregulation in methotrexate treated tumors (Schimke, 1984). No explicit definition or attempt to model MDR on a cellular basis is tried either.

In a more realistic model, Duc and Nickolls (1987) linked the pharmacokinetic profile of a single course of drug therapy to a tumor growth model. The distribution of drug was determined theoretically in three physiological compartments (the plasma, over the sensitive tumor, and over the resistant tumor). They assumed that each tissue compartment exhibited first order kinetics and that the sensitive and resistant tumor compartments were completely segregated and did not communicate biophysically. A standard set of ordinary differential equations was derived and solved.

They modeled cell growth within each compartment independently. The presence of one population did not affect the growth or loss of the other. The level of drug at each site was determined separately. A general growth-death model for cycle specific drugs took the following form:

$$\frac{dN}{dt} = F(N)N(t) - kf(t)m_x F(N)N(t) \tag{1}$$

For cycle nonspecific drugs the model looked like:

$$\frac{dN}{dt} = F(N)N(t) - kf(t)m_x N(t) \tag{2}$$

In these general models, F(N) may be used to represent a logistic growth term. If one were to define F(N) appropriately, as the size of the tumor, N, grows, F(N) would shrink. This would force the rate of growth, dN/dt, to decelerate and eventually slow to zero. kf(t) represents the time-dependent distribution of drug in a particular tissue compartment. The terms on the right of the minus sign in these equations represent loss, and the difference in the cell death rates is due entirely to the term m_x, which is large if the population is sensitive and small if it is resistant.

The model could be modified to implicitly represent MDR by representing the pharmacokinetics of the two (segregated) tissue compartments as asymmetric, e.g., by allowing for increased outflow from the resistant tissue compartment. Then kf(t), representing whole tissue and not intracellular concentrations, would be different for the two populations. However, the authors do not address this aspect of resistance explicitly, and one should note that this type of model does not distinguish between decreased drug accumulation in the tissue and decreased drug accumulation in the cell.

C. The hybrid

Michelson and Slate (1989, 1991) developed a mathematical model which describes drug resistance in a more mechanistic manner, by defining it as one or all of the physiologic pathways listed in the Introduction above. Transitions (e.g., growth, death, acquisition of resistance) are expressed stochastically for individual cells. However, the probability rates defining the transitions are dynamic, determined at each time step by the level of active drug at the target site of the average tumor cell. The model assumes that: 1) Each cell experiences a transition (or none at all!) with a given probability during a given time step independently, and does not depend on its neighbors for signals or controls. 2) The lifelength of each individual cell is a random variable that is identically and independently distributed throughout the cell population. 3) Drug is uniformly distributed throughout the cell population; no spatial hindrance in access of drug to any cell is encountered. 4) Cell death is due strictly to the cytotoxic effects of the drug, and the risk of cell death is proportional to the concentration of drug at the target site in the cell. The distribution of drug moieties throughout the average cell is modeled using standard concentration-dependent first order compartment kinetics to describe the interstitial spaces, the cytoplasm, and the target site (usually the nucleus).

While the variable expression of any of the resistance mediators listed in the Introduction may be due to any of a number of underlying processes (e.g, gene amplification, alterations in transcriptional or translational efficiency, etc.), the model only deals with their functional consequences (e.g, decreased uptake, increased efflux, differential target sensitivity, etc.) explicitly. The very simple assumption is made that cell death is strictly proportional to the concentration of drug at the target site: If the concentration of an active agent at its target site is high enough, the cell will very likely die, and the more drug present, the more likely the death.

Now cell death is a probabilistic event, based entirely upon the ability of a drug to penetrate, distribute, and accumulate in the cell. Using these models, we have shown that any cell which pumps out enough drug such that its concentration at the target site remains "low enough", significantly enhances its chances for survival (Michelson and Slate, 1989; Slate and Michelson, 1991). And though the molecular structure of the pump is not addressed explicitly in this model, a simple pump is an inherent component in the calculations of the drug distribution equations. Thus, this type of hybrid model brings the definition of drug resistance from the tumor-wide level down to the cellular level.

One should recognize the inherent limitations of this model. Clearly, in large, poorly vascularized tumors, cell-cell interactions (including "competition" for the drug) cannot be ignored. Any claim that drug is equally distributed across a large tumor mass is self-evidently erroneous. However, on the micro-pharmacological level, the level at which the P-glycoprotein pump works, the distribution of drug *within* the tumor cell may yield a useful insight into the dynamics of drug resistance and its reversal. It is this simple fact that motivates the development of the types of models described in the next section.

III. Specific models for multidrug resistance

A. Intracellular micropharmacology

Demant *et al.* (1990) developed a model of drug transport very similar to the one Michelson and Slate (1992) developed to describe the P-glycoprotein pump on the molecular level. Their questions were slightly different, however. Demant and colleagues asked: "Could endosomal transport of drug under varying levels of pH account for a major portion of drug efflux in MDR cell lines"? Their model described three compartments: the extracellular medium, the cytoplasm, and the endosomal vesicles. Within the cytoplasm the drug could be in three states: free, bound to low affinity membrane binding sites, or bound to high affinity nuclear binding sites. Based upon the theory of mass action, assuming that the amount of drug bound to the membrane sites is significantly smaller than the number of possible sites (i.e., ignoring saturation), and assuming that equilibrium is achieved rapidly (i.e., instantaneously) in the cytoplasm, they derive an equation for the dissociation constant for membrane binding. Active transport across the membrane is then defined by Michaelis-Menten kinetics. From their model they concluded that active transport is the primary efflux mechanism in MDR cell lines, and that diffusion and exocytosis are not fast enough to account for the rapid efflux observed experimentally.

Michelson and Slate (1992, 1993) have expanded upon this initial model to include diffusion and the energy dependence of the pump, and have modeled active transport as a facilitated diffusion process (see Section III.C below).

B. Calculation of pump K_m

All mathematical models used to describe MDR transport are designed within an experimental context to explain an observed dynamic. Today most researchers (all?) believe that the pump binds to a cytotoxic target drug before actively transporting it out of the cytoplasm. The binding and facilitation of the transport are reminiscent of enzyme kinetics. Therefore, most (all?) theoreticians have described the pump and its dynamics using a Michaelis-Menten rate equation. Horio *et al.* (1990) set up their model to mimic an experiment in which apical-to-basal and basal-to-apical flux across MDCK epithelial cells was measured. Based upon differential flux characteristics and relative diffusion rates, they derived a final equation for the apparent K_m algebraically. Transport observed experimentally for each direction was used to get a handle upon the intracellular concentration of drug so that the Eadie-Hoffstra plots could be employed.

In a similar study Spoelstra *et al.* (1992) based their model on an experimental flow through system. Flux, defined as a time derivative, d/dt, is the result of net diffusion and Michaelis-Menten transport. They derived net flux estimates from intracellular and extracellular concentrations, and by assuming equal membrane diffusability in both directions. Using Scatchard plots they derived estimates for the number of binding sites available to the target drug, and their affinity. However, they also observed that the Hill slopes of their Scatchard lines were greater than one, making strict interpretation of their data difficult.

C. Energy-dependent facilitated diffusion

Michelson and Slate (1992) developed a model of the MDR pump based upon a washout experiment: tumor cells are preloaded with radioactively tagged drug, removed from the drug-loaded medium, washed, and restored to a new drug-free medium. The intracellular drug concentrations are then monitored over time.

The P-glycoprotein associated with MDR is an energy-dependent pump which can be described by the following enzyme kinetics:

$$\begin{gathered} E+2ATP \rightleftharpoons E{\cdot}2ATP \\ E{\cdot}2ATP+S(\text{inside}) \rightleftharpoons E{\cdot}2ATP{\cdot}S \\ E{\cdot}2ATP{\cdot}S \rightleftharpoons E+2ADP+2P+S(\text{outside}) \end{gathered} \tag{3}$$

where E is the concentration of p170, and S is the concentration of the substrate (drug). The rate of the first reaction is given by the typical Michaelis-Menten formula

$$V_1=\frac{V_{ATP}\cdot ATP}{K_A+ATP} \tag{4}$$

The rate of the second reaction is given by

$$V_2=\frac{V_S\cdot S}{K_M+S} \tag{5}$$

Because the total flux of substrate out of the cell is limited by the slower of the two rates, V_1 and V_2, we define V_{MAX}, the minimum efflux rate, as the minimum of the V_1 and V_2. Thus, the overall flux is given by

$$FLUX=\frac{V_{MAX}\cdot ATP\cdot S}{(K_A+ATP)(K_M+S)} \tag{6}$$

The level of ATP in the resting cell is dynamically maintained by conversion to and from ADP and free phosphate. For the purposes of the model, Michelson and Slate assumed that the cell is a homogeneously-mixed compartment packed with enzyme processes which maintain this homeostatic condition. They further assumed (for numerical purposes) that these enzyme reactions could be conglomerated into a single energy maintenance process which follows Michaelis-Menten kinetics. Mathematically, then, the energy pools of the cell can be described by

$$\begin{aligned} \frac{dATP}{dt} &= -\frac{V'ATP}{K'+ATP} + \frac{V^{*}ADP}{K^{*}+ADP} \\ \frac{dADP}{dt} &= \frac{V'ATP}{K'+ATP} - \frac{V^{*}ADP}{K^{*}+ADP} \end{aligned} \tag{7}$$

where (V^{*},K^{*}) and (V',K') are the Michaelis-Menten constants for the ADP-ATP conversion process.

When a cell is challenged with a cytotoxic agent, the efflux pump begins operating, and the ATP pool is depleted by two molecules for each molecule of drug pumped from the cell. Consequently, for each molecule of drug pumped from the cell two molecules of ADP are produced. Therefore, Eq. 7 was modified to be:

$$\begin{aligned} \frac{dATP}{dt} &= -2FLUX - \frac{V'ATP}{K'+ATP} + \frac{V^{*}ADP}{K^{*}+ADP} \\ \frac{dADP}{dt} &= 2FLUX + \frac{V'ATP}{K'+ATP} - \frac{V^{*}ADP}{K^{*}+ADP} \end{aligned} \tag{8}$$

To represent the entire pump and its machinery, one can generalize the formalism developed by Joshi (1985) as follows:

$$\begin{aligned} \frac{\partial S}{\partial t} - \alpha\frac{\partial^2 S}{\partial x^2} - F &= 0 \\ F &= \frac{V_{MAX}\cdot ATP\cdot S}{(K_A+ATP)\,(K_M+S)} \\ \frac{dATP}{dt} &= -2F - \frac{V'ATP}{K'+ATP} + \frac{V^{*}ADP}{K^{*}+ADP} \\ \frac{dADP}{dt} &= 2F + \frac{V'ATP}{K'+ATP} - \frac{V^{*}ADP}{K^{*}+ADP} \end{aligned} \tag{9}$$

where α is the diffusion rate constant for drug through the membrane (from the cytoplasm to the interstitial space) and x represents the perpendicular radial distance through the membrane from inside to out.

The spatial characteristics of the model assume that substrate is distributed only *within* the pump (i.e., within the transmembrane channel of the pump), and that the concentration of substrate in the cytoplasm is derived from a well mixed compartment. Therefore, the concentration of substrate at the internal membrane surface is representative of the entire cytoplasmic concentration. A similar assumption is made about the external cell surface.

What this model says, is that drug would normally leave a preloaded cell via diffusion. This assumption is similar to that made by Spoelstra *et al.* (1992). Mathematically, simple diffusion over a spatial distance, x, (in our case, radially through a membrane) is described by the partial differential equation given in System 9 when $F = 0$. To facilitate this diffusion, something extra must be added. In this model, the flux term, F. The derivation of the flux term is crux of the model. It is this term which explicitly accounts for the energy dependence of the pump and its binding characteristics for the target drug. Suspected inhibitors would also be accounted for by the flux term. Therefore, characterization of flux is necessary if one is to describe MDR reversal in a mechanistic way (see Section IV, below).

Numerical simulations were run to confirm the reasonableness of the model. A drug washout experiment was mimicked. The preloaded cells were allowed to equilibrate in fresh medium, and the cytoplasmic concentration profiles for the target drug were plotted over time. The results were as expected. Without the pump, drug halflife in the cytoplasm was 12 h (defined by pure diffusion). With the pump, the halflife was about 30 min (similar to that observed in the laboratory experiments). Rate constants for the different pumping processes were derived from both laboratory measurements and from standard literature estimates.

IV. Model for reversal of MDR

One way to reverse MDR is to take advantage of the energy dependence of the pump. However, based upon their model, Michelson and Slate concluded that that strategy was fraught with peril. Mathematical analysis of the basic model (System 9) shows that since the ATP-ADP predrug steady state is stable, dynamic, and nonoscillatory, one could, in theory, exhaust the ATP pool in the local milieu of the pump. This could be achieved by increasing flux (a self-defeating strategy), blocking the efficiency of ADP to ATP synthesis, or increasing ATP hydrolysis.

The fact that the energy specific parameters represent potential targets in resistance reversal completely ignores the fact that manipulating the energy producing systems in a cell would have to be exquisitely precise for one to minimize toxicity *in vivo*. However, Michelson and Slate's objection to this means of MDR reversal is based upon the theoretical fact that, with respect to ATP levels, the pump is, in fact, a self-regulating mechanism.

Once a cytotoxic challenge begins, and if sufficient levels of ATP and target drug are around, the pump will saturate, forcing flux to some positive constant, F^*, less than or equal to V_{max}. During the transition phase from a completely inactive pump to one which is totally saturated, the steady state concentrations of ATP and ADP shift in their phase space towards the ADP axis. When the pump is fully saturated and flux equals F^*, a new steady state is established. If the level of drug in the cytoplasm is such that the energy control mechanisms can maintain high enough ATP levels for the pump to function, all the drug will be pumped out, flux will tend to zero, and the original steady state will be re-established. If the levels of drug in the cytoplasm are such that one begins to exhaust the ATP pool, then flux decreases, maybe even being forced to zero, i.e., the pump stops, until the ATP levels are re-established, and a new steady state (at some nonzero ATP level) is achieved. When that occurs, the flux is re-established at a new lower level. When all the drug is pumped out of the cell, the flux again tends to zero, and the original steady state is re-established.

Simulation studies (Michelson and Slate, 1992) have shown that under physiologically reasonable conditions, even in the presence of energy disruption, the ATP pool is replenished much more quickly than the pump diminishes it. At best, the efflux of drug is slowed only slightly. If a compound could be created which only attacks the energy conversion process for the pump in tumor cells, and thus remain nontoxic to other cells *in vivo*, it would, at best, act as an adjunct to one of the other therapeutic strategies.

Given that the energy pool is probably a suboptimal target for MDR reversal, how would one inhibit the P-glycoprotein pump? The inhibitor could:

1. Attack the outward diffusion kinetics through the membrane.
2. Attack the transport activity of the pump at the inner surface of the membrane (e.g., its ability to bind drug).
3. Attack the efflux efficiency of the pump by "clogging" the transmembrane pore complex extracellularly.

Strategy number 2 is the one presently being pursued most actively in the clinic. Compounds are sought which block the efflux action at the drug binding site(s) of the pump. Essentially, this is the motivation for using the calcium channel blocker, verapamil, and other P-glycoprotein inhibitors as MDR reversal agents. Michelson and Slate (1993) have extended their original model to include such theoretical reversal agents.

Suppose one introduces a competitive inhibitor, I, into the system. Then the V_2 equation above (Eq. 5), should be altered as follows:

$$V_2 = \frac{V_S \cdot S}{K_M(1 + (I/K_I)) + S} \tag{10}$$

where K_I is the dissociation constant of the P-glycoprotein-inhibitor complex. Efflux becomes

$$F = \frac{V_{MAX} \cdot ATP \cdot S}{(K_A + ATP)\,(K_M(1 + (I/K_I)) + S)} \tag{11}$$

Adding inhibitor diffusion and efflux to the original system results in:

$$\begin{aligned}
&\frac{\partial S}{\partial t} - \alpha\frac{\partial^2 S}{\partial x^2} - F_1 = 0\\
&\frac{\partial I}{\partial t} - \beta\frac{\partial^2 I}{\partial x^2} - F_2 = 0\\
&F_1 = \frac{V_{MAX}\cdot ATP\cdot S}{(K_A + ATP)\ (K_M(1 + (I/K_I)) + S)}\\
&F_2 = \frac{V_{MAX}\cdot ATP\cdot I}{(K_A + ATP)\ (K_I(1 + (S/K_S)) + I}\\
&\frac{dATP}{dt} = -(F_1 + F_2) + \frac{V^*ADP}{K^* + ADP}\\
&\frac{dADP}{dt} = (F_1 + F_2) - \frac{V^*ADP}{K^* + ADP}
\end{aligned} \tag{12}$$

where β is the diffusion constant for the inhibitor, F_1 is the efflux of the drug, and F_2 is the efflux of the inhibitor. Since the same fixed number of ATP molecules is required to transport each molecule out of the cytoplasm (whether S or I), and since the two substrates are competitive inhibitors, a total efflux term, $(F_1 + F_2)$, must be added to the energy conversion equations.

A similar analysis can be developed for a noncompetitive inhibitor. By replacing V_2 with

$$V_2 = \frac{V_1\cdot S}{(K_M + S)\ (1 + (I/K_I))} \tag{13}$$

one rederives the diffusion equation with new flux terms

$$\begin{aligned}
&F_1^* = \frac{V_{MAX}\cdot ATP\cdot S}{(K_A + ATP)\ ((K_M + S)(1 + (I/K_I)))}\\
&F_2^* = \frac{V_{MAX}\cdot ATP\cdot I}{(K_A + ATP)\ ((K_I + I)(1 + (S/K_S)))}
\end{aligned} \tag{14}$$

Simple inspection of Eqs. 10 and 13 shows that a competitive inhibitor increases K_m by a factor of $(1 + I/K_I)$, a normalized measure of the inhibitor activity. The noncompetitive inhibitor, on the other hand, decreases the transport (reaction) rate by a factor of $1/(1 + I/K_I)$, which again has I normalized to its own binding affinity. The meaning here is clear: In the first instance, as the inhibitor competes with the substrate, more substrate will eventually overcome any inhibition, while in the second, no amount of substrate can overcome the blockade, and the reaction is slowed as long as inhibitor is present.

The question Michelson and Slate asked is: "What should the initial loading of a test compound be given its 1) inhibitive character, and 2) its affinity for the binding site? Is any one kind of inhibitor better than any other? Does one need to worry about the timing and strategy for administration"?

Their simulation studies showed that over a fixed range of initial loading and inhibitor affinities, noncompetitive inhibitors are more efficacious than competitive inhibitors. But these advantages are only evident at the lower loading concentrations (suggesting an answer to their second question about the pharmacokinetics of the inhibitor). If enough competitive inhibitor can get into the tumor cells prior to drug therapy, and if it can be maintained there at high enough levels despite the pump's activity, then competitive inhibitors with proper pharmacokinetic profiles may be acceptable as clinically relevant MDR reversal agents.

Michelson and Slate also looked at the question of whether a one site noncompetitive inhibitor, one which is pumped out of cell along with the target drug, is significantly different in its effectivity from a two-site noncompetitive or allosteric hindering inhibitor which is not. They showed that if an inhibitor is not extruded by the pump, then its advantage over a similar compound which is forced from the cytoplasm is exerted only at high affinity-low concentration combinations. If both inhibitors are low affinity, then no real advantage was observed.

V. Implications of the theory

A cautionary word is in order before one ventures to interpret the theoretical results described above. First, note that the models of Spoelstra, Horio, and Michelson and Slate are all variations upon the Michaelis-Menten transport theme. Each, in its own way, describes transport as a saturable rate phenomenon depending upon the mass-action chemistry of the pump molecule. The differences between the three models reside in the detailed descriptions of diffusion, energy dependence, etc. Each is based upon the experimental design of choice used within their own laboratories to investigate transport.

Second, these models are in a state of evolutionary flux. For example, Michelson and Slate's earlier results (1990, 1992) did *not* adequately mimic the *inhibition* of p170 activity. When modeling inhibition, the stereochemistry of the ATP binding sites, the nature of the inhibitor (competitive, noncompetitive, or allosteric), the stoichiometry of the drug binding site(s), etc., were not accounted for. It was these added layers of complexity that required the development of a more sophisticated model like that presented above.

Therefore, one must accept these initial results with some degree of caution. For example, in the Michelson and Slate model (1993) outlined above in Section IV, they modeled diffusion as the only other mechanism of inhibitor loss. This is especially important when considering the case of a two-site allosteric noncompetitive compound. For example, they completely ignored the possibility that the inhibitor is the target of biochemical degradation, metabolism etc. But suppose it is. Then one should expect that the target drug efflux will accelerate with inhibitor degradation, and might even approach that observed when a one-site noncompetitive inhibitor is used. In a worst case scenario, if the degradation rate of the two-site inhibitor is faster than the efflux rate of the one-site inhibitor, the effectivity of the stable one-site inhibitor could even surpass that of the two-site one.

The models that have been developed thus far can be used to make simple predictions about how MDR reversal agents could be optimally employed to block pumping activity. However, in order to create a more realistic model, one must consider other complexities of P-glycoprotein function. For example, how many binding sites are there and what are their structures? How does binding affect ATPase activity? Are all P-glycoprotein molecules identical? How does posttranslational modification of the protein affect is transport and binding characteristics? It is the evolution of these questions, derived from the experimentalist, that drives the theorist to develop new mathematical models of MDR and its reversal.

References

Birkhead BG, Gregory WM, Slevin ML and Harvey VJ (1986) Evaluating and designing cancer chemotherapy treatment using mathematical models. Eur J Cancer Clin Oncol 22: 3–8.

Coldman AJ and Goldie JH (1983) A model for the resistance of tumor cells to cancer chemotherapeutic agents. Math Biosci 65: 291–307.

Coldman AJ, Goldie JH and Ng V (1985) The effects of cellular differentiation on the development of permanent drug resistance. Math Biosci 74: 177–198.

Croop JM, Gros P and Housman DE (1988) Genetics of multidrug resistance. J Clin Invest 81: 1303–1309.

Curt GA, Clendeninn NJ and Chabner BA (1984) Drug resistance in cancer. Cancer Treat Rep 68: 87–99.

Day RS (1986) Treatment sequencing, asymmetry, and uncertainty: Protocol strategies for combination chemotherapy. Cancer Res 46: 3876–3885.

Demant EJF, Sehested M and Jensen PB (1990) A model for the computer simulation of P-glycoprotein and transmembrane pH mediated anthracycline transport in multidrug-resistant tumor cells. Biochim Biophys Acta 1055: 117–125.

Duc HN and Nickolls PM (1987) Multicompartment models of cancer chemotherapy incorporating resistant cell populations. J Pharmacokin Biopharm 2: 145–177.

Goldie JH and Coldman AJ (1979) A mathematical model for relating the drug sensitivity of tumors to their spontaneous mutation rate. Cancer Treat Rep 63: 1727–1733.

Goldie JH and Coldman AJ (1984) The genetic origin of drug resistance in neoplasms: Implications for systemic therapy. Cancer Treat Rep 67: 923–931.

Goldie JH, Coldman AJ and Gudauskas GA (1982) Rationale for the use of alternating non-cross-resistant chemotherapy. Cancer Treat Rep 66: 439–449.

Gottesman M and Pastan I (1988). Resistance to multiple chemotherapeutic agents in human cancer cells. Tr Pharmacol Sci 9: 54–58.

Gregory WM, Birkhead BG and Souhami RL (1988) A mathematical model of drug resistance applied to treatment for small-cell lung cancer. J Clin Oncol 6: 457–461.

Harris AL (1984) Drug resistance to cancer chemotherapy. Drugs Today 20: 657–663.

Hokanson JA, Brown BW, Thompson JR, Jansson B and Drewinko B (1986) Mathematical model for human myeloma relating growth kinetics and drug resistance. Cell Tissue Kinet 19: 1–10.

Horio M, Pastan I, Gottesman M and Handler J (1990) Transepithelial transport of vinblastine by kidney-derived cell lines. Application of a new kinetic model to estimate In situ K_m of the pump. Biochim Biophys Acta 1027: 116–122.

Inaba M, Kobayashi H, Sakurai Y and Johnson RK (1979) Active efflux of daunorubicin and adriamycin in sensitive and resistant sublines of P388 leukemia. Cancer Res 42: 4730–4733.

Joshi RR (1985) On control of the function of the sodium-potassium pump in malignant cells. Bull Math Biol 47: 551–564.

Juliano RL and Ling V (1976) Cell surface P-glycoprotein modulating drug permeability in Chinese hamster ovary cell mutants. Biochim Biophys Acta 455: 152–162.

Kartner N, Riordan JR and Ling V (1983) Cell surface P-glycoprotein associated with multidrug resistance in mammalian cell lines. Science 221: 1285–1288.

Ling V and Thompson LH (1974) Reduced permeability in CHO cells as a mechanism of resistance to colchicine. J Cell Physiol 83: 103–116.

Marx JL (1986) Drug resistance of cancer cells probed. Science 234: 818–820.

Michelson S and Slate D (1989) Emergence of the drug-resistant phenotype in tumor subpopulations: A hybrid model. JNCI 81: 1392–1401.

Michelson S and Slate DL (1992) A mathematical model of the multi-drug resistant P-glycoprotein pump. Bull Math Biol 54: 1023–1038.

Michelson S and Slate DL (1993) A mathematical model for the inhibition of the multidrug resistance-associated P-glycoprotein pump. Bull Math Biol, in press.

Moscow JA and Cowan KH (1988) Multidrug resistance. JNCI 80: 14–20.

Pastan I and Gottesman M (1987) Multiple drug resistance in human cancer. New Engl J Med 316: 1388–1393.

Schimke RT (1984) Gene amplification, drug resistance, and cancer. Cancer Res 44: 1735–1742.

Slate DL and Michelson S (1991) Drug resistance reversal strategies: A comparison of experimental data with model predictions. JNCI 83: 1574–1580.

Spoelstra EC, Westerhoff HV, Dekker H and Lankelma J (1992) Kinetics of daunorubicin transport by P-glycoprotein of intact cancer cells. Eur J Biochem 207: 567–579.

Tsuruo T, Iida H, Tsukagoshi S and Sakurai Y (1982) Increased accumulation of vincristine and adriamycin in drug-resistant P388 tumor cells following incubation with calcium antagonists and calmodulin inhibitors. Cancer Res 42: 4730–4733.

Warr JR and Atkinson GF (1988) Genetic aspects of resistance to anticancer drugs. Physiol Rev 66: 1–26.

Address for offprints: S. Michelson, Department of Biomathematics, Syntex Discovery Research, MS A4–100, 3401 Hillview Avenue, Palo Alto, CA 94303, USA.

Cytotechnology **12**: 325–345, 1993.

The biology of radioresistance: similarities, differences and interactions with drug resistance

Simon N. Powell and Edward H. Abraham
Department of Radiation Oncology, Massachusetts General Hospital, Boston, MA 02114, USA

Key words: cell cycle arrest; cell membrane; DNA repair; oncogene; radiation resistance; signal transduction

Abstract

Cells and tissues have developed a variety of ways of responding to a hostile environment, be it from drugs (toxins) or radiation (summarized in Fig. 1). Three categories of radiation damage limitation are: (i) DNA repair (ii) changes in cellular metabolism (iii) changes in cell interaction (cell contact or tissue-based resistance; whole organism based resistance).

DNA repair has been evaluated predominantly by the study of repair-deficient mutants. The function of the repair genes they lack is not fully understood, but some of their important interactions are now characterized. For example, the interaction of transcription factors with nucleotide excision repair is made clear by the genetic syndromes of xeroderma-pigmentosum groups B, D and G. These diseases demonstrate ultraviolet light sensitivity and general impairment of transcription: they are linked by impaired unwinding of the DNA required for both transcription and repair. The transfer of DNA into cells is sometimes accompanied by a change in sensitivity to radiation, and this is of special interest when this is the same genetic change seen in tumors. DNA repair has a close relationship with the cell cycle and cell cycle arrest in response to damage may determine sensitivity to that damage. DNA repair mechanisms in response to a variety of drugs and types of radiation can be difficult to study because of the inability to target the damage to defined sequences *in vivo* and the lack of a satisfactory substrate for *in vitro* studies.

Changes in cellular metabolism as a result of ionizing radiation can impart radiation resistance, which is usually transient *in vitro*, but may be more significant *in vivo* for tissues or tumors. The mechanisms by which damage is sensed by cells is unknown. The detection of free radicals is thought likely, but distortion to DNA structure or strand breakage and a direct effect on membranes are other possibilities for which there is evidence. Changes in extracellular ATP occur in response to damage, and this could be a direct membrane effect. External purinergic receptors can then be involved in signal transduction pathways resulting in altered levels of thiol protection or triggering apoptosis. Changes in the functional level of proteins as a consequence of ionizing radiation include transcription factors, for example c-*jun* and c-*fos*; cell cycle arrest proteins such as GADD (growth arrest and DNA damage inducible proteins) and p53; growth factors such as FGF, PDGF; and other proteins leading to radioresistance. Mechanisms for intercellular resistance could be mediated by cell contact, such as gap junctions, which may help resistance to radiation in non-cycling cells. Paracrine response mechanisms, such as the release of angiogenic factors via membrane transport channels may account for tissue and tumor radiation resistance. Endocrine response mechanisms may also contribute to tissue or tumor resistance.

MECHANISMS OF RADIATION RESISTANCE

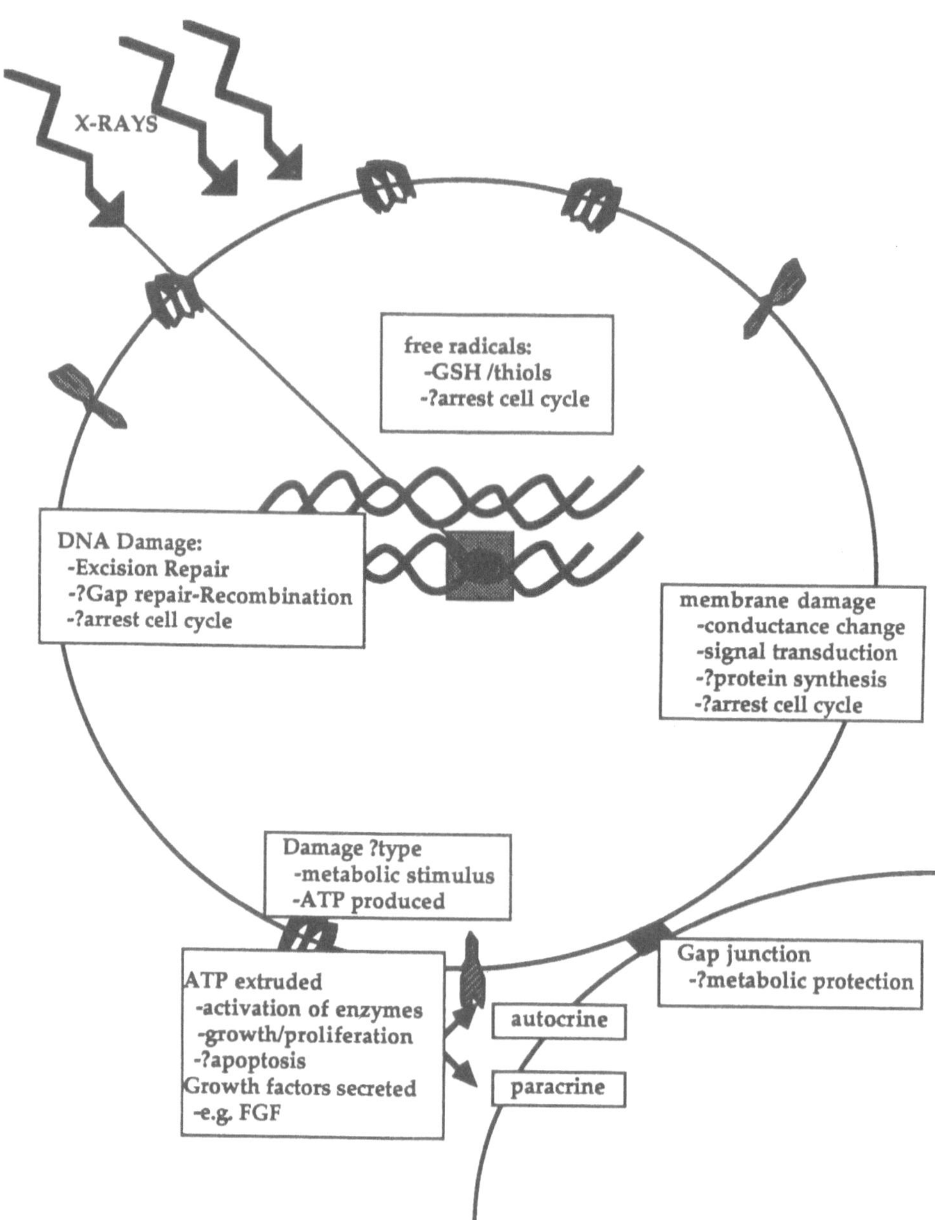

Fig. 1. The mechanisms of radiation resistance shown schematically. Ionizing radiation exposure to a cell leads to energy deposition throughout the cell. Free radicals are formed within the DNA, cytosol and membrane, and the biological consequences are shown. Resistance to the effects of ionizing radiation can be mediated by DNA repair, which may be assisted by increased cell cycle arrest; or by damage activated change in signal transduction pathways arising from either DNA damage or membrane damage.

DNA repair

Cell killing in mitosis induced by ionizing radiation is generally thought to be mediated by double-stranded DNA damage. The damage which triggers apoptosis may also be strand breakage, but the evidence is less clear and there is no known specificity to double-strand damage. Although there is some evidence that the amount of damage inflicted per unit of radiation may vary between cell or tumor types, this is not thought to be the major factor determining intrinsic radiation resistance (Ward, 1990). The processing of the inflicted damage appears to determine radiation resistance and sensitivity: this conclusion comes from studies limited to radiation sensitive mutants of normal tissue cell lines. However, in human tumor cells, the level of cell killing still correlates with double-stranded DNA damage and thus the molecular mechanisms involved in the processing of this type of damage are likely to be a major factor in determining intrinsic tumor cell radioresistance. The same is true for cytotoxic drugs which cause "functional" double stranded DNA damage (cisplatin, etoposide, bleomycin, for examples).

When double-stranded DNA damage is induced in cells, the first step in the removal of that damage is to "clean-up" the damaged region. As a minimum, this will involve the excision of at least one nucleotide from each strand, by the action of specific endonucleases (Lindahl, 1990). Double-stranded DNA damage is in close proximity on each strand (suggested by the clustering of ionizations) and nucleotide excision will lead to a loss of sequence. Furthermore, simple ligation of strand breaks is not going to restore the sequence. Although ligase can rejoin breaks, the lack of this enzyme in yeast leads to hypermutability rather than clear radiosensitivity. Bloom's syndrome is a rare human disease characterized by a early onset of cancer and hypermutability but cell killing from radiation is only at the sensitive end of the normal spectrum (Freidberg, 1985). The association of radiation sensitivity with a lack of DNA recombination function was initially made in a yeast mutant of *S. cerevisiae* (rad52) but this association has now extended to mammalian cells with the observation of a general radiation sensitivity in *scid* mice. Fibroblasts from *scid* mice were also found to be deficient in strand break repair using either of two damage assays, pulsed field gel electrophoresis (Biederman *et al.*, 1991) and neutral filter elution (Hendrickson *et al.*, 1991) thus associating double-strand damage processing with recombination.

All changes in radiosensitivity cannot be attributed to the efficiency of double-strand break rejoining, although this is characteristic of well described rodent cell mutants (see review: Jeggo, 1990). Many mutant cell lines which differ in radiosensitivity from their parent line demonstrate no differences in strand break rejoining (e.g., *irs*1, *irs*2, EM9, BLM-2). *Irs*1 shows low repair fidelity using a plasmid reconstitution assay (Debenham *et al.*, 1988b) and EM9 shows impaired recombination between plasmid molecules (Wahls and Moore, 1990) as well as increased induced sister chromatid exchanges. Human tumor cells have infrequently been found to differ in double-strand break rejoining, which in addition, correlated with radiation response (Schwartz *et al.*, 1988; Kelland *et al.*, 1988). Often tumors differing in sensitivity have no difference in double-strand break rejoining but do differ in repair fidelity (Powell *et al.*, 1992).

Repair-deficient mutants

It is thought that radiosensitive mutants lack one or more components of the repair process. If two different mutants lack different components of repair, a cell formed by their fusion will lack neither repair component. They will have "complemented" each other and the cells will be radioresistant. If the same component is missing from both mutants, complementation with cell fusion will not occur. Complementation groups have been described for a number of phenotypes: radiosensitive mutants of Chinese hamster cells, xeroderma-pigmentosum, and ataxia-telangiectasia. Many complementation groups exist for a single phenotype (at least 5 for ataxia-telangiectasia, and 7 for xeroderma-pigmentosum (X-P), with at least 3 other variants of X-P) and this implies that many genes are involved in DNA repair. The number of genes and enzymes involved in excision repair and recombination is considerable. For example, in bacteria, there are four genes required (uvr a,b,c and d) to complete the incision step of the excision repair process. In mammalian cells this process is more complex. Xeroderma pigmentosum appears to be characterized by an inability to incise the damaged DNA strand 5′ to the single-strand lesion. For this syndrome alone, there are at least 7 complementation groups (Fischer *et al.*, 1985; Bootsma and Hoeijmakers, 1993). Similarly, there are a range of complementation groups for uv-sensitive CHO cells, and the human genes which complement the CHO mutants are termed excision repair cross-complementing (ERCC) genes (Bootsma *et al.*, 1988). Initially, these groups appeared to be distinct from the genes deficient in xeroderma pigmentosum, but overlap has now been identified and homology with yeast mutants identified (ERCC2 and XP-D with RAD3, ERCC3 and XP-B with RAD25, ERCC5 and XP-G with RAD2; Bootsma and Hoeijmakers, 1993). These observations emphasize that DNA repair processes involve a complex sequence or cascade of enzymatic action and repair proteins are highly conserved.

The ionizing radiation sensitive syndrome, ataxia-telangiectasia, is characterized by five complementation groups (McKinnon, 1987; Jaspers *et al.*, 1985) although the precise nature of the repair defect has not yet been determined (see section on repair of double-strand damage). Ionizing radiation-sensitive mutants of CHO cells are documented (see reviews: Jones *et al.*, 1988; Jeggo *et al.*, 1991; Thacker and Wilkinson, 1991) and at least eight complementation groups have now been established. The human DNA repair gene which complements the EM9 mutant of AA8 CHO cells has been isolated, located at chromosome 19q 13.2/13.3 and termed XRCC-1 (Thompson *et al.*, 1990). Further genes are at various stages of isolation: XRCC-2 for the irs1 mutant (Tucker *et al.*, 1991) XRCC-4 (chromosome 5) for the XR-1 mutant (Giaccia *et al.*, 1990) and XRCC5 (chromosome 2) for the xrs mutants (Jeggo *et al.*, 1992). Little is known in detail of the function of these genes. Most detail is known for XRCC-1, which corrects the high frequency of sister-chromatid exchanges and single-strand break rejoining defect in EM9, and appears to have DNA ligase III activity. How this corrects the sensitivity to ionizing radiation is not yet clear.

For bacteria there has long been evidence that the repair deficiency is the cause of radiosensitivity. There is not only a series of radiation sensitive mutants which have specific repair defects, but correction of the defect by either gene transfer or microinjection of protein can fully reverse the sensitivity back to wild-type (Friedberg, 1985). Similarly, for RAD mutants of yeast (*Saccharomyces cerevisiae* and *Schizo-saccharomyces pombe*) radiation sensitivity falls into 3 main classes: ionizing radiation sensitive, uv sensitive, or both. Some of the genes which are defective in these mutants have been cloned, and DNA transfer of these candidate genes corrects both the repair defect and the sensitivity. For those genes more fully characterized, remarkable sequence homology exists between budding yeast and human genes. For example, the rad10 gene of *Saccharomyces cerevisiae* is homologous with rad103 of *Schizo-saccharomyces pombe* which is in turn homologous with the human gene ERCC-1 (Bootsma *et al.*, 1987). Furthermore, the development of temperature sensitive mutants in yeast, has shown that ionizing radiation sensitivity goes hand in hand with the double-strand break repair defect in rad54-3 (Frankenberg-Schwager and Frankenberg, 1990).

The functional defect of various ionizing radiation sensitive yeast mutants has been investigated using a plasmid repair assay in which there is a gap in the test gene sequence. Rad50, rad52 and rad54 show a complete repair deficit, rad55 had normal efficiency and accuracy of repair, and the rad57 mutant was capable of rejoining the gap but without restoration of the sequence. Rad53 showed partial inhibition of rejoining, with normal accuracy (Glazer *et al.*, 1989). This emphasizes that double-strand breaks or gaps have *at least* two components to their repair: rejoining (which is not a simple religation) and the accuracy of the rejoining. The yeast mutant, rad52, which is double-strand-break-repair and recombination-deficient, shows that once plasmid DNA is integrated then its stability is enhanced relative to normal cells. Similarly, the *xrs*

Chinese hamster cell sensitive mutant is reported to be hypomutable, suggesting a stability of its genome (Zdzienicka *et al.*, 1988). This may also apply to the *scid* phenotype (see section on recombination repair).

Radiation resistance in tumors and mediated by gene transfer of oncogenes

In contrast, tumors are characterized by the instability of their DNA, leading to aneuploidy with bizarre chromosomal content. This leads to a second line of inquiry linking recombination to the genetic basis of radioresistance. Tumorigenesis has been reported to be associated with the acquisition of radioresistance. This comes from the observation of the response to radiation of normal cells and their malignant counterparts (e.g., fibroblasts and fibrosarcoma; colonic epithelium and colon carcinoma) but more recently from models of tumorigenesis. The transfection of oncogenes (*ras* and *myc*) into embryonic fibroblasts has led to the acquisition of radioresistance (McKenna *et al.*, 1988; 1990a; 1990b). The mechanism for this is subject to further inquiry, and one suggestion has been that increased cell-cycle delay is found in the transformed and more resistant cell line (McKenna *et al.*, 1991) and that the effect is mediated by cyclin B. However, non-specific transfection can also transform cells and increase radioresistance (M.H.L. Green, personal communication; Pardo *et al.*, 1991) and the association between radiosensitivity and cell-cycle delay are not always linked genetically (Komatsu *et al.*, 1989).

These studies raise important gaps in our knowledge about the relationship between radiation resistance and tumor development. A hypothesis might be that tumor development is associated with relative genetic instability and an increased frequency of recombination, which in turn leads to more rapid repair, damage restriction and hence relative radioresistance. Other hypotheses involve cell cycle arrest, such as *ras* outlined above, a lack of apoptosis in tumors relative to normal cells, or tolerance of damage. Considerable interest exists in the role of tumor suppressor genes in human cancer, especially p53 and Rb (retinoblastoma) genes. Both genes have been found to have a role in cell-cycle control (Diller *et al.*, 1990; Kastan *et al.*, 1991; Goodrich *et al.*, 1991). P53 appears to be involved in the G1/S cell cycle checkpoint, and mutant, inactive or absent p53 allows unchecked progress into synthesis which may provoke an increase in mutational events. Whereas this would explain the high mutability in tumors, it would not explain the functional lack of p53 being associated with radioresistance (F. Pardo, personal communication). Information about the effect of lacking p53 on radiosensitivity is not available for tumors. The effect of p53 knockout upon thymocyte sensitivity was to cause radioresistance because of the failure to induce apoptosis following irradiation (Lowe *et al.*, 1993). Ascertaining which mechanism predominates in tumors *in vivo* at low radiation dose (repair, apoptosis or tolerance) is going to be a major goal in the coming years (see Fig. 2). The "switch" between triggering apoptosis and triggering cell cycle arrest is of critical importance.

Repair during the cell cycle and cell cycle arrest

Failure to repair may result from a reduced time available for repair. It has long been suggested that an important determinant of the outcome of irradiation is the competition between fixation and repair of the induced damage. Fixation is likely to involve a mis-repair process or the passage through a particular stage in the cell cycle where the damage is made irreversible. Studies using repair inhibitors have implicated the G1/S border and the G2/M border as the critical points in the cell cycle for damage fixation. If this competition hypothesis is correct, then any delay through the cell cycle following irradiation might allow more time for repair to take place before fixation occurs. Repair varies through the cell cycle. The initial evidence supporting this statement came from cell survival studies. Terasima and Tolmach (1963) showed that synchronized HeLa cells have biphasic changes in radiosensitivity through the cell cycle, with peaks of resistance in G1 and at the S/G2 border. A similar response was seen in the L5178Y murine lymphoma resistant strains, but in the sensitive strains the G1 peak was lost in the S/S variant or both peaks were lower in the S variant. Using the neutral filter elution method, initial damage was similar in the resistant and sensitive strains during the cell cycle, but a lower rate and extent of repair was found in the sensitive strain in the G1 and G2 phases of the cell cycle (Wlodek and Hittelman, 1988). Other evidence suggests that double-strand break repair in sensitive mutants is deficient in only one phase of the cell cycle (Giaccia *et al.*, 1985).

Damage repair is characterized by a decrease in the number of lesions following exposure to the DNA damaging agent. When describing this repair, it is important to distinguish differences in the rate of removal versus differences in the completeness of removal of lesions. These two endpoints may have quite different significance. The relationship of the rate of double-strand break repair and radiosensitivity is complex. It is likely to be dependent upon cell cycle time, the extent of radiation induced synthesis delay and thus the time it takes to reach a critical phase of the cell cycle.

Following irradiation, most cells show a transient delay in replicative DNA synthesis, measured using tritiated-thymidine incorporation. The extent of cell cycle delay induced by a genotoxic agent is likely to be an important factor in determining the significance of the rate of repair. Cells which show little induced cell cycle delay can be more sensitive to that damaging agent. This is most characteristic in the cells derived from patients with ataxia-telangiectasia (McKinnon, 1987). The mechanism of radiation-induced change in synthesis rate is now beginning to be understood in genetic terms. Fornace *et al.* (1989) demonstrated the induction of gene expression by DNA damaging agents and nutrient deprivation using subtraction hybridization in rodent cells. The genes were designated as *gadd* (growth arrest and DNA damage inducible) and were demonstrated to have a distinct pathway from the growth arrest produced by terminal differentiation. A separate line of study into the function of p53 discovered it to be both a transcription factor and important in cell cycle delay at the G1/S checkpoint. The N-terminus was found to have an acidic domain (Fields and Jang, 1990) like other transcription factors; a consensus binding sequence was discovered (Kern *et al.*, 1991) suggesting p53 was functional as a tetramer; p53 was not essential for the normal cell cycle as cells containing no p53 were viable; p53 appeared to be required for a "stress" response to DNA damage and its induction was temporally associated with the G1 arrest (Kastan *et al.*, 1991). Caffeine treatment blocked the induction of p53 and the G1 arrest (as well as G2 arrest). Cells lacking p53 did not show G1 arrest, but did show G2 arrest. More recently, the cell cycle checkpoint pathway controlling G1/S arrest in response to DNA damage has linked *GADD*45 to p53 and the A-T gene (Kastan *et al.*, 1992). Cells lacking p53 do not induce *GADD*45 in response to DNA damage, the *GADD*45 gene contains a p53 binding sequence (interestingly in its third intron, not in the promoter region) and thus p53 is thought to be upstream of *GADD*45. A-T cells do not show an increase in p53 after ionizing radiation, suggesting the A-T gene product is upstream from p53. Further unpublished work from Kastan's group has suggested that DNA strand breakage is the major trigger of p53 induction by studying a wide variety of DNA damaging agents and modifying conditions. No clear distinction was made between double and single strand breaks in this analysis: both restriction endonuclease and DNase I treatment were effective inducers of p53.

Changes in accessibility of DNA to enzyme attack either by exogenous nuclease or by DNase I is observed almost immediately after irradiation and this change returns to normal following irradiation with a half-time of several hours, comparable to the timecourse for the restoration of normal DNA synthesis. How these changes relate to the molecular pathways described above is not clear. It is likely that there is a similar pathway controlling the G2/M checkpoint which is genetically distinct from the G1/S checkpoint. Override of the checkpoint for synthesis does not significantly accelerate the subsequent mitosis (Downes *et al.*, 1990). However, the association between lack of synthesis delay and radiosensitivity is not complete: the extent of cell cycle delay does not correlate well with radiosensitivity; override of the cell cycle arrest by caffeine does not necessarily lead to increased cell kill (Musk, 1991); and cell fusions have been isolated where the radiosensitivity is lost, but the lack of radiation induced synthesis delay is maintained (Komatsu *et al.*, 1989). Whether a prime function of cell cycle arrest is to allow repair of DNA damage is not fully resolved (Fig. 2). It seems likely that the G1/S checkpoint is less closely associated with radiation resistance than G2/M (see section on signal transduction and cell cycle control).

Repair mechanisms

Cells have developed a wide variety of repair mechanisms for dealing with the many different types of DNA damage. The complexity of this organization is only partly understood. For the basis of this brief discussion, I will discuss repair mechanisms under three headings: 1) reversal of damage; 2) excision repair for damage confined to a single strand; and 3) repair by recombination for double-strand damage. These three mechanisms broadly represent a hierarchy from the repair of small simple lesions to the restoration of large complex lesions. If the function of one component is defective, repair defaults to the next stage in the

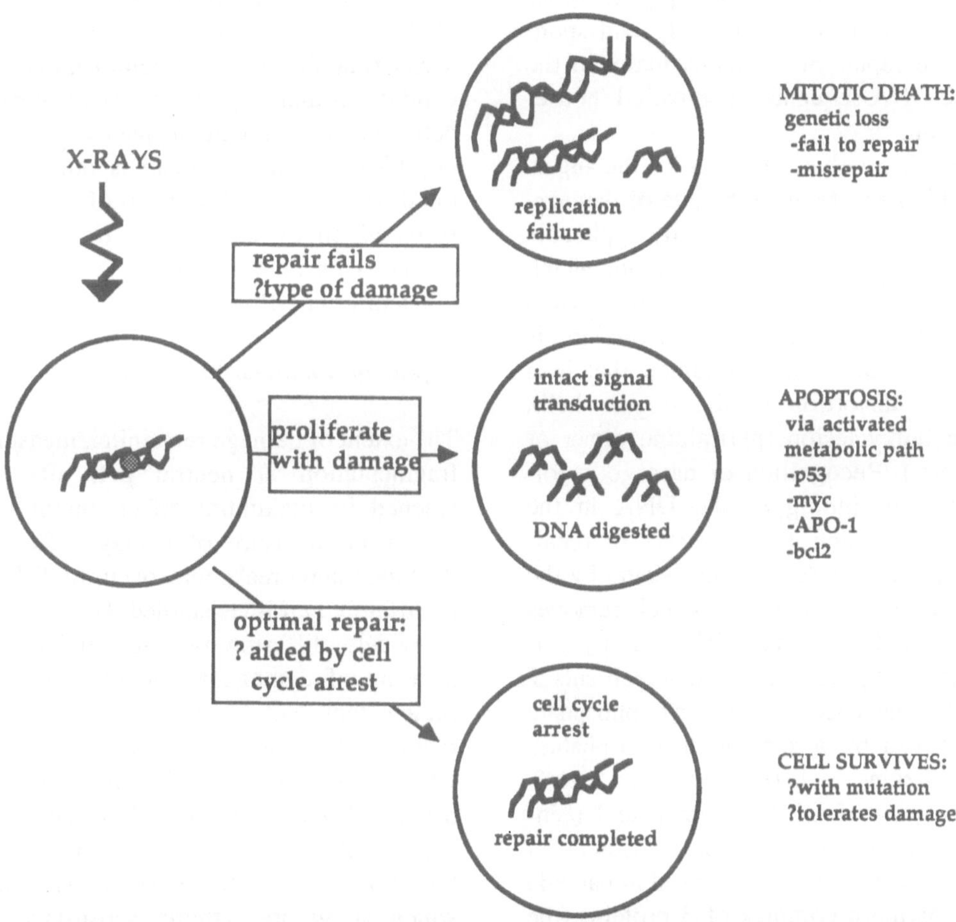

Fig. 2. The possible ways ionizing radiation leads to cell killing. DNA damage can lead to genetic loss and failure to survive mitosis; or can lead to activate apoptosis by an unknown damage signal. DNA damage may be correctly repaired or tolerated with the cell surviving.

hierarchy. The relative importance of recombination in controlling radioresistance will be emphasized.

The reversal of damage in DNA is the most direct mode of DNA repair. Examples of this mechanism of repair include: enzymatic photo-reactivation of pyrimidine dimers; repair of guanine alkylation; repair of single-strand breaks by direct rejoining; and the repair of sites of base loss by direct insertion (Friedberg, 1985). For most of these examples a single gene is required for a specific function. At first sight, this seems an economic use of genetic information compared with the multiple gene products required for the more complex repair processes. However, since there is a large variety of DNA lesions induced, this type of specific direct reversal of damage is likely to apply to those lesions which are produced frequently and spontaneously. These repair mechanisms maintain the integrity of the DNA sequence: a so-called 'housekeeping' function.

Excision repair involves cleavage of the sugar-phosphate backbone to remove the site of damage. This mechanism of repair is most easily applicable to lesions affecting a single DNA strand, which either requires single nucleotide excision or patch excision. The mechanism of damage recognition is unknown. Many authors have assumed that it is detected by the distortion to the double helix caused by the bulky lesion (pyrimidine dimer or drug adduct, etc.). Recognition of damage is followed by local unwinding of the DNA in the region of damage, which allows access of repair enzymes. Single nucleotide excision occurs by the following sequence: a glycosylase which removes the base damage and leaves an apurinic or apyrimidinic (a.p.) site; an a.p. endonuclease which cuts 5' to the a.p. site; the baseless deoxyribophosphate residue is removed by a specific 5' phosphatase enzyme. This enzyme is distinct from the 5' to 3' exonuclease activity of DNA polymerase I (Lindahl, 1990). Patch excision involves a different set of enzymes. This has been characterized in bacteria as initially involving a complex of 3 proteins (the UvrABC enzyme complex in E.coli). This complex binds to the DNA at the site of damage and incises the damaged strand at a distance 5' and then 3' to the lesion. Thus, the damaged patch is removed and the space is resynthesized by the action of DNA polymerase, and joined by ligase. The exonuclease activity of DNA polymerase makes the length of the resynthesized patch considerably larger than the size of the patch initially excised.

Excision repair is a multi-step process, and is less rapid in action than the single-step actions involved in the direct reversal of damage. The precise kinetics of excision repair are difficult to define. This is because, for pyrimidine dimer removal, the rate of removal depends on whether the sequence is actively transcribed (Bohr *et al.*, 1985) or even whether it is the sense or the anti-sense strand (Mellon *et al.*, 1987). There is some debate as to whether there are different mechanisms involved in the repair in active or inactive regions of DNA. This stems from the observation that xeroderma pigmentosum complementation group C exhibit normal repair of active genes, but are deficient in the repair of inactive DNA (Mayne *et al.*, 1988). However, there is other evidence that the different rates of removal of pyrimidine dimers is based on different delay times to initiate excision. This delay is partly cell cycle-dependent (Terleth *et al.*, 1990).

Repair of double-strand damage

The extent of damage remaining, measured as DNA fragmentation at neutral pH, after repair has reached its maximum effect, might be the most significant measure of damage. The removal of double-strand breaks may occur to different extents in different cell lines studied. However, many cells of widely differing radiosensitivity decrease the measure of damage at the same rate and to the same extent. Notable exceptions to this include the radiosensitive mutants of CHO cells (Kemp *et al.*, 1984; Giaccia *et al.*, 1985) the murine lymphoma cells, L5178Y (Wlodek and Hittelman, 1987) and normal fibroblasts from *scid* mice (see below). Fibroblasts from patients with ataxia-telangiectasia, which show an extreme sensitivity to ionizing radiation, are little different from normal fibroblasts either in the induction or repair of double-strand breaks using these techniques (Lehmann and Stevens, 1977).

In addition to whether the DNA remains fragmented, closure of double strand breaks can lead to errors. Measures of repair fidelity have necessitated the use of exogenous DNA, in which damage can be defined prior to repair and the accuracy of repair can be measured by the restoration of gene sequence or gene function. Repair fidelity experiments can be cell-mediated requiring DNA transfer into living cells, or mediated *in vitro* using cell extracts. For cell-mediated repair, DNA probes are based either on virus infection or plasmid transfection. It has been known since the mid-1950s that damaged viruses can still infect their host cells, where they are reactivated. The demonstration that host-cell reactivation of damaged virus was dependent upon the host cell's repair mechanisms was of fundamental importance for subsequent studies (see review by Defais and Hanawalt, 1983). For example, reduced host-cell reactivation was demonstrated in xeroderma pigmentosum cells infected with uv irradiated viruses. This result was repeated for a wide variety of different viruses, emphasizing that the result depended upon the type of DNA damage and host-cell repair rather than the virus or probe.

Ionizing radiation damage to virus DNA is dominated by single-strand breaks and base damage. This problem of specificity with irradiated virus has led to the study of specific double-strand breaks in plasmid DNA. The use of plasmid DNA repair probes allowed the study of functional double-strand break (dsb) repair. The double-strand break was modelled by restriction endonucleases. The value of this assay was demonstrated in ataxia-telangectasia (A-T). A-T cells have demonstrated, unexpectedly, an ability to rejoin dsb equal to normal cells in the damage assays outlined above. A plasmid was transfected either undamaged or with a restriction endonuclease-induced break through a selectable marker gene. Growth of cells transfected with the damaged plasmid in the selection medium requires the rejoining of the break and the restoration of gene function. With this method, it was found that A-T cells may not lack the ability to rejoin breaks in DNA, but the errors involved in this rejoining were greater than in normal cells (Debenham *et al.*, 1988a). This type of study has been applied to the different complementation groups of the radiosensitive Chinese hamster cell-lines. Line *irs1* showed a low rejoin fidelity, which was not seen for either *irs2* or *irs3* relative to their parent V79 line (Debenham *et al.*, 1988b). This further suggests that the functional defect in complementation groups is different, although the nature of the repair defect remains far from clear. Repair fidelity has been studied in human tumors of widely differing radiosensitivity. Using methods similar to those described above, a striking correlation between sensitivity and low repair fidelity was observed (Powell and McMillan, 1993). The observation was not universal, and one of nine cell lines studied was radioresistant and showed low repair fidelity. However, radiosensitive clones of two of these tumors were found to have lower repair fidelity without changes in repair using strand breakage assays (Powell and McMillan, 1990; Powell *et al.*, 1992). This emphasizes that strand break repair measures alone are inadequate to describe the full spectrum of radiation resistance.

A radiation-induced dsb is associated with the loss of at least one base from each strand. This is because ionizing radiation rarely involves simple cleavage of the phospho-diester bond, and excision repair is required. If the damage is in opposite bases, the sequence will be lost and simple religation is not possible. If there is separation of the strand breaks or base damage, then excision of the damaged patch of DNA will lead to loss of the sequence if the excision size is greater than the separation of the strand breaks.

It has often been difficult to demonstrate differences in DNA repair between cells of differing radiosensitivity with the methods available. Chromosomal aberrations, by contrast, consistently correlate well with cell survival. One of the main problems with techniques to measure DNA damage is that they measure rejoining of fragments and not full functional repair. That is, they measure that a break has been removed but they give no information as to whether the functional integrity of the DNA has been restored. Chromosome aberrations identify many lesions in which the fragmentation in DNA has resolved but clear abnormalities remain. Many of the exchange events reduce fragmentation, including the intrachromosomal translocation (e.g., inversion) and the interchromosomal translocation. Exchange events in response to DNA double-strand breakage are mediated by the process of recombination.

Recombination repair

If DNA damage has affected both strands then either it will lead directly to a double-strand break or indirectly as a result of excision repair processes acting on both strands. Since excision repair is based upon the integrity of the complementary strand, this process could lead to loss of genetic sequence. The question remains, therefore, when double-stranded DNA damage and loss of genetic sequence has occurred, how can this be repaired? Can the sequence be restored or does the cell try to limit the damage? With current knowledge, the only conceivable way to restore sequence is the process of recombination, but as we shall see recombination may well have a more universal role.

Damaged DNA will combine with undamaged DNA, which is presumed to require at least local sequence homology. Strand exchange follows, which can lead to the restoration of genetic sequence. The missing single-stranded patch can be restored to both DNA double-strands by the action of polymerase and ligase as described for excision repair. Conversely, the same mechanism is used in the deletion of specific sequences by the process of intrachromosome recombination. The precise mechanism of recombination will not be discussed here; the most widely accepted model of recombination is that proposed by Holliday, in which strand crossing leads to the formation of the Holliday structure. This is a junction between the four incoming sequences of double-stranded DNA. The resolution of the structure determines whether a patch of sequence is exchanged between strands (heteroduplex) or whether there is both a short patch of heteroduplex as well as full strand exchange. This latter mechanism accounts for the recombination events occurring in sister chromatid exchange as well as meiotic recombination.

A number of lines of evidence suggest that recombination may be an important process in the repair of ionizing radiation-induced damage. Most assays of a recombination are based on exchange of gene sequence within a single plasmid or between plasmids. The assays have often used plasmids which remain extrachromosomal, where precise sequence information can be readily obtained. Many ionizing radiation-sensitive mutants have been shown to be deficient in recombination. For example, the rad52 mutant of *Saccharomyces cerevisiae* is deficient in recombination, as judged by the ability to recombine plasmid DNA sequences separated either as fragments or with intervening sequence (Perera *et al.*, 1988). This same mutant is also deficient in the repair of double-strand breaks measured by velocity sedimentation (Brunborg *et al.*, 1980; Ho, 1975). The finding that haploid wild-type yeast can also be repair deficient suggested that the likely template for the restoration of sequence is the homologous undamaged chromosome. However, highly aneuploid mammalian cells, where no homologous chromosomes exist, are able to repair dsb and perhaps all that is required for recombination and hence repair to take place in this case is local homology of sequence. There has been some concern in using information from yeast recombination and applying this to mammalian cell recombination. Yeast recombination is dominated by recombination between highly homologous sequences. This does not appear to be the pattern for mammalian cells, where 'non-homologous' recombination predominates (Subramani, 1989). "Non-homologous" recombination in this context means strand-exchange between non-homologous regions, rather than no homology at the site of strand-exchange. In other words, the extent of homology can be limited but still allow recombination. For small gaps in sequence, non-reciprocal exchange of DNA may require only short regions of homology. The larger the deleted sequence the more likely this will require extensive sequence homology such as exchange between homologous chromosomes.

The importance of recombination in the processing of ionizing-radiation damage in mammalian cells was highlighted by the publication by Fulop and Phillips (1990). Mice, exhibiting severe combined immune deficiency (*scid*) have an inability to rearrange correctly the immunoglobulin and T-cell receptor genes. The inability of their defective recombinase system leads to the mice having no detectable B or T lymphocytes. However, the defect is not specific to lymphocyte development, and other myeloid cells and fibroblasts show the same deficiency. This deficiency has been characterized by ionizing radiation sensitivity. Furthermore, *scid* cells have been found to be double-strand break rejoining deficient (Biederman *et al.*, 1991; Hendrickson *et al.*, 1991) thus linking the VDJ rejoining process of *scid* to a general fragment rejoining process. Roth *et al.* (1992) presented evidence that the *scid* phenotype is associated with the persistence of DNA hairpins at the signal sequences of the immunoglobulin gene. It may be that double strand cleavage of DNA leads to "protected" hairpin intermediates which the *scid* phenotype cannot resolve. The precise nature of the link between the *scid* phenotype and ioninzing radiation sensitivity is likely to be instructive in understanding double-strand DNA processing.

There is evidence that persistent DNA damage provokes recombination events. DNA double-strand breaks are not repaired as rapidly as the processes of damage reversal and excision repair. Furthermore, if excision repair is defective and damage persists, this alone appears to provoke recombination (Nairn *et al.*, 1988). It is suggested, therefore, that recombination is a repair mechanism which occurs, when damage cannot be handled by the more rapid and specific repair mechanisms of damage reversal and excision repair. There is a wealth of data to suggest that recombination is enhanced by the presence of DNA double-strand breaks both in lower organisms (Thaler *et al.*, 1987) and mammalian cells (Hoy *et al.*, 1987; Hamilton and Thacker, 1987). In other words, for serious damage such as double strand DNA breaks or gaps, recombination may act as the default repair pathway to limit the genetic loss. It was recently demonstrated, using the plasmid repair fidelity assay, that A-T cells show errors in association with the integration of circular and linear plasmid DNA suggesting an error-prone recombination as its repair defect rather than a defect specific to cleaved DNA (Powell *et al.*, 1993). The observations for A-T and *scid* suggest that their x-ray sensitivity are related to an ability to rearrange DNA in general and not specific to cleaved DNA. This may well be true for other x-ray sensitive cells, as their functional defect is discovered, just as X-P cells types B,D and G appear to have defective helicase function leading to both excision repair defects and transcription failure.

Resistance to ionizing radiation and cytotoxic drugs compared

The mechanisms of resistance to chemotherapeutic drugs may result from decreased blood supply, from decreased absorption into tissues, from decreased cell uptake, from altered metabolism and finally from altered damage processing. Ionizing radiation resistance is dominated by damage processing detailed above, and altered cellular metabolism. The potential for inducing drug resistance is considerably greater than for radiation resistance. It is known that exposure of tumor cells to one drug may induce cross-resistance to a variety of unrelated drugs to which the cell has not been previously exposed. Similarly exposure to ionizing radiation may likewise induce drug resistance, but drug exposure rarely induces ionizing radiation resistance. Potential sites of overlap between radiation and drug resistance mechanisms will be discussed.

Cellular metabolic responses to ionizing radiation

In exponentially growing cells *in vitro*, ionizing radiation rarely leads to a permanent change in radiation resistance. Transient changes in resistance do occur, but they decay with a half-life of a few hours. One example of this would be the ability of cells to arrest progress through the cell cycle in response to DNA damage, detailed above. However, cells *in vivo* are subject to more complex growth control and proliferation signals, with the potential for ionizing radiation to lead to changes in growth factors and regulatory signal transduction pathways. The array of "damage response genes" has progressively enlarged and some of these will lead to resistance to ionizing radiation. Possible levels for interaction would include: the ability to arrest cell cycle progression; the ability to inhibit apoptosis; direct interaction with repair functions; and the ability to stimulate proliferation.

How does the cell monitor damage or stress?

Three possible sites of damage surveillance could be: changes in the structure of DNA; changes in the free radical chemistry of the cell; and changes in membrane configuration associated with altered membrane conductance. There are few firm data, but superficial evidence to support all three routes. The presence bulky lesions on a single strand of DNA is rapidly detected by the excision repair machinery whether treated within the cell or introduced into the cell after the damage was inflicted. These studies have not extended to evaluating the effect of transfected damaged DNA upon other patterns of stress response. The G1/S checkpoint via p53 appears to be activated by DNA strand breakage. Compelling evidence in support of this view, from Michael Kastan's group, is that restriction endonucleases introduced into cells by electroporation were a powerful inducer of p53, and that this effect was eliminated by heat inactivation of the enzyme. Other sources of strand breakage were equally able to induce p53. For example, the effect of ultraviolet radiation in cells unable to repair the pyrimidine dimer (such as xeroderma pigmentosum) is to produce strand breaks when the replication fork goes by the pyrimidine dimer. If replication is inhibited by hydroxyurea and arabinoside (Ara-A), then the p53 induction is lost.

Membrane response to ionizing radiation

The molecular processes mediating the cellular response to ionizing radiation may in part act via membrane changes which in turn activate signal transduction from cell surface receptors to the nucleus. Distinguishing whether membrane or cytosolic chemistry has the dominant role in signalling adaptive survival mechanisms depends on the identification of the initial x-ray induced changes. The answer may be an interaction of both. The recent work of Kuo *et al.* (1992) has demonstrated that potassium channel activation is an early response (within seconds) of irradiated cells responding to increased levels of hydrogen peroxides and agents giving rise to reactive oxygen intermediates. Using the technique of whole cell voltage clamping, potassium currents increased after treatment with ionizing radiation, hydrogen peroxide and xanthine/xanthine oxidase, but not after tritiated thymidine, suggesting a membrane effect. The effect was abolished by free radical scavengers, but was maintained in the presence of agents which deplete protein kinase C (PKC – see below).

The cell membrane acts to maintain the intracellular concentration of ions. Permitting changes in ion concentration result in induction of the so called "early response genes". Ion channels may have a role in the initiation of or response to signal transduction, utilizing their porperty of rapid conformational alterations in reponse to changes in the membane chemistry. A hypothesis under investigation is that opening of the MDR ATP-linked chloride channel in response to increased local concentrations of peroxides is one initial response to ionizing radiation. The acetylcholine receptor in muscle cells is directly responsible for passing an ionic signal across the membrane in the form of sodium ion influx which activates the intracellular signal pathway activating contraction. The MDR system may function in a different way, via the efflux of ATP to the extracellular surface, which allows interactions with specific purinergic receptors and provides ATP for ectokinases. Abraham has observed (unpubl.) the prompt release of ATP by cells irradiated to a low dose, 0.1 Gy. The specificity of the observation to MDR1 protein, rather than a non-specific membrane leak, was suggested by the magnitude of ATP release being related to the level of MDR1 expression.

The so-called ABC (ATP binding cassette) proteins have been shown to be transporters of drugs and channels. It appears that the removal of drug is an active phenomenon which results in reduced cellular cytotoxicity. Recently the p-glycoprotein (PGP, p170, MDR1) has been shown to function as both a volume regulated chloride channel (see review: Higgins, 1992) and an ATP channel (Abraham *et al.*, 1992). Morever, it was found that the MDR protein utilizes the transmembrane ATP gradient to remove drug from cells. Reversing the transmembrane ATP electrochemical gradient by depolarizing the membrane potential and reversing the ATP chemical gradient can reverse drug efflux performed by PGP.

Does the development of drug resistance as manifest by the presence of the mdr1 gene product lead to altered sensitivity to ionizing radiation? Increases in the amount of MDR protein present in the plasma membrane has been demonstrated by *in vitro* and *in vivo* experiments following fractionated ionizing radiation (Hill, 1991; McClean *et al.*, 1993). The increased expression of MDR1 can be mediated by p53 and *ras* (Chin *et al.*, 1992). Overexpressing MDR1 does not necessarily lead to radioresistance, but it has been observed to stimulate glutathione levels which could act as a radiation protector.

The extrusion of the ATP leads to an extracellular ATP cycle (see Fig. 3) which can result in extracellular phosphorylations (Skubitz and Goueli, 1991) and activated purinergic receptors (Burnstock, 1990). Binding to the purinergic receptor can activate intracelluar enzymes, such as the adenylate cyclase and membrane phospholipids. In a complex multicellular organism this response may be part of a system of communication between cells. ATP does not reenter the cell as ATP, but is salvaged in the form of adenosine after all phosphates have been removed.

Changes in the extracellular ATP concentration can modulate a variety of cellular responses including apoptosis, regulation of the cell cycle, signal transduction leading to thiol protection. Zheng *et al.* (1991) demonstrated that extracellular ATP bound to purinergic receptors led to the mobilization of intracellular calcium, presumably via the phospholipase C and inositol triphosphate pathway, leading to apoptosis. Rappaport (1983) showed that treatment of human tumor cells with ADP or ATP lead to growth arrest in the S phase of the cell cycle. Abraham (unpubl.) has evidence that ATP can lead to an increase in the intracellular glutathione level.

The role of extracellular ATP, for example in the control endothelial function, has recently been reported and reviewed (Boeynaems, 1988; 1989; 1990; Pirotton, 1990; 1992). The control of endothelial proliferation and modulation of its function is crucial to tumor growth. In addition to responding to extracellular ATP, endothelial cells synthesize and release two vasodilators: protstacyclin (PGI2) and nitric oxide (NO) originally called endothelial cell relaxing factor (EDRF). PGI2 is released rapidly following ATP stimulation of endothelial cells but returns to baseline within minutes of stimulation. Sources of ATP for this stimulation are speculated to be from platelet degranulation, red blood cell hemolysis and from endothelial cells following shear stress.

The role of intracellular GTP in the signal transduction pathway is well known. GTP-binding proteins are found on the intracellular surface of the cell membrane. In particular, two forms of the GTP-binding proteins that control cell proliferation are the monomeric products of *ras*, already noted to be involved in the upregulation of MDR expression, and the heterotrimeric G proteins (Gilman, 1987). The molecules that perform the hydrolysis of GTP are GAP (GTPase activating protein) in association with p21 *ras,* or a related family of proteins. They function in many signalling pathways including that of the rhodopsin, adrenergic and muscarinic receptors. The rhodpsin receptor is notable for its ability to respond to radiation, albeit in the visibile range.

Genes induced following exposure to ionizing radiation

Following ionizing radiation, certain genes are induced rapidly and in the presence of protein synthesis inhibition (Weischelbaum *et al.*, 1991). These early response genes initiate a cascade probable transcription factors including *fos, jun* and Egr-1. The triggering mechanism of this induction is not known between nuclear damage and membrane damage or an interaction of both. The levels of c-*fos* and junB mRNA increase after x-ray exposure in HL60 cells (Sherman *et al.*, 1990) but

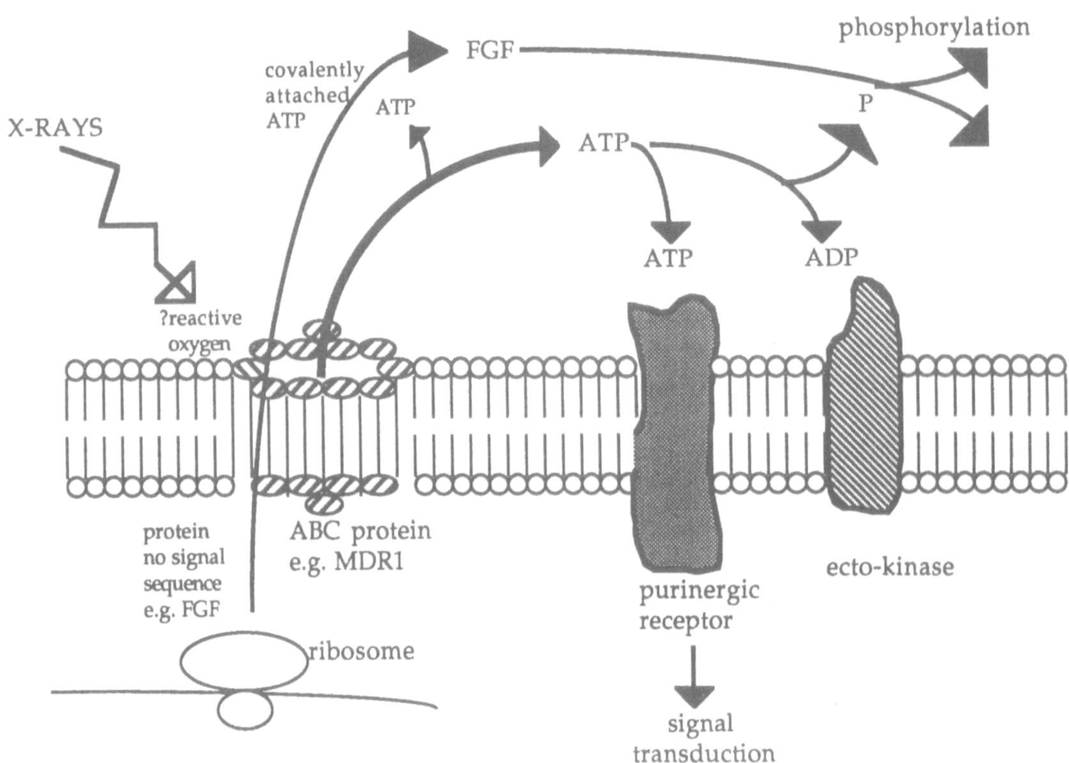

Fig. 3. The effects of ionizing radiation on membrane transport. Using the MDR1 protein as example, it is hypothesized that a direct effect of radiation is to alter the conductance of the channel and allow an increased efflux of ATP. This can act via an autorine loop to alter signal transduction pathways; or can phosphorylate extracellular proteins via an ecto-protein-kinase. Other proteins may be secreted via the MDR1 channel which can contribute to the tissue response to radiation.

in fibroblasts, sarcomas and epithelial tumor cells c-*jun* but not c-*fos* was induced (Hallahan *et al.*, 1991). Since fos-jun heterodimers are known to have higher affinity to the AP-1 DNA binding site than jun-jun homodimers, the cell lineage dependence of the different pattern of induction may have functional significance.

The induction of Egr-1, a nuclear phosphoprotein with a zinc-finger motif, has also been reported following ionizing radiation exposure (Hallahan *et al.*, 1991). This same protein has been reported to participate in the transition of cells from quiescent to proliferating states. How this observation relates to cell cycle arrest in response to ionizing radiation is unclear. The speed of gene induction is within 15 min, which makes it unlikely to be a consequence of the cell cycle arrest response which peaks after several hours.

It is not known whether these early response genes regulate the subsequent response to ionizing radiation. Another approach has been to analyse two dimensional gel electrophoresis before and after irradiation and compare the induced proteins with those induced by heat shock, hypoxia and alkylating agents. "x-ray specific induced proteins" (XIP) of various molecular weights have been described (Boothman *et al.*, 1989) but the type of protein or its function remain unknown. Transient induction in radiation resistance has been described for the same cells with initial x-ray doses of only 0.05 Gy and caffeine blocks the induction of XIPs (Hughes and Boothman, 1991).

If the early response genes make DNA binding proteins which act as transcription factors, what genes are transcribed? The answer seems to be a wide variety of genes, probably erroneously grouped under the heading of growth factors. PDGF and FGF are released from vascular endothelial cells after radiation (see section on secreted growth factors). Tumor necrosis factor (TNF) mRNA is increased after ionizing radiation (Hallahan *et al.*, 1989). TNF has been observed to sensitise the effects of radiation *in vitro* (Hallahan *et al.*, 1990) but how TNF induction *in vivo* functions to alter the effect of ionizing radiation is not known. The TNF gene promoter has an AP-1 binding site and an Egr-1 binding site suggesting induction via jun or Egr-1.

The relationship between x-ray induced gene expression and signal transduction

Many proto-oncogenes are thought to encode proteins involved in the cascade of events by which growth factors stimulate cell proliferation. Oncogenes have been identified that correspond with the growth factors (v-sis, c-*sis* = PDGF), the growth factor receptors (v-erb, v-fms, v-kit), transducers of the growth factor responses (v-src, v-ras, v-raf), and transcription factors that mediate growth factor-induced gene expression (v-jun, v-fos). The activation of phospholipase C, resulting in hydrolysis of membrane phospholipids to diacylglycerol (DAG) can be mediated via the epidermal growth factor receptor (EGFR) family of proteins which include the constitutively active, ligand binding independent v ErbB and c ErbB. DAG subsequently activates members of the phospholipid-dependent, serine-threonine specific PKC family. The c-*raf* proto-oncogene encodes a serine-threonine protein kinase which is activated following PKC activation. PKC plays a pivotal role in the molecular response to growth factors and mitogen, which control cell proliferation and differentiation. PKC is also activated by the tumor-promoting phorbol esters, which mimic the activating effect of DAG, by superoxide, and extracellular ATP. Extracellular ATP interacts with purine receptors to activate PKC. PKC once activated participates in the signal transduction to the nucleus followed by transcriptional activation of genes with phorbol ester-responsive cis elements within their promoter regions.

PKC may be "directly" activated by ionizing radiation. PKC activity extracted 15 s following x-ray treatment is 4.5 fold higher than control (Hallahan *et al.*, 1991). Peak PKC activity is seen at 30 s and returns to baseline after 60 s. Depletion of PKC by the phorbol ester TPA or the kinase inhibitor H7 abrogates the ability of x-rays to induce other stress response proteins such as c-*jun* and tumor necrosis factor (TNF). The speed of this response suggests a direct effect upon the membrane, but this is not proven: hydrolysis of phospholipids has been suggestesd (Nishizuka,1992). A different kinase inhibitor, selective for cyclic adenosine monophosphate depenedent protein kinase, does not block the induction of jun and Egr-1. Ionizing radiation is known to cause free radical production and subsequent oxidative damage within cells. One potential mechanism of PKC activation by ionizing radiation may involve degradation products of oxidized membrane phospholipids. Higly reactive hydroxyl radicals produced within the irradiated cell cause lipid peroxidations. These oxidized membrane phospholipids are hydrolyzed to form fatty acid byproducts such as DAG. Recent evidence has suggested that a tyrosine phosphorylation step is required in induction of PKC (Uckun *et al.*, 1993).

The relationship between signal transduction pathways and radiation resistance is complex. Many changes occur in response to ionizing radiation. The functional outcome, in terms of resistant or sensitive to radiation, is determined by multiple cellular metabolic factors which are cell lineage and tumor cell dependent. For example, the original observation by Sklar (1988) that transfection of the activated oncogene *ras* could confer radioresistance in NIH 3T3 cells has not been explained. The observation has been repeated for rat embryo cells (McKenna *et al.*, 1990) and further studies revealed no change in DNA damage and repair (using fragmentation assays) but did show an increased G2 delay. The molecular link between mutated *ras* and G2 delay is unknown. Transfection of mutated *ras* into already transformed cells (tumors not containing mutated *ras*) appears to have no effect on radiation resistance.

When DNA from a human laryngeal carcinoma was transfected into NIH 3T3 cells a c-raf transcript was identified in the transformants (Kasid *et al.*, 1987). The transformed cells had acquired radioresistance and antisense raf RNA reversed both the resistance and the tumorigenicity (Kasid *et al.*, 1989). How raf is conferring radioresistance is uncertain, but Burt *et al.* (1988) report that *ras* and raf increase MDR1 and glutathione S-transferase leading to chemoresistance.

Signal transduction and cell cycle control

There is no direct evidence linking the early response genes induced by ionizing radiation and cell cycle arrest induced by ionizing radiation. Kastan *et al.* (1992) have reported the G1/S checkpoint to be under the control of a p53 and GADD45 pathway following ionizing radiation, and that DNA strand breakage is a potent inducer of p53. The increased p53 expression is not mediated by increased transcription, but the mechanism of regulation is unknown. The relationship of the G1/S checkpoint and radioresistance may not be close: fibroblasts from p53 null mice show no altered sensitivity relative to heterozygotes and wild type (Powell, unpubl. observation; Kastan, personal communication). Paradoxically, thymocytes from p53 null mice show radioresistance, because of the inability to activate apoptosis (Lowe *et al.*, 1993). Again, the relationship to early reponse genes and signal transduction is not clear.

The G2/M checkpoint may be more important for radiation resistance. Although the final initiation of mitosis is mediated via cyclin B and p34 (cdc2 or cdk1 gene product) upstream control of this checkpoint in mammalian cells is not known. In the yeast *Saccharomyces cerevisiae* the RAD9 protein regulates the equivalent step. This protein has homology with part of the DNA binding domain of c-*jun* (Schiestl *et al.*, 1989) but there is no proof that c-*jun* is involved in cell cycle arrest.

Secreted growth factors

A class of proto-oncogenes encodes secreted proteins that function as growth factors. The MDR channel may function to release some of these proteins, and in particular for some of the important peptides which mediate the response to ionizing radiation and environmental stresses such as FGF, PD-ECGF, TNF, TGF-alpha (Vlodavsky, 1991a,b). Peptides such as FGF and PD-ECGF have the interesting property that they lack a signal peptide associated many secreted proteins (Mignatti *et al.*, 1992). In addition, FGF and PD-ECGF have covalent ATP and phosphate added extracellularly (Usuki *et al.*, 1989, 1991, 1992; Heldin *et al.*, 1991; Vilgrain and Baird, 1991; Feige and Baird, 1989, 1991). Whether these peptides may be secreted utilizing MDR type membrane proteins is under investigation. The covalent binding of ATP to the peptides may be the mechanism and driving force of the peptide secretion. After release, these factors can bind to the cells own receptors, which can then release internal stimulatory signals, with the potential for an autocrine stimulatory loop.

The use of beta-FGF in mice has been shown to protect against radiation induced lung injury (Fuks, unpubl.; Fuks and Weischelbaum, 1992). Radiation resistance can begin to have a broad basis: FGF in protecting against lung injury may be working to inhibit apoptosis in pneumocytes or to stimulate proliferation in macrophages. In other words, when tissue resistance is being evaluated, the whole functional unit must be taken into account involving autocrine, paracrine and endocrine signalling pathways. Changes in prostaglandin metabolism have long been known to impart tissue resistance to radiation. How these effects are linked to the signal transduction pathways described is not clear. Their actions may be tissue type dependent.

Cell-cell communication

The signalling systems which have developed for communication among cells and within cells under normal physiological conditions may alter at times of environmental stress. Much of this communication is known to involve hormonal signals, which often interact at the cell surface by not necessarily gaining entry into the cell. Some hormones cross the membrane and interact with intracellular receptors either in the cytosol or nucleus. The possible metabolic correlates of signal transduction include activation of protein kinases, alteration in phospholipid cycles, alteration in cellular calcium and ionic concentrations, and modulation of gene transcription and cell cycle.

The MDR1 glycoprotein has been hypothesized to function as an ATP channel and recent studies have shown that the transmembrane electrochemical gradient is capable of removing drugs. The release of ATP is capable of altering the extracellular microenvironment operating by means of a variety of surface receptors such as the purinergic and adenosine receptors, the P1, P2 and A1, A2 receptors respectively. Moreover, the extracellular ATP serves as a substrate for ecto-kinases, which have the potential to work in a paracrine fashion for cell-cell interaction. Similarly this may be the important role of FGF.

Gap junctions are ports of communication between cells. They are a normal feature of contact-controlled growth in normal epithelial cells, and often missing in tumors (e.g., Trosko *et al.*, 1990). It has been known for some time that radiation resistance can be acquired by cell contact (Durand and Sutherland, 1973). An interesting question is whether they provide an exploitable difference between normal and tumor cells for radiation resistance. Altered regulation of intercellular communication has been reported by various growth factors and peptide hormones (Madhukar *et al.*, 1989). Contact-inhibited cells are in a non-proliferative state which have been known for many years to have relative resistance to irradiation, via the functionally-defined increased potentially lethal damage repair. The molecular basis to this effect is not clear. Whether it is due to the quiecent state or an effect of cell-cell contact or a combination of both remains to be elucidated.

Summary of radiation resistance

Radiation resistance differs from drug resistance in lacking the changes in transport, uptake and extrusion. Damage processing and damage-induced changes in gene expression have a surprising degree of overlap. Repair genes for ionizing radiation are frequently cross resistant with drugs causing DNA strand breakage, in particular affecting both strands. Strand-breakage is a common theme for induced G1/S cell cycle arrest. The relationship between changes in signal transduction pathways and radiation resistance is not clear, but certain oncogenes transfected into normal cells involve these pathways and lead to resistance. G2/M arrest may be more closely associated with radioresistance than G1/S arrest.

References

Abraham EH, Prat AG, Gerweck L, Seneveratne T, Arceci RJ, Kramer R, Guidotti G and Cantiello HF (1993) The multidrug resistance (mdr1) gene product functions as an ATP channel. Proc Natl Acad Sci USA 90 (1): 312–316.

Biedermann KA, Sun J, Giaccia AJ, Tosto LM and Brown JM (1991) Scid mutation in mice confers hypersensitivity to ionizing radiation and a deficiency in DNA double-strand break repair. Proc Natl Acad Sci USA 88: 1394–1397.

Blöcher D and Pohlit W (1982) DNA double-strand breaks in Ehrlich ascites tumour cells at low doses of x-rays. II. Can cell death be attributed to double-strand breaks? Int J Radiat Biol 42: 329–338.

Boeynaems JM, Demolle D, Pirotton S, Raspe E, Lecomte M, Hepburn A, Van Coevorden A and Erneux C (1988) Control of prostacyclin production by vascular cells: role of adenine nucleotides and serotonin. Adv Exp Med Biol 243: 13–20.

Boeynaems JM and Pearson JD (1990) P2 purinoceptors on vascular endothelial cells: physiological significance and transduction mechanisms. Trends Pharmacol Sci 11: 34–37.

Boeynaems JM, Pirotton S, Van Coevorden A, Raspe E, Demolle D and Erneux C (1988) P2-purinergic receptors in vascular endothelial cells: from concept to reality. J Recept Res 8: 121–132.

Bohr VA, Smith CA, Okumoto DS and Hanawalt PC (1985) DNA repair in an active gene: removal of pyrimidine dimers from the DHFR gene of CHO cells is much more efficient than in the genome overall. Cell 40: 359–369.

Boothman DA, Bouvard I and Hughes EN (1989) Identification and characterization of x-ray-induced proteins in human cells. Cancer Res 49: 2871–2878.

Bootsma D, Koken MHM, van Duin M, Westerwald A, Yasui A, Prakash S and Hoeijmakers JHJ (1987) Homology of mammalian, drosophila, yeast and *E. coli* repair genes. In: Fielden EM, Fowler JF, Hendry JH and Scott D (eds.) Proceedings of the 8th International Congress of Radiation Research, Edinburgh, July 1987, Taylor and Francis Ltd., London, pp. 412–417.

Bootsma D, Westerveld A and Hoeijmakers JHJ (1988) DNA repair in human cells: from genetic complementation to isolation of genes. Cancer Surveys 7 (2): 303–316.

Bootsma D and Hoeijmakers JHJ (1993) Engagement with transcription. Nature 363: 114–115.

Bradley MO and Kohn KW (1979) x-ray induced DNA double strand break production and repair in mammalian cells as measured by neutral filter elution. Nucleic Acids Res 7: 793–804.

Brunborg G, Resnick MA and Williamson DH (1980) Cell-cycle specific repair of DNA double-strand breaks in *Saccharomyces cerevisiae*. Radiat Res 82: 547–558.

Burnstock G (1990) Overview: purinergic mechanisms. Ann NY Acad Sci 603: 1–18.

Burt RK, Garfield S, Johnson K, *et al.* (1988) Transformation of rat liver epithelial cells with v-H-ras or v-raf causes expression of MDR-1, glutathione-S-transferase-P and increased resistant to cytotoxic chemicals. Carcinogenesis 9: 2329–2332.

Chin KV, Ueda K, Pastan I and Gottesman MM (1992) Modulation of activity of the promoter of the human MDR1 gene by *Ras* and p53. Science 255: 459–462.

Debenham PG, Webb MBT, Stretch A and Thacker J (1988a) Examination of vectors with two dominant selectable genes for DNA repair and mutation studies in mammaliancells. Mutat Res 199: 145–158.

Debenham PG, Jones NJ and Webb MBT (1988b) Vector-mediated DNA double-strand break repair analysisin normal, and radiation-sensitive,Chinese hamster V79cells. Mutat Res 199: 1–9.

Defais MJ and Hanawalt PC (1983) Viral probes for DNA repair. Adv Radiat Biol 10: 1–32.

Diller L, Kassel J, Nelson CE, Gryka MA, Litwak G, Gebhardt M, Bressac B, Ozturk M, Baker S, Vogelstein B and Friend SH (1990) p53 functions as a cell cycle control protein in osteosarcomas. Mol Cell Biol 10 (11): 5772–5781.

Downes CS, Musk SR, Watson JV and Johnson RT (1990) Caffeine overcomes a restriction point associated with DNA replication, but does not accelerate mitosis. J Cell Biol 110: 1855–1859.

Durand RE and Sutherland RM (1973) Growth and radiation survival characteristics of V79-171b Chinese hamster cells: a possible influence of intercellular contact. Radiat Res 56: 513–527.

Fairman MP, Johnson AP and Thacker J (1992) Multiple components are involved in the efficient joining of double stranded DNA breaks in human cell. Nucleic Acids Research 20 (16): 4145–4152.

Fields S and Jang SK (1990) Presence of a potent transcription activating sequence in the p53 protein. Science 249: 1046–1049.

Fischer E, Keijzer W, Thielmann HW, Popanda O, Bohnert E, Edler L, Jung EG and Bootsma D (1985) A ninth complementation group in xeroderma pigmentosum, XP-I. Mutat Res 145: 217–225.

Fulop GM and Phillips RA (1990) The scid mutation in mice causes a general defect in DNA repair. Nature 347: 479–482.

Fornace AJ, Nebert DW, Hollander MC, Luethy JD, Papathanasiou M, Fargnoli J and Holbrook NJ (1989) Mammalian genes coordinately regulated by growth arrest signals and DNA damaging agents. Mol Cell Biol 9 (10): 4196–4203.

Frankenberg-Schwager M and Frankenberg D (1990) DNA double-strand breaks: their repair and relationship to cell killing in yeast. Int J Radiat Biol 58: 569–575.

Friedberg EC (1985) DNA repair. W.H. Freeman & Co.

Fuks Z and Weichselbaum RR (1992) Radiation tolerance and the new biology: growth factor involvement in radiation injury to the lung. Int J Radiat Oncol Biol Phys 24: 183–184.

Giaccia A, Weinstein R and Stamato TD (1985) Cell-cycle dependent repair of double-strand breaks in a gamma-ray sensitive Chinese hamster ovary cell. Somat Cell Genet 11: 485–491.

Giaccia AJ, Denko N, MacLaren R, Mirman D, Waldren C, Hart I and Stamato TD (1990) Human chromosome 5 complements the DNA double-strand break-repair deficiency and gamma-ray sensitivity of the XR-1 hamster variant. Am J Hum Genet 47: 459–469.

Glazer VM, Glazunov AV, Tevzadze GG and Koltovaia NA (1989) Repair of a double-stranded gap in plasmid DNA in radiosensitive mutants of *Saccharomyces cerevisiae*: effectiveness and precision. Mol Gen Mikrobiol Virusol 1989 Sep (9): 14–20.

Goodrich DW, Ping Wang N, Qian YW, Lee EY-HP and Lee WH (1991) The retinoblastoma gene product regulates progression through the G1 phase of the cell cycle. Cell 67: 293–302.

Hallahan DE, Sprigg DR, Beckett MA *et al.* (1989) Increased tumor necrosis factor mRNA following ionizing radiation exposure. Proc Natl Acad Sci USA 86: 1014–1017.

Hallahan DE, Sukhatme VP, Sherman ML, Virudachalam S, Kufe D and Weichselbaum RR (1991a) Protein kinase C mediates x-ray inducibility of nuclear signal transducers EGR1 and JUN. Proc Natl Acad Sci USA 88: 2156–2160.

Hallahan DE, Virudachalam S, Beckett M, Sherman ML, Kufe D and Weichselbaum RR (1991b) Mechanisms of x-ray-mediated protooncogene c-*jun* expression in radiation-induced human sarcoma cell lines. Int J Radiat Oncol Biol Phys 21: 1677–1681.

Hallahan DE, Virudachalam S, Schwartz JL, Panje N, Mustafi R and Weichselbaum RR (1992) Inhibition of protein kinases sensitizes human tumor cells to ionizing radiation. Radiat Res 129: 345–350.

Hallahan DE, Virudachalam S, Sherman ML, Huberman E, Kufe DW and Weichselbaum RR (1991) Tumor necrosis factor gene expression is mediated by protein kinase C following activation by ionizing radiation. Cancer Res 51: 4565—4569.

Hamilton AA and Thacker J (1987) Gene recombination in x-ray sensitive hamster cells. Mol Cell Biol 7 (4): 1409—1414.

Henderson EE and Long WK (1981) Host cell reactivation of uv and x-ray damaged Herpes simplex virus by Epstein-Barr virus transformed lymphoblastoid cell lines. Virology 115: 237—248.

Hendrickson EA, Qin X-Q, Bump EA, Schatz DG, Oettinger M and Weaver DT (1991) A link between double-strand break-related repair and V(D)J recombination: the scid mutation. Proc Natl Acad Sci USA 88: 4061—4065.

Higgins CF (1992) ABC transporters: from microorganisms to man. Annu Rev Cell Biol 8: 67—113.

Hill BT (1991) Interactions between antitumour agents and radiation and the expression of resistance. Cancer Treat Rev 18: 149—190.

Ho K (1975) Induction of DNA double-strand breaks by x-rays in a radioresistant strain of the yeast *Saccharomyces cerevisiae*. Mutat Res 30: 327—334.

Hoy CA, Fuscoe JC and Thompson LA (1987) Recombination and ligation of transfected DNA in CHO mutant EM9, which has high levels of sister chromatid exchange. Mol Cell Biol 7 (5): 2007—2011.

Hughes EN and Boothman DA (1991) Effect of caffeine on the expression of a major x-ray induced protein in human tumor cells. Radiat Res 125: 313—317.

Jaspers NGJ, Painter RB, Paterson MC, Kidson C and Inoue T (1985) Complementation analysis of ataxia-telangiectasia. In: Gatti RA and Swift M (eds) Ataxia-telangiectasia: genetics, neuropathology, and immunology of a degenerative disease of chilhood. Alan R. Liss, Inc., New York, pp. 147—162.

Jeggo PA (1990) Studies on mammalian mutants defective in rejoining double-strand breaks in DNA. Mutat Res 239 (1): 1—16.

Jeggo PA, Tesmer J and Chen DJ (1991) Genetic analysis of ionising-radiation sensitive mutants of cultured mammalian cell lines. Mutat Res 254: 125—133.

Jones NJ, Cox R and Thacker J (1988) Six complementation groups for ionizing radiation sensitivity in Chinese hamster cells. Mutat Res 193: 139—144.

Kasid U, Pfeifer A, Weichselbaum RR, Dritschilo A and Mark GE (1987) The raf oncolgene is associated with a radiation-resistant human laryngeal cancer. Science 237: 1039—1041.

Kasid U, Pfeifer A, Brennan T, Beckett M, Weischelbaum RR, Dritschilo A and Mark GE (1989) Effect of anti-sense c-raf-1 on tumorigenicity and radiation sensitivity of a human squamous cell carcinoma. Science 243: 1354—1356.

Kastan MB, Onyekwere O, Sidransky D, Vogelstein B and Craig RW (1991) Participation of p53 protein in the cellular response to DNA damage. Cancer Res 51: 6304—6311.

Kastan MB, Zhan Q, El-Deiry WS, Carrier F, Jacks T, Walsh WV, Plunkett BS, Vogelstein B and Fornace Jr. AJ (1992) A mammalian cell cycle checkpoint pathway utilizing p53 and GADD45 is defective in ataxia-telangiectasia. Cell 71: 587—597.

Kelland LR, Edwards SM and Steel GG (1987) Induction and rejoining of DNA double-strand breaks in human cervix carcinoma cell lines of differing radiosensitivity. Radiat Res 116: 526—538.

Kemp LH, Sedgwick SG and Jeggo PA (1984) X-ray sensitive mutants of Chinese Hamster ovary cells defective in double strand break rejoining. Mutat Res 132: 189—196.

Komatsu K, Okumura Y, Kodama S, Yoshida M and Miller RC (1989) Lack of correlation between radiosensitivity and inhibition of DNA synthesis in hybrids (A-T x HeLa). Int J Radiat Biol 56 (6): 863—867.

Kuo SS, Saad AH, Koong AC, Hahn GM and Giaccia AJ (1993) Potassium-channel activation in response to low doses of gamma-irradiation involves reactive oxygen intermediates in nonexcitatory cells. Proc Natl Acad Sci USA 90: 908—912.

Lehmann AR and Stevens S (1977) The production and repair of double-strand breaks in cells from normal humans and from patients with ataxiatelangectasia. Biochem Biophys Acta 474: 49—60.

Lindahl T (1990) Repair of intrinsic DNA lesions. Mutat Res 238: 305—311.

Lowe SW, Schmitt EM, Smith SW, Osborne BA and Jacks T (1993) p53 is required for radiation-induced apoptosis in mouse thymocytes. Nature 362: 847—849.

Madhukar BV, Oh SY, Chang CC, Wade M and Trosko JE (1989) Altered regulation of intercellular communication by epidermal growth factor, transforming growth factor-beta and peptide hormones in normal human keratinocytes. Carcinogenesis 10: 13—20.

Mayne LV, Mullenders LHF and Van Zeeland AA (1988b) Cockayne's syndrome: a uv sensitive disorder with a defect in the repair of transcribing DNA but normal overall excision repair. In: Friedberg EC and Hanawalt PC (eds) Mechanisms and Consequences of DNA Damage Processing, Liss, New York, pp. 263—266.

McClean S, Hosking LK and Hill BT (1993) Dominant expression of multiple drug resistance after *in vitro* X-irradiation exposure in intraspecific Chinese hamster ovary hybrid cells. J Natl Cancer Inst 85: 48—53.

McKenna WG, Nakahara K and Muschel RJ (1988) Site-specific integration of H-ras in transformed rat embryo cells. Science 241: 1325—1327.

McKenna WG, Weiss MC, Endlich B, Ling CC, Bakanauskas VJ, Kelsten ML and Muschel J (1990a) Synergistic effect of the v-myc oncogene with H-ras on radioresistance. Cancer Res 50: 97—102.

McKenna WG, Weiss MC, Bakanauskas V, Sandler H, Kelsten ML, Biaglow J, Tuttle S, Endlich B, Ling CC and Muschel J (1990b) The role of the H-ras oncogene in radiation resistance and metastasis. Int J Rad Oncol Biol Phys 18: 849—860.

McKenna WG, Iliakis G, Weiss MC, Bernhard EJ and Muschell RJ (1991) Increased G2 delay in radiation-resistant cells obtained by transformation of primary rat embryo cells with the oncogenes H-ras and v-myc. Radiat Res 125: 283—287.

McKinnon PJ (1987) Ataxia telangectasia: an inherited disorder of ionizingradiation sensitivity in man. Human Genet 75:

197–208.

Mellon I, Spivak G and Hanawalt PC (1987) Selective removal of transcription-blocking DNA damage from the transcribed strand of the mammalian DHFR gene. Cell 51: 241–249.

Mignatti P, Morimoto T and Rifkin DB (1992) Basic fibroblast growth factor, a protein devoid of secretory signal sequence, is released by cells via a pathway independent of the endoplasmic reticulum-Golgi complex. J Cell Physiol 151: 81–93.

Musk SRR (1991) Reduction of radiation-induced cell cycle block by caffeine does not necessarily lead to increased cell killing. Radiat Res 125: 262–266.

Nairn RS, Humphrey RM and Adair GM (1988) Transformation depending on intermolecular homologous recombination is stimulated by UV damage in transfected DNA. Mutat Res 208 (3–4): 137–141.

Nishizuka Y (1992) Intracellular signaling by hydrolysis of phospholipids and activation of protein kinase C. Science 258: 607–614.

North P, Ganesh A and Thacker J (1990) The rejoining of double-strand breaks in DNA by human cell extracts. Nucleic Acids Res 18 (21): 6205–6210.

Pardo FS, Ong A, Bristow RG, Taghian A and Borek C (1991) Role of transfection and clonal selection in mediating radioresistance. Proc Natl Acad Sci USA.

Perera JR, Glasunov AV, Glaser VM and Boreiko AV (1988) Repair of double-strand breaks in plasmid DNA in the yeast *Saccharomyces cerevisiae*. Mol Gen Genet 213 (2–3): 421–424.

Pirollo KF, Garner R, Yuan SY *et al.* (1989) Raf involvement in the simutaneous genetic transfer of the radioresistant and transforming phenotypes. Int J Radiat Biol 55: 783–796.

Pirotton S, Boutherin Falson O, Robaye B and Boeynaems JM (1992) Ecto-phosphorylation on aortic endothelial cells. Exquisite sensitivity to staurosporine. Biochem J 285: 585–591.

Pirotton S, Robaye B, Lagneau C and Boeynaems JM (1990) Adenine nucleotides modulate phosphatidylcholine metabolism in aortic endothelial cells. J Cell Physiol 142: 449–457.

Powell SN and McMillan TJ (1991) Clonal variation in DNA repair in a human glioma cell line. Radiother Oncol 21: 225–232.

Powell SN, Whitaker SJ, Edwards SM and McMillan TJ (1992) A DNA repair defect in a radiation-sensitive clone of a human bladder carcinoma cell line. Br J Cancer 65: 798–802.

Powell SN, Whitaker S, Peacock JH and McMillan TJ (1993) Ataxia-telangietasia: an investigation of the repair defect in the cell line AT5BIVA by plasmid reconstitution. Mutat Res 294: 9–20.

Powell SN and McMillan TJ (1993) Human tumor cell repair fidelity correlates with radioresistance. Int J Radiat Oncol Biol Phys.

Protic-Sabljic M and Kraemer KH (1986) Host cell reactivation by human cells of DNA expression vectors damaged by ultraviolet radiation or by acid-heat treatment. Carcinogenesis 7 (10): 1765–1770.

Rainbow AJ and Howes M (1981) Decreased repair of gamma-irradiated adenovirus in xeroderma pigmentosum fibroblasts. Int J Radiat Biol 36: 621–629.

Rapaport E (1983) Treatment of human tumor cells with ADP or ATP yields arrest of growth in the S phase of the cell cycle. J Cell Physiol 114: 279–283.

Roth DB, Menetski JP, Nakajima PB, Bosma MJ and Gellert M (199.) V(D)J Recombination: broken DNA molecules with covalently sealed (hairpin) coding ends in scid mouse. thymocytes. Cell 70 (6): 983–991.

Schiestl RH, Reynolds P and Prakash S (1989) Cloning and sequence analysis of *Sacharomyces cerevisiae* RAD9 gene and further evidence that its product is required for cell cycle arrest induced by DNA damage. Mol Cell Biol 9: 1882–1896.

Schwartz JL, Rotmensch J, Giovanazzi S, Cohen MB and Weichselbaum RR (1987) Faster repair of DNA double-strand breaks in radioresistant human tumor cells. Int J Radiat Oncol Biol Phys 15: 907–912.

Sherman ML, Datta R, Hallahan DE *et al.* (in press) Protein kinase-C mediates inducibility of nuclear signal transducers EGR-1 and c-*jun*. Proc Natl Acad Sci USA.

Singh SP and Lavin MF (1990) DNA-binding protein activated by gamma radiation in human cells. Mol Cell Biol 10: 5279–5285.

Sklar MD (1988) The *ras* oncogenes increase the intrinsic resistance of NIH 3T3 cells to ionizing radiation. Science 239: 645–647.

Skubitz KM and Goueli SA (1991) Basic fibroblast growth factor is a substrate for phosphorylation by human neutrophil ecto-protein kinase activity. Biochem Biophys Res Commun 174: 49–55.

Subramani S (1989) Analysis of recombination in mammalian cells using SV40 and SV40 derived vectors. Mutat Res 220: 221–234.

Terasima R and Tolmach LJ (1963) x-ray sensitivity and DNA synthesis in synchronous populations of HeLa cells. Science 140: 490–492.

Terleth C, Waters R, Brouwer J and van de Putte P (1990) Differential repair of UV damage in *Saccharomyces cerevisiae* is cell cycle dependent. EMBO J 9: 2899–2904.

Thacker J and Wilkinson RE (1991) The genetic basis of resistance to ionising radiation damage in cultured mammalian cells. Mutat Res 254: 135–142.

Thaler DS, Stahl MM and Stahl FW (1987) Tests of the double-strand-break repair model for red-mediated recombination of phage lambda and plasmid lambda dv. Genetics 116 (4): 501–511.

Thomas BJ and Rothstein R (1989) Elevated recombination rates in transcriptionally active DNA. Cell 56: 619–630.

Thompson LH, Brookman KW, Jones NJ, Allen SA and Carrano AV (1990) Molecular cloning of the human XRCC-1 gene, which corrects defective DNA strand break repair and sister chromatid exchange. Mol. Cell Biol. 10 (12): 6160–6171.

Trosko JE, Chang CC, Madhukar BV and Klaunig JE (1990) Chemical, oncogene and growth factor inhibition gap junctional intercellular communication: an integrative hypothesis of carcinogenesis. Pathobiology 58: 265–278.

Tucker JD, Jones NJ, Allen NA, Minkler JL, Stewart SA, Thompson LH and Carrano AV (1991) Cytogenetic characterization of the ionizing radiation sensitive Chinese hamster mutant irs1. Mutat Res 254 (2): 143–152.

Uckun FM, Tuel Ahlgren L, Song CW, Waddick K, Myers DE, Kirihara J, Ledbetter JA and Schieven GL (1992) Ionizing radiation stimulates unidentified tyrosine-specific protein kinases in human B-lymphocyte precursors, triggering apoptosis and clonogenic cell death. Proc Natl Acad Sci USA 89: 9005–9009.

Uckun FM, Schieven GL, Tuel-Ahlgren LM, Dibirdik I, Myers DE, Ledbetter JA and Song CW (1993) Tyrosine phosphorylation is a mandatory proximal step in radiation-induced activation of the protein kinase C Signaling pathway in human B-lymphocyte precursors. Proc Natl Acad Sci USA 90 (1): 252–256.

Usuki K, Heldin NE, Miyazono K, Ishikawa F, Takaku F, Westermark B and Heldin CH (1989) Production of platelet-derived endothelial cell growth factor by normal and transformed human cells in culture. Proc Natl Acad Sci USA 86: 7427–7431.

Usuki K, Miyazono K and Heldin CH (1991) Covalent linkage between nucleotides and platelet-derived endothelial cell growth factor. J Biol Chem 266: 20525–20531.

Usuki K, Saras J, Waltenberger J, Miyazono K, Pierce G, Thomason A and Heldin CH (1992) Platelet-derived endothelial cell growth factor has thymidine phosphorylase activity. Biochem Biophys Res Commun 184: 1311–1316.

Vilgrain I and Baird A (1991) Basic fibroblast growth factor is phosphorylated by an ecto-protein kinase associated with the surface of SK-Hep cells. Ann NY Acad Sci 638: 445–448.

Vlodavsky I, Bar Shavit R, Ishai Michaeli R, Bashkin P and Fuks Z (1991) Extracellular sequestration and release of fibroblast growth factor: a regulatory mechanism? Trends Biochem Sci 16: 268–271.

Vlodavsky I, Bashkin P, Ishai Michaeli R, Chajek Shaul T, Bar Shavit R, Haimovitz Friedman A, Klagsbrun M and Fuks Z (1991) Sequestration and release of basic fibroblast growth factor. Ann NY Acad Sci 638: 207–220.

Wahls WP and Moore PD (1990) Relative frequencies of homologous recombination between plasmids introduced into DNA repair-deficient and other mammalian somatic cell lines. Somat Cell Mol Genet 16 (4): 321–329.

Weichselbaum RR, Hallahan DE, Sukhatme V, Dritschilo A, Sherman ML and Kufe DW (1991) Biological consequences of gene regulation after ionizing radiation exposure. J Natl Cancer Inst 83: 480–484.

Wlodek D and Hittelman WN (1987) The repair of double-strand DNA breaks correlates with radiosensitivity of L5178Y-S and L5178Y-R cells. Radiat Res 112: 146–155.

Zdzienicka MZ, Tran Q, van der Schans CP and Simons JWIM (1988) Characterization of an x-ray sensitive mutant of V79 Chinese hamster cells. Mutat Res 194: 239–243.

Zheng LM, Zychlinsky A, Liu CC, Ojcius DM and Young JD (1991) Extracellular ATP as a trigger for apoptosis or programmed cell death. J Cell Biol 112: 279–288.

Address for offprints: Simon N. Powell, Department of Radiation Oncology, Massachusetts General Hospital, Boston, MA 02114, USA.

Cytotechnology **12**: 347–356, 1993.

Kinetic resistance to anticancer agents

Marie-Thérèse Dimanche-Boitrel, Carmen Garrido and Bruno Chauffert
INSERM 252, Faculty of Medicine, F-21033 Dijon, France

Key words: alkylating drugs, anthracycline, cell cycle, cisplatin, confluence, kinetic resistance, multidrug resistance, topoisomerase

Abstract

Adherent epithelial cancer cells, such as colon cancer cells, are much more resistant to anthracyclines and to many other major anticancer agents when the cell population reaches confluence. Our purpose is to analyze the mechanisms of this confluence dependent resistance (CDR) that is probably the major cause of the natural resistance of solid tumors to chemotherapy. Some drugs (anthracyclines, etoposide and vincristine) but not others (cisplatin, melphalan and 5-fluorouracil) accumulate less in confluent than in nonconfluent cells. A decrease of the passive transmembrane drug transport in confluent cells is associated to a reduced membrane fluidity. However, the predominant mechanism of CDR is an increase in the intrinsic resistance of the DNA to the drug-induced damage. This mechanism is now relatively well understood for anthracyclines and etoposide that act mainly through an inhibition of the topoisomerase II: as the enzyme level is low in slowly proliferating confluent cells, the number of drug-induced DNA strand breaks is lower than in rapidly growing nonconfluent cells which highly express the topoisomerase II gene. Mechanisms of CDR for the other drugs are less clear and could involve an increase in the ability to repair damaged DNA. Attempts to circumvent CDR could consist in the stimulation of the cell proliferation by hormones or growth factors, or in the recruitment of quiescent cells into the S and G2 phases by previous treatment of confluent cells with infratoxic concentration of DNA-damaging agents.

Introduction

In contrast to hematological malignancies that are frequently responsive to chemotherapy, most of the common solid tumors, such as digestive tract carcinomas, are inherently resistant to the usual anticancer drugs. Besides the low intrinsic drug sensitivity of the cancer cells [1], impaired drug distribution in tightly aggregated cell nodules, poor vascularization that results in poor oxygen supply and low pH in tumor depth could partly explain the chemoresistance of solid tumors [2–4]. An additional mechanism of the solid tumor resistance is suggested by the fact that confluent cancer cells in culture are more resistant to anticancer drugs than nonconfluent cells, even if incubation medium is regularly renewed and if the cells are incubated with drug as monolayer in the presence of a controlled pH and atmosphere. This confluence dependent resistance (CDR) has already been described and attributed either to a decrease of drug accumulation or to a reduction of the cell proliferation [5–11]. The purpose is to analyze the different mechanisms of the CDR in a well-defined model of cultured colon cancer cells and to propose new strategies to overcome this major cause of failure of chemotherapy in solid tumors.

Evidence of CDR to anthracyclines and to other anticancer agents in colon cancer cells

We used the HT 29 human colon cancer cell line, that lacks the P-glycoprotein 180 (PGP-180), to study the kinetic resistance of colon cancer cells to anthracyclines independently of the multidrug resistance (MDR) phenomenon. Sensitivity of HT 29 cells to two anthracyclines, doxorubicin (DXR) and deoxydoxorubicin (deoDXR), strongly decreased as the cell density increased (Fig. 1). DeoDXR was more active than DXR due to its better penetration through the plasma membrane [12]. As similar results were obtained when culture medium was changed every 12 h in the pre- and posttreatment period, reduction in glucose concentration, in oxygen supply or a drop in the pH of the culture medium could not be advocated, as for other observations in the literature [13—15], to explain CDR. Confluent HT 29 cells displayed a similar decrease in chemosensitivity to other major anticancer drugs when compared to nonconfluent cells (Fig. 2).

CDR was also observed in the DHD/K 12/PROb rat colon cancer cells which display the classical MDR phenotype characterized by the *mdr*-1 gene expression and an active efflux of anthracyclines [16,17]. Sensitivity of DHD/K12/PROb cells to DXR was potentiated by amiodarone, a potent MDR inhibitor [18], only in nonconfluent cells but not in confluent cells (Fig. 3). This experimental observation could explain the relatively disappointing results of several clinical trials aiming to circumvent MDR. Most of these trials were carried in patients with colon or renal carcinomas considering the number and the poor chemosensitivity of these tumors which display a MDR phenotype [19]. As CDR is a predominant mechanism of resistance in solid tumors, negative results of such trials cannot be attributed completely to the inefficacy of the MDR modifiers. We think that trials on MDR modifying drugs must only concern rapidly growing hematological malignancies where the influence of kinetic resistance is reduced.

Reduced drug accumulation in confluent cells

A significant decrease in drug cellular accumulation was measured in confluent cells for DXR, vincristine (VCR) and etoposide (VP-16) but not for cisplatin (CDDP), melphalan (L-PAM) or 5-fluorouracil (5-FU) (Fig. 4). Many anticancer drugs are

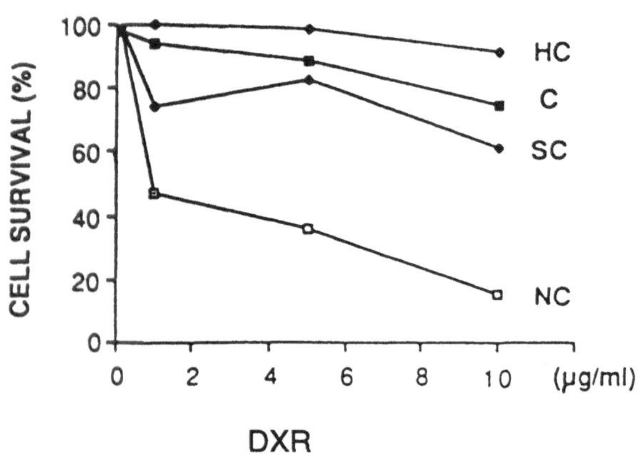

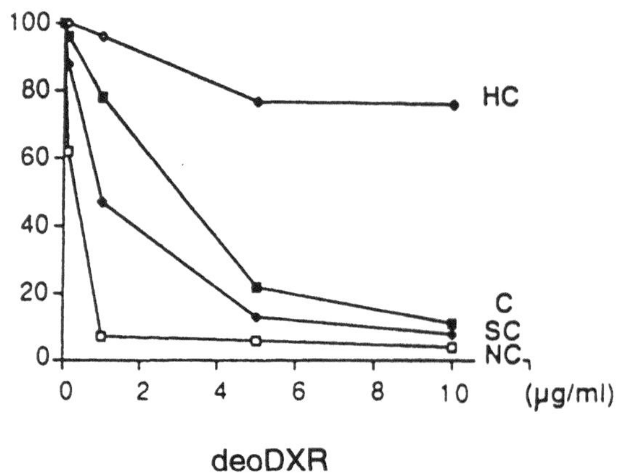

Fig. 1. Survival of HT 29 cells treated with DXR or deoDXR at various degrees of confluence. Cell density at the treatment time was 309 ± 94 cells/mm² for nonconfluent cells (NC), 1805 ± 206 cells/mm² for subconfluent cells (SC), 3478 ± 627 cells/mm² for confluent cells (C) and 7000 ± 851 cells/mm² for hyperconfluent cells (HC). Cells were seeded 48 h before treatment (1 h, 37°C), then washed twice and allowed to grow again in drug-free medium for 3 days. Percent of cell survival at the end of the experiment was measured by a methylene blue colorimetric assay [22].

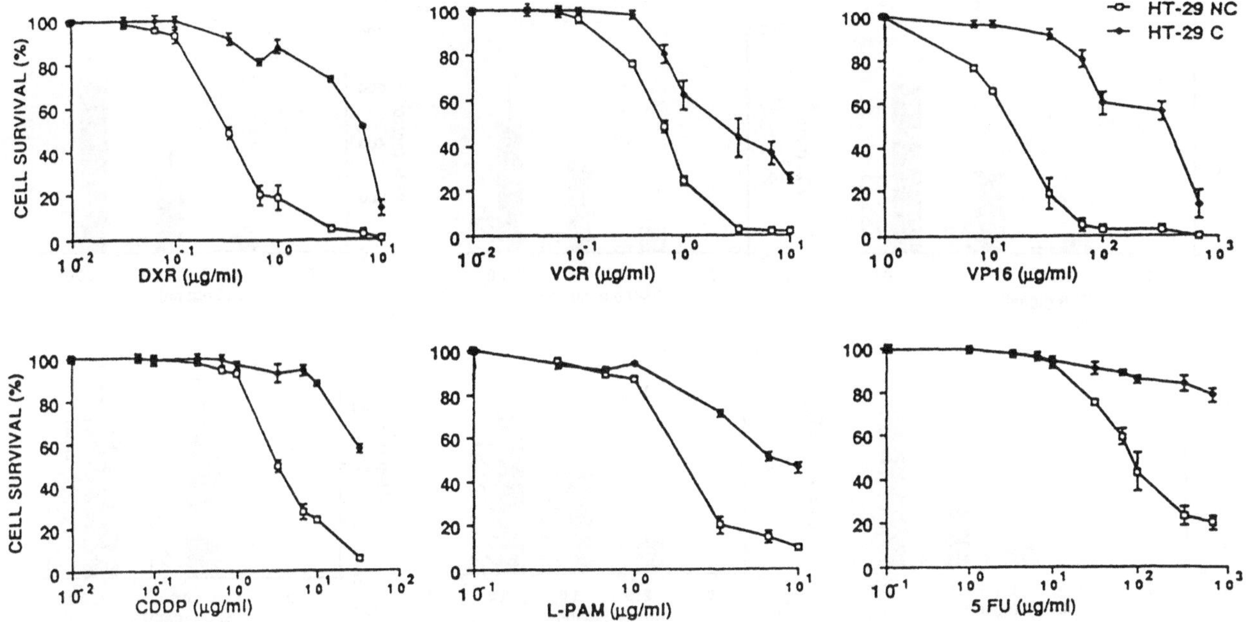

Fig. 2. Survival of confluent (dark symbols) or nonconfluent (open symbols) HT 29 cells after treatment with six major anticancer drugs. Cells were seeded at various density 48 h before a 4-h treatment with drugs, then cultivated for 6 additional days. Percent of cell survival was measured by a methylene blue colorimetric assay [24].

thought to passively diffuse through the plasma membrane with a transport rate depending on their lipophilicity [20,21]. Plasma membrane of confluent cells appears less permeable to passively diffusing molecules than that of nonconfluent cells. Low permeability for ions, nucleotides, sugars, and polyamines has been considered as a general characteristic of resting cells. For instance, passive transmembrane diffusion of D-alanine or sodium chromate is about two-fold reduced in confluent compared to nonconfluent cells [22]. Decreased membrane permeability in nonproliferating cells seems to be an adaptive response to the reduced need of metabolite supply. Not only passive diffusion but also active transmembrane transport of amino acids or some anticancer drugs such as melphalan are influenced by the rate of cell proliferation [23].

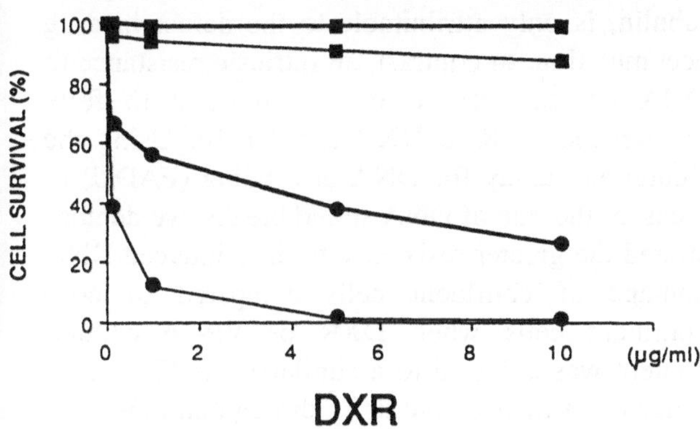

Fig. 3. Survival of the multidrug resistant rat colon cancer cells, DHD/K12/PROb after a treatment with DXR in presence (—) or not (▬) of amiodarone, a potent MDR inhibitor. Cells were seeded 48 h before treatment (1 h, 37°C). Cell survival was measured after a 72-h additional culture in drug-free medium by a methylene blue colorimetric assay. DXR toxicity was potentiated by amiodarone only on nonconfluent cells (●) but not on confluent cells (■).

Confluent cells grown in monolayer are tightly attached one to each other and are exposed to the incubation medium by a smaller surface of plasma membrane than nonconfluent cells. It could be hypothetized that geometric parameters are the only cause of the reduced drug accumulation in confluent cells. However, when confluent cells were

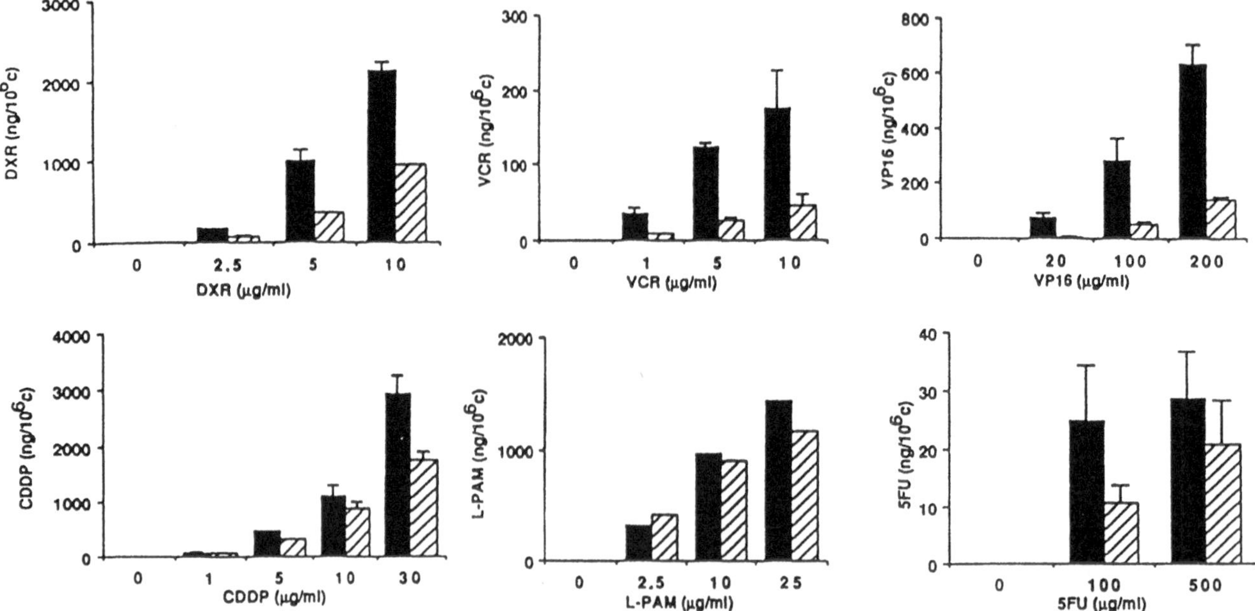

Fig. 4. Drug accumulation (ng/10^6 cells) in confluent (dashed bars) or in nonconfluent (dark bars) HT 29 cancer cells [24]. Cells were seeded 48 h before treatment (4 h, 37°C). Intracellular drug content was measured by an appropriate assay (radioactivity incorporation for DXR and VCR, HPLC for L-PAM and 5-FU, atomic absorption for CDDP).

trypsinized and immediately incubated in suspension with DXR or deoDXR , drug accumulation was still less than in nonconfluent cells for up to 12 h after trypsinization [22]. Using fluorescence polarization with a diphenylhexatriene derivative as a probe, we determined that membrane fluidity was significantly decreased in confluent cells even after the cell detachment by trypsin [24]. Relation between membrane permeability and fluidity has been well established [25] and could be due to a modification of the cholesterol and phospholipid content as well as a different organization and repartition of lipids in the basolateral or apical domains of polarized cells.

Intrinsic resistance to drug-induced DNA damage in confluent cells

To counterbalance the reduction of drug uptake with confluence, drug concentration in incubation medium was adjusted to produce equal intracellular drug content. Even in such conditions, confluent HT 29 cells were persistently more resistant than nonconfluent cells to DXR and VP-16 but not to VCR [24]. CDR of VCR, whose cellular target is tubulin, is only attributable to the defect in drug accumulation. In contrast, an intrinsic resistance to DNA damage has to be hypothetized to fully explain the CDR to DXR and VP-16. Using the fluorescent assay for DNA unwinding (FADU) to measure the rate of DNA strand breaks, we demonstrated the greater resistance to drug-induced DNA damage of confluent cells compared to nonconfluent cells when DXR or VP-16 cellular content was adjusted to a similar value [24]. Both drugs exert their cytotoxicity through an inhibition of the nuclear topoisomerase II enzyme that produces stable DNA strands breaks [26–30]. Topoisomerase II is highly expressed in rapidly proliferating cells as in the regenerating liver after partial hepatectomy [31] or in the lectin stimulated lymphocytes [32] whereas it is low expressed in quiescent cells [33–35]. Cells with diminished

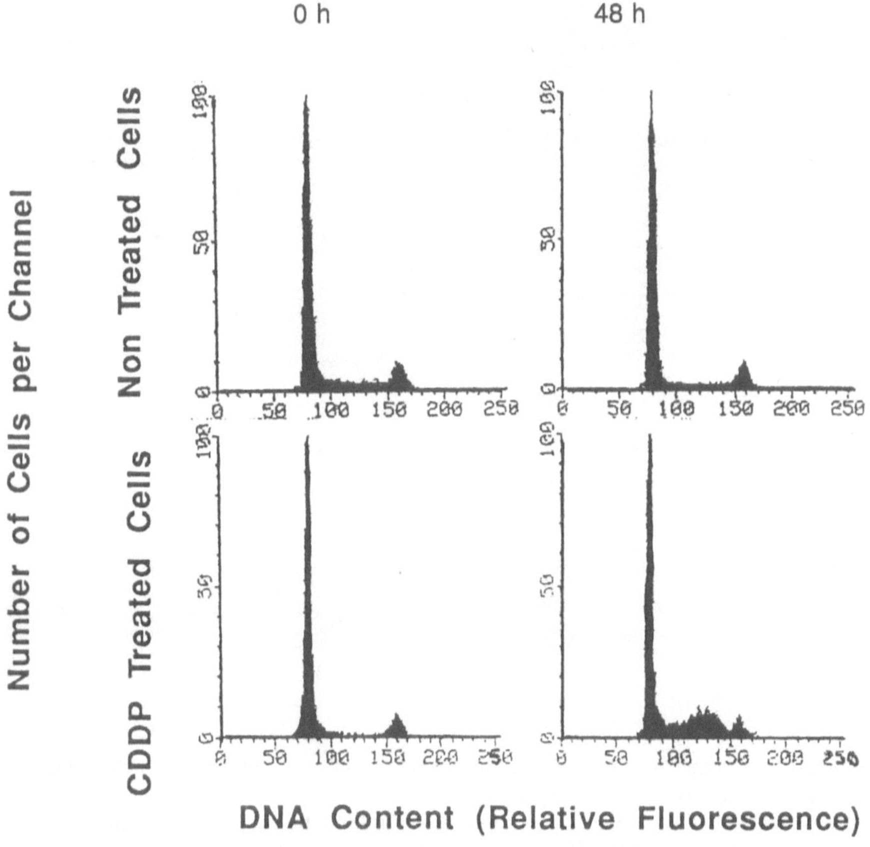

Fig. 5. Recruitment of confluent HT 29 cells into S-phase after a pretreatment with cisplatin (CDDP). Confluent HT 29 cells were treated for 4 h by CDDP (5 μg/ml), cultivated again in drug-free medium and recovered for cell cycle analysis 48 h after treatment. In a parallel experiment, incorporation of tritiated thymidine was not increased in S-phase recruited cells (data not shown). We conclude that the increased number of cells in S-phase in CDDP pretreated confluent cells is rather due to a longer time of passage through the S-phase than to a stimulation of the cell proliferation.

topoisomerase II content are functionally resistant to the DNA-damaging and cytotoxic effects of topoisomerase II inhibitors [36–41]. The decrease of the topoisomerase II level that is observed in confluent colon cancer cells (Fig. 6) could be the predominant biochemical explanation of the CDR to anthracyclines and etoposide.

Mechanisms of CDR to cisplatin, alkylating drugs and antimetabolites are far less known. DNA synthesis and passage through the whole cell cycle is often required for the anticancer drug cytotoxicity. For instance, cell killing induced by vincristine or aphidicolin requires the termination of the cell cycle and protein synthesis. Quiescent cells that do not traverse the cell cycle and mitosis do not die [42]. Binding of cisplatin to DNA has been shown to be insufficient by itself to cause cancer cell death. A passage through the S-phase and the transient accumulation of the cells in the G2/M phase was necessary before the activation of the programmed death process (apoptosis) that involves the activation of DNA-nucleases [43–47]. In addition, repair of DNA damage, that is considered as a major cause of resistance to alkylating agents, cisplatin or X-irradiation, could be more efficient in quiescent cells than in rapidly dividing cells [48–52].

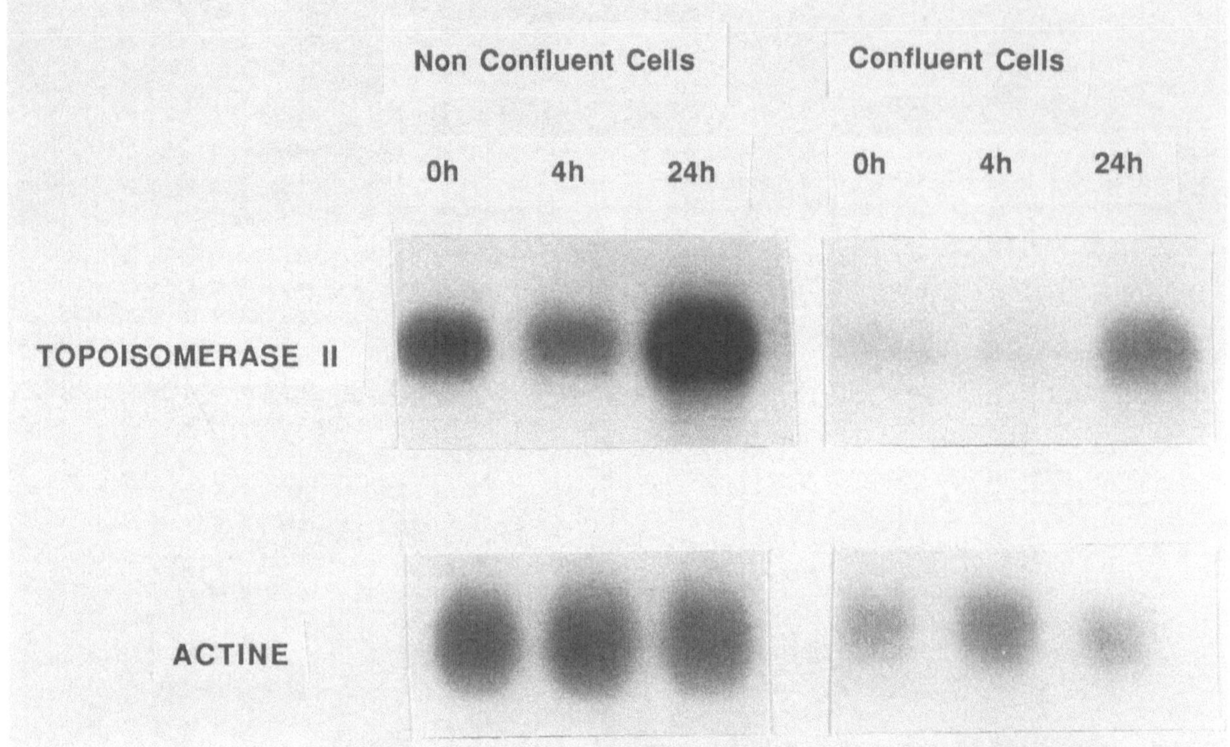

Fig 6 Increased expression of topoisomerase II mRNA in confluent HT 29 cells after a treatment with CDDP Confluent HT 29 cells were treated for 4 h by CDDP (5 μg/ml), cultivated again in drug-free medium and recovered for Northern blot 4 h or 24 h after treatment Topoisomerase II expression was higher in nonconfluent HT 29 cells compared to confluent cells Topoisomerase II expression was strongly increased in confluent HT 29 cells recovered 24 h after CDDP treatment, probably due to the recruitment of S-phase cells

Mechanisms of confluence-dependent growth arrest in confluent cancer cells

When colon cancer cells become crowded in culture, most of the cells exit the cell cycle and enter in a density-dependent quiescence (G0). Even if abnormal in their cell cycle regulation, some transformed epithelial cells such as colon cancer cells retain a partial contact-inhibition at confluency and a limitation in the saturation density of the cell population. In contrast of nontransformed fibroblasts which display a total inhibition of the cell proliferation at confluence, contact inhibition remains incomplete in colon cancer cells: about 5 to 15% of the cells are still in S- or G2/M phase, even in highly confluent population. All cells can be induced to resynthetize DNA following the cell trypsinization and by seeding at low density. Cell-quiescence that occurs despite the optimal nutrients and growth factors supply and the pH stability in the medium, appears to be only related to the cell-cell contacts. Mechanisms and molecules involved in the contact inhibition phenomenon are poorly understood. Cell adhesion molecules and membrane junctional complexes may influence regulation of cell growth [53], perhaps through the control of the intracellular pH [54]. We have observed that mouse mammary cells that express a cell-adhesion molecule, E-cadherin, are more resistant to DXR than control, nontransfected cells (unpublished results). Adhesion molecules could constraint the cancer cells to stay in the monolayer epithelial sheet instead to overlap the one on each other. More the culture confluent is, smaller the size and the protein content of the cells are. Recent evidence suggest that cell size is a key factor in control of cell proliferation through the synthesis of specialized cell-cycle control proteins as cyclins [55]. Molecular control of contact inhibition could also be due to a specialized membrane glycoprotein, contact-inhibin, which has been not only found in contact-inhibited untransformed fibroblasts but also in many cancer cells, including the HT 29 colon cancer cells [56]. Further studies are necessary to better understand the regulation of the cell proliferation by cell-cell contact because they could have a direct implication to explain the CDR of solid tumors.

Ways to circumvent CDR

Since the discovery of new very effective anticancer agents seems to be unlikely before the end of this century, breaking the cure barrier for solid tumors imply the use of the available drugs with the maximal efficacy and the integration of the few new agents, such as taxol or camptothecin analogs, in rational combination therapies. Basic principles of polychemotherapy in solid tumors have been empirically established with some success by clinical oncologists. The beneficial effect of tumor reduction mass has been established in animal models [57] and in the treatment of breast and ovarian carcinoma [58–60]. Maximal surgical removal of the tumor mass not only permit a better access of drugs to the residual malignant cells but also a transitory recruitment of proliferating cells which are better oxygenated and nutrient supplied. However, the small microscopic foci that persist after surgery, often contain quiescent cells that are resistant to adjuvant chemotherapy. Experimental studies on millimetric spheroids have shown that only the cells of the tumor periphery are proliferating whereas the inner cells are quiescent or necrotic. [61].

Hormonal manipulation has been proposed to overcome the quiescence related resistance. Estrogen have been used to recruit quiescent cells into the active cell cycle before chemotherapy of metastatic breast cancer [62]. The poorly convincing results of this study could be attributed to the tumor heterogeneity of the hormonal receptors, an insufficient cell synchronization or a stimulation of tumor growth. Moreover hormone stimulation of cell proliferation cannot be envisaged for many tumors, as digestive tract tumors, because they lack a well defined proliferation-inducing hormone or specific growth factors [63].

Recruitment of quiescent cells into the proliferative compartment of the cell cycle can be theoretically achieved after a first cytoreductive chemotherapy [64–72]. Theoretically, a second shot of cytotoxic drugs given at a appropriate time could be more efficient on a tumor mass containing reproliferating cells. However, these cytokinetic concepts on polychemotherapy have not impacted as heavily on clinical cancer therapy as originally hoped probably on account of the difficulty to obtain a synchronized recruitment of the tumor cells and to define the exact time of the second shot chemotherapy.

New approaches merit to be investigated for the rational combination of anticancer drugs. As topoisomerase II inhibitors exert a preferential cytotoxicity during the S- and G2/M phases of the cell cycle, cell recruitment in those stages could be a theoretical way to circumvent CDR of slowly growing tumors. DNA damaging drugs, or X-irradiation, are known to prolong the S- and G2 phases, most likely to give time to repair the DNA before mitosis, thus avoiding the transmission of unrepaired mutations to progeny cells [73,74]. We have recently found (manuscript in preparation) that a first treatment of confluent HT 29 colon cancer cells with infratoxic concentrations of cisplatin, melphalan or mitomycin C leads to an increased accumulation of the cells in the S- or G2 phases of the cell cycle (Fig. 5). Both these phases are associated to an increased expression of the topoisomerase II gene (Fig. 6) and, as a result, to an increased cytotoxicity of doxorubicin on confluent cells (Fig. 7). Similar results have been obtained by Weisberg using cisplatin [75] and Erba *et al.* using methotrexate and aphidicolin [76]. We think that an optimal schedule of polychemotherapy for clinical practice could consist in a first shot treatment with alkylating agents and/or cisplatin followed after a 48 or 72 h delay by topoisomerase II inhibitors, etoposide and/or anthracyclines. In order to maximize the synergistic antitumor efficacy of drug combination, chronology of the S phase onset and

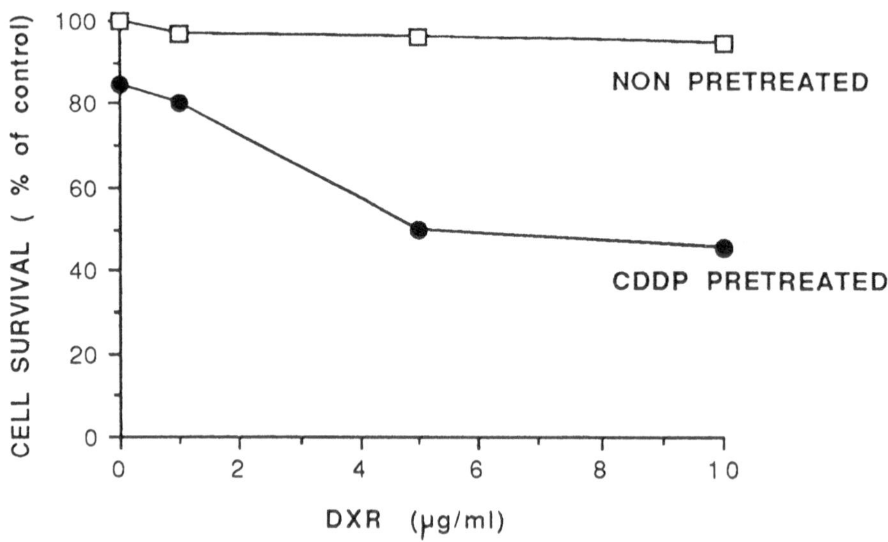

Fig. 7. Enhancement of DXR cytotoxicity on confluent HT 29 cells after a first shot treatment with CDDP. Confluent HT 29 cells were treated for 4 h by a low toxic concentration of CDDP (5 µg/ml), cultivated again in drug-free medium for 48 h then treated for 4 h by DXR. Cell survival was estimated 6 days after the DXR treatment by a methylene blue colorimetric assay. The divergence of the survival curves between CDDP pretreated and nonpretreated cells demonstrates the greater cytotoxicity of DXR on partially S-phase synchronized cells.

the way of administration of the first shot drugs, i.e. bolus or infusion, should be carefully determined in future clinical trials.

References

1. Salmon SE, Alberts DS, Meyskens FL, Durie BGM (1980) Clinical correlations of *in vitro* drug sensitivity. In: Salmon SE (ed) Cloning of Human Tumor Stem Cells. New York: AR Liss, 223–245.
2. Sutherland RM (1988) Cell and environmental interactions in tumor microregions: the multicell spheroid model. Science 240: 177–184.
3. Tannock IF (1968) The relation between cell proliferation and the vascular system in a transplanted mouse mammary tumour. Br J Cancer 22: 258–273.
4. Kerr DJ, Kaye SB (1987) Aspects of cytotoxic drug penetration, with particular reference to anthracyclines. Cancer Chemother Pharmacol 19: 1–5.
5. Barranco SC, Novak JK (1974) Survival responses of dividing and nondividing mammalian cells after treatment with hydroxyurea, arabinosylcytosine, or adriamycin. Cancer Res 34: 116–118.
6. Twentyman PR, Bleehen NM (1975) Changes in the sensitivity to cytotoxic agents occurring during the life history of monolayer cultures of a mouse tumour cell line. Br J Cancer 31: 417–423.
7. Chambers SH, Bleehen NM, Watson JV (1984) Effect of cell density on intracellular adriamycin concentration and cytotoxicity in exponential and plateau phase EMT6 cells. Br J Cancer 49: 301–306.
8. Drewinko B, Patchen M, Yan LY, Barlogie B (1981) Differential killing efficacy of twenty antitumor drugs on proliferating and nonproliferating human tumor cells. Cancer Res 41: 2328–2333.
9. Epifanova OI, Smolenskaya IN, Polunovsky VA (1978) Responses of proliferating and nonproliferating Chinese hamster cells to cytotoxic agents. Br J Cancer 37: 377–385.
10. Valeriote F, Van Putten L (1975) Proliferation-dependent cytotoxicity of anticancer agents: a review. Cancer Res. 35: 2619–2630.
11. Madoc-Jones H, Bruce WR (1967) Sensitivity of leukemic cells in exponential and stationary phase to 5-fluorouracil. Nature 215: 302–303.
12. Soranzo C, Ingrosso A (1988) A comparative study of the effects of anthracycline derivatives on human adenocarcinoma cell line (LoVo) grown as a monolayer and as spheroids. Anticancer Res 8: 369–373.
13. Born R, Eichholtz-Wirth H (1981) Effect of different physiological conditions on the action of adriamycin on Chinese hamster cells *in vitro*. Br J Cancer 44: 241–246.
14. Smith E, Stratford IJ, Adams GE (1980) Cytotoxicity of adriamycin on aerobic and hypoxic Chinese hamster V 79 cells *in vitro*. Br J Cancer 41: 568–573.
15. Duncan MR, Robinson MJ, Dell'Orco RT (1982) Exit of human diploid cells from low serum quiescence. Cell Biol

Int Rep 6: 369–377.
16. Chauffert B, Martin F, Caignard A, Jeannin JF, Leclerc A (1984) Cytofluorescence localization of adriamycin in resistant colon cancer cells. Cancer Chemother Pharmacol 13: 14–18.
17. Petit JM, Chauffert B, Dimanche-Boitrel MT, Genne P, Duchamp O, Martin F (1993) *mdr*-1 gene-expression and villin synthesis in a colon cancer cell line differentiated by sodium butyrate. Anticancer Res, in press.
18. Chauffert B, Martin MS, Hamman A, Michel MF, Martin F (1986) Amiodarone-induced enhancement of doxorubicin and 4′-deoxydoxorubicin cytotoxicity to rat colon cancer cells *in vitro* and *in vivo*. Cancer Res 46: 825–830.
19. Goldstein LJ, Gottesman MM, Pastan I (1991) Expression of the *mdr*-1 gene in human tumors. In: Ozols RF (ed) Molecular and Clinical Advances in Anticancer Drugs. New York: Kluwer Academic Publishers, 101–120.
20. Siegfried JM, Burke TG, Tritton TR (1985) Cellular transport of anthracyclines by passive diffusion: implications for drug resistance. Biochem Pharmacol 34: 593–598.
21. Epifanova OI (1984) Differential sensitivity of proliferating and resting cells to antitumour drugs and its probable causes. In: Lapis K and Jeney A (eds) Regulation of control cell proliferation. Budapest: Akademia Kiado, 434–457.
22. Pelletier H, Millot JM, Chauffert B, Manfait M, Genne P, Martin F (1990) Mechanisms of resistance of confluent human and rat colon cancer cells to anthracyclines: alteration of drug passive diffusion. Cancer Res 50: 6626–6631.
23. Goldenberg GJ, Lyons RM, Lepp JA, Vanstone CL (1971) Sensitivity to nitrogen mustard as a function of transport activity and proliferative rate in L 5178Y lymphoblasts. Cancer Res 31: 1616–1618.
24. Dimanche-Boitrel MT, Pelletier H, Genne P, Petit JM, Le Grimellec C, Canal P, Ardiet C, Bastian G, Chauffert B (1992) Confluence-dependent resistance in human colon cancer cells: role of reduced drug accumulation and low intrinsic chemosensitivity of resting cells. Int J Cancer 50: 677–682.
25. Magin RL, Niesman MR, Bacic G (1990) Influence of fluidity on membrane permeability: correspondence between studies of membrane models and simple biological systems. In: Aloia RC, Curtain CC and Gordon LM (eds) Membrane Transport and Information Storage. New York: AR Liss, 221–238.
26. Liu LF (1990) Topoisomerase poisons in cell killing and drug resistance. In: Mihich E (ed) Drug-resistance Mechanisms and Reversal. Rome: John Libbey CIC, 251–263.
27. Osheroff N (1989) Biochemical basis for the interactions of type I and II topoisomerase inhibitors with DNA. Pharmacol Ther 41: 223–241.
28. Osheroff N (1989) Effect of antineoplastic agents on the DNA cleavage/relegation reaction of eukaryotic topoisomerase II: inhibition of DNA relegation by etoposide. Biochemistry 28: 6157–6160.
29. Tewey KM, Rowe TC, Yang L, Halligan BD, Liu LF (1984) Adriamycin-induced DNA damage mediated by mammalian DNA topoisomerase II. Science. 266: 466–468.
30. Hancock R, Charron M, Lambert H, Lemieux M, Pankov R, Pepin N (1990) Topoisomerase II as a target of antitumor agents. Pharmacol Ther (suppl) 10: 119–137.
31. Duguet M, Lavenot C, Harper F, Mirambeau G, De Recondo AM (1983) DNA topoisomerase from rat liver: physiological variations. Nucleic Acid Res 11: 1059–1075.
32. Taudou G, Mirambeau G, Lavenot C, Vermeersch J, Duguet M (1984) DNA topoisomerase activities in concanavalin A-stimulated lymphocytes. FEBS 176: 431–435.
33. Heck MMS, Hittelman WN, Earnshaw WC (1988) Differential expression of DNA topoisomerase I and II during the eukaryotic cell cycle. Proc Natl Acad Sci USA 85: 1086–1088.
34. Zwelling LA, Estey E, Silberman I, Doyle S, Hittelman W (1987) Effect of cell proliferation and chromatin conformation on intercalation induced, protein associated DNA cleavage in human brain tumor cells and human fibroblasts. Cancer Res 47: 251–257.
35. Nelson WG, Liu LF, Coffey DS (1986) Newly replicated DNA is associated with DNA topoisomerase II in cultured rat prostatic adenocarcinoma cells. Nature 322: 187–189.
36. Chow KC, Ross WE (1987) Topoisomerase-specific drug sensitivity in relation to cell cycle progression. Mol Cell Biol 7: 3119–3123.
37. Estey E, Adlakaha RC, Hittelman WN, Zwelling LA (1987) Cell cycle stage dependent variations in drug-induced topoisomerase II mediated DNA cleavage and cytotoxicity. Biochemistry. 26: 4338–4344.
38. Sullivan DM, Glisson BS, Hodges PK, Smallwood-Kentro S, Ross WE (1986) Proliferation dependence of topoisomerase II mediated drug action. Biochemistry 25: 2248–2256.
39. Sullivan DM, Latham MD, Ross WE (1987) Proliferation-dependent topoisomerase II content as a determinant of anti-neoplastic drug action in human, mouse, and Chinese hamster ovary cells. Cancer Res 47: 3973–3979.
40. Robbie MA, Baguley BC, Denny WA, Gavin JB, Wilson WR (1988) Mechanism of resistance of noncycling mammalian cells to 4′-(9-acridinylamino) methanesulfon-*m* anisidide: comparison of uptake, metabolism and DNA-breakage in log- and plateau-phase Chinese hamster fibroblast culture. Cancer Res 48: 310–319.
41. Davies SM, Robson CN, Davies SL, Hickson ID (1988) Nuclear topoisomerase II levels correlates with the sensitivity of mammalian cells to intercalating agents and epipodophylotoxins. J Biol Chem 263: 17724–17729.
42. Kung AL, Zetterberg A, Sherwood SW, Schimke RT (1990) Cytotoxic effect of cell cycle specific agents: result of cell cycle perturbation. Cancer Res 50: 7307–7317.
43. Eastman A, Barry MA (1992) The origins of DNA breaks: a consequence of DNA damage, DNA repair or apoptosis. Cancer Investigation 10: 229–240.
44. Pera MF, Rawlings CJ, Roberts JJ (1981) The role of

DNA repair in the recovery of human cells from cisplatin-induced toxicity. Chem Biol Interact 37· 245–261

45 Fraval HNA, Roberts JJ (1979) Excision repair of *cis*-diamminedichloro-platinum(II)-induced damage of Chinese hamster cells Cancer Res 39: 1793–1797.

46 Sorenson CM, Eastman A (1988) The mechanism of *cis*-diamminedichloro-platinum(II)-induced cytotoxicity the role of G2 arrest and DNA double strand breaks Cancer Res 48: 4484–4488

47. Barry MA, Behnke CA, Eastman A (1990) Activation of programmed cell death (apoptosis) by cisplatin, other anticancer agents drugs, toxins and hyperthermia Biochem Pharmacol 40: 2353–2362.

48. Roberts JJ, Brent TP, Crathorn AR (1971) Evidence of the inactivation and repair of the mammalian DNA template after alkylation by Mustard gas. Eur J Cancer 7: 515–5524

49 Mendonca MS, Rodriguez A, Alpen EL (1990) Differential repair of potentially lethal damage in exponentially growing and quiescent cells Radiation Res 122: 38–43.

50 Dikomey E (1990) Induction and repair of DNA strand breaks in X-irradiated proliferating and quiescent cells. Int J Radiat Biol 57: 1169–1182.

51. Baral E, Auer G (1990) *In vitro* effect of doxorubicin on nonproliferating and proliferating epithelial cells. Int J Radiat Oncol Biol Phys 19: 963–965.

52. Geleziunas R, McQuillan A, Malapetsa A, Hutchinson M, Kopriva D, Wainberg MA, Hiscott J, Bramson J, Panasci L (1991) Increased DNA synthesis and repair-enzyme expression in lymphocytes from patients with chronic lymphocytic leukemia resistant to nitrogen mustards. J Natl Cancer Inst 83: 557–564

53 Cybulsky AV, Bonventre J, Quigg RJ, Wolfe LS, Samant DJ (1990) Extracellular matrix regulates proliferation and phospholipid turnover in glomerular epithelial cells. Am J Physiol 259: F326–F337.

54. Galkina SI, Sudina GF, Margolis LB (1992) Cell-cell contacts alter intracellular pH. Exp Cell Res 200: 211–214.

55 Ohtsubo M, Roberts JM (1993) Cyclin-dependent regulation of G1 in mammalian fibroblasts Science 259. 1908–1912.

56 Wieser RJ, Schutz S, Tschank G, Thoams H, Dienes HP, Oesch F (1990) Isolation and characterization of a 60 to 70 kDa plasma membrane glycoprotein involved in the contact-dependent inhibition of growth. J Cell Biol 111. 2681–2692

57 Skipper HE (1986) Experimental adjuvant chemotherapy an overview Recent result Cancer Res 103. 6–29

58 De Vita VT (1983) Implications for surgical adjuvant treatment of cancer. Cancer 51 1209–1220

59 Bonnadona G (1989) Conceptual and practical advances in the management of breast cancer J Clin Oncol 7: 1380–1397.

60. Cohen C (1985) Surgical considerations in ovarian cancer Semin Oncol 12. 53–56

61 Durand RE (1986) Chemosensitivity testing in V 79 spheroids role of drug delivery and cellular microenvironment J Natl Cancer Inst 77 247–242

62 Conte PF, Pronzato P, Rubagotti A, Alama A (1987) Conventional versus cytokinetic polychemotherapy with estrogenic recruitment in metastatic breast cancer: results of a randomized cooperative trial. J Clin Oncol 5: 339– 347

63. Morris DL, Watson SA, Durrant IG, Harrison JD (1989) Hormonal control of gastric and colorectal cancer in man. Gut 30: 425–429.

64. Schabel FM (1969) The use of tumor growth kinetics in planning "curative" chemotherapy of advanced solid tumors. Cancer Res 29: 2384–2389.

65. Kal W (1973) Proliferation behavior of proliferating and quiescent cells in a rat rhabdomyosarcoma after irradiation as determined by DNA measurements. Eur J Cancer 9· 753–756.

66. Braunschweiger PG, Schiffer LM (1980) Therapeutic implications of cell kinetic changes after cyclophosphamide treatment in spontaneous and transplantable mammary tumors. Cancer Treat Rep 62: 727–733.

67. Braunschweiger PG, Schiffer LM (1980) Cell kinetic-directed sequential chemotherapy with cyclophosphamide and adriamycin in T1699 mammary tumor. Cancer Res 40 737–733.

68. Burke PJ, Vaughan WP, Karp JE (1980) A rational for sequential high-dose chemotherapy of leukemia, timed to coincide with induced tumor proliferation. Blood 55: 960–968.

69. Aglietta M, Colly L (1979) Relevance of recruitment-synchronization in the scheduling of 1-β-D-arabinosyl-cytosine in a slow-growing acute myeloid leukemia of the rat. Cancer Res 39 2727–2732.

70. Karp JE, Humphrey RL, Burke PJ (1981) Timed sequential chemotherapy of cytoxan-refractory multiple myeloma with cytoxan and adriamycin based on induced proliferation Blood 57: 468–475.

71 Salmon SE (1975) Expansion of the growth fraction in multiple myeloma with alkylating agents. Blood 45: 119–123.

72 Rosso R, Alama A, Repetto L, Conte PF (1990) Timed sequential chemotherapy following ifosfamide-induced kinetic recruitment in refractory ovarian cancer. Cancer Chemother Pharmacol 26: 43–44.

73. Painter RB (1986) Inhibition of mammalian cell DNA synthesis by ionizing radiations Int J Radiat Biol 49 771–781.

74. Russev G, Boulikas T (1992) Repair of transcriptionally active and inactive genes during S and G2 phases of the cell cycle. Eur J Biochem 204: 267–272

75 Weisberg TF (1992) Cisplatin treatment of cultured tumor cells results in prolongation of S Phase. potential clinical implications. Abst 367, Proc ASCO.

76. Erba E, Sen S, Lorico A, D'Incalci M (1992) Potentiation of etoposide cytotoxicity against a human ovarian cancer cell line by pretreatment with nontoxic concentrations of methotrexate or aphidicolin. Eur J Cancer 28: 66–71

Address for correspondence. M -T Dimanche-Boitrel, INSERM 252, Faculty of Medicine, F-21033 Dijon, France

Cytotechnology **12**: 357–366, 1993.

Cytochromes P450 and drug resistance

Johannes Doehmer[1], Arnold R. Goeptar[2], Nico P.E. Vermeulen[2]
[1]*Institut für Toxikologie und Umwelthygiene, Technische Universität München, Lazarettstrasse 62, 80636 München, Germany;* [2]*Leiden/Amsterdam Center for Drug Research, Division of Molecular Toxicology, Free University Amsterdam, De Boelelaan 1083, 1081 HV Amsterdam, The Netherlands*

Key words: anticancer drugs, cytochrome P450, drug resistance, metabolic activation

Abstract

Cytochromes P450 are the key enzymes for activating and inactivating many drugs, in particular anticancer drugs. Therefore, individual expression levels of cytochromes P450 may play a crucial role in drug safety and drug efficacy. Overexpression of cytochrome P450 may yield rapid turnover and elimination of drugs before the target site was reached and any pharmacological effect is observed. Therefore, it may be vital to know the individual cytochrome P450 status in order to select the appropriate drug before drug resistance occurs. Expression levels and activity of cytochromes P450 depend on many different factors. These factors include tissue and organ specific expression, sex- and age-dependent expression, genetic differences yielding polymorphic forms, competitive inhibition or induction of cytochromes P450 due to multiple drug interaction, nutrition and diet. Genetically engineered test cells defined for cytochromes P450 are available for studying drugs for metabolic activation and for identifying the metabolically competent cytochrome P450 isoform.

Cytochrome P450 mediated drug resistance

Cytochromes P450 play a crucial role in the biotransformation process of many drugs and other xenobiotics (Fig. 1) (Cholerton *et al.*, 1992; Gonzalez, 1992). Upon oxidative or reductive metabolism by cytochromes P450, drugs may be changed into metabolites with altered pharmacological properties. In some cases, metabolites are as active as the initial compound. In other cases, drugs depend on metabolic activation by cytochromes P450 in order to become active. In most cases, oxidative metabolism via cytochromes P450 inactivate drugs. Inactivation may occur rapidly, before the drug had a chance to reach its target organ and to exert its pharmacological effect. Therefore, cytochrome P450 mediated drug resistance is a major concern.

The cytochrome P450 superfamily

Cytochromes P450 are extremely versatile enzymes with a broad capacity for handling drugs and other xenobiotics (Fig. 2) (Guengerich, 1993). They also play an important role in the metabolism of endogenous substrates such as fatty acids and steroid hormones. Cytochromes P450 are virtually present in any species looked at. They have been detected in bacteria, fungi, insects, plants, fish, amphibia and mammals. Protein and gene sequence alignments and comparisons have revealed relationships between the different cytochromes P450 (Gonzalez, 1989). Based on these studies, cytochromes P450 have been delineated from each other, starting with an ancestral cytochrome P450 gene, and resulting into a still growing evolutionary tree (Gonzalez and Nebert, 1990). The diversity of the cytochromes

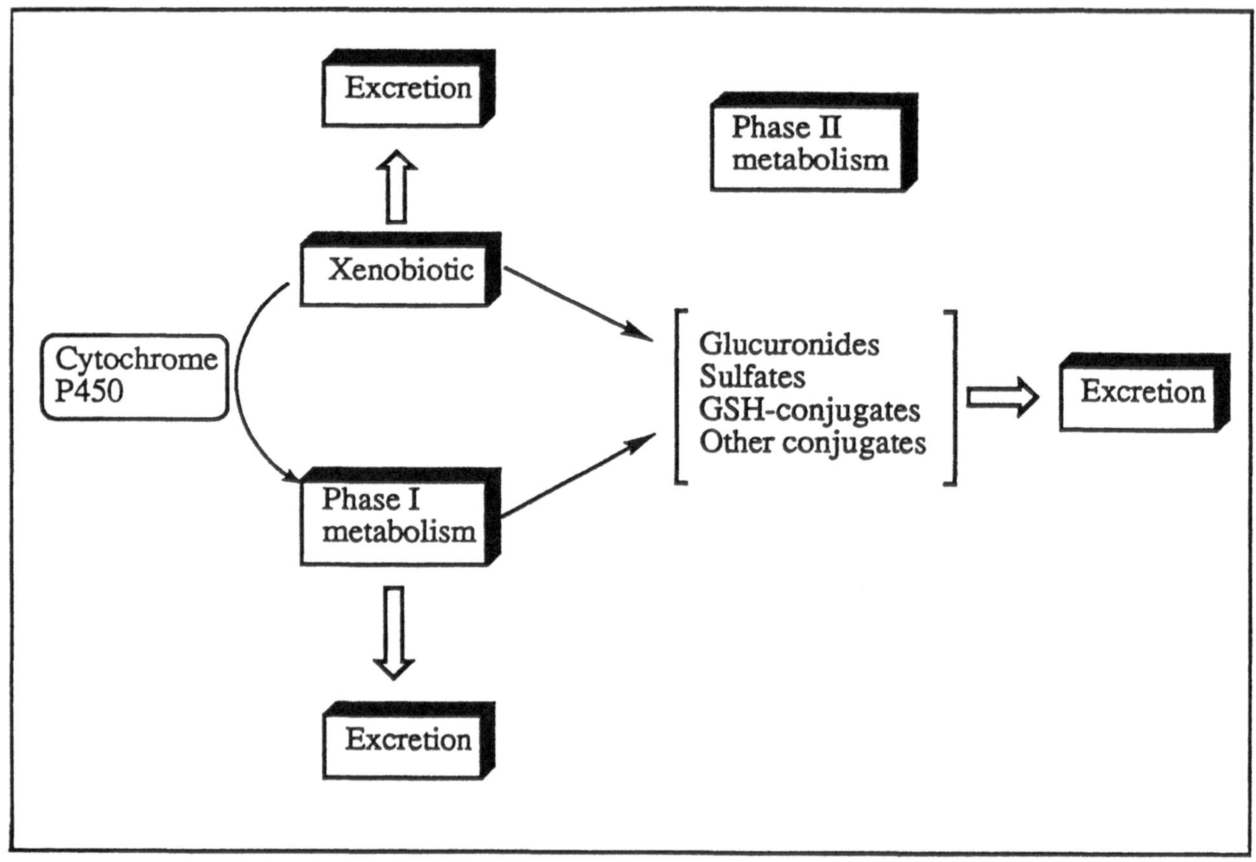

Fig. 1. Drug metabolism and excretion pathways for xenobiotic compounds.

P450, which has developed over billions of years, is amazing. It is therefore believed that cytochrome P450 gave selective advantage to organisms. Plants may synthesize toxic chemicals as a protective measure from being eaten by mammals or killed by fungi. Mammals or fungi in return may have gained cytochrome P450 activity, in order to degrade this toxic chemical. This interesting hypothesis was referred to as a silent warfare between plants and other organisms. Cytochromes P450 may also have been of advantage in order to survive toxic air pollutants. Therefore cytochromes P450 are considered to be important enzymes for protecting an organism against xenobiotics.

Today, 221 different cytochromes P450 have been listed (Nelson *et al.*, 1993). They have been classified into 36 gene families, of which 12 families exist in all mammals. These 12 families comprise 22 mammalian subfamilies. About 70 different human cytochromes P450 are known. A uniform nomenclature has been developed and recommended. According to this widely accepted nomenclature the cytochrome P450 genes and cDNAs are abbreviated as italicized *Cytochrome P450*, followed by an Arabic number denoting the family, a letter designating the subfamily, followed again by an Arabic number, denoting the individual gene.

All Cytochrome P450 have in common a heme group with iron as the central atom which can be loaded with molecular oxygen after receiving electrons from an oxidation of NADPH + H^+ mediated by a the adjacent Cytochrome P450 reductase. One atom of the oxygen is transferred to

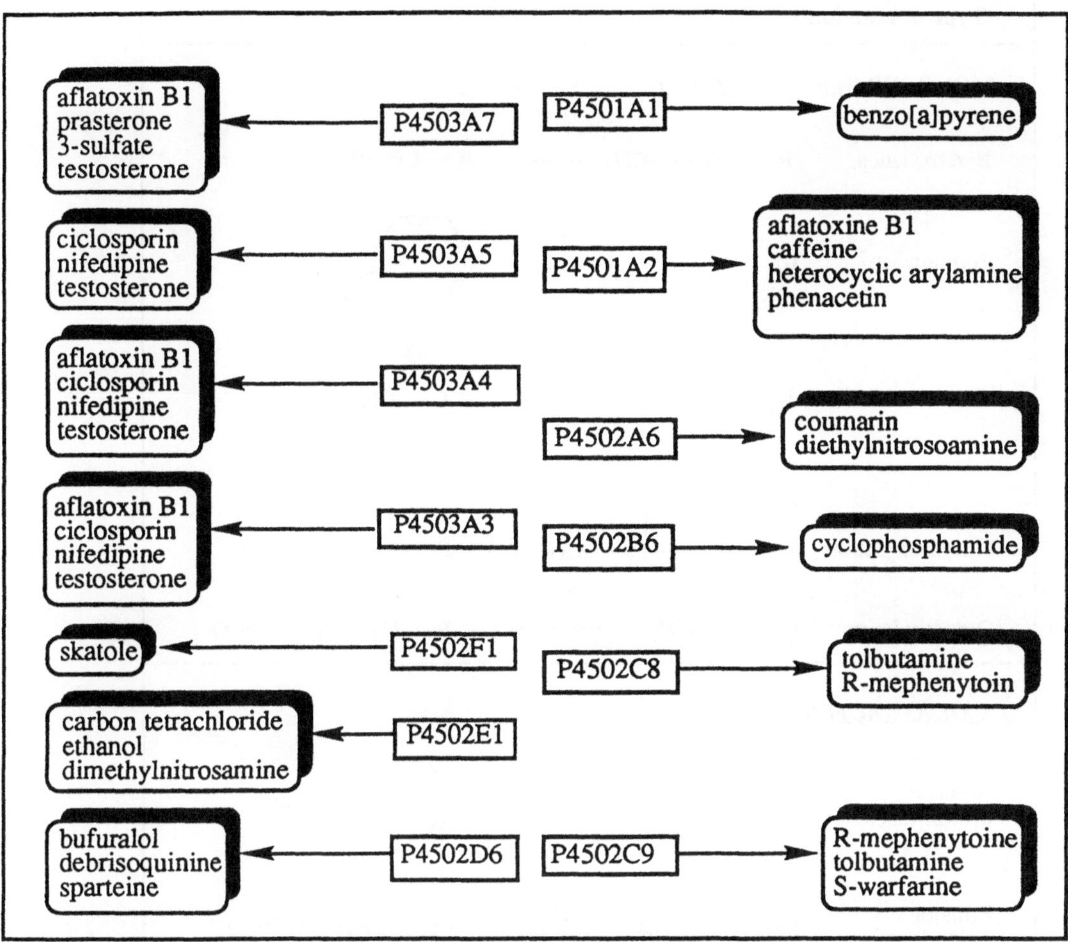

Fig. 2. Human cytochromes P450 and their substrates.

the substrate, the other is found in water. Cytochromes P450 differ by their substrate binding sites. Usually, substrate specificities overlap between the different Cytochrome P450 due to varying degrees of homology between Cytochrome P450. This is considered as an evolutionary wanted effect of plasticity covering a broad range of substrates, even man-made chemicals. Cytochrome P450 catalyze thousands of different reactions. Similarities in the chemistry have been used as criteria in order to group these oxidative reactions (Fig. 3) (Guengerich, 1993).

Factors influencing cytochrome mediated drug metabolism

Several factors may influence the individual Cytochrome P450 activity. The most important factors are: sex, age, organ tropism, inducibility, genetic polymorphism, nutrition and diet. Each of these factors may be sufficient in causing drug resistance either by rapid turnover to inactive products or by a lack of metabolic activation.

Sex-dependent expression of Cytochrome P450 is programmed neonatally, and has been described for several Cytochrome P450 (Ryan and Levin, 1993). For example, striking difference were found

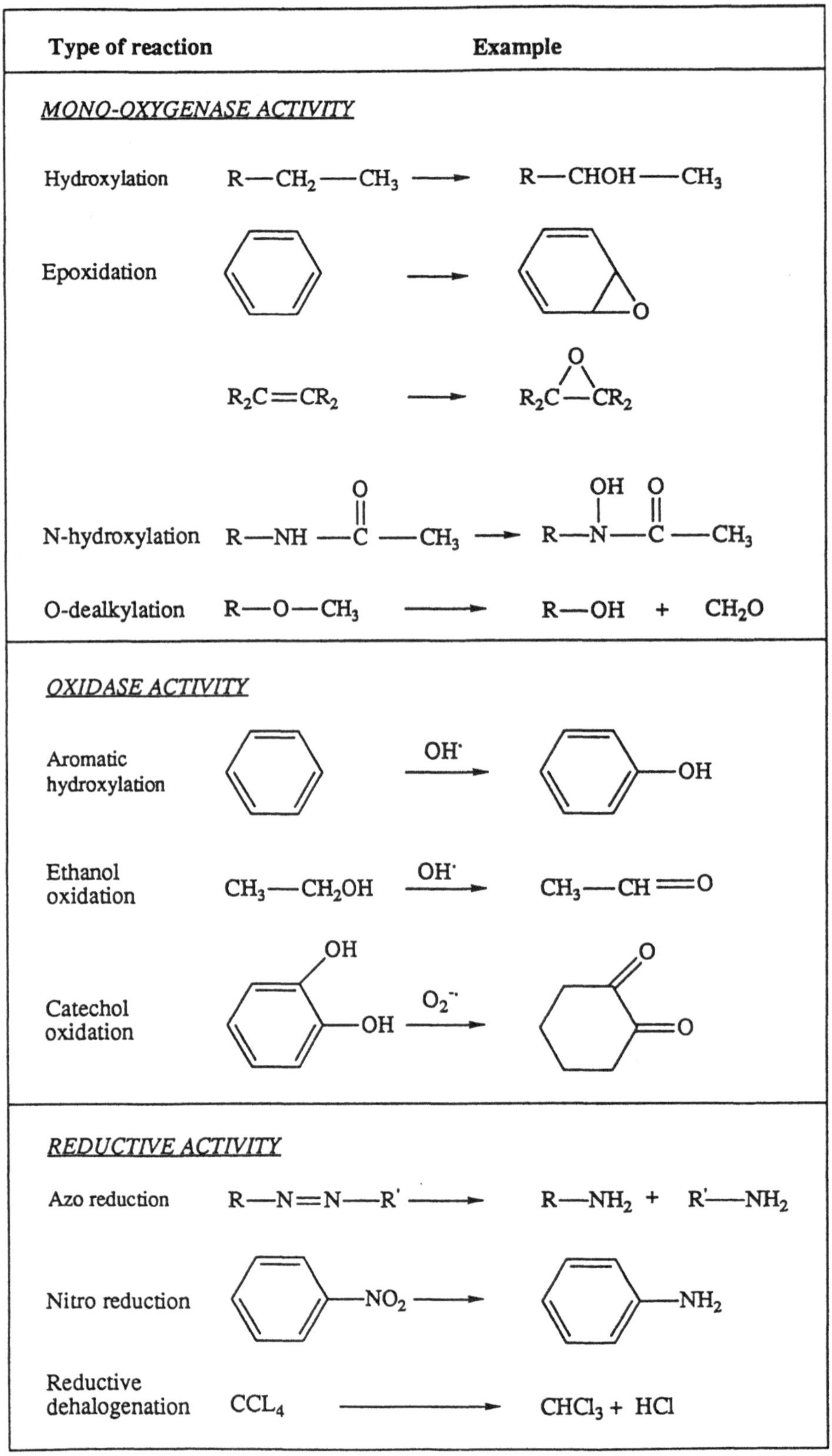

Type of reaction	Example
MONO-OXYGENASE ACTIVITY	
Hydroxylation	$R{-}CH_2{-}CH_3 \rightarrow R{-}CHOH{-}CH_3$
Epoxidation	$\rightarrow$ O
	$R_2C{=}CR_2 \rightarrow R_2C{-}CR_2$ (with O bridging)
N-hydroxylation	$R{-}NH{-}C(=O){-}CH_3 \rightarrow R{-}N(OH){-}C(=O){-}CH_3$
O-dealkylation	$R{-}O{-}CH_3 \rightarrow R{-}OH + CH_2O$
OXIDASE ACTIVITY	
Aromatic hydroxylation	$\xrightarrow{OH^{\cdot}}$ OH
Ethanol oxidation	$CH_3{-}CH_2OH \xrightarrow{OH^{\cdot}} CH_3{-}CH{=}O$
Catechol oxidation	OH, OH $\xrightarrow{O_2^{\cdot-}}$ O, O
REDUCTIVE ACTIVITY	
Azo reduction	$R{-}N{=}N{-}R' \rightarrow R{-}NH_2 + R'{-}NH_2$
Nitro reduction	$NO_2 \rightarrow NH_2$
Reductive dehalogenation	$CCL_4 \rightarrow CHCl_3 + HCl$

Fig. 3. Typical cytochrome P450 catalyzed reactions.

for cytochromes Cytochrome P450 2C11, expressed in male rats, but not in female rats; and Cytochrome P450 2C12, in female rats, but not in male rats. Cytochrome P450 2C11 is responsible for 2α- and 16α-hydroxylation of testosterone and progesterone. This Cytochrome P450 mediates oxidative degradation of a variety of different drugs (Ryan *et al.*, 1984). Cytochrome P450 2C12 mediates 15β-hydroxylation of 5α-androstane-3α,17β-diol-3,17 disulfate. Cytochrome P450 2C12 is responsible for a female specific hepatic 15β-hydroxylation of C21 steroids which upon are secreted as sulphate metabolites (Gustafsson and Ingelman-Sundberg, 1975). Sex-specific differences are also growth hormone/ growth hormone release factor mediated. In male rats growth hormone is pulse modulated and secreted every 3 to 4 h, whereas in female rats a sustained and continuous release is observed. The way growth hormone is secreted has an influence on the Cytochrome P450 isozyme pattern (Gustafsson *et al.*, 1983; Waxman *et al.*, 1985).

Genetic polymorphism may lead to different Cytochrome P450 activities. The classical example is the polymorphism described for Cytochrome P450 2D6 (Kimura *et al.*, 1989). Two distinct classes of fast and slow metabolizers among a population are observed, they are characterized by their efficiency to metabolize the drug debrisoquine. The same Cytochrome P450 is responsible for other drugs as well, such as cardiovascular agents, psychoactive drugs, or morphine derivatives. It is indicated to probe patients first with debrisoquine, before a drug can be given in adequate therapeutic dose.

Although most cytochromes P450 are expressed in the liver, some cytochromes P450 are expressed in other organs. The human Cytochrome P450 1A1 is almost exclusively expressed in lung tissue. Very low levels of cytochromes P450 have been detected in other tissues as well, such as brain and kidney.

Nutrition and diet may have an influence on cytochrome P450 expression. Starvation, for example, leads to a massive expression of the ethanol inducible Cytochrome P450 2E1, which in turn may degrade efficiently several drugs.

Many Cytochrome P450 are not constitutively expressed rather than upon induction by various drugs. Phenobarbital induces Cytochrome P450 2B1. Increased activity of particular cytochromes P450 may lead to rapid inactivation of other drugs. Overexpression of the phenobarbital inducible Cytochrome P450 leads to inactivation of steroid hormones as used for birth control drugs, degradation of lidocaine as used for anesthetics, increased turnover of coumarin as used as anticoagulant. Phenobarbital itself is inducer and substrate at the same time. For this reason, patients may develop drug tolerance against phenobarbital.

Cytochrome P450-mediated activation and inactivation of anticancer drugs

Cytochromes P450 play a decisive role in the metabolism of several anticancer drugs, such as cyclophosphamide (Clarkee and Waxman, 1989), procarbazine (Prough *et al.*, 1984), etoposide (Haim *et al.*, 1987) and 2-chloroethylnitrosoureas (Potter *et al.*, 1984). In particular, cyclophosphamide requires metabolic activation by Cytochrome P450 2B1-mediated hydroxylation at the 4-position to form 4- hydroxycyclophosphamide (Fig. 4) (Le-Blanc and Waxman, 1989). This hydroxylated metabolite serves as the precursor for the cytostatic acting metabolites acrolein and phosphoramide mustard. Inefficient expression of metabolically competent cytochromes P450 isoforms may result into inefficient bioactivation and drug resistance.

Quinones are chemical entities quite common in nature (Powis, 1989; Powis, 1987). Several of these quinones have been extensively studied for their use as therapeutics in chemotherapy of cancer. Among those quinones, mitomycin C and adriamycin are frequently used for the treatment of human tumours (Fig. 5). These quinone-containing antitumor drugs require reductive bioactivation to elicit their cytostatic activity (Fig. 6). Although Cytochrome P450 are generally classified as monooxygenases they can also catalyze a reductive reaction, which has been extensively studied for CCl_4 (Slater and Sawyer, 1971). Usually, Cytochrome P450-mediated reduction reactions proceed under low oxygen tension. Mitomycin C and

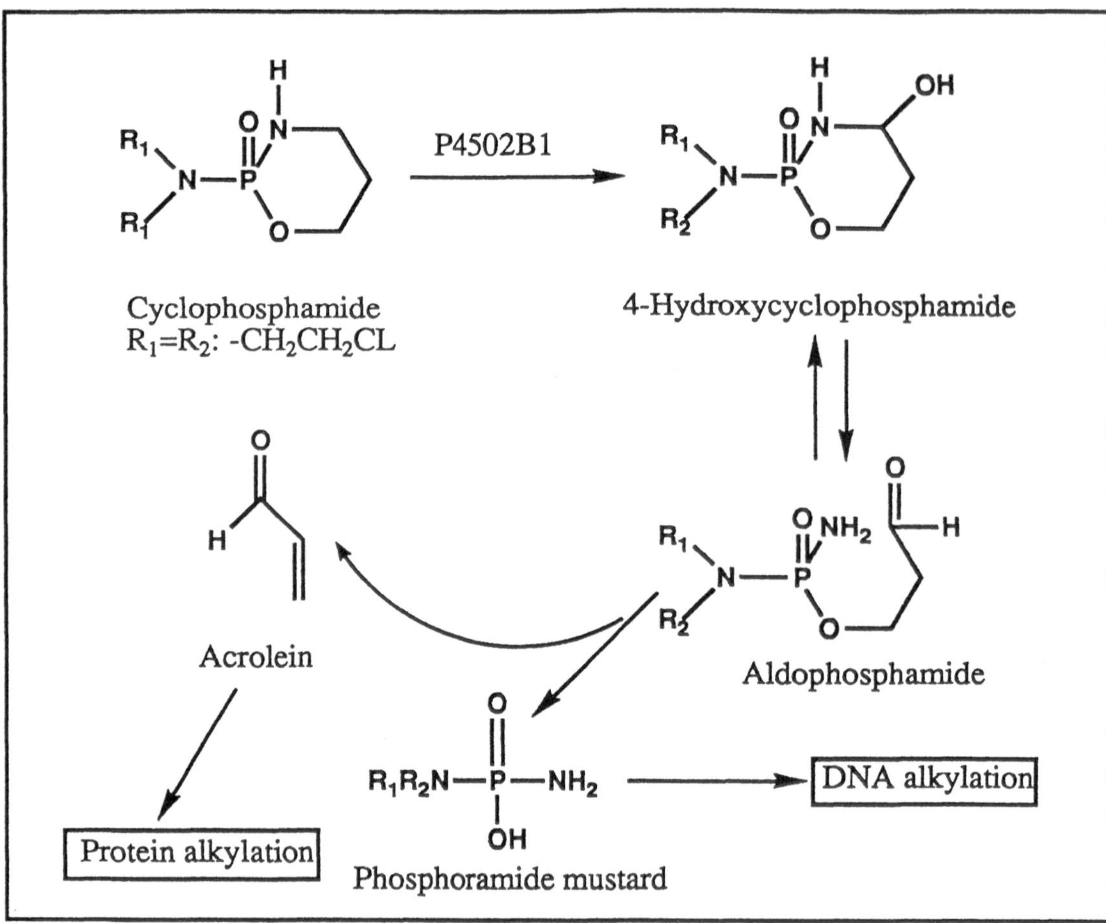

Fig. 4. Cytochrome P450 2B1 mediated metabolic activation of cyclophosphamide by hydroxylation.

adriamycin can undergo one-electron reduction to the corresponding semiquinone free radicals or by two-electron reduction to the corresponding hydroquinones. Several other enzymes, including xanthine oxidase and NADH-cytochrome b5 reductase have been implicated in the one-electron reduction of mitomycin C and adriamycin. However, Cytochromes P450 are in 20- to 30-fold excess of these enzymes in liver cells, and therefore play a key role in the one-electron reduction of quinones, such as mitomycin C. The molecular mechanism involved in quinone cytostatic activity is not known in detail. The unwanted cytotoxic properties of quinones are generally thought to be mediated through the one-electron reduction to the respective semiquinone free radicals (Fig. 6) (Powis, 1987). This one-electron reduction pathway is particularly sensitive to molecular oxygen due to reoxidation of the semiquinone free radical under concomitant formation of superoxide anion radicals, hydrogen peroxide, and the extremely reactive hydroxyl radicals. As a result of reductive bioactivation of quinones potentially cytotoxic free radicals intermediates are formed causing cell death (O'Brien, 1991).

Test systems for cytochrome P450-dependent drug metabolism

For drug safety and efficacy reasons it is of paramount importance to know the metabolic pathways,

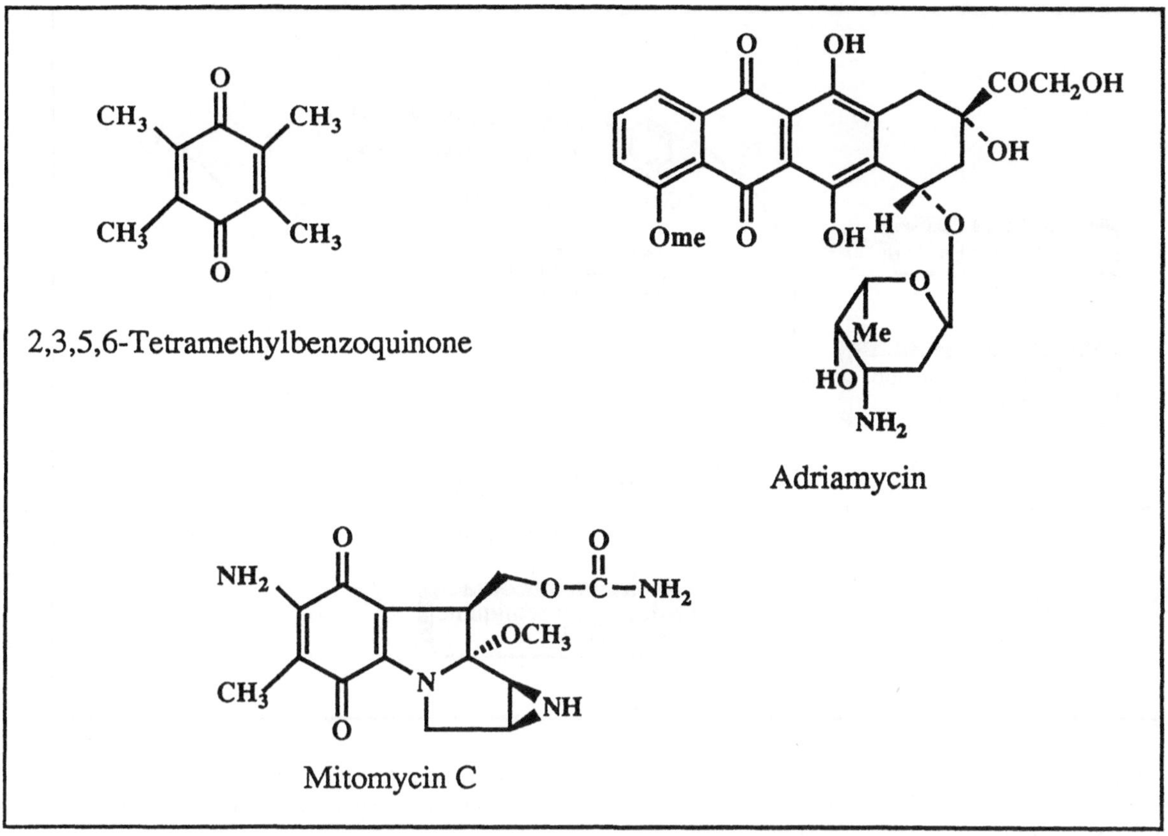

Fig. 5. Chemical structures of 2,3,5,6-tetramethylbenzoquinone, mitomycin C and adriamycin.

and in particular, to identify the metabolically competent cytochrome P450, and to identify the metabolites. Timely experimental systems are needed to solve those questions.

Most metabolism testings still take place in animals, as they are considered to the best predictors of the effect of drugs on the human organism. These studies have revealed problems associated with extrapolation to the human organism. Drug metabolism in humans and animals differ tremendously. In addition, complex organisms are not very suitable when detailed knowledge on a particular Cytochrome P450 is wanted. As always in science, in order to address specific questions, less complex and more analytical test systems are needed.

As an attempt to reduce complexity and as an alternative to testing drug metabolism in animals, metabolically competent somatic cells from animals and humans, e.g., hepatocytes, have been isolated and propagated in culture. There are several disadvantages in the use of primary hepatocytes. Primary cells rapidly loose tissue-specific functions upon cultivation within 2 or 3 days. This is particularly true for the expression of cytochromes P450. Loss of Cytochrome P450 activities occurs at different rates. Consequently, the Cytochrome P450 isoenzyme profile changes constantly in cultivated primary hepatocytes (Rogiers *et al.*, 1990). Therefore, primary hepatocytes cannot be standardized as is needed for quantitative studies, such as enzyme kinetics. At best, primary hepatocytes may be applied in more qualitative studies; in addition, especially when human hepatocytes are not readily available.

Recombinant DNA technology has significantly

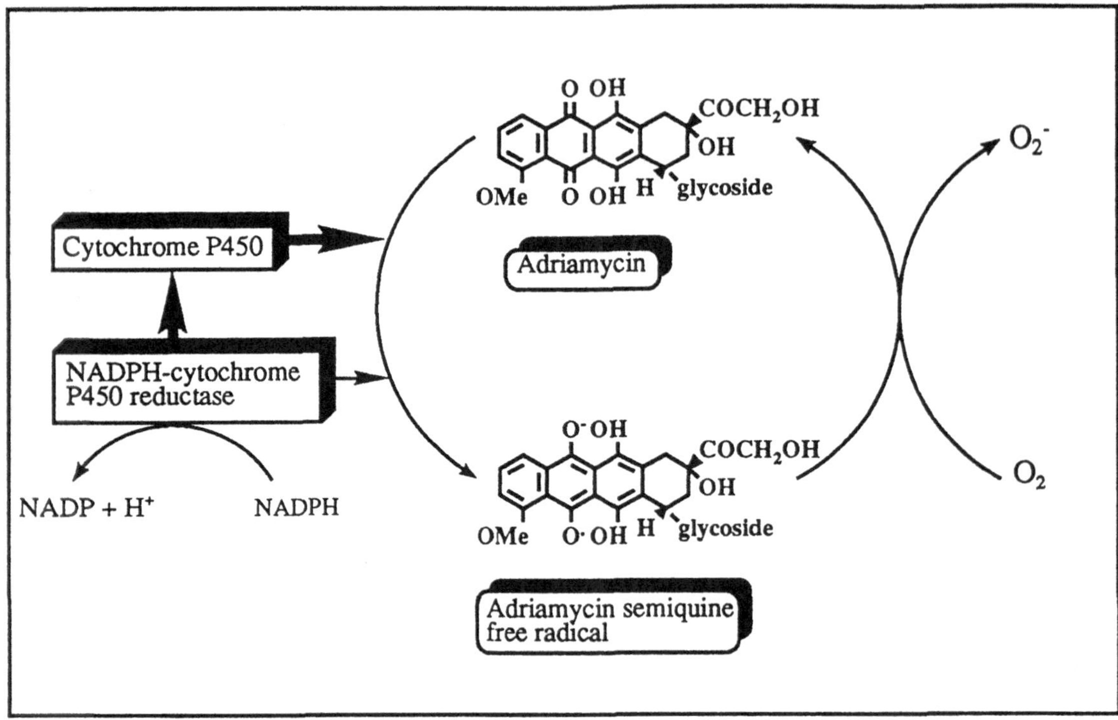

Fig. 6. Schematic presentation of the one-electron reductive bioactivation of the cytostatic quinone adriamycin. Under aerobic conditions the semiquinone free radical of adriamycin can redox cycle with molecular oxygen and produce reactive oxygen intermediates capable of initiating cell damage.

contributed to the advances in the area of drug metabolism in recent years. Several genes encoding animal and human cytochromes P450 have been cloned and successfully studied and applied by heterologous expression in well-established cell lines (Gonzalez *et al.*, 1991). Even cytochromes P450 become available, which are difficult to purify from native tissues due to low abundance. Heterologous expression of cytochromes P450 has been achieved in bacteria, yeast and mammalian cell lines (Doehmer and Greim, 1993). For example, the Chinese hamster V79 cell line is being engineered for expression of a variety of animal and human cytochromes P450. These cells have no endogenous Cytochrome P450 activity, and are therefore defined for the Cytochrome P450 activity acquired upon cDNA-mediated gene transfer. These cells have been repeatedly shown to be a perfect analytical tool for the study of Cytochrome P450 mediated drug metabolism. This maybe depicted by several applications of V79 cells genetically engineered for cytochromes P450 (Doehmer *et al.*, 1992). The cytostatic drug cyclophosphamide was shown to become activated upon Cytochrome P450 2B1-mediated hydroxylation (Doehmer *et al.*, 1990). Caffeine was dimethylated in Cytochrome P450 1A2 expressing cells (Fuhr *et al.*, 1992). Cytochrome P450 1A2 expressing cells were capable of producing paracetamol from phenacetin by O-deethylation (Jensen *et al.*, 1993). Cell lines genetically tailored for metabolic competence are state of the art in elucidating drug resistance due to metabolic effects. Of course, the same holds for other enzymes or membrane-bound proteins, causing drug resistance by overexpression as is the case for DHFR or the mdr-proteins.

Conclusion

Drug resistance may be caused by an array of factors, known and unknown. Drug metabolism related resistance is one of the most important factors involved. Drug resistance should be looked at as a naturally occurring event, because the enzymes implicated in drug metabolism have evolutionary evolved as a protective means against harmful effects by toxic agents in the environment and food. This may be in line with the evolutionary meaning of MDR proteins by protecting a cell against toxic chemicals. It needs an intelligent approach in drug design to override drug resistance, and to overcome these evolutionary implemented barriers against chemicals.

Acknowledgements

Johannes Doehmer is funded by the Bundesgesundheitsamt (BGA), Berlin, ZEBET.

References

Cholerton S, Daly AK, Idle JR (1992) The role of individual human cytochromes P450 in drug metabolism and clinical response. TiPS Reviews 13: 434–439.

Clarke L, Waxman DJ (1989) Oxidative metabolism of cyclophosphamide: identification of the hepatic mono-oxygenase catalysts of drug activation. Cancer Res 49: 2344–2350.

Doehmer J, Greim H (1993) Cytochromes P450 in genetically engineered cell cultures: the gene technological approach. In: Schenkman JB, Greim H (eds) Cytochrome P450. Heidelberg: Springer, 415–429.

Doehmer J, Seidel A, Oesch F, Glatt HR (1990) Genetically engineered V79 Chinese hamster cells metabolically activate the cytostatic drugs cyclophosphamide and ifosfamide. Env Health Persp 88: 63–65.

Fuhr U, Doehmer J, Battula N, Wölfel C, Kudla C, Keita Y, Staib AH (1992) Biotransformation of caffeine and theophylline in mammalian cell lines genetically engineered for expression of single cytochrome P450 isoforms. Bioch Pharmacol 43: 225–235.

Gonzalez FJ (1992) Human cytochromes P450: problems and prospects. TiPS Reviews 13: 346–352.

Gonzalez FJ (1989) The molecular biology of cytochrome P450s. Pharmacological Reviews 40: 243–288.

Gonzalez FJ, Crespi CL, Gelboin HV (1991) cDNA-expressed human cytochromes P450s: a new age of molecular toxicology and human risk assessment. Mutat Res 247: 113–127.

Gonzalez FJ, Nebert DW (1990) Evolution of the P450 superfamiiy: animal-plant "warfare", molecular drive and human genetic differences in drug oxidation. Trends Genet 6: 182–187.

Guengerich FP (1993) Metabolic reactions: types of reactions of cytochrome P450 enzymes. In: Schenkman JB, Greim H (eds) Cytochrome P450 Heidelberg: Springer, 89–103.

Gustafsson JA, Ingelman-Sundberg M (1975) Regulation and substrate specificity of a steroid sulfate-specific hydroxylase system in female rat liver microsomes. J Biol Chem 250: 3451–3458.

Gustafsson JA, Mode A, Norstedt G, Skett P (1983) Sex-steroid induced changes in hepatic enzymes. Annu Rev Physiol 45: 51–60.

Haim N, Nemec J, Sinha BK (1989) *In vitro* metabolism of etoposide (VP-16-213) by liver microsomes and irreversible binding of reactive intermediates to microsomal proteins. Biochem Pharmacol 36: 527–536.

Ryan DE, Dixon R, Evans RH, Ramanathan L, Thomas PE, Wood AW, Levin W (1984) Rat hepatic cytochrome P450 isozyme specificity for the metabolism of the steroid sulfate, 5α-androstane-3α, 17β-diol-3, 17-disulfate. Arch Biochem Biophys 233: 636–642.

Kimura S, Umeno M, Skoda RD, Meyer UA, Gonzalez FJ (1989) The human debrisoquine 4-hydroxylase (Cytochrome P450 2D) locus: sequence and identification of polymorphic Cytochrome P450 2D6 gene, a related gene and a pseudogene. Am J Hum Genet 45: 889–905.

LeBlanc GA, Waxman DJ (1989) Interaction of anticancer drugs with hepatic mono-oxygenase enzymes. Drug Met Rev 20: 395–439.

Nelson DR, Kamataki T, Waxman DJ, Guengerich FP, Estabrook RW, Feyereisen R, Gonzalez FJ, Coon MJ, Gunsalus IC, Gotoh O, Okuda K, Nebert DW (1993) The P450 superfamily: update on new sequences, gene mapping, accession numbers, early trivial names of enzymes, and nomenclature. DNA and Cell Biol. 12: 1–51.

O'Brien PJ (1991) Molecular mechanisms of quinone cytotoxicity. Chem Biol Interactions 80: 1–41.

Potter DW, Levin W, Ryan DE, Thomas PE, Reed DJ (1984) Stereoselective mono-oxygenation of carcinostatic 1-(2-chloropethyl)-3-(cyclohexyl)-1-nitrosourea by purified cytochrome P450 isoenzymes. Biochem Pharmacol 33: 609–613.

Prough RA, Brown MI, Dannan GA, Guengerich FP (1984) Major isozymes of rat liver microsomal cytochrome P450 involved in the N-oxidation of N-isopropyl-a-(2-methylazo)—p—toluamide, the azo derivative of procarbazine. Cancer Res 44: 543–548.

Powis G (1989) Free radical formation by antitumor quinones. Free Rad Biol Med 6: 63–101.

Powis G (1987) Metabolism and reactions of quinoid anticancer agents. Pharmac Ther 35: 57.

Rogiers V, Vandenberghe Y, Callaerts A, Verleye G, Cornet M, Mertens K, Sonck W, Vercruysse A (1990) Phase I and phase II xenobiotic biotransformation in cultures and co-

cultures of adult rat hepatocytes. Biochem Pharmacol 40: 1701–1706.
Ryan DE, Levin W (1993) Age- and gender-related expression of rat liver cytochrome P450. In: Schenkman JB, Greim H (eds) Cytochrome P450. Heidelberg: Springer, 461–476.
Slater TF, Sawyer BC (1971) The stimulatory effects of carbon tetrachloride on peroxidative reactions in rat liver fraction *in vitro*. Biochem J 123: 815–821.
Waxman DJ, Dannan GA, Guengerich FP (1985) Regulation of rat hepatic cytochrome P450: age-dependent expression, hormonal imprinting, and xenobiotic inducibility of sex-specific isoenzymes. Biochemistry 24: 4409-4417.

Address for offprints: Dr J. Doehmer, Institut für Toxikologie und Umwelthygiene, Technische Universität München, Lazarettstrasse 62, 80636 München, Germany.

Cytotechnology **12**: 367–384, 1993.

Role of matrix metalloproteinases in invasion and metastasis: biology, diagnosis and inhibitors

Susan McDonnell and Barbara Fingleton
School of Biological Sciences, Dublin City University, Dublin, Ireland

Key words: invasion, metalloproteinases, metastasis, TIMPS

Abstract

The processes of tumour invasion and subsequent metastasis are the most lethal aspects of cancer. Whilst many factors are involved, the matrix metalloproteinases (MMPs) have been implicated as key-rate limiting enzymes in the invasive process. This family consisting of eight members of similar structure, can be roughly divided into three groups based on substrate specificity. All are secreted in a latent form and require proteolytic cleavage for activation. The expression of these enzymes is regulated at the transcriptional level by a variety of growth factors and oncogenes. They are also regulated at the protein level by a family of specific inhibitors called the tissue inhibitors of metalloproteinases (TIMPs). Studies in human tumour samples have shown a positive correlation between metalloproteinase expression and metastatic potential. The levels of metalloproteinase expression have been manipulated using molecular biology techniques in several cell lines and shown a similar correlation. These results suggest that an understanding of metalloproteinase expression and proteolytic activity may lead to the development of effective therapeutic agents with the potential to reduce the incidence of metastatic cancer.

Abbreviations: MMP — matrix metalloproteinase; TIMP — tissue inhibitor of metalloproteinase; rTIMP — recombinant tissue inhibitor of metalloproteinase; TPA — 12-O-tetradecanoyl phorbol-13-acetate; TRE — TPA responsive element; EGF — epidermal growth factor; TGF-β — transforming growth factor β; ELISA — enzyme-linked immunosorbent assay; bp — base pair; PCR — polymerase chain reaction; RT-PCR — reverse transcription polymerase chain reaction.

Introduction

Over the past decade the fields of cancer biology, molecular biology and genetics have provided proof that cancer is, in essence, a genetic disease. At the molecular level we can now say that cancer for the most part is caused by an accumulation of somatic mutations. According to Vogelstein and Kinzler (1993), three to six mutations appear to be required to form a cancer. This 'multi-hit' hypothesis, involving both activation of an oncogene and loss of tumour suppressor gene activity, has been carefully documented by Vogelstein's group in colon cancer (Fearon and Vogelstein, 1990; Vogelstein and Kinzler, 1993). The development of colon tumours appears to be initiated by mutation of the APC tumour suppressor gene product which results in the formation of benign adenomas. Mutation of the RAS gene often occurs in one of these benign tumour cells, resulting in clonal expansion. Further mutations in the DCC and p53 tumour suppressor genes appear to result in the progression from a benign to a malignant state in which the resultant tumour cells are able to invade the surrounding tissues and metastasize to other organs.

Although tremendous progress has been made in the treatment and diagnosis of many different types of cancers, a therapeutic regime which would result in most cancers being cured has not yet been developed. Whilst the phenomenon of multidrug resistance obviously impedes chemotherapeutic treatment, it is the phenomenon of tumour invasion and subsequent metastasis that is the most lethal aspect of cancer. Indeed, multidrug resistance is generally associated with cancers that have metastasized as these are not readily treatable by surgery or radiation therapy (Gottesman, 1993). In many cases a tumour has already metastasized by the time a patient presents with a malignant tumour. The presence of undetectable metastases compounds the clinical problem as it is difficult to treat what cannot be seen. Since metastatic lesions may progress differently from their primary tumours, the effective treatment of metastatic disease presents a formidable problem. If researchers can learn what gives the metastatic cells this ability, then it may be possible to disarm them and develop new cancer therapies. A major challenge to cancer scientists is the development of improved methods to predict the metastatic potential of a patient's individual tumour and prevent local invasion.

Invasion and metastasis

Recent studies of metastasis have revealed it as an arduous, multistep process in which only a small proportion of the tumour cells survive to start a new tumour. First, cells detach from the primary tumour and move into the blood vessels that nourish it. They are carried in the general circulation until they lodge in a capillary bed. Cells that survive this trip may then penetrate the blood vessel cell wall, invade the surrounding tissues and begin to proliferate and form a new tumour mass. Purely on the basis of circulatory anatomy one can predict where a large percentage of metastases will arise. The lungs are a common site of metastasis, because the heart pumps all the blood through their capillaries. Similarly, metastases from the colon often arise in the liver, as the liver receives direct drainage from the large intestine. Metastases also arise in other organs due to the special 'soil' for survival (hormones or growth factors) present there.

As can be seen from the above scenario, tumour cells must manifest invasive properties at several points: to enter and escape the circulation and to penetrate into normal tissues. Liotta (1986) has postulated a three-step theory of invasion which can be summarized as follows: (1) adhesion of tumour cells to the basement membrane, a process which is mediated by specific cell surface receptors which specifically bind to components of the matrix such as laminin and fibronectin; (2) secretion of destructive enzymes which can locally degrade the matrix; (3) migration of tumour cell into the region of the matrix modified by proteolysis.

In order for tumour cells to invade the surrounding tissue, they must traverse a series of tissue compartments separated by the extracellular matrix which consists of the basement membrane and the underlying interstitial stroma (Liotta *et al.*, 1983). The extracellular matrix is composed of a network of various fibrous proteins such as collagens, laminin, fibronectin and proteoglycans. For many years the extracellular matrix was thought to be an inert structure whose only function was to provide structural support to organisms. We now know that extracellular matrix degradation has a profound influence on many normal and pathological processes, including embryonic development, tumour invasion and arthritis.

There are four main classes of proteases involved in proteolytic degradation of the extracellular matrix (Shi *et al.*, 1993): (1) serine proteases, e.g. plasminogen activators; (2) cysteine proteases, e.g. cathepsin B and cathepsin L; (3) aspartyl proteases, e.g. cathepsin D and (4) metalloproteinases, e.g. collagenases and stromelysins. It is generally thought that these enzymes may degrade local tissues through which tumour cells invade, and that they may also be involved in penetration through the basement membrane during the extravasation of tumour cells from blood vessels (Sloane and Honn, 1984; Dano *et al.*, 1985). Although all of these proteinases interact in a complex manner that ultimately leads to the process of invasion it has been suggested that initial extracellular matrix degradation is due to the metalloproteinases (Liotta, 1991a).

The metalloproteinases are believed to be the normal, physiological mediators of matrix degradation (usually referred to as matrix metalloproteinases, MMPs). They are secreted proteins, placing them in the proper location for extracellular matrix degradation, and their enzymatic activities are most potent at pH values close to neutrality (Matrisian, 1992). There is quite a large amount of evidence that supports a critical role for metalloproteinases in invasion and metastasis. Rifkin and colleagues (Mignatti *et al.*, 1986) demonstrated that a proteinase cascade is required for the invasion of melanoma cells and that the metalloproteinases play a major role in this process. Additional evidence has come from studies with the inhibitors of MMPs, the tissue inhibitor of metalloproteinases (TIMPs). An inverse correlation between TIMP levels and the invasive potential of murine and human cells has been observed (Hicks *et al.*, 1984; Halaka *et al.*, 1983). Denhardt and colleagues (Khokha *et al.*, 1989) have used an antisense approach to demonstrate a role for MMPs in tumour invasion. In these experiments cells that were not normally invasive became invasive when they no longer produced TIMP after transfection with an antisense TIMP construct. In addition, Alvarez *et al.*, (1990) have demonstrated that repeated injections of recombinant human TIMP inhibits lung metastases following intravenous injection of Ha-*ras* transfected rat embryo cells. It is presumed that the action of the TIMPs in these studies is to inhibit matrix degradation by metalloproteinases. The aim of this review is to examine the expression of members of this family of proteinases in human tumour samples and evaluate critically their role in tumour invasion and metastasis.

Matrix metalloproteinases (MMP) family

Structure and function

The MMPs are a multigene family of matrix degrading zinc metalloproteinases that degrade at least one component of the extracellular matrix. Molecular cloning of the various family members has revealed considerable amino acid conservation and raises the possibility that the genes arose by duplication of a single primordial gene. So far in humans, eight members of the MMP family have been identified and characterized (Table 1). The MMPs have been known by various names through isolation by different groups, but for the purposes of this review the designation recommended by the International Union of Biochemistry and Molecular Biology will be used. Therefore the 72 kDa and 92 kDa type IV collagenases will be referred to as gelatinase A and gelatinase B respectively, similarly PUMP-1, (putative metalloproteinase) will be referred to as matrilysin.

Analysis of the amino acid sequences reveals that these proteins contain several distinct domains that are conserved among the various members (Fig. 1). The first of these is the leader sequence (~17 amino acids) which targets the molecule for secretion, is subsequently removed and is therefore not present in the latent enzyme (Wilhelm *et al.*, 1987). The second domain is the propeptide (~80 amino acids), which has been shown to be cleaved when MMPs are activated and which contains the highly conserved sequence PRCGV/NPD (Grant *et al.*, 1987; Stetler-Stevenson *et al.*, 1989a; Nagase *et al.*, 1990) present in all members of the MMP family (see Fig. 1). The available data suggest that this region is involved in maintaining the enzyme in a latent state since mutations in this region result in an enzyme that no longer requires proteolytic activation (Matrisian *et al.*, 1991). *In vitro*, this activation can be achieved by a variety of agents, including organomercurials, oxidants, sulfhydryl alkylating agents and, in some cases, proteolytic cleavage by trypsin or plasmin (Wilhelm *et al.*, 1987; Stetler-Stevenson *et al.*, 1989a; He *et al.*, 1989; Springman *et al.*, 1990). This spectrum of activators suggests that a conformational change is

Table 1. Properties of human matrix metalloproteinase

MMP No.[a]	Name[b]	Latent	M_r (kDa) Active	Degrades
1	Interstitial collagenase	55	45	Fibrillar collagens Gelatin Proteoglycan
8	Neutrophil collagenase	75	58	As interstitial collagenase
2	Gelatinase-A	72	66	Denatured collagens Collagen IV, V, VII, X Elastin
9	Gelatinase-B	92	86	As gelatinase-A
3	Stromelysin-1	57	45	Proteoglycan Collagen II, IV, XI Gelatins, laminin, Fibronectin
10	Stromelysin-2	57	44	As stromelysin-1
7	Matrilysin	28	19	As stromelysin-1 Elastin Entactin
11	Stromelysin-3	51	44	Unknown

At present the family consists of eight members, their original MMP classification (a) is given, followed by the name recommended by the International Union of Biochemistry and Molecular Biology (b). The extracellular matrix substrates listed are representative and are not necessarily comprehensive.

required for activation and that a thiol bond is involved. The current view of MMP activation is that the N-terminal part of the molecule is folded around in the latent enzyme so that the cysteine residue in the conserved PRCGV/NPD region complexes with a zinc molecule. Activation results when a conformation change dissociates the cysteine from the zinc atom and replaces it with water, this has been referred to by Van Wart and colleagues as the 'cysteine switch' mechanism (Springman *et al.*, 1990). The activated metalloproteinase is then capable of autoproteolysis. *In vivo*, the endogenous activator is thought to be plasmin which has been shown to activate MMPs in a coculture system of keratinocytes and fibroblasts (He *et al.*, 1989). *In vitro*, other enzymes such as cathepsin G and neutrophil elastase have been shown to activate stromelysin-1 and gelatinase A (Okada and Nakanishi, 1989). In stromelysin-3 there is an additional insert of 10 amino acids between the propeptide and the catalytic domain (Basset *et al.*, 1990). The third domain is the catalytic domain (~170 amino acids) which contains the conserved sequence HEF/L/IGH, postulated to be the zinc-binding domain. The fourth domain which is present in all members of the MMPs except matrilysin shares homology to hemopexin and vitronectin (Hunt *et al.*, 1987) and is linked to the catalytic domain by a Pro-rich sequence of 5—10 amino acids. It was originally postulated to be involved in substrate specificity but, as can be seen from Table 1, matrilysin substrate specificity is similar to stromelysin-1 and -2. This domain is required for the specific binding and cleavage of fibrillar collagen in human interstitial collagenase but not stromelysin-1 (Clark and

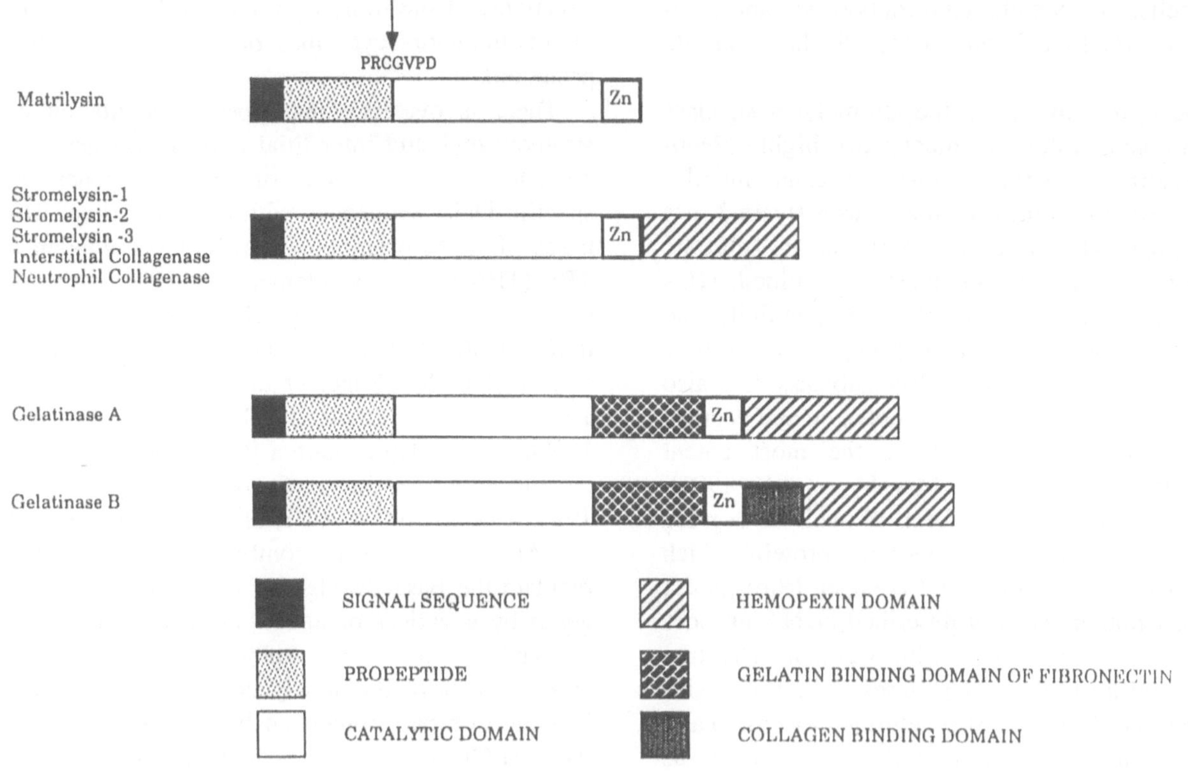

Fig 1 Domain structure of the MMP family members All members of the MMP family contain at least three protein domains: a pre domain encoding the leader sequence that targets the enzymes for secretion, a prodomain which is removed when the enzyme becomes activated, and the catalytic domain which contains the zinc binding region All members except matrilysin contain a domain that shares homology with hemopexin. Gelatinase A and B contain an additional fibronectin domain, and gelatinase B contains an additional collagen binding domain Additional information on their molecular weights and substrate specificities is given in Table 1 Each domain is designated by particular shading as indicated in the figure

Cawston 1989; Murphy *et al.*, 1992). The gelatinases A and B both contain an additional domain which contains three repeats of a sequence homologous to the gelatin binding domain of fibronectin (Collier *et al.*, 1988; Wilhelm *et al.*, 1989), this domain may mediate the sequestering of these enzymes within the extracellular matrix. Finally, at the C-terminal end of the gelatinase B catalytic domain there is an additional domain that has some homology to collagen (Wilhelm *et al.*, 1989).

Broadly speaking the MMPs may be divided into 3 subclasses with respect to their substrate specificity; the type I collagenases, the type IV collagenases, and the stromelysins (see Table 1 for additional details on molecular weights and substrates). The collagenase subclass has two members: interstitial collagenase and neutrophil collagenase. Both these enzymes cleave the alpha chains of types I, II and III collagen at a single site. Interstitial collagenase is produced by fibroblasts and macrophages in particular, while the expression of neutrophil collagenase is restricted to cells of the neutrophil lineage (Hasty *et al.*, 1990a). The collagenase type IV subclass contains two members: gelatinase A and gelatinase B. Both of these enzymes degrade denatured collagens (gelatins) and are specific for the degradation of type IV basement membrane collagen. Expression of gelatinase A is widespread and is frequently elevated in malignant tumours, gelatinase B was traditionally thought of as the macrophage gelatinase but its expression has been described in malignant cells, neutrophils, cytotrophoblasts and keratinocytes (Collier *et al.*, 1988; Wilhelm *et al.*, 1989).

The third subclass is the stromelysin subclass which contains three members: two highly homologous enzymes, stromelysin-1 and -2, and matrilysin. A possible fourth member, stromelysin-3, has been cloned (Bassat *et al.*, 1991) but its substrate specificity has not yet been determined. This subclass has the widest substrate specificity and degrades proteoglycans and glycoproteins such as fibronectin and laminin. This subclass has also been shown to cleave type IV collagen and degrade elastin with matrilysin being the most potent (Wilhelm *et al.*, 1987; Murphy *et al.*, 1991). Matrilysin has recently been shown to degrade entactin, a basement membrane protein which bridges laminin and type IV collagen (Sires *et al.*, 1993). Another recently described cDNA encodes a murine metalloelastase with a domain structure that is similar to the collagenases and stromelysins (Shapiro *et al.*, 1992). Stromelysin-1 was originally cloned using subtractive hybridization from a virus transformed rat cell line (Matrisian *et al.*, 1985), however, it is not widely expressed normally but can be induced readily by growth factors, tumour promoters and oncogenes in mesenchymal cells such as fibroblasts (Matrisian *et al.*, 1985; Kerr *et al.*, 1988, McDonnell *et al.*, 1990a). Stromelysin-2 and matrilysin were originally cloned from human tumour samples (Muller *et al.*, 1988), and have been found in a number of tumour types (Bassat *et al.*, 1990; Pajouh *et al.*, 1991; McDonnell *et al.*, 1991) and also in normal cells including macrophages (Busiek *et al.*, 1992) and the endometrium (Rodgers *et al.*, 1993).

Regulation of the MMPs

The expression of these enzymes is very highly regulated and occurs at many different levels. Regulation occurs at the transcriptional level by gene expression and at the protein level by proteolytic activation or by complexing to TIMPs (see section on inhibitors) which are also independently regulated. Studies over the past several years have shown that MMPs are regulated both positively and negatively at the transcriptional level by a variety of growth factors, cytokines, oncogenes and tumour promoters.

There is evidence that positive regulation of stromelysin-1 and interstitial collagenase genes is mediated by the action of protooncogenes on specific DNA sequences within the proximal 210 bases of these promoters. These genes contain a TRE (TPA responsive element) or AP-1 (activator protein-1) binding site which is recognized and transactivated by the product of the protooncogenes c-*fos* and c-*jun* (Angel *et al.*, 1987; Kerr *et al.*, 1988; Schonthal *et al.*, 1988; McDonnell *et al.*, 1990b). They also contain a PEA-3 element which recognizes the c-*ets* oncogene (Wasylyk *et al.*, 1990, 1991). Mutational analysis has shown that the AP-1 alone, or in combination with PEA-3 regulate the basal levels and inducibility of these genes by a variety of agents (Angel *et al.*, 1987; Wasylyk *et al.*, 1991; Auble and Brinckerhoff, 1991). We have previously demonstrated that the induction of rat stromelysin by epidermal growth factor (EGF) requires the induction of c-*fos* and c-*jun* protooncongene products and the activation of protein kinase C (McDonnell *et al.*, 1990b). Induction of interstitial collagenase by Ha-*ras*, v-*mos*, v-*src* and TPA also requires c-*fos* induction and the TRE (Schonthal *et al.*, 1988). Tumour necrosis factor-α (TNF-α) induction of interstitial collagenase also correlates with induction of c-*jun* and requires the TRE (Brenner *et al.*, 1989). Platelet derived growth factor (Kerr *et al.*, 1988), nerve growth factor (Machida *et al.*, 1989) and interleukin-1 (Frisch and Ruley, 1987; Sirum and Brinckerhoff, 1989) induction of stromelysin may also act through a similar pathway. The promoter region of the gelatinase A (Frisch *et al.*, 1990) is very different from those of stromelysin-1 and -2 and collagenase, most notably by the absence of an obvious TATA box and AP-1 site. The gelatinase B gene contains TATA and AP-1-like motifs but it is not clear that they are acting as true transcriptional elements (Huhtala *et al.*, 1990).

Recent studies (Kerr *et al.*, 1990) have shown that the transforming growth factor-β (TGF-β) inhibits the EGF induction of stromelysin gene expression through an upstream sequence referred to as the TGF-β inhibitory element (TIE). Interestingly, the Fos protein is present in the protein complex that recognizes the TIE element (Kerr *et al.*, 1990). Rahmsdorf and colleagues have also suggested that glucocorticoid inhibition of collagenase gene expression is mediated by the interaction of the glucocorticoid receptor and the Fos/Jun protein complexes (Jonat *et al.*, 1990). Thus it appears probable that protooncogene induction by growth factors can be involved in both positive and negative regulation of metalloproteinase gene expression.

Interestingly, TIMP-1 and metalloproteinases appear to be reciprocally regulated in some systems, TGF-β causes an increase in TIMP-1 mRNA in human fibroblasts (Overall *et al.*, 1989). Additionally the mouse TIMP promoter also contains an AP-1 site and can be transcriptionally activated by many of the same agents that activate collagenase or stromelysin (Campbell *et al.*, 1991). In conclusion we can say that the ultimate matrix-degrading activity of a metalloproteinase is therefore subject to regulation at multiple levels and must meet a number of requirements before proteolysis occurs: the presence of an inducer and absence of an inhibitor is required for transcription of the mRNA; enzyme must be activated and the balance between protease and inhibitor must favour proteolysis.

Tissue inhibitors of metalloproteinases (TIMPs)

The tissue inhibitors of metalloproteinases (TIMPs) are a multigene family (Stetler-Stevenson *et al.*, 1990) which at present consists of two members, TIMP-1 and TIMP-2. There is an overall 38% identity between the two human TIMPs at the amino acid sequence level, although the degree of similarity is higher at 68%. Both contain 12 cysteine residues at virtually identical positions clearly indicating that the two proteins are from the same family. These two proteins bind noncovalently to active metalloproteinases in a 1:1 molar ratio and specifically inhibit their enzymatic activity. TIMP-1 also forms a complex with the inactive proform of gelatinase B (Wilhelm *et al.*, 1989), while TIMP-2 forms a complex specifically with the proform of gelatinase A (Stetler-Stevenson *et al.*, 1989b).

TIMP-1 is a glycoprotein (M_r 30 Kda) and its cDNA and amino acid sequence has been reported (Docherty *et al.*, 1985; Carmichael *et al.*, 1986), although purification and characterization of the protein from rabbit bone tissue had earlier been achieved by Cawston (1981), and from various other sources also (Welgus and Stricklin, 1983; Murphy *et al.*, 1981). Early studies investigating the link between TIMP-1 expression and invasive ability in cell line models, showed that TIMP-1 levels were decreased 10–20 fold in highly invasive as compared to the level in normal or poorly invasive cells (Hicks *et al.*, 1984). The hypothesis that decreased TIMP levels resulted in increased invasive/metastatic properties, was further investigated using antisense RNA to induce reduction of TIMP levels in a non-invasive murine cell line. Using an amnion invasion assay, the cells which had reduced TIMP expression were shown to become invasive in comparison to control cells (Khokha *et al.*, 1989).

With the availability of recombinant TIMP (Carmichael *et al.*, 1986), numerous investigators have looked at the potential therapeutic use of TIMP-1 as an anti-invasive agent. Recombinant TIMP (rTIMP) has been shown to inhibit the invasive ability of tumour cells in both *in vivo* and *in vitro* assays. Addition of 5 μg/ml rTIMP resulted in complete inhibition of the invasive ability of a murine melanoma cell line B16-F10 in an amnion invasion assay. This assay measures the ability of cells to penetrate an amniotic membrane isolated from human placenta (Schultz *et al.*, 1988).

To investigate the *in vivo* effect of rTIMP, a highly metastatic rat embryo cell line (transfected with Ha-*ras*) known as HRAS1 was injected into the tail vein of nude mice. The ability of these cells to colonize the lungs was inhibited by up to 83% following repeated injections of rTIMP (Alvarez *et al.*, 1990). Systemic administration of rTIMP has also been used successfully in the treatment of collagen-induced arthritis in mice (Carmichael *et al.*, 1989).

The second member of the TIMP family, TIMP-2 is a nonglycosylated protein (M_r 23 kDa) which was isolated and cloned more recently and by several different groups (DeClerck *et al.*, 1989; Goldberg *et al.*, 1989; Stetler-Stevenson *et al.*, 1989b; Boone *et al.*, 1990). The activity of TIMP-2 has also been well characterized. It reacts stoichiometrically with active interstitial collagenase and also prevents the activation of this enzyme from its 52 kDa proform to 42 kDa active form (DeClerck *et al.*, 1991a). TIMP-2 has been purified from a complex it forms with the inactive form of gelatinase A (Ward *et al.*, 1991). The binding of TIMP-2 to this gelatinase does not prevent the gelatinase becoming activated thus suggesting that the TIMP-2 is bound to a site other than the active site. It is thought that the TIMP-2 prevents autocatalytic activation of the gelatinase (Howard *et al.*, 1991). Also the TIMP-2 while complexed to gelatinase A has been shown to retain inhibitory activity for other metalloproteinases (Kolkenbrock *et al.*, 1991). The inhibitory activity of TIMP-2 has been shown to be greater for the gelatinase A than that of TIMP-1 (Howard *et al.*, 1991) although both TIMP-1 and TIMP-2 can also inhibit all members of the other metalloproteinase subclasses (Matrisian, 1990).

The inhibitory activity of recombinant TIMP-2 (rTIMP-2) on the invasion of smooth muscle cell multilayers by human fibrosarcoma cells HT-1080 has been demonstrated by DeClerck and colleagues (DeClerck *et al.*, 1991b). Purified TIMP-2 has also been shown to inhibit the ability of HT-1080 cells to invade a matrigel membrane (Albini *et al.*, 1991). In this study anti-gelatinase A antibodies were used to show that it is specifically this proteinase that is responsible for the invasive properties of these cells.

While these studies using purified and recombinant TIMPs have been quite successful and are thought to be functioning through inhibition of metalloproteinase activity, they should be interpreted with some caution as TIMPs have been shown to have some growth-factor-like activity (Stetler-Stevenson *et al.*, 1992). Potentially, drugs mimicking the cysteine-containing peptide or propeptide region could be developed as therapeutic agents for the inhibition of invasion and metastasis. As yet none of the MMPs have been crystallized, once this has been achieved, it may be possible to design agents to bind to the active site and thereby inhibit their activity.

The existence of other inhibitors of metalloproteinases has also been demonstrated. α_2-macroglobulin, a nonspecific protease inhibitor has potent activity but due to its large size (750 kDa), lack of specificity and irreversible inhibitory mechanism, its activity is most likely confined to proteases in the bloodstream (Welgus and Stricklin, 1983). Other secreted inhibitors which are of smaller size than the TIMPs have been termed IMPs. They appear to share some sequence homology with TIMPs but are independently regulated (Banda *et al.*, 1992). Their significance *in vivo* has yet to be determined.

Expression and localization of MMPs in human tumours

The expression of various members of the metalloproteinase family by tumour cell lines has long been associated with metastatic potential. Some of the earliest experiments conducted by Liotta and his colleagues demonstrated a positive correlation between type IV collagenase activity and the invasive potential of variants of the murine melanoma cell line B16 (Liotta *et al.*, 1980). Recent experiments (Sreenath *et al.*, 1992) have shown a similar correlation in rat embryo cell lines transformed with ras. However, in these experiments correlation was seen with stromelysin-1 and -2, and not with gelatinase A or B. MMP expression is also associated with a range of normal events such as trophoblast invasion during pregnancy (Lala and Graham, 1990), endometrial changes during the reproductive cycle (Rodgers *et al.*, 1993), differentiation of mononuclear phagocytes (Welgus *et al.*, 1992) as well as other disease states, principally arthritis (Hasty *et al.*, 1990b).

Since the isolation and cloning of the human cDNAs for these proteinases (Muller *et al.*, 1988; Levy *et al.*, 1991), it has been possible for researchers to examine human tumour tissue for metalloproteinase expression. As can be seen from Table 2, an extensive number of studies have been carried out in the past few years examining a variety of tissues, including breast, colon, prostate, head and neck, pulmonary, and oesophageal. Table 2, summarizes most of these studies, and also shows the various techniques used.

The expression of metalloproteinases in human tumours has been be examined by a variety of techniques. Early studies used northern blotting (Muller *et al.*, 1991) which required a large amount of tissue and isolation of intact RNA. Recent studies have used *in situ* hybridization, a very sensitive and specific method which allows localization of the mRNA to specific cell types. Although, this technique can be technically difficult, the use of nonradioactive probes has made it possible to be more routinely used (Komminoth, 1992). However, it has been suggested by some investigators that for retrospective studies, protein-based techniques such as immunohistochemistry should be used as archival material is unlikely to have been collected in RNase free conditions. The mRNA of the metalloproteinase can also be quantitated using a modification of the PCR reaction, reverse transcription PCR (RT-PCR), this technique has been reported as being able to detect the mRNA from a single embryo (Rappolee *et al.*, 1988). One of the advantages of this method is that the RNA does not need to be fully intact and can be amplified directly from archival material (Shibata *et al.*, 1988).

Antibodies (both monoclonal and polyclonal) are now available to many of the metalloproteinases. Thus the technique of immunohistochemistry has made it possible to localize expression of the protein to specific cell types. A number of studies have used both immunohistochemistry to localize

Table 2. Expression of MMPs in human tumours

Location carcinoma	MMP	Method[a]	Reference
Breast	Stromelysin-3	NB, ISH	Basset *et al.*, 1990
	Gelatinase B	IHC	Monteagudo *et al.*, 1990
	Gelatinase B	ELISA	Zucker *et al.*, 1993
Colon	Matrilysin	NB, ISH, IHC	McDonnell *et al.*, 1991
	Gelatinase A	NB, IHC	Levy *et al.*, 1991
	Gelatinase A	IHC	Poulsom *et al.*, 1992
Gastric	Matrilysin	NB, ISH, IHC	McDonnell *et al.*, 1991
Head and Neck	Stromelysin-2	NB, ISH, IHC	Muller *et al.*, 1991[b], 1993
	Stromelysin-3	NB, ISH, IHC	Muller *et al.*, 1993
	Interstitial collagenase	NB, ISH, IHC	Muller *et al.*, 1991[b], 1993
	Interstitial collagenase	ISH	Gray *et al.*, 1992
Oesophageal	Gelatinase A	IHC	Shima *et al.*, 1992
	Stromelysin-1	IHC	Shima *et al.*, 1992
Prostate	Matrilysin	NB, ISH	Pajouh *et al.*, 1991
	Gelatinase A	NB	Pajouh *et al.*, 1991
Pulmonary	Stromelysin-3	NB, ISH	Urbanski *et al.*, 1992
	Gelatinase A and B	NB, ISH	Urbanski *et al.*, 1992
	Interstitial collagenase	NB, ISH	Urbanski *et al.*, 1992

This table summarizes some of the studies that have been undertaken in the past few years, examining the expression of MMPs in human tumour specimens. Various different techniques have been used to examine their expression. (a) Abbreviations used: NB — northern blotting, IHC — immunohistochemistry, and ISH — *in situ* hybridization. (b) The studies conducted in 1991 only used northern blotting for analysis.

the protein and *in situ* hybridization to localize the mRNA (McDonnell *et al.*, 1991; Muller *et al.*, 1993). Recently a number of ELISAs have been established which allow direct quantification of the levels of MMPs expressed. An ELISA for the detection of gelatinase B in plasma showed concentrations of gelatinase B from normal healthy subjects to be 9 ± 11 ng/μl and in patients with gastrointestinal cancers the plasma gelatinase B level was increased to 18 ± 23 ng/μl and to 21 ± 22 ng/μl in breast cancer patients (Zucker *et al.*, 1993). Another ELISA for the quantification of stromelysin has been developed and used for measuring the levels of stromelysin in arthritic patients (Obata *et al.*, 1992).

Unfortunately, it is beyond the scope of this review to examine in detail all of the studies listed in Table 2, instead we will focus on a few specific examples. The coexpression of several members of the MMP gene family appears to be a general characteristic of human carcinomas, and has been observed in breast (Basset *et al.*, 1990; Monteagudo *et al.*, 1990), colon (Levy *et al.*, 1991; McDonnell *et al.*, 1991) and prostate (Pajouh *et al.*, 1991). There are also a number of tissues where particular metalloproteinases are expressed in preference to others. Stromelysin-3 has been shown to be expressed in 100% (30 of 30) of all invasive breast cancers analysed so far (Basset *et al.*, 1990). The stromelysin-3 cDNA was cloned from a cDNA library constructed from a breast cancer surgical specimen rather than an established breast cancer cell line, in order to have both neoplastic and stromal cell components. Interestingly, in this study stromelysin-3 mRNA expression is restricted to the stromal cells immediately surrounding the islands of malignant epithelial cells of the invasive components of the tumours. It has been suggested that stromelysin-3 is one of the stroma-derived factors that have long been postulated to play an important role in progression of epithelial malignancies. In this study, the gelatinase A, stromelysin-1, stromelysin-2 and matrilysin were expressed in both malignant and benign tumours, while the gelatinase B and interstitial collagenase were detected in only the breast carcinomas but not to the same extent as stromelysin-3.

A more recent study (Muller *et al.*, 1993) has shown that stromelysin-3 is expressed in 95% (106 of 111) of head and neck squamous cell carcinomas examined. The stromelysin-3 mRNA and protein was specifically detected in stromal cells immediately surrounding invasive cancer cells (similar to breast carcinomas). They also observed that tumours with high levels of stromelysin-3 RNA were more likely to exhibit high local invasiveness, suggesting that stromelysin-3 may contribute to the progression of head and neck carcinomas. The levels of stromelysin-2 and interstitial collagenase were also elevated in head and neck squamous carcinomas (Muller *et al.*, 1991; Muller *et al.*, 1993). However, the stromelysin-2 mRNA was localized specifically to the neoplastic cells, and the interstitial collagenase mRNA was localized to both stromal and neoplastic cells (Muller *et al.*, 1993).

In contrast to the above studies, matrilysin expression has been found to be restricted to mainly neoplastic cells in colon, stomach (McDonnell *et al.*, 1991) and prostate (Pajouh *et al.*, 1991) carcinomas. We have recently extended our initial observations on matrilysin expression in colon carcinomas by investigating the stage of colorectal tumour progression associated with matrilysin gene expression. This was done by examining a series of colon lesions ranging from small adenomas to carcinomas. Preliminary data suggests that matrilysin expression arises in areas of benign polyps destined to become malignant carcinomas (Newell *et al*,. 1993). Similarly gelatinase A has been found by immunohistochemical methods to be expressed specifically in neoplastic cells in breast (Monteagudo *et al.*, 1990) and in colon (Levy *et al.*, 1991) carcinomas.

It appears from a number of studies that certain metalloproteinases may be expressed in a tissue specific manner, stromelysin-3 in stromal cells of breast carcinomas (Basset *et al.*, 1990) and matrilysin in epithelial cells of colon and prostate (Pajouh *et al.*, 1991; McDonnell *et al.*, 1991). A similar tissue specificity is also seen in cycling human endometrium, one of the few examples of normal tissue in which metalloproteinase expression has been observed (Rodgers *et al.*, 1993). Matrilysin is present in the epithelial cells of proliferative and menstrual endometrium, while stromelysin-1 is seen in the stromal component during the same cyclic interval. The possibility that tissue-specific elements may control the expression of metalloproteinase will be an interesting topic for future research.

As mentioned previously, the final proteolytic activity of metalloproteinase will be dependent not only on their expression but also on their activation which is largely controlled by their inhibitors, the TIMPs. A recent study (Urbanski *et al.*, 1992) has examined not only the expression of metalloproteinases in pulmonary carcinomas but also their inhibitors. Both TIMP genes were expressed in all normal and malignant tissue samples analysed. Two major classes of TIMP-2 transcripts of 3.5 kb and 1.0 kb were observed. In almost all cases the 3.5 kb species was the most abundant with the 1.0 kb form being barely visible for some tumour RNA. *In situ* hybridization showed TIMP-1 and TIMP-2 transcripts expressed predominately over tumour stromal cells. These findings support the idea that the TIMPs may play a role in human neoplasia due to their abilities to inhibit the active forms of metalloproteinases. It is important for future work to look not only at the expression of metalloproteinases but also their inhibitors.

In vitro models

Most of the data presented in the above section shows a correlation between MMP expression and tumour formation, some studies even show correlation with extent of metastasis (Muller *et al.*, 1993). The levels of MMPs have been manipulated by molecular biology techniques in a number of cell culture cell lines. These studies suggest that the expression of these proteins can contribute to the invasive and metastatic ability of malignant cells. In the following section we will look at 3 studies which have attempted to look directly at the effect of stromelysin and matrilysin on the invasive behaviour of tumour cell lines.

Studies using the classical two-stage model of chemical carcinogenesis in mouse skin demonstrated a positive correlation between stromelysin-1 mRNA expression and tumour progression (Matrisian *et al.*, 1986). This protocol consistently produces benign papillomas after treatment with a single dose of dimethylbenz(a)anthracene (DMBA), followed by repeated application of tumour promoter 12-O-tetradecanoyl phorbol-13-acetate (TPA). At 25—30 weeks following initial treatment, 5—7% of these animals develop squamous cell carcinomas. Analysis of the mRNA extracted from these tumours revealed that stromelysin-1 transcripts were detected in 73% of the carcinomas and in only 6% papillomas (Matrisian *et al.*, 1986). The murine papilloma cell line SP-1 was transfected with a vector expressing stromelysin cDNA (Matrisian *et al.*, 1991) under control of a strong viral promoter. For these studies, a cDNA containing a single base-pair mutation resulting in a valine to glycine change at position 92 was used. This mutation caused autoactivation of the stromelysin-1 protein, and resulted in over 90% of the protein being secreted in the active form (Park *et al.*, 1991). The transfected SP-1 cells were tested in an *in vitro* invasion assay and showed up to a 2.2-fold increase in their invasive potential. In this *in vitro* invasion assay cells are seeded onto matrigel-coated filters containing 10 μm pores in a Membrane Invasion Culture System (MICS) originally developed by Hendrix and colleagues (Hendrix *et al.*, 1987). Following 72-h incubation, cells that have completely invaded the matrigel-coated filters are collected by treatment with trypsin-EDTA, stained with hematoxylin and counted. These results show that expression of stromelysin-1 contributes to the invasive ability of these cells. Expression of stromelysin-1 is a late event in tumour progression which correlates with the conversion of a tumour from the benign to the malignant state.

Bowden and colleagues have observed that 14/18 prostate adenocarcinomas express matrilysin mRNA (Pajouh *et al.*, 1991). Using *in situ* hybridization this mRNA was localized to epithelial cells and not to stromal or inflammatory cells. This data suggests that matrilysin may be involved in the invasion and metastasis of prostate adenocarcinomas. To test this hypothesis, a human prostate cancer cell line DU-145 which does not express matrilysin and is nonmetastatic was transfected with a plasmid containing the full length matrilysin cDNA (Powell *et al.*, 1993). Using a severe combined immunodeficient (SCID) mouse model of tumour cell invasion the matrilysin transfected cell lines were observed to invade through the diaphragm more effectively than control transfected cell lines. These studies suggest a functional role for matrilysin in the initial invasion of prostate cancer.

In a previous study we have shown that matrilysin is expressed in 8/10 gastric carcinomas and 6/8 colon carcinomas examined (McDonnell *et al.*, 1991). Using *in situ* hybridization and an affinity purified antibody, the expression of matrilysin mRNA and protein was localized to tumour cells and was not detected in stromal or lymphocytic cells. We have conducted a series of experiments to determine whether matrilysin expression is causally related to colon tumour cell invasion and metastasis by genetically altering the expression of matrilysin in two colon derived cell lines. The SW480 and SW620 cell lines provide an ideal model system to test this hypothesis. The SW480 cells are derived from a primary carcinoma, do not express matrilysin RNA or protein, and are relatively noninvasive in an *in vitro* invasion assay. The SW620 cells are derived from a lymph node metastasis from the same patient as the SW480 cell line, express matrilysin RNA and protein, and are significantly more invasive in an *in vitro* invasion assay compared to the SW480 cells (Newell *et al.*, 1993).

The SW480 cells have been transfected with a plasmid containing the full length matrilysin cDNA and the neomycin resistance gene. This cDNA has been mutated in the highly conserved PRCG VPD region so that matrilysin protein is secreted in an active form. The transfected cells displayed various degrees of invasiveness, ranging from a 1.5 fold to 4 fold increase in invasive ability when compared to control cells transfected with neomycin only (Newell *et al.*, 1993). We now hope to transfect the SW620 cells with a plasmid containing the matrilysin cDNA in the antisense orientation, thus eliminating expression of matrilysin in this cell line. These experiments will determine whether matrilysin expression plays a functional role in the invasive and metastatic ability of colon carcinoma cell lines. This evidence, in combination with the studies of Vogelstein and coworkers who have characterized many of the genetic alterations which occur in colon carcinogenesis, may result in a more thorough understanding of the genetic events which lead to matrilysin expression and colon tumour progression.

Studies in which the levels of metalloproteinases have been manipulated by molecular biology techniques suggest that the expression of these proteins can contribute to the invasive and metastatic ability of malignant cells. However, evidence obtained from examining the expression in human tumor samples indicates that levels of more than one member of this family should be manipulated at any one time. For example, with current techniques it should be possible to transfect matrilysin-expressing cells with other metalloproteinases and then look to see if the invasive ability increases synergistically.

Anti-metastatic gene nm23

No review of invasion and metastasis would be complete without some discussion on metastasis suppressor genes. As mentioned briefly in the introduction, the malignant phenotype is the culmination of a series of genetic changes that involves both activation of oncogenes and loss of tumour suppressor genes. Initial evidence for the existence of genes with metastasis suppressing activities came from cell fusion experiments (Sidebottom and Clark, 1983). The loss of the metastatic phenotype from hybrids resulting from the fusion of a normal cell and a metastatic tumour cell indicated the existence of gene(s) in the normal cell that could reverse/overcome the metastatic properties of the tumour cells. The hybrids were tumourigenic indicating that the gene(s) linked to metastasis was independent of those involved in cell growth control.

A possible candidate for a metastasis suppressor gene was first isolated using differential hybridization with mRNAs extracted from high and low metastatic variants of a murine melanoma cell line K-1375 (Steeg *et al.*, 1988). The isolated cDNA was then used in Northern blots to show that its expression inversely correlates with metastatic potential. In further studies total RNA was isolated from infiltrating ductal carcinomas and fibroadenomas (benign tissue) of breast and probed for nm23 expression (Bevilacqua *et al.*, 1989). The result correlated with level of lymph node involvement, a commonly used marker in breast cancer. Significantly all patients with lymph node involvement showed lower levels of nm23 than the fibroadenoma patients. In samples from patients without lymph node involvement, a range of nm23 expression levels were seen and it was suggested that some of the higher levels may be due to contaminating normal tissue. Studies such as these and a similar one correlating low nm23 expression with reduced disease free interval and overall survival in breast cancer patients (Hennessy *et al.*, 1991), suggest the value of nm23 as a prognostic indicator. Before its usefulness can be truly ascertained, large scale prospective and retrospective studies are necessary. As noted earlier techniques such as RT-PCR, *in situ* hybridization and immunohistochemistry may be suitable for retrospective studies.

An important question concerning reduced nm23 levels is whether this causes metastasis or is associated with it. To answer this question, transfections of highly metastatic K-1375 murine melanoma cells with nm23 were performed (Leone *et al.*, 1991). Stable high-expressing clones were isolated and examined for tumourigenicity and metastatic potential by injection into mice. The percentage of mice that developed primary tumours after injection with nm23 clones was less when compared with mice injected with control cells. Also, tumour size was, on average 2.7 times smaller in the nm23 group than the control group. This study provided clear evidence that nm23 itself can suppress certain tumourigenic and metastatic processes in the systems examined.

A further study investigated nm23 allelic deletions in colorectal carcinoma patients (Cohn *et al.*, 1991). The DNA of 21 patients free of metastasis was analyzed and found to be heterozygous using Southern analysis. Of these, 11 were found to have an nm23 allelic deletion in tumour DNA. After a follow-up time of 25 months, the occurrence of metastases development was significantly higher in the group with nm23 deletion as compared to the nondeleted group. A more recent study has used RT-PCR to examine more closely the link between nm23 gene deletion and metastasis in colorectal cancer (Wang *et al.*, 1993). Using primers flanking the entire nm23 coding region, 533 base pair (bp) sequences were amplified. Of the 13 paired samples (normal and tumour), 11 gave the expected size fragment, one sample had a 533 bp band in both tumour and normal samples and a second smaller 470 bp band was also seen in the tumour sample. The final sample showed 533 bp band in the normal and 350 bp band in the tumour. Sequence analysis showed the 533 fragments in all cases to be identical to the 533 bp published data sequence. The 470 bp was equivalent to the nm23 sequence with a 64 bp deletion whereas the 350 bp fragment did not match any known sequence.

As yet no definitive proof of the normal biochemical function of nm23 is available. Postulations as to its possible function are based on its 77% homology to the Drosophila developmental gene *awd* and its >50% homology to NDP kinases (Wallet *et al.*, 1990). Mutations in the *awd* gene in Drosophila leads to a number of developmental alterations in tissue morphology and differentiation (Dearolf *et al.*, 1988). The similarity of nm23 to *awd* would suggest that it is involved in normal tissue development and that alterations in its expression would result in aberrant development perhaps contributing in the progression towards malignancy. The NDP kinases are a group of enzymes whose function is to catalyze the transfer of a phosphate group from nucleoside triphosphates (NTPs) to nucleoside diphosphates (NDPs). Two areas in which they are particularly important and where abnormal behaviour may lead to malignancy are in microtubule formation and G-protein function (Liotta *et al.*, 1991).

Microtubules are important amongst other things in relation to motility and are necessary for correct mitotic spindle formation — alterations in their assembly/disassembly due to incorrect NDP kinase activity could therefore influence the motile ability of a cell and/or mitosis.

G-proteins are active in signal transduction

relaying information from first messengers such as hormones or growth factors through effectors to second messengers within the cell. Their correct function is dependent on their ability to replace bound GDP with GTP and then to hydrolyze the GTP to GDP in a cyclic manner that is equivalent to a switching on/off mechanism (Linder and Gilman, 1992). NDP kinases (and possibly nm23) apparently have a role in the switching on event and alterations in NDP kinase activity could result in a particular G-protein being permanently switched on or off. This is important when it is realized that G proteins regulate either by stimulation or inhibition a wide range of cell functions. It is also possible that nm23 has other biochemical functions which have not yet been elucidated.

Conclusion

Evidence presented in this review strongly suggests that the expression of members of the metalloproteinase family of proteolytic enzymes is associated with the progression of a tumour to a malignant, invasive and metastatic state. The induction of the metalloproteinase genes may be as a result of the activation of oncogenes or the loss of tumour suppressor genes in the tumour cells. The specific metalloproteinase that is induced in a tumour cell may be dictated by the tissue in which the tumour arises. These results suggest that an understanding of metalloproteinase expression and proteolytic activity may lead to the development of effective therapeutic agents with the potential to reduce the incidence of metastatic cancer.

Acknowledgements

The authors would like to express their thanks to the Health Research Board and the Irish Science and Technology Agency (EOLAS) for their continued support and funding. Many thanks also to Geraldine Grant and Keara Hall who took time off busy schedules to read this manuscript and make helpful suggestions.

References

Angel P, Imagawa M, Chiu R, Stein B, Imbra RJ, Rahmsdorf HJ, Jonat C, Herrlich P & Karin M (1987) Phorbol ester-inducible genes contain a common cis element recognized by a TPA-modulated transacting factor. Cell 49: 729–739.

Albini A, Melchiori A, Santi L, Liotta LA, Brown PD and Stetler-Stevenson WG (1991) Tumor cell invasion inhibited by TIMP-2. J Natl Cancer Inst 83: 775–779.

Alvarez OA, Carmichael DF and De Clerck YA (1990) Inhibition of collagenolytic activity and metastasis of tumor cells by a recombinant human tissue inhibitor of metalloproteinases. J Natl Cancer Inst 82: 589–595.

Auble DT and Brinckerhoff CE (1991) The AP-1 sequence is necessary but not sufficient for phorbol induction of collagenase in fibroblasts. Biochemistry 30: 4629–4635.

Banda MJ, Howard EW, Herron GS and Apodeca G (1992) Secreted inhibitor of metalloproteinases (IMPs) that are distinct from TIMP. Matrix (Germany) Suppl 1: 294–298.

Basset P, Bellocq JP, Wolf C, Stoll I, Hutin P, Limacher JM, Podhajcer OL, Chenard MP, Rio MC and Chambon P (1990) A novel metalloproteinase gene specifically expressed in stromal cells of breast carcinomas. Nature (Lond) 348: 699–704.

Bevilacqua G, Sobel ME, Liotta LA and Steeg PS (1989) Association of low nm23 RNA levels in human primary infiltrating ductal breast carcinoma with lymph node involvement and other histopathological indicators of high metastatic potential. Cancer Res 49: 5185–5190.

Boone TC, Johnson MJ, DeClerck YA and Langley KE (1990) cDNA cloning and expression of a metalloproteinase inhibitor related to tissue inhibitor of metalloproteinases. Proc Natl Acad Sci (USA) 87: 2800–2804.

Brenner DA, O'Hara M, Chojkier M and Karin M (1989) Prolonged activation of c-*jun* and collagenase genes by tumour necrosis factor-α. Nature 337: 661–663.

Busiek DF, Ross FP, McDonnell S, Murphy G, Matrisian LM and Welgus HG (1992) The matrix metalloproteinase matrilysin (PUMP) is expressed in developing human mononuclear phagocytes. J Biol Chem 267: 9087–9092.

Campbell CE, Flenniken AM, Skup D and Williams BRG (1991) Identification of a serum- and phorbol ester-responsive element in the murine tissue inhibitor of metalloproteinase gene. J Biol Chem 266: 7199–7206.

Carmichael DF, Sommer A, Thompson RC, Anderson DC, Smith CG, Welgus HG and Stricklin GP (1986) Primary sequence and cDNA cloning of human fibroblast collagenase inhibitor. Proc Natl Acad Sci USA 83: 2407–2411.

Carmichael DF, Stricklin GP and Stuart JM (1989) Systemic administration of TIMP in the treatment of collagen-induced arthritis in mice. Agents Actions 3: 378–379.

Cawston TE, Galloway WA, Mercer E, Murphy G and Reynolds JJ (1981) Purification of rabbit bone inhibitor of collagenase. Biochem J 195: 159–165.

Clark IM and Cawston TE (1989) Fragments of human fibroblast collagenase. Purification and characterization. Biochem J 263: 201–206.

Cohn KH, Wang F, De Soto-LaPaix F, Solomon WB, Patterson LG, Arnold MR, Weimar J, Feldman JG, Levy AT, Leone A and Steeg PS (1991) Association of nm23 Hi allelic deletion with distant metastasis in colorectal carcinoma. Lancet 338: 722–724.

Collier IE, Wilhelm SM, Eisen AZ, Marmer BL, Grant GA, Seltzer JL, Kronherger A, He C, Bauer EA and Goldberg GI (1988) H-*ras* oncongene-transformed human bronchial epithelial cells (TBE-1) secrete a single metalloproteinase capable of degrading basement membrane collagen. J Biol Chem 263: 6579–6587.

Dano K, Andreasen PA, Hansen-Grondahl J, Kristensen P, Nielsen LS and Skriver L (1985) Plasminogen activators, tissue degradation and cancer. Adv Cancer Res 44: 139–266.

DeClerck YA, Yean TD, Ratzkin BJ, Lu HS and Langley KE (1989) Purification and characterization of two related but distinct metalloproteinase inhibitors secreted by bovine aortic endothetial cells. J Biol Chem 264: 17445–17453.

DeClerck YA, Yean T, Lu HS, Ting J and Langley KE (1991a) Inhibition of autoproteolytic activation of interstitial procollagenase by recombinant metalloproteinase inhibitor MI/TIMP-2. J Biol Chem 266: 3893–3899.

DeClerck YA, Yean TD, Chan D, Shimada H and Langley KE (1991b) Inhibition of tumor invasion of smooth muscle cell layers by recombinant human metalloproteinases inhibitor. Cancer Res 51: 2151–2157.

Docherty AJP, Lyons A, Smith BJ, Wright EM, Stephens PE, Harris TJR, Murphy G and Reynolds JJ (1985) Sequence of human tissue inhibitor of metalloproteinases and its identity to erythroid-potentiating activity. Nature 318: 66–69.

Fearon EJ and Vogelstein B (1990) A genetic model of colorectal tumorigenesis. Cell 61: 759–767.

Frisch SM and Ruley HE (1987) Transcription from the stromelysin promoter is induced by interleukin-1 and repressed by dexamethasone. J Biol Chem 262: 16300–16304.

Frisch SM, Reich R, Collier IE, Genrich LT, Martin G and Goldberg GI (1990) Adenovirus E1A represses protease gene expression and inhibits metastasis of human tumor cells. Oncogene 5: 75–83.

Goldberg GI, Marmer BL, Gregory GA, Eisen AZ, Wilhelm S and He C (1989) Human 72-kilodalton type IV collagenase forms a complex with a tissue inhibitor of metalloproteases designated TIMP-2. Proc Natl Acad Sci USA 86: 8207–8211.

Gottesman MM (1993) How cancer cells evade chemotherapy: Sixteenth Richard and Linda Rosenthal foundation award lecture. Cancer Res 53: 747–754.

Grant GA, Eisen AZ, Marmer BL, Roswit WT and Goldberg GI (1987) The activation of human skin fibroblast procollagenase. Sequence identification of the major conversion products. J Biol Chem 262: 5886–5889.

Gray ST, Wilkins RJ and Yun K (1992) Interstitial collagenase gene expression in oral squamous cell carcinoma. Am J Pathol 141: 301–306.

Halaka AN, Bunning RAD, Bird CC, Gibson M and Reynolds JJ (1983) Production of collagenase and inhibitor (TIMP) by intracranial tumours and dura *in vitro*. J Neurosurg 59: 461–466.

Hasty KA, Pourmotabbed TF, Goldberg GI, Thompson JP, Spinella DG, Stevens RM and Mainardi CL (1990a) Human neutrophil collagenase. A distinct gene product with homology to other matrix metalloproteinases. J Biol Chem 265: 11421–11424.

Hasty KA, Reife RA, Kang AH and Stuart JM (1990b) The role of stromelysin in the cartilage destruction that accompanies inflammatory arthritis. Arthritis Rheum 33: 388–397.

He C, Wilhelm SM, Pentland AP, Marmer BL, Grant GA, Eisen AZ and Goldberg GI (1989) Tissue cooperation in a proteolytic cascade activating human interstitial collagenase. Proc Natl Acad Sci USA 86: 2632–2636.

Hendrix MJC, Seftor EA, Seftor REB and Fidler I (1987) A simple quantitative assay for studying the invasive potential of high and low metastatic variants. Cancer Lett 38: 137–147.

Hennessy C, Henry JA, May FEB, Wesley BR, Angus B and Lennard TWJ (1991) Expression of the anti-metastatic gene nm23 in human breast cancer: association with good prognosis. J Natl Cancer Inst 83: 281–285.

Hicks NJ, Ward RV and Reynolds JJ (1984) A fibrosarcoma model derived from mouse embryo cells: growth properties and secretion of collagenase and metalloproteinase inhibitor (TIMP) by tumor cell lines. Int J Cancer 33: 835–844.

Howard EW, Bullen EC and Banda MJ (1991) Regulation of the autoactivation of human 72-kDa progelatinase by tissue inhibitor of metalloproteinases-2. J Biol Chem 266: 13064–13069.

Huhtala P, Chow LT and Tryggvason K (1990) Structure of the human type IV collagenase gene. J Biol Chem 265: 11077–11082.

Hunt LT, Barker WC and Chen HR (1987) A domain structure common to hemopexin, vitronectin, interstitial collagenase and a collagenase homolog. Protein Seq Data Anal 1: 21–26.

Jonat C, Rahmsdorf HJ, Park K-K, Cato ACB, Gebel S, Ponta H and Herrlich P (1990) Anti-tumor promotion and anti-inflammation: Down modulation of AP-1 (Fos/Jun) activity by glucocorticoid hormone. Cell 62: 1189–1204.

Kerr LD, Holt JT and Matrisian LM (1988) Growth factors regulate transin gene expression by c-*fos* dependent and c-*fos* independent pathways. Science 242: 1424–1427.

Kerr LD, Miller DB and Matrisian LM (1990) TGF-β1 inhibition of transin/stromelysin gene expression is mediated through a c-*fos* binding sequence. Cell 61: 267–278.

Khokha R, Waterhouse P, Yagel S, Lala PK, Overall CM, Norton G and Denhardt DT (1989) Antisense RNA-induced reduction in murine TIMP levels confers oncogenicity on Swiss 3T3 cells. Science 244: 947–950.

Kolkenbrock H, Orgel D, Hecker-Kia A, Noack W and Ulbrich N (1991) Complex between a tissue inhibitor of metalloproteinases (TIMP-2) and 72 kDa progelatinase is a metalloproteinase inhibitor. Eur J Biochem 198: 775–781.

Komminoth P (1992) Digoxigenin as an alternative probe labeling for *in situ* hybridization. Diagnostic Mol Pathol 1: 142–150.

Lala PK and Graham CH (1990) Mechanisms of trophoblast

invasiveness and their control: the role of proteases and proteases inhibitors. Cancer Met Rev 9: 369–379.

Leone A, Flaton U, Richter-King C, Sandeen MA, Margulies IMK, Liotta LA and Steeg PS (1991) Reduced tumor incidence, metastatic potential and cytokine responsiveness of nm23-transfected melanoma cells. Cell 65: 25–35.

Levy AT, Cioce V, Sobel ME, Garbisa S, Grigioni WF, Liotta LA and Stetler-Stevenson WG (1991) Increased expression of the Mr 72,000 type IV collagenase in human colonic adenocarcinoma. Cancer Res 51: 439–444.

Linder ME and Gilman AG (1992) G proteins. Sci Amer 267: 36–43.

Liotta LA, Tryggvason K, Garbisa S, Hart I, Foltz CM and Shafie S (1980) Metastatic potential correlates with enzymatic degradation of basement collagen. Nature 284: 67–68.

Liotta LA, Rao CN and Barsky SH (1983) Tumour invasion and the extracellular matrix. Lab Invest 49: 636–649.

Liotta LA (1986) Tumor invasion and metastases — role of the extracellular matrix. Cancer Res 46: 1–7.

Liotta LA, Steeg PS and Stetler-Stevenson WG (1991a) Cancer metastasis and angiogenesis: an imbalance of positive and negative regulation. Cell 64: 327–336.

Liotta LA, Stetler-Stevenson WG and Steeg PS (1991b) Metastasis suppressor genes. Import Adv Oncol 85–100.

Machida CM, Rodland KD, Matrisian LM, Magun BE and Ciment G (1989) NGF induction of the gene encoding the protease transin accompanies neuronal differentiation in PC 12 cells. Neuron 2: 1587–1596.

Matrisian LM, Glaichenhaus N, Gesnel MC and Breathnach R (1985) Epidermal growth factor and oncogenes induce transcription of the same cellular mRNA in rat fibroblasts. EMBO J 4: 1435–1440.

Matrisian LM, Bowden GT, Krieg P, Furstenberger G, Briand JP, Leroy P and Breathnach R (1986) The mRNA coding for the secreted protease transin is expressed more abundantly in malignant than in benign tumors. Proc Natl Acad Sci USA 83: 9413–9417.

Matrisian LM (1990) Metalloproteinases and their inhibitors in matrix remodeling. Trends Genet 6: 121–125.

Matrisian LM, McDonnell S, Miller DB, Navre M, Seftor EA and Hendrix MJC (1991) The role of the matrix metalloproteinase stromelysin in the progression of squamous cell careinomas. Am J Med Sci 302: 157–162.

Matrisian LM (1992) The matrix-degrading metalloproteinases. Bioessays 14: 455–463.

McDonnell S and Matrisian LM (1990a) Stromelysin in tumour progresion and metastasis. Cancer Metast Rev 9: 305–319.

McDonnell SE, Kerr LD and Matrisian LM (1990b) Epidermal growth factor stimulation of stromelysin mRNA in rat fibroblasts requires induction of proto-oncogenes c-*fos* and c-*jun* and activation of protein kinase C. Mol Cell Biol 10: 4284–4293.

McDonnell S, Navre M, Coffey RJ and Matrisian LM (1991) Expression and localization of the matrix metalloproteinase Pump-1 (MMP-7) in human gastric and colon carcinomas. Mol Carcinogenesis 4: 527–533.

Mignatti P, Robbins E and Rifkin DB (1986) Tumor invasion through the human amniotic membrane : requirement for a proteolytic cascade. Cell 47: 487–498.

Monteagudo C, Merino MJ, San-Juan J, Liotta LA and Stetler-Stevenson WG (1990) Immunohistochemical distribution of type IV collagenase in normal, benign and malignant breast tissue. Am J Pathol 136: 585–592.

Muller D, Quantin B, Gesnel MC, Millon-Collard R and Breathnach R (1988) The collagenase gene family in humans consists of at least four members. Biochem J 253: 187–192.

Muller D, Breathnach R, Engelmann A, Millon R, Broner G, Flesch H, Dumont P, Eber M and Abecassis J (1991) Expression of collagenase-related metalloproteinase genes in human lung or head and neck tumours. Int J Cancer 48: 550–556.

Murphy G, Cawston TE and Reynolds JJ (1981) An inhibitor of collagenase from human amniotic fluid. Biochem J 195: 167–170.

Murphy G, Cockett MI, Ward RV and Docherty AJP (1991) Matrix metalloproteinase degradation of elastin, type IV collagen and proteoglycan. A quantitative comparison of the activities of 95 kDa and 72 kDa gelatinases, stromelysin-1 and -2 and punctuated metalloproteinase (pump). Biochem J 277: 277–279.

Murphy G, Allan JA, Willenbrock F, Cockett ML, O'Connell JP and Docherty AJP (1992) The role of the C-terminal domain in collagenase and stromelysin specificity. J Biol Chem 267: 9612–9618.

Nagase H, Enghild JJ, Suzuki K and Salvesen G (1990) Stepwise activation mechanisms of the precursor of matrix metalloproteinase (stromelysin) by proteinases and (4-aminophenyl) mercuric acetate. Biochemistry 29: 5783–5789.

Newell K, McDonnell S, Witty JP, Gaire M, Rodgers WH and Matrisian LM (1993) The stromelysin subclass of metalloproteinases in tumor progression and metastasis. In: Barrett AJ and Bond J (eds) Proceedings of the 9th International ICOP Conference, Proteolysis and Protein Turnover.

Obata K, Iwata K, Okada Y, Kohrin Y, Ohuchi E, Yoshida S, Shinmei M and Hayakawa T (1992) A one step sandwich enzyme immunoassay for human matrix metalloproteinase 3 (stromelysin- 1) using monoclonal antibodies. Clin Chim Acta 211: 59–72.

Okada Y and Nakanishi I (1989) Activation of matrix metalloproteinase 3 (stromelysin) and matrix metalloproteinase 2 (gelatinase) by human neutrophil elastase and cathepsin G. FEBS Lett 249: 353–356.

Overall CM, Wrans JL and Sodek J (1989) Independent regulation of collagenase 72 kDa progelatinase and metalloendoproteinase inhibitor expression in human fibroblasts by transforming growth factor β. J Biol Chem 264: 1860–1869.

Pajouh MS, Nagle RB, Breathnach R, Finch JS, Brawer MK and Bowden GT (1991) Expression of metalloproteinase genes in human prostate cancer. J Cancer Res Clin Oncol 117: 144–150.

Park AJ, Matrisian LM, Kells AF, Pearson R, Yuan Z and Navre M (1991) Mutational analysis of the transin (rat stromelysin) auto-inhibitor region demonstrates a role for residues surrounding the "cysteine switch". J Biol Chem 266: 1584–1590.

Poulsom R, Pignatelli M, Stetler-Stevenson WG, Liotta LA, Wright PA, Jeffrey RE, Longcroft JA, Rogers L and Stamp GW (1992) Stromal expression of 72 kDa type IV collagenase (MMP-2) and TIMP-2 mRNAs in colorectal neoplasia. Am J Pathol 141: 389–396.

Powell WC, Knox JD, Navre M, Grogan TM, Kittelson J, Nagle RB and Bowden GT (1993) Expression of the metalloproteinase matrilysin in DU-145 cells increases their invasive potential in severe combined immunodeficient mice. Cancer Res 53: 417–422.

Rappolee DA, Wang A, Mark D and Werb Z (1989) Developmental expression of PDGF, TGF-α and TFG-β genes in preimplantation mouse embryos. Science 241: 1823–1825.

Rodgers WH, Osteen KG, Matrisian LM, Navre M, Giudice LC and Gorstein F (1993) Expression and localization of matrilysin, a matrix metalloproteinase in human endometrium during the reproductive cycle. Am J Obstet Gynecol 168: 253–260.

Schonthal A, Herrlich P, Rahmsdorf HJ and Ponta H (1988) Requirement for fos gene expression in the transcriptional activation of collagenase by other oncogenes and phorbol esters. Cell 35: 325–334.

Schultz RM, Silberman S, Persky B, Bajkowski WS and Carmichael DF (1988) Inhibition by human recombinant tissue inhibitor of metalloproteinases of human amnion invasion and lung colonization by murine B16-F10 melanoma cells. Cancer Res 48: 5539–5545.

Shapiro SD, Griffin GL, Gilbert DJ, Jenkins NA, Copeland NG, Senior RM and Ley TJ (1992) Molecular cloning, chromosomal location, and bacterial expression of a novel murine macrophage metalloelastase. J Biol Chem 267: 4664–4671.

Shi YE, Torri J, Yieh L, Wellstein A, Lippman M and Dickson RB (1993) Identification and characterization of a novel matrix-degrading protease from hormone-dependent human breast cancer cells. Cancer Res 53: 1409–1415.

Shibata DK, Arnhem M and Martin WJ (1988) Detection of human papillomavirus in paraffin embedded tissue using the polymerase chain reaction. J Exp Med 167: 225–230.

Shima I, Sasaguri Y, Kusukawa J, Yamana H, Fujita H, Kakegawa T and Morimatsu M (1992) Production of matrix metalloproteinase-2 and metalloproteinase-3 related to malignant behaviour of oesophageal carcinoma. A clinopathologic study. Cancer 70: 2747–2753.

Sidebottom E and Clark SR (1983) Cell fusion segregates progressive growth from metastasis. Br J Cancer 47: 399–405.

Sires UI, Griffin GL, Broekelmann TJ, Mecham RP, Murphy G, Chung AE, Welgus HG and Senior RM (1993) Degradation of entactin by matrix metalloproteinases, susceptibility to matrilysin and identification of cleavage sites. J Biol Chem 268: 2069–2074.

Sirum KL and Brinckerhoff CE (1989) Cloning of the genes for human stromelysin and stromelysin-2: differential expression in rheumatoid synovial fibroblasts. Biochemistry 28: 8691–8698.

Sloane BF and Honn KV (1984) Cysteine proteinases and metastasis. Cancer Metast Rev 3: 249–263.

Springman EB, Angleton EL, Birkedal-Hansen H and Van Wart H (1990) Multiple modes of activation of latent human fibroblast collagenase: evidence for the role of a Cys 73 active-site zinc complex in latency and a 'cysteine switch' mechanism for activation. Proc Natl Acad Sci USA 87: 364–368.

Sreenath T, Matrisian LM, Stetler-Stevenson WG, Gattoni-Celli S and Pozzatti RO (1992) Expression of matrix metalloproteinase genes in transformed rat cell lines of high and low metastatic potential. Cancer Res 52: 4942–4947.

Steeg PS, Bevilacqua G, Kopper L, Thorgeirsson UP, Talmadge JE, Liotta LA and Sobel MF (1988) Evidence for a novel gene associated with low tumor metastatic potential. J Natl Cancer Inst 80: 200–204.

Stetler-Stevenson WG, Krutzsch HC, Wacher MP, Margulies IMK and Liotta LA (1989a) The activation of type IV collagenase proenzyme: sequence identification of the major conversion product following organomercurial activation. J Biol Chem 264: 1353–1356.

Stetler-Stevenson WG, Krutzsch HC and Liotta LA (1989b) Tissue inhibitor of metalloproteinase (TIMP-2): a new member of the metalloproteinase inhibitor family. J Biol Chem 264: 17374–17378.

Stetler-Stevenson WG, Brown PD, Ohisto M, Levy AT and Liotta LA (1990) Tissue inhibitor of metalloproteinases-2 (TIMP-2) mRNA expresssion in tumor cell lines and human tumor tissues. J Biol Chem 265: 13933–13938.

Stetler-Stevenson WG, Bersch N and Golde DW (1992) Tissue inhibitor of metalloproteinase-2 (TIMP-2) has erythroid-potentiating activity. FEBS Lett 296: 231–234.

Urbanski SJ, Edwards DR, Maitland A, Leco KJ, Watson A and Kossakowska AE (1992) Expression of metalloproteinases and their inhibitors in primary pulmonary carcinomas. Br J Cancer 66: 1188–1194.

Vogelstein B and Kinzler KW (1993) The multistep nature of cancer. Trends Genet 9: 138–141.

Wallet V, Mutzel R, Troll H, Barzu O, Wurster B, Vernon M and Lacombe MA (1990) Dictyostelium nucleoside diphosphate kinase highly homologous in nm23 and *awd* proteins involved in mammalian tumour metastasis and Drosophila development. J Natl Cancer Inst 18: 1199–1202.

Wang F, Patel U, Ghosh L, Chen H and Banergee S (1993) Mutation in nm23 gene is associated with metastasis in colorectal cancer. Cancer Res 53: 717–720.

Wasylyk B, Wasylyk C, Flores P, Begue A, Leprince and Stehelin D (1990) The c-*ets* proto-oncogenes encode transcription factors that co-operate with c-*fos* and c-*jun* for transcriptional activation. Nature 346: 191–193.

Wasylyk C, Gutman A, Nicholson R and Wasylyk B (1991) The c-*ets* oncoprotein activates the stromelysin promoter through the same elements as several nonnuclear oncoproteins. EMBO J 10: 1127–1134.

Welgus HG and Stricklin GP (1983) Human skin fibroblast inhibitor. J Biol Chem 258: 12259– 12264.

Welgus HG, Senior PM, Parks WC, Kahn AJ, Ley TJ, Shapiro

SD and Campbell EJ (1992) Neutral proteinase expression by human mononuclear phagocytes: a prominent role of cellular differentiation. Matrix (Germany) Suppl 1: 363–367.

Wilhelm SM, Collier IE, Kronberger A, Eisen AZ, Marmer BL, Grant GA, Bauer EA and Goldberg GI (1987) Human skin fibroblast stromelysin: structure, glycosylation, substrate specificity, and differential expression in normal and tumorigenic cells. Proc Natl Acad Sci USA 84: 6725–6729.

Wilhelm SM, Collier IE, Marmer BL, Eisen AZ, Grant GA and Goldberg GI (1989) SV40-transformed human lung fibroblasts secrete a 94-kDa type IV collagenase which is identical to that secreted by normal human macrophages. J Biol Chem 264: 17213–17221.

Zucker S, Lysik RM, Zarrabi M and Moll T (1993) Mr 92,000 type IV collagenase is increased in plasma of patients with colon cancer and breast cancer: Cancer Res 53: 140–146.

Address for correspondence: Susan McDonnell, School of Biological Sciences, Dublin City University, Dublin 9, Ireland.

Cytotechnology **12**: 385–392, 1993.

Clinical significance of cellular resistance in tumours to cytotoxic chemotherapy and radiotherapy

Maeve Pomeroy[1] and Michael Moriarty[2]
[1]*Blackrock Clinic, Rock Road, Blackrock, Co. Dublin, Ireland;* [2]*Department of Radiotherapy and Clinical Oncology, Saint Luke's Hospital, Rathgar, Dublin 6, Ireland*

Key words: antineoplastic agents, drug resistance, neoplasm, radiation tolerance, radiotherapy

Abstract

Cellular resistance is a significant component of tumour treatment failure. More detailed understanding of resistance mechanisms has enabled us to plan circumvention strategies, though these are not yet in routine clinical use. Such resistance is, however, only one of several factors which render cure of advanced malignant disease difficult. It is important for researchers in this field to see not only therapeutic opportunities but also limitations of these approaches. It is hoped that increased cooperation between clinicians and scientists in the field of cellular resistance will yield further improvement in tumour response rates and cure.

Introduction

Currently available cancer treatments achieve a high cure rate for the two thirds of patients who present with localised tumour. While intensive laboratory and clinical research has resulted in improved treatment response rates for many tumours, only a small proportion of patients presenting with distant metastases are cured. Patients may fail to respond to therapy e.g., nonsmall cell lung carcinoma or may only show partial response e.g., breast carcinoma. Even among those patients who achieve a clinical, or even pathologically confirmed, complete remission e.g., ovarian carcinoma,there may be a high relapse rate. For the majority of locally advanced or metastatic solid tumours and for several hematological malignancies, we have not yet succeeded in converting treatment response to cure. It can be appreciated that novel approaches are needed to address these clinical situations. The causes are frequently multifactorial and include:

1. Failure to implement current optimal surgery, chemotherapy, radiotherapy or combined modality therapy;
2. Where optimal treatment is unknown, treating patients outside of large, well designed studies;
3. Host factors, such as multiple pathology or psychosocial factors, which necessitate modification of optimal therapy;
4. Problems of normal tissue tolerance and treatment dose distribution;
5. Kinetic factors such as the presence of tumour cells in relatively insensitive phases of the cell cycle e.g., G_0.
6. Cellular heterogeneity within tumours (Allalunis-Turner *et al.*, 1993)
7. Tumour cellular resistance, either primary or acquired secondary to exposure to chemotherapy and/or radiotherapy. Examples include the

primary chemotherapy resistance of pancreatic carcinoma and renal adenocarcinoma or the acquired resistance of the initially sensitive small cell lung carcinoma, breast carcinoma, ovarian carcinoma or myeloma.

Mechanisms of cellular resistance to treatment

The past decade has shown rapid progress in elucidating important mechanisms of cytotoxic drug resistance at the cellular level which include:

1. Decreased cytotoxic drug accumulation in tumour cells due to P-glycoprotein mediated multidrug resistance and similar mechanisms (Dano, 1973; Coley HM *et al.*, 1991);
2. Alterations of detoxification mechanisms such as glutathione-S-transferase in platinum resistance (Twentyman *et al.*, 1992);
3. Changes in DNA repair enzymes, topoisomerase I and II and polymerases in cisplatinum and doxorubicin resistance (Beck and Danks 1991);
4. Genomic alterations e.g., c-erb B2 over expression in breast carcinoma associated with resistance to several cytotoxic agents, excluding doxorubicin; (Paik, 1992).
5. Regrowth potential i.e., tumour cell regrowth between courses of chemotherapy, e.g., non-Hodgkins lymphoma or acute leukaemia (Larsson and Nygren, 1993).

Response to radiation treatment is known to depend on the well described four R's of radiotherapy, *viz.*, repair, redistribution, reoxygenation and repopulation but it is now increasingly apparent that individual cell radiosensitivity, the fifth R, is probably the most important factor in determining responsiveness/resistance(Steel and Peacock, 1989). It is appreciated that there is a high correlation between primary chemotherapy and radiotherapy resistance (Panis, 1990) though acquired resistance to radiotherapy, either by the use of previous radiotherapy, or by the use of chemotherapy does not appear to be an important event. Fractionated or single X-ray dose, however, has been shown to induce P-glycoprotein mediated drug resistance *in vitro* (McClean S *et al.*, 1993).

Interventional strategies to overcome multidrug resistance

These understandings have led to the discovery or design of resistance circumvention strategies:

A. Chemosensitisers such as calcium channel blockers e.g., verapamil and isomers, cyclosporins, including non-immunosuppressant analogues (Twentyman, 1992) or monoclonal antibody (Mechetner and Roninson 1992) or toxins targeted against P-glycoprotein or antisense cDNA for protein kinase C (Ahmad and Glazer, 1993) have shown promise. Such strategies have been extensively tested *in vitro* and in laboratory animals, but clinical use is as yet investigational, and several studies are in progress.

B. Cytotoxic drug analogues have been developed and tested which not only have a better toxicity profile but are theoretically less susceptible to the known mechanisms of cellular drug resistance. These include analogues of doxorubicin (Coley *et al.*, 1992) and cisplatinum (Twentyman *et al.*, 1992) and such strategies have considerable potential for clinical modification of resistance and are being actively researched. However, data derived from *in vitro* studies of specific tumour cell lines may not be replicated with other tumour cell lines from the same or other species(Goddard *et al.*, 1991).

C. DNA repair enzyme function may have importance in the development of tumour cell resistance. Techniques such as the administration of 5-fluorouracil 24 h before platinum can theoretically increase cell kill due to decreased cross linkage removal. It should be remembered, however, that DNA repair may be important for normal tissue tolerance of chemotherapy (Reed, 1991).

D. Gene therapy has been used *in vivo* to transfer drug sensitivity to tumour cells e.g., to the cells of a glioblastoma, using the herpes simplex virus enzyme thymidine kinase and a retroviral vector. Ganciclovir is an antiviral agent which was rendered cytotoxic by this means (Oldfield *et al.*, 1993). This innovative strategy shows exciting potential though targeting needs further refinement.

E. Modulation of chemotherapeutic effect by the addition of a drug which is not itself cytotoxic, e.g., the administration of folinic acid to potentiate 5-fluorouracil, is a technique which has shown improved response rates in the treatment of metastatic colorectal carcinoma. Folinic acid increases the binding of thymidylate synthetase to 5-fluoro-2′-deoxyuridine monophosphate (FdUMP) and thus increases 5-FU toxicity (Bleyer, 1989).

F. The addition of radiation to chemotherapy in resistant tumours has yielded disappointing results, at least, in part, due to the high correlation between radiation and chemoresistance. Various novel approaches, such as continuous infusional chemotherapy (Hannsen, 1989) or low dose daily chemotherapy e.g., oral vepesid (Clark, 1992) may show benefit. The use of radiation particles such as neutrons with the capacity for high linear energy transfer may overcome resistance to some tumours (e.g., chordoma, melanoma) relatively resistant to conventional radiotherapy (Schmitt and Wambersie, 1990).

Discussion

The hypothesis that the understanding of mechanisms of tumour cellular resistance can be translated into improved treatment results in the clinic remains unproven. Certainly, it has not been feasible to rapidly implement routine treatment sensitivity testing, as in microbiology, for a variety of reasons, including the considerable complexity of the topic.

Interpretation of *in vitro* studies may be difficult, as multiple mechanisms of drug/radiotherapy resistance exist and may even coexist in the same cell line (Morgan *et al.*, 1990). For even a single cytotoxic agent such as doxorubicin, there may be several mechanisms of resistance, only some of which may be reversible by currently identified circumvention agents (Redmond *et al.*, 1991). Data for cross-resistance may vary widely, depending on the tumour cell lines selected for evaluation (Goddard, 1991). Simulation of clinically achievable drug dose, radiation total dose and fractionation regimen with repair, redistribution, reoxygenation, and repopulation, are difficulties which may further confound results.

Similar problems exist with regard to the use of animal tumours and xenografts as well as interspecies differences in life cycle and metabolism. Such experiments do, however, allow *in vivo* modelling of the magnitude of expression and clinical significance of *in vitro* resistance mechanisms.

Moreover, in the clinical context, drug resistance will usually have arisen to combination chemotherapy and may be expected to be the result of several mechanisms of drug resistance. The patterns of resistance may also be expected to vary widely in different individuals, even with similar histopathology, stage at diagnosis, tumour burden and previous treatment.

Essential, for the regular clinical utilisation of cytotoxic drug sensitivity testing, is the availability of reliable, reproducible, inexpensive and preferably simple testing methods which rapidly yield results and can be performed in nonspecialist pathology laboratories. Methods of potential utility include:

MTT Assay;
DiSC Assay;
immunohistochemical stains;
flow cytometry;
genetic probes;
PCR techniques.

Some of these techniques have the advantage of being applicable to both fresh tumour specimens and also paraffin-wax embedded specimens. Imaging techniques *in vivo*, using ^{11}C colchicine and PET Scanning (Mehta *et al.*, 1992) could be adapted to detect MDR in a similar way to the *in vivo* dosimetry of π-meson radiotherapy (Shirato *et al.*, 1989).

Clinical perspective

Tumour resistance at a cellular level must be viewed in the context of several different clinical problems.

1. If optimal known chemotherapy agent selection is not delivered at the appropriate interval, dose, schedule and sequence, treatment results will be inferior. Continuing medical education and improved dissemination of information by technological means, international meetings and interdisciplinary co-operation seek to address this problem.
2. Ideally, where best treatment has not been identified, patients are most appropriately treated in the context of large clinical trials, a notion which is widely acknowledged but less frequently implemented, with less than 5% of all patients entered into clinical trials. Such approaches have been constructive in developing treatments for childhood leukaemia (Chessels *et al.*, 1992) and adult testicular germ cell tumours (Bajorin *et al.*, 1993).
3. Hopefully, a more highly organised approach to the evaluation of new drug treatments will allow identification of effective treatments, possibly unaffected by acquired treatment resistance mechanisms. New NCl, European and Japanese drug screening programs using *in vitro* techniques, with centralised cell data banks, offer prospects for improvements (Hellman and Carter, 1987).
4. A major continuing difficulty is a lack of specificity of chemotherapy and radiotherapy for tumour cells, leading to problems of normal tissue tolerance, frequently of bone marrow, requiring dose limitation. Circumvention techniques in this context include use of Colony Stimulating Factors (CSFs), Peripheral Stem cell Rescue (Kessinger and Armitage, 1993) after stimulation by chemotherapy and CSFs, and Bone Marrow Transplantation, autologous or allogeneic (Bensinger, 1992). Neutropenic sepsis is treated by multiple, systemic, highly potent antibiotics, to treat the most prevalent and most serious infections, including Pseudomonas septicaemia. Recent advances in antiemetic control, in particular the 5-HT3 receptor antagonists such as granisetron and ondansetron, have prevented nausea and vomiting being a significant problem for most patients. Analogues of cisplatinum, such as carboplatinum, have reduced problems of nephrotoxicity and peripheral neuropathy, as have newer anthracyclines and anthracenediones reduced cardiotoxicity encountered with doxorubicin.
5. High dose chemotherapy and/or high dose total body irradiation have been rendered feasible by the above methods of toxicity reduction, in particular those techniques which circumvent bone marrow toxicity.

   For chemotherapeutic agents which are phase specific or cell cycle stage specific and can only cause cell kill at very limited phases of the cell cycle e.g., antimetabolites such as 6-mercaptopurine and hydroxyurea, dose response curves show a plateau and the potential for benefit from dose escalation is limited. Duration of exposure, however, may yield improved response but may also increase toxicity.

   Cell-cycle stage nonspecific cytotoxic agents, such as alkylating agents and cisplatinum, act at several phases in the cell cycle and show a log-linear increase in cell kill with increased dose. In this group of drugs, there are clear theoretical advantages in dose escalation with rescue. Traditionally, this approach has been used after remission induction by more conventional regimens. Recent research has however employed the innovation of bringing high dose chemotherapy "Upfront" (Crown *et al.*, 1993), an approach which would be supported by the Goldie-Coldman Hypothesis of Drug Resistance (Goldie and Coldman, 1979). More clinical studies will be required before we can evaluate whether this approach can realise its undoubted potential benefit.
6. Improved supportive care has also facilitated the combined use of radiotherapy and chemotherapy and such combinations are being further evaluated from the point of view of exploiting mechanisms. It is suggested that the two modalities might lead to therapeutic gain in four main ways:

A. Spatial Cooperation, where chemotherapy and radiotherapy are used to treat neoplastic disease at different anatomical sites, e.g., chemotherapy as treatment of small cell lung carcinoma with cranial irradiation to treat putative microscopic involvement in the central nervous system.

B. Independent Cell Kill, where the modalities do not directly interact to achieve cell kill. Nonetheless, it is very important to select chemotherapeutic agents that do not exacerbate radiation effects to critical normal tissues within the field of radiation treatment, e.g., considerable care needs to be taken in the combined treatment of Hodgkins Disease with bulky mediastinal nodal enlargement where mediastinal irradiation may follow combined chemotherapy including bleomycin. A recognised toxicity of bleomycin is pulmonary fibrosis and here toxicityn may be additive.

C. Protection of Normal Tissues. This mechanism involves cytotoxic agents to protect against the effects of irradiation or indeed a second cytotoxic drug, e.g., stem cell mobilisation by cyclophosphamide in association with Colony Stimulating Factors, allowing dose intensification of another agent such as carboplatinum. Timing is critical in this interaction and its potential may be limited.

D. Enhancement of Tumour Response. Here the objective is to achieve a superadditive (i.e., synergistic) effect on tumour cell kill without incurring increased toxicity to normal tissues.

Overall, it is clear that therapeutic benefit can be obtained by simple addition of radiotherapy and chemotherapy but the benefits are probably proportional to the effectiveness of the individual modalities and detectable benefit of combined treatment is unlikely in already chemoresistant situations. It is possible that acquired chemoresistance can be prevented by the earlier use of combined modality therapy but this remains to be established though recent publications have suggested that, in certain circumstances, the reverse may be true (McClean *et al.*, 1993). Also problems of long-term toxicity, such as the increased induction of second neoplasms, need to be considered.

7. Better understanding of complex pharmacokinetics, e.g., 5-fluorouracil, presently being investigated *in vivo* by PET Scan, schedule dependency, e.g., low dose oral vepesid therapy vs. high dose intravenous therapy, chemotherapy combinations and evaluation of theoretical concepts such as alternating, noncross-resistant chemotherapy regimens or sequential cross-resistant regimens such as dacarbazine and fotemustine for malignant melanoma, will hopefully, in the future, clarify many areas of uncertainty.

8. Dose distribution is, of necessity, a well studied problem for the radiation oncologist and is also a continuing problem for the drug therapy of tumours. Inadequate blood supply to areas of tumour and failure of drug delivery to pharmacologically privileged or "Sanctuary" sites such as brain in leukaemia and lymphoma or testis in leukaemia continue to cause relapse at these sites. The use of combined modality therapy, routes of administration such as intrathecal delivery of drugs such as methotrexate and cytosine arabinoside, use of lipophilic drugs such as nitrosureas, platinum and epipodophyllotoxins and disruption of the bloodbrain barrier by methods to increase vascular permeability, have yielded limited succes.

9. Areas of tumour hypoxia and necrosis have traditionally been regarded as obstacles to tumour cure by radiotherapy, e.g., in glioblastoma and selection of high linear energy transfer radiation such as neutrons have been evaluated. Such treatment has sometimes been associated with not only increased tumour destruction but increased toxicity to normal tissues.

More recently, the presence of these hypoxic cells has been seen as a therapeutic target for bioreductive agents (Adams, 1992) or prodrugs which only become activated in the presence of hypoxia and cause cell death by ischaemic necrosis.

Tumour vasculature is another focus of

research and the use of techniques such as ARCON using accelerated radiotherapy, carbon dioxide and nicotinamide have shown benefit (Rojas, 1992) and further studies are in progress.

10. Better communication between health professionals and patients, increased access to information about diagnosis and treatment options, the availability of counselling services, support groups and selective use of psychological and psychiatric interventions may be expected to help patient compliance with challenging treatment regimens.

Many of the above problems are highly pertinent to treatment outcome for the individual cancer patient, as also is the issue of primary existence, or development on treatment, of cellular resistance to chemotherapy or radiotherapy. Recent reports suggest that lonidamine (Kiura, 1992) and histidinol (Warrington, 1992) may, under certain circumstances, be more cytotoxic to drug resistant than to parent cells. Combinations which include multidrug chemotherapy, biological therapy and tamoxifen for metastatic melanoma (Richards, 1992) have also improved response rates or for the same condition chemotherapy may be followed by "Mopping up" by liposome activated macrophage-therapy which is at present thought to be unassociated with the development of cellular drug resistance (Fidler, 1992).

Summary

Much has been achieved in recent years toward the better understanding of mechanisms of cellular drug and radiation resistance and design of circumvention strategies. Much work remains to be done, in multidisciplinary cooperation between scientists and clinicians, to allow us to utilise this information to increase cancer cure rates for our patients. It is hopefully possible to make real clinical progress in overcoming acquired resistance by understanding mechanisms and intervening. This should have benefit in the near future for improving treatment results in small cell lung, breast and ovarian carcinomas. Overcoming primary resistance is probably a longer term goal and may call for different strategies outlined in other chapters.

References

Adams GE (1992) Failla Memorial Lecture Redox, radiation and reductive bioactivation. Radiat Res (US) 132 (2): 129–139.

Ahmad S, Glazer RI (1993) Expression of the antisense cDNA for protein kinase Cα attenuates resistance of doxorubicin resistant MCF-7 breast carcinoma cells. Mol Pharmacol (US) 43 (6): 858–862.

Allalunis-Turner MJ, Barron GM, Day RS 3d, Dobler KD, Mirzayans R (1993) Isolation of two cell lines from a human malignant glioma specimen differing in sensitivity to radiation and chemotherapeutic drugs. Radiat Res (US), 134 (3): 349–354.

Bajorin DF, Sarosdy MF, Pfister DG, Mazumdar M, Motzer RJ, Scher HI, Geller, Fair WR, Herr Sogani P *et al.*, (1993) Randomized trial of etoposide and cisplatinum versus etoposide and carboplatinum in patients with good risk germ cell tumours, a multi-institutional study. J Oncol US 11 (4): 598–606.

Becks WT, Danks MK (1992) Mechanisms of resistance to drugs that inhibit DNA topoisomerases. Semin Cancer Biol (US) 2 (4): 235–244.

Bensinger WI Supportive care in marrow transplantation (1992) Curr Opin Oncol (US) 4 (4): 614–623.

Bleyer WA (1989) New vistas for Ieucovorin in cancer therapy. Cancer (US) 63: 995–1007.

Chessels JM, Bailey, CC, Richards S (1992) MRC UKALL X The U.K. protocol for childhood ALL 1985–90 The Medical Research Council Working Party on Childhood Leukaemia Leukaemia (England) 1992; 6 (Suppl 2): 157–161.

Clark PI (1992) Clinical pharmacology and schedule dependecy of the podophyllotoxin derivatives. Semin Oncol (US) 19 (2 suppl 6): 20–27.

Coley HM, Twentyman PR, Workman P (1992) Further examination of 9-alkyl and sugar modified anthraclines in the circumvention of multidrug resistance. Anticancer Drug Des (England), 7 (6): 471–481. (US) 41.

Coley HM, Workman P and Twentyman PR (1993) Retention of activity by selected anthracyclines in a multidrug resistant human large cell lung carcinoma line without P-glycoprotein hyperexpression. Br J Cancer 63: 351–357.

Crown J, Kritz A, Vahdat L, Reich L, Moore M, Hamilton N, Schneider J, Harrison M, Gilenski T, Hudis L *et al.* (1993) Rapid administration of multiple cycles high-dose myelosuppressive chemotherapy in patients with metastatic breast cancer. J Clin Oncol (US) 11 (6): 1144–1149.

Dano K (1973) Active outward transport of duanomycin in resistant Ehrlich Ascites tumour cells. Biochim Biophys Acta 324: 466–483.

Goddard PM, Valenti MR, Harrap KR (1991) The role of murine tumour models and their acquired platinum-resistant counterparts in the evaluation novel platinum antitumour agents: a cautionary note. Ann Oncol (Netherlands) 2 (8): 535–540.

Goldie JH and Coldman AJ (1979) A mathematical model for relating the drug sensitivity of tumours to the spontaneous mutation rate. Ca Tr Rep 63: 1927–1979.

Hansen RM, Quebbeman E, Anderson T (1989) 5-Fluorouracil by protracted venous infusion. A review of current progress Oncology (Switzerland) 46 (4): 245–250.

Hellman K and Carter SK (1987) Fundamentals of Cancer Chemotherapy (McGraw Hill).

Kessinger A and Armitage D (1993) The use of peripheral stem cell support of Hodgkins Disease chemotherapy. Import Adv Oncol (US) 167–175.

Kiura K, Ohnoshi T, Ueoka H, Takigawa N, Tabata M, Segawa Y, Shibayama T, Kimura I (1992) An adriamycin-resistant subline is more sensitive than the parent human small cell lung cancer cell line to lonidamine. Anticancer Drug Des (England) 7 (6): 463–470.

Larsson R and Nygren P (1993) Laboratory prediction of clinical chemotherapeutic drug resistance: a working model exemplified by acute leukaemia. Eur J Cancer 29A: 1208–1212.

McClean S, Hosking LK and Hill BT (1993) Expression of P-glycoprotein-mediated drug resistance in CHO cells surviving a single X-ray dose of 30 Gy. Int J Radiat Biol (England) 63 (6): 765–773.

Mc Clean S, Whelan RD, Hosking LK, Hodges GM, Thompson FH, Meyers MB, Schuurhuis GJ and Hill BT (1993) Characterisation of the P-glycoprotein over-expressing drug resistance phenotype exhibited by Chinese hamster ovary cells following their *in vitro* exposure to fractionated X-irradiation. Biochim Biophys Acta (Netherlands) 1177 (2): 117–126.

Mechetner EB, Roninson IB, (1992) Efficient inhibition of P-glycoprote-mediated multidrug resistance with a monoclonal antibody. Proc Natl Acad Sci USA 89 (13): 5824–5828.

Mehta BM, Rosa E, Fissekis JD, Bading JR, Biedler JL, Larson SM (1992) *In vivo* identification of tumour multidrug resistance with ^{3}H colchicine. J Nucl Med (US) 33 (7): 1373–1377.

Oldfield EH, Ram Z, Culver KW, Blaese RM, De Vroom HL, Anderson WF (1993) Gene therapy for the treatment of brain tumours using intra-tumoral transduction with the thymidine kinase gene and intravenous ganciclovir. Human Gene Ther (US) 4 (1): 36–39.

Paik S (1992) Clinical significance of erb-B-2 (HER-2/neu) protein. Cancer Invest (US) 10 (6): 575–579.

Panis X, Coninx P, Nguyen TD, Legros M (1990) Relation between responses to induction chemotherapy and subsequent radiotherapy in advanced or multicentric squamous cell carcinomas of the head and neck. Int J Radiat Oncol Biol Phys (US) 18 (6): 1315–1318.

Redmond SM, Joncourt F, Buser K, Ziemiecki A, Fey M, Margison G, Cerny T (1991) Assessment of P-glycoprotein, glutathione-based detoxifying enzymes and 06-alkylguanine-DNA alkyltransferase as potential indicators of constitutive drug resistance in human colorectal tumours. Cancer Res (U.S.) 51(8) 2092-7.

Reed E (1991) Alkylating agents and platinum: is clinical resistance simply a tumour cell phenomenon? Curr Opin Oncol (US) 3 (6): 1055–1059.

Richards JM, Mehta N, Ramming K and Skosey P (1992) Sequential chemoimmunotherapy in the treatment of metastatic melanoma. J Clin Oncol 10 (8): 1338–1443.

Rojas A (1992) ARCON: accelerated radiotherapy with carbogen and nicotinamide. BJR Suppl (England) 24: 174–178.

Schmitt G and Wambeisie A (1990) Review of the clinical results of fast neutron therapy. Radiother Oncol (Netherlands) 17 (1): 47–56.

Shirato H, Harrison R, Kornelsen RD, Lam GK, Gaffney CC, Goodman GB, Grochowski E and Pate B. (1989) Detection of pion induced radioactivity by autoradiography and positron emission tomography. Med Phys (US) 16 (3): 338–345.

Steel GG and Peacock JH (1989) Why are some tumours more radiosensitive than others? Radiother Oncol (Netherlands) 15 (1): 63–72.

Twentyman PR (1992) Cyclosporins as drug resistance modifiers. Biochem Pharmacol (England) 43: 109–117.

Twentyman PR, Wright KA, Mistry P, Kelland LR, Murrer BA (1992) Sensitivity to novel platinum compounds of panels of human lung cancer cell lines with acquired and inherent resistance to cisplatinum. Cancer Res (US) 52: 5674–5680.

Warrington RC (1992) L-histidinol in experimental cancer chemotherapy: improving the selectivity and efficacy of anticancer drugs, eliminating metastatic disease and reversing the multidrug-resistant phenotype. Biochem Cell Biol (Canada) 70 (5): 365–375.

Cytotechnology **12**: 393, 1993.

Key Word Index

adjuvant therapy 137
alkylating drugs 347
anthracycline 247
anticancer drugs 358
antineoplastic agents 385
antisense 231
ATP-binding cassette superfamily 33
ATPase 33
atypical multidrug resistance 137

C219 91
camptothecin 127
cancer 289
cancer chemotherapy 171, 231
carcinoma 231
cell cycle arrest 325
cell membrane 325
chemosensitizer 171
circumvention of MDR 213
cisplatin 347
clinical trials 213
clonal variation 231
confluence 347
cross-resistance patterns 265
cytochrome P450 358
cytogenetics 63
cytokines 231

diagnosis of multidrug resistance 91
DNA repair 325
doxorubucin 155
drug resistance 1, 358, 385
drug transport 33, 109
drugresistance 265

etoposide 155

facilitated diffusion 315
fractionated X-irradiation 265

gene amplification 63
glutathione 155
glutathione peroxidase glutathione reductase 155
glutathione S-transferase 155

haematological malignancies 213
human tumor cell lines 109

invasion 367

kinetic resistance cell cycle 347

low resistance 257

mathematical models 315
mdr-1 gene 231
MDR 1, 127, 213, 257, 315
membrane protein alterations 109
metabolic activation 358
metalloproteinases 367
metastasis 367
modulation 171
monoclonal antibody 91
MRK16 91
multidrug resistance (MDR) 231
multidrug resistance 1, 33, 63, 91, 155, 171, 213, 289, 347

neoplasm 385
non-P-glycoprotein-mediated multidrug resistance 265
non-P-glycoprotein multidrug resistance 109

oncogene 325

P-glycoprotein 1, 33, 91, 137, 155, 213, 231, 257
P-glycoprotein-mediated multidrug resistance 265
P-glycoprotein regulation 265
P-glycoproteins 63
physical mapping 63

quantitative PCR 289

radiation resistance 325
radiation tolerance 385
radiotherapy 385
reverse transcriptase polymerase chain reaction (RT-PCR) 289

sequestration of drug 109
signal transduction 325
solid tumour 231

therapy of multidrug resistance 91
TIMPs 367
topoisomerase 137
topoisomerase 347
topoisomerase I 127
topoisomerase II 265
verapamil 109
vincristine 155